The Illustrated Dictionary of
Southern African Plant Names

CARL LINNAEUS
(CARL VON LINNÉ)

'The Greek derivatives of the names of many plants are extremely difficult to discover, and therefore we must often be satisfied by conjecture.'
– CARL LINNAEUS, *PHILOSOPHIA BOTANICA*, 1751

Hugh Clarke & Michael Charters

EDITED BY
Eugene Moll, Department of Biodiversity & Conservation Biology, University of the Western Cape

First published by Jacana Media (Pty) Ltd in 2016

10 Orange Street
Sunnyside
Auckland Park 2092
South Africa
+2711 628 3200
www.jacana.co.za

© Hugh Clarke and Michael Charters, 2016

All rights reserved. No part of this publication may be reproduced, stored in a retrieval system or transmitted in any form or by any means – electronic, mechanical, photocopying, recording or otherwise – without the prior written permission of the copyright owner(s).

ISBN 978-1-4314-2443-6
ISBN 978-1-4314-2498-6 (hardcover)

Design by Shawn Paikin
Set in Arial
Printed and bound by ABC Press, Cape Town
Job no. 002817

See a complete list of Jacana titles at www.jacana.co.za

Disclaimer

The information contained in this book has been collected from many secondary sources and personal communications. Because this information was collected over a period of seven years, between 2009 and 2016, and many sources were accessed on several occasions, we have not included date of access for any web pages that appear in the bibliography. Apart from the research time it has taken to bring information from different sources together, very little of it represents our own original work. Most of the content here is, or at least is intended to be, scientific fact not traditionally covered by copyright. Much of our material is in the public domain. All images and text remain the intellectual property of their original owners. Where they can be found we have tried to contact the copyright holders for permission to use information contained in their publications. If you are the intellectual owner of anything presented in this book and are unhappy with the manner of its usage, please do not hesitate to contact the authors, who will rectify or remove entries in any future edition. Where entries in this book are original work by the authors, the authors would like to declare that this work is released into the public domain.

This book is dedicated to the many people who have given their lives to the conservation of Southern African fauna and flora, and to our wives, Fenja and Miriam, who have shown huge tenacity in supporting us throughout the many years it took to complete

Special acknowledgements

The authors would like to give special acknowledgement to four sources that have been paramount in making this publication unique, informative and readable.

Wikipedia – The Free Encyclopedia has provided a wealth of information, and without this source our task would have been all the more difficult. We would like thank anonymous Wikipedia contributors who, through their research, have paved the way for us in terms of historical biographical information for our short entries. We would also like to thank those individuals who provided Wikipedia with images of the various botanists, professors, plant collectors, explorers and others, which visually makes all the short biographies more interesting. In some cases information is so limited that our entries are very similar to Wikipedia, but our aim throughout has been to use other informational sources as well to provide a different presentation, although in a few cases this has been impossible.

Botanical Explorations of Southern Africa
These magnificent books by Mary Gunn and LE Codd (first edition) and HF Glen and G Germishuizen (second edition) have played a major role in helping us write short biographies for all those individuals who have visited, lived or worked in South Africa and been leading plant collectors. The depth of historical information in these books is astonishing and made our writing a short biography for all plant eponyms so much easier. Hugh Glen has been especially helpful in other ways, as indicated on the acknowledgements page. Many of our extracts for the biographies come directly from the *Botanical Explorations* text and again, where possible, we have tried to make the text different, but in 150 words, the maximum count for our entries, this was not always possible.

Checklist of families and genera of vascular plants – APGII version 2. The authors acknowledge the late Dr Koos Roux (1954–2013), who was a fern specialist, curator and collections manager of the Compton Herbarium from 2003 until his tragic death, for his work in the above-mentioned list. In 2012, we were struggling to establish what genus was associated with what family. The internet gave conflicting answers, as in some parts of the world the APGII system (Angiosperm Phylogeny Group II system) of flowering plant classification, which the South African National Biodiversity Institute uses, had been replaced by a later APGIII version, a mostly molecular-based system of plant taxonomy. So genera were sometimes placed in a different family from what is accepted in Southern Africa, which adheres to the APGII System. Dr Roux was drawing up a definitive list and was helpful and supportive of our project. On completion of his checklist he gave us a copy of this valuable guideline for Southern African plant classification.

Origins and Meanings of Names of South African Plant Genera. WPU Jackson's book published in 1990 was a forerunner to this book. He pioneered the idea that plant genera names should be meaningful, and his contribution to our book is given in the chapter on the history of this book.

BELOW: Beautiful post-fire flowering of *Watsonia* (CM)

Contents

Foreword	vi
Donors, sponsors and subscribers	viii
Introduction	x
The history of this book	xiii
A short guide to the many aspects of plant names	xvi
The plant eponym biographies	xix
A note regarding the biographies	xx
A brief note about plant taxonomy	xxi
Dictionary of Southern African plant names	xxii
Vascular plants	1
Bryophytes (mosses and liverworts)	377
Lichens	405
Glossary of botanical terms used in this book	428
Bibliography	435
Sources used to compile the biographies	437
Authors' acknowledgements	473
Contributors and photographers	477
Editor's endnote	479
About the authors and editor	482

Abbreviations

Gk. = Greek; La. = Latin; N.La. = New Latin; O.E. = Old English; O.La. = Old Latin; spp. = Author referring to all species in a given genus; IPNI = International Plant Names Index.

Foreword

It was a very great honour for me to be asked by Hugh Clarke and Michael Charters to write the foreword to this dictionary, which I believe is an historically significant contribution to regional botany *sensu lato*. I am, therefore, delighted to have the opportunity to personally inform users of just how important this dictionary is and to be able to say a few things about the people and effort involved in bringing this tome to you, the interested public.

Hugh and Michael, amateur botanical enthusiasts, have put together a dictionary that I truly believe is a 'once-in-a-lifetime' book and one you will not find anywhere else in the world. The book is both unique in terms of content and innovative in its presentation. In it you will find explanations of the meaning of botanical names that were all described in botanical Latin and incorporating other languages and, where applicable, the inference that the name has as far as the plant genus is concerned. But there is also so much more!

The book contains numerous beautiful full-colour illustrations of plants by botanical photographers to emphasise, and in many cases illustrate, the meaning of the names. It has been my privilege to be part of this photographic selection and other small educational additions that explain the meaning of botanical terms that may be unknown to users. Throughout the book you will find interesting narratives about the lives of the people after whom plants are named, and there are over 600 images of these people (which adds verve to this text). The entries for each of the 26 letters of the alphabet have been used to honour 24 of South Africa's top botanists, with many of them supplying their own images and biographies. There are also other features, not normally seen in plant name dictionaries, such as information on why plant generic names are changed. So the compiled information in this dictionary embraces a juxtaposition of the meaning of botanical names fused with the background of people in the history of botany.

In completing this *magnum opus* Hugh and Michael have had to navigate many obstacles to achieve the several milestones required for completion. Not least was that many authors did not explain why they gave their genera the names they did, so interpretation of the name can only be through logical deduction and sometimes a little intelligent guesswork. In writing up botanical histories there were some difficulties in uncovering centuries-old biographical information from limited available resources. Also, there was a lack of a network of local botanists interested in this field and Hugh and Michael struggled with translating 17th- to 19th-century Latin texts, and other constraints such as Hugh's limited understanding of botanical technical terms and the vagaries of plant taxonomy.

In the beginning of his research Hugh's business background, love of mountain walking and wildflower identification left him gloriously unaware of not only the enormity of the task ahead but also of the challenges he would face researching the content.

It was some years into the project before Hugh met Michael, initially through the latter's Eponym Dictionary of Southern African Plants website and, later, personally when Michael visited South Africa. It soon became evident that Michael would be an ideal co-author and their friendship and close working arrangement became an enduring partnership that has added great value to this dictionary.

Although Hugh initially thought that the book would take six months to complete (see the chapter on the history of the book) it actually took more than six years.

Over time I started to feel guilty of having suggested Hugh write this book. Fortunately, with his diligent work ethic and dogged persistence, he stuck to his task, and I was happy to help him towards the end with identifying and sourcing the colour illustrations of the genera.

Hugh mentions in the acknowledgements that a number of people helped him with the text. When it came to compiling the colour photographs I was able, through a wonderful lifetime network of botanical colleagues and friends, to obtain the pictures we needed. In a way this was some atonement for imposing six years of hard labour on Hugh's life – which, by the way, he actually enjoyed, as it was the kind of detective work he loved. And I was also thankful to be able to make a contribution towards this book, not as an author but as Hugh and Michael's botanical editor. I had fun in resurrecting images I had not studied for years, reading wonderful historical biographies about plant eponyms I knew by name but never knew the story behind, and the excitement of seeing a really good book develop.

Hugh's initial list of Southern African plant genera was provided by the South African National Biodiversity Institute (SANBI) in 2009, and he used it with other sources as his master list. Later in 2012 the late Dr Koos Roux of SANBI came to his assistance with an updated list. As Hugh worked through the list, it became evident that some genera and families in Roux's list, representing an APGII view, differed slightly from lists available from elsewhere – so there have been, and will continue to be, some questions about the inclusion or exclusion of one genus or another and maybe also about the current taxonomic validity of some of the genera. This is because plant nomenclature has always been, and will continue to be, in a state of flux because the classification of plants is a human construct (you will learn this as you use this dictionary). In fact, this tome is really about people, and how and why people have attributed the names to many of our plant genera.

An additional complicating factor is that systems of plant classification are not universally agreed on, nor are the philosophical foundations on which the schemes are based unanimously accepted.

Without doubt Hugh and Michael will have critics simply because they have had the courage to compile this dictionary, but what they have achieved is an excellent and beautifully illustrated book that will and can be enjoyed by all those who have an interest in plants – from amateur gardeners to professional botanists, from game rangers to historians and others at the cutting edge of research.

Thus, what we have here is an authoritative dictionary of the attribution of plant generic names taken from a SANBI list provided to Hugh a few years ago and Koos Roux's more up-to-date publication. Clearly, in a field that evolves continuously, some of the names may no longer be considered valid although they are still of historical interest because they are in the literature. However, the authors had to draw a line in the sand after which no more changes could be entertained or else the book would still be in manuscript form.

So well done Hugh Clarke and Michael Charters – you have given us a significant contribution for continual enjoyment by all who love our flora.

Eugene Moll, Cape Town

Donors, sponsors and subscribers

Our deep gratitude to all those who supported the publication of this dictionary – donors, sponsors, subscribers and the Flora & Fauna Publications Trust – because without their financial assistance this book would not have been published.

Donors

Botanical Education Trust

Botanical Society of South Africa – Kirstenbosch branch

E Oppenheimer & son

Richard and Margaret Clarke

Nicholson Educational Trust

Rod and Rachel Saunders

The Mapula Trust

Sponsors

Arnold, Miles & Anna-Marie
Botanical Society of South Africa – Kogelberg Branch
Botanical Society of South Africa – KZN Inland Branch
Bradshaw, John K
Burrows, John & Sandie
Buffelskloof Nature Reserve Trust
Cox, Jill & Graham
De Klerk, Dave
De Wet, Willemina Lunae
Friends of Cape of Good Hope
Gie, Susan & Richard
Gilfillan, Ted & Rem
GroundTruth
Loffler, Paul & Linda
Malan, Dorothy

Mauss, Joachim – Plettenberg Bay
Oliver, Marian Jean
Palmer, Tony & Tally
Sievers, Chrissie
Stewart, Iain
Thomas, Peter & Val
Thompson, Clive – Discovery Trails
Tree Society of Southern Africa
Van Wyk, Braam & Elsa
Visser, Henning
Woods, Ted

Subscribers

Albertyn-Cross, Lynda
AndBeyond
AWOL Tours
Bainbridge, Bill & Sheila
Barker, Jill & Walter

Donors, sponsors and subscribers

Barker, WR
Bashall, Frank
Berrisford, Peter & Merrilee
Buchel, Pat & Mike
Burton-Moore, PJM
Bytebier, Dr Benny
Campbell, Hazel
Collins, Amanda & Donald
Cook, HFG
Cowling, Richard
Delport, Mariana – Cape Eco-Tours
De Villiers, Jeanne & Kay
De Villiers, Joe
Drummond, Mark & Rosie
Esler, Karen J
Everard Read Gallery
Fletcher, Kelsea
Ford, Roger & Letitia
Francis, Henry & Suria
Friedman, Russel & Bonnie
Friends of Silvermine Nature Area
Gebhardt, Annemarie
Gill, Kevin
Giliomee, Jan
Grant, Marylynn
Grieve, G, K, A & J
Grobbelaar, Carl
Haarhoff, Philippa J
Haumann, Carel & Sumè
Haumann, Tielman and Ester
Henman-Weir, Felicity
Higginson, Di
Hyde, Mark
Izzard, Kelly
James, Alison & Douglas
James, Dave & Chris
Kewley, Doggy & Barbara

Khoury, Grace
Khoury, Joshua
Kremer-Köhne, Sylvie
Laidler, Gigi & Dennis
Landman, Nina
Leroy, John & Astri
Leyden, Murray
Lovelock, Heidi
Macdonald, Dr Ian AW
Mann, Graham & Mandy
Marquard, Cael
McCallum, Ian & Sharon
McClurg, Tim & Helen
McDonald, David J
McWade, Alistair
Meere CMF
Mengell, Marion
Mountain, Alan & Jenny
Nichols, Geoff & Lynne
Oliver, EGH (Ted)
Powrie, Les & Sal
Pratt, Duncan
Raitt, Lincoln & Rosemary
Raulstone, Dave
Smart, R
Smit, Dirk Jasper
Smuts, Stephen & Olwyn
Snow, Janet & Tim
The Sommervilles
Steyn, Christna
Swanepoel, Wessel
Tarboton, Warwick & Michèle
Timm, Heidi
Van Berkel, Nicola
Wood, Janet

* Ernest Oppenheimer & Son (EOS) is the investment holding company of the Oppenheimer Family; founders of the global mining company Anglo American, and managing shareholders of De Beers (the world's leading Diamond Company).

The Oppenheimer Family has a rich history and association with conservation work in South Africa, having long been supporters of biodiversity conservation, research in natural sciences and wildlife administration.

Through numerous programmes EOS remains a dedicated custodian of South Africa's unique ecological heritage.

Introduction

This first edition of the *Illustrated Dictionary of Southern African Plant Names* provides information that is for the most part unknown to the general public. It has been collected from a number of external sources, which has taken us, Hugh Clarke and Michael Charters, thousands of hours to research. Plant names can be traced back about 2 000 years to ancient writers including Theophrastus, Pliny the Elder and Pedanus Dioscorides. They observed and recorded the characteristics of the trees, shrubs, herbs and flowers that they knew and gave them new names that were not always the prevailing folk or common names. This is how the systematisation of our knowledge of plants began.

Our work is limited to Southern African generic names. Even these botanical names are not cast in stone. Sometimes, different botanists have assigned different names to the same plant. In these cases it has been agreed that the oldest name will be given preference. Sometimes names are invalidly published and must therefore be corrected. Sometimes new research reveals that plants in a particular genus actually belong in a different genus. Once again, a new name must be given. Taxonomy is an on-going process that continues at an ever-greater rate due to the advent of sophisticated molecular analysis based on plant genetic (DNA) analysis.

Humans have occupied the southern reaches of the African continent for many hundreds of thousands of years. They were well acquainted with the flora that surrounded them. They knew about the benefits of particular plants for sustenance, medicine and other uses and which plants to avoid. Africa's earliest inhabitants named plants in local languages, and these were passed down orally from generation to generation.

When in the latter part of the 17th century Europeans first landed at the foot of Africa because they desperately needed to resupply their vessels, they realised the flora of the Cape of Good Hope, as it came to be known, was so very different. These early settlers came upon some plants that seemed to be related to those already known to them, but many were unique and totally unknown. Up until the 17th century, botany and medicine were one and the same and at that time in Europe there was a special interest in collecting plants, not only for their beauty and unusual shapes but also for herbal properties and there was an intimate connection between what today we would call medicine, pharmacy and botany.

Naturalists such as Theophrastus, Dioscorides, Albertus Magnus, Andrea Caesalpino, John Gerard, Gaspard Bauhin, John Ray, and Joseph Pitton de Tournefort had already tried or were in the process of trying to organise the floristic world into an understandable structure, to describe and assign names to plants that would place them in a system that related similar groups of plants to each other. This process was taken to a new level by the Swedish botanist Carl Linnaeus, who built on the work of Gaspard Bauhin and established a system of binomial nomenclature according to which plants were given both generic and species names. Competing systems were developed by, among others, Michel Adanson, Antoine Laurent de Jussieu, Augustin Pyramus de Candolle and John Lindley.

The names given to these early plants are hardly ever alluded to today as a result of the European Renaissance, a cultural movement that lasted roughly from the 14th to the 17th century. It was during this period that classical Latin, which was derived from the Roman writers of the early first millennium, became and remained the sole language of learning throughout Europe for centuries. Indeed, it became a general-purpose language (as is English today in science) and was used especially by academics, diplomats, ecclesiastics, physicians and legal practitioners, and for domestic and international communications.

If it had not been for Latin as *lingua franca* of its day, we may not have had the uniform international system of botanical nomenclature of today. For centuries, botanists wrote about plants in Latin with thousands of publications being produced. Over time the practice changed from writing the names of plants in classical Latin to writing them in what is now called botanical Latin, which evolved from classical Latin as herbalists of the 16th century established the tradition, subsequently taken up by Linnaeus, that plants should be given Latin names. If there had been no common knowledge and usage of Latin, works of local botanists in the vernacular would have been unknown outside their own region. Linnaeus's work reflected the fact that a huge advance had been made in the knowledge of the complexities of structure and relationship of plants, and language had to evolve accordingly.

Botanical Latin has, as a result, grown far beyond its original form with the inclusion of Greek and other foreign words, and vast numbers of new words describing things that were essentially unknown in the ancient world. William T Stearn makes the point that Pliny the Elder would have well understood the Latin descriptions of plants in the 15th and 16th centuries but would have been lost by the divergent Latin of the 18th and 19th centuries. Proof of the giant leap in knowledge from the time of the ancients to the present day can be found in the fact that the early botanists such as Theophrastus in the third century BCE described about 500 plants, Pliny three centuries later described about 1 000, herbalists of the 15th and 16th centuries perhaps 4 000, Linnaeus in the 18th century around 7 300, and modern botanists some 250 000 to 300 000 species of flowering plants.

According to Stearn, the net result of this is that botanical Latin 'is now as unintelligible to classical scholars as Modern English would be to a Frenchman who had learned only Anglo-Saxon' (*Botanical Latin*, William T Stearn, 1987). The authors can vouch for this, not that we are Latin scholars but when we downloaded paragraphs of botanical Latin to try to find out what something meant, and used Google to translate (which provides reasonably good translations for some languages), we found we often couldn't make heads or tails of the text. The tragedy of this, in a way, is that one has to know Latin or botanical Latin in order to read these earlier publications. And as most people can't, whole centuries of work lie mostly unread except by a few scholars.

The majority of flower-lovers have limited or no knowledge of Latin. Fewer still, without instruction or training, can understand botanical Latin terminology. Statistics show that around the end of the 19th century, only 56 per cent of students at public schools were taught Latin. We were unable to establish the exact figures for this century, but some sources indicate that it is now down to between three and eight per cent. This was an important reason for writing this book, because there is no current Southern African botanical dictionary available that provides English translations of botanical names.

But there is a second issue that we feel needs to be addressed. Most botanical dictionaries give the botanical Latin name and English equivalent but do not explain their meaning and application. For example, take the genus *Adenandra*. A simple translation from the Greek is *aden* = gland, and *aner, andros* = man, male. Because this does not provide any insight into what the name *Adenandra* really means, we give, for each entry, the meaning and application to the best of our ability. In botanical Latin *aner, andros* = male, refers to the *male* reproductive parts of a flower, consisting of the *stamens*. Therefore, our entry is: 'Gk. *aden* = gland; *aner, andros* = man, a male, hence stamen. The stamens (gland-bearing anthers) are a striking feature of the flowers' – and we also illustrate them. See page 8.

While the translations of most botanical plant names are easy to determine, the meanings and implications of botanical Latin names are far more difficult to unravel for various reasons, not least their insignificance and meaninglessness at times. Henri Cassini (1781–1832) gave this reason for naming plants. 'I have given almost all my genres or subgenres insignificant names, and often mythological, because […] against common opinion, a generic name is even better, if it is more insignificant and less unpleasant to the ear' (translated from the French original, from *Bulletin des Sciences*, 1818).

Another difficulty is that many plant names are derived from languages other than English and are often recorded in journals that are housed in foreign libraries, academic institutions and botanical facilities that are not easily accessible.

There is also a third issue that we feel is of great importance. Plants are often named in honour of or to commemorate individuals who have made some distinctive contribution to society or who have stature or status in some way – a patron of botany, a respected botanist, educator or academic, a botanical collector or explorer, or a person who funded botanical explorations and the like. Sometimes these individuals were friends, colleagues, relatives or assistants to botanists or collectors, or farmers on whose lands particular plants were discovered. Nevertheless, they deserve to be remembered, and the names of plants commemorating them should be explained and recorded.

Our biographical dictionary entries aim to bring individuals to life, flesh out the usual basic details, place them in a historical context and reveal them as the interesting figures they were. These biographies are by no means exhaustive but at least give an idea about these lives.

Introduction

In short, we have tried to make this book useful in six new ways.

1. The text gives the Greek or Latin translation of the botanical (scientific) name and explains what this name implies or infers about the plant that bears the name.
2. Where 'old' genus names were included in the database provided to us by the South African National Biodiversity Institute, the current genus name is shown, e.g. *Acanthodium = Blepharis*. Our major source for this was Koos Roux's checklist, as mentioned on page iv.
3. Some 900 short biographies (plant eponyms) are given for all individuals for whom plants have been named, including a few histories never before recorded.
4. To make these biographies even more interesting, the book provides an image for more than 600 of these individuals and in addition there are more than 500 full-colour illustrations.
5. Vascular plants, bryophytes and lichens are individually divided into separate sections for the sake of easy access, because the main focus of this dictionary is on vascular plants.
6. Our asterisk key gives additional information relating to each genus's status.

Please inform Hugh Clarke (gascoyne@mweb.co.za) of any errors or omissions.

We wish you enjoyment, pleasure and edification in reading this book.

Hugh Clarke & Michael Charters

obartia flowering in spring (CM)

The history of this book

This book is an extensive revision of the late WPU Jackson's 1990 book *Origins and Meanings of Names of South African Plant Genera* (University of Cape Town). Jackson's work was founded on two major sources: *List of Species of South African Plants* edition 2 part 1 by GE Gibbs Russell, C Reid, J van Roy and L Smook of the Botanical Research Institute (in *Memoirs of the Botanical Survey of South Africa*, no. 51, 1984); and *List of Species of Southern African Plants* edition 2 part 2 by GE Gibbs Russell, WG Welman, E Retief, K Immelman, G Germishuizen, BJ Pienaar, M van Wyk and A Nicholas of the Botanical Research Institute (in *Memoirs of the Botanical Survey of South Africa*, no. 56, 1987).

In Jackson's 'booklet', as he calls it – which has 189 pages and contains about 2 800 entries – he writes: 'This list contains names considered to be correct as well as several other names now considered obsolete. All the extant names are included in the present booklet, but I have excluded a few obsolete names whose origins I could not trace.'

It is clear that Jackson thought his book contained a comprehensive list of all the South African genera currently in use and some older ones that had become obsolete. There is logic in having a database that contains the current genera and those that can be traced back, say, 60 years. For example, in October 1950, the first edition of Mary Maytham Kidd's *Wild Flowers of the Cape Peninsula* was published, and this classic book is still in print, albeit under the title of *Wild Flowers of the Table Mountain National Park* and with a revised text by Terry Trinder-Smith. But although the flowers therein remain the same, some plants have been renamed over the years as a result of new scientific evidence indicating under which genus they should fall.

Exactly to what degree Jackson expanded the work of the individuals mentioned above is unknown, but it is astonishing that he found so much information in an age before the internet. As an academic – he was a professor of medicine trained in the classical sense – he presumably had knowledge of Latin and Greek, which must have helped him interpret the scientific names. Also, by the time he wrote this book, he had been 'botanising' for some 40 years or so, as his interest in botany went back to the late 1940s, and he therefore had a good knowledge of plants. In addition, his life-long friend Dr Jim MacGregor, a Greek scholar, gave him enormous assistance (and refused co-authorship).

On his retirement Jackson registered for a BSc, majoring in botany, so he could further his knowledge. With this experience and knowledge base we assume that he could intuitively interpret the meaning of the various botanical expressions that he found in Latin publications. Maybe not known to many, he even wrote two wonderful coffee table books, *Wild Flowers of Table Mountain* and *Wild Flowers of the Fairest Cape*, for which he also took the photographs.

He was lucky too, that he was surrounded by highly esteemed botanists willing to help him, and he gives acknowledgement to, among others, Dr John Rourke and his staff at the Compton Herbarium, Diana Rex at the Bolus Herbarium, Dr Hugh Glen, Dr DJB Killick, E Potgieter, librarian at the Botanical Research Institute in Pretoria, and Ashley Nicholas, South African liaison botanist at Kew Gardens – all of whom provided information, help and support.

The above information is the background to our book. Jackson created an excellent framework on which our book was modelled in terms of layout and presentation, and of course there was a solid base of names that served to support further research.

So, how did this current book come about? After Hugh Clarke wrote *Common Wild Flowers of Table Mountain* (with co-author Bruce MacKenzie), he thought, 'Why not write a book like *A Guide to the Trees of Table Mountain*?' After all, this would be a good companion to his flower book, and the number of tree species on the mountain was very moderate, about 55 or so.

But Hugh learned that the respected botanist Eugene Moll was already working on a tree book for the Peninsula. The two met up, and it was at that meeting at Kirstenbosch in 2009 that Eugene suggested Hugh update WPU Jackson's *Origins and Meanings of Names of South African Plant Genera*, which was by then 20 years out of date.

The history of this book

Eugene never let on that he had been involved with Jackson's book, and Hugh only realised his involvement by chance a year later when he glanced at the title page and spotted Eugene's name.

The first task Hugh undertook to update Jackson's book was to request a current list of genera names from the South African National Biodiversity Institute (SANBI). Hugh guessed that 10 new genera – if that, and probably fewer – would be created each year, so that after a lapse of 20 years he would have to translate the names and find out the meanings of 200 generic names. In addition he would also have to update any genus name or family names that were now incorrect in Jackson's book. Easy. But the document that came from SANBI contained 4 766 genera, nearly 2 000 genera more than in Jackson's book.

Jackson's 1990 database had about 2 800 names; Hugh had about 4 766 based on the information Reuben Roberts of SANBI gave him. Koos Roux, who in 2012 drew up an APGII database, had some 9 000 names for Southern Africa, including invasives, while Les Powrie provided Hugh with a Southern African file of 14 174 generic names, some of which were 'old' names. Either Jackson did not have all the names he should have had, or maybe Hugh was given too many obsolete names. Whatever the case, the slog began.

Hugh admits that he could have worked more swiftly if he had been able to purchase the right books – such as Quattrocchi's exhaustive four-volume *CRC World Dictionary of Plant Names* (1999) – but they were often unavailable and prohibitively expensive. Fortunately some 'limited time' access to these books on the internet was available and authors such Quattrocchi generously gave permission for information in their books to be used.

University collections often contained books that were dated. Many books giving detailed descriptions of the various genera did not translate the meaning of the generic name, nor give its inference. Ironically, the specific epithets attached to the genus names were often explained. It seemed that many professional botanists had little interest in the meaning and inference of generic names, not only because of the translation difficulties from Latin, but because what was unravelled seemed vague. 'Many botanists don't care what the name means,' one of South Africa's top botanists commented in a casual conversation with Hugh.

But fortunately there are some who care. One of these is Dr Hugh Glen, who had previously populated the SANBI database with both the language translation and the description (inference) of what these names implied for about 500 names – and that was a great start.

Meanwhile, Michael Charters, an American living in California with a passion for Southern African flowers, was working away at biographies for his website *Eponym Dictionary of Southern African Plants.* This is undoubtedly the best website available relating to Southern African plant names and includes both generic and specific epithets with the individuals for whom they have been named.

Hugh only discovered this eponym dictionary when, after about two years' work, he decided to including short biographies in his database. Both Hugh and Michael found that their best source of easily accessible information was the internet, and this is how they met. Michael freely gave of his plant eponym information to Hugh, and this was reciprocated later by Hugh, as initially Michael did not include invasive genera – alien, exotic, non-indigenous plants from foreign countries – in his database. Despite sharing information, which involved over 400 emails to each other, both authors did studious independent research for their plant eponym biographies. In this way wrongful attributions were immediately corrected.

There was help from the botanical history researcher David Hollombe, one of Michael's acquaintances, who has the ability to ferret out obscure biographical information. Both Hugh and Michael had help from international academics (see the authors' acknowledgements) for which they are grateful.

Because of Michael's major contribution to this book, Hugh invited him to be a co-author. Michael also took on tasks of exceptional value such as proofreading an early version. Hugh wrote most of the short entries and biographies, the latter often based on Michael's work. Hugh collected more than 600 pictures of individuals after whom plants are named, some of them centuries old. Michael also contributed a number of his plant photographs.

With hundreds of pages assembled, Eugene stepped in, suggesting an illustrated book and committing himself to raising funds. His extensive network of friends and colleagues – made over 50 years of botanical activity such as lecturing

The history of this book

at various universities and in the NGO sector, as a public advocate for botanical education (particularly with a plant ecology and tree identification focus) – enabled him to raise the funds needed in only three months.

Eugene secured contributions from seven first-class botanical photographers (and Hugh added two more) and read early drafts of the manuscript in order to select the roughly 500 images that reflect the various plants that grow in different regions of Southern Africa and that matched the explanations of the genus names.

A further contribution from Eugene has been obtaining pictorial and text contributions from 24 top South African botanists, each a specialist in their own field, to open each of the alphabetical letter sections. He declined an offer for co-authorship of this dictionary, simply saying: 'This is your book. I enjoyed helping you.'

Nevillea flowering on the Cape Peninsula (CM)

Geranium flower (CM)

Terminalia twig and flowers (T)

A short guide to the many aspects of plant names

1. An introduction to scientific plant names

In the early years of botany, there was no systematic or standardised method for documenting plant names. Some scientific names are based on common ones. Going back a few centuries, a plant's Latin name was associated with its description, often a paragraph long. The plant known today as *Chrysanthemum indicum* was published by one author as *Chrysanthemum indicum: foliis simplicibus ovatis sinuatis angulatis serratis actis*, while another plant was known as *Chrysanthemum indicum: foliis subpalmatis sinuato-lobatis dentatis stipulatis, caule ramoso, ramis unifloris, calycis squamis rotundatis*. All of this was cumbersome and confusing, because the plants were not differentiated by their generic and specific names but by the added Latin descriptions. There was no uniform way of naming plants, and different people had different ideas about how to name them.

But in 1753 the Swedish botanist Carl Linnaeus expounded his system of binomial classification in the seminal *Species Plantarum*, using a two-part naming of plants that could be applied all over the world.

Linnaeus's system worked like this. The first name, the generic epithet, say *Disa*, designates the genus to which the species belongs; the second name, say *uniflora*, is the specific epithet that identifies the species. The two words, *Disa* and *uniflora*, when put together like this, *Disa uniflora*, make up a unique name. The genus name is always written first and starts with a capital, the second name is always in lower case – even if it is named after a person. Linnaeus called this second name the 'trivial' name. He believed the function of the name was simply to give the species a unique label; it did not have to have any meaning at all, although more often than not the specific epithet does have some meaning, hence, for example, *Disa lutea* (*luteus* = yellow, referring to the colour of the flower); *Disa tenuifolia* (*tenui* = thin, fine, slender; *folius* = leaf, referring to its thin, slender leaves).

The term 'genus' refers to a group of one or many closely related plants, called species, having common structural or genetic characteristics distinct from those of other groups. The many varieties of *Disa* found today may look very different from each other, but a few million years ago all of the plants of this particular genus had a common ancestor. The word 'genus' applies to a single group, but when used collectively the word 'genera' (plural) is employed.

A generic name can enfold as many species as are discovered, such as *Disa cornuta, Disa ferruginea, Disa rosea* and any others that are structurally, physically or genetically related. In listing a number of species names, the following shorthand is allowable – *D. cornuta, D. ferruginea, D. tenuifolia*. Note that the word 'species' is both singular and plural.

Going back even further in time, plant genera can be placed into even larger assemblages called families. For example, the *Disa* genus – together with a number of other plant genera such as *Disperis, Eulophia, Pterygodium, Satyrium* and others – falls within the orchid or Orchidaceae family. So the *Disa* genus has characteristics resembling other members of the orchid family. In the descriptions of the generic names, *Disa* is reflected with its family name like this: *Disa (Orchidaceae)*.

So to sum up, a group of species makes up a genus, and a group of genera makes up a family. Botanists are generally not concerned with groupings beyond this point. The term monotypic is applied to a grouping such as a genus that has only a single member.

And one final point is that by convention generic and species names are always written in italics.

2. The meanings of Greek and Latin words in this dictionary

In this *Illustrated Dictionary of Southern African Plant Names* you will find nearly 5 000 generic translated names together with descriptions of what these names may imply.

Usually the genus name is derived from Greek or Latin. Sometimes the derivation of the genus name is based on another language such as Arabic.

Latin prefixes and suffixes

The Greek letter 'u' is written as a 'y' in Latin. So the **Latin word 'hyper' meaning 'over, beyond, above' is rendered 'huper' in Greek.** Also, the Greek letter 'k' is almost always replaced by 'c' in Latin. In most cases we use the Latinised spelling, even for Greek words.

Some Latin words have prefixes, a verbal element placed at the beginning of the word to qualify meaning, such as bi-, as in bicycle, meaning two. Hence *biflorus* = La. *bi-* = two; *floris* = flower; meaning the plant has two flowers. The prefixes will present no problem.

The suffixes don't pose a problem either especially as we have translated them for you. The meanings of Latin words can change *slightly* depending on the suffix. So, for example, the Latin word *asperus* can end -*us*, -*a*, -*um* (depending on its gender) and means 'rough' whereas the Latin word *asperulum,* which has a somewhat different ending, means 'slightly rough' – the suffix -*ulus*, -*a*, -*um* meaning 'slightly,' or 'to a lesser degree.' This word might be used when describing the surfaces of the flowering plant's leaves.

Here is the best piece of advice: forget the adjectival endings – these are just anomalies. The meanings of the stems of words remain constant across all genera, and the various endings do not usually significantly change the meaning of the word.

3. Botanical translations and meanings

Many botanical dictionaries only give the Latin or Greek translation of the generic name but do not give any explanation for the naming. We have tried to explain both the translation and what this infers as far as the plant is concerned.

4. Generic names

Botanical names do not fall into any standardised categories of meaning. They reflect the wide range of interests of the authors who named the plants, and their choice of names is arbitrary, but overall the names can be grouped according to a number of categories. Examples of our categories are given below together with some examples for each category which are easy to understand.

The genus names can be associated with the following:

Descriptive features: morphological or structural parts of the plant such as roots, stems, buds, leaves, flowers parts, fruits, etc., which will give some idea as to its appearance. Descriptive features of plants refer to size, shape, the colour of flowers, leaves and stems, fragrances, scent or odours associated with some plants, the taste of fruits being sweet, bitter, poisonous, even the reaction of sensitive plants to touch, some of which shrink away when touched.

Habitat: the natural place or environment where a plant normally lives or grows. There are 24 habitats mentioned in our entries, ranging from sandy soils, mountain slopes and rocky areas to marshes, windy locations, high altitude mountains and even, surprisingly, arctic and subarctic habitats.

Geographical location: a specific point on the earth's surface such as a town, a broader area such as a country, or a place. Many geographic names change over time – Abyssinia became Ethiopia; Ceylon became Sri Lanka; South West Africa became Namibia. Early taxonomists were familiar with classical literature and often used ancient Greek and Latin names for localities. Only the more modern names are given in our text. An example of a geographical genus is name *Kogelbergia*, referring to Kogelberg Biosphere Reserve near Cape Town.

Mythological figures: The Roman and Greek worlds in particular were full of stories about gods and goddesses, fairies, sprites and nymphs, supernatural beings, satyrs, centaurs and humans who accomplished great tasks or suffered great tragedies. The floral world is often associated with mythological figures. Examples of genera named for such figures are *Achillea* (for Achilles), *Pandorea* (for Pandora and her famous box) and *Mercurialis* (for the war god Mercury).

Animal metaphors: Animals and imagery associated with them has often been used to describe a botanical feature. *Arctopus* for instance contains the Greek words for bear (*arktos*) and foot (*pous*), as author Carl Linnaeus thought the leaf of this species resembled the footprint of a bear's paw.

Anagrams: Referring to a word or name formed by rearranging the letters of another. This change sometimes occurs when an author discovers two genera names and gives their name to the first genus and 'disguises' their name with an anagram for the second. Sometimes it is just a matter of whimsy. Examples are *Achneria*, (an anagram of *Eriachne*) and *Bopusia* (from *Sopubia*).

A short guide to the many aspects of plant names

Miscellaneous: Some plant names fall into no category except this one. For example, the meanings can be humorous (such as *Ludorugbya*, referring to rugby) or have sexual connotations (*Juglans* which contains a reference to Jupiter's glans).

While these examples are straightforward, the interpretation of some plant names provided in this book are not as straightforward. To understand the meaning and interpretation of some plant names requires some botanical knowledge of plant structures, which can be put in terminology that is not commonly known. The glossary of terms at the end of the book will be of assistance in these instances.

Typical mopane veld (TW)

Wurmbea flower (CM)

The parasitic *Harveyi* in full bloom (CM)

The plant eponym biographies

During the course of our research we came across individuals who were surprised to know that a flower like *Begonia* was named after someone and that very few people are familiar with the history of plant eponym names. Our aim was, therefore, to fill in this knowledge gap in South African botanical history and we tried to track down all the Southern African plant eponym names on our database and record them in this book to ensure the details of the derivation of the generic names are not lost forever.

Many books – such as M Gunn and LE Codd's book *Botanical Explorers of Southern Africa* (1981), and the updated second edition written by HF Glen and G Germishuizen (2010) – contain biographies of explorers who contributed much to the botanical knowledge of Southern African flora through their search for new species. But they do not include the names of individuals who were not botanical explorers.

Many of our plant generic names are derived from individuals in other parts of the world who have never visited the region. In fact our database contains plant genera that are named after individuals born in 36 different countries.

It may not be generally known that most Southern African plants are named after individuals from overseas. Readers may be surprised that a statistical analysis of Southern African plant eponyms in this book shows that the most named individuals in our book come from Germany 256, the United Kingdom 175 and France 157, and a large number of them never even set foot in Africa, as huge numbers of never-before-seen plants were shipped to Europe, which became a collecting point for new exotic plants.

Contributions were also made by South Africa 58 and Italy 58, Sweden 57, Switzerland 30, the Netherlands 26, Denmark and Austria, each 19, Spain 18, Portugal and America each 16, Belgium 10 and Czech Republic 8 names. Twenty-one other countries contributed but not more than four genera or fewer names per country.

The Kingdom of Prussia, which no longer exists, contributed 11 names. Between 1701 and 1918 this kingdom included parts of present-day Germany, Poland, Russia, Lithuania, Denmark, Belgium and the Czech Republic. Many of the names seem to be German but we cannot accurately allocate these names to specific present day countries.

A note regarding the biographies

Because botanical history goes back centuries, some of the images we were able to procure for this book are not as clear as we would like them to be.

Also, many of the individuals you will read about in the biographies wrote books in a foreign language. To make some of the book titles easy for readers to understand we have in many cases provided an English translation.

Conserving botanical history

An invitation from Michael Charters

In addition to plant genera eponym names in this book, many species are named after individuals whose details are not recorded in this book. My website *The Eponym Dictionary of Southern African Plants* records both the names of individuals after whom a genus or species is named.

Should any readers wish to know more about these individuals, please refer to Michael Charters's website (http://www.calflora.net/southafrica/plantnames.html).

Should you know of an individual, a family member, friend, botanist, etc., whose name is not found in his comprehensive dictionary, please write to him at mmlcharters@calflora.net.

Nevillea in fynbos (CM)

A brief note about plant taxonomy

Right upfront we need to make it quite clear that this is not a book on plant taxonomy. Plant taxonomy is the science that finds, identifies, describes, classifies, and names plants. Within this book of plant names you will come across the names of many plant collectors who set out to foreign countries to find new plants; you'll learn about the hardships they endured, the courage they showed and sometimes read about the tragedy of their death. But our book does not describe how, over centuries, plant identification improved, how plant descriptions developed as a result of more discoveries about characteristics, neither will you read in detail about plant classification systems.

To address all the aspects of taxonomy relating to the Southern African plant names, even if this is only restricted to genera and excluded species, would be a monster problem requiring hundreds of pages more if done properly.

However, there is a positive in that our book focuses on 'finding plants', which is touched on in many of the biographies and also – our book's major challenge – to discover and clarify what the plant names mean. Our endeavour in particular has been not just to translate the botanical Latin, which many books have done, but to dig deeper and ask: 'so what?' What does this plant name infer or imply as far as the plant is concerned? Why did the author give the plant this name? Seemingly not much research has been done in this field. In many cases some answers have been found. But mainly, this question has not been answered.

So in this book our research aim was this: if plant names are a component of plant taxonomy, what attention has been paid to these names? Our challenge was to find out whether these names were meaningful or meaningless. It turns out that they are both. And even if knowing what a plant name means is not of great importance, in many instances the name is of help to identifying the plant or at least understanding what this hitherto unknown plant name means. This dictionary then is about extending our knowledge and having fun by knowing what the plant generic names mean.

Watsonia flowering in spring (CM)

Dictionary of Southern African plant names

The text on the following pages, listed alphabetically, is structured to provide information in two ways.

If the genus is named after an individual, the entry provides:

The genus name in bold italics: e.g. **Abelia**. The family name to which that genus belongs follows in brackets (Caprifoliaceae). The name of the author who described the genus is then given in accepted abbreviated form: e.g. R.Br. for Robert Brown. This is followed by the name of the individual after whom the plant is named: e.g. Clarke, Abel, together with a brief bibliography.

If the name does not relate to an individual, the entry provides:

The genus name in bold italics: e.g. **Abelmoschus**. The family name (Malvaceae). The abbreviated name of the author who described the genus: e.g. Medik. for Friedrich Kasimir Medikus. This is followed by the translation of the plant name – e.g. 'Arabic abu-l-mosk = father of musk' – and an explanation as to why it was given this name, the inference or implication.

It is not botanical practice to provide full author names, but should you wish to know the full names, a list of botanists by author abbreviation can be found at Wikipedia.org/wiki/List_of_botanists_by_author_abbreviation.

Please note that while the information provided below is believed to be accurate for the author describing the genus, there are occasionally some differences between major data sources, such as Tropicos and The Plant List. One of the reasons for this is that different people (botanists and taxonomists) have different ideas on how plants are related or how closely they are related. So there is not just one system of taxonomy but several. The South African National Biodiversity Institute (SANBI) adheres to the APGII system of taxonomy. Other countries, such as the United States, adhere to the APGIII system. In South Africa, if you look on the internet for the *Aloe* genus in Plant ZA to find out which family the genus *Aloe* belongs to, the answer is Asphodelaceae. In the United States, you will find it is linked to Xanthorrhoeaceae because the APGIII system has different allocations. Similarly, if you read older texts, you will discover that some of the generic names are put into completely different families from those of today as a result of new scientific findings.

Asterisk key indicating the genus status

* The genus is a 'naturalised alien': that is, an invasive or exotic genus now part of Southern African flora

** The genus name is unresolved: that is, neither an accepted name nor a synonym according to The Plant List

*** The genus name is no longer an accepted name and has been subsumed into another genus

**** The genus name could not be found, such as *Arthothelium,* so may be defunct

Note: finally, if genus name X is shown to equal genus name Y (Name X = Name Y), this means one genus or a part of it is now placed in another genus or, according to Koos Roux's APGII database, could be in the future.

The dictionary is divided into three sections: Vascular plants (such as flowers and trees) (pages 1–376), Bryophytes (mosses and liverworts) (pages 377–404) and Lichens (pages 405–427).

Vascular Plants

A

Acacia burkei, now called *Senegalia burkei* after the Australians, who have about 1 000 species of *Acacia* trees indigenous on their continent, and who approached the botanical naming committee of the 17th International Botanical Congress in Vienna (2005) with an appeal that won them the right to use the genus name *Acacia* even though Linnaeus in 1773 based it on the African species *Acacia nilotica*. This controversial decision was ratified by the 18th IBC in Melbourne (2011). (EJM)

Eugene John Moll (1941–) was born in Bulawayo and graduated from the University of KwaZulu-Natal Pietermartizburg with a PhD in plant ecology. He spent 10 years working in KwaZulu-Natal as a botanical survey officer, 20 years at the University of Cape Town, where he met and taught many excellent young people and encouraged them to become botanists, and then 10 years at the University of Queensland before retiring to Cape Town in 2003. He now holds an honary position at the University of the Western Cape in the department of biodiversity and conservation biology, where he still teaches an honours module in fynbos ecology.

Abelia* (Caprifoliaceae*) R.Br.
For Clarke Abel (1780–1826), British surgeon and naturalist. In 1816–1817, on the recommendation of Sir Joseph Banks, Abel went on a mission to China with Lord Amherst. He served in the British embassy at Canton as chief medical officer and naturalist, during which time he visited the well-known nursery gardens at Fa Tee (Fa Ti). For two years, he collected plants and seeds, and sent back to England a number of consignments of specimens and seeds – a major consignment was lost as a result of a pirate attack. They were shipwrecked on the way home resulting in the loss of all his specimens; however, some specimens left behind in Canton were later returned to him by Sir George Staunton. Abel kept a documented record of his travels, later published as *Narrative of a Journey in the Interior of China* (1818). In 1823, he became Lord Amherst's surgeon-in-chief when the earl was appointed governor-general in India. Abel was elected a Fellow of the Royal Society in 1819.

Abelmoschus* (Malvaceae) Medik.
Arabic *abu-l-mosk* = father of musk; referring to the musky smell of the seeds.

Aberia = ***Dovyalis*** (Salicaceae) Hochst.
For Mount Aber in the Semien mountains in Ethiopia (previously Abyssinia); referring to the location of the type species.

Abildgaardia (Cyperaceae) Vah.
For Peder Christian Abildgaard (1740–1801), Danish zoologist and veterinarian. He dropped out of secondary school for financial reasons and worked in a pharmacy for five years where he learned chemistry. Thereafter, he studied medicine at the University of Copenhagen obtaining his doctorate in 1768. In 1773 he started his own veterinary college and became the director of the Royal Veterinary College at Christianshavn and through his work obtained the deserving title of 'father of Danish veterinary science'. During his career he published many works on medicine and zoology with his main interest being natural history, mainly dealing with intestinal parasites, leeches and protozoans. He was the author of *Historia Brevis Regii Instituti Veterinarii Hafniensis* (1788), founder of The Natural History Society (1789) and was the first researcher to describe the mineral species Cryolite from specimens brought back from Greenland.

Abrodictyum (Hymenophyllaceae) C.Presl
Gk. (*h*)*abros* = soft, delicate; *diktyon* = a net, netted; referring to the lace-like appearance of the fronds of some species. The cell walls appear as a net.

Abroma* (Malvaceae) Jacq.
Gk. *a* = not; *broma* = food. The plant is mildly toxic and unfit to be eaten.

Abrus (Fabaceae) Adans.
Gk. (*h*)*abros* = soft, delicate. Smith and Stearn indicate of Arabic origin; an Egyptian name according to Prosper Alpini (1592); referring to the texture of the leaves.

Abryanthemum* = ***Carpobrotus***
(Aizoaceae) Neck.
Gk. *a* = intensive (very much); *embryon* = fruit; *anthemon* = flower; referring to its profuse flowering and fruiting.

Abutilon (Malvaceae) Mill.
Arabic *aubutilun*, Latinised as *abutilon,* first used by Ibn Sina Avicenna (Abd Allāh ibn Sīnā) (980–1037) for a mallow-like plant or mulberry tree. The leaves of some species in this genus resemble those of the mulberry tree.

Acacia* = ***Vachellia*** and ***Senegalia***
(Fabaceae) Mill.
Gk. *akakia* (*a-kakos*) = not good. Gk. *ake, akis* = a sharp point; probably refers to the spines (stipules) that are typical in many of these genera. The name given by early Greek botanist-physician Pedanius Dioscorides (c 40–90) to the Egyptian thorn, *Vachellia nilotica,* for its medicinal properties. However, this name remains controversial, and Southern African 'Acacias' are divided into two genera.

(MC)

Acaena (Rosaceae) L.
Gk. *akaina* or La. *acaena* = a spike, from *ake* = a point; alluding to the spiny calyces or to the prickles on the fruit.

Acalypha (Euphorbiaceae) L.
Gk. *akalephe* = a nettle or its sting. The leaves of some species are nettle-like, resembling those of *Urtica*, the true nettle.

Acampe (Orchidaceae) Lindl.
Gk. *akampe, akamptos* = not bent, stiff, straight; referring to the plant's little, brittle, inflexible flowers.

Acanthobotrya*** (Fabaceae) Eck. & Zeyh.
Gk. *akanthos* = thorn; *botrys* = bunch of grapes; referring to the seed pods.

Acanthodium = Blepharis (Acanthaceae) Delile.
Gk. *akanthos* = thorn; *-odes* = resembling *Acanthus* (q.v.).

Acanthonotus = Indigofera (Fabaceae) Benth.
Gk. *akanthos* = thorn; *notos* = the South; alluding to its distribution in many parts of the Southern Hemisphere.

Acanthopsis (Acanthaceae) Harv.
Acanthus (q.v.) Gk. *-opsis* = resembling.

Acanthosicyos (Cucurbitaceae) Welw. ex Hook.f.

Nara melon (EJM)

Gk. *akanthos* = thorn; *sikyos* or *sykios* = wild cucumber; referring to the thorny cucumber or gourd such as the !Nara and the Gemsbok cucumber.

Acanthospermum* (Asteraceae) Schrank
Gk. *akanthos* = thorn; *sperma* = a seed; referring to its spiny seeds or 'fruits'.

Acanthotheca*** **= Dimorphotheca** (Asteraceae) DC.
Gk. *akanthos* = thorn; *theke* = a case, capsule; referring to its thorny seeds or 'fruits'.

Acanthus* (Acanthaceae) L.
Gk. *akanthos* = thorn; referring to the plant's spiny leaves and flower spikes.

Acaulon = Aloinopsis (Aizoaceae) N.E.Br.
La. *acaulon* = without a stem. This name was changed, *inter alia*, because a moss had earlier been given this name.

Acca (Myrtaceae) O.Berg.
A Peruvian/South American vernacular name, also possibly linked to an ancient Hebrew word for hot sand; referring to its preferred habitat – it grows well in warm climates.

Acer* (Sapindaceae) L.
La. *acer* = sharp; referring to the characteristic points on maple leaves. It could also refer to the maple tree's hard wood, which was used, among other things, for spear hafts by Roman soldiers.

Acetosa*** (Polygonaceae) **= Rumex**
La. *acetosus* = vinegary (acetose); pre-Linnaean name for common sorrel and other plants with acidic leaves.

Acharia** (Achariaceae) Thunb.
For Erik Acharius (1757–1819), Swedish botanist and physician who pioneered the taxonomy of  lichens and is known as the 'father of lichenology'. In 1773 he entered Uppsala University as a pupil of Linnaeus, graduating from Lund University in 1782. He was appointed professor of botany at Wadstena Academy, Stockholm, in 1801 and devoted himself to the study of lichens. All his publications were connected with that class of plants, his *Lichenographia Universalis* (Göttingen, 1804) being the most important. He was appointed a provincial doctor at Vadstena in 1789 and did much to improve health care. He was particularly concerned about the treatment of venereal diseases and founded a special hospital for this at Vadstena in 1795 where he was director.

Achillea* (Asteraceae) L.
La. *achillea*, Gk. *achilleios* = Achilles, legendary hero of the *Iliad,* son of Peleus and Thetis, who supposedly learned the healing properties of plants of this genus from Chiron the Centaur and used them to staunch the wounds of his soldiers

at the siege of Troy. In Greek mythology Achilles was invulnerable, except for his heel. Since he died due to a poisonous arrow shot into his heel the term 'Achilles' heel' has come to mean a person's principal weak point.

Achimenes (Gesneriaceae) P.Br.
Gk. *akheimanos* = to suffer from the cold, literally *a* = without; *kheimon* = winter, hence 'tender or sensitive to the cold'; referring to the plant's need for warmth – a summer plant par excellence.

Achnatherum = Stipa (Poaceae) P.Beauv.
Gk. *achne* = chaff, glume, i.e. scale; *ather* = stalk, barb; referring to the characteristic awned lemma 'Needlegrass'.

Achneria*** = Pentaschistis (Poaceae) Munro ex. Benth.
Possibly Gk. *achne* = chaff, glume or a word play, anagram of *Eriachne*, (*erion* = wool, *achne* = a scale); alluding to cataphylls (or lemmas?).

Achyranthes* (Amaranthaceae) L.

(GN)

Gk. *achyron* (*akhuron*) = chaff, husk; *anthos* = flower. The floral parts are chaffy.

Achyrocline (Asteraceae) (Less.) DC.
Gk. *achyron* = chaff, husk, straw; *cline* = bed; referring to its appearance. In some areas, agglomerations of plants form floral mattresses.

Achyronia (Fabaceae) Kuntze
Gk. *achyron* = chaff claw or 'fingernail'; alluding to the chaffy calyx.

Achyropsis (Amaranthaceae) (Moq.) Hook.f.
Gk. *achyron* = chaff, husk: -*opsis* = resembling *Achyranthes* (q.v.).

Acidanthera = Gladiolus (Iridaceae) Hochst.
Gk. *acidos* = pointed object; *anthera* = anther; referring to the shape of the anthers.

Acmadenia (Rutaceae) Bart & H.Wend.
Gk. *akme* = highest point; *aden* = a gland; referring to the glands on the anthers.

Acmena* (Myrtaceae) DC.
Gk. *acmenae* = mythological name for the very beautiful nymphs attending Venus. The flowers and fruits are attractive.

Acokanthera (Apocynaceae) G.Don
Gk. *akris* = sharp point; *anthera* = anthers; referring to the sharp anthers borne within the flowers.

Aconitum (Ranunculaceae) L.
Gk. *Akiniton*, the ancient Gk. name used by Theophrastus for a very poisonous plant. Derived possibly from Gk. *akon* = dart or javelin, the tips of which were poisoned with the plant's deadly toxin, or *akonitos* = 'without struggle'. The plant's common name 'wolf's bane' refers to the deadly toxins extracted from the plant, which were in the past used to kill wolves 'without struggle'.

Acorus* (Acoraceae) L.
Gk. *akorus*, from *akoron*, a name used by Dioscorides, which in turn was derived from *coreon* meaning 'pupil of the eyes'. The plant was used in herbal medicine as a treatment for the inflammation of the eye.

Acrachne (Poaceae) Wight & Arn. ex Chiov.
Gk. *akra*, *akros* = the terminal point; *achne* = a scale, chaff; referring to the fine-pointed terminal glumes and lemmas.

Acranthemum*** = Tapinanthus (Loranthaceae) Tiegh.
Gk. *akros* = top, extremity; *anthemon* = flower; referring to the flowers which occur at the extremities.

Acridocarpus (Malpighiaceae) Guil & Perr.
Gk. *akris* = a locust; *karpon* = a fruit; alluding to the winged fruit. (Note: The Greek word *akris* for locust has many connotations not explained here.)

Acriulus*** = Scleria (Cyperaceae) Ridl.
Gk. *akr-* = sharp; -*ulus* = little, small; referring to the awns.

Acrocephalus*** = Haumaniastrum (Lamiaceae) Benth.
Gk. *akros* = at the tip, end; *kephale* = head. The flowers are at the top of the branches.

Acroceras (Poaceae) Stapf
Gk. *akros* = at the tip, end; *keras* = horn, spur. The upper subtending glume and sterile lemma each contract into a horn-like structure.

Acrodon (Aizoaceae) N.E.Br.
Gk. *akros* = at the tip, end; *odus, odontos* = tooth; referring to the apically dentate leaves.

Acrolepis*** = **Ficinia** (Cyperaceae) Schrad.
Gk. *akros* = at the tip, end; *lepis* = a scale; referring to the perianth reduced to scales.

Acrolophia (Orchidaceae) (Lindl) Pfitzer
Gk. *akros* = at the tip, end; *lophos* = crest; *akrolophia* = mountain ridge; referring to the raised edges (crests) on the flower's lip or habitat. 'Alluding either to the keels on the lip or, more probably, to the mountainous habitat of all the species.' (Pfitzer, 1888–1889).

Acropteris = **Asplenium** (Aspleniaceae) Link

(NC)

Gk. *akros* = at the tip, end; *pteris* = fern, from *pteron* = wing; alluding to the sori which are terminal on the veins.

Acrosanthes* (Aizoaceae) Eck. & Zeyh.
Gk. *akros* = at the tip, end; *anthos* = flower. The flowers appear, solitary, on a flower stalk.

Acrostachys*** = **Helixanthera** (Loranthaceae) (Benth. & Hook.f.) Tiegh.
Gk. *akros* = at the tip, end; *stachys* = spike. Possibly because the flowers appear at the end of the flower stalk.

Acrostemon (Ericaceae) Klotzsch
Gk. *akro* = end, extremity; *stemon* = a stamen; referring to the stamens which protrude well beyond the petals.

Acrostichum (Pteridaceae) L.
Gk. *akros* = at the tip, end, extremity; *stichos* = row or line; referring to the distal spore-bearing pinnae.

Acrotome (Lamiaceae) Benth. ex End.
Gk. *akros* = highest, terminal; *temnein* = to cut; possibly referring to the deeply incised corolla.

Actaea* (Ranunculaceae) L.
Gk. *aktiai* = one of the ancient names of *Sambucus* (*nigra*), 'the elder tree', which the *Actaea* is supposed to resemble. The leaves and habitats are said to be similar.

Actinanthella (Loranthaceae) Balle
Gk. *actis, actinos* = ray; *anthos* = flower; *-ella* (diminutive); referring to the ray-like flowers.

Actiniopteris (Pteridaceae) Link
Gk. *actis, actinos* = ray; *pteris* = a fern; referring to the radial or shaped lamina. Commonly known as the ray-fern.

Adansonia (Malvaceae) L.
For Michel Adanson (1727–1806), French botanist and naturalist of Scottish descent. He  studied at the University of Paris (1741–1746) and conducted research in the Canary Islands (1748). He also did research in Senegal (1749–1753) on its animals, plants, shells, people, commerce and languages as reflected in his *Histoire Naturelle du Sénégal* (1757). In 1763, in his *Familles Naturelles des Plantes*, he proposed a new method of plant classification based on a sum of their characteristics rather than Linnaeus's essentialist system (based on flower morphology). He also published monographs on the Baobab tree and catfish. In 1765, he began a massive encyclopaedia (27 volumes) entitled *l'Ordre Universel de la Nature*. This was rejected by the Académie des Sciences in 1774, as being too large.

Adelanthus (Adelanthaceae) Mitt.
Gk. *adelos* = obscure, uncertain; *adelo* = unknown, secret, or hidden; *anthos* = flower; referring to the initial confusion about the nature of the perianth.

Adenachaena*** (Asteraceae) DC.
Gk. *aden* = gland; *khaino* = to gape, open. The fruit, an achene, carries striae covered with papillae or glandular granules.

Adenandra (Rutaceae) Willd. (Images on following page)
Gk. *aden* = gland; *-andra* = male, stamens; alluding to the appearance of the stamens (gland-bearing anthers).

Adenandra (CM)

Adenanthellum (Asteraceae) B.Nord.
Gk. *aden* = gland; La. *anthellum* = for small flowers; referring to the glandular ray-florets.

Adenanthemum = Adenanthellum (Asteraceae) B.Nord.
Gk. *aden* = gland; *anthemon* = flower; referring to the glandular ray-florets.

Adenanthera (Fabaceae) L.
Gk. *aden* = gland; *anthera* = anther; referring to the gland-bearing anthers or gland(-tipped) anthers.

Adenanthos* (Proteaceae) Labill.
Gk. *aden* = gland; *anthos* = flower; referring to the prominent nectaries within the floral tube.

Adenia (Passifloraceae) Forssk.
Named after the city of Aden in Yemen where Forsskål came across this plant. While there are glands (Gk. *aden*) on the petiole and in the flowers of some species, this is not the reason for the name.

Adenium (Apocynaceae) Roem. & Schult.
Also named after the city of Aden in Yemen. The first known species of this genus, *A. obesum*, was collected there by Forsskål.

Adenocarpus (Fabaceae) DC.
Gk. *aden* = gland; *carpus* = fruit; referring to the sticky pods of the shrub.

Adenocline (Euphorbiaceae) Turcz.
Gk. *aden* = gland; *klino* = bed; referring to the glandular disk at the base of the flower.

Adenoglossa (Asteraceae) B.Nord.
Gk. *aden* = gland; *glossa* = tongue; referring to the gland-tipped appendages of the stamens.

Adenogonum = Engleria (Asteraceae) Welw.
Gk. *aden* = gland; *gonum* = knee, bend (*gonia* = angle); alluding to the rows of conspicuous oil glands on or along the ribs of the achenes.

Adenogramma (Molluginaceae) Rchb.
Gk. *aden* = gland; *gramma* = line; referring to the ridged or striped lines to the glands.

Adenolobus (Fabaceae) Torre & Hillc.
Gk. *aden* = gland; *lobos* = capsule or pod. The pods are sticky-glandular in this genus.

Adenopodia (Fabaceae) C.Presl
Gk. *aden* = a gland; *pous* = a foot. Each leaf has a conspicuous gland just above the base of the petiole.

Adenostemma (Asteraceae) J.R. & G.Forst.
Gk. *aden* = gland; *stemma* = garland, wreath. The corolla tubes carry glandular hairs.

Adenostyles* (Asteraceae) Cass.
Gk. *aden* = gland; *stylos* = style; referring to the glandular styles (the stigmatic arms).

Adhatoda (Acanthaceae) Mill.
Apparently Latinised from Tamil and Sinhalese *ada* = a goat; *thodai* = not-touch (untouchables?). The leaves are so bitter that not even goats will eat them.

Adiantopsis (Pteridaceae) Fée
Adiantum (q.v.); Gk. *-iopsis* = resembling, likeness.

Adiantum (Pteridaceae) L.
Gk. *a-* = not, without; *diantos* = wettable, hence dry, incapable of being wetted. The fronds of 'maiden hair' shed water and are reputed to remain dry even in a rain-shower or when plunged into water.

Adina (Rubiaceae) Salisb.
Gk. *adinos* = clustered, crowded. The plant has tightly clustered heads of flowers (Umber to Quattrocchi, 2000).

Adonis* (Ranunculaceae) L.
Adonis, in Greek mythology, was a beautiful youth, son of Cinyrus, king of Cyprus. One version of his death is that when Ares, the god of war, heard that Aphrodite, the goddess of love, beauty and sexual rapture was chasing Adonis to mate with him he became jealous and sought revenge. Aphrodite

begged Adonis to give up the dangerous sport of hunting because she could not bear to lose him, knowing that the young man would have a terrible end. Adonis ignored her advice and was killed by Ares disguised as a wild boar. Aphrodite heard his cries and went to him and he died in her arms. Upon his death she turned the blood drops that fell from his wounds onto the soil into windflowers (the short-lived anemone) as a memorial to their love. The petals of these flowers are blood-red.

Adoxa (Adoxaceae) L.
Gk. *a-* = without; *doxa* = glory, glorious; referring to its unattractive inflorescence, which is similar in colour to the leaves.

Adromischus (Crassulaceae) Lem.
Gk. *adros* = thick; *miskhos* = a stalk; referring to the thick stalks of the species.

(MC)

Adventina*** = Galinsoga (Asteraceae) Raf.
La. *adventicius* = foreign, strange, extraneous, coming from abroad or from outside. Rafinesque states: 'Named after its adventitious production', when growing in an unusual place. He described two species that did not appear to be native, one blooming in spring and the other in autumn, growing spontaneously in Bartram's garden.

Aeollanthus (Lamiaceae) Mart. ex Spreng.
Gk. *ailos* = the god of wind; *anthos* = flower; referring to its habitat – the plant is essentially a mountain- and cliff-dwelling species.

Aerangis (Orchidaceae) Rchb.f.
Gk. *aer* = air; *angos, aggeion, angeion* = vessel; probably 'alluding to the long (blood-)vessel-like spur of the lip in the type species' (Reichenbach, 1865).

Aeranthus (Orchidaceae) Lindl.
Gk. *aer* = air; *anthos* = flower; alluding either to the damp habits in which species of this genus usually grow ('mist-flowers') or to the epiphytic habit ('air-flowers') (Lindley, 1824).

Aerisilvaea*** (Euphorbiaceae) Radcl.-Sm.
For Herbert Kenneth Airy Shaw (1902–1985), English botanist and classicist. He studied classics  at Cambridge University but changed to natural sciences, obtaining his BA degree in 1925. He joined Kew Gardens and became acting principal scientific officer at the end of 1948. Among his achievements he authored nine new botanical families, conducted extensive research in the family Euphorbiaceae with papers relating to this family in Siam (1972), Borneo (1975), New Guinea and Australia (1980), Sumatra (1981) and Philippines (1983) and became an expert in Asian botany and entomology. He had a special interest in spermatophytes. He also amassed a compilation of Russian botanical terms, worked on Index Kewensis and the preparation of the seventh edition of *A Dictionary of the Flowering Plants and Ferns* by John Christopher Willis.

Aerva* (Amaranthaceae) Forssk.
This is the Latinised form of an Arabic name – the genus was first described in *F Aegypt Arab*.

Aeschynanthus* (Gesneriaceae) Jack
Gk. *aischynos* = shame; *anthos* = flowers; referring to the blushing red colour of the flowers.

Aeschynomene (Fabaceae) L.
Gk. *aischynos* = shame; *mene* = mind. The plant is sensitive and shrinks when touched.

Aesculus (Sapindaceae) L.
Originally this was a Latin name for a kind of oak with edible acorns, but Linnaeus transferred this name to the horse-chestnut whose seeds are not edible.

Aethephyllum = Cleretum (Aizoaceae) N.E.Br.
Gk. *aethes* = irregular; *phyllon* = leaf; alluding to the variable and irregular leaf shape.

Aethusa (Apiaceae) L.
Gk. *aethusa* = burning one; referring to its pungency. Poisoning from *A. cynapium*, 'fool's parsley', shows symptoms of heat in the mouth and throat.

Afrafzelia* = Afzelia** (Fabaceae) Pierre
Gk. *afra* = pertaining to Africa; Spanish *zelia* = sunshine; *Afzelia* (q.v.). The genus loves sunshine.

Aframomum* (Zingiberaceae) K.Schum.
Gk. *afra* = from Africa; *amomum* = a kind of Indian spice. The African genus is said to share some of the genus *Amomum*'s pungent and aromatic properties.

Afrocanthium (Rubiaceae) (Bridson) Lantz & B.Bremer
La. *Afro* = of Africa; *Canthium* (q.v.). A genus separated from *Canthium*, occurring only in East and Southern Africa.

Afrocarpus (Podocarpaceae) Gaussen

Ripening fruits (MC)

La. *Afro* = of Africa; *karpos* = fruit. The African *Podocarpus* (q.v.).

Afrocrocus (Iridaceae) J.C.Manning & Goldblatt
La. *Afro* = of Africa; *crocus* = saffron yellow. The name alludes to its African distribution and its resemblance to the genus *Crocus*.

Afroligusticum (Apiaceae) C.Norman
La. *Afro* = of Africa. *Ligusticum* is a Northern Hemisphere genus in the *Apiaceae* family believed to be from the Liguria region of Italy. This plant refers to an African variety of this genus.

Afrolimon (Plumbaginaceae) Lincz.
La. *Afro* = of Africa; Gk. *leimon* = meadow; referring to its natural grassland-type habitat.

Afroqueta (Turneraceae) Thulin & Razafim.
La. *Afro* = of Africa; the American genus *Piriqueta*. *Afroqueta capensis* is the only species in this genus, which was previously called 'Piriqueta', now renamed.

Afrosciadium (Apiaceae) P.J.D.Winter
La. *Afro* = of Africa; *skiadion* = parasol, hence umbel; referring to the prominent umbel at the end of the stem.

Afrotysonia (Boraginaceae) Rauschert
La. *Afro* = of Africa; *Tysonia* (q.v.).

Afzelia (Fabaceae) Sm.
For Adam Afzelius (1750–1837), Swedish botanist. He studied languages and botany (under Carl Linnaeus) at Uppsala University and became a senior lecturer in oriental languages at Uppsala University in 1777, and a botanical demonstrator in 1785. In 1792 and again in 1794–1796 he went to botanise in Sierra Leone. On his return voyage, a storm destroyed the live plants he hoped to bring to England, but his specimens and notes survived.  From 1797–1798 he acted as secretary of the Swedish embassy in London. In 1799, he took up his position again as botanical demonstrator at Uppsala. Three years later, he was elected president of the Zoophytolithic Society (later called the Linnaean Institute) and in 1812 he became professor of *materia medica* at the university. In addition to his various botanical papers, in 1823 Afzelius published Linnaeus's autobiography.

Agapanthus (Agapanthaceae) L'Hér. (Image on opposite page)
Gk. *agapē* = love; *anthos* = flower. Derivation unknown. *Agapeo* means 'to be contented with'. Perhaps the author, L'Héritier, was expressing his pleasure, i.e. 'flower with which I am well pleased'.

Agastache* (Lamiaceae) Gronov.
Gk. *agan* = much; *stachys* = a spike or 'an ear or grain of wheat'; referring to the numerous flower spikes.

Agapanthus (MC)

Agathosma (CM)

Agasyllis (Apiaceae) Spreng.
Old Gk. *aga-* = good, kind, pure, chaste; *ullus* = diminutive; the Greek name of the Ammoniac plant.

Agathaea (Asteraceae) Cass.
Gk. *agathos* = good; referring to the plant's beauty.

Agathelpis = Microdon (Scrophulariaceae) Choisy
Gk. *agathos* = good; *elpis* = hope; named after the Cape of Good Hope.

Agathisanthemum (Rubiaceae) Klotzsch
Gk. *agathis* = a ball of thread; *anthemum* = flower; possibly referring to the inflorescence sometimes in quite dense subcapitulate clusters.

Agathosma (Rutaceae) Willd.
Gk. *agathos* = good; *osmē* = smell, odour; referring to fragrant oils in the glands of the leaves.

Agave* (Agavaceae) L.
Gk. *agauos* = admirable, noble; referring to the spectacular appearance of the tree in flower.

Agelaea* (Connaraceae) So. ex Planch.
Gk. *agele* = herd, a crowd. These plants tend to cluster together along the margins of forests.

Agelanthus (Loranthaceae) Tiegh.
Gk. *agele* = a herd; *anthos* = flower. The flowers of these hemi-parasitic shrubs grow on trees in clusters.

Ageratina* (Asteraceae) Spach.
Gk. *ageratos, ageraton* = ageless; *ina* = diminutive, hardly ageing; referring to the non-fading flowers.

Ageratum* (Asteraceae) L.
Gk. plant *ageraton* = from *ageraos* = ageless; from *a-* = without; *geras* = old age; referring to the 'ever-lasting' flowers which retain their colour for a long time.

Agialid*** (Zygophyllaceae) Adans.
Gk. *agialid*, derived from the Arabic name for the tree; 'heglig'.

Agialida*** (Zygophyllaceae) Kuntze
See above.

Agiella = Balanites (Zygophyllaceae) Tiegh.
Possibly diminutive of *Agialid*.

Agnirictus*** = Stomatium (Aizoaceae) Schwantes
Gk. *agni, agnos* = pure, chaste; *rictus* = wide, open mouth; referring to the leaves which have a mouth-like appearance.

Agonis* (Myrtaceae) (DC.) Sweet
Gk. *agon* = gathering or clustering; referring to the multitude of seeds.

Agrimonia (Rosaceae) L.
Gk. *argema* = an eye disease, hence *argemos* = a white spot. The plant was believed to cure the 'white spots' caused by eye cataracts.

Agrocharis (Apiaceae) Hochst.
Gk. *agros* = field; *charis* = delight; referring to their preferred habitat.

Agropyron (Poaceae) Gaertn.
Gk. *agros* = field; *pyros* = wheat; referring to wheat-like field grass but not cultivated as a cereal; 'wheat grass'.

Agrostemma* (Caryophyllaceae)
Gk. *agros* = a field; *stemma* = wreath or garland. The flowers used to be woven into harvest festival wreaths.

Agrostis (Poaceae) L.
Gk. *agrostis* = grass, from *agros* = field; La. *agros* = couch-grass. Ancient name for a forage grass.

Ailanthus* (Sapindaceae) Desf.
Moluccan (Indonesian) name *ailanto* = sky tree or 'tree of the gods' or 'tree of heaven'; alluding, perhaps, to its resistance to pollution, freedom from insects and disease, and ability to grow in almost any soil (unfortunately, it is also an invasive).

Aidia* (Rubiaceae) Lour.
Gk. *aidios* = eternal, everlasting; referring to the durability of the wood.

Aidomene = Asclepias (Apocynaceae) Stopp
Gk. *aidomene* = respecting. Inference not clear.

Aira (Poaceae) L.
Gk. *aira* = an ancient Greek name for a weed; probably referring to a different European grass, *Lolium temulentum*, known as poison darnell.

Aistocaulon = Aloinopsis (Aizoaceae) Poelln.
Gk. *aistos* = unseen, hidden; *kaulos* = stalk, stem; referring to the succulent's stem which is hidden.

Aitonia = Nymania (Meliaceae) Thunb.
For William Aiton (1731–1793), English gardener at the botanical garden in Chelsea in 1754 and responsible for the management of the Royal Botanic Gardens at Kew from 1759. He substantially increased the Kew Gardens flower collections and was mainly instrumental in sending Francis Masson to the Cape in 1772; one of the earliest collectors. In 1783 he was given responsibility for the royal forcing and pleasure gardens at Kew and Richmond serving King George III. He published *Hortus Kewensis* (1789), a catalogue of plants at Kew, which describes 5 600 species in three volumes, with 13 plates. His son, William Townsend Aiton (1766–1849), who succeeded his father at Kew, edited the second edition, published in 1810–1813 in five volumes.

Aizoanthemum (Aizoaceae) Dinter ex Friedrich
Gk. *aei* = ever, always; *zoe* = life; *anthemon* = flower; referring to the almost perennial flowering season.

Aizoon (Aizoaceae) L.
Gk. *aei* = ever, always; *zoos, zoon* = alive, a living thing; alluding to the ability of the plant to live under difficult circumstances.

Ajuga (Lamiaceae) L.
Possibly Gk. *a-* = without; *zygo, zygon* (La. *jugum*) = yoke. Linnaeus imaginatively named it *Ajuga*, meaning 'has no ox's yoke', because the buds on the spike are not connected. This has been interpreted, variously, as an allusion to the fact that the calyx is not divided and is in fact a single petal, or that the sepals surrounding the buds are not connected, or that this is a reference to the apparently missing upper corolla lip. (Umberto Quattrocchi has suggested that this name could be a corruption of an old Latin name *Abiga*, applied by Pliny the Elder to another plant.)

Alafia* (Apocynaceae) Thouars
Alafia is a commune in the Timbuktu region of Mali. The word is also an African greeting = peace, tranquillity, wellness, calmness. Possibly related to where the type species was found. Meaning obscure.

Alatoseta (Asteraceae) Compton
La. *alatus* = winged; *seta* = a bristle; referring to the bristly and winged pappus.

Alberta (Rubiaceae) E.Mey.

Flowers of the Magnificant flame-bush (MC)

For Albertus Magnus (Albert of Bollstädt and other names) (c 1200–1280), German scholar, philosopher and author. He became a Dominican in 1223 and taught in various priories before being

sent to the University of Paris (c 1241). He received his doctorate in Theology (1245). From 1260–1262 he was Bishop of Regensburg. He paraphrased all of the known works of Aristotle and pseudo-Aristotle (17 volumes), wrote a botanical work *De vegetabilus et plantis* (seven volumes), and wrote on astronomy, chemistry, embryology, geography, mineralogy, music, physiology, palaeontology and zoology, etc. He was the first person to produce arsenic in a free form, recognised that the Milky Way was composed of stars, and experimented with photosensitive chemicals like silver nitrate.

Albertisia (Menispermaceae) Becc.
For Count Luigi Maria d'Albertis (1841–1901),

Italian ethnologist, naturalist and collector. In 1876–1877 he made three attempts to reach the source of the Fly River in Papua New Guinea. He failed because of shallow waters. On his last trip all but two of his crew deserted when he beat a Chinese crewman to death. He used unscrupulous collecting techniques, such as using dynamite to kill specimens in the water, keeping a python on board his boat to protect his stores, and firing rockets with dynamite at the hostile Papuans. On his return in 1878 he sent papers to the Royal Geographical Society, the Anthropological Institute and the Royal Colonial Institute. In 1880 he published his two-volume work *New Guinea: What I Did and What I Saw*.

Albizia (Fabaceae) Durazz.
For Filippo Degli Albizzi (1724–1789), an Italian nobleman, who introduced the silk tree (*Albizia julibrissin*) from Istanbul (then Constantinople) into Italy in about 1749. This tree then spread into the rest of Europe.

Albuca (Hyacinthaceae) L.
La. *albus* = white or *albicans* = becoming white; referring to the colouring of some *Albuca* flowers.

Alchemilla (Rosaceae) L.
Arabic *al-khimia* = alchemy. Alchemists in the Middle Ages attributed miraculous properties to water exuded by these plants by the process of guttation (the exudation of water from leaves as a result of root pressure).

Alchornea (Euphorbiaceae) Sw.
For Stanesby Alchorne (1727–1800), English

Albuca (MC)

botanist, a collector of British plants. From 1771–1772 he was praefectus horti (= head gardener) at the Chelsea Physic Garden, the second oldest botanical garden in England. The word 'Physic' here relates to the science of healing as this garden was originally established (1673) as an apothecaries' garden, i.e. for medicinal plants. Later he became assay master at the Royal Mint and amassed an important library of early printed books.

Alciope* = Capelio** (Asteraceae) DC.
The name of a nymph in Greek mythology, mother of Celmisius, whose name was given to the related New Zealand genus *Celmisia*.

Aldrovanda (Droseraceae) L.
For Ulysses Aldrovandus (Ulisse Aldrovandi) (1522–1605), Italian nobleman, botanist,

pharmacologist, naturalist and entomologist. He studied medicine and philosophy at the universities of Bologna (1539) and Padua (1547), became professor of natural sciences at the University of Bologna (1561) and director of the botanical garden of Bologna, (1568) one of the first in Europe. He was the author of many publications and was considered by Linnaeus as the 'father of natural history studies'. He developed one of the first museums of natural history, still at the university, containing 25 000

specimens, of which 7 000 were dried plants. He was the first person to describe insect parasitism. He coined the term 'geology' in 1603.

Alectra (Orobanchaceae) Thunb.
Gk. *alektor, alektruon* = a cock; an allusion to the resemblance of the flowers to a cock's comb.

Alepidea (Apiaceae) F.Delaroche
Gk. *a-* = without; *lepis* = a scale. Speculatively, meaning free of sap-sucking insects.

Aletris (Nartheciaceae*) L.
Gk. *aletris, aletridos* (*aletho, aleo* = to grind); allegedly after a female slave, more likely the name for female millers who ground the mealy texture of the perianth. The common names for this genus include colicroot, colic-weed, crow corn, unicorn root.

Aleurites* (Euphorbiaceae) J.R. & G.Forst.
Gk. *aleuron* = floury. The young growth appears to be dusted with flour – the pale pubescence gives a mealy appearance.

Aleuritopteris (Pteridaceae) Fée
Aleurites (q.v.); Gk. *pteris* = fern. The lower surfaces of the fronds seem to have been dusted with powder.

Alhagi (Fabaceae) Gagnep.
Arabic name for a pilgrim or nomad. Possibly meaning the plant was found everywhere, widespread.

Alinula (Cyperaceae) J.Rayna.
For Aline Marie Raynal-Roques (1937–), French professor, botanist, explorer and botanical collector in Europe  (France) and tropical Africa (Cameroon, Mali, Senegal) from 1960–1965. Wife of Jean Raynal, who named the genus after her. She worked extensively in the laboratory of phanerogams in the French National Museum of Natural History. She authored or co-authored a number of books, including *Le Botanique Redécouverte, Collection des Nouvelles Fleurs* (1994), *Un Amour d'Orchidée: le Mariage de la Fleur et de l'Insect* (2005) with Albert Roguenan and Yves Sell, and *Le Genie des Végétaux: des Conquérants Fragile* (2006) with Marcel Bournérias and Christian Block.

Alisma* (Alismataceae) L.
Gk. (*h*)*alisma* = salt-loving; name for a water plant used by Dioscorides (physician, pharmacologist, botanist, c 40–90), and adopted by Linnaeus.

Alistilus (Fabaceae) N.E.Br.
Gk. alistilus = winged style.

Allenia*** = **Radyera** (Malvaceae) E.Phillips
For Robert Allen Dyer (1900–1987), South African botanist. He was an assistant to Selmar  Schönland in the division of botany, curator of the Albany Museum Herbarium, and director of the Botanical Research Institute in Pretoria from 1944–1963. He was largely responsible for initiating the botanical survey section, founded the Pretoria National Botanical Garden, and began the *Flora of Southern Africa*. He edited *Flowering Plants of Africa*, has written extensively on South African flora, especially taxonomy, and has received numerous medals and awards from professional organisations.

Allium (Alliaceae) L.
La. garlic, used by Pliny the Elder. *All* was also Celtic for hot, pungent or burning, referring to the taste or the effect upon the eyes. The genus also includes various edible onions, chives, and leeks.

Allocassine (Celastraceae) N.Robson
Gk. *allos* = different or diverse; probably referring to the different *Cassine* (q.v.) in the same family.

Allophylus (Sapindaceae) L.
Gk. *allos* = different or diverse; *phylon* = tribe; hence *allophylus* meaning 'a stranger'; possibly alluding to belonging to a foreign or different (botanical) tribe.

Allosorus = Cheilanthes (Pteridaceae) Bernh.
Gk. *allos* = different or diverse; *sorus* = sorus or sori. The sori are not all the same – alluding to two different types of sori.

Alloteropsis (Poaceae) Host
Gk. *allotrios* = strange, belonging to another; *opsis* = appearance. The spikelets and inflorescences somewhat resemble that of *Panicum*.

Alnus* (Betulaceae) Mill.
Ancient Latin name for the alder tree. (Koos Roux places this under Casuarinaceae*).

Aloe (Asphodelaceae) L. (Images on opposite page)
Gk. *aloē* (from earlier Semitic word *alloeh*) = bitter. The liquid or dried juice found in the leaves is bitter.

Aloe plant in flower (above) and juicy leaf section (left) (EJM – Both)

Aloinopsis (Aizoaceae) Schwantes
Aloina (q.v.); Gk. *-iopsis* = resembling; because some original species, now moved to *Nananthus*, resemble *Aloe*.

Alonsoa (Scrophulariaceae) Ruíz & Pav.
The authors state that the '*genus nuncupatum D. Zenoni Alonso*' (the genus is named after D. Zenoni Alonso). Possibly this could be Cenón (or Zenón) Alonso Acosta Zorilla y Dávila (1756–?), a Spanish government official, and some sources record it as honouring Alonso Zanoni, a Spanish soldier in Bogotá, Columbia, but these details are unsubstantiated.

Alpinia* (Zingiberaceae) Roxb.
For Prospero Alpino (1553–1617), Italian physician and botanist. He took a doctorate at Padua in 1578, and shortly thereafter went to Egypt for three years (1580–1583) as physician to the Venetian Consul in Cairo. Here, he perceived the sexual differences of plants: 'the female date-trees or palms do not bear fruit […] unless the dust found in the male sheath or male flowers is sprinkled over the female flowers', a preconception of the sexual system in Linnaeus's *Systema Naturae* (1735). Alpino's work *De Medecina Egyptiorum* (1591) and *De Plantis Aegyptii Liber* (1592) introduced the coffee plant, banana and baobab tree to European readers. In 1593 he was appointed professor of botany at Padua, where he also established a botanical garden.

Alphitonia* (Rhamnaceae) Reissek ex End.
Gk. *alphiton* = barley meal; referring to the dry, mealy quality of the mesocarp in the fruits.

Alsine*** = Stellaria (Caryophyllaceae) L.
Latin name for a plant derived from Gk. *alsinē* (*Parietaria lusitanica*), possibly 'chickweed'. It could also be a combination of *Stellaria* and *Arenaria* of other classification (Merriam Webster Dictionary).

Alsophila = Cyathea (Cyatheaceae) R.Br.
Gk. *alsos* = grove, glade forest; *philos* = beloved (*-phila* = -loving); referring to the plant's preferred habitat – the plant loves shade.

Alstonia* (Apocynaceae) R.Br.
For Charles Alston (1683–1760), Scottish botanist. He studied initially at Edinburgh until 1703, when his father, a physician, died. But later under the patronage of the duchess of Hamilton studied at Leiden University for two years (1715–1716) under Herman Boerhaave. On his return to Scotland he became a lecturer and later king's botanist and regius professor of botany for the Royal Gardens at Edinburgh University. He had a special interest in *materia medica* – the therapeutic properties of any substance used for healing (i.e. medicines) – and had a special interest in opium, among other drugs. He was critical of Linnaeus's 'sexual system' and wrote a book *Tirocinium Botanicum Edinburgense* (1753) in which he attacked this botanical theory.

Alstroemeria* (Colchicaceae) L.
For Baron Klas von Alstroemer (Clas Alströmer, Claus von Alstroemer) (1736–1796), Swedish naturalist, chancellor of Gothenburg and close friend and pupil of Linnaeus. From 1760–1764 Alstroemer travelled throughout southern Europe, collecting plants for Linnaeus. Linnaeus named this South American plant after him.

Alternanthera* (Amaranthaceae) Forssk.
La. *alternus* = alternate; *anthera* = an anther; alluding to the alternation of stamens and staminoids of androecium: the stamens in some species being alternately sterile and fertile.

Althaea (Malvaceae) L.
Gk. *althaino* = to cure, *althos* = medicine. The

powdered root of the plant has been used for medicinal purposes.

Althenia (Potamogetonaceae) F. Petit.

For Jean Althen (Hovhannes Althounian) (1709–1774), author of *Mémoire sur la Culture de la Garance*, an Armenian/Persian agronomist who developed the cultivation of madder in France. Although the plant had been present in the region before his arrival, it was Althen who developed its cultivation, turning it into an industry. In 1754 he arrived in Avignon where he started experimenting with the cultivation of madder. After some early teething problems cultivation started in St Chamond, then continued around Avignon in 1763. By 1772 they produced enough madder to supply factories in India. He died in poverty in 1774 as a result of poor business decisions. The industry he founded reached its climax in the 1860s with over 50 mills (*CRC World Dictionary of Plant Names*).

Alvesia (Lamiaceae) Welw.
For Bento Antonio Alves, possibly a horticulturalist at the Portuguese Royal Botanical Garden, which was managed by Friedrich Martin Joseph Welwitsch, director of the garden. Welwitsch did botanical exploration in the Canary Islands and Madeira, possibly with Alves, who became a friend and companion of the Austrian botanist. From 1853–1861 they botanised in Angola, then a Portuguese colony, during which they discovered *Welwitschia mirabilis*. Welwitsch named the genus Alvesia after Alves in 1869.

Alysicarpus (Fabaceae) Desv.
Gk. *alysis* = a chain; *karpos* = fruit. The pod is constricted between each seed (moniliforme). The pods are chain-like.

Alyssum* (Brassicaceae) L.
Gk. *alyssos* = without madness, rage. The plant was believed to cure 'madness' caused by rabies.

Amaranthus (Amaranthaceae) L.
Gk. *a-* = against; *marainein* = to fade, waste away; alluding to the lasting quality of the flower.

Amaryllis (Amaryllidaceae) L.
Gk. *Amaryllis* (La. *amarysso* = to sparkle): the name of a pretty shepherdess in Virgil's pastoral *Eclogues*. This attractive flower is commonly known as 'Belladonna Lily'.

Amaryllis, also known as the 'March-lily' (CM)

Amauropelta = Thelypteris
(Thelypteridaceae) Kunze
Gk. *amauros* = dark; La. *pelta* = small shield; referring to the small, often obscure indusia characteristic of this genus.

Amblygonocarpus (Fabaceae) Harms
Gk. *amblys* = blunt; *gonia* = angle; *karpos* = fruit. In section the fruits are roughly diamond-shaped, with very blunt angles.

Ambraria*** = Nenax** (Rubiaceae) Cruse
A name formerly applied to the genus *Anthospermum* by Heister, probably derived from Cape Ambra, in northern Madagascar. As indicated, *Ambraria* has been sunk into synonymy with *Nenax*.

Ambrosia* (Asteraceae) L.
Gk. *ambrotus* = immortal (Ambrosia was the food of the Olympian gods, conferring immortality). A Greek and Latin name for a plant that gave off a strong odour when the leaves are rubbed.

Ambulia*** = Limnophila** (Plantaginaceae) Lam.
Modified from *Ambul* (Adans.), said to be the Sanskrit name or possibly a modification of the Malabar *Ambuli*. Possibly from La. *ambulo* = to walk; referring to the plants 'mobility' in that it is an aquatic plant and because of this ability to move around freely is a widely spread invasive plant.

Amelanchier (Rosaceae) Medik.
An ancient French name probably derived from *amalenquièr, amelanchièr*, early French names for the European *Amelanchier ovalis*, referring to Canada.

Amellus (Asteraceae) L.
Gk. *amellus* = star. Possibly this is the name given by Virgil to a blue daisy that grew on the banks of the river Mella, perhaps similar to *Aster amellus,* the European michaelmas daisy or Italian aster.

Ammannia (Lythraceae) L.
For Paul Ammann (1634–1691), German physician and botanist, who studied medicine at the University of Leipzig and received his doctorate in physics in (1662). He held three chairs in his career, becoming professor of medicine (1664), of botany (1674) (when he was also director of the medical garden), and of physiology (1682). He authored *Supellex Botanica* in 1675, an enumeration of the medical plants in the garden and others in the vicinity. He also produced *Medecina Critica* (1670), *Paraenesis ad Docentes Occupata Circa Institutionum Medicarum Emendationem* (1673), *Irenicum Numae Pompilii cum Hippocrate* (1689), and *Character Naturalis Plantarum* (1676).

Ammi* (Apiaceae) L.
Ancient Latin name for an umbelliferous plant, possibly from the Gk. *ammos* = sand; referring to its preferred habitat.

Ammocharis (Amaryllidaceae) Herb.
Gk. *ammos* = sand; *charis* = delight. It grows well in sandy soils.

Ammophila* (Poaceae) Host
Gk. *ammos* = sand; *-philos* = loving. It thrives in sandy soils.

Amoebophyllum* = **Mesembryanthemum** (Aizoaceae) N.E.Br.
Gk. *amoib-* = to change; *phyllon* = a leaf; referring to the leaves being alternate (unusual among mesembs).

Amorphophallus (Araceae) Blume ex Decne.
Gk. *amorphos* = deformed, shapeless; *phallos,* hence 'formless phallus'; alluding to the interesting spadix.

Amorphospermum* = **Niemeyera** (Sapotaceae*) F.Muell.
Gk. *amorphos* = deformed; *spermum* = seed; referring to the absence of obvious cotyledons (seed leaves) in the seeds.

Ampelocissus (Vitaceae) Planch.
Gk. *ampelos* = a vine, especially a grape-vine; *kissos* = ivy; a plant with fruit-like grapes that climbs like ivy.

Ampelopteris = **Cyclosorus** (Thelypteridaceae) Kunze
Gk. *ampelos* = a climbing plant; *pteris* = a fern; alluding to the production of proliferus vine-like shoots in the axils of the pinnae.

Ampelopsis (Vitaceae) Michx.
Gk. *ampelos* = a vine. The plant is a climbing shrub.

Amperea (Euphorbiaceae) A.Juss.
For André-Marie Ampère (1775–1836), French physicist, mathematician and polymath. As  a youngster he taught himself mathematics and became a mathematics teacher (1799). Later he was professor of physics and chemistry at the École Centrale in Bourg-en-Bress (1802), tutor then professor of mathematics at the École Polytechnique (1803, 1809) and chair of experimental physics at the Collège de France (1824). He started investigating the relationship between electricity and magnetism in 1820 and, in 1826, wrote *Mémoire sur la Théorie Mathématique des Phénomènes Électrodynamiques Uniquement Déduite de l'Experience,* which showed that electromagnetic phenomena were not only empirically demonstrable but also mathematically predictive. He coined the term 'electrodynamics' and one of his discoveries, 'Ampère's law', was given his name to honour him.

Amphiasma* (Rubiaceae) Bremek.
Gk. *amphi* = both sides, around, on all sides; *asma* = song (a word that Jackson says 'does not seem to make sense'). The Arabic word *asma* = excellent. Therefore, possibly, the meaning is: the genus is excellent all round.

Amphibolia (Aizoaceae) Bolus ex A.G.J.Herre
Gk. *amphiboles* = ambiguous, doubtful, uncertain. It can mean 'attacked from both sides'; possibly relating to the habitat; the type species was on a sea cliff going down to the high tide line.

Amphicosmia = **Alsophila** (Cyatheaceae) Gardner
Gk. *amphi* = both sides; *cosmos* = world. This genus is found in both the Old and New Worlds, it is a native of both South Africa and Brazil.

Amphidoxa = **Gnaphalium** (Asteraceae) DC.
Gk. *amphi-* = around, on both sides; *doxa* = glory. The all-round glory refers to both the plant's seasonal flowering and its foliage.

Amphigena*** = ***Disa*** (Orchidaceae) Rolfe
Gk. *amphigenes* = of doubtful parentage. This may refer to its resemblance to '*Herschelia* in habitat and to *Monadenia* in the single pollinary gland, but distinct from both ... and not agreeing with any other member of the *Disa* group' (Rolfe, 1913).

Amphiglossa (Asteraceae) DC.
Gk. *amphi* = around; *glossa* = tongue; referring to the long tongue-like leaves.

Amphilophium* (Bignoniaceae) Kunth
Gk. *amphi* = around, on both sides; *lophium* = a little crest. The plant is crested on all sides.

Amphisiphon = ***Daubenya*** (Hyacinthaceae) W.F. Barker
Gk. *amphi* = around; *siphon* = tube; referring to the long staminal columns, which have filaments attached to the perianth tube.

Amphithalea (Fabaceae) Eck. & Zeyh.
Gk. *amphi* = around; *thallos* = a green stalk; hence flowering all around the stem.

The thick green stalk hidden by a mass of flowers (CM)

Amphoranthus = ***Phaeoptilum***
(Nyctaginaceae) S.Moore
Gk. *amphora* = a two-handled jar with a narrow neck; *anthos* = flower; referring to the companulate (bell-shaped) perianth. The flower is shaped like an amphora.

Amsinckia* (Boraginaceae) Lehm.
For Wilhelm Amsinck (1752–1831), German businessman, senator and first Bürgermeister (mayor) of Hamburg, patron of botany and benefactor of the botanical garden in Hamburg. He attended the Johanneum and academic high school in Hamburg, studied in Leipzig and Göttingen (1771–1774) and obtained a licentiate qualifying to take a doctorate. In 1786 he became a town councillor (alderman) managing various public offices, and was elected Mayor in 1802. He took office during the French occupation of Hamburg and was particularly active in the negotiations with the French Republic. He made many improvements to Hamburg relating to land reclamation, educational improvement, lighthouse construction, and island requisition.

Amsonia* (Apocynaceae) Walter
For John Amson, (1698–1763) American physician, mayor of Williamsburg, Virginia from 1750–1751, and an associate of the famous botanist John Clayton. The genus is most often said to have been named for an 18th-century Virginia physician and traveller, Charles Amson, but research in 2004 by James Pringle, who traced the history of the first collected species of the genus and the variety of the names given to the plant, clearly establishes that Charles probably never existed. Very little is known about John Amson.

Amygdalus = ***Prunus*** (Rosaceae) L.
The Greek word for almond. Gk. *amysso* = to lacerate; alluding to the fissured shell.

Anacampseros (Anacampserotaceae) L.
Gk. *anakampto* = to cause the return of; *eros* = love. The plant was supposed to be able to restore love.

Anacardium* (Anacardiaceae) L.
Gk. *ana* = upwards; *kardia* = heart. Originally applied by 16th-century apothecaries to the 'heart-shaped' fruit of an Indian fruit (*Semecarpus anacardium*); later used by Linnaeus as a generic name for the cashew (Hugh Glen).

Anaclanthe*** = ***Babiana*** (Iridaceae) N.E.Br.
Gk. *anaklao* = to bend back; *anthe* = flower; probably referring to the flower's petals, which open widely.

Anacyclus* (Asteraceae) L.
Gk. *an* = without; (*anthos* = flower); *kyklos* = ring; referring to the arrangement of the outer flowers on the disk.

Anagallis (Primulaceae) L.

(CM)

Gk. *ana* = again; *agallein* = to delight in or *anagelis, anagalao* = to laugh. Possibly because these flowers open whenever the sun strikes them or alluding to its herbal-medicinal value said to alleviate depression and many other conditions. Common name 'pimpernel'.

Anaglypha* = *Gibbaria (Asteraceae) DC.
La. *anaglyphus* = embossed, Gk. *anaglyphein* = to emboss (*glyphein,* to carve); perhaps referring to some parts of the inflorescence being glandular. Perhaps referring to striate-sulcate leaves (Hollombe).

Anapalina* = *Tritoniopsis (Iridaceae) N.E.Br.
Gk. *anapalin* = reversed order; referring to the bracteoles being longer than the bracts.

Anapausia (Lomariopsidaceae) C.Presl
Gk. *ana* = upward; *pausis* = stop, halt. Allusion unknown. (Stewart humorously suggested that Presl rested after describing the genus.)

Anapeltis* = *Phlebodium (Polypodiaceae) J.Sm.
Gk. *ano* = without; *peltis* = small, round shield; referring to the naked sori in opposition to *Pleopeltis.*

Anaphalis (Asteraceae) DC.
Gk. *ana* = upwards, above, on, upon, up, high; *falos* (*phalos*) = shining, white; referring to its white woolly foliage and flowers. (It was the Greek name for a similar plant.)

Anaphrenium** = *Heeria*** (Anacardiaceae) E.Mey.
Gk. *ana* = like or upward; *phrin* = a heart; referring to the heart-shaped fruits.

Anarrhinum* = *Diclis (Scrophulariaceae) Desf.
Gk. *ana-* = like or resembling; *rhis, rhinos* = nose or snout; referring to the corolla which looks like a calf's snout. (Some sources place it in *Plantaginaceae.*)

Anarthria* (Anarthriaceae*) R.Br.
Gk. *anarthros* = lacking vigour; referring to its appearance.

Anastrabe (Stilbaceae) E.Mey. ex Benth.
Gk. *ana* = upwards, again; *strabos* = squint or distorted; probably referring to the larger stamens curving around the smaller.

Anaxeton (Asteraceae) Gaertn.
Gk. *anaxein* = to polish; alluding to the shiny upper surface of the leaves.

Anchusa (Boraginaceae) L.
Gk. *anchousa,* La. *anchusa* = paint. Some species of this genus, such as Anchusa *tinctoria,* are used for cosmetics (rouge) or as an emollient to soothe and soften the skin.

Ancistrum* = *Acaena (Rosaceae) J.R. & G.Forst.
Gk. *ankistron* = fish-hook; alluding to the hook-like spicules.

Ancylanthos** = *Lagynias*** (Rubiaceae) Desf.
Gk. *ankylos* = bent; *anthos* = a flower. The flower tube is slightly curved, which is unusual in Rubiaceae.

Ancylobotrys (Apocynaceae) Pierre
Gk. *ankylos* = crooked, bent, curved; *botrys* = a grape or bunch of grapes; referring to the clusters of flowers carried on bent flower stalks at the tips of the stem.

Anderbergia (Asteraceae) B.Nord.
For Arne Alfred Anderberg (1954–), Swedish botanist and taxonomist, and professor of botany at the Swedish Museum of Natural History. He obtained a PhD from the University of Stockholm (1985), lectured there from 1990 and has been head of phanerogamic botany at the museum since 2001, of botany since 2013. He has worked extensively with the APG classification systems.

His special interests include *Asteraceae* and *Ericales*. He has authored, inter alia, *Taxonomy and Phylogeny of the Tribe Gnaphalieae* (1991), *Asteraceae: Cladistics and Classification* (1994) with Kåre Bremer, *Phylogenetic Relationships in the Order Ericales* (2002), the entry on *Asteraceae* for *The Families and Genera of Flowering plants* (2007) with Joachim Kadereit and Charles Jeffrey, and *Systematics, Evolution, and Biogeography of the Compositae* (2009) with Vicki Funk, Alfonso Susanna, Tod Stuessy and Randall Bayer.

Andrachne (Phyllanthaceae) L.
Gk. *andrachne* = ancient Greek name for an evergreen shrub. An ancient Greek name for 'purslane' (*Portulaca*).

Andradia** = ***Dialium*** (Fabaceae) Sim
For Alfredo Augusto Freire de Andrade (1859–1929), Portuguese politician and colonial administrator, a lieutenant-general in the Portuguese army (engineering corps), governor-general of Mozambique 1906–1910, and author of *Relatorios Sobre Moçambique*. He made a geological reconnaissance of the Portuguese territories between Lourenço Marques and the Zambezi River. Other positions he held include lecturer at the school of the Army, acting governor of Mozambique in 1892 and 1895, director general of the colonies from 1911–1913, secretary general of the Ministry of Education, chairman of the Board of Public Instruction, president of the Board of Public Instruction and minister for Foreign Affairs.

Androcymbium = ***Colchicum*** (Colchicaceae) Willd.
Gk. *andro-* = male (in botanical language, stamen); *kymbium* = cup or saucer (La. *cymba*). The petal limbs enfold the stamens. Common names 'men-in-a-boat', 'cup-and-saucer'.

Andropogon (Poaceae) L.
Gk. *andros* = a man, male; *pogon* = beard, hence 'bearded male'; alluding to the awns or to the long hairs on the raceme internodes and pedicels of sterile and male-only spikelets.

Androsace (Primulaceae) L.
Gk. *andros* = a man, male; *sakos* = a shield. There seem to be many interpretations of what this name refers to. Some sources refer it to the shape of the anther, others link to a calyx, or the exposed stamens of heterostyled species, or to its round hollow leaf.

Androsiphon = ***Daubenya*** (Asparagaceae) Schltr.
Gk. *andro-* = male; *siphon* = tube. Meaning unclear.

Androstachys (Picrodendraceae) Prain
Gk. *andros* = male; *stachys* = spike. Perhaps alluding to the male and female flowers which are borne on separate inflorescences.

Aneilema (Commelinaceae) R.Br.
Gk. *a-* = without; *eilema* = veil, covering; referring to the absence of spathes or sheaths (from the involucre).

Anemia (Anemiaceae) Sw.
Gk. *aneimon* = unclothed, naked; referring to the uncovered sporangia on the sporophylls.

Anemone (Ranunculaceae) L.
Gk. *anemos* = wind. Named after the nymph Anemone who was turned into a flower by a jealous goddess, and forever buffeted by the north wind. Anemones thrive in exposed windy places, their preferred habitat.

Angelonia (Plantaginaceae) Humb. & Bonpl.
From *Angelon,* the South American name for the plant, a snapdragon.

Anginon (Apiaceae) Raf.
Anginon = one of the names of hemlock given by Dioscorides; referring to the plant's toxicity.

Angiopteris* (Marattiaceae) Kaulf.
Gk. *aggeion* (*angion*) = a vessel; *pteris* = winged, hence 'winged vessel'; referring to the contracted fertile pinnules that enclose the sporangia – with reference to the open (aggregated) sporangia.

Angkalanthus (Acanthaceae) Balf.f.
Gk. *ankalis* = bent arm; *anthos* = flower; possibly referring to the highly reflexed flower petals.

Angylocalyx (Fabaceae) Taub.K.
La. *ankylos* = curved; calyx; referring to the bent calyx.

Angophora (Myrtaceae) Cav.
Gk. *Aggeion* = a vessel; *phorein* = to bear; referring to the urn-shaped fruits.

Angolluma*** = ***Orbea*** (Apocynaceae) R.Munster
La. *Angola* = the country called Angola; *luma* = a vernacular name used by the botanist A Gray as the generic name for new species.

Anigozanthos* (Pontederiaceae) Labill.
Gk. *anoigos-anthos* = open flower; possibly referring to the division of the floral tube into six unequal parts, or possibly 'with tube-like flowers'.

Angraecum (Orchidaceae) Bory
Latinised name of Malay *angurek* for orchids of this type.

Anila (Fabaceae) Kuntze
Derived from the Arabic *an-nI,* the indigo plant, having its linguistic roots in Persian: *al-* + *nI* (from

Sanskrit *nl,* from *nla-* = dark blue). The colour of some species is dark blue or indigo.

Aniseia (Convolvulaceae) Choisy
La. *aniso-* = unequal; referring to the sizes of the two outermost sepals or to the calyx.

Aniserica*** = **Erica** (Ericaceae) N.E.Br.
La. *aniso-* = unequal; the genus *Erica* (q.v.); perhaps referring to the oblique corolla.

Anisochaeta (Asteraceae) DC.
La. *aniso-* = unequal, different; *chaete* = mane, bristle; referring to the varying sized scales of the pappus.

Anisodontea (Malvaceae) C.Presl
La. *aniso-* = unequal; *-odontos* = toothed; alluding to the irregularly dentate leaves.

Anisogonium = **Diplazium** (Athyriaceae) C.Presl
La. a*niso-* = unequal; *gonia* = an angle; alluding to the unequal anastomosing (the reconnection of two streams that previously branched out, such as blood vessels or leaf veins) of the veins.

Anisopappus (Asteraceae) Hook. & Arn.
La. *aniso-* = unequal; *pappos* = *pappus,* a modified calyx. The bristles of the pappus surrounding the achenes are unequal, not all the same length.

Anisophyllea* (Anisophylleaceae) R.Br. ex Sabine.
La. *aniso-* = unequal; *phyllos* = leaf, therefore unequal leaf; referring to the dimorphism of the leaves (the plant has two leaf forms).

Anisoramphus = **Crepis** (Asteraceae) DC.
La. *aniso-* = unequal; *ramphos* = beak; referring to the achenes, the inner ones having longer beaks than the outer.

Anisostigma** = **Tetragonia** (Aizoaceae) Schinz
La. *aniso-* = unequal; stigma; alluding to the two long, papillose styles and two short epapillose styles.

Anisotes (Acanthaceae) Nees
La. *an-* = not; *iso-* = equal; *otis* = ear. The upper and lower corolla lips are unequal.

Anisothrix = **Pentatrichia** (Asteraceae) O.Hoffm.
La. a*niso-* = unequal; *thrix* = hair; alluding to the pappus with 10 short outer bristles and five long inner ones.

Anisotoma (Apocynaceae) Fenzl
La. a*niso-* = unequal; *tome* = cut; hence, *aniso-tome* = 'unequally cut'; possibly referring to the leaves or lobes.

Annesorhiza (Apiaceae) Cham. & Schltdl.
Gk. *anison* = anise, an aromatic seed, dill; *rhiza* = root; referring to the flavour of the tuberous rootstock, much used as a food.

Annona (Annonaceae) L.
La. *annona* = harvestable food, such as wheat, yearly produce; from La. *annus* = year. It was the name of a wheat allotment given to the people of Rome, when necessary, by the government to stave off famine. This allotment (started by Gaius Gracchus in 123 BCE), of 6.5 bushels of corn at a reduced price was given to up to 300 000 Roman householders and was in effect, with some eligibility variations, until the end of the Roman Empire; referring to the edible fruit of this species. Possibly after Annona, the Roman goddess (often depicted on coins), responsible for produce grown in a year.

Anochilus*** = **Pterygodium** (Orchidaceae) Rolfe
Gk. *ano-* = upwards; *-cheilos* = lip; referring to 'the uppermost position of the lip in flower which is non-resupinate' (Rolfe, 1913).

Anoda* (Malvaceae) Cav.
Gk. *anodia* = without joint; from La. *nodus* = joint, node (Cavendish); Gk. *an-* = without; *odon* = tooth (Jackson). 'The flowering stem lacks nodes, unlike other related genera' (Cavendish). Jackson states the second option, saying this is more consistent as 'the leaves do (not bear) teeth as do other genera'.

Anonidium (Annonaceae) Eng. & Diels.
Gk. *anona* or *Annona* (q.v.); *idium* = a little; referring to yearly produce.

Anogramma (Pteridaceae) Link
Gk. *ano-* = upwards; *gramma* = line; referring to the elongated (or line-shaped) sori that mature first near the tips or *an* = without; *gramma* = line without indusia (W.B.G. Jacobsen).

Anogeissus (Combretaceae) (DC.) Wall.
Gk. *ano* = above, towards-the-top; *geissos* = covering; referring to the scale-like fruiting heads; the fruits are covered with scales on top.

Anoiganthus*** = **Cyrtanthus** (Amaryllidaceae) Baker
Gk. *anoigo* = to open; *anthos* = flower; referring to the perianth.

Anomalanthus (Ericaceae) Klotzsch
Gk. *anomalos-* = unusual, abnormal, unequal, strange; *anthos* = flower; referring to its unusual morphology.

Anomalesia*** = **Gladiolus** (Iridaceae) N.E.Br.
Gk. *anomalos* = unusual, irregular, unequal, strange. The perianth lobes are very unequal.

Anomatheca*** = **Freesia** (Iridaceae) Ker Gawl.
Gk. *anomos* = irregular, abnormal; *theca* = a case, capsule; alluding to the papillose wall of the capsule in the type species.

Anopteris = **Pteris** (Pyeridaceae) Prantl. ex Diels.
Gk. *ana* = upwards; *pteris* = fern; alluding to the erect rhizome.

Anosporum*** = **Cyperus** (Cyperaceae) Nees
Gk. *ano-* = irregular; *sporum* = spore; possibly referring to the irregular shapes of the seeds.

Anredera* (Basellaceae) Juss.
Not explained by author. Possibly a misspelling of Spanish *enredadera* = creeping or climbing plant. The Madeira vine, a South American evergreen climber.

Ansellia (Orchidaceae) Lindl.

The region's largest epiphytic orchid (HS)
(MC)

For John Ansell (?–1847), British plant collector, gardener and assistant botanist. He joined a three-ship expedition to the Niger in 1841. He travelled on the *Wilberforce* with botanist Dr Theodor Vogel and naturalist Lewis Frazer. The expedition should have reached the Niger in March, the dry season, but due to delays, only reached the Niger in August. Deaths from malaria and other illnesses increased rapidly once they were on the river. As a result, of the 145 Europeans who took part in the expedition, 53 ultimately perished mostly of fever. The ships lost too many crew to function properly. With difficulty they reached the island of Fernando Po where Ansell found his plant. He returned to England, very ill with fever, and died 5 years later.

Antegibbaeum (Aizoaceae) Schwantes ex C.Weber
La. *ante-* = before; *gibba* = hump or swelling; referring to the shape of the leaf. Jackson writes 'the stage before gibbaeum', i.e. possibly becoming hunch-back like.

Anthemis* (Asteraceae) L.
Gk. *anthemon* = flower; referring to its profuse flowering.

Anthephora (Poaceae) Schreb.
Gk. *anthos* = flower; *-phero* = to bear, carrying; referring to the involucre which may look like a calyx.

Anthericum (Agavaceae) L.
Gk. *anthera* = an anther; *kerkos* = hedge; Linnaean name from the Gk. *antherikos*, asphodel (St Bernard's lily); apparently alluding to the tallness of the flower stems.

Antherothamnus (Scrophulariaceae) N.E.Br.
Gk. *anthera* = an anther; *thamnos* = a bush; perhaps referring to the unequal and irregular stamens. 'The author does not explain this' (Jackson).

Antherotoma (Melastomataceae) (Naudin) Hook.f.
Gk. *anthera* = an anther; *tomos* = slice. The anthers open by a terminal pore.

Anthochortus (Restionaceae) Nees
Gk. *anthos* = flower; *khortos* = pasturage, fodder, grass; possibly referring to its habitat or usage.

Anthocleista (Gentianaceae) Afze ex R.Br.
Gk. *anthos* = flower; *kleistos* = closed; refers to the shape of the corolla in some species.

Antholyza*** = **Gladiolus** (Iridaceae) L.
Gk. *anthos* = flower; *lyssa* = rage, the 'rage flower'. A poetic allusion to the open-mouthed

flowers that are meant to look like an angry animal.

Anthospermum* (Rubiaceae) L.
Gk. *anthos* = flower; *spermum* = seed. Although the flowers are usually dioecious – unisexual, male and female – 'male' flowers sometimes have ovaries capable of ripening seed.

Anthoxanthum (Poaceae) L.
Gk. *anthos* = flower; *xanthos* = yellow; referring to the colour of the panicles (spikelets), which are yellow-green.

Anthriscus* (Apiaceae) Pers.
Gk. *anthriskon* = a name for chervil; from *anthos* = flower; from Ita. *risco* = possibly danger = a hot spice. *Anthriscus cerefolium* is an Old World herb used by the ancient Greeks to spice up dishes. Not to be confused with poisonous lookalikes.

Anthyllis (Fabaceae) L.
Gk. *anthos* = flower; *ioulos* = down, hence 'downy flower'. An ancient name used by Dioscorides and Pliny the Elder. The downy flower provides an allusion of first beard growth, hence the common name, 'Jupiter's beard'.

Anticharis (Scrophulariaceae) Endl.
La. *anti-* = without, lacking; *charis* = grace or charm, hence 'lacking charm'; referring to its appearance, especially when non-flowering.

Antidesma (Phyllanthaceae) L.
Gk. *anti* = against; *desma* = Johannes Burman's term for poison; an oblique reference to one species supposedly used as an anti-venom against snake bite. (Burman was a friend of Linnaeus.)

Antigonon* (Polygonaceae) Endl.
Gk. *anti* = opposite; *gonon* = an angle; referring to the jointed flower stalks and the sharp angles of the zigzag stems.

Antimima (Aizoaceae) N.E.Br. emend. Dehn.
Gk. *antimimos* = imitating. The first species looked like *Argyroderma*.

Antiphiona (Asteraceae) Merxm.
La. *anti-* = resembling; *Iphiona*, previous name of the genus, origin unknown.

Antirrhinum (Plantaginaceae) L.
Gk. *anti-* = resembling or similar to; *rhis, rhinos* = nose, snout; hence 'nose-like'; referring to the resemblance of the flower's corolla to a calf's snout.

Antithrixia (Asteraceae) DC.
La. *anti-* = resembling; the genus *Athrixia* (q.v.); particularly the pappus, as commented by William Henry Harvey.

Antizoma (Menispermaceae) Miers
La. *anti-* = resembling; *zoma* = a covering; referring to the woolly calyx.

Antonia (Loganiaceae) Pohl
For Anton Victor (1779–1835), archduke of Austria and grand master of the Teutonic Knights, and  viceroy of Lombardy-Venetia 1816–1818. Although elected as prince-bishop of Münster and archbishop and prince-elector of Cologne, due to the French occupation of those cities he never assumed his powers, and those territories were secularised in the reorganisation of the Holy Roman Empire, thus technically his predecessors were the last holders of those offices. He is described as a 'high botany lover and promoter'. His reign was a short one. The Bishopric of Münster was occupied by Prussia in 1802 and the electorate of Cologne was partly occupied by the Hessians in the same year leaving him with little power. He never married and left no heir.

Antrophyum* (Pteridaceae) Kaulf.
Gk. *antron* = cave, cavern, hole; *phyo* = to grow; referring to places for this fern to vegetate.

Antunesia = Distephanus (Asteraceae) O.Hoffm.
For Father José Maria Antunes (1856–1928), plant collector in Angola in 1895, who collected with Eugène Dekindt (1865–1905), German plant collector in Angola from 1899–1902. He collected in South Africa and Angola from 1889–1903.

Apatesia (Aizoaceae) N.E.Br.
Gk. *apatē, apatesis* = deceit, deception; alluding to the likeness of this genus to *Hymenogyne*, before the flower has opened (Hans Herre).

Aphania*** (Sapindaceae) Blume
Gk. *a* = not; *phanes* = visible, seen, noticed; hence 'inconspicuous'; referring to the tree's flowers which are 'obscure, inconspicuous' and hardly open.

Aphelexis = Edmondia (Asteraceae) D.Don.
Gk. *apheles* = simple; *exis* = habit; referring to the habit of the species (*Paxton's Botanical Dictionary*).

Aphloia (Aphloiaceae) (DC.) Benn.
Gk. *a-* = without; *phloios* = bark; without, or possibly with only the thinnest of, bark.

Aphyllocalpa = Osmunda (Osmundaceae) Cav.

(NC)

Gk. *a-* = without; *phyllon* = leaf; *calpa* = urn. The urn-shaped sporangia are not borne on leafy frond tissue as the lamina is reduced or lacking in the fertile portions of the leaves.

Aphyteia*** (Hydnoraceae) L.
Gk. *a-* = without; *phyton* = a plant; referring to a plant which has neither root, stem, nor leaves.

Apicra*** = Haworthia (Asphodelaceae) Willd.
Gk. *a-* = without; *picros* = sharp, bitter in taste. The sap of this plant is not bitter, like most aloes.

Apium (Apiaceae) L.
La. parsley, celery; possibly derived from Celtic *apon* = water or Sanskrit *ap-* = relating to water. The plants grow in wet ground, marshes and salt marshes.

Aplanodes (Brassicaceae) Marais
Gk. *a-* = without, negative; *planodes* = wandering about. The author, Marais, says, 'Not an introduced plant as I first thought it might be'.

Apochaete*** = Tristachya (Poaceae) (C.E. Hubb.) J.B.Phipps.
Gk. *apo-* = away from; *chaete* = flowing hair, mane, bristle; referring to the lemma lobes which bear minute aristulate bristles.

Apocynum = Eustegia (Apocynaceae) L.
Gk. *apokynon* = against dogs (*apo-* = against; *kynon* = dogs). *Apocynum venetum* is toxic and in ancient days was used as a poison to kill dogs, hence its common name, dogbane.

Apodolirion (Amaryllidaceae) Baker
Gk. *-apod* = without a foot; *lirion* = a white lily; referring to the stemless, *Gethyllis*-like 'lily'.

Apodynomene*** = Tephrosia (Fabaceae) E.Mey.
Gk. *apodynomo* = to lose strength, weaken. Meaning unclear.

Apodytes (Icacinaceae) E.Mey. ex Arn.
Gk. *apoduein* = to strip off; referring to the tiny sepals which cover nothing leaving the corolla uncovered.

Aponogeton (Aponogetonaceae) L.f.

The flowers and fruits of the 'Water blommetjie' are a key ingredient in bredie. (CM)

Celtic *apon* = water, Gk. *geiton* = near, neighbour; referring to an aquatic plant that was found in the neighbourhood of healing Aquae Aponi hot springs near Padua, Italy.

Apoxyanthera = Raphionacme (Apocynaceae) Hochst.
Gk. a*poxys* = tapering; *anthera* = anther. The

name derived from the extremely sharp anthers: 'nomen derivat um ab antheris acuminatissimis'.

Aptenia = Mesembryanthemum (Aizoaceae) N.E.Br.
Gk. *apten* = wingless (*ptenos* = winged); referring to the wingless seed (capsule).

Aptosimum (Scrophulariaceae) Burch. ex Benth.
Gk. *Aptosimum* = not falling (*a-* = not; *ptosis* = fall); referring to the capsules which remain attached to the stems long after the seeds have been shed (William John Burchell, 1822).

Aquilegia* (Ranunculaceae) L.
La. *aquila* = eagle or *aqua* = water; *legere* = to collect or draw; referring to the shape of the flower petals, which can look somewhat like an eagle's claw in some species or referring to the nectar at the base of the spurs.

Arabidopsis* (Brassicaceae) Heynh.
Gk. *Arabis* (q.v.); *-opsis* = resembling (*Arabis*).

Arabis* (Brassicaceae) L.
Gk. *Arabis* = genus from Arabia; possibly referring to its ability to grow on rocky or sandy soil.

Arachis* (Fabaceae) L.
Gk. *a-* = without; *rachis* = a branch or rachis. Pliny the Elder's name for a vetch-like plant; in modern times refers to plants whose fruits are usually underground.

Arachniodes (Dryopteridaceae) Blume
Gk. *arachne, arachnion* = the spider or its web; *-odes* = resembling; alluding to the finely divided pentagonal outline of the lamina of some species (*Flora of Zimbabwe*) or to the indumentum of the fern. (It has also been suggested that Blume (the author) saw fungal hyphae or spider webs on his original material.)

Arachnocalyx = Erica (Ericaceae) Compton
Gk. *arachnise* = spider; calyx.

Araujia* (Apocynaceae) Brot.
For Antonio de Araujo de Azevedo, count of Barca (1754–1817), Portuguese diplomat and politician, and patron of botany. He served under John VI (1767–1826) in Portugal and Brazil (1807) as minister plenipotentiary at the Hague in 1787 and St Petersburg, minister of Foreign Affairs and War (1804), of the Interior (1806 and 1814), of

the Navy (1815), and president of the Royal Board of Trade and as chairman of the Royal Treasury. He taught the Brazilians how to manufacture porcelain, had his own splendid botanical garden, introduced a Chinese colony to cultivate tea in Brazil (1812) and tried to establish a royal school of sciences, arts and crafts. On his death, his archive of 7 000 documents on different topics was integrated into the Royal Library. *Araulia* is the Brazilian name for this plant.

Arbutus* (Ericaceae) L.
The Latin name for the so-called 'strawberry tree' in Europe; alluding to the tree's red berries which somewhat resemble a strawberry.

Arctium* (Asteraceae) L.
Gk. *arktion* = bear-like, from *arktos*, or Celtic *arth* = bear; referring perhaps to the rough prickly heads of these plants called 'burrs' noted for easily catching on to fur and clothing.

Arctopus (Apiaceae) L.

Male plant left and female right (CM)

Gk. *arktos* = a bear; *pous* = a foot. Linnaeus thought the leaf of this species resembled a bear's paw.

Arctotheca (Asteraceae) J.C.Wendl.
Gk. *arktos* = a bear; *theke* = a case, capsule; referring to rough densely woolly fruit.

Arctotis (Asteraceae) L.
Gk. *arkto-* = brown bear; *-otis* = an ear. The bear-like ears have been linked, variously, to the earlike pappus scales, outer involucral bracts or the shaggy fruit.

Arenaria* (Caryophyllaceae) L.
La *arena* = sand; *-aria* = belonging to, connected with. The species is found in sandy areas.

Arenifera (Aizoaceae) A.G.J.Herre
La. *arena* = sand; *-fera* = carrying. The sand adheres to the rough leaves.

Arethusa (Orchidaceae) L.
For a mythological wood (fountain) nymph named Arethusa; possibly because of the wet habitat usually preferred by most species.

Argemone* (Papaveraceae) L.
Gk. *argemos* = a white spot. The plant was believed to cure the 'white spots' caused by eye cataracts.

Argeta = Gibbaeum (Aizoaceae) N.E.Br.
Gk. *argeta* = chalky; alluding to the whitish colour of the plant.

Argyreia (Convolvulaceae) Lour.
Gk. *argyro* = silver; alluding to the whitish colour of the plant.

Argyroderma (Aizoaceae) N.E.Br.
Gk. *argyros* = silver; *derma* = skin; referring to the fat leaves.

Argyrolobium (Fabaceae) Eck. & Zeyh.
Gk. *argyros* = silver; *lobion* = diminutive of *lobos* = lobe, pod, capsule. The fruits are covered in silvery hairs.

Aridaria = Mesembryanthemum (Aizoaceae) N.E.Br.
La. *aridus* = dry; -*aria* = connected with; referring to the arid habitat of the Mesembryanthemum.

Aristea (Iridaceae) Aiton
La. *arista* = a point or beard. The leaves are sharp-pointed.

(CM)

Aristida (Poaceae) L.
Gk. *aristos* = bristle, or awn, from an 'ear' of corn; La. *arista* = awn; *aristida* = beard of grain; referring to the bristle-like appendage at the tip of the grass floret's awned lemma.

Aristolochia* (Aristolochiaceae) L.
Gk. *aristos* = best; *lochia, locheia* = childbirth. The flowers of some species are supposed to resemble the human foetus in the best position for easy delivery. Classical authors thought that it eased delivery, hence the common name 'birthwort' applied to some species.

Arnica (Asteraceae) L.
Probably Gk. *arnakis* = sheep's skin; referring to the soft, woolly leaves.

Arrhenatherum (Poaceae) P.Beauv.
Gk. *arrhen* = male; *ather* = a bristle. The upper floret in each spikelet is male and awned.

Arrowsmithia (Asteraceae) DC.
For John Arrowsmith (1790–1873), English cartographer and map publisher, the nephew of  Aaron Arrowsmith (1750–1823), an acclaimed map publisher. He worked for his uncle from 1810–1823. After his uncle's death he commenced his own business and, for 38 years, until he retired in 1861, he produced many meticulously detailed quality maps of foreign countries. For his acclaimed *London Atlas of Universal Geography*, he states that he examined more than 10 000 sheets of private maps, charts and plans. In 1839 he bought Aaron's existing business from his uncle's sons. He was a founding and council member of the Royal Geographical Society and received the Society's Patron's Gold Medal (1863) for services rendered to geographical science.

Artabotrys (Annonaceae) R.Br.
Gk. *artao* = to support; *botrys* = a bunch of grapes; alluding to tendrils which supports the fruit as it ripens.

Artemisia (Asteraceae) L.
Middle English *artemesie* = mugwort, from Old French, from La. *artemisia*, from Gk. *artemisiā* = wormwood, after Artemis, the Greek goddess of hunting and chastity. Linnaeus's reason for naming this plant as such is unknown.

Artemisiopsis (Asteraceae) S.Moore
Artemisia (q.v.); -*iopsis* = resembling this genus.

Arthraerua (Amaranthaceae) (Kuntze) Schinz
Gk. *arthron* = a joint; -*urus* = tailed; alluding to the jointed stems.

Arthratherum* (Poaceae) P.Beauv.
Gk. *arthron* = a joint; *ather* = beard of grain, barb;

referring to the jointed connection of the awn with the apex of the lemma.

Arthraxon (Poaceae) P.Beauv.
Gk. *arthron* = joint; *axon* = axis. At maturity the inflorescence axis breaks into segments.

Arthrocnemum (Amaranthaceae) Moq.
Gk. *arthron* = jointed; *kneme* = shin, limb, thus 'jointed limb'. The plants are shortly branched (Jackson), possibly referring to the woody bases branching into fleshy, jointed green stems.

Arthropodium (Agavaceae) R.Br.
Gk. *arthron* = a joint; *pous, pod* = foot; referring to the jointed flower stalks.

Arthropteris (Oleandraceae) J.Sm.
Gk. *arthron* = a joint; *pteris* = fern; referring to the joint, or articulation, at the base of the pinnae.

Arthrosolen* = Gnidia** (Thymelaeaceae) C.A.Mey.
Gk. *arthros* = a joint; *solen* = a tube. Possibly refers to the fact that the filament of some species can be seen but is jointed to the tube.

Aruncus* (Rosaceae) L.
Gk. *aryngos* = goat's beard; referring to the straggling white plumes in *A. dioicus*.

Arundinaria (Poaceae) Michx.
A plant that looks like a reed or a genus closely related to *Arundo* (q.v.).

Arundinella (Poaceae) Raddi
Arundo (q.v.); Gk. *-iella* = diminutive. Like *Arundo*, but a smaller reed.

Arundo* (Poaceae) L.
La. = reed or cane; Celtic *aru* = water; referring to this reed's preferred habitat (wetlands).

Asaemia* = Athanasia** (Asteraceae) (Harv.) Harv. ex Benth. & Hook.
Gk. *a* = without; *semos* = mark, sign; relating to a comparison, seemingly, with the genus Stilpnophyton.

Asarina (Plantaginaceae) Mill.
Spanish *antirrhinium* = Snapdragon.

Asarum* (Aristolochiaceae*) L.
Gk. *asaron, a* = without; *saron* = sweeping. The Greek name for this genus used by Dioscorides. Wild ginger.

Asclepias (Apocynaceae) L.
Named after Asklepios, the ancient Greek god of medicine (Latin: Aesculapius); possibly because of the many folk medicinal uses for the milkweed plants.

Ascolepis (Cyperaceae) Steud.
Gk. *askos* = bladder, belly, wineskin; *lepis* = scale. The hypogynous scale enclosing the achene in some species (David Gledhill).

Askidiosperma (Restionaceae) Steud.
Gk. *askidion* = a little bag, from *askos* = bag; *sperma* = seed; possibly referring to the seeds being bagged for conservation reasons.

Aspalathus (Fabaceae) L.
Gk. *aspalathos* = a scented bush that grew in Greece. Named after a plant known to classical Greek writers as *Aspalathos*.

Aspalthium* = Bituminaria** (Fabaceae) Medik.
Gk. *aspalathos* = a scented bush.

Asparagopsis* = Asparagus** (Asparagaceae) Kunth
Asparagus (q.v.); Gk. *-iopsis* = resembling, i.e. resembling this plant.

Asparagus (Asparagaceae) L.

'Spears' can appear days after a fire and young plants within a week. Mature plants flower weeks after fire. (EJM)

(EJM)

Gk. *asparagos*, derived from the Persian *asparag* = sprout or shoot, or *spargan* = to swell. Georg Christian Wittstein suggested some earlier possible derivations including *a-* = intensive;

sparassa = to tear; referring to the many South African species that have sharp spines, some curved like hooks.

Aspasia (Orchidaceae) Lindl.
Gk. *aspasio* = glad, delightful. Named in honour of Aspasia, the beautiful mistress of the Athenian statesman Pericles; alluding to its pretty flowers.

Aspazoma = *Mesembryanthemum* (Aizoaceae) N.E.Br.
Gk. *aspazomai* = to clasp; referring to the leaf sheath clasping the stem (Hans Herre).

Asphodelus* (Asphodelaceae) L.
Gk. *asphodelos* = a sceptre: a decorated stick that is carried by a queen or king, hence the common name 'king's spear', of unknown origin. A mythological reference to asphodelus, a flower that connected with the dead, perhaps named for the greyish colour of its leaves and yellowish flowers indicating the pallor of death (Wiki).

Aspidium = *Tectaria* (Tectariaceae) Sw.
Gk. *aspis* = shield; *aspidium* = small shield; alluding to the shape of the fern's indusium.

Aspidixia*** = *Viscum*** (Santalaceae) Tiegh.
Gk. *aspidos* = small shield; *ixos* = mistletoe, birdlime. The 'birdlime' refers to the viscous sap.

Aspidoglossum (Apocynaceae) E.Mey.
Gk. *aspidos* = small shield; *glossa* = tongue; referring to the deeply lobed corolla resembling a forked tongue.

Aspidonepsis (Apocynaceae) Nicholas & Goyder
Gk. *aspidos* = shield; *anepsia* = cousin; referring to the close relationship of this genus with Aspidoglossum.

Aspidosperma (Apocynaceae) Mart. & Zucc
Gk. *aspidos* = small shield; *spermum* = seed. The genus name refers to the shielded seed.

Aspilia (Asteraceae) Thouars
Gk. *a-* = without; *spilos* = a spot or blemish. Hence, without a blemish, spotless.

Asplenium (Aspleniaceae) L.
Gk. *a-* = without; *splenon*, from *splen* = spleen. Commonly known as 'spleenwort'. Herbalists, going back to the time of Pliny the Elder and Dioscorides believed this fern could cure spleen and liver problems.

Assonia*** = *Dombeya*** (Malvaceae) Cav.
For Claudio Ignacio Jordan de Asso y del Rio (Ignatius de Asso) (1742–1814), Spanish scholar,

recognised expert in law, languages, economics, history and science (geology, botany, zoology). He obtained his doctorate in 1764, worked as a lawyer (1771–1775) publishing works on civil rights, became a diplomatic counsel in Amsterdam in 1776, and conducted many botanical explorations (1778–1784), compiling a complete inventory of the geology, flora and fauna of the Kingdom of Aragon in northeastern Spain. He translated several books from Arabic (1782), wrote books on economics, agriculture, botany, fish and natural sciences, and held the chairs of chemistry and botany at Real Sociedad Económica Aragonesa (1797–1802).

Astephanus (Apocynaceae) R.Br.
Gk. *a-* = without; *stephanos* = a crown; alluding to the absence of a corona (crown).

Aster (Asteraceae) L.
Gk. *astron* = a star; referring to the radiate heads of the flowers.

Asteracantha = *Hygrophila* (Acanthaceae) Nees
Gk. *astron* = a star; *Acanthus* (q.v.). The inference is not clear – the leaves splay out somewhat star-like in some species.

Asterochaete*** = *Carpha*** (Cyperaceae) Nees
Gk. *astron* = a star; *chaete* = flowing hair, mane, bristle; referring to the fruits and sexual organs (Umberto Quattrocchi).

Asteropterus*** = *Leysera*** (Asteraceae) L.
Gk. *astron* = a star; *pteron, a* wing.

Asthenatherum*** = *Centropodia*** (Poaceae) Nevski.
Gk. *astheneo* = to become weak; *ather* = barb or spine; referring to the awns, which are relatively weak compared to *Danthonia* from which genus it was separated.

Astragalus (Fabaceae) L.
Gk. *astragalos*, probably from *astron* = a star; *gala* = milk or Gk. *astragalos* = ankle bone. The first interpretation is that the plant known as 'milk vetch' is a plant in pasturage which was supposed to increase milk yield. The second interpretation is that its seed shape is similar to the ankle bone.

Astridia (Aizoaceae) Dinter
For Astrid Elise Wilberg (1887–1960), wife of the German botanist Martin Heinrich Gustav

Schwantes (1891–1960), archaeologist and professor of pre-history at Kiel University, and mesemb specialist, and author of *Flowering Stones and Mid-day Flowers* (1957).

Astripomoea (Convolvulaceae) A.Meeuse
Probably Gk. *astron* = a star, and the genus *Ipomoea* (q.v.); a stellately pubescent plant.

Astrochlaena * = *Astripomoea*
(Convolvulaceae) Hallier.f.
Gk. *astron* = a star; *chlaina* = a cloak, blanket. The plant is stellately pubescent.

Astroloba (Asphodelaceae) Uitewaa.
Gk. *astros* = star; *lobos* = lobe; refers to the star-like shape of the petals.

Asystasia (Acanthaceae) Blume
Gk. *a-* = without; *systasia* = consistency, hence 'inconsistency'. Unlike most corollas in the family Acanthaceae, the corollas are almost regular, which is unusual.

Atalaya (Sapindaceae) Blume
From the Timorese name *atalai* for a species of this genus.

Athamanta (Apiaceae) L.
Gk. = of or from Mount Athamas in Sicily, where some of these plants are found.

Athanasia (Asteraceae) L.
Gk. *a-* = without; *thanatos* = death, hence 'immortal'; referring to the persistent dry involucral bracts.

Athrixia (Asteraceae) Ker Gawl.
Gk. *a-* = intensive; *thrixia* = hair. The lower surface of the leaves of some species such as *Athrixia phylicoides* is densely felted by a mass of fine woolly hair. Alternately, in the *Encyclopedia of Plants* (1828), co-author John Claudius Loudon alludes to the 'absence of hairs upon the receptacle and the stigmas of the ray'. So there may be two interpretations.

Athyriopsis = *Deparia** (Athyriaceae) Ching
Gk. *a-* = without; *thurium* = shield; *opsis* = resembling *Athyrium* (q.v.).

Athyrium (Athyriaceae) Roth
Gk. *athryos* = door-less. The sporangia just push back the outer edge of the indusium (Stewart, Johnson and Michel); referring to the enclosed sori (Davesgarden.com); the indusium is hinged on one side (*Hardy Fern Home*).

Atractylis (Asteraceae) L.
Gk. *atractos* = spindle; *atraktylis* = a thistle-like plant used for making spindles; refers to the long spines of the outer bracts.

Atraphaxis = ***Polygonum*** (Polygonaceae) L.
Gk. *a-* = without; *trepho* = to nourish; name for the goat wheat (La. *atriplex*). The ancient name for the herb Orache.

Atrichantha (Asteraceae) Hilliard and B.Burtt
Gk. *a-* = without; *thrichos* = hair; *anthos* = plant; alluding to the hairless corolla lobes.

Atriplex (Amaranthaceae) L.
La. *atri* = black, dark; *plexus* = interwoven, plaited; possibly referring to the black seeds.

Atropa (Solanaceae) L.
In Greek mythology, Atropos was one of the three fates, female deities who supervised fate rather than determined it. She dealt with death by cutting the cord of life. Her sisters were Clotho who spun the cord and Lachesis who determined its length. The species *Atropa belladonna*, commonly known as 'deadly nightshade', has foliage and berries that are extremely toxic. The pharmacologically active ingredients of *Atropa* include atropine and scopolamine.

Audouinia (Bruniaceae) Brongn.

One of the genera in a Cape endemic family (CM)

For Jean Victoire Audouin (1797–1841), French naturalist, entomologist and ornithologist. He was a student of medicine and an assistant to Pierre André Latreille (1762–1833), professor of entomology at the Muséum National d'Histoire Naturelle, and his successor. His principal work *Histoire des insectes nuisibles à la vigne* (1842) was completed after his death by Henri Milne-Edwards and Émile Blanchard. He also published *Dictionaire classique d'histoire naturelle* (1822) with Bory de Saint-Vincent, and other publications. He was one of the founders of the Société Entomologique de France and twice its chairman (1834, 1837). He was also a member of the French and Swedish Academy of Sciences.

Augea (Zygophyllaceae) Thunb.
For Johann Andreas Auge (1711–1805), German gardener, naturalist, and botanical collector, who worked at the botanical garden at Leiden under the direction of Herman Boerhaave. He arrived at the Cape in 1747 and went to work as a gardener for the governor. Subsequently the next governor, Ryk Tulbagh, promoted him to superintendent of the garden. He was a participant in a land expedition from the Cape Colony to Namibia from July 1761 to April 1762. He acted as a guide to the Swedish botanist Carl Peter Thunberg in 1772. From about 1781 on he was completely blind. Sometime in the late 1790s a band of Xhosa warriors attacked his farm and he lost all his possessions including his botanical books and dried plant specimens.

Aulacorhynchus = Tetraria (Cyperaceae) Nees
Gk. *aulax, aulaco* = a furrow or groove; *rhynchus* = beak, nose; referring to the grooved and paired beak-like fruit.

Aulax (Proteaceae) P.J.Bergius
Gk. *aulax* = a furrow. The leaves of *A. cancellata* are inconspicuously channelled, while some of the floral parts are microscopically grooved.

Aulaya*** = Harveya (Orobanchaceae) Harv.
Gk. *aulaia*, La. *aulaeum* = a curtain (especially in the theatre); referring to the lirellae-like ascomata which appear as furrows on the thallus.

Aulojusticia = Justicia (Acanthaceae) Lindau
Gk. *aulos* = tube, alternately *aule* = abode, and the genus *Justicia* (q.v.). The 'long-tubed *Justicia*' shares its habitat or lives in close proximity with plants of the *Justicia* genus.

Aulostephanus** = Brachystelma (Apocynaceae) Schltr.
Gk. *aulos* = tube; *stephanus* = garland, wreath. Possibly referring to the shape of the inflorescence.

Australina* (Urticaceae) Gaudich.
La. *australis* = southern. Species occur in the Southern Hemisphere, e.g. Africa, Australia and New Zealand.

Australluma*** = Caralluma (Apocynaceae) Plowes
La. *australis* = southern; *luma* used by the botanist A Gray as the generic name for a new species.

Austrocylindropuntia = Opuntia (Cactaceae) Backeb.
La. *australis* = southern; *cylindrus* = cylinder; *opuntia* = the genus from which this segregate was removed (initially referring to the Greek city, Opus). Austrocylindropuntia is the South American version of *Cylindropuntia* and differs from it in that the spines lack papery sheaths (CactiGuide.com).

Austroderia (Restionaceae) N.P.Barker and H.P.Linder
La. *austro-* = southern; possibly because it is found in the Southern Hemisphere.

Austromimusops*** = Vitellariopsis (Sapotaceae) A.Meeuse
La. *austro-* = southern; and the genus *Mimusops* (q.v.).

Avena* (Poaceae) L.
Gk. a*vena* = an ancient name for oats, which has been traced back to Virgil (90–19 B.C) and further.

Avenastrum*** = Helictotrichon (Poaceae) Jess.
Avena (q.v.); *astrum* = incomplete resemblance. The plant is somewhat similar to *Avena*.

Avicennia (Acanthaceae) L.

One of our mangroves with pencil roots (EJM)

For Ibn Sina Avicenna (Abd Allāh ibn Sīnā) (980–1037), Arabian philosopher, physician and polymath. He had an extraordinary intelligence and is said to have memorised the entire *Qur'an* by the age of 10, overtaken his teachers at the age of 14, and become a qualified physician by the age of 18. His interests included medicine, psychology, pharmacology, geology, physics, mathematics, astronomy, chemistry and philosophy. He was also a poet, Islamic scholar and theologian. He is said to have written 450 treatises of which 240 survive. His most famous works are Kitab Al-Shifa (*The Book of Healing*), a vast philosophical and scientific encyclopaedia, and *al-Qanun* (*The Canon of Medicine*) which was a standard medical text comprising the entire medical knowledge available from ancient and Muslim sources.

Aviceps* = *Satyrium (Orchidaceae) Lindl.
La. *avis* = bird; *caput* = head; alluding to the flower which 'in a way simulates the head of a bird' (Lindley, 1838).

Avonia* = *Anacampseros
(Anacampserotaceae) (E.Mey. ex Fenzl) G.D.Rowley
Possibly from *avus* = grandfather; referring to the mass of miniature branching stems covered by the white, scale-like papery stipules. Derivation is uncertain.

Axonopus* (Poaceae) P.Beauv.
La. *axon, axonos* = axis, stem, axle; *pous, podos* = foot; referring to the digitate inflorescence which branches out somewhat like the spokes of a wheel (the racemes around the upper part of the rachis).

Azalea* = *Rhododendron (Ericaceae) L.
Gk. *azaleos* = dry; referring to the plant's dry habitat or its brittle dry wood.

Azanza* = *Thespesia (Malvaceae) Alef.
For Miguel José de Azanza (1746–1826), Spanish military officer, colonial administrator and  diplomat. He served in the military early in his career but switched to a diplomatic career. He became secretary of the Spanish embassy in St Petersburg; chargé d'affaires in Berlin (1784–1786); minister of War (1793–1796) during the war with France; and viceroy of New Spain (1796–1800). At this time, the Spanish botanist Martin Sessé y Lacasta and the Mexican physician and naturalist Jose Mariano Mozinò, supported by Spain, were conducting an ambitious survey of the natural history of the colonies of New Spain. They named a plant *Azanza insignis* (unpublished), possibly recorded later as Augustin Pyramus de Candolle's *Hibiscus Azanzae*. Back in Spain, Azanza became secretary of the Treasury (1808) for King Ferdinand VIII, then worked for Napoleon's brother, Joseph Bonaparte, who was appointed king of Spain. Azanza worked for him and was made duke of Sante Fe. With the defeat of the French, Azanza was forced into exile and died in poverty in France.

Azima (Salvadoraceae) Lam.

(EJM)

Arabic *azim* = defender; *azima* = mighty; referring to the straight spines, often in whorls of four, which 'resemble' the swords of soldiers of Azimus, a small but martial city of Thrace (Palmer).

Azolla (Azollaceae) Lam.
Gk. *azo* = to dry; *ollyo* = to destroy, kill or perish. The plants are killed by drought.

Azorella (Apiaceae) Lam.
Diminutive form of Azores (islands in the eastern Atlantic Ocean).

B

A collage of *Babiana* flowers of which there are about 80 species (MC – All)

Michael Charters (1946–) is a non-professional botanist and wildflower photographer who has been documenting the flora of Southern California photographically for the past 20 years, taking hundreds of fields trips and some 200 000 photographs. He initially became interested in the flora of the Cape region of South Africa because of the affinity to that of Southern California, and he has visited South Africa four times, spending weeks at a time photographing plants in the field in both the Eastern and Western Cape. His work has appeared in dozens of books and scholarly articles.

Babiana (Iridaceae) Ker Gawl.
Dutch *baviaantje,* Afrikaans *bobbejaantjie* or its Cape corruption *babiaantjie* = baboon. Baboons relish the corms.

Baccharis (Asteraceae) L.
The derivation of this name is unclear as Linnaeus gives no explanation of it. Possibilities include Gk. *bakkaris, bakkaridos* = the name for a plant which provides unguent, a soothing or healing ointment, or La. *Bakchos, Bacchus* = named after the god of fertility, wine, revelry, and sacred drama (La. *bacca* = a fruit or berry) or *Bakcharis* = an ancient Greek name used by Dioscorides for sowbread (Umberto Quadrocchi).

Baccharoides (Asteraceae) Moench
Baccharis (q.v.); Gk. *-oides* = resembling.

Bachmannia (Capparaceae) Pax
For Franz Ewald Bachmann (1856–c 1916), German naturalist and physician. He studied  at Breslau, becoming a doctor in 1883, and came to the Cape that year. He practiced medicine in Darling and Hopefield from 1883–1887 and during this time collected species on the Sandveld along the West Coast (Malmesbury–Clanwilliam) including some mosses, lichens and fungi. From 1887–1888 he travelled extensively around Natal and collected in Pondoland while an agent for the Berlinische Pondo Gesellingschaft, reporting on natural resources before returning to Berlin. His travels are described in *Reisen, Erlebnisse und Beobachtungen in der Kapolonie, Natal und Pondoland* (1901) but with little botanical detail.

Bacopa (Scrophulariaceae) Aubl.
An aboriginal name used for a plant of this genus by natives of French Guiana. In the Indian subcontinent *Bacopa monniera* is also called Brahmi, a name derived from Brahma, the creator god of the Hindu pantheon of deities. In the Ayurvedic *Materia Medica, Bacopa* has been recognised for its brain-enhancement characteristics.

Baeobotrys*** = ***Choristylis*** (Iteaceae) J.R. & G.Forst.
Gk. *baios, baeo-* = small, little; *botrys* = a bunch (as of grapes); referring to the small rounded unopened flowers in the inflorescence.

Baeometra (Colchicaceae) Salisb. ex Endl.
Gk. *baios* = *little*, small; *metron* = measure; referring to its size.

Baikiaea (Fabaceae) Benth.
For William Balfour Baikie (1825–1864), Scottish naval surgeon, naturalist, explorer and linguist.  He was commander of the Niger expedition of 1854 proving the Niger and Binue river could be navigated by steam ship as described in his *Narrative of an Exploring Voyage up the Rivers Kwora and Binue* (1856). Further expeditions up the Niger in 1857 and 1859 failed as on both occasions the steamer was wrecked. Baikie established a settlement near the confluence of the Niger and Benue at Lokoja where he was an unofficial British consul and agent. He remained there studying the country, its peoples and languages, and translated the Bible into some of the Central African languages. In 1864 he left Lokoja to return home but died en route in Sierra Leone.

Baillonella (Sapotaceae) Pierre
For Henri Ernest Baillon (1827–1895), French botanist and physician. He was a professor of  natural history at the Faculté de Médecine, Paris, director of the Paris botanical garden, and author of numerous botanical works such as the *Dictionnaire de botanique* (1876–1892) (four volumes), *Histoire des plantes* (1866–1895) (13 volumes) and *Histoire naturelle des plantes* of Madagascar (three volumes). In 1867 he won the high state award *Légion d'honneur* and just before his death in 1894, he was elected a foreign member of the Royal Society in London.

Baissea (Apocynaceae) A.DC.
For Nicholas Sarrabat (pseudonym, 'de la Baisse') (1698–1739), French Jesuit priest and scientist, and professor of mathematics at Marseilles. He discovered the comet of 1729 without the aid of a telescope, conducted experiments on magnetism and the causes and variations of wind patterns (1730), traced the circulation of sap in plants (1733) by immersing living plant roots in red pokeweed berry juice, and went on an archaeological expedition (1735–1736) to the islands of Milos and Malta. He won three science competitions organised by the Académie Royale. Although the rules did not allow further entries, he

submitted an entry under the pseudonym 'de la Baisse', to which he later confessed, hence 'de la Baisse, alias Sarrabat'.

Balanites (Zygophyllaceae) Delile.

The fluted trunks are outstanding (EJM)

Gk. *balanos* = an acorn, *balanites* = acorn-having; referring to the acorn-shaped fruit.

Ballota (Lamiaceae) L.
Gk. *ballote* (Dioscorides) from *ballein* or *ballo* = to reject. (Some sources state *ballota* = little ball); alluding to the plant's offensive odour.

Ballya = Aneilema (Commelinaceae) Brenan
For Peter René Oscar Bally (1895–1980), Swiss botanist and taxonomist. He was a plant collector (1943–1975) both in Europe and India but mainly in tropical Africa (Ethiopia, Ghana, Kenya, Somalia, Sudan, Tanzania, Uganda and Zimbabwe). He was head of the herbarium of Coryndon National Museum, Nairobi, Kenya (1938–1958), and authored *East African Succulents* volumes 1–6 (*Journal of East African Natural History*, 1940–1946). In 1960, he moved to Swaziland and worked on the genus *Aloe*, collecting and

corresponding with Gilbert W Reynolds, author of *Aloes of Tropical Africa and Madagascar* (1966). In 1961, Bally published *The Genus Monadenium: A Monographic Study* (1961).

Baloskion* (Restionaceae) Raf.
Meaning unknown. Gk. *balos* = to throw; *skion* = skin, hide; possibly referring to the capsule falling with glume and perianth attached. Perhaps a Rafinesque contrived word.

Balsamodendrum*** = Commiphora (Burseraceae) Kunth
Gk. *balsamon* = aromatic; *dendron* = tree; Balsam-tree; myrrh is made from a species of this genus.

Bambusa* (Poaceae) Schreb.
17th-century Dutch colonial sources indicate the word is the Latinised form of the Indo-Malayan vernacular name *bambu* or *bamboe*. Earlier sources show that it should be 'Mambu' but was spelled incorrectly (D Ohrnberger, *The Bamboos of the World*, 1999).

Banksia* (Proteaceae) f.
For Joseph Banks (1743–1829), English naturalist, explorer, plant collector and patron of the natural sciences, an interest which developed from visits to the Chelsea Physic Garden and British Museum. He explored parts of Newfoundland and Labrador with Constantine John Phipps (1766), South America, Tahiti, New Zealand and Australia with Captain James Cook (1768–1771), and Scotland and Iceland a year later. He visited the Cape, briefly, in 1771 and explored the Cape flora with Swedish naturalist Daniel Solander, and a year later sent out Francis Masson, a gardener at Kew, to collect botanical specimens. In 1778 he became president of the Royal Society, a position he was to hold for 41 years. During this time he substantially enlarged the Kew Gardens and the British Museum collections by sending botanists and explorers to different parts of the world.

Baphia (Fabaceae) Afzel. ex Lodd.
Gk. *baphé* = a dye; referring to the brilliant red dye given by the heartwood of the tropical African *B. nitida*, used to colour British bandana handkerchiefs.

Baphiastrum (Fabaceae) Harms
Gk. *baphé* = a dye; *-astrum* = inferior or incomplete resemblance to *Baphia* (q.v.).

Baptisia (Fabaceae) Vent.
Gk. *baptisis* = to dip, bathe, from *baptein*. The plant was formerly used to make a dye or colouring agent.

Barbacenia (Velloziaceae) Vand.
Possibly after Luis Antonio Furtado de Castro do Rio de Mendonca and Faro (1754– 1830),  Brazilian sixth viscount and first earl of Barbacena. From 1772 he studied law and philosophy at the University of Coimbra, becoming the first graduate to receive a doctorate in philosophy. While completing his law degree he competently conducted classes in natural history for the Italian Domingo Vandelli (1735–1816), professor of chemistry and natural history (and the genus author) when Vandelli was unable to give lectures. After his return to Lisbon he became one of the founders of the Royal Academy of Sciences of Lisbon (1779) and its secretary until he left for Brazil upon his appointment as governor and captain-general of Minas Gerais, one of the largest states in Brazil.

Barbarea (Brassicaceae) R.Br.
 For Saint Barbara, patron of artillerymen and miners, and an early Christian martyr. According to legend Saint Barbara was beheaded by her own father, a wealthy heathen named Dioscorus, for expressing a belief in Christ. One species of 'water cress' was dedicated to Saint Barbara.

Barberetta (Haemodoraceae) Harv.
For Mary Elizabeth Barber (née Bowker) (1818–1899), British botanist, entomologist,  painter, and poet, daughter of Miles Bowker, an 1820 settler. When young she read *The Genera of South African Plants* (1838) by William Henry Harvey and corresponded with him and also Joseph and William Hooker and Charles Darwin. Over the years she became a noted authority on South African flora and made a significant contribution to science in written papers and specimen collections which she sent to the Royal Botanic Gardens at Kew; including the aloe which was named by Robert Allen Dyer (1874) in her honour. Although she had no formal training, she painted natural history and landscape subjects; many of which are in the Albany museum, Grahamstown.

Barleria (Acanthaceae) L.

Excellent border plant in full sun (EJM)

 For Jacques Barrelier (1606–1673), French Dominican monk, biologist, botanist, and physician. He spent 23 years in Rome where he established a botanical garden at the Saint-Xyste convent. During this time he visited France, Spain and Italy to study flowers. He spent years working on his book *Orbis Mundi Hortus Botanicus*, describing the species he collected in his travels and making in excess of 300 detailed copper engravings of these plants. Before this project was completed, Barrelier died of asthma and fire destroyed all his notes. The plates survived and the French botanist Antoine de Jussieu (1656–1758), director of Jardin des Plantes in Paris, published his book posthumously in 1714 as *Plantae per Galliam, Hispaniam et Italiam observata*. Linnaeus spelled his name incorrectly.

Barnardiella = Moraea (Iridaceae) Goldblatt
For Thomas Theodore Barnard (1898–1983), English professor of social anthropology at the University of Cape Town (1926–1933) and authority of the Linnaean period. Soon after moving to South Africa he became interested in South African flora, particularly Iridaceae, and began to cultivate them. He grew a number of

 species and hybrids of *Gladiolus* on an extensive scale in England. He collected some 250 specimens of plants, mainly in the southwestern Cape. With GJ Lewis and AA Obermeyer he wrote *A Revision of the South African Species of Gladiolus*. He was awarded the Harry Bolus Medal in 1977 by the South African Botanical Society and was a Fellow of the Linnaean Society.

Barosma *** = ***Agathosma*** (Rutaceae) Willd.
Gk. *baro* = heavy; *osma* = fragrance, scent; referring to the powerful fragrance of these shrubs.

Barringtonia (Lecythidaceae) J.R. & G.Forst.

Thickets were a characteristic of small East Coast estuaries. (EJM)   (GN)

 For Daines Barrington (1727–1800), English lawyer, then Welsh judge (1757), naturalist, author and antiquarian, friend of Sir Joseph Banks, Fellow of the Royal Society, an author on diverse subjects such as: legal tracts – *Observations on the Statutes, Chiefly the More Ancient, from Magna Charta to 21st James I* (1766); exploration – *Probability of Reaching the North Pole* (1775); history and antiquities (particularly north Wales); music composers – *Mozart's Visit to London* (Aged Eight); Anglo-Saxon translations – of Alfred's *Orosius*; accounts of noted prodigies – *William Crotch*, *Charles and Samuel Wesley*, and *Garret Wellesley*; birdsong – *Essay on the Language of Birds*. He was elected a Fellow of the Society of Antiquaries in 1768.

Barrowia = ***Orthanthera*** (Apocynaceae) Decne.
For John Barrow (1764–1848), English statesman, geographer and prolific author. He served as  secretary to Lord Macartney and governor of the Cape of Good Hope (1797–1804). He made two major treks as described in his *Travels Into the Interior of South Africa* (1806), and drew up a large detailed map; improved 10 years later by the naturalist/explorer William John Burchell. In 1804, when the colony was returned to the Batavian Republic, Barrow returned to England. He became second secretary to the Admiralty, a position he held for 40 years. He promoted many Arctic expeditions, notably those of John Ross and William Parry. He was a Fellow of the Royal Society and a founding member of what became the Royal Geographical Society in 1830.

Bartholina (Orchidaceae) R.Br.
For Thomas Bartholin (1616–1680), Danish doctor, theologian, mathematician, and professor  of anatomy at Copenhagen, and physician to King Christian V of Denmark. He discovered the lymph vessel system in humans, following William Henry Harvey's work on the circulation of blood. Bartholin was first to describe the thoracic duct in humans, shortly after Jean Pecquet recognised this duct in animals. During his lifetime, Bartholin achieved fame both as a teacher and for his many textbooks. His illustrated revision of a seminal work by his father Caspar Bartholin, in which the work of two contemporary anatomists Gasparo Aselli and Harvey was recognised, became the standard reference work on anatomy.

Bartschia *** = ***Bartsia*** (Orobanchaceae) L.
Orthographic variant of *Bartsia*.

Bartsia * (Orobanchaceae) L.
For Johann Bartsch (Latinised as Johannes Bartsius) (1709–1738), German physician and naturalist, born in Königsberg, East Prussia (now Kaliningrad, Russia). He graduated in medicine at the University of Leiden, where in 1735 he met Linnaeus. Linnaeus was assisted and supported by Bartsch with his *Flora Lapponica* (1737), a botanical account of the five months he spent in Lapland in 1732 collecting plant and other specimens. When in 1737 the post of medical officer of the Dutch East India Company in Dutch

Guiana was offered to Linnaeus by Herman Boerhaave, he suggested Johan Bartsch be given the job. Bartsch died six months after arriving in Dutch Guiana, now Surinam. He was the author of *Thesis de Calore Corporis Humani Hygraulico*.

Basananthe (Passifloraceae) Peyr.
Gk. *basanos* = used in the sense of 'agony' or 'torture'; *anthos* = flower. Some species have thorny shrublets.

Basella (Basellaceae) L.
The local Malabar (Indian) name, a region in southern India, for vine spinach (*Basella alba*). Wikipedia says the type species was found in Madagascar so the seeds may have been brought across by monsoon winds.

Basilicum (Lamiaceae) Moench
La. *basilicas*, Gk. *basilikos* = princely, royal. Although not named directly after royalty, Basil is one of the 'kings of herbs' referring to its aromatic, culinary, antioxidant, antiviral, antimicrobial, and other healing properties.

Bassia (Amaranthaceae) Al.
For Ferdinando Bassi (1710–1774), Italian botanist and naturalist, prefect (director) of the garden of exotic plants in the Bologna Botanical Garden from 1763 to 1774, author of *Ambrosina, novum plantae genus* and contributor to the *Istituto delle Scienze*. As a young man he studied the natural sciences and became an assistant to botanist Giuseppi Monti who facilitated his contacts with Italian and other European naturalists. He corresponded with the scientists of his time such as Albrecht von Haller and Linnaeus with whom, for 10 years from 1763–1773, he provided descriptions and accurate drawings of new botanical species and information about other natural science matters.

Basteria* = Cullumia** (Asteraceae) Houtt.
For Job Baster (1711–75), Dutch physician, biologist, zoologist, botanist, a student of Boerhaave (1668–1738), professor of botany and chemistry at Leiden University. Baster published many works on natural history including a treatise on the classification of plants and animals (1768) and *Opuscula Subseciva* (1759–1765), miscellaneous observation on plants, animals, seeds and embryos and *Natuurlijke Uitspanningen Behelzende Eenige Waarnemingen over Sommige Zee-planten en Zee-insecten*

(*Natural Accounts about Some Observations of Some Marine Plants and Insects*). He was a Fellow of the Royal Society in London. The journal 'Basteria' of the Dutch Malacological Society is named after him.

Basutica* = Gnidia** (Thymelaeaceae) E.Phillips
From Basutoland; referring now to the independent Kingdom of Lesotho within South Africa's borders; but when discovered in 1944 it was a British crown colony called Basutoland.

Bauhinia (Fabaceae) L.

Flowers and characteristic bi-lobed leaves (EJM)

For Gaspard (or Caspar) Bauhin (1560–1624) and Johan Bauhin (1541–1613), Swiss-French botanists and herbalists. Gaspard became professor of anatomy and botany at the University of Basel, as well as dean of the medical faculty and university rector. He authored *Pinax Theatri Botanici* (1596) describing and classifying some 6 000 species of plants using a binomial nomenclature, a precursor of Linnaeus's classification system. Johann became professor of rhetoric at the University of Basel (1566) and physician to Duke Frederick I of Württemberg at Montbéliard (1570). With his son-in-law, Jean Henri Cherler, he wrote *Historia Plantarum Universalis,* a compilation of everything that was known at that time about botany. This was published posthumously. He also established botanical gardens in Montéliard and Stuttgart.

Baynesia (Apocynaceae) Bruyns
For Maudsley Baynes (1881–1971), English explorer, possibly studied at Oxford but seemingly absconded with funds from the social club. He went to South Africa to 're-start' his life. He was

commissioned in 1911 by the Kaokoveld Land and Mining Company of London (to which the Kaokoveld was sold by the German Colonial Government) to explore the lower reaches of the Kunene River from the Ruacana Falls to the Coastal Desert. He was the first European to investigate this area. The journey involved extreme hardships and he only just survived it. The Germans of Zessfontein were so impressed with his feat that they named the great and unknown mountainous range on the Kunene River in Namibia in his honour.

Beaumontia* (Apocynaceae) Wall.
For Diana Beaumont (née Wordworth) (1765–1831) of Bretton Hall, Yorkshire, described in the Curtis Botanical Magazine Volume 7 (New Series) in 1833 as 'an ardent lover and munificent patroness of Horticulture who oversaw the work done in Bretton Hall Garden for three decades'. A wealthy woman and wife of Colonel TR Beaumont (1758–1829). The two developed two conservatories and a botanical garden between 1792 and 1831 as part of their 20 000 hectare estate. Diana amassed a large mineral collection and was an obsessive gardener. She built a massive bell-shaped glass house 70 feet in height and 100 feet in diameter to house her exotic plants. Unfortunately, the conservatories and glass house were destroyed by her son, Thomas Wentworth, when she died.

Becium = Ocimum (Lamiaceae) Lindl.
Gk. *bekion* = an ancient name for sage.

Beckeropsis = Pennisetum (Poaceae) K. Schum.
For Johannes Becker (1769–1833), German botanist. He was one of the founding members of the Senckenbergische Naturforschende Gesellschaft (Senckenberg Nature Research Society) in 1817, an initiative of Germany's beloved poet Johann Wolfgang von Goethe (1749–1832). Becker was the Society's first curator for botany, also author of the *Flora der Gegend um Frankfurt am Main*. The Herbarium Senckenbergianum comprises his personal collection. He was at the same time Sitfsbotanikus at the botanical garden and enriched the herbarium with many specimens from plants cultivated there.

Begonia (Begoniaceae) L.
For Michel Bégon (1638–1710), French patron of botany, a veteran French colonial administrator who served in Canada in the 1670s and early 1680s, then in the French colonies in the Caribbean for another decade. He became involved in the spice trade around 1684, when he became governor-intendant of Saint-Domingue (Santo Domingo). He was a dedicated plant collector and met Charles Plumier (who named the Begonia after him) in the Antilles. In 1688 he became intendant of the port of Rochefort, which he considerably transformed and developed. He gathered much data on French commerce and actions overseas, and advised the government extensively on matters of trade and commerce until his death in 1710.

Behnia (Asparagaceae) Didr.
For Wilhelm Friedrich Georg Behn (1808–1878), German anatomist and zoologist, professor of anatomy and physiology at the University of Kiel, director of the zoological museum at Kiel, friend and companion of Danish botanist Didrik Ferdinand Didrichsen (1814–1887). He received his doctorate from the University of Kiel in 1832 and later became director of the Institute of Anatomy and was appointed to the Zoological Museum. In 1845–47 he participated in a circumnavigation of the globe aboard the Danish ship 'Galathea' collecting natural history material. From 1869–1878 he was elected president of the Leopoldina-Carolina Akademie der Naturforscher, one of the oldest scientific academies in Germany.

Beilschmiedia (Lauraceae) Nees
For Carl (Karl) Traugott Beilschmied (1793–1848), German pharmacist, botanist, bryologist, and author of books on plant phytogeography. He studied pharmacy at the University of Bonn (1820–1822), running his own business from 1826 onwards. He combined his work with field botany and editing the German version of Wikstrom's botanical annual bibliographies. He obtained his PhD in 1837. His major work was *Pflanzengeographie, nach Alexander van Homboldt's Werke Ueber die Geographische Vertheilung der Gewächse* (1831) (*Plant Geography, According to Alexander von Humboldt's Work on the Geographical Distribution of Plants*), in which he discusses plant geographic distribution over different climatic zones and different altitudes.

Belamcanda* = **Iris** (Iridaceae) Adans.
A vernacular Asian name, probably from western India; possibly this name refers to the 'leopard lily', because of the 'leopard' markings on the petals.

Bellardia* = ***Bartsia*** (Orobanchaceae) All.
For Carlo Antonio Lodovico Bellardi (1741–1826), Italian botanist and physician. He was a pupil of Carlo Allioni (1728–1804) at the botanical garden of Turin, professor of botany at the University of Turin and director of the Turin Agricultural Institute. His herbarium was damaged with the loss of many specimens when he placed this herbarium under the management of French agriculturist M Bonafous. Bellardi was the author of *Stirpe Nova vel Minus Note Pedemontii Descriptae et Iconibus Illustratae* (1808).

Bellidiastrum** = ***Osmitopsis*** (Asteraceae) Less.
Bellis (q.v.), a name used by Pliny the Elder; *astrum* = somewhat resembling.

Bellis* (Asteraceae) L.
Gk. *bellis* = pretty, handsome, a name used by Pliny the Elder.

Belmontia = ***Sebaea*** (Gentianaceae) E.Mey.
Named after some member of the Belmonte family, specific details unknown, whose garden supplied plants to Albertus Seba (See Sebaea).

Bequaertiodendron** = ***Englerophytum*** (Sapotaceae) De Wild.
For Joseph Charles Corneille Bequaert (1886–1982), Belgian-born American botanist and entomologist, who obtained a PhD from the State University of Ghent (1908), spent seven years working in the Belgian Congo (1910–1916), was appointed research associate in Congo zoology at the American Museum of Natural History (1917), became a naturalised US citizen (1923), and was assistant professor in medical entomology at Harvard Medical School until 1945. He accepted the curatorship of recent insects at the Museum of Comparative Zoology at Harvard until his retirement in 1956. He was president of the US Malacological Union and produced more than 250 publications on medical entomology, molluscs, botany, and systematics of insect families.

Berardia** = ***Nebelia*** (Bruniaceae) Sond.
For Etienne Bérard (1764–1839), a former professor, proprietor of an industrial chemical factory in Montpellier, and treasurer of the Ecole de Pharmacie from 1813–1839.

Berberis (Berberidaceae) L.
Latinised form of the Arabic name *barbaris* for the fruit.

Berchemia (Rhamnaceae) Neck. ex DC.

The leaves have distinct, deeply sunk side-veins. (EJM)

The fruits are very tasty. (AC)

For Jacob Peter Berthout van Berchem (1763–1832), also called Jacob-Pierre von Berchem, Dutch naturalist, born in The Netherlands, died in France. He published various notes on the fauna and flora of Switzerland. He was secretary of the Society of Physical Sciences in Lausanne and he wrote several works on mineralogy including *Principes de Minéralogie* (1795) with Henry Struve. He also wrote one of the earliest Swiss guidebooks describing various routes that might be followed in the neighbourhood of Chamonix and Mont Blanc. He was one of the contributing authors – with T Pennant, Sprungli, Wyttenbach, Van Berchem, and Studer – responsible for the three-volume work *Travels in Switzerland* (1789) edited by Pulteney.

Bergenia* (Saxifragaceae*) Moench
For Karl August von Bergen (1704– 1759), German anatomist and botanist. He studied medicine at Viadrina University (the University of Frankfurt) from 1727 where he was taught by his father, Johann Georg von Bergen (died 1738) and Andreas Ottomar Goelicke (1670–1744), both anatomists, and undertook further studies at the University of Leiden under professor Herman Boerhaave (1668–1738) and others. He also studied in Paris and Strasbourg. After he obtained his doctorate in medicine (1732) he became a

professor taking the Chair of Anatomy and Botany when his father died. Among his duties was responsibility for the university's botanical garden – his major work being *Flora Francofurtana*, Frankfurt (Oder), published in 1750. He also wrote on topics ranging from the general distribution of membranes in animals, anatomy, pathology and therapy, to an essay on the rhinoceros.

Bergeranthus (Aizoaceae) Schwantes
For Alwin Berger (1871–1931), German botanist and horticulturist, superintendent of the Hanbury Garden in La Mortola, Italy, and authority on Cactaceae. He worked at the botanical gardens of Dresden and Frankfurt. He was also director of the department of botany of the Natural History Museum in Stuttgart, and was the author of *Die Agaven* published in 1915 which described 274 species of agaves. He provided a comprehensive account of *Crassulaceae* in Adolf Engler and Karl von Prantl's *Die Natürlichen Pflanzenfamilien* and first described a few genera, including *Encephalocarpus* and *Epiphyllanthus*.

Bergia (Elatinaceae) L.
For Peter Jonas Bergius of Stockholm (1730–1790), Swedish physician and botanist, plant collector, and a pupil of Linnaeus. He was appointed professor of natural history and pharmacy at the Collegium Medicum in Stockholm in 1761. He authored several books between 1757–1780 including *Descriptiones Plantarum ex Bona Spei Capita* (1767), a book on the plants and the mosses of the Cape of Good Hope (Cape Peninsula), *Materia Medica* (1780), a treatise on fruit trees (1780), and a historical work on the city of Stockholm in the 15th–16th centuries. Bergius never left Sweden; the specimens were sent to him by Johann Andreas Auge, superintendent of the Company's Garden. He was made a member of the Royal Society in 1770.

Berkheya (Asteraceae) Ehrh.

For Jan (Johannes) le Francq van Berkhey (1729–1812), Dutch botanist, physician, naturalist, poet, professor of natural history at Leiden University (1773–1795), author of the *Natural History of Holland*, (seven volumes, 1769–1778), a reference work on Dutch botany, geology, topography and cultural history in the 18th century, a *Natural History of Cattle in Holland*, (six volumes, 1805–1811), botanical works like *Expositio Characteristica Structurae Florum qui Dicuntuv Composite*, and polemic, satirical and rhetoric poems. His *In Praise of Gratitude* (1773) received the gold medal from the Dutch poetry society, but many of his poems, which he read out in the streets and which kept the crowd spellbound, angered and upset his enemies. In 2010 an asteroid was named after him.

Berkheyopsis*** = **Hirpicium** (Asteraceae) O.Hoffm.
Berkheya (q.v.); Gk. *opsis* = resembling; but unlike *Berkheya* this genus does not have the spines.

Berlinia (Fabaceae) Sol.
For Andreas Berlin (1746–1773), Swedish botanist, who studied at Uppsala University as a pupil of Linnaeus. He travelled to London with a letter of introduction from Linnaeus to find a botanical expedition he could join. According to Jarvis (2007), he became Sir Joseph Banks's secretary from 1770–1773. In 1773 he travelled with the English naturalist Henry Smeathman on an expedition to Bance (now Bunce) Island off the coast of Sierra Leone. The expedition's purpose was to explore the central parts of Africa. Prior to reaching the mainland, Berlin died of a stomach illness while on Isles de Los, an island group lying off Conakry in Guinea. Before his death Berlin managed to send one of the three specimens to Linnaeus from which the genus was defined.

Bernhardia*** = **Psilotum** (Psilotaceae) Bernh.
For Johann Jakob Bernhardi (1774–1850), German botanist, physician and horticulturist. He graduated with degrees in philosophy and medicine from the University of Erfurt (1799), became extraordinary professor of medicine at the university (1805) and director of the city's botanical garden, and became an advisor on medical matters to the state of Prussia (1819). He was an important author on the flora of Germany and respected researcher making plant anatomy discoveries relating to the annular thickenings in vessels previously unrecognised. He assembled a herbarium of 60 000 plants from North and South America, Asia and Africa now in the Missouri Botanical Garden and forming the nucleus of its herbarium collection.

Berrisfordia**** = **Conophytum** (Aizoaceae) L.Bolus
For Francis ('Frank') Berrisford (1898–1973), an accountant, mountain climber and keen naturalist,

and member of the Mountain Club of South Africa for many years. Among others, he climbed with the prolific plant collector Elsie Esterhuysen (q.v. *Elsiea*), and also with Dr Keppel Barnard (1887–1964), director of the South African Museum, who had a particular interest in Colophon Stag beetles (family Lucanidae) which are endemic to high altitude mountainous areas of the Western Cape in South Africa. Barnard asked Berrisford to collect anything that seemed unusual at the top of high peaks and was responsible for forwarding the plant specimen discovered by Berrisford to South African botanist Harriet Margaret Louisa Bolus. The only species in this genus, *B. khamiesbergensis*, was later transferred to *Conophytum* by Martin Schwantes.

Bersama (Melianthaceae) Fresen.

The knobbly fruits open to reveal the red seeds. (EJM)

Bersama refers to the Ethiopian name for this genus. Original meaning unknown.

Bertilia (Asteraceae) B.Nord.
For Bertil Nordenstam (1936–), Swedish botanist, botanical historian and academic. He studied at Lund University where he became an associate professor after obtaining a PhD in  1968. In 1980 he became a professor in the department of phanerogamic botany at the Swedish Museum of Natural History. He has done extensive research work in Greece, Australia, Mongolia, China, South Africa, Namibia, Ecuador, Peru and the Caribbean especially relating to *Senecioneae* and *Calenduleae* tribes of the family *Asteraceae,* and he is the author of 50 genera plus hundreds of new species. He was editor of the *Compositae Newsletter* for 25 years. His historical interests have led him to publishing books and papers on the life and times of Carl Peter Thunberg, Olof Swartz and Erik Ekman. He was elected a member of the Royal Swedish Academy of Sciences in 1984, where he has served as vice-president and chairman of biological sciences.

Bertya (Euphorbiaceae) Planch.
For Leonce Auguste Marie Compte de Lambertye (1810–1877), French botanist, horticulturist, author of *Catalog Raisonné des Plantes Vasculaires qui Croissent Spontanément dans le Département de la Marne*, popularised viticulture and fruit tree pruning. The author writes that this genus should have been named after M. le Compte Léonce de Lambertya.

Berula (Apiaceae) W.D.J.Koch
La. *berula, ae,* derived from the French vernacular name *berle,* Latin name for an aquatic plant similar to watercress. The generic name is apparently an ortho-variant of another well-known Apiaceae genus, *Ferula* (ferule) = giant fennel.

Berzelia (Bruniaceae) Brongn.

One of the most colourful species – most others have whitish flowers (EGHO)

For Jöns Jacob Berzelius (1779–1848), renowned Swedish chemist. He was appointed professor of chemistry and pharmacy at the Karolinski Institute (1807), where he developed a chemical  'shorthand' in which chemical elements were written in symbols. He discovered the law of constant proportions which states that every pure substance always contains the same elements combined in the same proportions. He and his students identified many new chemical elements such as Silicon, Lithium, Selenium and others and he coined many new terms such as catalysis, polymer, isomer and allotrope. He was secretary of the Royal Swedish Academy of Sciences from 1818–1848.

Beschorneria* (Agavaceae) Kunth
For Friedrich Wilhelm Christian Beschorner (1806–1873), German physician, psychiatrist and amateur botanist. He studied medicine at the University of Bonn and the University of Wroclaw and in 1830 became a doctor of medicine. He worked as second-in-charge in a psychiatric institution in Lubiaz, Poland until 1835. In 1838, he was the first director and head of the Psychiatry Department in Owińska. Carl Sigismund Kunth (1788–1850) named a succulent after him, but no details could be found about Beschorner's 'vivid interest in botany'.

Besenna = Feuilleea (Cucurbitaceae) A.Rich.
A term used to describe the bark of an Abyssinian (= Ethiopian) species, *Albizia anthelmintica*, known locally as 'mesenna' or 'besenna'. The meaning of 'besenna' or 'messena' probably relates to the use of the bark as a remedy to expel tapeworms.

Beta* (Amaranthaceae) L.
Derivation uncertain. Possibly Celtic *bett* = red or Gk. *beta* = beet; alluding to the red colour of the roots, such as beetroot, spinach and chard.

Betula* (Betulaceae) L.
Latin name for a birch tree.

Bewsia (Poaceae) Gooss.
 For John William Bews (1884–1938), Scottish ecologist, and professor of botany at Natal University College in South Africa, of which he became principal in 1931. He was a prolific author of books and papers, including such publications as *The Grasses* and *Grasslands of South Africa* (1918), *An Introduction to the Flora of Natal and Zululand* (1921), *Researches on the vegetation of Natal* (1923), *Plant Forms and their Evolution in South Africa* (1925), *Studies in the Ecological Evolution of the Angiosperms* (1927), *The World's Grasses* (1929), *Human Ecology* (1935), and *Life as a Whole* (1937). He was a Fellow of the Linnaean Society.

Bidens (Asteraceae) L.

Flowers and buds (EJM)   Fruits (MC)

La. *bi* = two; dens = teeth; referring to the two teeth or bristles on the dry one-seeded fruit, an achene, which, possibly, hooks onto any passing animal, thus dispersing the seed or relating to the two barb-toothed pappi of the original species.

Bignonia* (Bignoniaceae) L.
For Abbé Jean-Paul Bignon (1662–1743), French ecclesiastic, statesman and scholar, writer and  preacher, member of the Académie Française and the Academie Royale des Inscriptions et Belles-Lettres, and librarian to King Louis XV at the Bibliothèque nationale de France from 1718–1741, at that time the leading library in Europe.
One of Bignon's innovations was to open the library to the public, albeit for only a brief period each week. He began a major catalogue of the library, which was completed 10 years after his death. Bignon also contributed to the *Médailles du règne de Louis le Grand*, *Sacre de Louis XV*, and

Journal des Savants, and authored *Life of Francis Levesque, Priest of the Oratory* (1684).

Bijlia (Aizoaceae) N.E.Br.
For Deborah Susanna Malan (Mrs William van der Bijl) (1872–1942), South African naturalist, field collector and nurserywoman in the Western Cape, founder and first president of the South African Succulent Society in 1931. She was very interested in Karoo plants and sent many specimens to NE Brown at Kew Gardens, author of the name *Bijlia*. She also corresponded with Harry Bolus.

Bilabrella*** = **Habenaria** (Orchidaceae) Lindl.
La. *bi-* = two; *labrellum* = upper lip; referring to the 'conspicuous petals which, with the labellum situated beneath, suggest an upper lip' (Lindley, 1834).

Bilderdykia*** = **Fallopia** (Polygonaceae) Dumort.
For Willem Bilderdyk (Bilderdijk) (1756–1831), Dutch poet, prolific author and polyglot; familiar with 25–30 languages, modern, ancient and exotic. He studied at Leiden, obtained his doctorate in law in 1782 and practised as an advocate at the Hague. From 1795–1806 he was forced into exile because of his strong monarchical and political beliefs, living in Germany and London. On his return Louis-Napoléon made him his librarian and later president of the Royal Institute (1809–1811); a job he lost on Napoleon's abdication. He endured great poverty and ended his career a history tutor at Leiden. He published 300 000 lines of poetry and books on geology, national history, literary perspectives, architecture and 30 volumes on linguistics.

Biophytum (Oxalidaceae) DC.
Gk. *bios* = life; *phytum* = plant, hence 'plant of life'; referring to the reactive 'irritable' nature of the foliage. The leaves and pods are sensitive to touch.

Bischofia* (Phyllanthaceae) Blume
For Gottlieb Wilhelm Bischoff (1797–1854), German botanist, lexicographer and university professor. He studied at the Academy of Fine Arts, Munich (1819), and botany at the Friedrich-Alexander University in Erlangen (1821). He became a professor of botany at the University of Heidelberg in 1825 and later professor of botany at Erlangen (1833) and director of the university's botanical garden (1839). His main interest was the study of cryptogamic plants. Among his publications he wrote (in German) a *Handbook of Botanical Terminology* (1833), in which he coined the terms 'antheridium' and 'archegonium'; *Textbook of General Botany* (1834); *A Dictionary of Descriptive Botany* (1839); *Medical and Pharmaceutical Botany* (1843) and *Contributions to the Flora of Germany and Switzerland* (1851).

Bistella = **Vahlia** (Vahliaceae) Adans.
La. *bi-* = two; *stella* = a star. Name not explained by the author. Possibly referring to the 'twining' of flowers.

Bituminaria (Fabaceae) Heist. ex Fabr.
La. *bituminaria* = concerning bitumen. The leaves when rubbed give off a bituminous, asphalt smell.

Blackiella*** = **Atriplex** (Amaranthaceae) Aellen
For John McConnell Black (1855–1951), Scottish botanist, linguist, author and illustrator who emigrated to Australia in 1877. His major publications were *The Naturalised Flora of South Australia* (1909) and *The Flora of South Australia* (published in four parts from 1922–1929) describing 2 430 species, both indigenous and naturalised. He began a revision of this book in 1939 and was working on part three in 1951 when he died. Already fluent in six languages he developed an interest in Australian Aboriginal languages and published papers on them. He received many distinctions for his botanical work, including being made an associate *honoris causa* of the Linnaean Society, London (1930) and being awarded an MBE (1945). He was president of the Royal Society of South Australia (1933–1934).

Blackwellia*** = **Homalium** (Salicaceae) Warb.
For Elizabeth Blackwell (1707–1758), Scottish botanist, botanical illustrator and engraver, the first British woman to produce a herbal book. *A Curious Herbal* was published in 1735 as a reference work designed for physicians. It illustrates in colour some 500 plants and herbs that were considered to have medicinal qualities.

She drew specimens of plants from the Chelsea Physic Garden. Sir Hans Sloane provided financial support for this project, which took six years to complete. The funds she raised from this project were used to secure the release of her botanist husband Alexander Blackwell from debtor's prison. Later, he left his family, relocated to Sweden, got involved in a political conspiracy, was arrested and hanged for treason.

Blaeria*** = **Erica** (Ericaceae) L.
For Patrick Blair, MD, FRS (1666–1728), Scottish physician, surgeon and medico-botanical author, who practiced in Dundee, London and Boston. His three major publications were *Osteographia Elephantina* (1713), in which he describes his dissection of an elephant and anatomical findings; *An Account of the Dissection of a Child* (1717), in which he described a child who had pyloric stenosis; and *Practice of Physick, Anatomy, and Surgery* (1718). In 1720 he gave a lecture on the sexual characteristics and fertilisation of the plant, which caught the attention of Linnaeus. He was elected a member of the Royal Society. During the Jacobite Rising of 1715, he was arrested, thrown into prison and sentenced to death. As a result of the intercession of Hans Sloane, Richard Mead and others, he was eventually pardoned.

Blainvillea (Asteraceae) Cass.
For Henri Marie Ducrotay de Blainville (1777–1850), French zoologist and anatomist. Initially, he  planned to study painting but switched his career choice and became professor of anatomy and zoology in the Faculty of Sciences at Paris (1812–1830), a member of the French Academy of Sciences (1825) and the Royal Swedish Academy of Sciences (1837). His publications cover a wide range of subjects such as the distribution of the animal kingdom (1816), the skeletal and dental system of recent and fossil mammals (1839–1864), French fauna (1821–1830), general and comparative physiology (1833), a manual of conchology and malacology (1825–1827), and a history of the science of organisms (1845).

Blastania*** = **Ctenolepis** (Cucurbitaceae) Kotschy & Peyr.
Gk. *blastano* = to bring forth; *blastos* = bud, sprout, embryo; possibly referring to the plant's rapid growth and ability to spread.

Blaxium = **Osteospermum** (Asteraceae) Cass.
Meaning unknown to authors.

Blechnum (Blechnaceae) L.
Gk. *blekhnon* = an ancient name used for ferns in general.

Blepharanthera** = **Brachystelma** (Apocynaceae) Schltr.
Gk. *blepharon* = eyelid; *anthera* = anthers; referring to the fringed anthers.

Blepharis (Acanthaceae) Juss.

The petals are fused. (TD)

Gk. *blepharis* = eyelash, hence fringed; referring to the delicate fringe of hairs that adorn the bracts and calyx below the flowers.

Blepharolepis*** = **Scirpus** (Cyperaceae) Nees
Gk. *Blepharis* (q.v.); *lepis* = scale. This grass-like species can appear scruffy.

Blepharophyllum** (Ericaceae) Klotzsch
Gk. *Blepharis* (q.v.); *phyllon* = leaf. The leaves are fringed with 'eyelashes'.

Blighia* (Sapindaceae) K.D.Koenig
For William Bligh (1754–1817), British mariner, navigator and colonial governor. He joined the  navy in 1770, was appointed master on *Resolution* on James Cook's third voyage (1776–1780), saw action against France in 1780–1783 and was promoted to lieutenant, became commander of *HMS Bounty* (1787) during which the crew mutinied and set Bligh and 18 others adrift in an open boat seven metres long. He skilfully navigated 3 618 miles (5 822 km) to Timor in 47 days. In 1791 the Admiralty appointed him a captain, later rear-admiral then vice admiral (1814). He received a gold medal from the Royal Society of Arts (1793)

and was elected a Fellow of the Royal Society (1801). Fifteen years after the *Bounty* mutiny he was appointed governor of New South Wales in Australia.

Blindia (Seligeriaceae) Bruch & Schimp.
For Jean-Jacques Blind (1806–1867), German Pastor and plant collector in Münster, Germany from 1834–1848. He contributed to the collection of plants from Alsace and the Vosges made by JB Mougeot associated with CG Nestler, now preserved in the herbarium at the Museum of Grenoble.

Blotiella (Dennstaedtiaceae) R.M.Tryon
For Marie-Laure Tardieu-Blot (1902–1998), French botanist and physician, pteridologist and plant collector, who obtained a doctorate in natural science from the Faculty of Sciences in Paris, 1943, was deputy director of the National Museum of Natural History from 1964, and co-author of floras of New Caledonia, Madagascar and the Comoros, Cambodia, Laos and Vietnam, Cameroon, including *General Flora of Indo-China* (1932–1950), and *Pteridophytes of French intertropical Africa* (1953). In 1932, he was appointed a member of the Société Botanique de France.

Blumea (Asteraceae) DC.
For Karl Ludwig von Blume (Karel Lodewijk Blume) (1796–1862), German botanist, studied  medicine at Leiden University and went to Java to study native medicines (plants) in 1817. He became director of the Buitenzorg Botanical Gardens, now Bogor (1722), returned to The Netherlands in 1726 for health reasons and became director of the state herbarium at Leiden. He authored, *inter alia*, the four-volume *Rumphia* (1835–1848), in line with the Linnaean classification system, and *Flora Javae nec non Insularum Adjacentium* (1829), considered among the finest works among the literature of the Old World orchids. He was elected a foreign member of the Royal Swedish Academy of Sciences in 1855.

Bobartia (Iridaceae) L.
For Jacob Bobart (1599–1680), German botanist and the first horti praefectus (superintendent, head gardener) of the Oxford Physic Garden; which cultivated medical herbs; the first garden of its kind in England. He was the author of *Catalogus Plantarum Horti Medici Oxoniensis, sci Latino-Anglicus et Anglico-Latinus* (1648); a

Usually most common on overgrazed pastures (CM)

catalogue of 1600 plants that were in the garden. His son, Jacob Bobart the Younger (1641–1719), succeeded his father as horti praefectus and became acting professor of botany at Oxford.

Bobgunnia (Fabaceae) J.H.Kirkbr & Wiersema
For Charles R 'Bob' Gunn (1927–), research botanist at the Systematic Botany and Mycology Laboratory, Agricultural Research Service, US Department of Agriculture, and an authority on the *Fabaceae*. He participated in the First International Legume Conference at the Royal Botanic Gardens, Kew, in 1978. The aim of the conference was to arrive at a consensus on the tribal and generic classification of legumes. As a first step he surveyed legume seed characters, then prepared a nomenclature of legume genera for use in his databases. He described the subfamilies *Caesalpinioideae* (Gunn, 1991) and *Mimosoideae* (Gunn, 1984). He was the contributing editor of a newsletter called *The Drifting Seed*, and author of *Legume Fruit and Seeds*.

Boeckeleria = Tetraria (Cyperaceae) T.Durand
For Johann Otto Boeckeler (1803–1899), German apothecary-botanist and algologist of Oldenburg, specialist in sedges, author of *Flora oder allgemaine Botanische Zeitung* and *Die Cyperaceen des Königlichen Herbariums zu Berlin* (1868).

Boeckhia *** = ***Hypodiscus*** (Restionaceae) Kunth
For Philipp August Böckh (Boeckh) (1785–1867), German classical scholar and antiquarian. In

1803, he studied theology at the University of Halle-Wittenberg but switched to philology when he met Friedrich August Wolf, a classical scholar who taught at Halle. Böckh became his best student. After graduating he moved into academia and in 1811 became professor of rhetoric and classical literature at Humboldt University, Berlin, later dean and rector. Böckh extended the meaning of philology to higher levels to include not only the study of languages and literature, but all documents and their contextual setting – history, beliefs, morality, religion, art and antiquity. He was an appointed member of the Royal Prussian Academy of Sciences in 1814.

Boehmeria * (Urticaceae) Jacq.
For Georg(e) Rudolf Boehmer (Böhmer) (1723–1803), German physician and botanist, who studied at the University of Leipzig before becoming professor of botany and anatomy at the University of Wittenberg in 1752. Later he became professor of therapy. While working at the university he had part-time duties as city physician for Wittenberg and later at Kemberg. He authored many works, systematically recording the history of botany including his five-volume *Library of Natural History* (1785–1789), *Lexicon of Things in Herbarium, Technical History of Plants, Systematic and Literary Handbook of Natural History,* and *Economics and Related Sciences and Arts*.

Boerhavia (Nyctaginaceae) L.
For Herman Boerhaave (1668–1739), Dutch anatomical physician, botanist and chemist, who

held three professorships at the University of Leiden, that of professor of botany and medicine (1709), of practical medicine (1714, a year in which he also became rector), and of chemistry (1718). His reputation was such that he tutored Peter the Great in maritime affairs and knew Linnaeus and Voltaire well. During his career he introduced the modern system of clinical instruction, described 'Boerhaave's syndrome', a condition involving a torn oesophagus, described plant species, and published numerous works in medicine and chemistry. In 1728 he was elected to the French Academy of Sciences.

Boivinella *** = ***Englerophytum*** (Sapotaceae) (Pierre ex Baill.) Aubrév. & Pellegr.
For Louis Hyacinthe Boivin (1808–1852), French botanist, plant collector for the French National Museum of Natural History in Paris. He visited the islands of the Indian Ocean (Comoros, Seychelles and Réunion) as botanist on the Oise expedition (1846–1852) and visited Mauritius in 1847–1849 as well as the coasts of Africa, the Canary Islands and Madagascar. He gathered a vast and important collection of specimens that was deposited in the Museum of Natural History for identification and classification of the new species. Unfortunately, shortly after he returned from his mission, exhausted and suffering from malaria, he died in a hospital in Brest.

Bojeria *** = ***Euphorbia*** (Euphorbiaceae) DC.
For Wenceslas Bojer (1797–1856), Czech-born naturalist, horticulturist, botanist and plant

collector. He worked in the Imperial Museum at Vienna, studied and collected the flora of Mauritius, Madagascar, the Comoro Islands and the east coast of Africa extensively, and was the author of *Hortus mauritianus: ou enumeration des plantes, exotiques et indigènes, qui a l'Ile Maurice croissent, disposées d'après la méthode naturelle* (1837) and *Nouvelles espèces de plantes à Madagascar et îles Comores* (1841). He was one of the founders of the Société royale des Arts et des Sciences de l'île Maurice, the first scientific association of Mauritius, was curator at the Museum Desjardins (1842) and director of the Jardin des Pamplemousses (1848).

Bokkeveldia = ***Strumaria*** (Amaryllidaceae) U. & D.Mül-Doblies
Afrikaans *bokkeveld* = lit. 'goat field'; referring to a veld type that occurs in the northwest Cape that includes Cape fynbos, mountain renosterveld and succulent Karoo vegetation types.

Bolandia (Asteraceae) Cron
Afrikaans *boland* = lit. 'up country'. A region in the Western Cape.

Bolbitis (Dryopteridaceae) Schott.
La. *bolbos* = bulb; *-itis* = likeness; referring to the small gemmea borne on fronds in some cases or to the vegetative buds which occur on the apical portion of the frond in the type species.

Bolborchis = ***Nervilia*** (Orchidaceae) Zol. & Moritzi.
La. *bolbos* = bulb; *orchis* = orchid; alluding to the conspicuous ovoid pseudobulbs.

Bolboschoenus (Cyperaceae) (Asch.) Palla
Gk. *bolbos* = swelling or bulb; La. *schoenus,* from Gk. *skhoinos* = a rush or reed. *Boluschoenus* is a rhizomatous perennial, the name of which alludes to the presence of corms whereas the genus *Schoenus* 'has no tubers', perhaps alluding to the bulbous base of these rush-like plants.

Bolusafra (Fabaceae) Kuntze
For Harry Bolus (1834–1911), English-born South African botanist, botanical artist, stockbroker, and plant collector who arrived in South Africa in 1850. He started collecting in 1865 and organised six major collecting expeditions within South Africa, Mozambique and Swaziland during 1883–1904. He had a passion for orchids, writing, *inter alia*, the three-volume work *Icones Orchidearum Austro-Africanum Extra-Tropicarum* (1893–1913), the third volume published posthumously. He founded the Bolus Herbarium and bequeathed his library to the (now) University of Cape Town, as well as sufficient funds to develop a chair of botany. He was a Fellow of the Linnaean Society, and member and president of the South African Philosophical Society (later the Royal Society of South Africa).

Bolusanthemum = ***Bijlia*** (Aizoaceae) Schwantes
For Harriet Margaret Louisa Bolus (*née* Kensit) (1877–1970), South African analytical botanist,

taxonomist and daughter-in-law of Harry Bolus, whose son, Frank, she married. She obtained a BA honours from the South African College, Cape Town (now the University of Cape Town) (1902) and an honorary DSc from the University of Stellenbosch (1936). In 1903 she was made curator of Harry Bolus's private herbarium, bequeathed to the University of Cape Town in 1911, but retained her position until 1955 when she retired aged 77. She had a vast knowledge of Cape flora with a special interest in *Mesembryanthemaceae*. In her career she described some 1 700 new species of succulent plants. She wrote popular and scientific works including *Notes on Mesembryanthem and Some Allied Genera,* A *Book of South African Flowers,* and *A Second Book of South African Flowers.*

Bolusanthus (Fabaceae) Harms
Bolusafra (q.v.); *anthos* = flower.

Pendulous flowers can be spectacular in spring. (EJM)

Bolusia (Fabaceae) Benth.
Bolusafra (q.v.).

Bombax* (Malvaceae) L.
Gk. *bombyx* = silk (or cotton); referring to the silken fibres or referring to the wool in the pods.

Bolusiella (Orchidaceae) Schltr.
See *Bolusafra*; La. *-iella* = (diminutive).

Bonamia (Convolvulaceae) Thouars
For François Bonamy (1710–1786), French physician and botanist, from a long line of

apothecaries. He received a doctorate in medicine in 1735 from the Royal University of Nantes and became professor of botany where he taught for over 40 years, during which time he became university president (rector) and dean of the faculty. He was also a member of the Royal Society of Medicine. He established a garden of apothecaries (medicinal plants) at Nantes and taught botany for free or next to nothing to those who visited the garden. He was author of *Florae Nannetensis prodromus*

(*Flora around Nantes*) (1782) and founded the first society of agriculture in France.

Bonatea (Orchidaceae) Willd.
For Guiseppe Antonio Bonato (1753–1836), Italian botanist who was professor of botany at  Padua and praefectus of the Botanical Garden of Padua, author of *Pisaura automorpha e Coreopsis formosa*. His personal herbarium, together with that of his predecessor Giovanni Marsili, formed the basis of the Padua Herbarium, now housing in excess of a half million specimens.

Bonnaya*** = **Lindernia** (Linderniaceae) Link & Otto
For Charles François, Marquis de Bonnay (1750–1825), French military officer, magistrate,  diplomat and statesman, twice president of the National Assembly, French ambassador in Copenhagen (1814), royal French envoy to Prussia, envoy extraordinary and minister plenipotentiary in Berlin (1815), minister of state and member of the king's Privy Council (1820) and governor of the royal chateau of Fontainebleau (1821). He authored *Prospectus d'un nouveau Journal* (1789), *La Prise des Annonciades* (1790) and a translation of *Tristram Shandy* (1785).

Bonyunia (Loganiaceae) M.R.Schomb. ex Progel.
Possibly a Portuguese name as it occurs in lowland regions of the Amazon River watershed and elsewhere in South America. Meaning obscure.

Boophone (Amaryllidaceae) Herb.
Gk. *bous* = ox; *phonos* = murder, slaughter; referring to the bulb sap which is poisonous, quite capable of killing oxen; hence one of the common names is 'Oxbane'. *Boopis* means 'ox-eyed', a result of induced hallucinations from *Boophone disticha*, a traditional Zulu medicine.

Bopusia*** = **Graderia** (Orobanchaceae) C.Presl
Possibly an anagram of *Sopubia*.

Boraginella*** = **Trichodesma** (Boraginaceae) Kuntze
La. *borago*, possibly from La. *burra* = a heavy garment; *-iella* = diminutive; referring to the hairy leaves and stem.

Borago* (Boraginaceae) L.
The derivation of this name is uncertain. One suggestion is that the name is a corruption of La. *corrago* = courage, from *cor* = the heart; *ago* = I bring (food), named because the plant had a reputation since Roman times 'to give one courage' as a result of its food value. The herb can be eaten raw as salad greens or cooked like spinach, use as tea, etc. Other possibilities have been suggested.

Borassus (Arecaceae) L.
Gk. *borassos* = the immature spadix of the female date palm.

Borbonia = **Aspalathus** (Fabaceae) L.
For Jean-Baptiste Gaston de Bourbon, Duke of Orléans (1608–1660), third son of Henri IV  of France, and patron of botany. His life is not related here. David Hollombe, researcher, advises 'Gaston of Orleans, brother of Louis XIII, established a botanical garden at his palace of Blois, which had acquired much celebrity from the works of Morison, and by drawings of the most remarkable plants. Gaston of Orléans, not satisfied with the mere collection of plants of every country in his garden at Blois, had them described by learned botanists, and the most remarkable species drawn on vellum, by the painter Robert, eminent for his skill in that branch of the art.'

Borraginoides*** = **Trichodesma** (Boraginaceae) Boehm
Borago (q.v.); Gk. *-oides* = resembling Borage. The double 'r' in 'Borrago', also found in 'Borraginaceae' is probably an ancient version of the family name or typographic error. The only place where this is found is in Philip Miller's *Gardeners Dictionary,* 1754. This name is no longer listed on The Plant List.

Borreria*** = **Spermacoce** (Rubiaceae) G.Mey.
Borrera (q.v.). (This can be found in the Lichens section.)

Boschia (Malvaceae) Korth.
For Johannes van den Bosch (1780–1844), Dutch lieutenant who became Governor-General of the Dutch East Indies (1830–1833). He ruled

the colony jointly with an advisory board at a time when the Netherlands was near bankruptcy. To redress this situation, he introduced his 'cultivation system', mainly in Java. He ordered farmers to devote 20% of their land for government crops for export and peasants had to work in government-owned plantations for 60 days of the year. This policy allowed the Dutch East Indies budget to be balanced. In 1834, Bosch returned to the Netherlands, was appointed minister of colonies and demanded ever-higher financial results from the colonies at the detriment of farmers and slaves. In 1839 he was criticised for the opaqueness of his policy on loans between the government and Netherlands Trading Society and he stepped down in 1840 with the honourable title of Minister of State and was also made a 'count' by royal decree. Korth(als), author, was an official botanist in Java (1831–1836).

Boscia (Capparaceae) Lam.

Trees often show a sharp browse-line as they are most palatable to large herbivores. (EJM)

For Louis Augustin Guillaume Bosc (1759–1828), French botanist, invertebrate zoologist, entomologist and horticulturist. He studied botany under Antoine Laurent de Jussieu at the Jardins des plantes. In 1797 he moved to America, due to the unrest in France, where he explored its natural riches. Much of what he discovered is included in the works of BGE de Lacépède (fish), PA Latreille (reptiles) and others. In 1803, a few years after his return to France, he was made inspector of the gardens and the public nurseries at Versailles. His copious contributions to scientific literature include *Dictionnaire d'histoire naturelle*, the *Encyclopedic methodique*, and he was one of the editors of the *Annales de l'agriculture francaise*. In 1806 he was elected a member of the Academy of Science and in 1825 took the Chair of Plant Culture at the National Museum of Natural History.

Bosqueia*** = Trilepisium (Moraceae) Thouars ex Baill.
Portuguese *bosqueia* = forest, woods. The genera consist mainly of trees, shrubs or vines.

Bothriochloa (Poaceae) Kuntze
Gk. *bothrion* = a little pit; *chloa* = grass; referring to the pitted glumes.

Bothriocline (Asteraceae) Oliv. ex Benth.
Gk. *bothros* = hollow; *cline* = bed. The flowers are in hollows in the receptacle.

Botryceras** = Laurophyllus
(Anacardiaceae) Willd.
Gk. *botrys* = bunch, cluster; *keras* = a horn. The female inflorescence resembles a cluster of much-branched antlers.

Botrychium (Ophioglossaceae) Sw.
Gk. *botrys* = bunch, cluster (of grapes) or raceme; alluding to the grape-like clusters of sporangia on the fertile segments of the fronds.

Botryopteris*** = Helminthostachys
(Ophioglossaceae) C.Presl
Gk. *botrys* = bunch, cluster; *pteris* = fern; a fern with sporangia borne in grape-like clusters.

Boucerosia*** = Caralluma (Apocynaceae) Wight & Arn.
Gk. *boukeros* = like the horns of an ox; referring to the shape of the sections of the corona.

Bouchea (Verbenaceae) Cham.
For Peter Carl (Karl) Bouché (1783–1856) and Peter Friedrich Bouché (1785–1856), German horticulturists and botanists, members of the Bouché family of nurserymen. Peter Carl was a founding member of the Society of Gardeners of the Royal Government of Prussia in 1822 and a student of Carl Willdenow, with whom he went on collecting trips. He also corresponded with other contemporaries in the botanical field such as Carl Sigismund Kunth, Diederich Franz Leonhard von Schlechtendahl and Adelbert von Chamisso. Peter Friedrich Bouché was also an entomologist whose collection is in the German Entomological Institute.

Bougainvillea* (Nyctaginaceae) Comm. ex Juss.
For Louis-Antoine, Comte de Bougainville (1729–1811), French admiral and explorer. He

joined the army in 1753 and took part in the Seven Year War with Britain (1754–1756), obtaining the rank of colonel. In 1763 the French tried to colonise the 'Isles Malouines' (Falkland Islands/Malvinas). Bougainville made three expeditions to the island, landing 130 settlers, but when the Spanish government claimed the islands were theirs, for political reasons, the French withdrew. In 1766, Louis XV gave Bougainville permission to circumnavigate the world, the first Frenchman to do so, a three year journey he described in his *Voyage autour du monde*. Between 1779 and 1782 Bougainville saw active service during the American War of Independence, which led to Britain's defeat. He obtained the rank of vice-admiral (1791), was made a senator by Napoleon (1799), was given the Légion d'honneur (1804), and was made the comte de Bougainville by Napoleon (1808).

Bowiea (Hyacinthaceae) Harv. ex Hook.f.
For James Bowie (1789–1869), English botanist and plant collector. He joined the staff of the Royal Botanic Gardens, Kew, in 1810, and after being trained in horticulture and plant collecting was sent to Brazil in 1814 by Joseph Banks. Two years later, he was instructed to go to South Africa where he spent six years (1818–1822) collecting plants mainly in the Eastern Cape region. The Irish botanist William Henry Harvey, who spent a considerable amount of time in South Africa considered Bowie to have provided Europe's gardens with more succulents than any other individual. He was recalled to England in 1823 and lost his job, but returned to South Africa in 1827 and worked as garden superintendent and plant collector for Baron von Ludwig, a prominent and wealthy Capetonian until around 1842, then for Ralph Henry Arderne (1802–1885) at what is today Arderne Gardens in Claremont. He wrote the earliest guide to Cape flora printed in South Africa (1829). He died an alcoholic and in poverty.

Bowkeria (Stilbaceae) Harv.
For James Henry Bowker (1822–1900), South African naturalist and army colonel who took part in the seventh and eighth Cape Frontier Wars (1846–1853). He became an authority on butterflies collected in the Eastern Cape and elsewhere (including plants) together with his

artist sister, Mary Elizabeth Barber née Bowker (1818–1899), a noted botanist and entomologist who corresponded regularly with Joseph and William Hooker at the Royal Botanic Gardens, Kew, and Charles Darwin. She sent many plant specimens, many previously unknown, to the herbarium of Trinity College, Dublin and Kew Gardens. She authored scientific papers on botany and entomology and general articles in many fields, including stone-age implements. The Bowkeria name commemorates them both.

Brabejum (Proteaceae) L.

(EJM)

Derivation uncertain, possibly Gk. *brabeion* = sceptre; referring vaguely to the shape of the star-like leaves which can be made into a 'crown' in the same way that bay or laurel leaves were made into a 'crown' awarded to winners at the Pythian games held at Delphi.

Brachiaria (Poaceae) (Trin.) Griseb. (Image on following page)
La. *brachium* = forearm; *-aria* = related to; referring to the branched inflorescence that extends like arms almost at right angles along the main axis.

Brachyachne (Poaceae) (Benth.) Stapf
Gk. *brachys* = short; *achne* = scale, glume; alluding to the lemma or flowering glumes being shorter than the two outer glumes.

Brachycarpaea = Heliophila (Brassicaceae) DC.
Gk. *brachys* = short; *karpos* = fruit, hence 'short fruits'; possibly referring to the short-stalked fruits.

Brachiaria (EJM)

Brachychiton* (Malvaceae) Schott & Endl.
Gk. *brachys* = short; *chiton* = a tunic; referring to the overlapping scales or the bristles surrounding the seed in fruit.

Brachychloa (Poaceae) S.M.Phillips
Gk. *brachys* = short; *chloa* = young corn or grass; referring to the racemes, which are short in comparison to those of *Leptochloa*.

Brachycorythis (Orchidaceae) Lindl.
Gr. *brachys* = short; *kopus, korys* = helmet; referring to the strongly cucullate perianth, which may sometimes vaguely resemble a helmet (Lindley, 1838).

Brachylaena (Asteraceae) R.Br.
Gk. *brachys* = short; *chlaina* = a cloak; refers to the short floral bracts surrounding the disk, which are shorter than the florets.

Brachymeris*** = **Phymaspermum** (Asteraceae) DC.
Gk. *brachys* = short; *meris* = a part. The leaves and parts of the flower are small.

Brachypodium (Poaceae) P.Beauv.
Gk. *brachys* = short; *podion* = little foot; referring to the short-stalked (subsessile) spikelets.

Brachyrhynchos*** = **Senecio** (Asteraceae) Less.
Gk. *brachys* = short; *rhynchos* = beak, bill; possibly referring to the narrow or prolonged tip on the fruit or seeds.

Brachysiphon (Penaeaceae) A.Juss.
Gk. *brachys* = short; *siphon* = tube; referring to the floral tube, which is short.

Brachystegia (Fabaceae) Benth.
Gk. *brachys* = short; *stegia* = a roof or covering. The buds are protected by a pair of bracteoles.

Brachystelma (Apocynaceae) R.Br.
Gk. *brachys* = short; *stelma* = crown, garland, wreath; alluding to the short staminal corona of some species.

Brachystelmaria = Brachystelma (Apocynaceae) Schltr.
Gk. *brachys* = short; *stelma* = crown, garland, wreath; *-aria* = related to; the same as above.

Brackenridgea (Ochnaceae) A.Gray
For William Dunlop Brackenridge (1810–1893), Scottish-born American nurseryman, landscape architect and horticulturist who came to the United States in 1837. He was the naturalist and assistant botanist on the US exploring expedition of 1838–1842 led by commodore Charles Wilkes, and reportedly  did all the collecting and serious botany work. For several years on the US ship *Vincennes*, he was responsible for collecting plants during its exploration of New South Wales, and the Oregon and California coasts. He also collected in the Philippines, Fiji and Tahiti, the Canary and Cape Verde Islands, New Zealand, Madeira and Hawaii.

Bracteolaria = Baphia (Fabaceae) Hochst.
La. relating to the bracteoles. Possibly referring to the plant's distinctive bracteoles.

Brasenia (Cabombaceae) Schreb.
For Christoph Brasen (1738–1774), a Danish surgeon. He first met the Moravian missionaries in 1758 and later became the leader of the 1771 missionary expedition that established the Moravian mission of Nain on the coast of Labrador, the purpose of which was to convert the Inuit residents there to Christianity. He served as its first superintendent, his administrative skills, strong leadership, surgical expertise and scientific curiosity being highly valued. In 1774, while exploring the northern Labrador coast to establish a second mission post, he drowned during a fierce storm. The genus was named in 1789 by German naturalist and professor of natural history, Johann Christian Daniel von Schreber, who frequently

received collected plant specimens from the Moravians, including those from Brasen.

Brasilettia (Fabaceae) DC.
Spanish *brasilete* = the Brazil-wood tree.

Brassica* (Brassicaceae) L.
Latin name for the cabbage plant, used by Plautus.

Braunsia (Aizoaceae) Schwantes
For Hans Heinrich Justus Carl Ernst Brauns (1857–1929), German entomologist and physician. He qualified with a PhD from Schwerin University, got his MD from Leipzig, and worked for the German Medical Service before emigrating to South Africa for health reasons in 1895. He practiced in Port Elizabeth, Hoopstad (in the then Orange Free State, now the Free State) and eventually settled in Willowmore in the Karoo in 1899. He collected insects extensively, especially *Hymenoptera,* and took field trips to the Karoo, Cape Province, Orange Free State and Transvaal. He provided specimens to the South African Museum, the Hungarian Natural History Museum, and Natal Museum. In 1924 he sold his collection to the Transvaal Museum. He was a member of the Royal Society of South Africa and an honorary member of the Societe des Sciences Naturelles, Musee du Congo Belge, Tervuren.

Brayulinea* = **Guilleminea**
(Amaranthaceae) Small
For William Bray (1865–1953) and Edwin Burton Uline (1867–1933), US botanists, both students of the Amaranthaceae.

Bray received his PhD in 1898, worked for 10 years at the University of Texas before moving to Syracuse University, where he became professor of botany and first dean of the New York State College of Forestry at Syracuse University. He pursued a course of research in the areas of phytogeography, forest resources in Texas and New York, plant adaptive strategies, pest species, and plant ecology. In 1915, he was one of the co-founders of the Ecological Society of America. Uline was a New York high school principal with a number of botanical publications to his credit. Further details unknown.

Breonadia* (Rubiaceae) Ridsdale
For Jean Nicolas Bréon (1785–1864), French horticulturalist and botanist. He was admitted to the Muséum National d'Histoire Naturelle in Paris as a gardener-student, where he was trained by A Thouin and A de Jussieu (1809). He was sent to Ajaccio, Corrsica in 1813 and became the first director (1817–1831) of the Jardin du Roy (now the Jardin de l'Etat) on Île Bourbon (today Réunion Island). He compiled the first catalogue of plants on Réunion Island called *Catalogue des Plantes Cultivées au Jardin Botanique et de Naturalisation de île Bourbon* (1822). He also explored the east coast of Madagascar (1818) and went to the Persian Gulf, Maldives and Mauritius (1821). His health deteriorated in 1827 and he retired in 1829. Bréon returned to France in June 1833 and later became a founder (1841) and president of the Cercle d'Horticulture, which became the Société Nationale d'Horticulture. He also collected plants in the Mascarene Islands.

Breweria (Convolvulaceae) R.Br.
For Samuel Brewer (1670–1743), amateur British botanist and bryologist, collector of plants, insects and birds. He came from a family of affluent textile merchants and was involved in wool manufacture, and botanised in northern Wales with Johann Jakob Dillen (Dillenius), a German botanist who moved to England and who became Sherardian professor of potany at Oxford University. He also provided him with some plants for the third edition of John Ray's (1607–1725) *Synopsis Methodica Stirpium Britannicarum* published in 1724. He seems to have been unfortunate in business with troubled family relationships. He became head gardener to the duke of Beaufort at Badminton.

Brexia* (Celastraceae) Thouars
La. *brexia* = rain; referring to the leaves which protect against rain.

Brianhuntleya (Aizoaceae) Chess., S.A.Hammer & I.Oliv.
For Brian Huntley (1944–), South African biodiversity expert and professor of botany at

the University of Cape Town. He studied at the universities of Natal, Pretoria and the Orange Free State, obtaining an MSc for research on Marion Island. He spent several years with the Transvaal Department of Nature Conservation and four years in Angola. In the 1970s he joined the CSIR (Council for Scientific and Industrial Research) and coordinated its savanna ecosystem project and later all national ecosystem projects. In 1989 he became director of the National Botanical Institute (later, the South African National Biodiversity Institute), and was responsible for eight national botanical gardens, three research centres, four major bio-regional programmes and more than 100 school-based environmental greening projects.

Bridelia (Phyllanthaceae) Willd.
For Samuel Elisée von Bridel (1761–1828), Swiss botanist, bryologist, poet, muscologist and librarian, and author of *Bryologia universa*. He studied at the University of Lausanne and later went to Gotha (Thuringia, Germany), where he taught the sovereign's children, princes August and Friedrich von Sachsen-Gotha. He was one of the foremost bryological leaders of his time, and also published the two-volume work entitled *Muscologia recentiorum*. Most of his moss herbarium was acquired by the Botanical Museum of Berlin and fortunately escaped destruction during an air raid in World War II.

Brillantaisia* (Acanthaceae) P.Beauv.
For M Louis-Marie Brillantais-Marion, 19th-century French timber merchant, shipbuilder, and founder with Captain Jean-François Landolphe of a company that had exclusive trading privileges in Benin granted by King Louis XVI. He apparently provided services that helped to facilitate the researches of the French botanist and entomologist Ambroise Marie François Joseph Palisot, baron de Beauvois, who collected in Benin and neighbouring Nigeria.

Briza* (Poaceae) L.
Gk. *brizo, brezein* = to vibrate; alluding to the flowers and seedheads which shake on their stalks in the slightest wind.

Brizopyrum*** = *Tribolium* (Poaceae) Stapf
Gk. *brizo, brezein* = to vibrate; *pyros* = wheat; possibly referring to the drooping branches of this genus.

Bromus (Poaceae) L.
Gk. *bromos* = a kind of oat, which provided edible grain.

Brousonnetia* (Moraceae) L'Her. ex Vent.
For Pierre Marie Auguste Broussonet (1761–1807), French physician, naturalist and politician. His career was chequered: doctorate in medicine,

University of Montpellier (1799); moved to London (1780) becoming a member of the Royal Society; published a work on fish *Ichthyologiae Deca 1*; returned to Paris (1782) and did some botanising before becoming permanent secretary of the Agricultural Society of Paris (1785). He was elected to the National Assembly (1789) and welcomed the French Revolution. But after seeing his friend Louis Bénigne François Berthier de Sauvigny accused of causing the famine and lynched on the street, and with extremists threatening his own life, he fled to Spain (1794) and then Morocco where he became French consul (1797) but had to flee again (1799) as a result of the plague. He became a commissioner of French government trade relations in Tenerife (1803), took up the chair of botany at Montpellier with responsibility for the city's botanical garden (1804), and published a catalogue of the garden under the title *Elenchus plantarum horti botanici monspeliensis*. He was again appointed a member of the legislature (1805).

Browallia* (Solanaceae) L.
For Johannes Browallius (1707–1755), Finnish and Swedish Lutheran theologian, physicist and

botanist. He was a professor of physics (1737–1746) at the Åbo Akademi, now the University of Helsinki, then professor of theology (1746–1749) and bishop of Turku (Åbo) (a diocese of the Church of Sweden). He became vice-chancellor of the Royal Academy of Turku from 1749 to 1755. Linnaeus's friendship with him was confirmed when Browallius defended Linnaeus's *Systema naturae* (1735) against Johan Georg Siegesbeck's vicious attack on his theory that stamens and pistils could be likened to male and female organs by which plants reproduce themselves. This friendship turned into hatred when Browallius advised Linnaeus to finish his studies abroad and marry a rich girl, even though he was already engaged to Sara Lisa Morea. In Linnaeus's absence abroad, he courted Morea.

Brownanthus = *Mesembryanthemum* (Aizoaceae) Schwantes
For Nicholas Edward Brown (1849–1934), British botanist at Kew Gardens from 1873. Although he

never visited South Africa, he became an expert on African plants, particularly succulents. In 1921 he was awarded the Captain Scott Memorial medal in recognition of his work on South African flora and in 1932 an honorary DSc was conferred on him by the University of the Witwatersrand. His papers appeared mainly in the *Kew Bulletin* and in *Flora Capensis*.

Brownleea (Orchidaceae) Harv. ex Lindl.
For Rev. John Brownlee (1791–1871), British botanist who was a gardener, theologian, Xhosa linguist and missionary in South Africa.

He arrived in Cape Town in 1817 and established a mission station on the site that would become King William's Town. He was a well regarded botanist and had an extensive garden of local plants, and sent local specimens to William Henry Harvey, an Irish botanist who came out to South Africa in 1835 and who wrote *Flora Capensis*.

Bruguiera (Rhizophoraceae) Lam.
For Jean Guillaume Bruguière (1750–1798), French physician, zoologist, botanical artist and plant collector. He studied medicine at the University of Montpellier before turning to the natural sciences.

He went to the Antarctic with Kerguelen Trémarac (1773), visiting the Cape, Mauritius, Rodrigues and Kerguelen islands, and Madagascar where he collected specimens. He was mainly interested in life forms such as molluscs and other invertebrates. He also accompanied the entomologist Guillaume-Antoine Oliver to Persia (1790) but had to abort the trip because of poor health. In 1792, he made another trip with Oliver to the Greek islands and Middle East, but died on the way home (1798). Bruguière partly completed a *Natural History of Worms,* Volume 1 (1792), contributed to the French encyclopaedia (the invertebrates section), and described various taxa, published posthumously.

Brunella = Prunella ** (Lamiaceae) L.
German *brunella* = name for Quinsey. Linnaeus incorrectly spelled the genus *Prunella*, now renamed. The genus allegedly has curative medicinal properties. Quinsey is a tonsillar abscess at the back of the throat caused by streptococcus, and is often a complication of tonsillitis.

Brunia (Bruniaceae) Lam.
For Alexander Brown (f 1692–1698), a naval surgeon and plant collector who worked for the East India Company around 1690 and collected in India, the Cape, Spain and Arabia, etc. sending specimens to Plukenet (1641–1706), an English botanist, royal professor of botany and gardener to Queen Mary; James Petiver (c 1665–1718) a London apothecary; Jacob Bobart (c 1665–1718) in Oxford and to Charles du Bois (1656–1740), an English merchant and botanist, treasurer of the East India Company. He amassed a vast herbarium of East Indian plants. No further details are known.

Brunsvigia (Amaryllidaceae) Heist.

The dried inflorescences form large 'balls' that roll through the veld on windy days scattering their seeds. (TD)

For Karl (Carl) Wilhelm Ferdinand (1713–1780), also known as Charles I or Karl I, duke of Brunswick-Lüneburg, able military commander, ruler of Brunswick-Lüneburg from 1773, patron of the arts and sciences who promoted the study of plants, including the beautiful Cape species *B. orientalis*. He also founded the *Collegium Carolinum*, an institute of higher education, which is today known as the Technical University of Brunswick. The name Brunswick is the Latin translation of Braunschweig, a town in Germany that was a sovereign duchy of northern Germany between the 10th and 19th centuries.

Bryomorphe (Asteraceae) Harv.
Gk. *bryon* = moss*; morphe* = form. An 'alpine' daisy with moss-like characteristics and preferring a moss-like habitat.

Bryonia* (Cucurbitaceae) L.
Gk. *bryo* = to grow rapidly; *bryon* = oyster-green. *Bryon* normally means moss, but *bryro* can mean 'luxuriant growth' as would be the case in leafy hedgerows, a use to which the species *B. dioica* is put.

Bryophyllum* = *Kalanchoe (Crassulaceae) Salisb.
Gk. *bryo* = to blossom, come into bud; moss; *phyllon* = a leaf; alluding to new plantlets sprouting from notches in the leaves.

Bubon * = *Notobubon*** (Apiaceae) L.
Gk. *bubon* = inflammation of a node; *boubon* = the groin; alluding to its medicinal qualities.

Bucco * = *Agathosma*** (Rutaceae) H.Wendl.
A word of Khoisan origin, possibly becoming the Zulu word *bucu,* and transliterated *boegoe* (Afrikaans), *buchu* (English). A genus of small shrubs, some of which are highly aromatic and used in herbal remedies and for commercial purposes such as flavouring and perfume.

Bucculina* = *Holothrix (Orchidaceae) Lindl.
La. *buccula, bucculina* = little cheek; alluding to the 'fleshy broad-toothed concave petals, which stand erect, and apparently adhere by their edges into a sort of vaulted pointed arch, giving the flower the appearance of having two little cheeks (bucculae)' (Lindley, 1836).

Buchenroedera* = *Lotononis (Fabaceae) Eck. & Zeyh.
For Wilhelm Ludwig von Buchenröder (1783–1841), South African botanist, lieutenant in the Hottentot Light Infantry, and later merchant in the Uitenhage District where he bought two farms. He was a friend of Danish botanical collector and apothecary Christian Friedrich Ecklon and German botanist Carl Ludwig Philipp Zeyher, and the genus was named in 1836 by them.

Buchnera (Orobanchaceae) L.
For Andreas Elias Buchner (1701–1769), a German physician and naturalist. He studied medicine at the universities of Halle and Leipzig, graduating in 1721. He did his housemanship (apprenticeship) in Erfurt, before becoming a physician in Rudolstadt. In 1729, he accepted an offer to become an associate professor and later full professor at Erfurt (1736–1745) and later at the University of Halle, serving as Vice-President on three occasions. For many years he was a member of Leopoldina, the German Academy of Natural Sciences and, in 1735 he became its president. He was the editor of *Miscellanea physico-medico-mathematica* (1731–1734) and author of *An Easy and Very Practicable Method to Enable Deaf Persons to Hear* (1759), printed in English (1770).

Bucranion * = *Utricularia*** (Lentibulariaceae) Raf.
La. *bucrania* = an ox head. A sculptured ornament, representing an ox skull adorned with wreaths, etc.

Buddleja (Scrophulariaceae) L.
For Adam Buddle (1660–1715), English amateur botanist, vicar of Farnham, Essex, and collector of British plants. He was educated at Cambridge University, obtaining a BA degree in 1681 and an MA degree in 1685 and was ordained into the Church of England in 1703. Further details of his life are obscure. He is credited with creating Britain's first herbarium and he compiled a new English flora, completed in 1708, but it was never published; the original manuscript is preserved at the Natural History Museum, London. He established a reputation as an authority on bryophytes.

Buglossoides (Boraginaceae) Moench
The genus *Buglossa*; Gk. *-oides* = resembling. Bugloss is a European plant of the genus *Anchusa.*

Bulbine (Asphodelaceae) Wolf
La. *bulbus* = an onion or bulb. A misnomer in that the plants do not have a bulbous base.

Bulbinella (Asphodelaceae) Kunth (Image on following page)
Bulbine (q.v.); Gk. *-ellus, -ella* = diminutive.

Bulbophyllum (Orchidaceae) Thouars (Image on following page)
Gk. *bolbos* = bulb; *phyllum* = leaf; referring to the thick, fleshy leaves – 'bulb-leaf'. These pseudobulbs bear one or two leaves at the apex.

Bulbostylis (Cyperaceae) Kunth
Gk. *bolbos* = bulb; *stylos* = a pillar, hence style. The deciduous style has a button-like base, which it leaves on the ovary when falling.

Bulliarda* = *Crassula (Crassulaceae) DC.
For Jean Baptiste François Bulliard (1752–1793), French physician, botanist and naturalist. He studied medicine in Langres, worked in hospices in Clairvaux and Paris, where he set up his own practice. He began his botanical studies at the Abbey of Clairvaux and was a pupil of Jean Jacques Rousseau. He authored *Flora parisiensis*, *Herbier de la France* (*Guide to the Herbs of France*) in 13 volumes, *Elémentaire Dictionnaire de botanique* (which consolidated botanical terminology and Linnaeus's system), *Histoire des plantes vénéneuses Suspectes et de la France* (which was reported to the police as

Bulbinella flowers resemble onion flowers but lack the bulbouos base. (CM)

Bulbophyllum (HS)

a dangerous book – being a guide to poisoning relatives and friends – and then removed from bookstores and confiscated) and *Histoire des Champignons de la France* (*Natural History of the Mushrooms of France*).

Bunburia = Cynanchum (Apocynaceae) Harv. Named after Charles Bunbury by his friend, the botanical author William Henry Harvey.

Bunburya = Tricalysia (Rubiaceae) Meisn. For Charles James Fox Bunbury (1809–1886), English baronet, naturalist, geologist with  a particular interest in paleobotany, botanist, traveller and plant collector. He was educated at Trinity College, Cambridge. He collected specimens on expeditions to Argentina, Uruguay and Brazil (1833–1835), visited South Africa for three months in 1838, botanising around Cape Town, often with William Henry Harvey, and then up to the Eastern Cape as far as the Keiskamma River, and went to Madeira (1853–1854) with Sir Charles Lyell the geologist. He authored *A Journal of a Residence at the Cape of Good Hope* (1848); *Notes on the Vegetation of Buenos Aires and the Neighbouring Districts* (1855); *Brazil, Account of a Journey in Brazil in 1833–35*; *Botanical Fragments* (1883) and numerous papers on geographical botany and on fossil plants. He was elected to the Royal Society in 1851.

Bunium (Apiaceae) L.
Gk. *bounion* = the name of an earth-nut; referring to plants, which have tuberous roots that are edible. Early Europeans called the potato an earth-nut.

Bupleurum (Apiaceae) L.
Gk. *bous* = ox; *pléuron* = rib; hence *bupleuros* = ox-rib. Various reasons suggested for name choice: an inference to ridges on the fruit or the cattle swelling after eating the leaves or the parallel veins of some leaves of the *Bupleurum* species.

Burchellia (Rubiaceae) R.Br.
For William John Burchell (1782–1863), English explorer, naturalist, traveller, artist and author.  The son of a botanist and nursery owner, he served his apprenticeship at Kew Gardens. In 1805 he travelled to St Helena where in 1807 he became a schoolmaster and acting botanist. In 1810 he travelled to Cape Town. Between 1810 and 1815, he travelled some 7 000km, some in uncharted territory, and collected 50 000 specimens which he took back to England. He described this journey in the two-volume *Travels in the Interior of Southern Africa* (1822). Apart from flowers, his name is also associated with Burchell's zebra and Burchell's coucal. Between 1825 and 1830, he travelled to Brazil collecting many specimens, including over 20 000 insects. His diaries of this trip are lost but his field notes survived.

Burkea (Fabaceae) Benth.
For Joseph Burke (1812–1873), British 'under-gardener' (botanist) and collector for the earl of Derby, Edward Smith Stanley (1775–1851), who was sent to South Africa arriving March, 1840. For the next two-and-a-half years, he explored the country together with the South African botanist Carl P Zeyher (1799–1858). By mid-June 1842, they had an 'immense collection of living and dead animals and dried plants, seeds, bulbs etc.' These were taken to England, the plant specimens finding their way to Kew Gardens. Shortly after his return, in December 1842, Burke married and settled in the United States where he went on a joint expedition with collectors from Kew to Hudson Bay and California.

Burmannia (Burmanniaceae) L.
For Johannes Burman (1707–1780), Dutch botanist and physician, a professor of botany at Amsterdam University who  studied at Leiden under Herman Boerhaave (1722), qualifying as a doctor in 1728, and who was a close friend of Linnaeus. He was the author of *Thesaurus zeylanicus* (Ceylon plants), *Rariorum Africanarum plantarum* (1738–1739), the first work devoted to plants of the Cape Colony, based on a collection by Nicolaes Witsen and the work of Caspar Commelin (and possibly Simon van der Stel), and *Flora malabarica* (dealing with plants on Ambon Island, part of the Maluku Islands of Indonesia).

Burnatia (Alismataceae) Micheli
For Emile Burnat (1828–1920), Swiss engineer, industrialist, magistrate and botanist. He worked at the *Conservatoire et Jardin botaniques* in Geneva. He was the author of *Flore des Alpes maritimes,* written in seven volumes with John Isaac Briquet and François Cavillier, and co-author with August Gremli of *Catalogue raisonné des Hieracium des Alpes Maritimes.*

Butomopsis (Alismataceae) Kunth
Gk. *bous* = ox; *tomos* = cutting; *-opsis* = resembling; referring to the sharp leaf margins.

Buttonia (Orobanchaceae) McKen ex Benth.
For Edward Button (1836–1900), British geologist, explorer, gold prospector, mining  pioneer, mine manager and amateur plant collector. He prospected in Natal from 1860–1864 as well as in the Transvaal from 1869–1871, discovering the Eersteling Gold Reef near Marabastad but mining operations were not successful because of difficulties with transport, labour and water supplies. In later years he was connected with the East Rand Gold Mines, Johannesburg. While in Natal in 1968, he collected plants, especially terrestrial orchids and seeds which were sent to Kew Gardens by Mark Johnstone McKen (1823–1872), horticulturalist, plant collector and sometime curator of the Natal Botanical Gardens.

Buxella*** = **Buxus** (Buxaceae) Tiegh.
Buxus (q.v.); Gk. *-ella* (diminutive).

Buxus (Buxaceae) L.
Classical Latin name for boxwood.

Byrsocarpus* = **Rourea** (Connaraceae) Thonn. ex Schumach.
Gk. *bursa* = a hide; *karpos* = fruit; referring to the fruit's hard skin.

C

Darrel Charles Herbert Plowes (1925–) grew up in various parts of South Africa and developed a lifelong interest in ecological and environmental science and a wide variety of natural history topics, especially birds, succulents, and butterflies. He has collected and studied succulent stapeliads (Asclepiadaceae) for more than 75 years and has established 20 new genera and published 51 new species, including five co-authored species. He obtained a BSc in agriculture at the University of the Witwatersrand in 1948 and worked in the agricultural sector in Zimbabwe. From 1956–1982 he was the head of African agriculture department in Manicaland Province in Zimbabwe. During his career he made comprehensive botanical and faunal surveys, resulting in more than 6 000 collected plant specimens and 30 000 photographs in various African countries. His film *Black Eagle Fly Free* received first prize at the South African Documentary Film Festival in 1975. He also published a book on the wildflowers of Zimbabwe in the same year. After his enforced retirement in 1982, he undertook many environmental impact assessments in Zimbabwe, Botswana, Sudan and Chad.

Illustration by **Auriol Batten** (née **Taylor** – 1918–2015) of a species of *Ceropegia*.

Batten obtained her degree in Botany and a teaching Diploma from UKZN. She taught Darrel at Estcourt High School (1941) and later he suggested she produce a book of 100 South African flowers painted in colour at natural size with their habitats drawn in pencil behind (a technique she pioneered and many others copied). This prestigious book won her a gold medal from the RHS in UK and an honorary doctorate from Rhodes. Her paintings have appeared in ten books on SA wildflowers, a series of stamps for the former Transkei, as well as on calendars for the BotSoc.

Cabomba* (Cabombaceae) Aubl.
Latinised from the vernacular American-Spanish name for 'water lily'; first found in French Guiana, South America.

Cacalia** (Asteraceae) L.
Gk. *kakos* = destructive; *lian* = exceedingly; hence 'very damaging'. An ancient name of uncertain meaning used by Dioscorides and Plinius for some similar plant.

Cadaba (Capparaceae) Forssk.

(TD)

La. *cadaba* = without leaves, from *kadhab*, the Arabic name of *C. farininosa*. This plant, 'desert broom', is leafless and survives in the harshest conditions.

Cadia (Fabaceae) Forssk.
La. *cado* = possibly to fall into decay. From the Arabic vernacular name *kadi*. Meaning very obscure. Arabic *kadi* = judge.

Cadiscus* = *Senecio (Asteraceae) E.Mey. ex DC.
La. *cadiscus* = a voting-urn, resembling a barrel; referring to the ribbed and furrowed achenes (Georg Christian Wittstein).

Caenopteris* = *Asplenium (Aspleniaceae) P.J.Bergius
Gk. *kainos* = new, recent; *pteris* = fern; referring to a newly found fern by Peter Jonas Bergius.

Caesalpinia (Fabaceae) L.
For Andrea Cesalpino (Latinised as Andreas Cæsalpinus) (1519–1603), noted Italian botanist, philosopher and physician to Pope Clement VIII. He was a professor of medicine and botany at Pisa and Rome, praefectus of the first botanical garden of Pisa and founder of the second. He was a forerunner of Linnaeus, having recognised the sexual aspect of plants, which he classified

by their fruits or seed, rather than alphabetically or by medicinal properties. He also did some physiological work and wrongly envisioned a 'chemical circulation' consisting of repeated condensation and evaporation of blood. His work predated that of the English physician William Harvey (1578–1657), who discovered the concept of the 'physical circulation' of blood.

Caesia (Hemerocalliaceae) R.Br.
For Federico Cesi (Fridericus Caesius) (1585–1630), Italian scientist, botanist, microscopist and

supporter of Galileo. When he was around 17 (1603), he founded the Accademia dei Lincei (academy of the lynx-eyed) in Rome. This first international scientific society existed in great secrecy in its early years. Galileo joined the Accademia in 1611 and donated his microscope to Cesi and the Accademia. After the condemnation of Copernicus, followed by Cesi's death and Galileo's condemnation, the Accademia ceased to exist. He was the first person to discover that ferns have spores. The words microscope and telescope were first used by Johannes Faber (1574–1629) in a letter of 13 April 1625 to Duke Federico Cesi (1585–1630).

Cailliea* = *Dichrostachys (Fabaceae) Guil. & Perr.
For René Caillié (1799–1838), French explorer. The orphaned son of a prison convict,

uneducated, frail and thin, he read *Robinson Crusoe* which kindled in him a love of travel and adventure. In 1818, aged 16, he made a voyage to Senegal, and again in 1824. That year the Paris-based Société de Géographie offered a 10 000 franc reward to the first European to see and return alive from Timbuktu. Caillié learned Arabic, the laws and customs of Islam, and started his exploration in 1827. He reached Timbuktu, 'a small, unimportant, and poor village', in 1828 and won the 10 000 francs prize, was awarded the Legion of Honour and a pension. His three-volume book *Travels through Central Africa to Timbuctoo; and across the Great Desert, to*

Morocco (1824–1828) was published at public expense.

Cajanus* (Fabaceae) DC.
Latinised from the Malay vernacular name *katchang* = bean, pea.

Calamagrostis (Poaceae) Adans.
Gk. *kalamos* = reed (Arabic, *kalom*); *agrostis* = grass.

Calamaria = Isoetes (Isoetaceae) Boehm
Gk. *calamosarius* = resembling a reed; referring to a reed-like quillwort.

Calamophyllum (Aizoaceae) Schwantes
Gk. *kalamos* = reed or reed pen (quill); *phyllon* = leaf; referring to the slender cylindrical leaves.

Calandrinia (Montiaceae) Kunth
For Jean Louis Calandrini (1703–1758), Swiss scientist and professor of mathematics and  philosophy at Geneva (1734–1750). He obtained a PhD in physics from the academy in Geneva in 1722, wrote on such subjects as the aurora borealis, comets, the effects of lightning, and flat and spherical trigonometry as well as a commentary on Isaac Newton's *Principia*.

Calanthe (Orchidaceae) R.Br.
Gk. *kalos* = beautiful; *anthe* = bloom; *anthos* = flower; alluding to 'the beautiful flowers of most of the species' (R Brown, 1821).

Caldesia (Alismataceae) Parl.
For Ludovico Caldesi (1822–1884), Italian botanist, politician, mycologist, naturalist, member of parliament and professor of botany at Faenza. He was the author of *Florae Faventinae Tentamen*. He was a student of the Italian botanists Filippo Parlatore and Giuseppe De Notaris. His collections of exsiccate, collected in Emilia-Romagna (Northern Italy) in 1860, are preserved in Florence, but there is other material at Wrocklaw (Poland), Vienna and in the herbarium of the department of biology at the University of Milan.

Calendula = Osteospermum (Asteraceae) L.
La. *kalendae* = calendar; *-ula* = little; probably referring to the almost year-round flowering season of some species.

Calicorema (Amaranthaceae) Hook.f.
Gk. *kalos* = beautiful; *korema* = broom; referring to the rigid and twiggy, much-branched shrub.

Calla = Zantedeschia (Araceae) L.
Gk. *kallos* = beautiful; referring to the plant's inflorescence, which is arum-like.

Calliandra* (Fabaceae) Benth.
Gk. *kallos* = beautiful; *aner, andros* = a male, hence stamen. The stamens are a striking feature of the flowers.

Callichilia (Apocynaceae) Stapf
Gk. *kallos* = beautiful; *-chilus* = a lip. The lobes of the corolla are conspicuous.

Callicysthus** (Fabaceae) Endl.
Gk. *kallos* = beautiful; *kysthos, kysthis* = a bladder; probably referring to the seed pod.

Callilepis (Asteraceae) DC.
Gk. *kallos* = beautiful; *lepis* = scale, hence 'beautiful scale'; referring either to the paleae on the receptacle or to the beaks produced from the ovary (Plantzafrica.com).

Callipteris = Diplazium (Athyriaceae) Bory
Gk. *kallos* = beautiful; *pteris* = fern.

Callistemon* (Myrtaceae) R.Br.
Gk. *kallos* = beautiful; *stemon* = stamen; referring to these beautiful flowers with their conspicuous stamens.

Callistigma = Mesembryanthemum (Aizoaceae) Dinter & Schwantes
Gk. *kallos* = beautiful; *stigma* = stigma.

Callitriche (Plantaginaceae) L.
Gk. *kallos* = beauty; *trichos* = hair, hairy, hence 'beautifully-haired'. The beautiful, thin stems of the aquatic plant inspired this name.

Callitris (Cupressaceae) Vent.
Gk. *kallios* = beautiful; *treis* = three; referring to the beauty of the three-whorled leaves and often cone scales, and columella lobes.

Calobota**** (Fabaceae) Eck. & Zeyh.
Gk. *kalos* = beautiful; *bota, botane* = plant.

Calodendrum (Rutaceae) Thunb. (Image on opposite page)
Gk. *kalos* = beautiful; *dendron* = tree. The tree's flowers are spectacular.

Calomeria (Asteraceae) Vent.
Gk. *kalos* = beautiful; *meris* = a part, portion. In 1804, the year of Napoleon Bonaparte's and Joséphine de Beauharnais' coronation, Joséphine asked French botanist, Étienne Pierre Ventenat (1757–1808) to name a flower in Napoleon's

honour. Ventenet could not name the flower 'Napoleona' as Pallissot de Beauvois used that name in his *Flora of Oware and Benin* (1802), and 'Bonapartea' was used by Messrs. Ruiz and Pavon in their *Flora of Peru*. Ventenat named a newly found plant *Calomeria,* which is thought to be 'an oblique, complimentary reference to Napoleon Bonaparte' (Davesgarden.com). *Calomeria* can be translated as 'the Greek equivalent of *bon partie,*' (source: *Botanophilia in 18th Century France*) which is a word play on 'Buonaparte' and French for 'good part' or *bon en partie* = good in parts. Ventenat is alleged to have said the name would give, loosely translated, '[T]he Emperor some small show of gratitude that he is entitled to expect from all those who cultivate the arts and sciences.'

Calodendrum in full bloom (EJM)

Calonyction* = *Ipomoea* (Convolvulaceae) Choisy
Gk. *kalos* = beautiful; *nyctios* = nocturnal; from *nyx, nycto-* = night. A night-blooming species, now renamed 'morning glory'.

Calophanes* = *Dyschoriste* (Acanthaceae) D.Don.
Gk. *kalos* = beautiful; *phanos, phaino* = to appear; referring to the flowers when they emerge.

Calophyllum (Calophyllaceae) L.
Gk. *kalos* = beautiful; *phyllon* = leaf; referring to the slender oppositely arranged leaves with leathery blades.

Calopsis = **Restio** (Restionaceae) P.Beauv. ex Desv.
Gk. *kalos* = beautiful; *-opsis* = sight. The greenery of *Calopsis* is especially attractive when used in flower arrangements and as a backdrop to a water garden.

Calorophus (Restionaceae) Labill.
Gk. *kalostrophos* = a rope maker, rope twister, from *kalos* = beautiful; *strophe* = turning; referring to the making of ropes from the reeds.

Calostemma* (Amaryllidaceae) R.Br.
Gk. *kalos* = beautiful; *stemma* = garland, wreath, crown; referring to the conspicuous golden staminal corona of the type species.

Calostephane (Asteraceae) Benth.
Gk. *kalos* = beauty; *stephanos* = crown; referring to the flowers.

Calotesta (Asteraceae) P.O.Karis
Gk. *kalos* = beauty; *testa* = outer coat of seed. The plant is characterised by its persistent, thick and strongly cutinised testa epidermis (Perola Karis).

Calothamnus (Myrtaceae) Labill.
Gk. *kalos* = beauty; *thamnos* = bush; referring to its appearance.

Calotheca*** = *Briza* (Poaceae) Desv.
Gk. *kalos* = beauty; *theka* = cup or container, case, capsule; referring to the lemma margins which extend as lateral wings (*Etymological Dictionary of Grasses*).

Calotes (Asteraceae) R.Br.
Gk. *kalos* = beautiful; *-otis* = an ear; a reference to the elaborate pappus scales of the first-named species (Davesgarden.com).

Calotropis* (Apocynaceae) R.Br.
Gk. *kalos* = beautiful; *tropis* = the keel of a ship; refers to the keeled 'corona leaflets' (FloraBase of Western Australian Flora).

Calpidisca*** = *Utricularia* (Lentibulariaceae) Barnhart
Gk. *kalpis* = urn; *diskos* = fleshy disk; possibly referring to the corolla which is urceolate: pitcher-like, hollow and contracted near the disk-shaped mouth.

Calpurnia (Fabaceae) E.Mey.

The pendulous yellow blossoms make a good show in spring. (TW)

For Calpurnius, Roman first-century CE poet, whose full name was Titus Calpurnius Siculus, a second-rate imitator of the poet Virgil. The trees resemble *Virgilia*.

Caltha* (Ranunculaceae) L.
La. *caltha* = cup, the Latin name for a marigold; referring to the cup-shaped flowers.

Calvaria** = **Sideroxylon** (Sapotaceae) Comm.
La. *calva* = bald, skull; *aria* = pertaining to. Named for a tree found in 1973, over 300 years old, endemic to Madagascar. Only 13 were found at that time, in poor condition and dying, hence the imagery.

Calycanthemum* = *Ipomoea
(Convolvulaceae) Klotzsch
Gk. *kalyx* = cup, hence calyx; *anthemon* = flower.

Calymella** = **Gleichenia** (Gleicheniaceae) C.Presl
Gk. *kalumma, kalymma* = veil, hood; *-ella* = diminutive. The ultimate pinnules are small and hood-like.

Calystegia (Convolvulaceae) R.Br.
Gk. *kalux* = cup, hence calyx; *stege* = covering; referring to the large persistent bracteoles which clasp the calyx.

Camelina* (Brassicaceae) Crantz
Gk. *khamai* = dwarf or on the ground; *linon* = flax. 'Dwarf', in this case means 'false', hence the common name 'false flax'. The plant is a weed that suppresses the vigour of flax crops.

Camellia* (Myrsinaceae) (Theaceae) L.
For Georg Joseph Kamel (La. Camellus) (1661–1706), Moravian Jesuit missionary, apothecary, artist and botanist. He was sent to the Mariana Islands (1683) and Manila, Philippines (1688) where he established a pharmacy, providing poor people with remedies for free. He botanised on Luzon island, north of Manila, collecting some 360 varieties of plants and herbs which he sent to the British botanists, Rev. John Ray, and James Petiver, who published *Herbarium aliarumque stirpium in insula Luzone Philippinarum* (*Herbs and Medicinal Plants in the Island of Luzon, Philippines*). Further specimens came from Chinese gardens at Manila. His first shipment of botanical drawings failed to reach England as a result of piracy. Kamel also co-wrote the first account of Philippine birds, *Observationes de Avibus Philippensibus* (1702), published by the Royal Society.

Campanopsis** = ***Wahlenbergia***
(Campanulaceae) (R.Br.) Kuntze
Campanula (q.v.); *opsis* = resembling.

Campanula* (Campanulaceae) L.
La. *campanula* = little bell; referring to the shape of the flower's corolla, 'bell-flower'.

Campium** = **Bolbitis** (Dryopteridaceae) C.Presl
Gk. *kampe* = bend, curve; alluding to the curving veins.

Campsis* (Bignoniaceae) Lour.
Gk. *kampe* = curved, bent; referring to the bent stamens.

Camptoloma (Scrophulariaceae) Benth.
Gk. *kamptos* = bent, flexible; *loma* = edge or fringe; possibly referring to the bent leaf edges in some species.

Camptorrhiza (Colchicaceae) E.Phillips
Gk. *kamptos* = curved; *rhiza* = root; referring to the plant's curved roots.

Campuloclinium* (Asteraceae) DC.
Gk. *kampulos* = a little bent; *clinium* = prostrate; possibly referring to the light green stems, erect in summer and spreading but prostate in winter.

Campylostachys (Stilbaceae) Kunth
Gk. *kampulos-* = bent, curved; *stachys* = spike. The inflorescence is a terminal, slightly drooping, globose spike.

Canahia = Kanahia (Apocynaceae) Steud.
Variant spelling of *Kanahia,* an Indian name meaning 'darling'.

Canavalia (Fabaceae) DC.
Latin form of the Malabar vernacular name *kanavali,* meaning 'forest climber'. The genus is a tropical vine.

Canna* (Cannaceae*) L.
Gk. *kanna, canna* = a small reed.

Cannabis* (Cannabaceae) L.
La. Derived from Scythian, an ancient Indo-European language, translated into ancient Greek *kánnabis* (c 440 BCE) referring to its recreational usage and later into La. *cannabis* = hemp; the plant name is commonly known as marijuana and can have adverse effects but also has medical uses.

Cannomois (Restionaceae) P.Beauv. ex Desv.

Note the large, deciduous, papery bracts on the cane-like stems. (EJM)

Gk. *kanna, canna* = reed, cane; *omoios* = similar; possibly referring to the huge number or species in the *Restionaceae* (about 400), many of which look similar.

Canscora* (Gentianaceae) Lam.
Cansjan-cora, Malayalam name recorded by Rheede.

Canthium (Rubiaceae) Lam.
From the Malabar name *canti* for a species of this genus. 'Turkey-berry' tree.

Cantuffa = Pterolobium (Fabaceae) J.F.Gmel.
Cantuffa = Kantuffa, the vernacular name for a plant from Kantuffa, Abyssinia (Ethiopia).

Capassa = Lonchocarpus (Fabaceae) Klotzsch
Vernacular name *Capassa* = 'no good' (wood is useless). The author writes (translated from the German) 'A tree with odourless and useless wood … hence his native name, 'capassa' = 'no good'.'

Capelio (Asteraceae) B.Nord.
Capelio = an anagram for *Alciope.*

Capeobolus (Cyperaceae) Browning
Cape- = collected at the Cape of Good Hope (Cape Peninsula); *bolos* = a lump, a round mass. 'Also commemorates the surname Bolus: associated with the botany department of the University of Cape Town and Bolus Herbarium' (*South African Journal of Botany* 65: 218, 1999).

Capeochloa (Poaceae) H.P.Linder & N.P.Barker
Cape = of the Cape Peninsula, South Africa; *chloa* = grass; referring to its habitat.

Caperonia (Euphorbiaceae) A.St.-Hil.
For Noël Caperon or Capperon (died 1572), French apothecary, discoverer of *Fritillaria meleagris* in the Orléans forest; other sources say they were found in meadows by the Loire. He sent this to Carolus Clusius (1526–1609), Flemish doctor and pioneering botanist, perhaps the most influential of all 16th-century scientific horticulturalists. Caperon's death was untimely. He was a Protestant, but not a French Calvinist Protestant. In 1572, French soldiers and Roman Catholic clergy attacked Huguenots (French Calvinist Protestants), a campaign that lasted some weeks and resulted in the deaths of thousands of Protestants. Caperon died during the St Bartholomew Massacre.

Capnitis** = ***Lotononis*** (Fabaceae) E.Mey.
Gk. *kapnos* = smoke*;-* = *itis* = likeness; referring to its smoky grey appearance or habitat.

Capnophyllum (Apiaceae) Gaertn.
Gk. *kapnos* = smoke; *phyllon* = leaf; referring to the colour of the leaves.

Capparis (Capparaceae) L.
Gk. *kapparis* (La. *capparis* = caper). The ancient Greek name for this genus; probably from the Arabic *kabar* = 'caper-bush'.

Capraria (Scrophulariaceae) L.
La. *caper* = goat; *-aria* = connected or related to; 'goatweed'. Goats are especially fond of *C. biflora*. Named by Paul Hermann in 1698, who reported the Portuguese name for the plant as 'cavritta' which he Latinised as *Capraria*.

Capsella* (Brassicaceae) Medik.
La. *capsa* = box, case; *-ella* = diminutive; referring to the small size of the seed pods.

Capsicum* (Solanaceae) L.
Gk. *kapto* = to bite; referring to the fiery flavour found in many peppers.

Caralluma (Apocynaceae) R.Br.
Caralluma is named from the Arabian *qarh alluhum* = a flesh wound or abscess; perhaps referring to the erect, fleshy, diffusely branched stem.

Cardamine (Brassicaceae) L.
Gk. *kardamon*, probably from *kardia* = heart; *damso* = I subdue; referring to its medicinal heart-strengthening properties.

Cardaria* = *Lepidium (Brassicaceae) Desv.
Gk. *kardia* = heart; *-aria* = pertaining to; referring to the heart-shaped fruit.

Cardiogyne* = ***Maclura*** (Moraceae) Bureau
Gk. *kardia* = a heart; *gyne* = a woman. The perianth of the female flower is roughly heart-shaped.

Cardiospermum (Sapindaceae) L.
Gk. *kardia* = a heart; *sperma* = a seed. The seeds have a conspicuous heart-shaped spot.

Carduus* (Asteraceae) L.
Latin name for thistle; Celtic *ard* = a point. The plants are mostly spiny.

Carex (Cyperaceae) L.
La. *caricis, carex* = reeds, sedges, rushes. Possibly also from La. *carere* = absent, having flaws. In some species the upper florescent spikes are without seeds because the spikes are male.

Carissa (Apocynaceae) L.
Sanskrit name for an Indian species of this genus.

Carlina (Asteraceae) L.
Named for the Holy Roman Emperor Carolus Magnus, otherwise known as Charlemagne (742–814). The preface to the *Codex Epistolaris Carolinus* states that the sick soldiers in King Charlemagne's army, suffering from the plague, were cured by a species of 'carline thistle', *Carlina acaulis*.

Carex (CM)

Caroxylon** = ***Salsola*** (Amaranthaceae) Thunb.
Gk. *karoxylon* = wood; from Karla, ancient province in Asia Minor.

Carpacoce (Rubiaceae) Sond.
Gk. *karpos* = fruit; *akoke* = a point. The fruit ends in a sharp point.

Carpanthea (Aizoaceae) N.E.Br.
Gk. *karpos* = fruit; *anthe, anthos* = flower; referring to the open fruit looking like a star-shaped flower when open (H.E.K. Hartmann).

Carpha (Cyperaceae) R.Br.
Gk. *karphos* = straw, twig or other small dry body; referring to its appearance.

Carphalea (Rubiaceae) Juss.
Gk. *karphos* = dry stick; referring to the desiccated look of the plant.

Carphopappus** = ***Pegolettia*** (Asteraceae) Sch.Bip.
Gk. *karphos* = a dry stick; *pappus* = a modified calyx. Referring to its 'dry stick' appearance.

Carpobrotus (Aizoaceae) N.E.Br. (Images on opposite page)
Gk. *karpos* = fruit; *brotos* = edible. The fruit can be eaten.

Carpolobia (Polygalaceae) G.Don
Gk. *karpos* = fruit; *lobos* = a pod, lobe; referring to the fruit's shape.

Carpolyza** = ***Strumaria*** (Amaryllidaceae) Salisb.
Gk. *karpos* = fruit; *lysis* = parting; referring to the irregular opening of the capsule (cf. *Antholyza*, derived from *lyssa* = rage).

Carpobrotus (EJM)

Carpobrotus: The fruits of some species are valued for jam-making (Suurvygie konfyt). (LEFT: CM, RIGHT: EJM)

Carpopodium*** = **Heliophila** (Brassicaceae) Eck. & Zeyh.
Gk. *karpos* = fruit; *podion* = little foot. Meaning unclear.

Carrpos = **Monocarpus** (Monocarpaceae) Prosk.
For Denis John Carr (1915–2008), English botanist and professor in the Department of Developmental Biology, Research School of Biological Sciences, Australian National University (1968–1980). He obtained his PhD from Manchester University and taught there (1958–1960) and was professor of botany at Queen's University, Belfast (1960–1967). He and his wife Stella Grace M (Maisie) (1912–1988), an ecologist, wrote a number of books together including *People and Plants in Australia* (1981) and *Plants and Man in Australia* (1983). They

were noted for their work in the alpine regions of New South Wales and Victoria and named several plant species, including the bloodwood, *Eucalyptus dampieri,* in 1987.

Carruanthus (Aizoaceae) Schwantes
Latinised Karoo = *Carru*; *anthos* = flower; referring to the Karoo desert, its habitat.

Carthamus* (Asteraceae) L.
From the Arabic *qurtom* = to paint. The seeds of this plant, *C. tinctorius,* were used for making yellow dye and the red florets provided red dye which was used for colouring, flavouring food (although it had little taste) and in medicines.

Carum = **Chamarea** (Apiaceae) Sond.
From Latin *caria* (Gk. *karia*), Asia Minor, according to Pliny the Elder, where it was discovered. Gk. *Karon*, 'caraway'.

Caruncularia*** = **Stapelia** (Apocynaceae) Haw.
La. *carunculatus* = with a prominent caruncle; *atus* = possession, connected or related to; referring to the seed coat outgrowth, usually obscuring the micropyle.

Carya* (Myricaceae) Nutt.
La. *karyon* = nut; referring to the walnut tree.

Caryotophora (Aizoaceae) Leistner.
Gk. *karyotos* = nut-shaped, *karyotis* = the cultivated date-palm, from *karyon* = a nut; *phorein* = to carry; referring to the fruits.

Casearia (Salicaceae) Jacq.
For the Dutch clergyman Johannes Casearius (1642–1678), a missionary, minister of the Dutch East India Company, and co-author of the first two volumes of Hendrik A Van Rheede's *Hortus Indicus Malabaricus.*

Casimiroa* (Rutaceae) La Llave
For Casimiro Gómez (unrecorded–1815), an Otomi Indian (an indigenous ethnic group) from

the town of Cardonal in Hidalgo, Mexico (not for the similarly named Casimiro Gomez de Ortega (1740–1818)). As a little boy he was adopted by a wealthy Spanish merchant, Pedro Marcos Gutierrez of Mexico City. When a citizen revolution took place in 1810, against the European-born Spaniards' appalling treatment of Mexican citizens, Casmiro returned to his homeland and gathered a group of Otomi Indians who became mercenaries and allies during the war. He showed his talent by managing the citizen revolutionaries, even though they lacked training and were poorly armed, and had some successes in various skirmishes with enemy forces. After many years of fighting, on 2 November 1815, he was taken by surprise by Antonio Castro at the Hacienda de Tenango and after the battle fled but was caught and taken to Tulancingo and killed.

Cassebeera (Adiantaceae) Kaulf.
For Johann Heinrich Cassebeer (1784–1850), German botanist, geologist, agriculture and wine expert, politician and pharmacist. He worked for most of his life as a pharmacist, at one time owning two pharmacies. He also had an interest in civic matters and, from 1814 onwards, served as a city councillor, as honorary mayor of Geinhausen, and also as state representative in the Kassell district. As a student he studied, in particular, chemistry and botany. In addition to his scientific activity in the fields of mineralogy and geology, he had a particular interest in cryptogams, wrote several articles and published a book on the evolution of mosses. He was awarded an honorary doctorate of pharmacy from the University of Marburg for his research.

Cassia (Fabaceae) emend. Gaertn.
Ancient Gk. name Kasia used by Dioscorides, Latinised as Cassia. Cassia can mean cinnamon, but that does not apply to this genus, mainly trees, but to the family Lauraceae. Meaning unknown, possibly of Chinese origin.

Cassine (Celastraceae) L.
The Latin word *cassine* was taken from the word *cassina*, which means 'tea' or 'black drink' (made from the leaves and stems of *Ilex vomitoria*, commonly called yaupon holly) in the Timucua language spoken in northern and central Florida, US. It appears to come from an earlier derivation, *assi*, also meaning 'tea' or 'black drink' in the Hitchiti language spoken by the eponymous tribe of Native Americans in Georgia, US. Timucua Native Americans used this brew for male-only purification and unity rituals.

Cassinia* (Asteraceae) R.Br.
For Alexandre Henri Gabriel Comte de Cassini (1781–1832), French botanist and naturalist who came from a family of astronomers and cartographers. In addition to his botanical interests he was a magistrate and an advisor at the supreme court of appeal with a particular

interest in teaching. He named many flowering plants and new genera in the sunflower family (*Asteraceae*), many of them from North America. He published 65 papers and 11 reviews in the *(Nouveau) Bulletin des Sciences par la Société Philomatique de Paris* between 1812 and 1821. After his first collection, *Opuscules phytologiques* (1826), he was elected a member of the Academy of Sciences and became an associate in 1831. He died of cholera in 1832. His great-great-grandfather discovered Jupiter's Great Red Spot and the Cassini division in Saturn's rings.

Cassinopsis (Icacinaceae) Sond.
Cassine (q.v.); Gk. -*opsis* = resemblance. The genus resembles *Cassine*.

Cassipourea (Rhizophoraceae) Aubl.
The name used for this tree species in Guiana; apparently a Guyanese vernacular name.

Cassytha (Lauraceae) L.

A dodder-like parasite (EJM)   (CM)

Arabic or Aramaic *kesatha* = a tangled wisp of hair. This describes the masses of matted stems that cover the host parasitic plant known as 'dodder laurel' and by other names. This plant closely resembles in looks 'dodder' in the genus *Cuscuta*, but is not related to it (*Florida Ethnobotany*, Daniel F Austin).

Castalis*** = Dimorphotheca (Asteraceae) Cass.
Perhaps Gk. *kasta* = chaste; or from Kastalla, a spring sacred to the muses, similar to Nymphia.

Castanea* (Fagaceae*) Mill.
The Latin name for a chestnut.

Casuarina* (Casuarinaceae*) L.
La. *Casuarinus* = a cassowary bird. The drooping branches of the tree are supposed to resemble the feathers of a cassowary.

Catalepis (Poaceae) Stapf & Stent
Gk. *kata* = under, downward; *lepis* = scale. The lower glume is small or reduced to a small linear scale.

Catalpa* (Bignoniaceae) Scop.
American Indian name *kutuhlpa* = head with wings; referring to the flower lobes or seeds. Another name for this tree is catawba, misspelled as catalpa by the describing botanist (Scopoli). In terms of botanical naming rules, the incorrect name remains in force.

Catapodium* (Poaceae) Link
Gk. *kata* = below, under, downward; *podion* = little foot; allusion obscure.

Catevala** = **Haworthia** (Asphodelaceae?) Medik.
Malayalam *kattarvala* = *Aloe vera*, spelled Catevala by Rheede.

Catha (Celastraceae) Forssk. ex Scop.
Latinised from the Arabic name *khat* = the name of a headcloth worn by Ancient Egyptian Pharaohs. The chewing of the leaves of khat as a social custom is well established as this can induce mild euphoria, reduce hunger, be used for 'Bushman's tea' (South African common name) as well as have harmful narcotic effects.

Catharanthus* (Apocynaceae) G.Don
Gk. *katharos* = pure, unspotted; *anthos* = a flower; possibly referring to the purity of the flowers, which are the same shade of pink or white throughout.

Cathastrum* = **Pleurostylia** (Celastraceae) Turcz.
Catha (q.v.); *-astrum* = inferiority or partial resemblance.

Catophractes (Bignoniaceae) D.Don
Gk. *kata* = below; *fraktes* = an enclosure or thorn hedge; alluding to the leaves and often flowers arising below the thorns.

Catunaregam (Rubiaceae) Wolf
Tamil *kattu* = wild; *naregam* = edible citrus; the name of a wild organic fruit from Sudan.

Caucalis* (Apiaceae)
Name used by Hippocrates and Theophrastus. Meaning obscure. Linnaeus used this word for some umbelliferous plants.

Caulipsolon (Aizoaceae) Klak
Gk. *kaulos* = stem; *psilos* = bare. The genus name *Caulipsolon* is an anagram of *Psilocaulon*, the genus in which it was previously placed. The stems and leaves have smooth (bare) surfaces.

Cautleya* (Zingiberaceae) (Benth.) Hook.f.
For Proby Thomas Cautley (1802–1871), English engineer and palaeontologist. In 1819 he joined the Bengal artillery. By 1825, he was assistant to captain Robert Smith, the engineer in charge of the construction of the 560km Eastern Yamuna Canal, and in 1836 he was appointed general superintendent for northwest Indian canals. He was actively involved in Dr Hugh Falconer's fossil expeditions in the Siwaliks, Southern Himalayas, and wrote numerous scientific papers on the geology and fossils of the Siwaliks (some with Hugh Falconer) which appeared in the proceedings of the Bengal Asiatic Society and Geological Society of London. In 1837 he was awarded the Geological Society's Wollaston medal and on his return to England in 1858 was awarded the KCB (Knight Commander of the Order of Bath).

Cavacoa (Euphorbiaceae) J.Léonard

Female flower (male flower on following page) (EJM)

For Alberto Judice Leote Cavaco (1916–?), Portuguese botany professor and plant collector who made botanical expeditions to France, Mozambique, and possibly Madagascar, as well as conducting research in Portugal. He taught for many years at the Agricultural Institute in the Ecole Polytechnique (Switzerland) and at the

Cavacoa: Male flowers (EJM)

University of Lisbon. He ended his academic career by studying phanerogams at the National Museum of Natural History of France. He published a number of books, in French, on the flora of different countries, focusing on a few families. These include *The Flora of Madagascar and the Comoros* (1959), *Flora of Gabon* (1963) and *Flora of Cameroon* (1974).

Cayratia (Vitaceae) Juss.
Possibly an Indian vernacular name. Meaning unknown.

Ceanothus (Rhamnaceae)
Gk. *keanothus* = a spiny shrub; Latin for thistle. Many species have spiny, holly-like leaves.

Cecropia (Urticaceae) Loefl.
Named after *Cecrops* or *Kekrops*, the mythical King of Attica and founder of Athens.

Cedrela* (Meliaceae) P.Browne
Spanish *cedrelo*, diminutive of *cedro*, Gk. *Kedros*, La. *cedrus* = cedar. The diminutive of *Cedrus*, whose wood appearance and aroma is similar, but a different family.

Cedronella* (Lamiaceae) Moench
La. *cedrus* (Gk. *kedros*) = cedar; Gk. *-ella* = diminutive. The scent of the leaves is supposed to resemble the cedar.

Ceiba* (Malvaceae) Mill.
Spanish, probably from Taino *ceíba* = name for a silk-cotton tree.

Celastrus* (Celastraceae)
Gk. *kelastros* = a name for an evergreen tree. The fruits are retained on the tree throughout winter.

Celeri*** (Apiaceae) Adans.
Fr. *céleri*, from Italian (Lombardy) dialect *selleri*, from Gk. *selinon* = parsley. Celery = a plant, *Apium graveolens*.

Celastrus* (Asteraceae) Cass.
Named for Celmis, a priestess in Greek mythology. The Dactyls were three mythical beings who developed working with iron, including Celmis (the smelter), Damnameneus (the hammer), and Acmon (the anvil). Significance obscure.

Celosia (Amaranthaceae) L.
Gk. *keleos* = burning; *kelos* = dry, burnt; referring to the brightly coloured flame-like flower heads.

Celsia = Verbascum (Scrophulariaceae) L.
For Olof Celsius the Elder (1670–1756), Swedish theologian, botanist, philologist, plant collector, teacher, professor of botany at Uppsala University, author of a book on biblical plants, *Hierobotanicos*, written in 1745–1747. It was Celsius who 'discovered' Linnaeus when he 'noticed a hungry-looking student in the botanic garden and, presumably surprised to find a student studying, enquired what he was doing. He invited him home and gave him a free room and meals. As a new year's gift, Linnaeus presented Celsius with a manuscript dissertation, *Praeludia Sponsaliorum Plantarum*, on the marriage of plants and their sexual analogies with animals' (*The Linnaean Correspondence*). In 1739 Celsius became a member of the Royal Swedish Academy of Sciences.

Celtis (Ulmaceae) L.
Latin name used by Pliny the Elder referring to the unrelated *Ziziphus lotus*. This name now refers to a genus of deciduous trees. The 'new' family names seem to be *Celtidaceae* or *Cannabaceae* (formerly *Ulmaceae*).

Cenchrus (Poaceae) L.
Gk. *kenchros* = millet (seeds).

Cenia* = *Cotula (Asteraceae) Comm.
Gk. *kenos* = empty; referring to the hollow receptacle.

Centaurea* (Asteraceae) L.
Gk. *kentaurieon* = in Greek mythology the centaur Chiron (*Kheiron*) was known for his knowledge and skill with medicine. The common name 'Centaury' is sometime incorrectly used for this genus, although this refers to the unrelated plant genus *Centaurium*.

Centella (Apiaceae) L.
Gk. *kentron* = a spur or sharp point; *-ella*= diminutive; probably alluding to the small, pointed styles.

Centema (Amaranthaceae) Hook.f.
Gk. *centema* = sting, perhaps from *kentema* = point of a weapon; referring to the spike-like bracteate thyrses.

Centranthus* (Valerianaceae) DC.
Gk. *kentron* = spur, sharp point; *anthos* = flower; referring to the flower having a spur-like base.

Centrapalus (Asteraceae) Cass.
Gk. *kentron* = spur, sharp-pointed; (*se*)*palus* = sepals or *palus* = a stake; possibly referring to the flower having spur-like sepals.

Centratherum* = *Vernonia (Asteraceae) Cass.
La. *centrum* = centre; *atherum* = prickle or awn; perhaps alluding to the spine-tipped middle phyllaries of the original species.

Centropodia (Poaceae) Rchb.
Gk. *kentron* = spur, sharp-pointed; *podion* = a small foot; possibly alluding to the short awn.

Centrosis* = *Calanthe (Orchidaceae) Thouars
Gk. *kentron* = spur, sharp-pointed. The lip is conspicuously spurred.

Centrostachys (Amaranthaceae) Wall.
Gk. *kentron* = spur, sharp-pointed; *stachys* = spike; referring to the many spurred spikes.

Centrostigma (Orchidaceae) Schltr.
La. *centrum* = centre; *stigma* = stigma; referring to part of the style, that ovary-surface that receives pollen in impregnation. 'The stigmatic concavity has a central position on the flower' (Schlechter, 1915).

Cephalandra* = *Coccinia (Cucurbitaceae) Schrad. ex Eck. & Zeyh.
Gk. *kephale* = head; *andros* = man; referring to the stamens.

Cephalanthus (Rubiaceae) L.
Gk. *kephale* = head; *anthos* = flower. The flowers are borne in compact rounded heads.

Cephalaria (Dipsacaceae) Schrad.
Gk. *kephale* = head; *-aria* = pertaining to. The flowers are aggregated into pedunculate heads.

Cephalocroton (Euphorbiaceae) Hochst.
Gk. *kephale* = a head; *Croton* (q.v.). Plants look like *Croton* (q.v.), but with flowers in heads.

Cephalomanes (Hymenophyllaceae) C.Presl
Gk. *kephale* = a head; *manes* = a kind of cup; *manos* = slender or loose; probably alluding to the slender receptacles in the type species.

Cephalophyllum (Aizoaceae) N.E.Br.
Gk. *kephale* = a head; *phyllon* = leaf; referring to the compact heads of leaves in some species.

Cephalostigma* = *Wahlenbergia (Campanulaceae) A.DC.
Gk. *kephale* = a head; *stigma* = stigma; referring to the shape of the stigma.

Ceradenia* (Polypodiaceae) E.Bishop
Gk. *cera-* = wax, from Gk. *keros*; *aden* = gland; referring to the waxy-glandular laminar hairs.

Ceradia* = *Othonna (Asteraceae) Lindl.
Gk. *keras* = horn; referring to the horned appearance of the branches.

Ceraria (Didiereaceae) H.Pearson & Stephens
Gk. *keras* = a horn. The branches of *C. namaquensis* look like horns.

Cerastium (Caryophyllaceae) L.
Gk. *keras* = horn; referring to the shape of the 'horned' seed capsules.

Ceratandra (Orchidaceae) Eck ex F.A.Bauer
Gk. *keration* = little horn; *andros* = male (stamen); referring to the rostellum arms that carry the pollinia (Fernkloof Nature Reserve).

Ceratandropsis* = *Ceratandra (Orchidaceae) Rolfe
Ceratandra (q.v.); *opsis* = resembling.

Ceratiosicyos** (Achariaceae) Nees
Gk. *keration* = little horn; *sikyos* = cucumber; referring to the fruit which has the common name 'creeping pod cucumber'.

Ceratocaryum (Restionaceae) Nees
Gk. *kerato* = horned;-*karyon* = nut, walnut; referring to the fruit, a large woody nut.

Ceratogonum = Oxygonum (Polygonaceae) C.A.Mey.
Gk. *kerato* = horned; *gonium* = angle; possibly referring to the prickly fruit.

Ceratonia* (Fabaceae) L.
Gk. *kerato* = horned; referring to the shape of the seed pods of *Ceratonia siliqua,* commonly called the carob or locust-tree.

Ceratophorus = Suregada (Euphorbiaceae) Sond.
Gk. *kerato* = horned; *phorus* = bearing, hence 'horn bearing'; referring to the corolla spurs.

Ceratophyllum (Ceratophyllaceae) L.
Gk. *kerato* = horned; *phyllum* = leaves; referring to the forked divisions of fronds, which resemble a stag's horn.

Ceratopteris (Pteridaceae) Brongn.
Gk. *kerato* = horned; *pteris* = a fern; alluding to the antler-like fertile fronds.

Ceratotheca (Pedaliaceae) Endl.

The dry fruit splits open from apex to base when ripe. (EJM)

Gk. *kerato* = horned; *theke* = a case, capsule; referring to the horned fruit (capsule).

Cereus* (Cactaceae) Mill.
La. *cereus* = torch, waxy, and hence 'candle-like'.

Tabernaemontanus first used the word 'cereus' in 1625 to refer to cacti with elongated (candle-like) bodies such as *Cereus hexagonus.*

Ceriops (Rhizophoraceae) Arn.
Gk. *keros* = wax; *-ops* = resembling; referring to the glands at the base of the stipules, which exude a waxy substance.

Cerochlamys (Aizoaceae) N.E.Br.
Gk. *keros* = wax; *chlamys* = cloak, mantle; referring to the leaves, which are coated with wax (Hans Herre).

Ceropegia (Apocynaceae) L.
Gk. *keros* = wax; *pege* = fountain. Linnaeus described this genus in volume 1 of his *Species plantarum*, which appeared in 1753. Linnaeus thought that the flowers looked like candles and *Ceropegia* literally means a fountain of wax.

Ceropteris = Pityrogramma (Adiantaceae) Link
Gk. *keros* = wax, beeswax; *pteris* = a fern; referring to the wax-like farina, which is on the underside of the fronds.

Cervia* = **Convolvulus*** (Convolvulaceae) Rodr.
La. *cervus, cervi-* = deer. This genus, now in *Convolvulus,* was probably named as such because it is largely resistant to being eaten by deer (in America).

Cervicina = Wahlenbergia (Campanulaceae) Delile.
La. *Cervia* (q.v.); *-ina* = resembling.

Cestrum* (Solanaceae) L.
Gk. *kestron* = point, sting. The leaves of *Cestrum* are lance-shaped and some berries are toxic (poisonous). The name was originally used by Dioscorides for a different plant; probably, betony.

Ceterach = Asplenium (Aspleniaceae) DC.
Possibly from the Gk. *sjetrak*, or the Persian *chetrak,* both ancient names for 'fern'. The name was used by Persian physicians.

Chaenostoma = Sutera (Scrophulariaceae) Benth.
Gk. *khaino* = gape, split, open; *stoma* = mouth; referring to the corolla tube, which opens widely.

Chaetacanthus = Dychoriste (Acanthaceae) Nees
Gk. *khaite* = spine, bristle; *akanthos* = thorn. Resembles the genus *Acanthus* with hairy bracts and calyx.

Chaetacme (Ulmaceae) Planch.

(EJM)

Gk. *khaite* = spine, bristle; *akmi* = highest point. The leaves are pointed at the tip and the branches covered with spines.

Chaetaria = *Aristida* (Poaceae) P.Beauv.
Gk. *khaite* = a bristle; *aria* = pertaining to; referring to the long and bristle-like lateral lemma lobes.

Chaetobromus (Poaceae) Nees
Gk. *khaite* = a bristle; *bromos* = a kind of oat; referring to the bristly hairs at the internodes.

Chaetospora = *Schoenus* (Cyperaceae) R.Br.
Gk. *khaite* = a bristle; *spora* = seed; referring to the hairy seeds.

Chamaealoe = *Aloe* (Asphodelaceae) A.Berger
Gk. *khamai* = on the ground; *Aloe*; referring to the the plant's low-growing, dwarf-like size.

Chamaecrista (Fabaceae) Moench
Gk. *khamai* = dwarf; *christa* = the earlier generic name.

*Chamaecrypta*** (Scrophulariaceae) Schltr. & Diels.
Gk. *khamai* = on the ground, low, small, dwarf; *kryptos* = hidden; alluding to the difficulty in seeing this plant.

Chamaefilix = *Asplenium* (Aspleniaceae) Hil.
Gk. *khamai* = on the ground; *filix* = fern. A terrestrial fern rather than an epiphyte.

*Chamaegigas** (Scrophulariaceae) Dinter
Gk. *khamai* = growing on the ground; *gigas* = giant; referring to the tiny plant's ability to thrive even in the desert.

Chamaesyce = *Euphorbia* (Euphorbiaceae) Gray
Gk. *khamai* = growing on the ground; *skyon* = fig; referring to the plant's prostrate growth pattern.

Chamarea (Apiaceae) Eck. & Zeyh.
Gk. *chamai* = growing on the ground. Possibly this is from the Khoisan name *chamare* (Adamson), a kind of umbellifera.

Chamelaucium (Myrtaceae) Desf.
Gk. *khamai* = low, small, dwarf; *leucos* = white. The derivation of the name is unclear.

*Chamira** (Brassicaceae) Thunb.
Gk. *khamira* = the name of a monster in Greek mythology. Probably referring to the plant's prostrate habit, Gk. *Chamai* = on the ground, low, dwarf.

Chapmanolirion = *Pancratium* (Amaryllidaceae) Dinter
For James Chapman (1831–1872), South African explorer, hunter, trader and photographer. He started exploring, going across the Limpopo River, and reached the Chobe River (1852) to within 110km of Victoria Falls, almost beating David Livingstone in his discovery (1853). In 1854 he went with Samuel H Edwards to parts of Lake Ngami, Northern Bechuanaland (now Botswana), the Okavango River, Damaraland, returning to Walvis Bay in 1855. From 1860–1864, he tried to explore the Zambezi from Victoria Falls to its delta with his brother, Henry, and Thomas Baines (1820–1875) but failed because of misfortune and sickness. He wrote *Travels in the Interior of South Africa* (1868), published shortly before his death. One of his sons was lost on the Titanic.

Charadrophila (Stilbaceae) Marloth
La. *charada* = a ravine; *philos* = loving; referring to its habitat.

Charieis = *Felicia* (Asteraceae) Cass.
La. *charis* = graceful, elegant; referring to its appearance.

Chascanum (Verbenaceae) E.Mey.
Gk. *khasme* = gaping, wide open; *kanum* = off-white, ash coloured; referring to the wide-open flower tube. *Chascanon* was the name given by Dioscorides for *Xanthium strumarium*.

Chascolytrum = *Briza* (Poaceae) Desv.
Gk. *khasko* = open, gape; *elytron* = sheath, cover;

referring to the glumes, which open to reveal grain as it matures.

Chasmanthe (Iridaceae) N.E.Br.

(CM)

Gk. *khasme* = wide open, gaping; *anthos* = flower; alluding to the shape of the corolla.

Chasmatophyllum (Aizoaceae) Dinter & Schwantes
Gk. *kasme* = gaping, open mouth; *phyllon* = leaf. As the young toothed leaves separate, they resemble the jaws of an animal.

Chasmone = Argyrolobium (Fabaceae) E.Mey.
Gk. *khasme* = gaping wide open; alluding to the gaping calyx.

Cheilanthes (Pteridaceae) Sw.
Gk. *keilo* = lip; *anthos* = flower; referring to the lip-like false indusium that covers the sporangia. The sorus is marginal in these ferns.

Cheiranthus = Erysimum (Brassicaceae) L.
Gk. *kheir* = a hand, possibly from Arabic *kheyri* = red-flowered; *anthos* = flower, thus 'hand-flower'; perhaps a reference to the custom of carrying these fragrant flowers in the hand as a bouquet.

Cheiridopsis (Aizoaceae) N.E.Br.
Gk. *kheiris* = a sleeve, -*opsis* = resembling;

referring to the old, dry leaves which form a protective sheath around new emerging green leaves.

Cheirostylis (Orchidaceae) Blume
Gk. *kheir* = hand; *stylus* = style; alluding to the lobed clinandrium, which bears a fanciful resemblance to a hand.

Chelidonium* (Papaveraceae) Tourn. ex l.
Gk. *khelidon* = swallow; allegedly because it comes into flower when the swallows arrive and fades at their departure.

Chenolea = Bassia (Amaranthaceae) Thunb.
Gk. *kheno* = goose; *olea* = olive. The allusion is obscure.

Chenoleoides = Bassia (Amaranthaceae) (Ulbr.) Botsch.
Gk. *Chenolea* (q.v.); -*oides* = resembling.

Chenopodiopsis (Scrophulariaceae) Hilliard
Gk. *Chenopodium* (q.v.); -*opsis* = resembling.

Chenopodium (Amaranthaceae) L.
Gk. *kheno* = goose; *podion* = little foot; hence 'goose-foot'; referring to the shape of the leaves which, in some species, resemble the webbed feet of geese.

Chibaca = Warburgia (Canellaceae) Berto.
Originally discovered in the Maputo District of Mozambique known as Chibaha (Xibahá) hence Bertolini's original generic name.

Chilianthus* = **Buddleja** (Scrophulariaceae) Burch.
Gk. *khilioi, chili-* = thousand; *anthus* = flower, hence 'thousand-flowered'. According to David Gledhill's *The Names of Plants*, 'Chiliarchus was in charge of 1 000 men', hence 'thousand-flowered' – but the spelling of the name is incorrect.

Chlidanthus* (Amaryllidaceae) Herb.
Gk. *khlide* = delicate (or luxury); *anthus* = flower; referring to the delicate funnel-shaped flowers.

Chilopsis* (Bignoniaceae) D.Don
Gk. *kheilos* = lip; *opsis* = resembling; possibly referring to the flaring, trumpet-shaped flowers, or alluding to the two-lipped corolla.

Chimaerochloa (Poaceae) H.P.Linder
Gk. *chima,* from *cheimon* = winter; *eros* = loving (passionate desire for); *chloa* = grass. The grass loves the winter season.

Chimonanthus* (Calycanthaceae) Lindl.
Gk. *kheimon* = winter; *anthos* = flower; referring to its flowering time.

Chionanthus (Oleaceae) L.
Gk. *chion* = snow; *anthos* = flower. The flowers of most species are white.

Chionochloa (Poaceae) Zotov
Gk. *chion* = snow; *chloa* = grass; referring to the preferred high altitude habitat of the grass, 'snow grass'.

Chirocalyx = Erythrina (Fabaceae) Meisn.
Gk. *chiro* = hand, calyx; referring to the spathaceous calyx, its slender limb-segments spreading like the fingers of a hand.

Chironia (Gentianaceae) L.
For Chiron, the good 'civilised' centaur of Greek mythology who studied medicine, astronomy, music and other arts. When he was accidentally shot, legend has it that Zeus, the god of sky and thunder, put him in the south at Alpha and Beta Centauri, pointers to the Southern Cross; referring to the plant's medicinal properties.

Chloridion = Stereochlaena (Poaceae) Stapf
Gk. *Chloris* (q.v.); *-idion* = diminutive.

Chloris (Poaceae) Sw.
Named after Chloris, the Greek goddess of flowers and the personification of spring or Gk. *chloros* = yellowish-green; *chloros* = verdant; alluding to the luxuriant, green leaves and pale yellow-greenish flowers in some species.

Chlorocodon = Mondia (Apocynaceae) Hook.f.
Gk. *chloros* = (yellow-)green; *kodon* = a bell. The flowers resemble greenish bells.

Chlorocyathus = Raphionacme (Apocynaceae) Oliv.
Gk. *chloro-* = green; *kyathos* = cup, ladle. The flowers are greenish and cup-shaped.

Chlorophytum (Agavaceae (Asparagaceae)) Ker Gawl.
Gk. *chloros* = yellow-green; *phyton* = a plant; referring to the green colour of the leaves and the flowers.

Chomelia (Rubiaceae) Kuntze
For Pierre Jean Baptiste Chomel (1671–1740), French physician and botanist. He studied botany and medicine from an early age, took botanising lessons with French botanist Joseph Pitton de Tournefort, and obtained a doctorate in Paris in 1697. In 1700 he undertook a research project

in the mountainous areas around Auvergne looking for medicinal plants and also evaluating the snowmelt water quality. He found many unknown plant species, which he sent to the Jardin du Roi. Between 1703–1720 he continued this research, communicating his findings with the Academy of Sciences. He became a partner in this Academy in 1707 and was elected dean of the faculty in 1738. He authored *Abrege de l'Histoire des Plantes Usuelles* (first edition 1712, last published 1803), containing herbal plants that had 'proven' medicinal value and in which he provided French and Latin names.

Chlorophytum: the Latin name for a green plant (CM)

Chondrilla* (Asteraceae) L.
Gk. *condrilla* = endive or chicory.

Chondropetalum = Elegia (Restionaceae) Rottb.
Gk. *chondros* = grain, corn, cartilage; *petalum* = petal. The flowers, while small, are abundant.

Chordiflex* (Restionaceae) B.G.Briggs & A.S.Johnson
La. *chorda* = a rope or twine; *-fex* = a maker. The plant is used for making ropes.

Chorisochora* (Acanthaceae) Vollesen.
Gk. *chori-* = separate, distinct, apart; *chora* = countryside; presumably referring to its habitat and that it is widely dispersed.

Choristosoria** = **Pellaea (Pteridaceae) Mett. ex Kuhn.
Gk. *choristo-* = separated, apart; *soros* = a heap, mound, sorus. There is a discrete indusium for each sorus.

Choristylis (Escalloniaceae) Harv.
Gk. *choris* = separate; *stylis* = style. Each flower has two separate styles.

Choritaenia (Apiaceae) Benth.
Gk. *chori-* = separate; *taenia* = ribbon. The two styles are linear (Robert Allen Dyer).

Chorizema (Fabaceae) Labill.
Gk. *choros* = a dance; *zema* = food; perhaps referring to the discoverers on finding this genus and water at the same location (Australian Native Plant society); alternatively, *chori-* = separate, distinct, apart; *nema* = thread referring to the free filaments of the flower (San Marcos Growers).

Chortolirion (Asphodelaceae) A.Berger
Gk. *chortos* = feeding place, pasture; *leirion* = lily. The plant grows in grasslands.

Christella** = **Cyclosorus (Thelypteridaceae) H.Lév.
For Konrad Hermann Heinrich Christ (1833–1933), Swiss jurist, pteridologist, naturalist, botanist and paleobotanist. He studied law at Basel University and Humboldt University of Berlin, obtaining his doctorate in 1856, and graduated as a notary a year later. He served as lawyer and notary from 1869 to 1908 in Basel. Besides his legal and political work, he supported missionary work in Africa and became an activist against the mistreatment of people in the Congo region, forming the Swiss League for the Protection of the Natives of the Congo State. He had a special interest in systematics, plant geography, the history of botany and ferns. Some 60 species were named in his honour. He authored at least four books, three in German, and *Ferns of the Maritime Alps* (1900) in English. In 1912 he was elected a foreign member of the Linnaean Society.

Chromolaena* (Asteraceae) DC.
Gk. *chroma* = colour; *laina* = a cloak; referring to the tips of the involucral bracts.

Chrozophora (Euphorbiaceae) Neck. ex A.Juss.
Gk. *chrozoa* = relates to colour, colouring; *phora* = bearing; referring to the blue-purple colouring that can be obtained from this plant used for food colouring. It was also used to illuminate ancient manuscripts.

Chrysanthellum* (Asteraceae) Rich.
La. *Chrysanthemum* (q.v.); *-ellum* = diminutive. The two species described by Rich. are considered to be illegitimate.

Chrysanthemoides** = **Osteospermum (Asteraceae) Fabr.
Chrysanthemum (q.v.); Gk. *-oides* = resembling.

Chrysanthemum* (Asteraceae) L.
Gk. *chrysos* = gold, golden; *anthemon* = flower, hence 'golden-flowered'; alluding to the golden colour of the flower, although 'most species are actually silver' (Jackson).

Chrysitrix (Cyperaceae) L.
Gk. *chrysos* = gold, golden; *tricho, trix* = hair; alluding to the golden hairs on the inflorescence.

Chrysocoma (Asteraceae) L.

(GAN)

Gk. *chrysos* = gold; *kome* = hair, locks; referring to the golden terminal heads.

Chrysodium** = **Acrostichum (Pteridaceae) Fée
Gk. *chrysos* = gold; *-odes, odium* = resembling; referring to the flower head.

Chrysophyllum (Sapotaceae*) L.
Gk. *chrysos* = golden; *phyllon* = leaf; referring to the leaves of some species that are often covered with golden hairs underneath.

Chrysopogon (Poaceae) Trin.
Gk. *chrysos* = golden; *-pogon* = beard; referring to the golden-coloured hairs under the spikelet.

Chrysopteris = Phlebodium (Polypodiaceae) Link
Gk. *chryso-* = golden; *pteris* = fern; referring to the golden-coloured fronds of the type species.

Chrysoscias = Rynchosia?*** (Fabaceae) E.Mey.
Gk. *chryso-* = golden; *skias* = cloak; referring to the golden flowers and in some species the large shade leaves.

Chymococca = Passerina (Thymelaeaceae) Meisn.
Gk. *chymos* = juice; *kokkos* = berry, seed; referring to the shining scarlet globose fleshy fruits.

Cibotium (Dicksoniaceae*) Kaulf.
Gk. *kibotion* = a small box or chest; alludes to the shape of the indusium.

Cichorium* (Asteraceae) L.
Ancient Arabic name for chicory.

Ciclospermum** (Apiaceae) Lag.
Gk. *ciclospermum* = circular seed. The seeds are rounded (the fruits of these plants are almost globular).

Cicuta (Apiaceae) L.
Latin word for hemlock, a very poisonous plant.

Cienfuegosia (Malvaceae) Cav.
For Bernardo de Cienfuegos (c 1580–1640), Spanish physician and botanist who wrote seven hand-written bound volumes, kept at the Spanish National Library, containing some 1 000 drawings of plants, most of them in colour. This monumental work contains a great deal of original data about plants and their application, especially in the realm of medicine. Cienfuegos also translated at least one book from Latin to Spanish, relating to the life of Father Gonzalo de Silveira, a priest of the Society of Jesus, martyred at Monomatapa, a city in Caffraria (1614).

Cincinalis = Pteridium (Dennstaedtiaceae) Gled.
La. *cincinalis* = with curls; possibly referring to curly ferns or referring to the false indusium.

Cineraria (Asteraceae) L.
La. *cinereus* = ash-coloured. Most species of this genus have leaves with an ashen-grey hair-covering on at least the lower surface.

Cinnamomum* (Lauraceae) Schaeff
Gk. *kinnamomon* = cinnamon; alluding to a spice obtained from the inner bark of several trees from the genus *Cinnamomum* that is used in both sweet and savoury foods.

Circandra (Aizoaceae) N.E.Br.
Gk. *kirkos* = circle, ring; *andros* = male; referring to short stamens (anthers) arranged in a ring around the top of the ovary.

Cirrhopetalum = Bulbophyllum (Orchidaceae) Lindl.
La. *cirrhos* = a curled or wavy tendril; *petalum* = petals; referring to the tendril-like petals of this orchid species.

Cirsium* (Asteraceae) Mill. emend. Scop.
Gk. *kirsion* = a kind of thistle, derived from *cirsos* = a swollen vein or welt. In ancient times, thistles were used as a remedy against swollen veins.

Cissampelos (Menispermaceae)

(GG)

Gk. *kissos* = ivy; *ampelos* = vine. The plant scrambles like ivy and bears fleshy fruits like grapes.

Cissus (Vitaceae) L.
Gk. *kissos* = ivy. These plants often climb like ivy.

Cistus* (Cistaceae) L.
Gk. *kistos* = name used in ancient Greece for an evergreen shrub.

Citropsis* (Rutaceae) (Eng) Swingle & M.Kellerm.
Gk. *kitron* = citrus; *-iopsis* = resembling, likeness. These trees look like species of *Citrus*.

Citrullus (Cucurbitaceae) Eck. & Zeyh.
Gk. *kitron* = citrus; *-ullus* = diminutive. The fruit is globose or ellipsoid (Robert Allen Dyer).

Cladium (Cyperaceae) P.Browne
Gk. *klados* = a branchlet such as twig or shoot; *-ium* = resembling; referring to the panicles.

Cladoraphis (Poaceae) Franch.
Gk. *klados* = stem, twig or shoot; *raphis* = needle; referring to the central axis of the inflorescence, which has a sharp-pointed tip.

Cladostemon (Capparaceae) A.Braun & Vatke

(EJM)

Gk. *klados* = a branch; *stemon* = stamen. The stamens branch off a conspicuous androphore.

Claoxylon (Euphorbiaceae) Juss.
Gk. *klaein* = to break; *xylon* = wood. This genus is noted for its brittle wood.

Clausena* (Rutaceae) Burm.f.
For Peder Claussen Friis (1545–1614), Norwegian parish priest and naturalist with an interest in Norway's geography, history and ancient languages. During his lifetime he wrote on a wide range of subjects – topogeographical works, a chronology of Norwegian kings, and translations of various kinds. He collected texts of the ancient laws and ordinances which he translated, clerical works, old Norse sagas, and so on. It was only after his death that his writings received recognition. A considerable amount of Norway's history and sagas was captured by Peder Claussen and in some cases is the only known record. He authored a natural history account, *Description of Norway and Adjacent Islands*, published posthumously in 1632.

Cleistachne (Poaceae) Benth.
Gk. *kleistos* = closed, locked away; *achne* = glume; referring to the uppermost of the two bracts, the palea, which is often surrounded by the lemma, hence closed.

Cleistanthus (Phyllanthaceae) Hook.f. ex Planch.
Gk. *kleistos* = closed; *anthos* = flower, hence 'hidden flower'. The flowers are hidden by the prominent, hairy bracts.

Cleistochlamys* (Annonaceae) Oliv.
Gk. *kleistos* = closed; *chlamys* = cloak. The calyx is closed in a bud.

Cleistopholis* (Annonaceae) Pierre ex Eng.
Gk. *kleistos* = closed; *pholis* = horny scale, scale. The plant has closed scales referring to the inner arrangement of the petals.

Clematis (Ranunculaceae) L.
Gk. *klematis* = from *klēma* = a vine branch, twig or tendril. The name of a climbing plant that can be readily cut and grafted.

Clematopsis* (Ranunculaceae) Bojer ex Hutch.
Clematis (q.v.); *-opsis* = resembling.

Cleome (Cleomaceae) L.
Cleonia was a plant name used by Dioscorides for a mustard-like plant; possibly from Gk. *kleio* = I shut. Also used by Theophrastus. The 'spider-flower'. The original source of this name is unknown.

Cleretum (Aizoaceae) N.E.Br.
Gk. *kleros* = fate, chance, lot (as in 'drawing lots'); *-etum* = the place dominated by a given plant. In early Sparta public land was apportioned to citizens by drawing lots, with the best land given to those who, by chance, drew a winning lot.

Clerodendrum (Lamiaceae) L.
Gk. *kleros* = chance or fate; *dendron* = tree. The Ceylonese 'chance-tree'; alludes to a native legend that the trees possessed medicinal properties, and also to the uncertainty of the medicinal properties of the different species.

Cliffortia (Rosaceae) L.
For George Clifford (1685–1760), Dutch merchant

and banker, amateur botanist and zoologist. He was a director of the Dutch East India Company and owned a magnificent garden at Hartecamp, Netherlands, as well as a private zoo in Amsterdam. George Clifford is best known as a patron of the Swedish naturalist Linnaeus, whom he employed as 'hortulanus' and who catalogued the family's unique collection of plants, herbarium and library. The result was Linnaeus's 530-page book *Hortus Cliffortianus* (1738), his first important work, in which he described many species from Clifford's garden. The publication was paid for by George Clifford as a private edition.

Clinopodium (Lamiaceae) L.
Gk. *klino-* = sloping, inclining; *podium* = little foot, hence 'bed-flowers'. A Dioscorides name. The knob-shaped appearance of the inflorescence resembles bed casters.

Clitoria* (Fabaceae) L.
Gk. *kleio* = I shut. Linnaean name for *Clitoria mariana*, the butterfly pea, from a supposed resemblance to the female clitoris or referring to the seed forming inside the flower before petal-drop (Plowden) or by analogy the legume in the persistent flower parts (David Gledhill).

Clivia (Amaryllidaceae) Lindl.
For Lady Charlotte Florentia Clive, Duchess of Northumberland (1787–1866), the grand-daughter of Baron Robert Clive (who founded the British Empire in India), and who for some time was the governess of the future Queen Victoria. Clivias were first discovered in 1815 by the naturalist and explorer William Burchell in the forests of Eastern Cape Province of South Africa.

Additional plants were collected in the same area in the early 1820s by plant collector and Kew gardener James Bowie, under the direction of Kew botanist James Lindley who named the plant 'Clivia' in 1828.

Chloanthes (Lamiaceae) R.Br.
Gk. *chloros* = grass green; *anthos* = flower; referring to the greenish-yellow flowers of the type species, the original specimens of which dried green.

Clutia (Euphorbiaceae) L.
For Theodorus Augerius Clutius (Outgers Cluyt) (1577–1636), Dutch botanist, horticulturalist, beekeeper and pharmacist, eldest son of Dirck Outgaertszoon Cluyt (Clutius) (1550–1598) from Delft, an apothecary, curator of the Leiden botanical garden, and an authority on medicinal herbs. Outgers studied and worked with his father in the garden. After his father died he hoped to become his successor, but failed in the attempt. Thereafter, he studied at the University of Montpellier for several years. Between 1602–1608 he travelled to France, Germany and Spain, and also, later, on three occasions to the desert of Barbary in North Africa to increase his knowledge and collect plants for the Leiden botanical garden. Leiden University rewarded him handsomely for his efforts. On his return to the Netherlands (1618), he worked as a physician and during that time worked hard to promote the Amsterdam Hortus Botanicus where he obtained a job against strong opposition. Herman Boerhaave honoured Outgers (and his father) by naming *Clutia pulchella* after them.

Cluytia* = *Clutia (Euphorbiaceae) L.
As for *Clutia*. *Clutia* was renamed *Cluytia* by a Mr Dryander in Aiton's *Hortis Kewensis* according to Curtis's *Botanical Magazine*, Volume 45, 1818, now reverted.

Clypea* = *Stephania (Menispermaceae) Blume
Gk. *clypea, clypeus* = a small circular shield used by the Romans held by a handle or worn on the forearm, also known as a buckler. Alluding to the buckler-formed filament.

Cnestis (Connaraceae) Juss.

(RB)

Gk. *knesiao* = to itch; *knestis* = a rasp; referring to the hairs on the fruits which cause itching.

Cnicus* (Asteraceae) L.
Gk. *knekos* = thistle, which in turn may be derived from *chnizein* = to injure; referring to the plant's extreme prickliness.

Cnidium (Apiaceae) (L.) Cusson
Ancient name for a potherb, a kind of nettle, 'orach'.

Coccinia (Cucurbitaceae) Wight & Arn.
La. *coccinus* = scarlet; Gk. *kokkos* = berry. The fruits are bright red when ripe.

Coccosperma = Erica (Ericaceae) Klotzsch
Gk. *kokkos* = scarlet-berried; *sperma* = seed; referring to the cochineal-like, scarlet seeds.

Cocculus (Menispermaceae) DC.
Gk. *kokkos* = berry; *-ulus* = small; referring to the fruit.

Cochlearia* (Brassicaceae)
Gk. *kochliarion* = spoon-shaped; alluding to the leaf shape of some species.

Cochlidium = Monogramma (Polypodiaceae) Kaulf.
Gk. *cochlea* = spoon; *eidos* = like. The tip of the frond is spoon-like.

Coddia (Rubiaceae) Verdc.
For Leslie Edward Wastell Codd (1908–1999), South African botanist and agriculturalist. He

studied at the universities of Natal, Cambridge, Imperial College of Tropical Africa, Trinidad, and UNISA. He became director of the Botanical Research Institute in Pretoria from 1963–1973, published *Trees and Shrubs of Kruger National Park* (1951), authored over 160 publications, edited the journal *Bothalia* (1958–1974), described many new taxa, helped to found and became president of the South African Association of Botanists, amassed a collection of over 11 000 plant specimens, and co-authored with Mary Gunn the major biographical work *Botanical Exploration of Southern Africa* (1981).

Codiaeum (Euphorbiaceae) A.Juss.
Malayan vernacular name, *kodiho* or *codebo*. Meaning unknown.

Codon (Boraginaceae) L.
Gk. *kodon* = a bell; referring to the shape of the corolla – especially the flowers of *C. royenii* which are deeply cup-shaped, even if they do not hang down (Plantza.com).

Codonanthemum* = Anomalanthus** (Ericaceae) Klotzsch
Gk. *kodon* = a bell; *anthemum* = flower; hence 'bell-flower'; referring to the bell-shaped flowers.

Codonostigma** (Ericaceae) Klotzsch
Gk. *kodono* = bell-shaped; stigma. The stigma is bell-shaped.

Coelachyrum (Poaceae) Hochst. & Nees
Gk. *koilos* = hollow; *achyron* = glume, chaff, husk, hence 'hollow-chaff'. The glumes are hollow.

Coelanthum (Molluginaceae) E.Mey. ex Fenzl
Gk. *koilos* = hollow; *anthos* = flower. The calyx is bell- or funnel-shaped.

Coelidium (Fabaceae) Vogel ex Walp.
Gk. *koilos* = hollow; *-idium* = diminutive; alluding to the frequency of the concave leaves (William Henry Harvey, *Flora Capensis*).

Coelorachis (Poaceae) Brongn.
Gk. *koilos* = hollow, *r(h)achis* = the main axis of a structure, such as a leaf or an inflorescence. The main axis of the inflorescence is hollow.

Coffea (Rubiaceae) L.

A twig in full flower (NC)

Turkish *kahve* from Arabic *qahwa* (1598) = coffee. There are many other claims to the source of this word.

Coilostigma = Erica (Ericaceae) Klotzsch
Gk. *koilos* = a hollow; stigma. The stigma is 'large, peltate or crater-like' (Brown in *Flora Capensis*).

Coix* (Poaceae) L.
Ancient Gk. *koix* = name of a palm (*Hyphaene coriacea,* in the works of Theophrastus). The genus has false 'fruits' which resemble the fruits of this palm.

Cola* (Malvaceae) Schott & End.
An African vernacular name for a species of this genus; akin to *k'ola* = the cola-nut.

Colchicum (Colchicaceae) L.
Gk. From Colchis, an ancient region of Georgia, near the north-eastern Black Sea, in the Caucasus.

Coldenia (Boraginaceae) L.
For Cadwallader Colden (1688–1776), Irish-born Scottish scientist, physician and politician. He studied medicine in London, emigrated to America in 1710, and became surveyor-general of the colony in 1718. In 1739 he moved to his farm in New York where he studied history and botany. He became one of the first men in Europe or America to completely master the new Linnaean system of plant classification but was not entirely satisfied with this system. He wrote papers on gravity and yellow fever, but his scientific work was generally ignored because it was not based upon grounded theory, i.e. observations, but only deductive reasoning. He was acting governor of New York from 1760–1762 and 1763–1765 and governor from 1769–1771.

Coleochloa (Cyperaceae) Gilly
Gk. *koleos* = a sheath; *chloe* = a grass. In this genus of *Cyperaceae,* the stems arise from a grass-like sheath.

Coleonema (Rutaceae) Bart. & H.Wend.
Gk. *koleos* = a sheath; *nema* = a thread; alluding to the filaments of the sterile stamens (staminoides) enfolded in the channel of petals.

Coleotrype (Commelinaceae) C.B.Clarke
Gk. *koleus* = sheath; *trypa* = a hole. The young shoots break through the leaf-sheath rather than emerging from the top of the sheath.

Coleus = Plectranthus (Lamiaceae) Lour.
Gk. *koleos* = a sheath; alluding to the filaments being connected at the base, which sheaths the style. The stamens are enclosed in the tubular corolla.

Colina = Mohria (Schizaeaceae) Greene.
Possibly after Eugene Jean Baptiste Colin (1845–1919), French pharmacist and microscopist, with no further details. 'The name commemorates the French Professor Colin' (E Greene).

Colophospermum (following page): Mopane woodlands cover large areas of the drier central parts of Southern Africa. (TW)

Seeds showing resin droplets (EJM)

Colocynthis *** = **Citrullus** (Cucurbitaceae) Mill.
Gk. *kolokunthis* = round gourd, such as a pumpkin.

Colophospermum = ***Copaiba*** (Fabaceae) J.Kirk ex J.Léonard. (Image on previous page)
Gk. *kolophonios* = resin; *sperma* = seed; referring to the oily resinous seeds.

Colpias (Scrophulariaceae) E.Mey. ex Benth.
Gk. *kolpos* = breast or womb; probably referring to the two pouches in the corolla tube which resemble breasts.

Colpodium (Poaceae) Trin.
Gk. *kolpos* = fold, cleft or hollow; referring to the glume apices which are often indented.

Colpoon (Santalaceae) P.J.Bergius
Gk. *kolpos* = breast or uterus; La. *colpus* = groove. Possibly because stamens pass through grooves in the ovary. 'Sinubus quatuor excavatis quos Stamina permeant' (Peter Jonas Bergius).

Colubrina (Rhamnaceae) Rich. ex Brongn.
La. *colubra* = snake; *-ina* = like; referring to the snake-like stems or stamens.

Columellea (Asteraceae) Jacq.
For Lucius Junius Moderatus Columella (4–c 70), Roman soldier, farmer and agricultural writer, author of *De Re Rustica*, preserved in 12 volumes, the most important source and complete description of Roman agriculture, farming methods and management. He also wrote about trees in *De Arboribus*.

Columnea* Gesneriaceae L.
For Fabio Colonna (Fabius Columnus) (1567–1640). Italian philologist, antique dealer, naturalist and writer on plants. He studied law at the University of Naples (1589) but epilepsy prevented him from practising. His interests switched to ancient authors of medicine, botany and natural history with a special interest in fossils which he researched from 1606–1616. In 1592 he wrote *Phytobasanos* (translated as *A Critical Examination of Plants*) and *Ekphrasis,* coining the word 'petal' previously known as 'floral leaves'. His books include *Opusculum de purpurea* (1675), *Apiaro* (1635) and *Yesor Messicano* (1628). He

had an interest in the microscope and telescope, and he invented a 50-stringed instrument, the pentecontachordon. He was an early member of the *Accademia dei Lincei.*

Colutea (Fabaceae) L.
Probably Gk. *kolutea*, a name used by Theophrastus for the 'bladder senna'. This Greek name has been used for these shrubs at least since Theophrastus.

Colysis (Polypodiaceae) C.Presl
Gk. *kolysis* = a separation or interruption; alluding to the discontinuous rows of sori.

Comborhiza (Asteraceae) Anderb. & K.Bremer.
Gk. *combos* = knot, tuber; *rhiza* = root; referring to the stunted basal parts of the plants (Marinda Koekemoer, SANBI).

Combretum (Combretaceae) Loef.
Latin name used by Pliny the Elder for a climbing plant of another genus.

Commelina (Commelinaceae) L.
For Jan or Johannes Commelin (1629–1692) and his nephew Caspar Commelin (1667–1731), Dutch botanists, who wrote the two-volume work *Horti medici amstelodamensis rariorum plantarum … descriptio et icones.* Johannes Commelin opened the popular gardens in Amsterdam in 1682, initially called Hortus Medicus, but later Hortus Botanicus. He was a spice merchant who used his wealth and connections to build the Amsterdam botanical gardens into Europe's leading centre for the study of botany in the late 17th century especially of exotic plants discovered by Europeans in the East Indies, Africa and the Americas. Commelin wrote much of the text for the aforementioned work and it was completed by his nephew Caspar Commelin (1667–1731) after this death.

Commicarpus* *** = **Boerhavia** (Nyctaginaceae) Stand.
Gk. *kommi* = gum; *karpos* = fruit; referring to the gummy-glandular fruit. The outer surface of the fruit is sticky.

Commidendrum (Asteraceae) DC.
Gk. *kommi* = gum*; dendron* = tree. 'Gumwood', presumably the tree is also resinous.

Commiphora (Burseraceae) Jacq.
Gk. *kommi* = gum; *phoros* = bearing. The plant

exudes an aromatic resin or sap with healing properties.

The orange ball is exuding resin. (EJM)

Comptonanthus* = **Lasiopogon*** (Asteraceae) B.Nord.
For professor Robert Harold Compton (1886–1979), the second director of the National  Botanical Gardens of South Africa (1919–1953). He started the *Journal of South African Botany* in 1935. During his 34 years as director of Kirstenbosch National Botanical Garden, he was also professor of botany at Cape Town University. His interests were mainly in taxonomy. After his retirement he settled in Swaziland where he produced *An Annotated Checklist of the Flora of Swaziland*. A prolific collector, he collected in excess of 35 000 specimens in his career. He was president of the South African Association for the Advancement of Science and received many awards and medals for his service to science, including an honorary DSc from the University of Cape Town.

Conchium* = **Hakea*** (Proteaceae) Sm.
La. *concha* = shell; *-ium* = little; alluding to the close resemblance of the fruit to a bivalve shell.

Congea (Lamiaceae) Roxb.
East Indian vernacular name for this plant.

Coniandra** = **Kedrostis (Cucurbitaceae) Schrad. ex Eck. & Zeyh.
Gk. *konos* = cone; *aner, andr-* = male; referring to the stamen, stamina.

Conicosia (Aizoaceae) N.E.Br.
Gk. *konikos* = conical, cone-shaped; referring to the cone-shaped capsule.

Coniogramme* (Pteridaceae) Fée
Gk. *konios* = dusty; *gramme* = line; referring to the spore cases arranged along the veins. The sporangia are not in round sori but in distinct brown lines following the veins.

Conium (Apiaceae) L.
Gk. *konas* = to whirl; referring to vertigo, one of the symptoms of ingesting this plant, some species of which are highly poisonous.

Conopharyngia* = **Tabernaemontana*** (Apocynaceae) G.Don
Gk. *konos* = cone; *pharynx* = throat. The tube of the flower is somewhat conical.

Conophyllum** = **Mitrophyllum (Aizoaceae) Schwantes
Gk. *konos* = cone; *phyllon* = leaf; referring to the leaf shape.

Conophytum (Aizoaceae) N.E.Br.
Gk. *konos* = cone; *phytum* = plant; alluding to the inverted cone shape of the plant.

Conostomium (Rubiaceae) (Stapf) Cufod.
Gk. *konos* = cone; *stoma* = mouth; *-ium* = diminutive. *Paxton's Botanical Dictionary* says this refers to the toothed theca (a pollen sac or cell of the anther).

Convallaria* (Ruscaceae) L.
La. *convallis* = of the valley; but typically grows in wooded moist forests.

Convolvulus (Convolvulaceae) L.

(CM)

La. *convolvere* = to bind around. The name describes the twining habit of this genus.

Conyza (Asteraceae) Less.
Gk. *konyza* = a fleabane. The plant was believed to drive away gnats (William Henry Harvey, *Flora Capensis*). *Paxton's Botanical Dictionary* derives the name from *konis* = dust; the powdered leaves

of some species drives away flies, etc., hence the trivial English name 'flea-bane.'

Copaiba = Copaifera (Fabaceae) Adans.
The Portuguese or Spanish name, *Copaiba,* for a tree from the native Tupi, South America.

Copaifera (Fabaceae) L.
Brazilian *copaiba* = balsam; *fero* = I bear. A species of this genus yields copal gum, used in the manufacture of varnish.

Copaiva = Copaifera (Fabaceae) Jacq.
The Portuguese or Spanish name, *Copaiba,* for a tree from the native Tupi, South America.

Copisma*** **= Rhynchosia** (Fabaceae) E.Mey.
Gk. *kopis* = a broad curved knife; possibly referring to the shape of the seed pod.

Coprosma* (Rubiaceae) J.R.Forst. & G.Forst.
Gk. *kopros* = dung, manure; *osme* = smell; referring to the foul smell of the species.

Coptosperma (Rubiaceae) Hook.f.
Gk. *koptos* = bruised, cut; *sperma* = seed. The seeds of this genus have a hilar cavity, and so appear bruised.

Corallocarpus (Cucurbitaceae) Hook.f.
La. *corallium* = coral; *karpos* = fruit; alluding to the operculate fruit.

Corbichonia (Mulliginaceae) Scop.
For Jean Corbichon (or Corbechon), a French writer, lived about 1350. He was an Augustinian monk, chaplain of King Charles V, and made himself known by a translation of a Latin encyclopaedic treatise authored by Barthélemy l'Anglais around 1240, entitled *De Proprietatibus Rerum* (*On the Properties of Things*). This work was commissioned by Charles V in 1372 as part of his royal programme to replace Latin with French as the language of learning, reviewed and corrected by Pierre Ferget, another monk of the order, and published under the title *Le grand propriétaire de toutes choses* (Lyons, 1482, 1485, 1491, 1500; Paris, 1510; Rouen, 1556).

Corchorus (Malvaceae) L.
Gk. *korchoros* = applied to this or a similar plant, from *koreo* = to purge; because of the laxative properties of some species.

Cordia (Boraginaceae) L.
For Euricius Cordus (1486–1535), German humanist, prolific poet, botanist, professor of medicine at the gymnasium in Bremen, author of *Botanologicon,* and for his son Valerius Cordus

(1515–1544), botanist, pharmacist, physician, and botanical collector, who found many new and rare species, mainly in central and southern Germany and Italy. The son, Valerius, received a bachelor's degree of medicine at the University of Marburg. He was one of the fathers of pharmacognostics (a subfield of pharmacology which studies natural drugs, including the study of their biological and chemical components, botanical sources, and other characteristics), and died in Rome.

Cordyla (Fabaceae) Lour.
Gk. *kordyle* = club-like; referring to the shape of the bud and fruit.

Cordyline* (Asparagaceae) Comm. ex R.Br.
Gk. *kordyline* = club-like; referring to the enlarged underground stems or rhizomes.

Cordylogyne (Apocynaceae) E.Mey.
Gk. *kordyle* = a club; *gyne* = woman, female. The style is produced above the anther appendages into a club-shaped structure (Robert Allen Dyer).

Coreopsis (Asteraceae) L.
Gk. *koris* = a bug, tick; *-opsis* = appearance. The achenes resemble ticks.

Coriandrum* (Apiaceae) L.
Gk. *koris* = bedbug; *-andra* = male, the stamens; referring to the smell, but the bad odour emanates mainly from the leaves and unripe fruit (Cavendish).

Cormophyllum = Cyathea (Cyatheaceae) Newman.
Gk. *kormos* = tree trunk, log; *phyllon* = leaf. The leaves grow from a tree-like trunk.

Cornus (Cornaceae) L.
La. *cornu* = a horn; referring to the hardness of the wood in some species. Latin name of the dogwood tree is cornelian cherry.

Coronilla (Fabaceae) L.
La. *corona* = crown; *illa* = diminutive; referring to the crown-shape food plant (crown vetch).

Coronopus*** **= Lepidium** (Brassicaceae) Zinn
Gk. *koronopous,* from *korona* = a crow; *pous* = foot, hence 'crow-foot'; referring to the deeply cleft leaves like the point of a crown.

Corpuscularia (Aizoaceae) Schwantes
La. *corpusculum* = a little body; *-aria* = pertaining to; referring to the leaves that form their body fusion and cluster formation (translation from French).

Corrigiola (Caryophyllaceae) L.
La. diminutive of the *corrigia* = shoelace; perhaps alluding to the slender stems.

Cortaderia (Poaceae) Stapf
Spanish *corta* = to cut; an Argentinian name for 'pampas grass'; referring to the sharp-edged leaves.

Corycium (Orchidaceae) Sw.

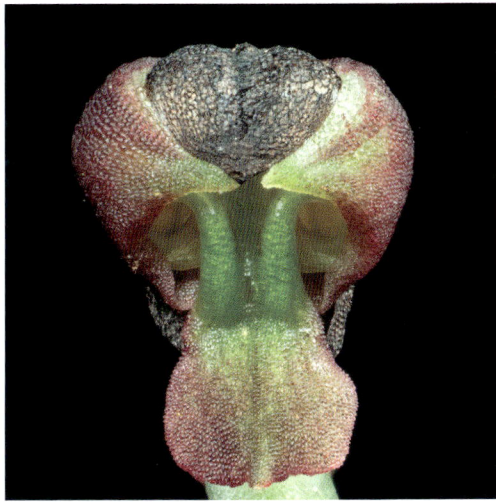

(HS)

Gk. *korys* = helmet or helmet-like structure; *korykion*, diminutive of *korykos* = leather bag. The uppermost three tepals of this coarse, globose flower converge and touch, but not fused together into a hood (shape).

Corydalis (Papaveraceae) Vent.
Gk. *korydalis* = a crested lark. The spur of the flower resembles the lark's crest.

Corylus* (Betulaceae*) L.
Gk. *korylos* = hazel, from Gk. *korus* = helmet; alluding to the shape and hardness of the nut shells or possibly the involucre.

Corymbis*** = Corymborkis (Orchidaceae) Thouars
Gk. *korymbos* = a cluster, corymb; referring to the corymbose inflorescence.

Corymbium (Asteraceae) L.
Gk. *korymbos* = a cluster; referring to a flat-topped or rounded clustered inflorescence with the lower petals longer that the upper (i.e. a corymb).

Corymborkis (Orchidaceae) Thouars
Gk. *korymbos* = a cluster, corymb; *orchis* = orchid; referring to the inflorescence, which is in a corymb or cluster.

Corynephorus* (Poaceae) P.Beauv.
Gk. *koryne* = a club; *phorus, phero* = bearing; referring to the tips of the awns, which swell and become club-shaped.

Corynocarpus (Urticaceae) J.R.Forst. & G.Forst.
Gk. *koryne* = a club; *karpus* = fruit; referring to the shape of the fruit.

Cosmos* (Asteraceae) Cav.

Beautiful in spring along Highveld roads – introduced from South America (EJM)

Gk. *kosmos* = ornament, beautiful; referring to the plant's elegant flowers.

Costularia (Cyperaceae) C.B.Clarke
Gk. *costulatus* = with fine ribs or veins; referring to the veins in the leaves.

Cotoneaster* (Rosaceae) Medik.
O.La. *cotone* = quince; *-aster* = an inferior imitation. Not the real thing. The leaves of some species resemble those of a quince, but the fruits of all species are smaller.

Cotula (Asteraceae) L.
Probably Gk. *kotule* = a little cup or hollow-shaped receptacle; referring to the shape of the involucre or of the flower head. Another source says 'or the concave base of the stem clasping leaves'.

Cotyledon (Crassulaceae) L.
Gk. *kotyledon* = seed leaf, from *kotyle* = cup, bowl; referring to the bowl- or spoon-shape of the broad seed leaves.

Coulteria (Fabaceae) Humb., Bonpl. & Kunth
For Thomas C Coulter (1793–1846), Irish physician, botanist and explorer, served as a physician in Mexico where he collected plants, best known for his exploration of Central Mexico (1825–1834), Arizona and Southern California (1831–1833). In 1834 he returned to Ireland. His collection is said to have contained over 50 000 specimens, representing between 1 500 and 2 000 species. Unfortunately, he lost his manuscripts between London and Dublin. He became curator of the herbarium at Trinity College, Dublin. The herbarium's nucleus was his personal collection. He was a member of the Royal Irish Academy and a Fellow of Trinity College.

Courbonia *** = ***Maerua*** (Capparaceae) Brongn.
For Alfred Courbon (1829–1895), French professor at the medical school of Toulouse and first-class surgeon in the French navy expedition which, on the orders of Louis-Napoléon Bonaparte, emperor of France, explored the Red Sea. Courbon investigated the medical geography, taking special notice of the diseases peculiar to each district, as well as of the relative frequency, severity, differences, etc., of the diseases common to various countries. He collected plants in Eritrea in 1859–1860 and authored two works relating to the expedition – a book, *Flore de l'île de Dissée (Mer Rouge)* (1863) and a paper, *Observations Topographiques et Medicales Recuillies dans le Voyage a l'Isthme de Suez, sur le Littoral de la Mer Rouge et en Abyssinie* (1861).

Couroupita (Lecythidaceae) Aubl.
A Latinised form of the French Guiana vernacular name, *kouroupitoumo* (Davesgarden.com).

Courtoisia *** = ***Courtoisina*** (Cyperaceae) Nees
For Richard Joseph Courtois (1806–1835), Belgian physician and botanist. At age five he met botanist Alexandre Louis Simone Lejeune who sparked his interest in plant collecting. He obtained his doctorate in medicine *cum summa lauda* in 1825 from the University of Ghent and that year was appointed deputy director of the Botanical Garden of the University of Liège at a paltry salary; so he was forced to translate German medical works to provide for his family. Between1825–1830 Lejeune and Courtois edited the *Choix de plantes de la Belgique* (many issues) and compiled the *Compendium Florae Belgicae,* 1828–1836. He was offered a botanical appointment at the University of Liège but died, probably of tuberculosis, before he could take it up.

Courtoisina *** = ***Cyperus*** (Cyperaceae) Soják.
See above.

Crabbea (Acanthaceae) Harv.
For George Crabbe (1754–1832), English doctor, minister, poet and amateur botanist. He served his apprenticeship and became a surgeon-apothecary in 1775; subsequently, he moved to London to become a writer. In 1782 he was ordained a priest, finishing his career as rector of Trowbridge, Wiltshire. Crabbe was a highly prolific poet and author (admired by Thomas Hardy for his realistic, unsentimental views) but destroyed much of his work. According to his son George Jr in *The Life of George Crabbe* (1834), his father destroyed his *Essay on Botany* in English because an academic criticised him for compiling it in a modern language. But for this, he wrote, '[M]y father might perhaps have had the honour of being considered as the first discoverer of more than one addition to the British flora.'

Cracca = ***Tephrosia*** (Fabaceae) L.
Latin name for a leguminous plant. Possibly from *cracca* = a kind of wild vetch.

Craibia (Fabaceae) Harms & Dunn
For William Grant Craib (1882–1933), British botanist who studied at Aberdeen University obtaining an MA in 1907. In 1908 he was appointed acting curator of the herbarium at the botanic garden in Calcutta (now Kolkata), and of the library of the Royal Botanical Society, but returned to Kew Gardens in the following year as assistant for India at Kew. During his early years, he lectured on forest botany (1915–1917) and on forest botany and Indian forest trees (1917–1920). In 1920 he became a professor of botany at Aberdeen University, a position he held for 13 years. He was an authority on Asiatic flora, especially *Primula Indigofera* and *Enkianthus*, and authored *Contributions to the Flora of Siam* (1912) and *Florae siamensis enumeratio* (1925). He was also an authority on systematic botany and

previously had published a book on the *Flora of Banffshire* (Scotland), his native county.

Crambe* (Brassicaceae) L.
La. *crambe*, Gk. *krambe* = a kind of cabbage.

Craspedolepis*** = Restio (Restionaceae) Steud.
Gk. *kraspedon* = edge; *lepis* = scale. Possibly referring to the leaf-sheaths which have membranous edges, in many cases showing a woolly scale within.

Craspedorhachis (Poaceae) Benth.
Gk. *kraspedon* = edge, fringe; *rhachis* = spine, axis; referring to spikelets which have been flattened on the main axis to form a wing-shaped structure.

Crassocephalum (Asteraceae) Moench
Gk. *krasson* = larger; *kephale* = a head. The flower heads are grouped in relatively large, robust inflorescences.

Crassula (Crassulaceae) L.

(CM)

La. *crassus* = thick; *-ula* = diminutive; referring to the fleshy succulent leaves.

Crataegus* (Rosaceae) L.
Classical Gk. name for this genus, derived from *kratos* = strength; referring to the hard wood.

Craterocapsa (Campanulaceae) Hilliard & B.Burtt
Gk. *krateros* = strong; *capsa* = a chest (*capsula*, a little box); referring to the hard capsule.

Craterostigma (Scrophulariaceae) Hochst.
Gk. *krater, crater* = mixing bowl, mouth of volcano, stigma. The style is 'shortly bilamellate at the somewhat cup-shaped apex' (*Flora Capensis*).

Crepidomanes = Hymenophyllum (Hymenophyllaceae) (C.Presl) C.Presl
Gk. *krepis* = slipper; *manes* = a kind of cup; alluding to the shape of the sorus.

Crepidorhopalon (Linderniaceae) Eb. Fisch.
Gk. *kepis, crepido* = pedestal, base, foundation; *rhopolos* = cudgel, club. The plant has a multicellular base and club-shaped end cell (*Flora of Zimbabwe*).

Crepis* (Asteraceae) L.
Latin name of a plant used by Pliny the Elder, Gk. *krepis* = slipper, sandal; possibly in reference to the shape of the fruit (*Flora of Northern America*).

Crinipes (Poaceae) Hochst.
La. *crinum* = hair; *pes* = foot; referring to the awn on the lower glume.

Crinum (Amaryllidaceae) L. (Image on following page)
Gk. *krinon* = lily (*crinon, crinum*).

Crocanthus*** = Malephora (Aizoaceae) Bolus
Gk. *krokeos* = saffron-coloured; *anthos* = flower; possibly referring to the flower colour.

Crocanthus*** = Crocosmia (Iridaceae) Klotzsch ex Klatt
Gk. *krokos* = thread; *anthos* = flower; referring to the red threads (stigmas) and yellow styles.

Crocodilodes*** = Berkheya (Asteraceae) Kuntze
Crocodia (q.v.); *-odes* = resembling.

Crocosmia (Iridaceae) Planch.
Gk. *krokos* = saffron; *osme* = smell; referring to the scent of the dried flowers when immersed in water. *Krokos* probably of Semitic origin (e.g. Arabic *kurkum*).

Crocoxylon*** = Cassine (Celastraceae) Eck. & Zeyh.
Gk. *krokos* = crocus, saffron; *xylon* = wood. The commonly known Afrikaans name, *Saffraanhout*, is a translation from the Greek.

Crocyllis = Gaillonia (Rubiaceae) E.Mey. & Hook.f.
Gk. *krokos* = crocus, saffron; *-ullus* = diminutive. The author (Meyer) gives no explanation.

Crinum (on previous page): A mass display of flowers (TD)

Cromidon (Scrophulariaceae) Compton
Anagram of the related genus *Microdon* (Compton).

Crossandra (Acanthaceae) Salisb.
Gk. *krossos* = fringe; *aner* = a man. The stamens are fringed.

Crossopteryx (Rubiaceae) Fenzl
Gk. *krossos* = fringe; *pteron* = a wing. The wing-like seeds are fringed.

Crossyne (Amaryllidaceae) Salisb.
Derivation obscure. Gk. *krossos* = fimbria; *ynis* = vomer (a ploughshare); by implication, throwing out from the earth something with a fimbriate leaf margin (Kesting).

Crotalaria (Fabaceae) L.
Gk. *krotalos* = rattle. The dry seed pods rattle when shaken.

Croton (Euphorbiaceae) L.
Gk. *kroton* = a tick. The seeds look like ticks.

Crudia (Fabaceae) Schreb.
From a Guyana vernacular name. Meaning unknown.

Cryophytum = Mesembryanthemum (Aizoaceae) N.E.Br.
Gk. *kruos* = frost, hence *cryo* = extreme cold; *phyton* = plant; alluding to cold-weather plant-life, occurring at high altitudes.

Cryphiantha ** = **Amphithalea** (Fabaceae) Eck. & Zeyh.
Gk. *kryphios* = concealed; *anthos* = flower; possibly in some species the small flowers are concealed by the large leaves.

Cryptadenia = Lachnaea (Thymelaeaceae) Meisn.
Gk. *kryptos* = hidden, concealed; *aden* = gland; alluding to the fact that the floral scales are enclosed within the hypanthium.

Cryptocarya (Lauraceae) R.Br.
Gk. *kryptos* = hidden, concealed; *karya* = a nut. The succulent fruit is partly hidden in a deepish cup.

Cryptogramma (Pteridaceae) R.Br.
Gk. *kryptos* = hidden, concealed; *gramme* = line. The reflexed leaf margin hides the line of sori.

Cryptolepis (Apocynaceae) R.Br.
Gk. *kryptos* = hidden, concealed; *lepis* = scale. The corona lobes are scale-like and deeply hidden inside the flower.

Cryptomeria* (Cupressaceae) D.Don
Gk. *kryptos* = hidden, concealed; *meris* = portion, hence 'hidden parts'. The cones (or flowers) of these trees are well hidden by the dense foliage.

Cryptostegia* (Apocynaceae) R.Br.
Gk. *crypto* = hidden; *stegios* = shelter; alluding to the stamens being concealed within the corolla tube.

Cryptostemma (Asteraceae) R.Br.
Gk. *kryptos* = hidden, concealed; *stemma* = a crown, garland; referring to the way the woolly seeds shroud the pappus.

Cryptostephanus (Amaryllidaceae) Welw. ex Baker
Gk. *kryptos* = hidden, concealed; *stephanus* = garland, wreath; alluding to the partially hidden inflorescence.

Ctenitis* (Dryopteridaceae) (C. Chr.) C.Chr.
Gk. *ktena* = comb; *-itis* = likeness; alluding to the comb-like appearance of some of the bipinnate species of the genus.

Ctenium (Poaceae) Panz.

(EJM)

Gk. *ktenion-* = a small comb; referring to the appearance of the inflorescence, which is usually a one-sided spike.

Ctenolepis (Cucurbitaceae) Hook.f.
Gk. *kteis, ktenos-* = a comb; *lepis* = scale; referring to the comb-like arrangement of scales.

Ctenomeria = ***Tragia*** (Euphorbiaceae) Harv.
Gk. *kteis, kten-* = a comb; *meros* = a part, portion. The sepals of the female flower are particularly lobulate.

Ctenopterella* (Polypodiaceae) Parris.
Ctenopteris (q.v.); Gk. *-ella* = diminutive.

Ctenopteris (Polypodiaceae) Blume ex Kunze
Gk. *kteis, kten-* = a comb; *pteris* = fern; alludes to the pinnae, which are pectinately lobed so as to be comb-like.

Cucumella = ***Cucumis*** (Cucurbitaceae) Chiov.
The genus *Cucumis* (q.v.); Gk. *-ella* = diminutive.

Cucumis (Cucurbitaceae) L.
La. *cucumis* = cucumber.

Cucurbita* (Cucurbitaceae) L.
La. *cucurbita* = a gourd.

Culcita* (Culcitaceae*) C.Pres.
La. *culcita* = a large cushion; alluding to the arched, cushion-like indusium or perhaps the soft hairy crown.

Cullen (Fabaceae) Medik.
For William Cullen (1710–1790). Scottish physician, apothecary, chemist and academic.

He studied pharmacology, medicine and chemistry at the universities of Edinburgh and Glasgow. After a short period of medical practice, he held the positions of lecturer in chemisty (1747) and professor of the practice of medicine (1751) both at Glasgow University, and professor of chemistry and medicine at the University of Edinburgh (1755). During his career, he was appointed president of the Faculty of Physicians and Surgeons of Glasgow, and helped found the Royal Medical Society and the Royal Society of Edinburgh. He wrote a number of books including *First Lines of the Practice of Physick, for the Use of Students* (1777) and *A Treatise on Materia Medica* (1789).

Cullumia (Asteraceae) R.Br. ex Aiton
For Sir John Cullum (1733–1785), British botanist, geneologist, antiquarian, cleric and scholar, and author of *History and Antiquities of Hawstead* (1785), and his brother, Sir Thomas Gery Cullum (1741–1831), a medical practitioner, surgeon and botanist, member of the Linnaean Society, and author of *Floræ Anglicæ Specimen imperfectum et ineditum* (1774). Both became fellows of the Royal Society.

Sir Thomas Gery Cullum   Sir John Cullum

Cunonia (Cunoniaceae) L.

For Johann (Joan, Johannes) Christian Cuno (1708–1780), German poet, botanist and merchant. He was conscripted into military service in 1724 and only left the army in 1740 during which time he travelled to Croatia, Slavonia, Hungary and Italy as a recruiting officer. In 1741, he met and married a widow Völkers whose husband had been a merchant. He developed this business and settled in Amsterdam, and also created a botanical garden and grew exotic plants. Cuno wrote poetry and also translated Dutch poetry into German. He wrote a poem about his garden *Ode über seinen Garten* (1749) to which David Sigismundus Augustus Büttner (1724–1768), professor of medicine and botany at the Collegium medico-chirurgicum, Berlin added a species list. Cuno corresponded with Linnaeus who named the genus after him.

Cupressus* (Cupressaceae) L.
The Latin name for the cypress tree.

Curio*** = Senecio (Asteraceae) P.V.Heath.
La. *curio* = lean, emaciated; referring to its appearance.

Curculigo* (Hypoxidaceae) Gaertn.
La. *curculi* = a kind of weevil. Referring to the shape of the ovary.

Curroria*** = Cryptolepis (Apocynaceae) Planch. ex Hook.f. & Benth.
For Andrew Beveridge Curror (1811–1843), Scottish surgeon, served in the royal navy on HMS Water-Witch, plant collector in Angola in 1839–1843. Records show that he travelled to the rarely visited island of Annobón in the Gulf of Guinea about 400km from Gabon where he collected two specimens sometime between 1839–1842. Little else is known about him.

Curtisia (Curtisiaceae) Aiton
For William Curtis (1746–1799), nurseryman, entomologist, and founder of *Curtis's Botanical Magazine,* first published in 1786 and still going today. He started his career as an apothecary but switched to botany and natural history. At the age of 25 he produced *Instructions for Collecting and Preserving Insects; Particularly Moths and Butterflies.* He was a demonstrator of plants and praefectus horti at the Chelsea Physic Garden from 1771 to 1777 and then established his own London Botanic Garden at Lambeth in 1779. He was the author of *Flora Londinensis* in six volumes, a work that was published over the period 1777–1798 and was devoted to urban nature.

Curtogyne*** = Crassula (Crassulaceae) Haw.
La. *curto* = short; *kyrtos* = curved, arched, bent;
gyne = woman, female; referring to the shape of the pistil.

Curtonus = Crocosmia (Iridaceae) N.E.Br.
Gk. *kyrtos* = curved, arched, bent; *onos* = axis; referring to the axis (flower-bearing portion of an inflorescence), which can bend almost horizontally.

Cuscuta (Convolvulaceae) L.

Real dodder (CM)   (CM)

Arabic *kashuta*, which became *cuscuta,* a mediaeval Latin name for the parasite Dodder, a member of this genus.

Cuspidia (Asteraceae) Gaertn.
La. *cuspidata* = tapering to a rigid point; referring to the leaves which are 'spinose-toothed' (Robert Allen Dyer).

Cussonia (Araliaceae) Thunb.
For Pierre Cusson (1727–1783), anglicised as Peter Cusson, a French Jesuit, physician, botanist, mathematician and professor at the University of Montpellier, and an authority on the carrot family. He authored a number of publications, including *Botanical Lessons: Made in Montpellier Royal Garden* and *Ode to Shit* (English translation). He had travelled extensively throughout Majorca, Spain and the Pyrenees, and amassed an excellent collection of specimens, which were regrettably disposed of by an elderly female relative with whom he lived, who cleaned his study in his absence.

Cuviera (Rubiaceae) DC.
For Georges Léopold Chrétien Frédéric Dagobert, Baron Cuvier (1769–1832), French naturalist and zoologist. He succeeded Jean-Baptiste Lamarck in the chair of comparative

anatomy at the Jardin des Plantes. He was instrumental in establishing the fields of comparative anatomy and palaeontology through his work in comparing living animals with fossils. He founded vertebrate palaeontology as a scientific discipline and established for a fact that extinctions take place, some catastrophic, a major new view of history. He was a prolific writer on natural history, comparative anatomy, molluscs, palaeontological and geological investigations, etc. His most famous work is the *Le Règne Animal* (*The Animal Kingdom*) (1817). In 1819, he was created a life peer in honour of his scientific contributions.

Cyamopsis (Fabaceae) DC.
Gk. *kyamos* = a bean; *-opsis* = resembling. The plant is a bean.

Cyanella (Tecophilaeaceae) L.
La. *cyaneus* = greenish-blue, Gk. *kyanos* = dark blue; *-ella* = diminutive; referring to the blue flowers in some species, although some species have white, pink and purple flowers.

Cyanotis (Commelinaceae) D.Don
Gk. *kyanos* = dark blue; *ous, ot* = ear; alluding to the shape of the blue tepals.

Cyanthillium* (Asteraceae) Blume
Gk. *kyanos* = blue; *anthyllion* = little flower; alluding to colour of the corollas.

Cyanus* (Asteraceae) (Juss.) DC.
Gk. *kyanos* = blue; *anthyllion* = little flower; referring to the bright blue flower head.

Cyathea (Cyatheaceae) Sm.

A shade-loving tree-fern (EJM)

Showing circinate vernation which is the neat 'unwinding' of the fronds that characterises all ferns, (EJM)

Gk. *kyathos* = a wine cup; referring to the cup-shaped sori on the underside of the fronds.

Cyathocoma (Cyperaceae) Nees
Gk. *kyathos* = cup, ladle; *kome* = hair, locks; possibly referring to the cup-like flowers, aristate glumes or nutlet that has a hairy beak.

Cyathopappus = *Elytropappus***
(Asteraceae) Sch.Bip.
Gk. *kyathos* = cup, ladle; *pappus* = a modified calyx; referring to the cup-shaped outer pappus.

Cyathula (Amaranthaceae) Blume
Gk. *kyathos* = cup, ladle; *-ula* = diminutive. The stamens may be united into a membranous cup at the base.

Cybistetes = *Ammocharis***
(Amaryllidaceae) Milne-Redh. & Schweick.
Gk. *kybisteter* = an acrobatic tumbler, from *kybistan* = to somersault; referring to the tumbling of fruiting inflorescences being blown about by the wind.

Cyclantheropsis (Cucurbitaceae) Harms
Gk. *kyklos* = circle; *anthera* = anther; *opsis* = resembling (the genus *Cyclanthera*); referring to the arrangement of circular anther cells (*CRC World Dictionary of Plant Names*).

Cyclonema = *Rotheca*** (Lamiaceae) Hochst.
Gk. *kyklos* = circle, circular; *nema* = a thread, filament; perhaps referring to the long exerted stamens in some species.

Cyclophorus = *Pyrrosia* (Polypodiaceae) Desv.

(NC)

Gk. *kyklos* = circle, circular; *phorein* = bearing. The sori cover the lamina surface.

Cyclopia (Fabaceae) Vent.
Gk. *kyklos* = circle, round, circular; *ops* = eye; referring to the circular base of the calyx; a more likely derivation than from Kyklops, the one-eyed monster of Homer's Odyssey.

Cyclopteris (Cystopteridaceae) Gray
Gk. *kyklos* = circle circular; *pteris* = fern; referring to the circular wing surrounding each seed.

Cycloptychis *** = ***Heliophila*** (Brassicaceae) E.Mey. ex Sond.
Gk. *kyklos* = circle, circular; *ptyx* = a ridge; referring to the radiating ridges of the fruit (*Flora Capensis*).

Cyclosorus = ***Thelypteris*** (Thelypteridaceae) Link
Gk. *kyklos* = circle, circular; *sorus* = sorus. The leaves bear round the sori.

Cyclospermum (Apiaceae) Lag.
Gk. *kyklos* = round; *sperma* = seed; referring to the (semi-)orbicular fruits.

Cyclostemon = ***Drypetes*** (Euphorbiaceae) Blume
Gk. *kyklos* = round; *stemon* = stamen; referring to the arrangement of the stamens.

Cycnium (Orobanchaceae) E.Mey. ex Benth. emend. Eng.
Gk. *kyknos* = swan; probably referring to the slender, elongated corolla tube like a swan's neck.

Cydonia * (Rosaceae) Mill.
Latin name for quince possibly named after the ancient town of Cydonia in northwest Crete.

Cylindrophyllum (Aizoaceae) Schwantes
Gk. *kylindros* = cylinder; *phyllon* = leaf; referring to the plant's cylindrical leaves.

Cylindropuntia = ***Opuntia*** (Cactaceae) (Engelm.) F.M.Knuth.
Gk. *kylindros* = cylinder; the genus *Opuntia* from which this genus was segregated; referring to the general form of the cactus. Opuntia was a town in Greece.

Cylista *** = ***Rhynchosia*** (Fabaceae) Aiton
Gk. *kylix* = a cup; *kylistos* = a large round stone or a circular entwined garland; alluding to the calyx being very large.

Cymation *** = ***Ornithoglossum*** (Colchicaceae) Spreng.
La. *cymation* = molding. Derivation unknown.

Cymbalaria * (Plantaginaceae) Hill
La. *cymbalum* = cymbal, from Gk. *kymbe* = boat; *-aria* = pertaining to; referring to the 'boat-like' shape of the leaves.

Cymbidium (Orchidaceae) Sw.
Gk. *cymbidium* = boat-like; referring to the hollow (boat-shaped) recess in the lip.

Cymbopappus (Asteraceae) B.Nord.
Gk. *kymbe* = boat; *pappus* = a modified calyx; referring to the boat-shaped pappus.

Cymbonotus (Asteraceae) Cass.
Gk. *kymbe* = hollowed vessel or boat; *notos* = back; alluding to the convex back of the achenes.

Cymbopogon (Poaceae) Spreng.

(EJM)

Gk. *kumbe* = boat; *pogon* = bearded; referring to the boat-shaped spatheoles which subtend the hairy racemes.

Cymbosetaria = ***Setaria*** (Poaceae) Schweick.
La. *cymba* = boat or cup (Gk. *kymbe*); *seta* = bristle; *-aria* = pertaining to; possibly referring to the fertile lemma which has a keel shape.

Cymodocea (Cymodoceaceae) K.D.Koenig
Gk. *kymodoke* (Cymodoce) one of the Nereids (goddesses) of 'steadying the waves' who, with her sisters Amphitrite and Kymatolege, possessed the power to still the winds and calm the sea. (Hesiod, Homer, Hyginus, Virgil). Inference unclear.

Cynanchum (Apocynaceae) L.

The flowers have a complicated structure and the plants exude a milky sap when damaged. (CM)

Gk. *kynos* = a dog; *anchein* = to strangle or throttle. Some species of this genus are poisonous, and several species have climbers or twiners, hence the common name 'dogbane', a dog-strangling vine.

Cynara (Asteraceae) L.
Gk. *kynara* = artichoke, probably from *kyon* = a dog; alluding to the involucral bracts resembling dogs' teeth. The phyllaries likened to dogs' teeth (Davesgarden.com).

Cynoctonum *** = ***Cynanchum***
(Apocynaceae) E.Mey.
Gk. *kynos* or *kyon* = dog; *ctonos* = murderer. The plant is poisonous.

Cynodon (Poaceae) Rich.
Gk. *kynos, kyon* = dog; *-odon* = toothed; referring to the sheathed tips of the sharp, hard scales of the plant's rhizomes and stolons or the tooth-like buds of the rhizome.

Cynoglossum (Boraginaceae) L.
Gk. *kynos* or *kyon* = dog; *glossa* = tongue; referring to the leaf-shape, 'hound's tongue'.

Cynorchis *** = ***Cynorkis*** (Orchidaceae)
Thouars
Gk. *kynos* or *kyon* = dog; *orchis* = orchid, testicle; referring to the small testiculate tubers.

Cynorhiza (Apiaceae) Eck. & Zeyh.
Gk. *kynos* or *kyon* = dog; *rhiza* = root of a plant. Derived from the Afrikaans vernacular name, *hondewortel* ('dog-root') – possibly because dogs try to pull the prominent roots up.

Cynorkis (Orchidaceae) Thouars
Gk. *kynos* or *kyon* = dog; *orchis* = orchid; referring to the tubers.

Cynosorchis *** = ***Cynorkis*** (Orchidaceae)
Thouars
Gk. *kynos* or *kyon* = dog; *orchis* = testis, orchid; referring to the tubers.

Cynosurus * (Poaceae) L.
Gk. *kynosoura* = dog's tail (Latin *cynosoura* = Ursa Minor); referring to the shape of the inflorescence.

Cyperus (Cyperaceae) L.

The compound flowering heads of papyrus are characteristic. (EJM)

La. *cuperos,* Gk. *kypeiros* = sedge or rush. The genus name for rush or sedge.

Cyphia (Campanulaceae) P.J.Bergius
Gk. *kyphos* = bent; referring to the shape of the style and stigma.

Cyphocalyx *** = ***Aspalathus*** (Fabaceae)
C.Presl
Gk. *kyphos* = bent, hunchbacked; calyx; referring to the bent or curved calyx.

Cypholepis = *Coelachyrum* (Poaceae) Chiov.
Gk. *kyphos* = bent; *lepis* = scale; referring to the concave shape of the lemma.

Cyphomandra* (Solanaceae) Mart.
Gk. *kyphos* = tumour, hump; *aner, andra* = male, stamen; refers to a thickening area on the connective tissue of the anther.

Cyphostemma (Vitaceae) (Planch.) Alston.
Gk. *kyphos* = bent; *stemma* = wreath or garland; apparently referring to the leaves arising at grotesque angles from the swollen stem (Hugh Glen).

Cypselodontia = *Dicoma* (Asteraceae) DC.
Gk. *kypsele* = hollow vessel, box; *odontos* = tooth. The receptacle is honeycombed, with the margins of the pits toothed.

Cyrtanthus (Amaryllidaceae) Aiton

Also known as Fire lily (CM)

Gk. *kyrtos* = curved; *anthos* = flower; referring to the curved perianth tube.

Cyrtomium (Dryopteridaceae) C.Presl
Gk. *kyrtoma* = curved, arched; referring to the leaflets – the veins rejoin forming an arch.

Cyrtopera*** = *Eulophia* (Orchidaceae) Lindl.
Gk. *kyrtos* = swelling; *pera* = pouch; referring to pouch-like spur of the lip.

Cyrtorchis (Orchidaceae) Schltr.
Gk. *kyrtos* = arched, curved; *orchis* = orchid. 'Curved orchid', all perianth segments are strongly recurved.

(HS)

Cystea = *Cystopteris* (Cystopteridaceae) Sm.
Gk. *kyste* = bladder; alluding to the sorus shape.

Cysticapnos (Papaveraceae) Mill.
Gk. *kystis* = bladder; *kapnos* = smoke, smoky; referring to the distinctively inflated, bladder-like fruit, dangling – like lanterns. The fruit in bud can look smoky pink-grey.

Cystopteris (Cystopteridaceae) Bernh.
Gk. *kystos* = bladder; *pteris* = fern; alludes to the inflated indusium.

Cytinus (Cytinaceae) L.
La. *cytinus* = the Greek word for the flower of pomegranate, referring to the calyx of the pomegranate blossom.

Cytisus* (Fabaceae) L.
Gk. *kytisos* = an old name for woody legumes; a kind of clover. A Greek name used by Linnaeus for *Medicago arborea*.

D

With more than 250 species of trees, shrubs and lianas. *Dalbergia melanoxylon* has been used for woodcarving and most notably for tone-woods (for musical instruments). So what does the future hold for the conservation of this remarkable genus that brings us some of the world's finest music? That is up to us. But if we lose the species that are threatened by unmanaged trade, the world will certainly be a poorer place.

An example of 'Ujamaa'-style carving by the Makonde people, an ethnic group in southeast Tanzania and northern Mozambique (AC)

(AC)

Born in Pietermaritzburg in KwaZulu-Natal in South Africa, **AB (Tony) Cunningham**'s (1957–) PhD focused on the values of plants to people in the Maputaland area of Southern Africa. He completed his PhD through the University of Cape Town and was supervised by Eugene Moll. He is an adjunct professor at the school of plant biology at the University of Western Australia and was awarded the title of distinguished economic botanist by the Society for Economic Botany in 2016 for his lifetime work on people and plants mainly in Africa, Southeast Asia and China.

Dactyliandra* (Cucurbitaceae) Hook.f.
Gk. *daktylos* = finger; *andros* = male; referring to the stamens, which are finger-like.

Dactylis* (Poaceae) L.
Gk. *daktylos* = finger; alluding to the finger-like appearance of the inflorescence.

Dactyloctenium (Poaceae) Willd. & L.

Berea-grass is an excellent lawn species for shady, sub-tropical gardens. (EJM)

Gk. *daktylos* = finger; *ktenos* = a comb; alluding to the finger-like inflorescence in which the spikelets resemble small combs.

Dactylopsis = Mesembryanthemum (Aizoaceae) N.E.Br.
Gk. *daktylos* = a finger; *-opsis* = resembling; referring to the finger-like leaves found on the branches.

Daemia = Pergularia (Apocynaceae) R.Br.
Gk. *daimon* = a natural spirit (which could be depicted as a devil – with claws); possibly referring to the stems of this vine, which climbs by means of sharp, recurved hooks.

Dahlgrenodendron = Cryptocarya (Lauraceae) J.J.M. van der Merwe & A.E. van Wyk
For Rolf Martin Theodor Dahlgren (1932–1987), a Swedish-born Danish botanist, plant collector and professor of botany at the University of Copenhagen from 1973–1987. In the 1950s and 1960s he visited South Africa, mainly the Cape, Natal, Transvaal and Zimbabwe, collecting over 5 000 species. He wrote extensively on plant systematics and cladistics. He developed a new, widely accepted system of Angiosperm classification, which is based upon many more characteristics than previous systems. It has instructive diagrams, called Dahlgrenograms. He authored *The Families of the Monocotyledons: Structure, Evolution, and Taxonomy* (1985) with Harold Trevor Clifford and Peter Yeo. In 1986 he was elected a member of the Royal Swedish Academy of Sciences.

Dahlia (Asteraceae) Thunb.
For Andreas (Anders) Dahl (1751–1789), Swedish botanist, physician and student of Linnaeus. He studied for a bachelor's degree at Uppsala University from 1770, after which he worked in Gothenburg as curator of the private natural history museum and botanical garden belonging to Claes Alströmer, a Swedish buyer and naturalist (see *Alstroemeria*). In 1786 he received an honorary medical doctor's degree at the University of Kiel, Germany. In 1787 he became associate professor and botanical director at the Academy of Turku (called 'Åbo' by Swedes), today Helsinki University. Parts of his collection, much of which was lost in a fire in 1827, are kept in the Helsinki Botanical Museum and in the Giseke's herbarium, Edinburgh. Dahl published *Observationes botanicæ circa Systema vegetabilium divi a Linné* (1787).

Dais (Thymelaeaceae) L.

The Pom-pom tree is aptly named. (EJM)

Gk. *dais* = a pine-torch. The flower-head on its pedicel resembles an unlit torch (Hugh Glen).

Dalbergia (Fabaceae) L.f.
For Carl Gustav Dahlberg (1721–1781), Swedish mercenary lieutenant-colonel, then plantation

owner when he married a Dutch widow (1746) through whom he obtained this resource, as well as supreme court judge in Surinam. From here he compiled zoological and botanical collections, some for Linnaeus. Named also for his brother Nils Ericsson Dahlberg (1736–1820), botanist and physician, student of Linnaeus in 1755, personal physician of Gustav III from 1768, and twice president of the Swedish Academy of Sciences.

Dalechampia (Euphorbiaceae) L.
For Jacques Daléchamps (1513–1588), French botanist, physician and professor of surgery, philologist, humanist and naturalist who lived in Lyon. His most important work was *Historia plantarum generalis* (1586–1587), a compilation of botanical knowledge of his time. He also made many translations of important works into French and Latin. His name is given variously as Daléchamps, Dalechamp, Dalechampius, or D'Aléchamps. The spelling of names was not as fixed at that time as it is now.

Danthonia (Poaceae) DC.
For Étienne Danthoine (1739–1794), French botanist and agrostologist from Marseilles, student of the grasses of Provence. At the time of his death he was the pharmacist in the military hospital in Grasse. He was a member of the Academy of Sciences of Marseilles and wrote articles on grasses, bedstraws and (published posthumously) gall wasps.

Danthoniopsis (Poaceae) Stapf
Danthonia (q.v.); Gk. *-iopsis* = resembling, likeness.

Darea* = *Asplenium (Aspleniaceae) Juss.
For George Dare (f 1680s–1690s), English apothecary who introduced foreign hymenophyllums into English horticulture. No further information found.

Dasispermum (Apiaceae) Raf.
Gk. *dasys* = hairy, shaggy, thick-haired; *spermum* = seed, hence 'thickly-haired seed'. The seeds have thick spines.

Dasystachys = *Chlorophytum***
(Agavaceae) Baker
Gk. *dasys* = thick, hairy; *stachys* = spike; referring to the hairy spikes.

Dasystemon* = *Crassula (Crassulaceae) DC.
Gk. dasys = thick, hairy; *stemon* = stamens; alluding to the very hairy stamens.

Datura* (Solanaceae) L.
Maybe from the Arabic *tatorah* but also said to be from the Hindi *dhatūrā* or the Sanskrit *dhattūrāh*, both meaning 'thorn apple'. The plant is poisonous.

Daubenya (Hyacinthaceae) Lindl.
For Charles Giles Bridle Daubeny (1795–1867),

English botanist, geologist and physician, professor of chemistry and botany at Oxford, Fellow of the Linnaean Society, and plant collector in the US, West Indies and Europe. He was the author of *On the Action of Light upon Plants, and of Plants upon the Atmosphere* (1836), *Sketch of the Geology of North America* (1839), *Lectures on Roman Husbandry* (1857), *Climate: an Inquiry into the Causes of its Differences and into its Influence on Vegetable Life* (1863), *Essay on the Trees and Shrubs of the Ancients,* and a *Catalogue of the Trees and Shrubs indigenous to Greece and Italy* (1865). He conducted plant experiments at the Oxford Botanic Garden. He was a Fellow of the College of Physicians and the Royal Society.

Daucus* (Apiaceae) L.
Gk. *daukos* = wild carrot or parsnip; classical name used by Theophastrus for a carrot-related plant.

Dauresia* (Asteraceae) B.Nord. & Pelser
Daures is the vernacular name for Brandberg, Namibia.

Davallia (Davalliaceae) Sm.
For Edmund Davall (1763–1798), English-born botanist who resided in Switzerland most of his life. He was a plant collector and a Fellow of the Linnaean Society and a friend and correspondent of Sir James Edward Smith (1759–1828), one of the founders of the Linnaean Society in 1788. Davall established a botanical garden at Orbe, Switzerland, and compiled a collection of Swiss plants. On his death, he left his herbarium and botanical papers to Sir James, including an unfinished work, *Illustrations of Swiss Plants*.

Decabelone = *Tavaresia*** (Apocynaceae) Decne.
Gk. *deka* = ten; *belone* = arrowhead, needle; referring to the ribbed stems which are covered in prickles – 10 or more.

Decaceras = *Brachystelma***
(Apocynaceae) Harv.
Gk. *deka* = ten; *keras* = horn, spur. *Brachystelma huttonii* (Harv.) is the synonym genus. A look

at this plant suggests that the 'horns' refer to segements of the flower that are united at the tips or the spindle-like roots.

Decalepis (Cyperaceae) Boeck. non Wight & Arn.
Gk. *deka* = ten; *lepis* = scale; referring to the corona scales and staminal filaments.

Decaneurum*** **= Vernonia** (Asteraceae) DC.
Gk. *deka* = ten; *neurum* = veins; referring to the leaf venation, which has a midrib plus more or less equal lateral nerves or veins.

Decorsea (Fabaceae) R. Vig.
For Gaston-Jules Decorse (1873–1907), a French army physician with an interest in ethnography and linguistics. He qualified with an MD from the University of Paris in 1898. From 1898–1902 he collected plants, insects and fossils in Madagascar for the Natural History Museum in Paris. He joined a French expedition to Chad in 1902–1904 and explored parts of Tunisia and Sudan in 1905. Arising from these explorations he wrote *Rabah et les arabes du Chari: documents arabes et vocabulaire* (1905) with Maurice Gaudefroy-Demombynes, followed by his book *Du Congo au Lac Tchad la brousse telle qu'elle est, les gens tels qu'ils sont : carnet de route* (*From Congo to Lake Chad, the Bush as it is, People as they are*) (1906). Decorse collected a wide range of specimens from terrestrial molluscs to snakes such as the holotype of the snake *Typhlops decorsei*. Four reptiles were named after him.

Decussocarpus = Nageia (Podocarpaceae) de Laub.
La. *decussos* = shaped like an X; *karpos* = fruit; a genus similar to and formerly included in *Podocarpus* (q.v.), in which each pair of opposite leaves is at right angles to the one below, i.e. decussate.

Deguelia (Fabaceae) Aubl.
From the Galibi (French Guiana) name for *Derris* = a tall, woody, climbing plant.

Deilanthe (Aizoaceae) N.E.Br.
Gk. *deile* = evening; *anthe* = flower; night-flowering.

Deinbollia (Sapindaceae) Schumach. & Thonn.

For Peter Vogelius Deinboll (1783–1874), Danish clergyman, member of parliament, entomologist and collector. He was born in Denmark but lived in Norway. His collection of insects found in the Natural History Museum in Oslo is one of the oldest collections in the museum.

Delairea (Asteraceae) Lem.
For Eugene Delaire (1810–1856), head gardener at the botanical gardens in Orléans from 1836 to 1856. No further information found.

Delonix (Fabaceae) Raf.
Gk. *delos* = evident; *onyx* = claw, from Gk. *onux* = nail; referring to the long-clawed petals.

Delosperma (Aizoaceae) N.E.Br. emend. Lavis.
Gk. *delos* = visible, open, transparent; *sperma* = seed; referring to seeds that are easily visible as they are in an unenclosed chamber of the capsule which has no covering membrane.

Delphinium (Ranuculaceae) L.
La. *delphinion* = larkspur, diminutive of *delphinos* or *delphis* = dolphin; referring to the flower's nectary in some species which is seen to resemble the shape of a dolphin.

Dendrobium (Orchidaceae) Sw.
Gk. *dendron* = tree; *bios* = life; referring to the genus being an arboreal epiphyte, the genus name meaning 'living on a tree'.

Dendrophthoe = Septulina (Loranthaceae) Mart.
Gk. *dendron* = tree; *phthoe* = to waste away; referring to the often deleterious effect of this hemi-parasitic plant on trees.

Denekia*** (Asteraceae) Thunb.
For Carl Henry Deneke (1735–1803), German botanist and surgeon who studied in Stockholm, Uppsala, Germany, France and England. He was a member of the Surgical Society of 1766, regimental surgeon to Suomenlinna, a group of six islands now part of the city of Helsinki, Finland, from 1768–1772, and occupied himself with ophthalmology, and military medicine in Finland from 1789–1790. He was a friend of the genus's author, Swedish naturalist and botanist Carl Peter Thunberg.

Deparia* (Athyriaceae) Hook. & Grev.
Gk. *depas* = cup, goblet, beaker; referring to the shape of the indusium in some of the species.

Derenbergia*** **= Conophytum** (Aizoaceae) Schwantes
For Julius Derenberg (1873–1928), German physician and succulent plant collector.

Derenbergiella = Mesembryanthemum (Aizoaceae) Schwantes
Derenbergia (q.v.); Gk. *-iella* = diminutive.

Dermatobotrys (Scrophulariaceae) Bolus
Gk. *derma* = skin, bark; *botrys* = bunch of grapes. The plant is an epiphyte that bears berries in threes.

Deroemeria* = **Holothrix** (Orchidaceae) Rchb.f.
For Rudolf Benno von Römer (Roemer) of Neumark and Löthain (1803–1870), a German botanist. He is remembered for his massive book collection of 2 600 botanical works going back to the 15th century, which he bequeathed to the University of Leipzig. For some considerable time it was thought this genus was named after Johann Jacob Roemer (1763–1819), German physician, entomologist and professor of botany at the University of Zurich, but Heinrich Gustav Reichenbach's original description of Deroemera in his publication *De Pollinis Orchidearum* states 'Dicavi nobilissimo Do Roemer, Lothainensi ac Neumarkensi ...' Deroemeria is an orthographic variant of Deroemera.

Derris (Fabaceae) Lour.
Gk. *deros* = skin, hide, leather covering; referring to the tough seed pods.

Deschampsia* (Poaceae) P.Beauv.
For Louis Auguste Deschamps (1765–1842) a French botanist, naturalist, physician and plant collector, mainly in Java. He went on an expedition (1791–1793) in search of the explorer Jean-François de Galaup, comte de Lapérouse. Bad weather and disease resulted in 89 of the 119 of his crew dying. The French expedition became stranded in Java and Deschamps was briefly interned by the Dutch. The Dutch governor Van Overstraten persuaded Deschamps to conduct research in Java, which he did. Further misfortune followed when the ship in which he was travelling home was captured by the British. His diary, drawings, manuscripts and specimens were taken from him and donated to the British Museum much later, in 1861. Afterwards he settled in Batavia as a physician until 1802, but eventually settled at Saint Omer in France.

Descurainia* (Brassicaceae) Webb & Berthe.
For François Déscourain (1658–1740), French pharmacist and botanist, and friend of Antoine, a teacher, and Bernard de Jussieu, botanical demonstrator at the Jardin du Roi in Paris. He wrote a flora of his native town Etampes, which he did not publish, but which his grandson, Jean-Etienne Guettard (1715–1786), turned into a book entitled *Observations sur les plantes* (two volumes, Paris, 1747).

Desmanthus* (Fabaceae) Willd.
Gk. *desme* = a bundle; *anthos* = flower; referring to the flowers which, for some species, serve as lucerne or fodder for donkeys, hence possibly collected in a bundle.

Desmazeria (Poaceae) Dumort.
For Jean Baptiste Henri Joseph Desmazieres (1786–1862), French pharmacist, botanist, horticulturist, and author of *Plantes cryptogames de Nord de la France*. He devoted himself to mycology, agrostology and horticulture. He published numerous studies on the plants of northern France as well as fungal flowering plants. He was the editor of the journals *Annales des sciences naturelles* and the *Bulletin de la société des sciences de Lille*. Among his important publications were *Recherches microscopiques et physiologique sur le genre Mycoderma* (*Microscopic and Physiological Research on the Genus Mycoderma*). He was a member of the Botanical Society of France, the Imperial Society of Science and the Botanical Society of Brussels.

Desmochaeta* = **Cyathula** (Amaranthaceae) DC.
Gk. *desmos* = a band or chain; *chaite* = a bristle. Possibly alluding to the long bristles on the inflorescence in some species.

Desmodium (Fabaceae) Desv.
Gk. *desmodion* = a small chain; referring to the jointed seed pods.

Desmonema* = **Euphorbia** (Euphorbiaceae) Miers
Gk. *desmos* = band, bond; *nema* = a thread, filament. The lower halves of the filaments are usually connate.

Detris* = **Felicia** (Asteraceae) Adans.
La. *detri*, stem of *detere* = to wear away.

Deverra (Apiaceae) DC.
La. *deverro* = sweep away; referring to the stems which are slim, thin, almost leafless, broom-like.

Devia (Iridaceae) Goldblatt & J.C.Manning
For Miriam Phoebe de Vos (1912–2005), South African botanist, plant collector and botany professor. She obtained a DSc (1940) from Stellenbosch University where she worked for most of her career, during which time she collected some 2 400 specimens and had a

particular interest in cyto-taxonomy and embryology, especially of Iridaceae. She is acknowledged for her scientific contributions to the morphology and anatomy of South African plants, especially her revisions of *Syringodea, Romulea, Ixia, Crocosmia, Duthiastrum* and *Chasmanthe*. A talented artist, she illustrated three colour plates of *Romulea* for *The Flowering Plants of Africa* (Volume 29, 1952). In 1974 she was awarded the Havenga prize for Biology from the South African Academy of Arts and Sciences.

Dewinterella *** = ***Hessea*** (Amaryllidaceae) D. & U.Müll-Doblies
Dewinteria (q.v.); Gk. *-iella* = diminutive.

Dewinteria (Pedaliaceae) Van Jaarsv. & A.E.van Wyk
For Dr Bernard de Winter (1924–?), South African botanist and plant collector. He obtained a DSc at the University of Pretoria, taught for a while, then joined the staff of the National Herbarium at the Botanical Research Institute of which he became director in 1973. He collected some 10 500 specimens, including extensive collections in South Africa, Namibia, Botswana and Zimbabwe. He contributed substantially to the botanical literature and was author of *Sixty-Six Transvaal Trees* with his wife Mayda and botanist DJB Killick. He was president of Section B of the South African Association of Advanced Science (1968) and executive president (1974). He was also president of the South African Biological Society (1969) and South African Association of Botanists (1976).

Diacarpa = ***Atalaya*** (Sapindaceae) Sim
Gk. *dia-* = through; *karpos* = fruit. The fruit consists of winged samarae.

Dialium (Fabaceae) L.
Possibly Gk. *dialion* = destroy; referring to the petals which fall off soon.

Diandrochloa *** = ***Eragrostis*** (Poaceae) De Winter
Gk. *di-* = two; *aner, andr-* = male, hence stamen; *chloa* = grass. The plant has two anthers and two stamens.

Dianthera (Acanthaceae) L.
Gk. *di-* = two; *anthera* = anthers. The term 'double anther' refers to the separated anther-cells.

Dianthera = ***Cleome*** (Capparaceae) Pax
Gk. *dis-* = two; *anthera* = anthers. Only two anthers are fertile in this genus.

Dianthus (Caryophyllaceae) L.

Delicate single Carnation-like flowers (CM)

Gk. *dios* = godlike, divine; *anthos* = flower; probably referring to the scent.

Diaphananthe (Orchidaceae) Schltr.

(HS)

Gk. *diaphanes* = from *dia-* and *phainein* = to show through, transparent; *anthos* = flower. The membranaceous texture of the perianth parts are delicate and somewhat transparent, diaphanous.

Diascia (Scrophulariaceae) BalleLink & Otto

(EJM)

Gk. *di-* = two; *askion* = wineskin, bladder, belly; referring to the two lateral corolla pouches.

Diastella (Proteaceae) Salisb. ex Knight
Gk. *diastellein* = to separate or expand; referring to the deeply divided perianth segments.

Dicerocaryum (Pedaliaceae) Bojer
Gk. *di-* = two; *keras* = horn; *karyon* = nut. The hard fruit carries two conical spines on its upper face.

Dicerothamnus (Asteraceae) Koekemoer
Gk. *di-* = two; *keras* = horn; *thamnus* = bush or shrub. Two-horned bush – referring indirectly to the Renosterveld, the habitat of Renosterbos.

Dichaelia*** = **Brachystelma** (Apocynaceae) Harv.
Gk. *dichaea* = having two parts; *elia* = little.

Dichanthium (Poaceae) Willemet.
Gk. *dicha-* = divided in two, bifid; *anthos* = flower; *-ium* = diminutive; alluding to the two kinds of spikelet pairs in the raceme, genderless (sterile) and male.

Dichapetalum (Dichapetalaceae) Thouars
Gk. *dicha-* = divided in two; *petalum* = petals. The petals are two-lobed or bifid.

Dichilus (Fabaceae) DC.
Gk. *di-* = two; *cheilos* = lip; alluding to the two-lipped calyx.

Dichondra* (Convolvulaceae) J.R. & G.Forst.
Gk. *di-* = two; *khondros* = grain. The fruit consists of two membranous capsules.

Dichosma** (Rutaceae) DC. ex Loudon
Gk. *dicha-* = in two, split, divided in two; *osma* = scent, smell, fragrance.

Dichrocephala (Asteraceae) L'Hér. ex DC.
Gk. *dichroos* = two-coloured; *kephale* = head. The marginal and disk flowers are different colours in one head.

Dichrostachys (Fabaceae) (A.DC.) Wight & Arn.

(EJM)

Gk. *dis-* = two; *chroos* = colour; *stachys* = spike. The flower spikes are yellow and lilac (Hugh Glen); the upper bisexual flowers are yellow, the lower neuter flowers rosy-purple (Jackson); flowers have coloured staminodes (Don Perrin).

Dicksonia* (Dicksoniaceae*) L'Hér.
For James Dickson (1738–1822), British botanist, mycologist and nurseryman. He came from a family of nurserymen and in 1772 set up a business as a nurseryman and seedsman in Covent Garden. By 1781 he became interested in cryptogams. Between 1785 and 1801 he published

his *Fasciculus plantarum cryptogamicarum Britanniae*, a four-volume work in which he published over 400 species of algae and fungi that occur in the British Isles. He is also the author of *Collection of Dried Plants, Named on the Authority of the Linnaean Herbarium and Other Original Collections*. He was a Fellow of the Royal Society, a founding member of the Linnaean Society, and a founding member of the Royal Horticultural Society.

Dicliptera (Acanthaceae) Juss.
Gk. *diclis* = double-folding (of doors); *pteron* = a wing; referring to the two-celled wing caps.

Diclis (Scrophulariaceae) Benth.
Gk. *diclis* = twice-folded (of a door). The capsule is bipartite and the upper corolla lobe is bilobed.

Dicoma (Asteraceae) Cass.
Gk. *di-* = two; *kome* = tuft of hairs; referring to the double pappus of the first described species (*Flora Capensis*).

Dicraeia*** (Podostemaceae) Thouars
Gk. *dikraios* = cleft, forked from *dis* = twice; *keraia* = a horn, projecting beam, branch.

Dicranopteris (Gleicheniaceae) Bernh.

(NC)

Gk. *dikranos* = two-branched (forked), headed; *pteris* = fern; referring to the fronds which branch repeatedly in two.

Dicrocaulon (Aizoaceae) N.E.Br.
Gk. *dikros* = forked; *kaulon* = stem; alluding to the two long lateral shoots, resembling a fork, which produce terminal flowers. At the end of the flowering season they often overreach the flowers.

Didelta (Asteraceae) L'Hér.
Gk. *di-* = double, twice; *delta* = the letter D, symbolised in Greek by a triangle. The outer bracts are in two rows and are triangular in shape.

Didiclis = Selaginella (Selaginellaceae) P.Beauv.
Gk. *di-* = two; *diklis* = double-folding, two-doored; possibly referring to a folded pinnule margin protecting the sori. The sporangia are two-valved.

Didymaotus (Aizoaceae) N.E.Br.
Gk. *didymos* = double; *aotos* = flower. The plant bears two flowers (Plantza.com).

Didymocarpus (Gesneriaceae) Wall.
Gk. *didymos* = in pairs; *karpos* = fruit; alluding to the twin fruit – the fruit are formed in pairs.

Didymochlaena (Hypodematiaceae) Desv.
Gk. *didymos* = twin; *khlainos* = cloak; referring to two sori which share an indusium.

Didymodoxa (Urticaceae) Wedd.
Gk. *didymos* = twin, paired; *doxa* = glory, glorious; possibly referring to the double fruit.

Didymoglossum (Hymenophyllaceae) Desv.
Gk. *didymos* = twin, paired; *-glossum* = tongue. The cup-like indusium is bilabiate.

Didymoplexis (Orchidaceae) Griff.
Gk. *didymos* = twin; *plexis* = interwoven. The sepals and petals connate at the base and form a short tube.

Dielsia* (Restionaceae) Gilg
For Friedrich Ludwig Emil Diels (1874–1945),

German botanist, director of Berlin-Dahlem botanical garden and museum, professor of botany at the University of Berlin, and plant collector in South Africa, Java, Eastern and Western Australia and New Zealand, later New Guinea and Ecuador, author of *Fragmenta Phytographiae Australiae Occidentalis* (1905) with Ernst Pritze. His collections were destroyed in an air raid in 1943.

Diectomis*** = Andropogon (Poaceae) Kunth
Gk. *dis-* = twice; *ektemnein, ektome* = to cut out, castrate; referring to the likelihood that two out of every three spikelets in each cluster will be sterile.

Dierama (Iridaceae) K.Koch
Gk. *diorama* = a funnel; alluding to the shape of the perianth.

Dietes (Iridaceae) Salisb. ex Klatt
Gk. *dis-* = twice; *etes* = an associate. Twice – the flowering stalks last two years or more; an associate – the genus closely resembles both *Iris* and *Moraea*.

Dietrichia (Crassulaceae) Tratt.
Probably for Johann Friedrich Gottlieb Dietrich (1765–1850), German botanist and later professor of botany at Jena University. At just 20, while on a botanical excursion, he met Goethe who, together with Duke Karl August of Weimar, gave him a scientific education in Jena. They also sponsored trips abroad to Kew Gardens, Huntington and Chelsea Gardens. From 1792–1801 he worked for Duke Licher Gartner (becoming his gardener from 1794) in Weimar. From 1801–1845 he was appointed the duke's services inspector (director) of botanical gardens in Wilhelmsthal in Eisenach. Dietrich was awarded the title of grand ducal council, he obtained his PhD and became professor of botany. He authored a number of books on botany and horticulture.

Digitalis (Plantaginaceae) L.
Gk. *digitalis* = finger-like; referring to the corolla shape. *D. purpurea* ('foxglove') can be fitted over a human fingertip with ease.

Digitaria (Poaceae) Haller
La. *digitum* = a finger; *-aria* = pertaining to. The inflorescences attached to the main stalks are splayed out like the fingers of a hand.

Digitariella = Digitaria (Poaceae) De Winter
La. *Digitaria* (q.v.); *-iella* = diminutive.

Diheteropogon (Poaceae) (Hack.) Stapf
Gk. *di-* = two; *Heteropogon* (q.v.); referring to the racemes of this genus which are paired, unlike *Heteropogon*, which has solitary racemes.

Dilanthes = Trachyandra (Asphodelaceae) Salisb.
Gk. *deile* = evening; *anthos* = flower; night-flowering.

Dilatris (Haemodoraceae) P.J.Bergius
Gk. *di-* = two; *latris* = servant; alluding to the two small anthers, 'servants' of the single large one.

Dimorphotheca (Asteraceae) Vail ex Moench
Gk. *di-* = two; *morphe* = form; *theke* =a fruit (a case or container); referring to the two different forms of cypselae (fruit) produced by the ray and disk flowers: those of the ray flowers wingless, three-cornered; those of the disk flattened and two-winged.

Up the sandy West Coast these flowers can form a stunning white mass in a wet spring. (CM)

Dinacria *** = ***Crassula*** (Crassulaceae) Harv.
Gk. *dis-* = two; *akros* = a point, hence 'forked'; 'referring to the apparently forked apex of the carpel, by which this little plant is known from all others of the order' (William Henry Harvey, *Flora Capensis*).

Dinebra (Poaceae) Jacq.
Arabic *danaiba* = a long tail; *denab* = queue; referring to the acuminate glumes and extended apices.

Dinocanthium = Pyrostria (Rubiaceae) Bremek.
Gk. *dino-* = terrible (as in dinosaur), and the genus *Canthium* (q.v.); alluding to the stiff, spine-tipped branchlets.

Dintera (Plantaginaceae) Stapf
For Moritz Kurt Dinter (1868–1945), government botanist in South-West Africa (now Namibia), explorer, and plant collector.

He studied botany and horticulture at the botanical gardens of Dresden and Strasbourg. Later he took charge of Sir Thomas Hanbury's La Mortola garden. He then spent six months at Kew Gardens learning English, before leaving for South-West Africa (1897) where he relied on the sale of botanical specimens for his livelihood. His collecting career spanned 38 years and his pressed collection was over 8 400 specimens, excluding large quantities of living plants and seeds, and his wife's collections, which were never numbered.

He authored *Deutsch-Südwest-Afrika: Flora-forst- und landwirtschaftliche Fragmenta* (Leipzig, 1909) and *Die vegetabilische Veldkost Deutsch-Südwest-Afrikas* (Okahandja, 1914).

Dinteracanthus*** = ***Ruellia*** (Acanthaceae)
C.B.Clarke ex Schinz
Dintera (q.v.); Gk. *akanthos* = thorn; genus *Acanthus* (q.v.).

Dinteranthus (Aizoaceae) Schwantes
Dintera (q.v.); Gk. *anthos* = flower.

Diodia* (Rubiaceae) L.
Gk. *diodos* = thoroughfare, passage; alluding to the fact that many species grow as weeds by the wayside.

Dioscorea (Dioscoreaceae)
For Pedanius Dioscorides, (c 40–90) Greek physician, whose *Materia Medica*, circulated in Latin, Greek, and Arabic, was the leading work of its kind throughout the Middle Ages and remains a major source of historical information relating to medicines used by the Greeks, Romans and other cultures.

Diosma (Rutaceae) L.
Gk. *dios* = divine; *osme* = fragrance; referring to the fragrant leaves, especially when crushed.

Diospyros (Ebenaceae) L.
Gk. *dios* = divine; *pyros* = literally, a 'grain of wheat' but in this instance fruit. The fruits are 'divine' – edible and very tasty. This name was originally applied to the Caucasion persimmon, *Diospyros lotus*.

Dipcadi (Hyacinthaceae) Medik.
A Turkish name originally for the grape hyacinth, *Muscari*.

Dipera*** = ***Disperis*** (Orchidaceae) Spreng.
Gk. *di-* = two, twice; *pera* = pouch, sac. The pouches or sacs are formed by the lateral sepals.

Dipetalia*** = ***Oligomeris*** (Resedaceae) Raf.
Gk. *di-* = two; *petalon* = petal; referring to the flower structure which, in addition to the sepals, usually includes two whitish petals that may be partially fused together.

Diphaca = ***Ormocarpum*** (Fabaceae) L.
Gk. *dis-* = two; *phaca* = a bract. Each flower is surrounded by two bracts.

Diphasiastrum = ***Lycopodium***
(Lycopodiaceae) Holub
Gk. *di-* = two; *phasis* = appearances; *astrum* = inferiority or partial resemblance; hence, 'false *Diphasium*'.

Diphasium = ***Lycopodium*** (Lycopodiaceae)
C.Presl ex Rothm.
Gk. *di-* = two; *phasis* = appearances. The indusium sometimes lies on both sides of the vein.

Dipidax*** = ***Wurbea*** (Colchicaceae) Lawson ex Salisb.
Gk. *di-* = double; *pidax* = spring, fountain; referring to the pair of nectaries at the base of the perianth-segments.

Diplachne*** = ***Leptochloa*** (Poaceae)
P.Beauv.
Gk. *diplax* = double-folded or *diploos* = double; *achne* = glume, lobe; referring to the minutely bilobed lemma.

Diplacorchis*** = ***Brachycorythis***
(Orchidaceae) Schltr.
Gk. *diplax* = double-folded; *orchis* = orchid; alluding to the presence of two parallel plate-like keels on the lip.

Diplazium (Athyriaceae) Sw.
Gk. *diplasios* = two-fold, double; referring to the frequently double sori in the ferns – double covering over the spores, two indusia on the same receptacle.

Diplecthrum*** = ***Satyrium*** (Orchidaceae)
Persl.
Gk. *di-* = twice; *plektron* = sting, spur; alluding to the two-spurred lip of the flowers, which occurs in most *Satyrium* species.

Diplocyatha** = ***Orbea*** (Apocynaceae)
(Thunb.) N.E.Br.
Gk. *diplo-* = double; *kyathos* = a cup; referring to the corolla tube campanulate with another tube arising from its base within, reaching to the mouth and there thickened into a recurved rim.

Diplocyclos* (Cucurbitaceae) (End) Post. & Kuntze
Gk. *diplo-* = double; *cyclos* = circle, ring; referring to the double groove around the edge of the seed.

Diplogastra = ***Platylepis*** (Orchidaceae)
Rchb.f.
Gk. *diplo-* = double; *gaster* = pouch, belly, stomach; referring to the basally bisaccate lip.

Diplolophium (Apiaceae) Turcz.
Gk. *diplo-* = double; *lophium* = little bundle, hence 'bunched'; possibly referring to the umbellate inflorescence at the top of the stem well separated from the bunched leaves at the base.

Diplopappus = Erigeron (Asteraceae) Cass.
Gk. *diplo-* = double; *pappus* = a modified calyx. The pappus bristles are in two rows in both ray and disk florets. However, *pappos* originally meant down, fluff, so the name could apply to the minutely downy tips of the style branches.

Diplorhynchus (Apocynaceae) Welw. ex Ficalho. & Hiern
Gk. *diplo-* = double; *rhynchos* = beak, bill; referring to the paired beak-like fruit.

Diplosoma (Aizoaceae) Schwantes
Gk. *diplo-* = double; *soma* = body; named because each plant of this succulent has only two fleshy leaves, which lie opposite each other on the ground open like a book.

Diplospora (Rubiaceae) DC.
Gk. *diplo-* = double; *spora* = seed; in reference to the cells of the fruit being two-seeded.

Diplostemma *= **Geigeria*** (Asteraceae) Steud.
Gk. *diplo-* = double; *stemma* = garland, crown.

Diplostephium (Asteraceae) Kunth
Gk. *diplo-* = double; *stephos* = wreath, crown; referring to the two-rowed pappus present on the achenes.

Diplotaxis* (Brassicaceae) DC.
Gk. *diplo-* = double; *taxis* = arrangement; referring to the double row of seeds in the pod.

Dipogon = Sorghastrum (Poaceae) Liebm.
Gk. *di-* = two; *pogon* = beard; referring to the double-bearded style.

Dirichletia (Rubiaceae) Klotzsch
For Johann Peter Gustav Lejeune Dirichlet (1805–1859), German mathematicial prodigy.

He studied under the French mathematician Jean Nicolas Pierre Hachette at the Faculté des Sciences de Paris (1816–1822). When just 22 he provided a partial proof for Fermat's Last Theorem and was immediately offered a post at the University of Breslau (1827), then later at the Prussian Military Academy, the University of Berlin (1931), and the University of Göttingen (1855). His main research interest was number theory. He pioneered many new concepts, publishing *Dirichlet's Theorem on Arithmetic Progressions, Dirichlet Character and Functions,* and proved the Dirichlet unit theorem, and more. He became a member of the Prussian Academy of Sciences in 1805 aged 27.

Disa (Orchidaceae) P.J.Bergius (Images on following page)
Origin obscure. Börge Pettison believes the plant was named after Queen Disa who occurs in a Swedish legendary saga. The author, Peter Jonas Bergius, was a Swedish botanist.

Dischisma (Scrophulariaceae) Choisy
Gk. *di-* = two; *schizein* = to split; referring to the divided corolla tube.

Discocapnos (Papaveraceae) Cham. & Schltdl.
Gk. *diskos* = a disk; *kapnos* = smoke; possibly alluding to the disk-shaped seed and smoky-greyish fruit nodules.

Discocarpus (Phyllanthaceae) Klotzsch
Gk. *diskos* = a disk; *karpos* = fruit; possibly referring to the circular and flattened shape of the seed (fruit).

Disparago (Asteraceae) Gaertn.
La. *dispar* = unlike, dissimilar; *-ago* = resemblance or connection; referring to the different sorts of floret in each tiny capitulum.

Disperis (Orchidaceae) Sw.

(HS)

Gk. *dis* = twice; *pera* = a pouch, sac; alluding to the pouches formed by the lateral sepals.

ALL: *Disa* species (HS)

Disphyma (Aizoaceae) N.E.Br.
Gk. *dis-* = two; *phyma* = tubercule; possibly relating to the nodules, swellings or growths at the ovary apex (on the capsule structure).

Dissotis (Melastomataceae) Benth.
Gk. *dissotis* = of two kinds; referring to the stamens or anthers of which there are two kinds that differ in length and structure.

Distegia*** = Didelta (Asteraceae) Klatt
Gk. *di-* = two; *stegos* = covering, roof; possibly alluding to the bract subtending the berries.

Distephanus (Asteraceae) Cass.
Gk. *dis-* = two; *stefanos* = a wreath, crown. The pappus of the seeds is composed of two different kinds of bristles, each in its own whorl.

Dittrichia* (Asteraceae) Greuter
For Manfred Dittrich (1934–), German botanist, specialist in the Asteraceae at the herbarium of the Geneva Conservatory and Botanical Garden, and director of the herbarium of the Berlin-Dahlem Botanical Garden and Botanical Museum. Author of *Flora Iranica* (1979).

Dobrowskya* = *Monopsis* (Lobeliaceae) C.Presl
For Joseph Dobrowsky (1753–1829), Hungarian theologian and philologist who lived in Bohemia (part of the Czech Republic), rector in the general seminary at Hradisch. His fame rests mainly on his slavonic studies – he researched Biblical fragments, wrote a history of the Bohemian and Moravian literature (1818), a German-Bohemian dictionary in two volumes (1802, 1821), and other works, but his botanical contributions were also noteworthy. He authored *Entwurf eines Pflanzensystems nach Zahlen und Verhältnissen* (*Design of a Plant System of Numbers and Ratios*) (1802). In 1795 he was temporarily confined to a mental hospital but recovered in 1803 and authored at least another seven publications before his death.

Dodonaea (Sapindaceae) Mill.
For Rembert Dodoens (or Rembertus Dodonaeus) (1517–1585), Flemish physician and herbalist. He studied medicine, cosmography and geography at the University of Leuven, worked mainly as a physician, but was court physician to the Austrian emperor Rudolph II in Vienna from 1575–1578 and professor of medicine at Leiden University in 1582. He was a prolific writer and one of the foremost botanists of his day, with 12 major publications to his name. Dodoens's *Cruydeboeck,* a reference book about herbs with 715 images (1554), was the most translated book of that time after the Bible. It was translated into French, English and Latin, and became a work of worldwide renown used as a reference book for two centuries.

Doellia (Asteraceae) Sch.Bip. emend. Anderb.
For Johann(es) Christoph (Christian) Doell or Döll (1808–1885), German botanist. He studied natural sciences, theology and philosophy at the University of Heidelberg from 1827. In 1831, he was ordained a priest and became a tutor of languages at the Mannheim Gymnasium. In 1840, he became a teacher of botany and zoology at the Bürgerschule Mannheim. From 1843–1872 he was a librarian at the Royal Library of Karlsruhe. He wrote several books for learning Latin and English and a number of botanical publications including *Rheinische Flora* (1843) and a three-volume work *Flora des Grossherzogthums Baden* (*Flora of the Duchy of Baden*) (1857–1862) featuring flowering plants and cryptogams. He also wrote a chapter on *Poaceae* for Carl Friedrich Philipp von Martius's work *Flora Brasiliensis*.

Dolichandra (Bignoniaceae) Cham.
Gk. *dolichos* = long; *aner, andr-* = male, stamen. The plant has long anthers.

Dolichandrone (Bignoniaceae) Fenzl
Gk. *dolichos* = long; *andron* = men's quarters; referring to the relatively long anthers in flowers of this genus.

Dolichochaete* = *Tristachya* (Poaceae) (C.E. Hubb.) J.B.Phipps.
Gk. *dolichos* = long; *khaite* = flowing hair, mane, bristle; referring to the long narrow bristle-like awns which emerge from the lobes of the upper lemmas.

Dolichos (Fabaceae) L.
Gk. *dolikhos* = A Latin and Greek name used by Theophrastus and Pliny the Elder for any long-podded bean; referring to this leguminous plant's long shoots.

Dolichothrix (Asteraceae) Hilliard & B.Burtt
Gk. *dolichos* = long; *thrix* = hair; referring to the long silky twin-hairs on the cypselas.

Dombeya (Malvaceae) Cav.
For Joseph Dombey (1742–1794), French botanist, physician, naturalist and traveller with Spanish botanists Hipólito Ruiz and José Pavón in Chile and Peru. He gathered much valuable information relating to the cinchona plant from which quinine was derived. His special interest was spermatophytes. He authored *Flore Péruvienne, L'Herbier de Dombey explique,* and *Observations de Dombey faites au Chili et au Pérou,* all of which were published posthumously. His career was sullied by misfortune such as his collections being captured by the British (specimens sent to the British Museum) or

confiscated by zealous officials. In 1793 he undertook a mission to the United States but never arrived there as a result of a storm. Eventually, he was captured by British privateers and imprisoned for ransom at the British colony in Montserrat, West Indies, where he died. His main work and collection is housed in the Jardin des Plantes in Paris.

Donella* = *Chrysophyllum (Sapotaceae*)
Pierre ex Baill.
La. *donna* = lady; *-ella* = diminutive, hence 'little lady' or *donum* = gift, hence 'little gift'. Like many botanical authors, Jean Baptiste Louis Pierre described the genus but did not indicate the reason for giving it this name.

Doodia* = *Bechnum (Blechnaceae) R.Br.
For Samuel Doody (1656–1706), British botanist, pharmacist, curator of the Chelsea Physic Garden, authority on cryptogams, Fellow of the Royal Society, and plant collector for the Rev. Adam Buddle. Although he did not publish any botanical works during his lifetime, he drew up a draft publication on mosses now in the British Library. His work was acknowledged by John Ray in his *Historia Plantarum* Volume 2 (1686) where he gave a long list of plants Doody found. Doody also shared his knowledge with other English botanists of the day such as Leonard Pluket, James Petiver, Hans Sloane and others.

Dopatrium* (Plantaginaceae) Buch.-Ham. ex Benth.
Hindi *dopatta* = silk scarf with golden threads. Perhaps referring to the 'scarf-like' look of some petals such as those of *Dopatrium longidens*.

Doratanthera = *Anticharis***
(Scrophulariaceae) Benth.
Gk. *dora* = hide, skin; *anthera* = anther. Meaning obscure.

Doria* = *Othonna*** (Asteraceae) Less.
For Andrea Doria (Dorea) (Andreas Dorias) (1468–1560), an Italian nautical pioneer, admiral, and statesman after whom many ships were named. He is credited with being the first mariner to devise a method for sailing a ship against the wind. As a statesman he was involved in many bloody wars, feuds and conspiracies. As captain-general of the Genoese navy from 1512 onward he fought against the French and Barbary pirates, helped Charles V capture Tunis, and saved the Holy Roman Empire from a disastrous expedition against the North African city of Algiers in 1541. He retired to a monastery at age 87, but when the French tried to annex the nearby island of Corsica he once again succeeded in leading the fleet to victory. Andrea Doria is often referred to as 'the father of his country'.

Dorotheanthus* = *Cleretum (Aizoaceae)
Schwantes
For Dorothea Schwantes (1849–?) (née Meyer), wife of farmer Jürgen Meyer and mother of German professor and botanist Gustav Martin Heinrich Gustav Schwantes, who published the genus in her honour in 1927.

Dortmannia* = *Lobelia*** (Lobeliaceae)
Steud.
For Jan Dortmann, a Dutch apothecary of Groningen, according to Clusius (1526–1609). No details found.

Doryanthes (Doryanthaceae) Correa
Gk. *doryos* = lance, spear; *anthos* = flower; referring to the long spear-like stem.

Doryopteris (Sinopteridaceae) J.Sm.
Gk. *doryos* = spear; referring to the shape of the fronds in some species.

Dovea* = *Elegia (Restionaceae) Kunth
For Heinrich Wilhelm Dove (1803–1879), Prussian meteorologist and physicist, a professor at the University of Königsberg, director of the Prussian Meteorological Institute. He was an associate professor at the University of Königsberg, an associate professor at the University of Berlin, and then a full professor at the Friedrich-Wilhelms-Universität. He was also the director of the Prussian Meteorological Institute and studied the effect of climate on the growth of plants. He was a Fellow of the Royal Society and a member of the Prussian Academy of Sciences; and has a crater on the moon named after him.

Dovyalis (Salicaceae) E.Mey. ex Arn.
Gk. *dovyalis* = spear; referring to the sharp spines that are typical of the genus.

Doxantha* = *Macfadyena*** (Bignoniaceae)
Miers
Gk. *doxa* = glory, glorious; *dox* = praise; *anthos* = flower; referring to the appearance of its bright yellow flowers.

Dracaena (Dracaenaceae) L.
Gk. *drakaina* = a female dragon. Various meanings are ascribed to the name, mainly centred around the red, resinous, blood-like sap of *D. draco* – dragon's blood, but there is no precise explanation for the name.

Dracocephalum (Lamiaceae) L.
Gk. *drakon* = dragon; *cephalum* = head; referring to the shape of the corollas.

Dracomonticola (Orchidaceae) H.P.Linder & Kurzweil
Gk. *drakon* = dragon; *monticolus* = inhabiting mountains, mountain dweller; referring to its habitat.

Dracophilus (Aizoaceae) (Schwantes) Dinter & Schwantes
Gk. *drakon* = dragon; *philos* = friend, lover; referring to one species growing on the Drachenfels ('Dragon's Rock') in Namibia.

Dracosciadium (Apiaceae) Hilliard & B.Burtt
Gk. *drakon* = serpent, dragon; *skiadeion* = parasol, umbrella or canopy; referring to the shape of the inflorescence.

Dracoscirpoides *** (Cyperaceae) Muasya
Gk. *drakon* = serpent, dragon; *Scirpoides* (q.v.); referring to the bull rush genus.

Dregea (Apocynaceae) E.Mey.
For Johann Franz (Jean François) Drège (1794–1881), German horticulturalist at Riga, Munich, Berlin and St Petersburg, botanical collector and traveller.  He collected at the Cape from 1826–1834 with his brother Carl Friedrich; Johann collecting botanical specimens and Carl collecting zoological and ethnological specimens. Johann explored more of South Africa than any previous collector. He kept meticulous records of collection sites, altitude, environmental features and vegetation. In 1903, Rudolf Marloth described him as 'the father of South African phytogeography.' He returned to Europe with an enormous collection of some 200 000 specimens of some 8 000 species. These were sold to the major European herbaria.

Dregeochloa (Poaceae) Conert
Dregea (q.v.); Gk. *chloe* = grass.

Drimia (Hyacinthaceae) Jacq. ex Willd.

A *Drimia* species in full flower (CM)

Gk. *drimys* = acrid, pungent; referring to the sap which is considered irritating or even toxic in many species.

Drimiopsis = Ledebouria (Hyacinthaceae) Lindl. & Paxton.
Drimia (q.v.); Gk. *-opsis* = resembling.

Droguetia (Urticaceae) Gaudich.
Possibly for Marc-Julien Droguet (1769–1836), chief physician in the French Navy, author of *Propositions sur les préceptes d'hygiène navale à observer sous les Tropiques du Cancer, du Capricorne* … (1806). Origin uncertain.

Drosanthemopsis = Jacobsenia (Aizoaceae) Rauschert
Drosanthemum (q.v.); Gk. *-opsis* = resembling.

Drosanthemum (Aizoaceae) Schwantes
Gk. *drosos* = dew; *anthemon* = flower; referring to the glittering leaf papillae.

Drosera (Droseraceae) L.

(EJM)

Gk. *droseros* = dewy; alluding to the dewy glistening leaf-glands.

Drymaria (Caryophyllaceae) Willd. ex Schult.
Gk. *drymos* = forest, (oak)wood; referring to the habitat of at least one species.

Drymoglossum = Pyrrosia (Polypodiaceae) C.Presl
Gk. *drymos* = forest wood; *-glossum* = tongue; an epiphyte with fertile tongue-like fronds.

Drynaria* (Polypodiaceae) (Bory) J.Sm.
Gk. *Dryads* = mythological tree nymphs; *-aria* = pertaining to. The sterile fronds resemble leaves of the oak, which was sacred to the Dryads.

Dryopteris (Dryopteridaceae) Adans.
Gk. *dryos-* = (oak) tree; *pteris* = fern; referring to the fern's habitat – it grows among oak trees.

Drypetes (Putranjivaceae) Vahl
Gk. *drypetes* = ready to fall from the tree; referring to the relatively large and heavy fruits.

Duchesnea = Potentilla (Rosaceae) Sm.
For Antoine Nicholas Duchesne (1747–1827), French horticulturist, agronomist, pioneer in hybridisation at the Royal Garden in Versailles; author of *L'Histoire des Fraisiers* (strawberries). At age 17 he crossed large Chilean strawberries *Fragaria chiloensis*, brought to Europe by the French military spy Amédée François Frézier, which being all female would  not fruit with male pollen from other strawberry species, thus winning favour with King Louis XV. He was the first botanist to conduct an in-depth taxonomic study of the genus *Cucurbita* and produced 258 paintings of gourds of that genu*s*.

Dufourea = Tristicha (Podostemaceae) Bory
For Léon Jean Marie (or Jean-Marie Léon) Dufour (1780–1865), French physician, botanist, mycologist and naturalist. He was an army doctor during the Peninsular War (1807–1814) between France and the allied powers of Spain, Portugal and the United Kingdom. He was a foreign member of the Royal Swedish Academy of Sciences and the author of *Recherches anatomiques sur les Carabiques et sur plusieurs autres Coléoptères* and 232 articles on arthropods.

Dumasia (Fabaceae) DC.
For Jean Baptiste André Dumas (1800–1884), a French chemist, member of the legislative assembly, minister of agriculture, a senator, and president of the Paris municipality, husband of Hermine Brongniart and son-in-law of French chemist, minerologist and zoologist Alexandre Brongniart (1770–1847). He is best known for his works on organic analysis  and synthesis, as well as the determination of atomic weights (relative atomic masses) and molecular weights by measuring vapour densities. He was the co-founder and founder-editor of *Annales des sciences naturelles* in 1824 with Adolphe Theodore Brongniart (his brother-in-law) and Jean Victoire Audouin, and authored *Die Philosophie der Chemie: Vorlesungen, geh. im Collège de France* in 1839.

Dumortiera (Marchantiaceae) Nees
For Barthélemy Charles Joseph Dumortier (1797–1878), Belgian politician and bryologist, president of the Chamber of Deputies, member of parliament. He became chairman of the *Société Royale de Botanique de Belgique*. He authored *Observations sur les graminées de la flore de Belgique* (1823), *Flora Belgica* (1827) and *Analyse des familles des plantes, avec l'indication des principaux genres qui s'y rattachent* (1829), as well as historical works.  From 1821–1825, he explored much of Europe with Alexander Lejeune. He became a member

of the Academie Royale des Sciences et Belle-lettres de Bruxelles in 1929. In 1831 he entered parliament, where he served until his death and played a key role in the creation of the state botanical garden.

Duosperma (Acanthaceae) Dayton
La. *duo-* = two; *-sperma* (also Gk.) = seeded; referring to the flattened-ellipsoid fruits with two seeds only.

Duranta* (Verbenaceae) L.
For Castor Durantes (1529–1590), Italian physician, botanist-herbalist and poet. He studied medicine at Perugia in 1567 and became a doctor to Gualdo Tadino and subsequently became physician to Pope Sixtus V. He authored *De bonitate et vitio alimentorum centuria* (*The Treasure of Health*) (1565) and his book *Herbario Novo*, focusing on preserving health and prolonging life, went through 11 editions and was still reprinted 130 years after his death. He started one of the earliest herbaria for medicinal plants from Europe and East and West Indies in Venice in 1584.

Duthiastrum (Iridaceae) M.P.de Vos
For Augusta Vera Duthie (1881–1963), South African botanist and plant collector, born in Knysna, studied at the universities of Cape Town (MA, 1910) and Stellenbosch (DSc,1929). She spent a year at Cambridge (1912) and a year in Australia (1920). She was a lecturer in botany and head of department at Victoria College (later to become Stellenbosch University), and she retired as senior lecturer in 1939. She established the Stellenbosch Herbarium, specialising in the flora of the Stellenbosch district, especially the sand flats, and on retirement collected extensively around Knysna. Eight species were named after her.

Duthiella = Duthiastrum (Iridaceae) M.P.de Vos
Duthiastrum (q.v) La. *-iella* = diminutive.

Duvalia (Apocynaceae) Haw.
For Henri Auguste Duval (1774–1814), French physician and botanist. He studied medicine at the University of Paris, was famous for his catalogue *Plantae succulenta in horto Alenconio* (1809) and was the author of *Enumeratio plantarum succulentum in horto Alenconio*. He was the first person to describe the plant genera *Gasteria, Haworthia* and *Ligulari,* and authored a book on all the species found naturally in the environs of Paris (1813).

Duvaliaranthus = X Duvaliaranthus (Apocynaceae) Bruyns
Duvalia (q.v.); *anthos* = flowers.

Duvernoia*** = Justicia (Acanthaceae) E.Mey. ex Nees
For Johann Georg Duvernoy (1692–1759), German professor of anatomy and surgery, and a botanist. A student in Paris (1708) of Joseph Pitton de Tournefort (who formulated the concept of genus), he established that certain large bones found in Siberia belonged to the mammoth, not the elephant, and was the author of *Designatio Plantarum Circa Tubingensem Arcem Florentium* (1722) about the native flora of the Tübingen area.

Dyerophytum (Plumbaginaceae) Kuntze
For Sir William Turner Thiselton-Dyer (1843–1928), British botanist and third director of the Royal Botanic Gardens, Kew. He studied mathematics and natural sciences at Oxford, and thereafter held various professorships before taking up a professorship at the Royal Horticultural Society in London in 1872. He became assistant director at Kew in 1875, director in 1885 and retired in 1905. During his tenure he created an international research laboratory at Kew, introduced a new rock garden, and helped Sri Lanka and Malaya develop rubber plantations. He authored an English edition of Julius Sachs's *Text-Book of Botany* (1875), editions of the *Flora Capensis* and of the *Flora of Tropical Africa,* and *Index Kewensis* (1905). Gk. *phyton* = plant.

Dymondia (Asteraceae) Compton
For Margaret E. Dryden-Dymond (1909–1952), a member of the horticultural staff at Kirstenbosch National Botanical Garden, who discovered and obtained plant material of this monotypic genus (*Dymondia margaretae*) in the Bredasdorp district on a Kirstenbosch expedition in 1933. She was honoured with both the specific and generic epithet for this taxon.

Dyschoriste (Acanthaceae) Nees
Gk. *dys* = poorly; *khoristos* = separated. The stigma is only weakly bilobed.

E

Myriad *Erica* of which there are at least 850 species (EGHO – All)

Cape botanist **Edward 'Ted' George Hudson Oliver** (1938–), a world authority on Ericas, has been doing research on *Erica* since 1958 and from the 1980s with his late wife Inge, mainly at the Stellenbosch Herbarium, with stints as SA botanist at the Royal Botanic Gardens, Kew, and as curator of the National Herbarium in Pretoria, and then from 1996 at the Compton Herbarium, Kirstenbosch. He retired in 2003 but continues with this research.

Eberlanzia (Aizoaceae) Schwantes
For Friedrich Gustav Eberlanz (1879–1966), German decorator, botanist, amateur naturalist and succulent collector. He came to South-West Africa (now Namibia) in 1914 and settled in Lüderitz. For nearly 50 years he explored much of the Namib desert on foot, collecting plants, animals, minerals and stone age tools, which he donated to the municipality in 1960. These items formed the nucleus of the Lüderitz Museum, opened in 1966.

Ebracteola (Aizoaceae) Dinter & Schwantes
La. *e-* = without, lacking; *bracteole* = small bract; suggesting, albeit incorrectly, that the flowers do not have bracteoles.

Ecbolium (Acanthaceae) Kurz
Gk. *ekbole* = expulsion; from *ekballein* = to cast out. The seeds are expelled violently when the fruits open.

Echidnopsis = Notechidnopsis (Apocynaceae) Hook.f.
Gk. *echidna* = viper; *-opsis* = resembling; referring to the snake-like stems or branches.

Echinochloa (Poaceae) P.Beauv.
Gk. *echinos* = hedgehog, hence prickly; *chloe* = grass; alluding to the bristly spines or hairs on the spikelets.

Echinolaena (Poaceae) Desv.
Gk. *echinos* = prickly, spiny (hedgehog, sea-urchin); *laina* (cloak); referring to the appearance of the plants, which have long, stiff, widely spreading leaves.

Echinopsilon* = Bassia** (Amaranthaceae) Moq.
Gk. *echinos* = hedgehog; *psilon* = smooth. The seeds are prickly; the nuts smooth and shining.

Echinopsis* (Cactaceae) Zucc.
Echinus (q.v.); Gk. *-opsis* = resembling.

Echinospermum = Lappula (Boraginaceae) Sw. ex Lehm.
Gk. *echinos* = hedgehog; *sperma* = seed. The nutlets are often bordered by one or more rows of hooked prickles (Robert Allen Dyer).

Echinostachys* = Pycnostachys** (Lamiaceae) E.Mey.
Gk. *echinos* = hedgehog; *stachys* = spike. Possibly because the flowers are densely spiked.

Echinus* = Braunsia** (Aizoaceae) Bolus
La. *echinos* = hedgehog and sea urchin; alluding to the spiny nature of the plants.

Echiostachys (Boraginaceae) Levyns
Gk. *echis* = viper; *stachys* = spike. The spike resembles that of *Echium* (q.v.). (Not explained by author).

Echium* (Boraginaceae) L.
Gk. *echion* = from *echis* = viper; the nutlets appearing to represent a viper's head. A name used by Dioscorides for a plant used to cure snake bites.

Ecklonea* = Trianoptiles** (Cyperaceae) Steud.
For Christian Friedrich Ecklon (1795–1868), a Danish pharmacist, botanist and plant collector, and one of the early botanical explorers of the Cape. He moved to South Africa in 1823, working first as an apothecary's apprentice and then as a pharmacist, and collected plants from 1823 to 1833, returning to Europe in 1828 with vast amounts of collected material, which were distributed to German and Danish botanists. According to INPI, Ecklon named a total of 1 974 different genera or species. During part of this time he worked with Karl Ludwig Philipp Zeyher with whom he published a catalogue of South African plants (1835–1837). From 1833–1838 he was in Hamburg working on revising his collection, later returning to South Africa where he eventually died.

Eclipta* (Asteraceae) L.
Gk. *ekleipo, ekleipsis* = to be lacking or leaving out, deficient; referring to the minute or even lack of a pappus.

Eclopes* = Relhania** (Asteraceae) Gaertn.
Gk. *eclopes* = to cut; referring to the sharp, pointed leaves.

Ectadiopsis* = Cryptolepis** (Apocynaceae) Benth.
Ectadium (q.v.); *iopsis* = resembling.

Ectadium (Apocynaceae) E.Mey.
Gk. *ektad* = outwards; *ektadios* = outstretched. Referring to a bushy shrub really spreading outwards.

Ectotropis = Delosperma (Aizoaceae) N.E.Br.
Gk. *ectos* = outside; *tropis* = keel of a ship; alluding to the capsule perfectly pendulous when old, supported on a long seta, curved only at its apex.

Edmondia (Asteraceae) Cass.
The previous belief that this genus might honour James W. Edmond, Scottish botanist (died 1815) has been discredited as new evidence indicates

he died in 1875. Possibly named after English-born Edmund Davall (*Davallia*. q.v.) It is possible that the French botanist Alexandre Henri Gabriel de Cassini (1781–1832) mis-spelled Devall's first name 'Edmund' as this English word 'Edmund' is spelled 'Edmond' in French, hence the genus *Edmondia*.

Eenia = ***Anisopappus*** (Asteraceae) Hiern
For Ture (Thure) Johan Gustaf Een (1837–1883), Swedish mariner, trader and plant collector.

He travelled and traded in Damaraland and Ovamboland (Namibia) between 1866 and 1871 as described in his *Minnen från en flerårig vistelse i Sydvästra Afrika* (*Memories of a Multi-Year Stay in South Africa*) (1872). In 1878 he again returned to Namibia trading and collecting natural history material and returning to Sweden in 1880 with a valuable collection of animal skins, ethnological specimens, seeds, bulbs and 300 herbarium specimens, which he catalogued. For a short time he went to Tristan da Cunha as captain of the telegraph, trading seal skins in the South Atlantic. In 1883 he joined an international Congo expedition under Henry Morton Stanley. He died while on board the SS *Biafra* at the mouth of the Congo River.

Egeria* (Hydrocharitaceae) Planch.
A mythological water nymph and wife of the legendary second king of Rome, Numa Pompilius. The plant is aquatic, hence its name.

Ehretia (Boraginaceae) P.Browne
For George Dionysius Ehret (1708–1770),

German botanical artist. He worked for two years as a gardener at the Jardin des Plantes in Paris before going to England. His unique style and clarity of plant illustration was sought out by specialists such as Sir Joseph Banks. During 1735–1736 he worked with Linnaeus and for George Clifford, a wealthy Dutch banker in the Netherlands, producing the *Hortus Cliffortianus* in 1738, a masterpiece of early botanical literature. Over 3 000 of his illustrations survive in collections such as Trew's published work, *Plantae Selectae* (1750) and in the Natural History Museum, England. His patroness was Margaret Cavendish Harley Bentinck, Duchess of Portland. He was elected a Fellow of the Royal Society in 1758.

Ehrharta (Poaceae) Thunb.
For Jakob Friedrich Ehrhart (1742–1795), Swiss-born German botanist and naturalist. He

studied botany at Uppsala University (1773–1775) under the guidance of Linnaeus, before returning to Hanover where Linnaeus's son was director of the Botanical Garden. Ehrhart was one of the first botanists to publish exsiccatae (= plant collections, precisely identified, named and labeled), for various botanists or institutions. From 1780–1793 he produced seven series of these exsiccatae, each of about 1 620 species. He also had his own general collection (*Hortus siccus*) of about 3 300 plant species. A set of these collections are kept in the herbarium of Moscow University. He was supposedly the first person to use the rank of subspecies in botanical nomenclature.

Eichhornia* (Pontederiaceae) Kunth

Introduced from South America for ponds and aquaria, this weed can be a huge problem. (EJM)

For Johann Albrecht Friedrich von Eichorn (1779–1856), Prussian lawyer and politician. In his early career he studied law at the University of Göttingen and for a short while was involved with the military during the Napoleonic wars.

After the war he resumed his judicial service before becoming a diplomat in Paris where he was a privy counsellor in the ministry of Foreign Affairs, and later became second director of the Foreign Affairs Office. He became a minister of Education and Public Welfare, a court advisor, and although he had a number of setbacks where he disagreed with the king, he became a member of the Prussian State Council (1817–1848) and a member of the State House at the Erfurt Union Parliament, including being interim President.

Einadia (Amaranthaceae) Raf.
Gk. *ein* = closed within; *andros* = males, stamen; referring to the solitary stamen.

Ekebergia (Meliaceae) Sparrm.
For Carl Gustav (Gustaf) Ekeberg (1716–1784), Swedish ship's captain for the Dutch East India  Company from 1750 who was also a trained chemist, physician, navigator and excellent cartographer. He made 10 trips to India and China between 1742 and 1778 and brought back numerous natural history specimens from his voyages for Linnaeus, with whom he had a close friendship. He wrote a number of books about his travels such as *Voyages aux Grandes-Indes dans les années* (1773) and *Ostindische Reise in den Jahren 1770 und 1771* (1785), as well as a medical book on inoculation. He was elected a Fellow of the Swedish Academy of Sciences and Knight of the Order of Vasa in 1777. He met Swedish botanist Anders Sparrman, who named Ekebergia *capensis* after him.

Elachyptera* (Celastraceae) A.C. Sm.
Gk. *elachys* = short; *pteron* = a wing. The original description states that the name was given 'referring to the short basal wing of the seed'.

Elaeagnus* (Elaegnaceae) L.
Gk. *elaia* = olive; *agnos* = pure, chaste; referring to the chaste-tree *Vitex agnus-castus*.

Elaeocarpus* (Elaeocarpaceae) L.
Gk. *elaia* = olive; *karpus* = fruit; referring to the fruit's likeness to real olives.

Elaeodendron (Celastraceae) Jacq.
Gk. *elaia* = olive; *dendron* = tree. The fruit of this tree looks oily and resembles the true olive.

Elaphoglossum (Dryopteridaceae) Schott ex J.Sm.
Gk. *elaphos* = stag, deer; *glossa* = tongue. The shape of the fronds resembles a deer's tongue.

Elatine (Elatinaceae) L.
Gk. *elate* = the 'silver fir'. While most plants of this genus are acquatic herbs, a few are not, and species such as *Elatine alsinastrum* have leaves that somewhat resemble the European silver fir *Albies alba*, but they are more green than silver.

Elatostema* (Urticaceae) J.R.& G.Forst.
Gk. *elatos* = elastic; *stemon* = a thread; alluding to the manner in which the staminal filaments spring out at anthesis.

Elegia (Restionaceae) L.

The rosette of infertile branches from a node make it an attractive garden subject. (EJM)

Gk. *elegeia*, *elegia* = song of lamentation; possibly referring to the sound restios make while they are moving in the breeze.

Eleocharis (Cyperaceae) R.Br.
Gk. *eleios* (*heleios*) = marsh, low ground; *charis* = grace, beauty; referring to the humid marshy zone which is the plant's preferred habitat.

Eleogiton** = ***Isolepis*** (Cyperaceae) Link
Gk. *eleios* (*heleios*) = marsh-neighbour; in analogy with *Potamogeton* (David Gledhill).

Elephantorrhiza (Fabaceae) Benth.
Gk. *elephas* = an elephant; *rhiza* = root. The rootstock is enormous.

Eleusine (Poaceae) Gaertn.
From Eleusis, a Greek city where the goddess of grain, Demeter, was worshipped. The Roman equivalent is Ceres.

Elionurus (Poaceae) Kunth ex Willd.
Gk. *eleios* = doormouse; *oura* = tail; alluding to the spike-like racemes curling strongly when old, somewhat like a mouse's tail.

Elphegea *** = ***Zyrphelis*** (Asteraceae) Less.
An historical name of unknown etymology. Some sources say a name used by boys meaning 'discerning friend', but this is not supported by the dictionaries consulted.

Elsiea *** = ***Ornithogalum*** (Hyacinthaceae) (Baker) F.M. Leight.
For Elizabeth 'Elsie' Esterhuysen (1912–2006), South African botanist and plant collector. She studied at the University of Cape Town, obtaining an MSc, worked in the McGregor Museum, Kimberley (1936–1937), and from 1938 in the Bolus Herbarium in Cape Town. She has been called 'the most outstanding collector ever of South African flora' (Karel Bremer) amassing some 36 000 herbarium specimens, many high-altitude species. She was an authority on Cape flora, particularly Restionaceae, Ericaceae and Rutaceae. She was also a talented illustrator, producing mainly black and white line drawings.

Elymandra (Poaceae) Stapf
Gk. *Elymus* (q.v.) = ancient Greek cereal, millet; *andros* = male (stamen).

Elymus (Poaceae) L.
Gk. *elumos, elymus* = millet. An ancient Greek name for wild rye or lime grass or similar.

Elynanthus *** = ***Tetraria*** (Cyperaceae) T. Lestib.
Gk. *elytron* = sheath, cover; *anthos* = flower; referring to the shape of the corolla.

Elytraria (Acanthaceae) Michx.
Gk. *elytron* = sheath, cover; perhaps alluding to the scale-like bracts.

Elytrigia = ***Elymus*** (Poaceae) Desv.
The name is derived from a combination of the genera *Elymus* (q.v.) and *Triticum*.

Elytropappus (Asteraceae) Cass.
Gk. *elytron* = a sheath, covering; *pappos* = down, fluff; referring to the small cup-like rim around the base of the feathery pappus (Plantzafrica.com); a second source says referring to the fluffy bracts.

Elytrophorus (Poaceae) P.Beauv.
Gk. *elytron* = a sheath, covering; *-phorus* = bearing, carrying; alluding to the sheathing glumes.

Embelia (Myrsinaceae) Burm.f.
Latinised from the Sri Lankan name *welembilla*, still in use for *Embelia ribes*.

Emelianthe (Loranthaceae) Danser.
Possibly for Emily Pauline Reitz Ferguson (1872–?), a plant collector of Riversdale in South Africa who collected in the Riversdale and Swellendam areas in the 1920s and 1930s; *anthos* = flower. No further details found.

Emex (Polygonaceae) Campd.
Derived from *ex* = out of, and *Rumex* = referring to the genus that was segregated from that genus.

Emilia (Asteraceae) Cass.
Not stated by the author, Cassini, but probably after the province of Emilia in Italy where his ancestor, Domenico Cassini (1625–1715), was professor of astronomy (1650–1669); or named after some unknown individual.

Eminia (Fabaceae) Taub.
For Mehmed Emin Pasha (originally Isaak Eduard Schnitzer) (1840–1892), Ottoman-German physician, explorer, naturalist and linguist. He graduated at the universities of Breslau, Königsberg, and Berlin becoming a doctor. In central Africa he collected plants, animals and birds, many of which he sent to museums in Europe and he published geographical and anthropological papers. After

serving under general Charles Gordon in Sudan as a district medical officer (1876–78) he became governor of Equatoria. In 1881 an Arab uprising forced him to withdraw his military force to the southernmost part of Equatoria. Henry Morton Stanley organised a relief expedition along the Congo, which proved to be unnecessary. Stanley persuaded Emin to accompany him to the coast reached in 1890. While working for the German East Africa Company he was murdered by two Arab slave traders while exploring in the region of Lake Tanganyika.

Emorya (Scrophulariaceae) Torr.

For William Hemsley Emory (1811–1887), United States Army officer, surveyor, civil engineer and explorer. He graduated from the United States Military Academy at West Point, New York, became a second Lieutenant in the Forth Artillery, but resigned in 1836 to go into civil engineering. Two years later he returned to the military and worked his way up the ranks to that of major. He specialised in mapping, which included the Canadian border (1844–1846), the Mexican-American border (1848–1853), and the Gadsen Purchase (1854–1857); then part of Mexico. His mapping skills were outstanding but he also made notes about the plant life and people who inhabited the sparsely populated Southwest. John Torrey (1796–1873) helped identify many of the plants his reconnaissance expedition collected.

Emplectanthus (Apocynaceae) N.E.Br.
Gk. *empleko* = to interweave, be entangled, render perplexed; *anthos* = flower; alluding to the combination of characters of different genera in one plant.

Empleuridium (Celastraceae) Sond.
Gk. *Empleurum* (q.v.); *-idion* = diminutive.

Empleurum (Rutaceae) Sond. & Harv.
Gk. *em-* = in; *pleuron, pleura* = membrane; referring to 'the seeds (which are) contained within the membranous inner hull of the ripe capsule' (Sonder, *Flora Capensis*).

Empodium (Hypoxidaceae) Salisb.
Gk. *em-* = within; *pous, pod-* = foot; alluding to the underground ovary.

Empogona (Rubiaceae) Hook.f.
Gk. *em-* = within; *pogon*= bearded; referring to the funnel-shaped corolla tube – rather short, densely bearded at the throat.

Enarganthe (Aizoaceae) N.E.Br.
Gk. *enarges* = brilliant, shining; *anthe* = flower; referring to its attractive pink flowers.

Encephalartos (Zamiaceae) Lehm.
Gk. *en-* = in; *kephale* = head; *artos* = bread; referring to the inner parts of the trunks of these trees, which are starchy and edible (Hugh Glen); or the farinaceaous pith from the upper part of the stem, which was kneaded into bread (Jackson).

Endonema (Penaeaceae) A.Juss.
Gk. *endo-* = within; *nema* = a thread, filament. The filament and style are largely contained inside the perianth tube.

Endostemon (Lamiaceae) N.E.Br.
Gk. *endo-* = within, inside; *-stemon* = stamen. The stamens are included within the corolla tube.

Englerastrum*** = *Plectranthus* (Lamiaceae) Briq.
Engleria (q.v.); Gk. *-astrum* = star; implying the star shape of some characteristic, but this word could also mean incomplete likeness or inferior in some way. The implication of this name could not be found.

Encephalartos: Typical cliff habitat (EJM)

Encephalartos: Female cone (EJM)

Engleria (Asteraceae) O.Hoffm.
For Heinrich Gustav Adolf (Adolf) Engler (1844–1930), German botanist, professor at the University of Berlin from 1889–1921 and director of the Berlin Botanical Gardens. After completing his PhD at the University of Breslau, he held various positions, including professorships at the universities of Kiel (1878–1884) and Breslau (1884–1888). He is recognised for his work on plant taxonomy and phytogeography; his major work being the 23-volume *Die Natürlichen Pflanzenfamilien* (*The Natural Plant Families*) (1887–1915), edited by Karl von Prantl, which he wrote with some notable experts, and which contained some 6 000 illustrations by Joseph Pohl. He was also founder and editor of the periodical *Botanische Jahrbücher*. He received the Linnaean Medal in 1913.

Englerodaphne = *Gnidia* (Thymelaeaceae) Gilg
For Adolf Engler (see *Engleria*); Gk. *daphne* = laurel; Engler's laurel.

Englerophytum (Sapotaceae*) K. Krause.
For Adolf Engler (see *Engleria*); Gk. *phyton* = plant; Engler's plant.

Engysiphon* = **Geissorhiza** (Iridaceae)
G.J.Lewis
Gk. *enguos* = narrow; *siphon* = tube; referring to the slender perianth tube.

Enicostema (Gentianaceae) Blume
Gk. e*n*- = in, inside; *icos* = twenty; *stema* = wreath, circle; alluding to the many flowers arranged in circles in the leaf axils along the stem.

Enneapogon (Poaceae) P.Beauv.
Gk. *ennea* = nine; *pogon* = beard; referring to the nine hairy awns of the lemma of the fertile floret.

Ensete (Musaceae) Horan.
A transcription of the vernacular name Amharic *ensat,* for this plant in Ethiopia.

Entada (Fabaceae) Adans.
The native Malabar (Indian) name for *E. scandens* (William Henry Harvey, *Flora Capensis*) the 'sea bean'. The large beans have narcotic properties and are commonly known as the African Dream Herb or Snuff Box Sea Bean.

Entadopsis = ***Entada*** (Fabaceae) Britton
Entada (q.v.); Gk. -*opsis* = resembling; a genus of plants looking like *Entada*.

Entandrophragma (Meliaceae) C.DC.
Gk. e*n*- = in; *andros* = male, stamen; *phragma* = partition (within the male membrane). The stamen filaments are joined to form a tube (Hugh Glen); referring to the connate filaments producing an urceolate tube (Jackson).

Enteropogon (Poaceae) Nees
Gk. *entero*- = internal; *enteron* = intestine; *pogon* = beard; alluding to the awned lemmas or bearded callus.

Enterosora* (Polyodiaceae) Baker
Gk. *entero*- = internal; *sora* = sorus; referring to the sori which are not superficial but embedded in the frond.

Enterospermum* = **Coptosperma**
(Rubiaceae) Hiern
Gk. *enteron* = intestine; *sperma* = seed. A cross-section through the seed resembles a similar section through an animal's intestine (Hugh Glen); referring to the 'thin testa deeply intruded into the horny albumen' (Robert Allen Dyer).

Entolasia (Poaceae) Stapf
Gk. *entos* = within; *lasios* = hairy, shaggy; referring to the hairy fertile lemma covered by the upper glume.

Entoplocamia (Poaceae) Stapf
Gk. *entos* = inwards; *plokamus* = lock of hair (in plural, curling locks); referring to the gynoecium which has long styles.

Enydra (Asteraceae) Lour.
Gk. *enydros* = living in or near water; referring to the plant's habitat.

Epaltes* (Asteraceae) Cass.
Gk. word meaning *healing*. The root of the Indian species *E. divaricata* has medicinal properties.

Epimedium* (Berberidaceae) L.
Gk. *epimedion* = literally 'about the middle', hence 'not this, not that', from *epi*- = almost, near; *medion* = middle; referring to the variability of the genus. Some species have single stems, others multiple stems; some have several leaves per stem, others solitary leaves; some are deciduous, others evergreen, etc. Linnaeus used this name for a different plant.

Ephippiocarpa = ***Callichilia*** (Apocynaceae) Markgr.
Gk. *efippion* = a saddle; *karpon* = a fruit. The fruits are saddle-shaped.

Epidendrum (Orchidaceae) L.
Gk. *epi-* = on, on top of, beside, near; *dendron* = tree; referring to the plant's epiphytic growth habit. It grows on trees.

Epilobium (Onagraceae) L.
Gk. *epi-* = upon; *lobion* = diminutive of *lobos* = pod, capsule. The flower and capsule appear together, the flowers rest on the pod-like fruit, the corolla being borne on the end of the ovary.

Epinetrum* = ***Albertisia*** (Menispermaceae) Hiern.
Gk. *epinetron* = a curved cover used for weaving. Possibly so named because some species are twining lianas, growing in secondary forests.

Epiphora* = ***Polystachya*** (Orchidaceae) Lindl.
Gk. *epi-* = upon; *phora* = bearing; from *pherein* = to carry; referring to its epiphytic habit, it bears down on other plants.

Epiphyllum = ***Schlumbergera*** (Cactaceae) Pfeiff.
Gk. *epi-* = upon*; phyllum* = leaf; referring to flowers which appear to bloom from a leaf, which are leaf-like stems (*Phylloclades*).

Epischoenus (Cyperaceae) C.B.Clarke
Gk. *epi-* = upon usually, here meaning 'akin to' the rush genus *Schoenus* (q.v.).

Epistemum ** (Fabaceae) Walp.
Gk. *episteme* = science, knowledge. Meaning obscure.

Equisetum (Equisetaceae) L.

Plants are filled with silica bodies, so very sandpapery to the touch. (GN)

La. *equus* = horse; *seta* = bristle or hair. The vertically bristle-branched spike is said to somewhat resemble a horse's tail, 'horsetail'.

Eragrostis (Poaceae) Wolf
Gk. *eros* = love; *agrostis* = grass; allegedly referring to the 'graceful heart-shaped spikelets', but also the 'graceful dancing spikelet' and 'female aroma of the inflorescences of many species'; 'love grass'.

Eranthemum (Acanthaceae) L.
Gk. *er* = spring; *anthemon* = flower. This alludes to a spring flower but some species occur in terminal spikes during the early winter season.

Eranthis * (Ranuculaceae) Salisb.
Gk. *er* = spring; *anthos* = flower; referring to the early-blooming nature of this spring flower.

Eratobotrys *** = ***Ledebouria*** (Hyacinthaceae) Fenzl ex Endl.
La. *eratos* = lovely, beloved; Gk. *botrys* = bunch of grapes. Meaning obscure.

Eremia = ***Erica*** (Ericaceae) D.Don.
Gk. *eremia* = solitude, desert, possibly from *eremic* = deserts or sandy regions; referring to the plant's preferred habitat.

Eremiella = ***Erica*** (Ericaceae) Compton
Eremia (q.v.); La. *-iella* = diminutive.

Eremiolirion (Tecophilaeaceae) J.C.Manning & F.Forest
Gk. *eremios* = desert or wilderness; *lirion* = white lily; referring to the plant's habitat – desert and xeric shrubland.

Eremiopsis = ***Erica*** (Ericaceae) N.E.Br.
Eremia (q.v.); Gk. *-opsis* = resembling.

Eremocarpus = ***Croton*** (Euphorbiaceae) Benth.
Gk. *eremos* = lonely, solitude, deserted; *karpos* = fruit; alluding to the solitary carpel of the pistillate flower.

Eremothamnus (Asteraceae) O.Hoffm.
Gk. *eremos* = a lonely place, solitude; *thamnos* = shrub, bush. The plant is native to the coastal desert of Namibia.

Erepsia (Aizoaceae) N.E.Br.
Gk. *erepo* = to cover (with a roof) or *erepso* = I shall hide; referring to the staminodes covering and hiding the stamens.

Eriachne = ***Digitaria*** (Poaceae) R.Br.
Gk. *eri(on)* = wool; *achne* = glume. The first species described had hairy glumes.

Erianthemum (Loranthaceae) Tiegh.
Gk. *eri(on)* = wool; *anthemon* = flower. The flowers are woolly-haired.

Erianthus = ***Saccharum*** (Poaceae) Rich.
Gk. *eri(on)* = wool; *anthus* = flower; referring to the lower glumes, which are hairy.

Erica (Ericaceae) L.
Gk. *ereike* = to break. The name used for a heath by Theophrastus (372–287 BCE) and Pliny the Elder. The stems are brittle and break easily (Lindsay); or possibly but less likely because of the ability of the plant to break up bladder stones (*Paxton's Botanical Dictionary*).

Ericinella *** = ***Erica*** (Ericaceae) Klotzsch
Erica. (q.v.). Gk. *-ella* = diminutive. The flowers are minute.

Erigeron (Asteraceae) L.
Gk. *eri* = early (*er* = spring); *geron* = old man; referring to the fluffy, white seed heads and to the flowers which occur in spring but wilt early and turn grey.

Erinus * (Plantaginaceae) L.
Gk. *er* = spring; *-inus* = indicating possession; so-called because the plants flower early.

Eriocarpha = ***Lasiospermum*** (Asteraceae) Lag. ex DC.
Gk. *erion* = wool; *carphon* = fruit; woolly fruit.

Eriocaulon (Eriocaulaceae) L.
Gk. *erion* = wool; *caulos* = stem; woolly stem.

Eriocephalus (Asteraceae) L.

(CM)

Gk. *erion* = wool; *kephale* = head. The fruiting capitula (fruiting heads) are woolly.

Eriochloa (Poaceae) Kunth
Gk. *erion* = wool; *chloe* = grass, hence 'woolly grass'; referring to the woolly spikelets and hairy glumes.

Eriochrysis (Poaceae) P.Beauv.
Gk. *erion* = wool; *chrysos* = golden; referring to the golden-yellow hairs of the inflorescence.

Eriosema (Fabaceae) (DC.) G.Don
Gk. *erion* = wool; *sema* = a banner. The standard petal of the flower is glandular-hairy.

Eriosemopsis (Rubiaceae)
Eriosema (q.v.); Gk. *-opsis* = resembling.

Eriospermum (Ruscaceae) Jacq. ex Willd.
Gk. *erion* = wool; *sperma* = seed. The seed is covered with white hairs.

Eriosphaera = ***Lasiospermum*** (Asteraceae) F. Dietr.
Gk. *erion* = wool; *sphaera* = globe. The heads are aggregated into woolly clusters.

Eriospora* = ***Coleochloa*** (Cyperaceae) Hochst.
Gk. *erion* = wool; *sphora* = seed; referring to the woolly seeds.

Eriudaphus* = ***Solopia*** (Salicaceae) Nees
Gk. *eri* = woolly; *edaphos* = bottom; referring to the tubular perianth, the bottom of which is densely hairy.

Erlangea (Asteraceae) Sch.Bip.
In honour of the University of Erlangen in Bavaria (Germany), where the *Asteraceae* specialist Carl (Karl) Heinrich Schultz (1805–1867), known as Schultz Bipontinus, studied botany from 1825–1826.

Erodium * (Geraniaceae) L'Hér.
Gk. *erodios* = a heron; *-odes, odium* = resembling; referring to the long beak on the fruit, the seed pod of which resembles the head and long beak of a heron.

Eroeda* = ***Oedera*** (Asteraceae) Levyns
Anagram of *Oedera,* the previous name for the genus, after Georg Oeder of Copenhagen, author of *Flora Danica,* a massive work initially designed to cover all plant species in the crown lands of the Danish King. The name was changed because Crantz had already used it for a genus in *Liliaceae* (1768), thus antedating Linnaeus (1771).

Eruca * (Brassicaceae) Mill.
Classical Latin name used by Pliny the Elder; *eruca* = caterpillar. Probably named as such as the leaves are used as a food by the larvae of some moth species.

Erucastrum (Brassicaceae) (DC.) C.Presl
Eruca (q.v.); *-astrum* = partially resembling, inferior, wild.

Erysimum (Brassicaceae) L.
Gk. *eryomai* = to help or save; referring to the centuries-old belief that the plant had medicinal properties.

Erythrina (Fabaceae) L. (Image on opposite page)
Gk. *erythros* = red. The flowers of some species are red.

Erythrococca (Euphorbiaceae) Benth.
Gk. *erythros* = red; *kokkos* = berry; the fruits are red.

Erythrina (EJM)

Erythrophleum (Fabaceae) Afze ex G.Don
Gk. *erythros* = red; *phloios* = bark; referring to the red sap of *E. suaveolens*.

Erythrophysa (Sapindaceae) E.Mey. ex Arn.
Gk. *erythros* = red; *physos* = bladder; referring to the red, balloon-like fruits.

Erythroxylum (Erythroxylaceae) P.Browne
Gk. *erythros* = red; *xylon* = wood; referring to the reddish colour of the wood in some species.

Escallonia* (Escallionaceae) Mutis ex F.
For Antonio Escallón y Flórez (1739–1819), physician, explorer, student and botanical associate and friend of Spanish botanist Jóse Celestino Mutís in Colombia who named the genus *Escallonia* in his honour in 1821. He was also an advisor to the viceroy Pedro de la Cerda Massia. He was from Spain and travelled and collected plants in South America, then settled in New Granada (present-day Colombia).

Eschscholzia* (Papaveraceae) Cham.
For Johann Friedrich Gustav von Eschscholtz (1793–1831), an Estonian surgeon, zoologist, entomologist and botanist.

He was professor of anatomy at the University of Dorpat and director of the Zoological Cabinet from 1819–1828. Eschscholtz served as physician and naturalist aboard the Russian circumnavigational expeditionary ship *Rurik* under the command of Otto von Kotzebue on two voyages, 1815–1818 and 1823–1826, exploring in the Pacific, Alaska and California. On the first he became a close friend of the botanist Adelbert von Chamisso. He was the author of *System der Akalephen* (1829) and the *Zoologischer Atlas* (1829–1833). His insect collections are in the zoological museums of Moscow, Tartu and Helsinki.

Esterhuysenia (Aizoaceae) Bolus
For Elizabeth 'Elsie' Esterhuysen, (1912–2006), South African botanist and plant collector (q.v. *Elsiea*).

Ethulia* (Asteraceae) L.f.
Linnaeus gave no explanation of this name. Perhaps Gk. *aithon* = fiery, sparkling; *oulios* = baneful, destructive; referring to the purple-red flowers of the first described species, *E. conyzoides*.

Euadenia = **Cladostemon** (Capparaceae) Oliv.
Gk. *eu-* = well, good; *aden* = a gland; referring to the well-marked conspicuous glands below the ovary at the end of the gynophore.

Eucalyptus* (Myrtaceae) L'Hér.
Gk. *eu-* = well; *kalyptos* = to cover. The operculum of the calyx conceals the floral parts at first.

Euchaetis (Rutaceae) Bart. & H.Wendl.
Gk. *eu-* = well, fine; *khaite* = long flowing hair, mane. The petals are bearded on the inner surface.

Euclea (Ebenaceae) Murray
Gk. *eukleia* = fame, glory, from *eu-* = good; *kleos* = report; perhaps referring to the good quality ebony-type wood of some species or its beautiful evergreen foliage.

Eucomis (Hyacinthaceae) L'Hér. (Image on following page)
Gk. *eukomes* = beautifully-haired, from *eu-* = well; *kome* = hair of the head; referring to the crown of leaves atop the inflorescence.

Eugenia (Myrtaceae) L.
For prince François Eugène of Savoy (1663–1736), French-born book collector, patron of the arts and botany, one of the most successful military commanders in modern European history. For six decades he took part in many campaigns and battles. It is a tribute to his courage that during his long

military career, while serving the three Habsburg Holy Roman Emperors, Leopold I, Joseph I, and Charles VI, he was wounded no less than 13 times. He distinguished himself with many victories in campaigns to save the Habsburg Empire from French conquest including the battles of Blenheim (1704), Turin (1706), Oudenarde (1708), and Malplaquet (1709), and he broke the westward thrust of the Turkish Ottomans at Petrovaradin (1716) and Belgrade (1717), thus liberating central Europe from 150 years of Turkish occupation.

Eucomis (MC)

Eulalia (Poaceae) Kunth
For Eulalie Delile (1800–1840), botanical artist and the sister of French botanist Alire Raffeneau Delile (1778–1850). Eulalie illustrated the work of the French naturalist Victor Jacquemont as well as Carl Sigismund Kunth's (1788–1850) *Révision des Graminées* (1829–1834). Perhaps Kunth named the genus *Eulalia* in gratitude for her support and friendship.

Eulophia (Orchidaceae) R.Br. ex Lindl.
Gk. *eu-* = good, well; *lophos* = crest; alluding to the crest on the lip.

Eulophidium = Oeceoclades (Orchidaceae) Pfitzer
Eulophia (q.v.); Gk. *-idion* = diminutive.

Eumorphia (Asteraceae) DC.
Gk. *eu-* = good, well; *morphe* = form; alluding to the neatness of the foliage.

Eulophia (HS)

Euonymus* (Celastraceae) L.
Gk. *eu-* = good; *onoma, onuma* = name. The auspicious 'of good name' for this plant is ironic, as it is poisonous to animals.

Eupatorium* (Asteraceae) L.
For Mithridates VI Eupator (132–63 BC), King of Pontus, who is said to have used a species of this genus as an antidote to poison.

Euphorbia (Euphorbiaceae) L.
Gk. *eu-* = well; *phorbe* = pasture or fodder; probably after Euphorbus, Greek physician to Juba II, King of Mauretania. Juba was educated in Rome and married the daughter of Antony and Cleopatra. He was apparently interested in botany and had written about an African cactus-like plant from the slopes of Mount Atlas, which he had found or knew about, which was  used as a powerful laxative. That plant may have been *Euphorbia resinifera,* and like all *Euphorbias* had a latexy exudate (milky emulsion from certain plants). Euphorbus had a brother named Antonius Musa who was the physician to Augustus Caesar in Rome. When Juba heard that Caesar had honoured his physician with a statue, he decided to honour his own physician by naming the plant he had written about after him.

Euptelea* (Eupteleaceae) Siebold & Zucc
Gk. *eu* = good; *ptelea* = elm; referring to the edible fruit.

*Euremia***** (Ericaceae) D.Don
Orthographic variant of *Eremia*.

Eurylobium = *Stilbe* (Stilbaceae) Hochst.
Gk. *eurys* = large, broad; *lobion* = diminutive of *lobos* = lobe, pod, capsule. The corolla lobes are broader than those in *Stilbe*.

Euryops (Asteraceae) Cass.

(CM)

Gk. *eurys* = large or broad; *ops* = eye or face; referring to the large showy capitula or flower head.

Eurystigma = *Mesembryanthemum* (Aizoaceae) Bolus
Gk. *eury* = wide; *stigma* = stigma; alluding to the plant's five broad, thick stigmas.

Eustachys (Poaceae) Desv.
Gk. *eu-* = good, well; *stachys* = spike; alluding to the fact that the inflorescence branches are true spikes.

Eustegia (Apocynaceae) R.Br.
Gk. *eu-* = good, well; *stege* = roof, cover; alluding to the triple corolla.

Euthystachys (Stilbaceae) A.DC.
Gk. *euthys* = straight; *stachys* = spike; referring to the upright, straight inflorescent spikes.

*Euxolus**** (Amaranthaceae) Raf.
Misspelling of *eukolos* = contented, peaceable; 'well shut' according to Rafinesque.

Evolvulus (Convolvulaceae) L.
La. *e-* = without; *volvulere* = to roll, twist, hence unroll, untwist; referring to its non-climbing habit, unusual among members of the 'morning glory' (*Convolvulus*) family.

*Evota**** = *Ceratandra* (Orchidaceae) Rolfe
Gk. *eu-* = well; *ous, otis* = ear; referring to the dilated (wide) arms of the rostellum.

Evotella (Orchidaceae) Kurzweil & H.P.Linder
Evota (q.v.); La. *-ella* = diminutive.

Exacum (Gentianaceae) L.
La. *exacum*, from *ex* = out, *agere* = to drive (out) or *exacon* = the Gaelic name for centaurium (an old genus name); alluding to the supposed property of the plant to expel poison.

Excoecaria (Euphorbiaceae) L.
La. *excaecare* = to blind; referring to the acrid, milky, poisonous juice which irritates the eyes and the skin.

*Exochaenium**** = *Sebaea* (Gentianaceae) Griseb.
Gk. *exo-* = out, outside; *khainein* = to gape; referring to an indehiscent pericarpium, or fruit, with a pericarp contiguous to the seeds.

*Exohebea**** = *Tritoniopsis* (Iridaceae) R.C.Foster
Gk. *exo-* = out, outside; *hebe* = pubescent, hairy; possibly after Hebe, the Greek goddess of youth, daughter of Zeus and Hera.

Exomis = *Chen* (Amaranthaceae) Fenzl ex Moq.
Gk. e*x-* = without; *omos* = shoulder; referring to a Greek tunic or vest worn by soldiers and workers without sleeves, leaving the shoulders bare. The flowers of some *Exomis* species grow in leaf axils near stem tips, half-hidden among the leaves. Possibly 'off-the-shoulder'. Meaning obscure.

Ezoloba (Fabaceae) B.-E.van Wyk & Boatwr.
Named after two botanical collectors who first collected the species (EZ = Eckl. & Zey.); Gk. *lobos* = pod, capsule; *lobium* when referring to a legume (member of Fabaceae), such as *Melolobium, Schizolobium*, etc.

Ezosciadium (Apiaceae) B.Burtt
See above – EZ = Eckl. & Zey.; *skiadeion* = parasol, umbrella, or canopy. The inflorescences are in units of umbels.

F

Flagelleria: a genus which has long stems, somewhat bamboo-like, often reaching the tree canopy; used for the making of baskets and similar woven products

Sizwe Cawe (1958–) hails from Mthatha in the Eastern Cape, South Africa. He studied plant ecology under Bruce McKenzie at the erstwhile University of Transkei after having completed a postgraduate diploma in rural survey at the International Institute for Aerospace Survey in Enschede, the Netherlands. He has studied the ecology of both the Afromontane and coastal forests of the then Transkei region, South Africa. He has been involved with Coert Geldenhuys in several projects relating to the use and management of the forests of that region by the local people. He has developed an interest in the use of plants by local people in the Eastern Cape (ethnobotany) after participating in the People & Plants Initiative that was run by Tony Cunningham. He currently lectures ecology, conservation biology and ethnobotany at Walter Sisulu University in the Eastern Cape.

Facelis* (Asteraceae) Cass.
Probably a reference to Facelis (Facelina), a goddess in Greek mythology usually depicted with a torch in her hand. La. *fax* = torch; referring to the inflorescence which appears at the top of the plant – like a hand-held flaming torch.

Fadogia (Rubiaceae) Schweinf.
Derivation uncertain. Possibly a Nigerian name, after Fadoga, Sudan (not traced in modern gazetteers).

Fagara*** = **Zanthoxylum** (Rutaceae) L.
An ancient name for an aromatic plant, possibly of Arabic origin, which can be traced back to the 1020s, first described by Ibn Sina (Avicenna) and which is associated with Sichuan pepper.

Fagelia = **Bolusafra** (Fabaceae) Neck.
For Hendrik Fagel the Elder (1706–1790), Dutch greffier or official scribe to the Staten-Generaal (States-General of Parliament) from 1744–1790 and book collector. Together with his son Hendrik Fagel the Younger and other members of the family, they collected the largest and most important Dutch private library collection of books anywhere in the world. When the French Revolution broke out in 1789, the Fagel family took the collection over to London where it was acquired by the University of Dublin in 1802. This massive collection takes up a mile of shelf-space and comprises some 20 000 volumes of miscellany – maps, atlases, statistics, politics, religion, accounts of expeditions, etc.

Fagonia (Zygophyllaceae) L.
For Guy-Crescent Fagon (1638–1718), French physician and botanist, professor of botany and chemistry at the Royal Gardens from 1671–1708 and from 1699–1718 its director. He was personal physician to Louis XIV of France. In 1669 he was made an honorary member of the French Academy of Sciences.

Fagopyrum* (Polygonaceae) Mill.
La. *fagus* = beech tree, but perhaps originally from Gk. *phago* = I eat (the edible mast of an old tree); Gk. *pyros* = grain, wheat. The fruit is a large triquetrous nut.

Fagraea (Loganiaceae) Thunb.
For Jonas Theodor Fagraeus (1729–1797), Swedish physician and botanist. He obtained an MA degree from the University of Lund (1751) and became a doctor of medicine from Uppsala (1758). Linnaeus, in a letter to Abraham Brook, wrote: 'None of all, who studied medicine in Uppsala in my time, have had [a] stronger head, and sharper *Acumen ingenii* (acuteness) than Doctor Fagræus.' From 1759–1760 he was librarian and curator of the scientific collections of Jonas Alströmer in Alingsås. He was a prolific writer on various topics and even wrote a manuscript on zoological systematics at Linnaeus's request. Linnaeus offered him a professorship in botany at St Petersburg, Russia, which he declined. In October 1779 a devastating fire burnt most of the precious part of Alströmerska library, costly instruments, collections, and Fagraeus's own entire library and unpublished manuscripts. He never really recovered from this set-back.

Faidherbia (Fabaceae) A. Chev.
For Louis Léon César Faidherbe (1818–1889), French general and governor of Senegal. For his military services in Algeria, Guadeloupe and France, he was decorated with the grand cross, and made chancellor of the order of the Legion of Honour. In 1872 he went on a scientific mission to Upper Egypt, where he studied the monuments and inscriptions. An enthusiastic geographer, philologist and archaeologist, he wrote numerous works, including *Collection des inscriptions numidiques* (1870), *Epigraphie phenicienne* (1873), *Essai sur la langue poul* (1875) and *Le Znaga des tribes sénégalaises* (1877), the last a study of the Berber language. He also wrote on the geography and history of Senegal and the Sahara, and *La Campagne de l'armée du Nord* (1872).

Falkia (Convolvulaceae) Thunb.
For Johan Peter Falck (Falk) (1733–1774), Swedish botanist and doctor, traveller, professor of botany at St Petersburg, and pupil of Linnaeus. He accompanied Linnaeus on his expedition to the island province of Gotland and tutored Carl Linnaeus the Younger. He undertook an expedition at the behest of the Russian Academy of Sciences to explore a vast area of Siberia, during which he collected a great deal of information about plants, animals and local peoples and customs. He committed suicide in Kazan after having become addicted to opium and enduring long spells of depression.

Fallopia* (Polygonaceae) Adans.
For Gabriele Falloppio (1523–1562), Italian physician and surgeon and professor of anatomy, surgery and botany at the University of Padua

(1551–1562). He was also superintendent of the university's botanical gardens. A pioneering anatomist, he was the first to describe, among many discoveries, the semicircular canals (*chorda tympani*) of the ear, the circular folds of the small intestine, and the inguinal band, later called Poupart's ligament, the clitoris and what are now known as the Fallopian tubes, as well as the *arteria profunda* of the penis. Among his publications was *Observationes anatomicae* (1561), which contained many descriptions of his anatomical research. His collected works, *Opera omnia,* were published after his death, in Venice (1584) and Frankfurt (1600).

Fanninia (Apocynaceae) Harv.
For George Fox Fannin (1832–1865), Irish botanist, plant collector and farmer who died at an early age in Natal (now KwaZulu-Natal). After moving to South Africa he became interested in the local plants and collected many which he sent to Irish botanist William Henry Harvey at Dublin. His sister Marianne Edwardine Fannin (later ME Roberts), who also lived in South Africa, painted and pressed many of these collected specimens.

Fatsia* (Araliaceae) Decne. & Planch.
Ancient Japanese, *Yatsude* = eight (present-day Japanese, *hach.*); referring to the lobes (the leaves have seven to nine broad lobes).

Faucaria (Aizoaceae) Schwantes
La. *fauces* = jaws, throat; *-aria* = denotes possession or relation to. The leaf-pairs resemble the open jaws of some mythical beast.

Faurea (Proteaceae) Harv.
For William Caldwell Faure (1822–1844), South African botanist, soldier and naturalist, teacher of mathematics at the South African College, Cape Town (now the University of Cape Town), made a special study of the genus *Oxalis,* went to India for the East India Company and became an ensign in the Second European Light Infantry. He died in an ambush at the early age of 22. He was praised by William Henry Harvey, who predicted that had he lived he would have become a fine botanist.

Felicia (Asteraceae) Cass.
Origin uncertain. La. *felix* = happy, cheerful, though in the neuter plural form *felicia* = happy things; possibly a reference to the bright flowers. Other sources vaguely refer to a mysterious German official in Regensburg called Felix who died in 1846 but speculatively and more

(CM)

 probably for the Italian Fortunato Bartolomeo de Felice (1723–1789), an Italian scholar established in Yverdon who led the European team that wrote the *Yverdon Encyclopedia,* published between 1770 and 1780 in 58 quarto volumes. This superseded the Parisian *Encyclopedie* of Diderot and d'Alembert published between 1751 and 1772.

Fenestraria (Aizoaceae) N.E.Br.
La. *fenestra* = window; *-aria* = pertaining to; referring to the window-like leaf apices (leaf ends).

Feretia* (Rubiaceae) Delile.
For Pierre Victor Adolphe Ferret (1814–1882), French brigadier-general and plant collector who led an adventurous expedition to Ethiopia (1839–1843). He went to Cairo for eight months to learn Arabic, took the pilgrim's route to Mecca and eventually reached Abyssinia. He authored the three-volume work *Voyage en Abyssinie dans les provinces du Tigré dú Samen et de l'Amhara* (1847) with the French plant collector Joseph Germain Galinier (1814–1888), in which they describe their journey, what Abyssinia (Ethiopia) was like in the 19th century, and its geology and natural history. Later he was chief of staff in Oran province, Algeria (1866).

Fernandoa* (Bignoniaceae) Welw. ex Seem.
For Dom Fernando II (1816–1885), German-born King-consort of Portugal and husband of Queen

Dona Maria II. In terms of Portuguese law he obtained this title, aged 21, on the birth of their son in 1837 and held it until the death of his wife in 1853. Then he assumed the title of regent of Portugal which he held for two years (1853–1855) during the minority of his son King Pedro V. During Dona Maria's pregnancies (they had 11 children), he took over her role as King of Portugal, which he did competently. They worked well as a team. Later in life, in 1869, he was invited to become King of Spain but declined the offer. Among his other functions he was President of the Royal Academy of Sciences and the Arts and Lord-Protector of the University of Coimbra.

Ferraria (Iridaceae) Burm. ex Mill.

There are many species and all smell of carion as they are fly pollinated. (MC)

For Giovanni Batista Ferrari (1584–1655), Italian Jesuit, professor of Hebrew and rhetoric at the Jesuit College in Rome, horticultural advisor to the Pope, and author of many illustrated botanical books, including *De Florum Cultura* in four volumes (1633), a horticultural book emphasising the planning and planting of gardens, and *Hesperides sive de Malorum Aureorum cultura* (1646), a 'citrus encyclopedia'. He also wrote a Latin-Syrian dictionary, a series of *Orations* – treatises on rhetoric, which emphasised good Latin usage, and a book on Sienese saints. He was the first scientist to provide a complete description of the limes, lemons and pomegranates, and their use in preventing scurvy.

Ferula (Apiaceae) L.
Latin name for fennel.

Festuca (Poaceae) L.
Latin name for stalk or straw; referring to a weedy grass, festucine, straw-coloured.

Feuilleea (Fabaceae) Kunze
For Louis Éconches Feuillée (Feuillet) (1660–1732), French explorer, botanist, astronomer and geographer. He studied at the Minim convent of Mane, was taught botany by the renowned Charles Plumier, and astronomy and cartography by Jean Mathieu de Chazelles. During his career he journeyed to the Levant (1699), Antilles (1703–1706), and western South America (1707–1711). He compiled an inventory of his observations in three volumes (1714–1725). His publications include *Journal des observations physiques, mathématiques, et botaniques* (1714) and *Suite du Journal* (1725). As a result of his achievements he was made a member of the Order of the Minims, and received the title of 'Royal Mathematician' from Louis XIV of France who also built an observatory for him on the Michaelmas Plain at Marseilles.

Ficinia (Cyperaceae) Schrad.
For Heinrich David August Ficinus (1782–1857), German physician, naturalist, botanist, professor of physics and chemistry at the medical-surgical academy in Dresden (1814), then was professor of natural history (1817). From 1822 onwards he worked in his father's pharmacy but also taught chemistry, technology and physics at the Technical Training Institute in Dresden (1828–1833). He wrote several literary works, textbooks and papers in the fields of botany, optics and mineral chemistry. They include *Flora of the Area around Dresden* (1807), *Optics or Attempts to Follow the Right Outline of the Whole Theory of Light* (1828), *Foundations of Medical Physics, Foundations of Medicinal Chemistry* (1815), and *General Natural History* (1839) (titles translated from German).

Ficus (Moraceae) L. (Image on following page)
La. = fig probably derived from an older tongue, Hebrew, *fag;* or Persian, *fica.* An old name for the edible fig, *Ficus carica.*

Fimbristylis (Cyperaceae) Vahl
Gk. *fimbria* = a fringe; *stylus* = style; referring to the fimbriated style.

Ficus: Typical figs in stages of ripening (EJM)

Finckea *** = **Acrostemon*** (Ericaceae) Klotzsch
For August Fincke (1805–1873), Polish pharmacist and botanist of Silesia who took over as apothecary in Krappitz in 1836. He was interested in researching the flora of upper Silesia.

Fingerhuthia (Poaceae) Nees
For Carl (Karl) Anton Fingerhuth (1798–1876), German botanist and physician, author of *Monographia Generis Capsici* (1832) and *Tentamen florulae lichenum Eiffliacae* (1829), and co-author with Matthias Joseph Bluff and Karl Friedrich Wilhelm of *Compendium florae Germaniae*.

Fintelmannia *** = **Trilepis*** (Cyperaceae) Kunth
For Joachim Anton Ferdinand Fintelmann (1774–1863), Prussian royal gardener. In 1795 he became a gardener's assistant at the Charlottenburg Royal Gardens. In 1804 Frederick William III appointed him as head gardener on the Peacock Island. This acclaimed park, which he designed with Peter Joseph Lenné from 1816–1834, had over 9 000 shrub roses, greenhouses, a palm house with 42 large palm trees, etc. In 1834, aged 60 years old, he was appointed head gardener at Charlottenburg, where he worked until he was 89 years old. He was the author of *Über Nutzbaumpflanzungen* (1856). He was awarded the Red Eagle Order third class with ribbon from Friedrich Wilhelm IV in 1853 and given the title *Oberhofgärtner auszeichnete* (*Oberhof excellent gardener*) in 1854.

Firmiana (Malvaceae) Marsili.
For Karl Joseph von Firmian (proper name Karl Gotthard von Firmian) (1716–1782), Austrian noble, politician and art collector. He studied at the University of Leiden and travelled extensively through France and Italy. In 1753, he was recruited by Francis I (1708–1765), king of the Holy Roman Empire, to become its ambassador to Naples and in 1756 minister plenipotentiary and governor-general of Lombardy, an area in northern Italy under Austro-Hungarian rule. He was an avid supporter of the arts and sciences. When he died he left a legacy of a library of 40 000 volumes and precious art collections.

Flacourtia (Achariaceae) L'Hér.
For Étienne de Flacourt (1607–1660), French botanist and traveller, who was appointed governor of Madagascar by the French East India Company in 1648 where he successfully pacified the mutinous French troops on the island, but failed to develop good relations with the indigenous peoples. Upon his return to France in 1655, de Flacourt wrote his *Histoire de la Grande Isle Madagascar* (1658), the first comprehensive account of the island. Not long after he was appointed director-general of the company, he again returned to Madagascar, but drowned on his voyage home in 1660. He was one of the first Europeans to describe the now extinct flightless elephant bird, *Aepyornis maximus*.

Flagellaria (Flagellariaceae*) L.
La. *flagellum* = long, tapering, supple, whip-like; -*aria* = possessing (this characteristic); Linnaean name. The leaf-tips carry tendrils, which might also be considered flagelloid.

Flanagania = ***Cynanchum*** (Apocynaceae) Schltr.
For Henry George Flanagan (1861–1919), South African citrus farmer with an interest in botany who collected actively around Komga and Kei Mouth of the Eastern Cape from 1889 onwards and later in other parts the Cape Province and Orange Free State. He developed, on his farm

in the Komga District, a noteworthy garden containing rare exotics as well as South African trees and shrubs. His herbarium was bequeathed to the National Herbarium in Pretoria.

Flaveria * (Asteraceae) Juss.
La. *flavus* = yellow. The plant is used in Chile to produce yellow dye.

Flemingia (Fabaceae) Roxb. ex Aiton f.
For John Fleming (1747–1829), English botanist and physician, member of the Indian Medical Service in Bengal, physician-general and president of the Bengal Medical Board, Fellow of the Royal and Linnaean Societies, the Royal Society of Edinburgh and the Horticultural Society, and author of *Catalogue of Indian Medical Plants and Drugs* (1810). He collected a large body of drawings done by native artists of Indian plants and sent specimens to Sir Joseph Banks for his collection. He worked for the Indian Medical Service from 1768 until his retirement in 1813, when he returned to England. He became a member of the Medico-Botanical Society of London and in 1818 was elected a member of parliament, a position he held for two years.

Fleurya *** = ***Laportea*** (Urticaceae) Gaudich.
For Camile Fleury, who was a merchant service apprentice on the *Uranie*, one of the two ships involved in the circumglobal expedition of 1817–1820 under Commander de Freycinet. Charles Gaudichaud-Beaupré, botanist and author of the Fleurya genus was on board the *Uranie*. The results of this voyage were published under Freycinet's supervision as the three-volume work *Voyage autour du monde sur les corvettes Uranie et la Physicienne* (1824–1844). Fleury was listed in this work as a member of the crew and was elevated to the rank of ensign during the course of the expedition. Some sources state after a JF Fleury, French botanist, author of *Orchids of the environs of Rennes* (1819), but this seems unlikely.

Floscopa (Commelinaceae) Lour.
La. *flos* = flower; *scope* = broom; referring to the broom-like flower spike at the top of the stem in some species.

Flueggea (Phyllanthaceae) Willd.
For Johannes (Johann) Flüggé (Fluegge) (1775–1816), German physician, cryptogamic botanist, and university lecturer. He studied medicine and natural history at the universities of Jena, Vienna, and Göttingen, and obtained a doctorate from the University of Erlangen in 1800. While studying in Jena he collected the oldest bryophyte sample in the Herbarium Hamburgense in 1798, and established the first botanical garden in Hamburg in 1810. He is remembered for his research into grasses (genus *Paspalum*). He was the author of *Graminum monographiae* (1810).

Fockea (Apocynaceae) Endl.
For Gustav Waldemar (Woldemar) Focke (1810–1877), German physician, plant physiologist and amateur microscopist. He studied at the University of Heidelberg, obtaining a PhD in 1833 and did post-doctoral studies under Stephan Ladislaus Finite (1804–1829), professor and director of the botanical garden at the University of Vienna, and under professor Christian Gottfried Erenberg (1795–1876) at the University of Berlin; the founder of the science of micropaleontology and microbiology, and also spent some time at the University of Halle. Despite all his training he did not publish many papers concerning his research, although he delivered a lot of lectures. He was highly involved in scientific societies and a member of the German Academy of Sciences Leopoldina among others. His major works were *De respiratione vegetabilium* (1833) and *Physiologische Studien* (1847).

Foeniculum * (Apiaceae) Mill.
La. *foenum* = hay (fennel); -*ulum* = little, hence 'little hay or hay-like'; alluding to the smell of the plant being similar to hay.

Forbesia *** = ***Empodium*** (Hypoxidaceae) Eckl.
For John Forbes (1799–1823), English plant collector and naturalist who visited the Cape in 1822 having been appointed by the Horticultural Society of London to collect as many plants, seeds, flowers, insects, minerals, bird skins and any other objects of natural history as possible. He travelled with a Royal Navy expedition under the command of Captain William Owen whose task was to survey the coasts of Africa using chronometers, since the available maps were either incomplete or wrong. Forbes's other task was to write a journal of the voyage. He died from malaria on the Zambesi River in Mozambique. Almost a third of captain William Owen's two-volume *Narrative of Voyages to the Shore of Africa, Arabia and Madagascar* (1931) is taken from the journals of his officers or from John Forbes.

Forficaria = Disa (Orchidaceae) Lindl.
La. *forfic* = scissors, pincers; *-aria* = possessing; referring to the deeply incised leaves and possibly the paired petal shape.

Forsskaolea (Urticaceae) L.
For Pehr Forsskål (Peter Forskaol, Petrus Forskål or Pehr Forsskåhl) (1732–1763), Finnish-born

Swedish botanist, zoologist, traveller and philosopher. Born in Helsinki, he studied botany at the University of Uppsala under Linnaeus and obtained a doctorate in philosophy and studied oriental languages at Göttingen. He was a naturalist on the Royal Danish expedition to Egypt and Yemen 1761–1763 where he died from malaria. Forsskål's notes, *Flora Aegyptiaco-Arabica*, were published posthumously in 1775. The Danish botanist Carsten Niebuhr, who went on this expedition as geographer, was the sole survivor.

Foveolina (Asteraceae) Källersjö.
La. *foveo* = keeps warm; *linum* = flax; referring to the properties of flax.

Fragaria* (Rosaceae) L.
La. *fragum* = strawberry; *-aria* = pertaining to; referring to the fruit.

Frankenia (Frankeniaceae) L.
For Johan Frankenius (Frank, Franke) (1590–1661), professor of anatomy, medicine and botany at Uppsala, Sweden, and the first writer on Swedish plants, author of *Speculum botanicum*, and a colleague of Linnaeus. In 1638 he made the first inventory of Swedish plants.

Fraxinus* (Oleaceae) Tourn. ex L.
La. *fraxinus* = spear; referring to the hardwood ash tree. The wood was also extensively used for making bows.

Freesia (Iridaceae) Klatt
For Friedrich Heinrich Theodor Freese (Vries) (1795–1876), a German physician and botanist from Kiel who learned much about South African plants from his contemporary Christian Friedrich Ecklon (1795–1868), and who, like his teacher, studied South African plants. This beautiful plant Freese discovered was named after him by the German botanist Klatt in 1866.

Fresenia*** = Felicia (Asteraceae) DC.
For Johann Baptist Georg Wolfgang Fresenius (1808–1866), German physician and botanist; he

Freesia: These geophytes have been cultivated because their flowers are sweetly scented. (CM)

studied medicine at the universities of Heidelberg, Würzburg and Giessen from which he obtained his doctorate in 1829. From 1831–1866 Fresenius was curator of the Senckenberg Herbarium and a teacher at the Senckenberg Research Institute and Medical Society. His special interest was phycology (study of algae).

Freylinia (Scrophulariaceae) Colla.
For Pietro Lorenzo, count of Freylino (Freilino) (1754–1820), Italian botanist and owner of a famous garden at Buttigliera d'Asti near Marengo in Italy in the early 19th century. He compiled a catalogue of the plants growing in his garden.

Freyliniopsis** (Scrophulariaceae) Engl.
Freylinia (q.v.); Gk. *-iopsis* = resembling, likeness.

Fridericia (Bignoniaceae) Mart.
For Frederick William III (Friedrich Wilhelm) (1770–1840), King of Prussia. After an early education from tutors, this 'soldier king' undertook

military training from age 14, becoming a colonel aged 20. In 1797 he became King of Prussia and sovereign prince of the canton of Neuchâtel in French-speaking Switzerland. In 1806 Friedrich Wilhelm decided to go to war against France while the army of Russia, a Prussian ally, was too far away to assist, resulting in a disastrous defeat. He lost his territories. The royal family fled to Russia. Years later, after Napoleon's defeat in Russia (1812), Prussia joined the coalition forces, which resulted in Napoleon's defeat (1813–1815). Friedrich Wilhelm was a patron of botany, though to what degree is unknown. The Berlin Botanical Garden's herbarium under professor C Willdenow, which housed an important collection of exotic plants, was financed by order of Friedrich Wilhelm.

*Friedrichsthalia**** = *Trichodesma*
(Boraginaceae) Fenzl
For Emanuel Ritter von Friedrichstal (1809–1842), Czech-born explorer, botanist, archaeologist and daguerreotypist who undertook several scientific journeys for the Austrian government recording his observations. He travelled to southeastern Europe from 1834–1836 (Greece, Turkey, Serbia), authoring two books about his observations (including the flora) and and was the first person to take daguerreotypes of the Mayan ruins in Central America (1837–1841). While travelling he became ill, probably with malaria, and returned to Vienna, where he died. La. *thalla* = abundance, from *thallein* = to sprout.

Friesodielsia (Annonaceae) Steenis
For Elias Magnus Fries (1794–1878), Swedish botanist, one of the founders of taxonomic mycology, professor of botany and applied economics at Uppsala University. He was elected an escort for the Royal Swedish Academy of Sciences, and was the author of the three-volume *Systema mycologicum* (1821–1832), *Elenchus fungorum* (1828), the two-volume *Monographia hymenomycetum Sueciae* (1857–1863), and *Hymenomycetes Europaei* (1874); and for Friedrich Ludwig Emil Diels (1874–1945), German botanist, director of Berlin-Dahlem Botanical Garden and Museum, plant collector in South Africa, Java, eastern and western Australia, and New Zealand, co-author of *Fragmenta Phytographiae Australiae Occidentalis* (1905) with Ernst Pritze.

Frithia (Aizoaceae) N.E.Br.
For Frank Frith (1872–1954), English-born horticulturalist and succulent plant collector. In 1900 he came to South Africa with the Royal Army Medical Corps (RAMC) and did service during the South African War. He remained in South Africa after the war and became the first horticulturist to join the South African Railways. His main interest was succulents, and a special coach was put at his disposal for the collection and succulents and other aloes throughout South Africa and South-West Africa (Namibia). In 1925 his garden design at the Wembley Empire Exhibition won the bronze Lindley Medal of the Royal Horticultural Society. While there, he took specimens to the British botanist Nicholas Edward Brown at Kew Gardens, who later published the genus after him.

*Fritillaria** (Colchicaceae) L.
La. *fritillus* = dice box; referring to the markings on the flowers of some species.

Fuchsia (Onagraceae) L.
For Leonhart (Leonhard) Fuchs (1501–1566), German physician and botanist. He obtained an MA and qualified as a medical doctor in 1524. After practising medicine for two years he turned to academia, and for the last 31 years of his life was a professor of medicine at the University of Tübingen (closed 1800); where he also served as chancellor on seven occasions. While there, he created a botanical garden, one of the oldest in the world. His main work was *De historia stirpium commentarii insignes* (*Notable Commentaries on the History of Plants*) (1542), featuring around 400 wild plants and 100 ornamental plants, accurately drawn, and with detailed illustrations made from woodcuts.

*Fugosia**** = *Cienfuegosia* (Malvaceae) Juss.
The name *Fugosia,* published in 1789, was abridged by Jussieu from *Cienfuegosia* (q.v.) referring to Bernardo de Cienfuegos (c 1580–1640), Spanish physician and botanist, and is considered an illegitimate name.

Fuirena (Cyperaceae) Rottb.
For Jørgen (Georg) Fuiren (1581–1628), Danish physician and botanist. He studied medicine, botany and mathematics at the University of Leiden (1598–1602), art at the University of Padua (1602), and obtained a medical doctorate (Basel and Strasbourg). He started practising as a physician in Copenhagen (1610) but gave up the practice (1623) to focus on botany, especially on Danish plants. He travelled throughout Scandinavia, and was a pupil of one of the Bauhin brothers. He came from a very wealthy merchant family, known for their generosity, and he did not need to worry about his medical practice, but could be freely occupied with botany. His last years were concerned with theological studies.

Fumaria (Papaveraceae) L.
La. *fumus* (*terrae*) = smoke of the earth; -*aria* = associated with, hence 'fumitory'; referring to the disagreeable odour of the plant's roots.

Furcaria* = *Hibiscus* (Malvaceae) Kostel.
La. *furcatus* = forked; *-aria* = pertaining to; referring to the leaves of the involucel which are divaricately forked.

Furcraea* (Agavaceae) Vent.
For Antoine François, Comte de Fourcroy (1755 – 1809), French chemist. Although he obtained a doctor's diploma in 1780 from the medical school in Paris, Fourcroy pursued a career in chemistry as a result of Professor JBM Bucquet's (1746–1780) influence. He became a popular lecturer in chemistry at the College of the Jardin du Roi. He worked with Antoine-Laurent de Lavoisier (1743–1794), the 'father of chemistry', and Guyton de Morveau and Claude Berthollet on the *Méthode de nomenclature chimique,* a work that helped standardise chemical nomenclature. He wrote many scientific memoirs for the Royal Society, a book on systematic entomology and, under Napoleon I, took a leading part in the establishment of schools for both primary and secondary education and scientific studies. In 1801 he was elected a foreign member of the Royal Swedish Academy of Sciences. (The botanical name was originally misspelled by the author Étienne Pierre Ventenat.)

Fusanus* = *Colpoon* (Santalaceae) Steenis
La. *fusus* or *fusanus* = a spindle; referring to so-called 'spindle tree' from whose wood fine charcoal is made in stick form, hence 'spindle'.

G

Gymnosporia is an Old World genus of plants that comprise suffrutices, shrubs and trees.

Marie Jordaan (1948–) obtained a BSc at Stellenbosch University and started to work in the National Herbarium, Pretoria in 1970. She was transferred to the Stellenbosch Botanical Research Unit in 1972 and worked there until 1975. In 1981 she joined the Durban Botanical Research Unit and returned to the National Herbarium, Pretoria in 1987 and worked there until her retirement in 2013. She also obtained an MSc and PhD in taxonomy from the University of Pretoria. Her main interests are in the tree families of Southern Africa, and her main research was done on Celastraceae (most notably *Gymnosporia*). She has published many articles and made substantial contributions to many books on trees.

Gagea * (Liliaceae) Salisb.
For Sir Thomas Gage (1781–1820), English botanist, seventh baronet of Hengrave Hall, Suffolk. Gage collected plants in Ireland and the Iberian Peninsula and contributed to *English Botany* by James Sowerby and JE Smith and was also a pioneer of lichens.

Gaillardia * (Asteraceae) Foug.
For Antoine René Gaillard de Charentonneau (1719–1789), French magistrate, patron of botany, naturalist, amateur botanist and member of the Académie des Sciences. Antoine René was an officer of the courts from 1740–1771 and from 1774–1779. He also received seeds of plants from the French colonies that he both cultivated himself and shared with other botanists. At one stage he owned the castle of Charentonneau but was seemingly forced to sell it around the time of the French Revolution.

Gaillonia (Rubiaceae) A.Rich. ex DC.
For François Benjamin Gaillon (1782–1839), French botanist and specialist in marine plants (phycologist). He sent many papers to many learned societies of which he was a correspondent, including the Linnaean Society in Paris, Lyon, Bordeaux, Normandy and the Academy of Rouen. Among his findings was the reason why oyster parks take a green colour at certain times of the year (1821). He contributed to the *Flore générale de France ou Iconographie, description et histoire de toutes les plantes* (*Flora of France and General Iconography, description and History of all the plants*) (1828–1829) undertaken by Jean-Deslongchamps Loiseleur, with Christiaan Hendrik Persoon, Jean-Baptiste Alphonse Dechaffour de Boisduval and Louis Alphonse de Brébisson.

Galactia (Fabaceae) P.Browne
Gk. *gala* = milk. Some species in the genus have a 'milky' sap containing latex, which is rare in the family.

Galaxia *** = ***Moraea*** (Iridaceae) Thunb.
La. *galaxia, galaxias* = the milky way; Gk. *gala* = milky or *galactos* = milky fluid; referring to the white or creamy flowers that some species in *Galaxia* have.

Galbanon = ***Notobubon*** (Apiaceae) Adans.
La. *galbanum* = the gum resin (sap) of an umbelliferous plant in Persia. The new name, *Notobubon galbanum,* is the notorious 'blister bush' of the Western Cape mountains.

Galeandra (Orchidaceae) Lindl.
Gk. *galea* = helmut; *andros* = male; referring to the helmet-shaped cap over the stamen.

Galega (Fabaceae) L.
Gk. *gale* = milk; *ega* = to bring on; referring to the belief that if animals consumed the plant, this would stimulate or increase their milk production.

Galenia (Aizoaceae) L.
For Greco-Roman Claudius Galen (130–201), famed physician, philosopher of repute, and the most prolific writer in antiquity with over 600 treatises (on anatomy, physiology, medicine, logic and philosophy). His surviving work runs to some three million words, which is thought to be only about one-third of his total output. Described by Emperor Marcus Aurelius as 'first among doctors and unique among philosophers', his 'medical bible' continued to exert an important influence over the theory and practice of medicine until the mid-17th century in the Byzantine and Arabic worlds and Europe.

Galeomma (Asteraceae) Rauschert
Gk. *galeos* = shark, dogfish; *-omma* = eyed; possibly referring to the many floral 'eyes' staring from this plant.

Galinsoga * (Asteraceae) Ruíz & Pav.
For Ignacio Mariano Martinez de Galinsoga (1766–1797), physician and botanist who held many important positions, including being a member of the Spanish Court and advisor on medical matters, and physician to Maria Luisa of Parma (1751–1819), queen consort and wife of King Charles IV of Spain. He was director of the Botanical Garden of Madrid and Chemistry Laboratory, president of the Royal College of Medicine (Madrid), director of real practice medicine studies, first general of the Royal Medical Armies, and an inspector of the Royal Practice Medicine Academy of Barcelona, the Royal College of Physicians and Surgeons of Zaragoza, and the Valladolid Academy of Surgery. He was a knight of the Royal and Distinguished Spanish Order of Charles III.

Galium (Rubiaceae) L.
Gk. *galion* = bedstraw, from *gala* = milk; referring to the flowers of *G. verum,* lady's bedstraw. This plant was, in the past, used to curdle milk and is still used to colour cheese (Don Perrin).

Galopina (Rubiaceae) Thunb.
La. *galopina* = a little errand boy, page or a low fellow, hence ragamuffin. Some fish species are also called '*galopin*' = swift leaper. Meaning unclear but could perhaps allude to the fact that the plant grows quickly.

Galpinia (Lythraceae) N.E.Br.

This small bushveld tree can be massed in white spring flowers that are shown off against the persistant autumn leaves. (EvJ)

For Ernest Edward Galpin (1858–1941), a South African botanist and banker. He left some 16 000 sheets to the National Herbarium in Pretoria, which formed the nucleus of its collection, and was dubbed 'the Prince of Collectors' by General Smuts. Galpin discovered half a dozen genera and many hundreds of new species. Numerous species are named after him, and his farm, called 'Mosdene', is commemorated in the genus *Mosdenia*. He sent many specimens to botanists such as Harry Bolus, John Medley Wood and Peter MacOwan. Galpin was a life member of the Linnaean Society, and Volume 13 of *Flowering Plants of South Africa* was dedicated to him.

Galtonia *** = ***Ornithogalum*** (Hyacinthaceae) Decne.
For Sir Francis Galton (1822–1911), British anthropologist, traveller and explorer, geographer, meteorologist, inventor, psychometrician and statistician, cousin of Charles Darwin, founder

of the science of eugenics and pioneer of fingerprinting. His many interests included psychology, genetics and heredity, and biology. He coined the phrase 'nature versus nurture', devised the first weather map, founded the biometric approach to genetics, explored much of Central Africa, and explored unknown areas of South-West Africa (now Namibia) in 1850. He authored *Narrative of an Explorer in Tropical South Africa, The Art of Travel, Hereditary Genius* and others. He published over 340 papers and books, was a Fellow of the Royal Society and was knighted in 1909.

Gamochaeta * (Asteraceae) Wedd.
Gk. *gamos* = united, joined, union; *khaite* = bristle, loose and flowing hair; alluding to basally connate pappus bristles.

Gamolepis (Asteraceae) Less.
Gk. *gamo-* = united, joined; *lepis* = scale. In some species the involucral bracts are fused into a toothed cup.

Garcinia (Clusiaceae) L.
For Laurent Garcin (1683–1751/1752), French-born army physician and botanist who worked as a doctor, although not fully qualified, in a Dutch regiment for 16 years visiting Spain, Portugal and Flanders (Bridel). He became chief surgeon of a ship of the Dutch East India Company and made three voyages to the East Indies between 1720–1729, visiting Arabia, Persia, India, Ceylon and Malaysia, collecting for Herman Boerhaave, and for García de Orta on the way. Back in Europe he qualified as a physician at the Reims Academy and practised medicine in Neuchâtel until the end of his life. He wrote a miscellany of papers, was a correspondent of the Academy of Sciences of Paris (1731) and kept up correspondence with some of the greatest scholars of his time. The *CRC World Dictionary of Plant Names* also attributes this epithet to Garcia de Orta, Portuguese physician who died in 1570. Many sources give his death date as 1751, but Tropicos and the Harvard Herbaria database give it as 1752.

Gardenia (Rubiaceae) J.Ellis.
For Alexander Garden (1730–1791), Scottish doctor, botanist and zoologist who lived in Charleston, South Carolina, the United States, where he practised for over 30 years. He used his spare time to study plants and living creatures, sending parcels of birds, fish, reptiles, amphibia,

insects and plants to Linnaeus and others. His support of England in the American Revolutionary War resulted in the confiscation of his property.  The plant name assigned to him by Linnaeus is a South African genus, although he never visited South Africa. He was a Fellow of the Royal Society of London (1773), a founder Fellow of the Royal Society of Edinburgh (1783) and on his return to England became the Royal Society's vice-president.

Garuleum (Asteraceae) Cass.
Possibly a misspelling of La. *caeruleum* = dark blue; referring to the deep blue colour of the marginal (ray), female florets.

Gasoul*** = **Mesembryanthemum** (Aizoaceae) Adans.
Possibly derived from the Arabic, *ghasul* = possibly a form of clay. Forrskål states that several plants were burnt for the *barilla* = soda ash (sodium carbonate) they contained, such as *Mesembryanthemum geiticulatim* and *nodiflorum*, both of which are called ghasul (rhassoul, ghassoul).

Gasteria (Asphodelaceae) Duval

(GAN)

Gk. *gaster* = abdomen, belly. The plant is named for its stomach-shaped flowers with swollen stems or the swollen base of the perianth tube (WPU Jackson) or the spikelets (Davesgarden.com).

Gastridium* (Poaceae) P.Beauv.
Gk. *gaster* = belly; *-idion* = diminutive; referring to the basally swollen glumes.

Gastrodia (Orchidaceae) R.Br.
Gk. *gastrodes* = pot-bellied; referring to the bell-shaped flowers (WPU Jackson) or swollen tuber of this orchid (Davesgarden.com).

Gaura* (Onagraceae) L.
Gk. *gauros* = majestic, splendid, superb; referring to the strikingly beautiful flowers of some species.

Gazachloa*** = **Danthoniopsis** (Poaceae) J.B.Phipps.
Gaza, short for Gazaland, the historical name for the region in southeastern Africa, Mozambique and Zimbabwe; *chloa* = grass; referring to the area where this grass was found.

Gazania (Asteraceae) Gaertn. (Images on opposite page)
Gk. *gaze, gaza* = riches, royal treasure; *chloe* = grass; or possibly after Theodorus Gaza (many spellings of this name) (1398–1478), a Greek scholar who moved to Italy in 1430. He became professor in Greek at the University of Ferrara (1447) and a Greek-Latin translator for Pope Nicholas V (1450–1455). He worked for King Alfonso V of Aragon (Alphonso the Magnanimous) (1456–1458) and subsequently for Cardinal Bessarion. He translated many works including Aristotle's *Problemata, De Partibus Animalium,* and *De Generatione Animalium* and Theophrastus' *Historia Plantarum,* works by noted Greek authors, and a Greek grammar (four books). He is regarded as one of the greatest classical scholars and humanists of the Renaissance.

Geigeria (Asteraceae) Griess.
For Philipp Lorenz Geiger (1785–1836), German chemist, pharmacist and professor of pharmacy  at the University of Heidelberg. In 1835 he discovered the poisonous alkaloid coniine in hemlock (Conium). He also isolated atropine, an alkaloid found in nightshade (*Atropa belladona*), jimsonweed (*Datura stramonium*) and mandrake (*Mandragora officinarum*), and the related alkaloids aconatine, daturine, hyoscyamine and atropine. From 1824–1836 he edited the *Magazin der Pharmazie*. His major works were the *Pharmaco-poeia Universalis* and his *Handbuch der Pharmacie*.

Gelsemium (Loganiaceae) Juss.
Latinised form of the Italian word for Jasmine =

Gazania: These stunning daisies make great borders in sunny, water-wise gardens. (MC)

gelsomino. The reason for calling the plant this name is not clear.

Geissoloma (Geissolomataceae) Lindl. ex Kunth
Gk. *gelsson* = eaves of a roof, tiles, or hem of garment; *loma* = fringe. The leaves may be minutely 'fringed' when young, and the four perianth segments are imbricate, like tiles on a roof.

Geissorhiza (Iridaceae) Ker Gawl.
Gk. *geisson* = title; *rhiza* = root; alluding to the regular overlapping of the corm tunics in some species.

Gelonium = Suregada (Euphorbiaceae) Willd.
Etymology uncertain. According to Georg Christian Wittstein, this is an East Indian name. Backer derives it from Gk. *gelon* = sparkling; referring to the glossy green leaves.

Gelseminum*** = Tecoma (Bignoniaceae) Kuntze
La. *gelsemino* = Italian for Jessamine, Jasmine. The name is derived from the Persian name 'Yasmin' meaning 'God's gift' or 'Gift from God'.

Gemmaria = Strumaria (Amaryllidaceae) U. & D.Mül-Doblies
La. *gemma* = bud or eye of a plant, or a jewel; *-aria* = pertaining to; referring to the beauty of some of these species such as the (now) *Strumaria ammocharis*.

Gendarussa*** = Justicia (Acanthaceae) Nees
Ambonese Malay *rusa* = deer. The meaning of the name is unclear. Perhaps the deer fed on this plant.

Geniosporum*** = Platystoma (Lamiaceae) Wall. ex Benth.
Gk. *genio* = related to chin or beard; *spora* = seed; alluding to the corolla protruding from the throat of the calyx.

Genista (Fabaceae) L.
La. *genista* = the Latin name for the broom shrub, possibly from Celtic *gen* or French *genet* = bush. Invasive plants with wide-branching root systems and thick tenacious stems. Because of their brush-like foliage, many of this species were used for sweeping in the 15th century.

Genlisea (Lentibulariaceae) A.St.-Hil.
For Stéphanie Félicité Ducrest de Saint-Aubin, Comtesse de Genlis (1746–1830), author of more than 80 works – historical novels, romances, and prose and poetical compositions. Her *La botanique Historique et Literature* (1810) discusses the traits, anecdotes, superstitions, celebrities, ceremony, etc. surrounding flowers. She lived through the French Revolution (1789) and was sympathetic to it but, with the fall of the Girondins (a political faction) in 1793, was compelled to take refuge in Switzerland, then Berlin and Hamburg. Her husband, from whom she separated in 1782, was guillotined. She only returned to France in 1799. Napoleon gave her a pension, but Louis XVIII withdrew it, and she had to live on the earnings from her writing.

Geochloa (Poaceae) H.P.Linder & N.P.Barker
Gk. *geo-* = earth; *chloa* = grass; possibly referring to its widespread occurrence.

Geophila* (Rubiaceae) D.Don
La. *geo* = earth; *philo* = loving, therefore earth-lover, ground-lover; referring to its creeping habit.

Georgeantha* (Anarthriaceae*) B.G.Briggs & A.S.Johnson
For Alexander Segger George (1939–), Australian

botanist and botanical historian. He started his career as a laboratory assistant in 1959 at the Western Australian Herbarium and studied botany at the University of Western Australia. In 1968, he was seconded as Australian botanical liaison officer at the Royal Botanical Gardens, London. An authority on the genera *Banksia* and *Dryandra,* he wrote, *inter alia, The Banksia Book* (1984) and *An Introduction to the Proteaceae of Western Australia* (1985) as well as popular wildflower books on Western Australia and historical biographies. From 1981–1993 he was executive editor for the *Flora of Australia* series. He also served as an adjunct associate professor with the School of Biological Sciences, Murdoch University. In 2012 he was awarded the Order of Australia for 'service to conservation and the environment as a botanist, historian and author'.

Geosiris* (Iridaceae) Baillon.
Gk. *geos* = earth; Iris (q.v.) referring to the Iris family of plants.

Geraniospermum* = Pelargonium** (Geraniaceae) Kuntze
Gk. *geranos* = a crane; *sperma* = seed. See *Geranium* below.

Geranium (Geraniaceae) L.
Gk. *geranos* = a crane. The seed pod resembles a crane's head and beak.

Gerardia (Orobanchaceae) L.
For John Gerard (1545–1607), English botanist, surgeon and author. For some 20 years he supervised gardens of wealthy aristocrats such as those of Lord Burleigh in the Strand, and at Theobalds in Hertfordshire. He also developed his own famous garden in Holborn, then the most fashionable district of London. In 1596, he published a catalogue, *Catalogus Arborum, Fruticumi*, of 1 033 plants and trees that he cultivated in his garden, the first complete catalogue ever published of the contents of a single garden. He is best known as the author of *The Herball or Generall Historie of Plantes* (1597), including the first illustration of a potato.

Gerardiina (Orobanchaceae) Engl.
Gerardia (q.v.); *-ina* = resemblance.

Gerardiopsis = Anticharis (Scrophulariaceae) Engl.
Gerardia (q.v.); Gk. *-iopsis* = resembling, likeness.

ABOVE: *Geranium* (EJM)

LEFT: *Geranium*: The beak-like column looks like the bill of a crane. (EJM)

Gerbera (Asteraceae) L.
For Traugott Gerber (1710–1743), German medical doctor, naturalist and explorer. He registered as a medical student at the University of Leipzig in 1730 and obtained a doctorate for his thesis, *De Thoracibus*, in 1735. Between 1739 and 1741 he led several expeditions on the Don and Volga rivers to search for medicinal plants and herbs and served as curator of the oldest (medical-pharmaceutical) botanical garden in Moscow from 1735–1742. He served in the Russian army in Finland in 1742. He was the author of *Dissertationem Physicam de Plantarum Transpiratione* and was a close friend of Swedish botanist Linnaeus, who published the genus *Gerbera* in 1758. Some sources also include his brother Fr. Gerber, who collected plants in the West Indies, in the commemoration.

Gerdaria = Sopubia (Orobanchaceae) C.Presl
Derivation unknown, but probably an anagram of of *Gerardia* (q.v.) as is *Graderia* (q.v.).

Germanea* = Solenostemon** (Lamiaceae) Lam.
For Jean-Joseph de Saint-Germain (1719–1791), French bronze caster and clock maker, artist, amateur botanist and distinguished cultivator of plants who grew a number of very rare plants in his garden. Jean-Baptiste Lamarck named this genus after Saint-Germain in 1788, shortly after the plant had bloomed for the first time. Later its name was changed to *Plectranthus*. Saint-Germain published *Manuel des Végétaux ou*

Catalogue Latin et François de Toutes les Plantes, Arbres et Arbrisseaux Connus sur le Globe de la Terre Jusqu'à ce Jour (1784). This book was arranged according to the Linnaean system, by classes, orders, genera and species, naming places where (some of them) grew around Paris, presumably in botanical gardens. The catalogue is in Latin and French.

Geropogon (Asteraceae) L.
Gk. *geron* = old man; *pogon* = beard; alluding to the long hoary down of the seed.

Gerrardanthus (Cucurbitaceae) Harv. ex Hook.f.
For William Tyrer Gerrard (c 1831–1866), English naturalist and traveller, plant, bird and insect collector with a special interest in ferns. He came to Natal, South Africa, in 1860/1861 and collected alone or with Mark Johnston McKen (1823–1872), horticulturalist and collector, along the south and north coast of Natal and inland to Ladysmith, Greytown and Zululand. He found many new genera and over 150 new species. He left Natal for Madagascar in 1865, having already suffered bouts of yellow fever, collapsed shortly after arriving with repeated attacks of fever and died late 1865 or early 1866. With McKen he published *Synopsis Filicum Capensium* (published 1870). He also collected in Australia.

Gerrardina (Achariaceae) Oliv.
Gerrardanthus (q.v.); *-ina* = resembling or possessing.

Gethyllis (Amaryllidaceae) L.
Possibly Gk. *getheo* = I rejoice; *ullus* = diminutive, but most sources say from *gethyon* = a bulb, onion or species of leek. The bulbs of this genus are somewhat similar to those of the leek.

Geum (Rosaceae) L.
Gk. *geum* = to taste well, possibly derived from *geyo* (*geuo*) = to impart a pleasant flavour; referring to the roots of the species.

Gibbaeum (Aizoaceae) (Haw.) N.E.Br.
La. *gibbus* = a hump on the back; referring to the thick lumpy shape of the succulent leaves.

Gibbaria = *Osteospermum* (Asteraceae) Cass.
La. *gibbus* = a hump; and *-aria* = pertaining to; perhaps referring to the keeled, kidney-shaped achenes (small, dry fruit).

Gibbaeum (CM)

Gigasiphon (Fabaceae) Drake
Gk. *gigas* = giant; *siphon* = tube. The tube of this flower is very large.

Girardinia (Urticaceae) Gaudich.
For Jean Pierre Louis Girardin (1803–1884), French chemist and agronomist and professor of chemistry at the School of Agriculture at Reun (1828–1857). He was dean and professor of chemistry at the Faculty of Lille (1857–1862), author of numerous works on the application of chemistry in agriculture such as *Note on the Use of Bones Crushed or Ground as Fertilizer* (1831), *Memoirs of Chemistry Applied to Industry, Agriculture, Medicine and Home Economics* (1839), *Considered Manure as Fertilizer* (1844) and *Short Instruction on the Use of Salt in Agriculture* (1849). He wrote an early work on volcanoes and other studies, was director of the School of Applied Sciences and the School of Industrial Arts and Mines, and was a member of the Academy of Medicine.

Gisekia (Gisekiaceae) L.
For Paul Dietrich Giseke (1741–1796), German botanist, physician, librarian and academic. He studied at the Academic Gymnasium in Hamburg and at the University of Göttingen and graduated in 1764. He went for an extended trip through France and Sweden, where he met Linnaeus, becoming his student and a close friend in Linnaeus's later years. Back from abroad he settled in Hamburg and started his practice as a physician but in 1771 started became professor

of physics and discourse at the Academic Gymnasium in Hamburg (later the University of Hamburg). He authored a number of books and was a contributor/editor of Linnaeus's *Praelectiones in Ordines Naturales Plantarum* (*Lectures in the Natural Order of Plants*) (1792).

Gladiolus (Iridaceae) L. (Images on opposite page)
La. *gladiolus* = a small sword; referring to the sword-like shape of the leaves.

Glandularia (Verbenaceae) J.F.Gmel.
La. *glandula* = having small glands; *-aria* = pertaining to, possession.

Glaucium* (Papaveraceae) Mill.
Gk. *glaukos* = grey-green, glaucous; referring to the colour of the leaves.

Gleditsia* (Fabaceae) L.
For Johann Gottlieb Gleditsch (Gleditsius) (1714–1786), German botanist and author on forestry. He obtained his PhD from the University of Frankfurt/Oder (1742) and taught botany, physiology and medical botany. In 1746 he moved to Berlin and became a professor of forestry and botany at the Collegium Medico-chirurgicum and became director of the Berlin botanical garden (1746).

He was the author of *Systematische Einleitung in der Neuere Forstwissenschaft* (*Systematic Introduction to the New Forestry*) (1774–1775). He belonged to the German Anti-Linnaean 'school' and developed a 'natural' system to describe plants, *Système des Plantes, Fondé sur la Situation et la Liaison des Etamines* (*System of Plants, Based on the Situation and the Binding of Stamens*) (1751).

Gleichenia (Gleicheniaceae) Sm.
For Wilhelm Friedrich Freiherr von Gleichen-Russwurm (sometimes recorded as Wilhelm Friedrich von Gleichen) (1717–1783), German lieutenant-colonel, botanist and amateur naturalist. He resigned from the military and court service in 1756 and devoted himself to microscopic examinations. One of his works, *New Discoveries in the Vegetable Kingdom from Microscopical Observations*, reflects the growing interest in microscopy as a tool of investigation and a means of categorising the natural world. Among his books (abbreviated titles) were *The Latest from the Kingdom of Plants*, and *Microscopic Studies and Observations* (1764), *History of the Common House Fly* (1764), *Microscopic Observations of the Seeds of Animals and Different Infusions* (1778) and *Exquisite Microscopic Discoveries in Plants, Flowers and Insects* (1777).

Glekia (Scrophulariaceae) Hilliard
For Georg Ludwig Engelhard Krebs (1792–1844), German apothecary, one of the most prolific botanical and zoological collectors of his era, naturalist to the King of Prussia. He came to South Africa in 1817. During the next 21 years he went on 14 expeditions, mainly in the Eastern Cape, and to the north as far as the Orange River, and also to Natal. He also went to Madagascar, Mauritania and Réunion.

He sent his collections to Martin Heinrich Karl Lichtenstein, director of the Berlin Natural History Museum – his 12th consignment contained 7 245 dried plant specimens, a quagga (now extinct), a rhinoceros, an elephant, 900 birds, over 7 000 insects, and a preserved 'Bushman' (Khoisan). He was recommended to receive the Prussian Order of the Red Eagle for meritorious contributions to science but died before the award could be made.

Glia* (Apiaceae) Sond.
Name from *gli*, the plant is so called by the 'Hottentots' (Sonder) or Khoisan, who prepared from the roots an inebriating liquor providing great glee (WPU Jackson's pun).

Glinus (Molluginaceae) L.
Gk. *glinos* = sweet juice, maple, used by Theophrastus. Some species are eaten for food; others have herbal properties.

Glischrocolla (Penaeaceae) (End) A.DC.
Gk. *glischro* = gluey, sticky; *kola* = a glue. Possibly because the plant exudes a natural gummy adhesive.

Globba* (Zingiberaceae) L.
Derived from an Indonesian word for the plants, 'galoba'.

Globularia (Plantaginaceae) L.
La. *globulus*, diminitive form of *globus* = round head, sphere; *-aria* = possessing; referring to the rounded shape of the flower heads.

Globulariopsis (Scrophulariaceae) Compton
Globularia (q.v.); Gk. *-iopsis* = resembling, likeness. The flowers are in dense, more or less globose (globe-shaped) spikes.

Gladiolus species have been used in horticulture and the cut-flower trade for many years. (CM, except where indicated)

Globulea* = **Crassula** (Crassulaceae) Haw.
La. *globulus* = a globule or small globe; alluding to the waxy globules with which the petals are tipped.

Gloriosa = **Littonia** (Colchicaceae) L.

The Flame lily is the national flower of Zimbabwe. (EJM)

La. *gloriosus* = glorious; referring to the colours and shape of the flowers.

Glossochilus (Acanthaceae) Nees
Gk. *glossa* = tongue; *cheilos, chilos* = lip; referring to the bilabiate corolla.

Glossostephanus = **Oncinema** (Apocynaceae) E.Mey.
Gk. *glossa* = tongue; *stephanus* = garland, wreath; referring to the inflorescence, specifically the five-lobed, tongue-shaped petals.

Glossostylis* = **Alectra** (Orobanchaceae) Cham. & Schltdl.
Gk. *glossa* = tongue; *stylis* = style. 'The style is produced [...] into a long beak-like point acutely bifid at the apex' (Brown in *Flora Capensis*).

Glottiphyllum (Aizoaceae) Haw. ex N.E.Br.
Gk. *glotta* = tongue; *phyllon,* leaf; referring to the tongue-like leaf.

Gloveria (Celastraceae) Jordaan
For Ruth Glover (later Ruth Wordsworth) (f 1908–1925), who worked on the staff of the Bolus Herbarium for Harry Bolus from 1908–1914 and did work for the South African museum, including joining the Percy Sladen memorial expedition of 1910–1911, which visited

Khamiesberg, Giftberg and Oliphants River Mountains. She and Miss E Stephens worked in the valley of the Oliphants River in the vicinity of the Warm Baths springs, and the latter wrote up the species found in that area. She moved to Bulawayo, Southern Rhodesia (now Zimbabwe) after her marriage and became Ruth Wordsworth.

Glumicalyx (Scrophulariaceae) Hiern
La. *gluma* = husk. The author writes that the calyx is glumaceous (resembling a glume).

Glyceria (Poaceae) R.Br.
Gk. *glykeros* = sweet (sweet grass); alluding to the sweet herbage and seeds of *G. fluitans*.

Glycine* (Fabaceae) L.
Gk. *glykys* = sweet; referring to the sweet roots and leaves of some species of the soya bean genus.

Glycyrrhiza* (Fabaceae) L.
Gk. *glykys* = sweet; *rhiza* = root; referring to the sweet root (licorice) that has been used in food and medicine for thousands of years.

Gmelina (Lamiaceae) L.
For Johann Georg Gmelin (1709–1755), German naturalist, botanist and geographer who obtained

a medical degree at age 18, a Fellowship at the Academy of Sciences at 19 and was appointed professor of chemistry and natural history at the University of St Petersburg at 21. From 1731–1742 he explored much of the Urals and Western Siberia. His *Flora Sibirica* describes 1 178 species, 294 illustrated.
In 1747 he became professor of medicine at the University of Tübingen and director of the university's botanical gardens. He was elected a foreign member of the Royal Swedish Academy of Sciences in 1749.

Gnaphalium (Asteraceae) L.
Gk. *gnaphalion* = a flock of wool, downy-like. The plant is hairy all over, like a plant whose soft white leaves are used as cushion stuffing.

Gnidia (Thymelaeaceae) L.
Derivation uncertain. Linnaeus only states 'habitat in Aethiopa', Africa, where it is widely

Usually some of the first plants to flower post grassland fires (CM)

distributed. Possibly *Gnidia* was named after a Greek city, Knidos, where a kind of laurel grew, or Cnidus in Caria (modern Turkey) (Hugh Glen). Another possibility is that it could be a Greek word for Daphne or laurel; in Greek mythology, Daphne was a pretty nymph who was turned into a laurel bush (WPU Jackson). It might also have been named after Knossos in Crete (spelled Knidiossos in one version), with the G being substituted for K.

Gomphocarpus (Apocynaceae) R.Br.
Gk. *gomphos* = a peg, nail, club; *karpos* = fruit; referring to the fruits that are swollen and egg-shaped with tapering tips (John Manning).

Gompholobium* (Fabaceae) Sm.
Gk. *gomphos* = a peg, club; *lobion* = diminutive of *lobos* = lobe, pod, capsule; referring to the club-like seed pod.

Gomphostigma (Buddlejaceae) Turcz.
Gk. *gomphos* = club; *stigma*; referring to the club-shaped stigma.

Gomphrena* (Amaranthaceae) L.
Gk. *gomphos* = a peg, club. An ancient classical name used by Pliny the Elder for *Amaranthus tricolor*, a plant similar to *Gomphrena*; probably in allusion to the shape of the flowers.

Gonatopus (Araceae) Hook.f. ex Engl.
Gk. *gonion* = angle, knee joint; *-pus* = foot (*pous* = a foot); alluding to the 'knee-like swelling on the petiole' (Hooker).

Gongrothamnus = Distephanus (Asteraceae) Steetz.
Gk. *goggros* = a knob; *thamnos* = a shrub. Lumpy glands at the base of the petiole remain when the leaves fall, giving leafless stems a knobbly aspect.

Gonioma (Apocynaceae) E.Mey.
Gk. *gonio* = angular, an angle; La. *-ma* = a suffix that may indicate the result of an action (Stearn). The fruits stick out at right angles to the stalk.

Goniopteris = Cyclosorus (Thelypteridaceae) C.Presl
Gk. *gonia* = an angle; *pteris* = fern; referring to the veins meeting at sharp angles.

Gonostemon** (Apocynaceae) Haw.
Gk. *gonia* = an angle; *stemon* = stamen; possibly referring to the positioning of the stamen at an acute angle.

Gorskia = Capaiba (Fabaceae) Bolle.
For Stanislaw Batys Górski (1802–1864), Polish botanist, entomologist, pharmacist and doctor.

He studied medicine and natural sciences at Vilnius University, graduating in 1825, conducted research on flora in the Polish Bialowieza Forest (1826) and became head of the Botanical Garden of Vilnius University (1829–1832), during which time he compiled a catalogue of plants in the garden. From 1832–1841 he was assistant professor of botany and pharmacology at the Medical-Surgical Academy in Vilnius. He authored works on botany, entomology and zoology. In 1842 Vilnius University was closed by the Russian authorities, and the bulk of Górski's natural history collections was transferred to Kiev; he saved a part of his private collection of insects and herbals.

Gorteria (Asteraceae) L.
For David de Gorter (1717–1783), Dutch botanist, physician, plant collector, professor of medicine at

the University of Harderwijk, where he also studied, and possibly also for his father, Johannes de Gorter (1689–1762), professor of medicine at Harderwijk from 1725–1754. De Gorter Jr graduated in 1734. In 1735 Linnaeus came to Harderwijk where he obtained his PhD (under De Gorter Sr). A friendship sprang up between the younger De Gorter and Linnaeus. Together they made collecting trips around Harderwijk, with De Gorter subsequently authoring one of the first floras to use Linnaeus's form of

binomial nomenclature, *Flora Belgica* (1767). After retiring from the university, De Gorter became chief physician at the court of Elizabeth (Elizaveta Petrovna), Empress of Russia, succeeded by his son.

Gosela (Scrophulariaceae) Choisy
Derivation not stated in original publication, presumably an anagram of *Selago*.

Gossypioides (Malvaceae) Skovst. ex J.B.Hutch.
La. *gossypium* = cotton; Gk. *-oides* = resembling (from the Arabic *goz,* a soft substance). Plants of this genus resemble and are related to cotton (*Gossypium* spp.).

Gossypium (Malvaceae) L.

(EJM)

La. *gossypium* = cotton, possibly from the Arabic *goz* = silky. A name used by Pliny the Elder to describe cotton.

Graderia (Orobanchaceae) Benth.
Apparently an anagram of *Gerardia* (q.v.). Another such anagram is *Gerdaria*.

Grammanthes*** = Crassula (Crassulaceae) DC.
Gk. *gramma* = a line; *anthos* = a flower; referring to the V-shaped mark on the petals (does not occur in all species).

Grammatotheca (Lobeliaceae) C.Presl
Gk. *gramma* = a line; *theka* = a case, capsule; referring to the elongate cylindrical capsule.

Grammitis (Polypodaceae) Sw.
La. *gramma* = a line; *-itis* = likeness, close connection; alluding to the parallel lines of sori on each side of the midrib – the sori are often elongate or coalescing.

Grangea (Asteraceae) Adans.
Possibly for Nicholas Tourtechot-Granger (c 1680–1734), French physician and botanist-gardener who worked at a Christian hospital in Tunis for several years and who had an interest in natural history. He returned to France in 1728, then in 1731 accompanied his friend Jean Pierre Pignon to Egypt, where the latter became French consul in Cairo. He travelled up the Nile to Aswan and documented monuments, plants, animals and minerals. In 1732 he was commissioned by King Louis XV to travel around the Middle East to gain natural history information. He visited Crete, Egypt again, Cyprus, Palestine and Syria but died en route two days' march from Basra, either in 1734 or 1737. He was the author of *Relation du Voyage Fait en Egypte, par le Sieur Granger, en L'année 1730*, published in 1745. Derivation uncertain. He was also honoured with the genus *Grangeria* in the family Chrysobalanaceae.

Graphorchis = Eulophia (Orchidaceae) Thouars
Gk. *graphis* = brush, pencil; *orkis* = orchid; referring to the 'pencil' or 'brush' strokes on the inner petals of some species.

Gratiola (Plantaginaceae) L.
Diminutive of *gratia* = grace; referring to the plant's supposed healing properties.

Greenwayodendron* (Annonaceae) Verdc.
For Percy James Greenway (1897–1980), South African-born botanist who became an East African botanical expert. He was educated at the East Anglican Institute of Agriculture and Royal Horticultural Society (1912–1916). He worked for Kew Gardens in various capacities from 1919–1924 and then moved to the Imperial Forest Institute, Oxford. In 1928, he was appointed botanist in charge of the East African Agricultural Research

Station at Amani, Tanganyika (Tanzania) (1927–1950) and later at Nairobi, where he retired in 1958. During his career he collected in Kenya, Zambia, and Zimbabwe and was said to be able to distinguish over 2 000 Serengeti plants. He was awarded an OBE in 1951 for his work at the East African High Commission and an honorary DSc degree from the University of the Witwatersrand in 1951. He co-authored *Kenya Trees and Shrubs* (1961) with Ivan Robert Dale (1904–1963).

Grevea (Montiniaceae) Baill.
For a Monsieur H Grevé (?–1895), a French plant collector who discovered the type species on the banks of the Morondava River in Madagascar, probably in 1883, and gave it to Henry Ernest Baillon, French botanist and professor of natural history, who named it *Grevea madagascariensis* in his honour in 1884. For more than 20 years, Grevé collected plants, fossils and other natural history objects for Alfred Grandidier, French naturalist and explorer. Grevé settled at Morondava and married the daughter of a Sakalava chief, raised cattle and crops and served as representative for the Messageries Maritimes, a French merchant shipping company. In 1895, during the second Franco-Hova War against the island's Merina aristocrats, seeking to restore property confiscated from French residents, Grevé was taken prisoner by Hova soldiers and shot on orders of the Hova military government.

Grevillea* (Proteaceae) R.Br. ex Knight
For Charles Francis Greville (1749–1809), British antiquarian and collector with a special interest in minerals and precious stones. He was a keen horticulturist, Fellow and vice-president of the Royal Society, a Fellow of the Linnaean Society, and a co-founder in 1804 of the Horticultural Society of London (now called the Royal Horticultural Society)

with his close friend Sir Joseph Banks and five others. He introduced and grew many rare plants, 14 of which were illustrated in *Curtis's Botanical Magazine*. When his father died in 1773, his brother became Earl of Warwick, and Charles Greville inherited his seat of Warwick in the House of Commons and became a lord of the Admiralty.

Grewia (Malvaceae) L.
For Nehemiah Grew (1641–1712), British botanist and physiologist, physician, microscopist, known as 'the father of plant physiology'. He graduated from the University of Cambridge in 1661 and read for a medical degree from Leiden University

in 1671. From 1672 he practised as a physician. Grew published a number of works, including *Idea of a Phytological History* (1673), and his major work, *Anatomy of Plants* (1682). Much of Grew's pioneering physiological work was done with the microscope, as was the research by his Italian contemporary Marcello Malpighi, who sent botanical and zoological works to the Royal Society in 1675 and 1679, so there was a likely exchange of information. Grew was a Fellow of the Royal Society and at one time its secretary.

Greyia (Melianthaceae) Hook. & Harv.

The bright scarlet flowers appear before the leaves, making this a stunning small tree. (EJM)

For Sir George Grey (1812–1898), English soldier, explorer, governor, politician and patron of botany. He entered the Royal Military College, Sandhurst, in 1826, becoming a captain in 1839. From 1837–1839 he explored part of Australia, as described in his two-volume *Journal of Two Expeditions of Discovery in North-West and Western Australia.* Grey became governor of South Australia (1841–1845), of New Zealand (1845–1853) and (1861–1868), and of the Cape Colony (1854–1861). Grey was a good linguist and keen naturalist who ceaselessly collected geological and biological specimens for the Kensington Museum and Kew Gardens. He strongly supported William Henry Harvey's project to describe the flora of South Africa, and Volume 1 of *Flora Capensis* is dedicated to him.

Grielum (Neuradaceae) L.
Gk. *grelos* = old; referring to the hoary leaves (William Henry Harvey, *Flora Capensis*).

Grisebachia = ***Erica*** (Ericaceae) Klotzsch
For August Heinrich Rudolph Grisebach (1814–1879), German botanist, phytogeographer  (i.e. a person who studies the geography of plant distribution), professor on the university medical faculty at Göttingen, pioneer in plant systematics, plant collector and taxonomist, Fellow of the Linnaean Society, author of *Genera et Species Gentianearum* (1838), *Die Vegetation der Erde* (1872) and *Flora of the British West Indian Islands* (1864), and director of the Botanical Garden of Göttingen. He followed in the footsteps of his uncle (a professor of botany), explored the island of Cuba, travelled extensively throughout the Balkan region and worked on the botany of South America.

Grossera (Euphorbiaceae) Pax
For Wilhelm Carl Heinrich Grosser (1869–1942), German botanist. He contributed to *Das Pflanzenreich* by Adolf Engler. La. *grossus* = thick.

Grubbia (Grubbiaceae) P.J.Bergius
For Michael (Mikael) Grubb (af Grubbens) (1728–1808), Swedish botanist, mineralogist, merchant, botanical collector.  He graduated with a PhD from Åbo Academy (later Helsinki University) (1748) and worked in Guangzhou, Canton (1749–1755), founding a branch there of the Swedish East India Company. He visited the Cape in 1764 and collected specimens, many bought from Johann Andreas Auge (q.v. *Augea*) and others, which he presented to Peter Jonas Bergius (1730–1790). This collection formed the basis of Bergius's *Descriptiones Plantarum ex Capita Bonae Spei Plantae Capenses* (1767), a flora of the Cape Province. He became a director of the Swedish East India Company (1766–1769) and was elected a member of the Swedish Royal Academy of Sciences in 1767 and knighted in 1768, when he took the name Af Grubbens.

Grumilea *** = ***Psychotria*** (Rubiaceae) Gaertn.
La. *grumula* = a small heap. The albumen of the seed is grainy.

Guajacum ** (Fabaceae) L.
Spanish *guayaco* = yielding guaiac resin and guaiacol (orthomethoxyphenol), which are used in medicinal products.

Guatteria (Annonaceae) Ruíz & Pav.
For Giovanni Battista Guatteri (1739–1793), Italian botanist, Jesuit abbot, professor of botany, founder of the New Botanical Garden of Parma. After 10 years' study at the universities of Parma and Padua, he became professor of botany at Parma (1769). Guatteri revitalised botanical teaching by focusing on the direct study of plants instead of Greek and Arabic texts, used the most modern botanical texts by Gómez de Ortega (which he translated) and Bodoni, adopted the new classification system of Linnaeus and, over years, completely renovated the university's botanical garden to make it among the best in the world. He was also inspector of mines and fossils for the Duchy, gave advice on scientific land management and promoted the artificial cultivation of mushrooms.

Guettarda (Rubiaceae) L.
For Jean-Étienne Guettard (1715–1786), French physician, naturalist, botanist and geologist (mineralogist). He was  curator of the natural history collection of the French scientist René de Réaumur (1741), a member of the Faculty of Medicine of Paris (1742) and one time *médecin botaniste* to the French Prince Louis, Duc d'Orléans (1747–1752). He was the first to map France geologically, his *Atlas et Description Minéralogiques de la France* (1780) showing the mineralogical distributions of much of Europe. He was one of the first scientists to notice the relationship between the distribution of plants and the soils and subsoils. During his research he discovered (1765) Kaolin clay in Alençon, which resulted in the production of Sèvres porcelain.

Guibourtia (Fabaceae) Benn.
For Nicholas Jean Baptiste Gaston Guibourt (1790–1861), French pharmacologist and  professor of pharmacology at the Ecole de Pharmacie, Paris from 1832, and member of the editorial board of the *Journal of Pharmacy and Chemistry* from 1846. He authored *Histoire abrégée des Drogues Simples* (*A History*

of Plants Used for Extracting Drugs) (1869) and *Manuel Légal des Pharmaciens et des Élèves en Pharmacie* (*Legal Handbook for Pharmacists and Pharmacy Students*) (1852). He was a founder and populariser of the study of *materia medica* – that is, homeopathic medicines – in France. He published extensively on exotic woods and natural dyes and was appointed a member of the National Academy of Medicine in 1824.

Guilandina (Fabaceae) L.
For Melchior Wieland (Latinised as Guilandinus) (c 1520–1589), Prussian naturalist, traveller and lecturer (professor) and demonstrator of medicinal herbs at Padua, and *praefectus* (director) of the Botanical Garden of Padua. After studying in Königsberg and Rome, he was commissioned by the University of Padua to travel to Greece, Syria, Asia, Palestine and Egypt to collect plants. During his travels he was captured by pirates – Gabriele Fallopio, chair of botany at Padua paid the 200 scudi ransom – and later shipwrecked before returning to Italy. In 1561, he succeeded Anguillara as second director of the Botanical Garden at Padua. He featured in Alpino Prospero's *De Medecina Aegyptiorum* and wrote *Epistolae de Stirpibus* (1558), *Papyrus* (1572), and *Synonyma Piantarum* (published posthumously in 1608).

Guilleminea* (Amaranthaceae) Kunth
For Jean Baptiste Antoine Guillemin (1796–1842), French botanist, traveller and author. He studied with Jean Pierre Étienne Vaucher (1763–1841) and Augustin Pyramus de Candolle (1778–1841) and later became curator of the herbarium and library of botanist Jules Paul Benjamin Delessert (1773–1784). In 1827 he worked as an *aide-préparateur* at the Muséum National d'Histoire Naturelle and in 1834 he succeeded Adolphe Brongniart (1801–1876) as an assistant naturalist to the chair of botany. He was co-author with Achille Richard and George Samuel Perrottet of a work on the flora of Senegambia (present-day Senegal and Gambia) titled *Florae Senegambiae Tentamen* (1830–1833).

Guizotia* (Asteraceae) Cass.
For François Pierre Guillaume Guizot (1787–1874), French politician, statesman and historian.

He studied law in Paris but had a fascination with history. During his career he held many senior positions that reflect his versatility and also the volatility of the times: chair of modern history, University of Paris (1812), secretary general of the Ministry of Justice (1815–1816), director in the Ministry of the Interior (1819–1820) and (1830–1832), minister of public instruction (1832), London ambassador, then foreign minister (1840), and premier (1847). He fled to England during the French Revolution (1848), and for the next 20 years devoted himself to historical writing. A scholar of the highest order, he wrote eight major historical works (29 volumes).

Gunillaea (Campanulaceae) Thulin
For Gunilla Thulin, wife of Swedish botanist Mats Thulin (1948–), professor of systematic

botany at Uppsala University and author of this genus published in 1974. Thulin has devoted more than 40 years to the study of African plants, with a particular focus on the flora of the Horn of Africa region. In 1988–2006 he was editor of the four-volume work *Flora of Somalia*, and his discovery in 2006 of the new species *Acacia fumosa*, a tree that dominates the vegetation over large areas in eastern Ethiopia, got much attention. He has authored more than 200 papers, including works on the genus *Wahlenbergia* and *Fabaceae* of Ethiopia and has co-authored, with Kåre and Birgitta Bremer, *Introduction to Phylogeny and Systematics of Flowering Plants* (2003).

Gunnera (Gunneraceae) L.
For Johan Ernst Gunnerus (1718–1773),

Norwegian clergyman, naturalist collector and renowned scholar. He studied at the universities of Copenhagen (1737), Halle (1742) and Jena (1745), became professor of theology at the Copenhagen University (1754) and bishop of Trondheim (1758). He founded Norway's first scientific institution, Det Trondhiemske Selskab (The Trondheim Society), which became the Royal Norwegian Society of Sciences and Letters in 1766. He was the author of *Flora Norvegica* (1766–1776) and was elected a foreign member of the Royal Swedish Academy of Sciences. He discovered many plants, animals, fish and birds not known to science and had extensive correspondence with the Swedish naturalist Carl Linnaeus about them.

Gussonea*** = Solenangis (Orchidaceae) A.Rich.
For Giovanni Gussone (1787–1866), Italian

botanist and physician. He studied medicine at the University of Naples (1811) but abandoned his medical career and instead devoted his life to botany. He became director of the Botanical Garden of Boccadifalco, near Palermo (1817–1827), later director of the Botanical Garden of Caserta, Naples, and in 1861 he became emeritus professor of botany at the University of Naples. He made many expeditions in Italy, collecting for his herbarium some 14 000 Italian plant species, now housed in the Royal Botanical Gardens at Naples. He authored *Catalogus Plantarum* (1821), *Plantae Rariores* (1826), *Florae Siculae Prodromus* (1827–1828), *Flora Sicula* (1829) and *Florae Siculae Synopsis* (1842–1844).

Gutenbergia* (Asteraceae) Sch.Bip.
For Johann (Johannes) Gutenberg (Johann Gensfleisch zur Laden zum Gutenberg) (1400–1468), German goldsmith, printer, publisher and craftsman. He was the first European to develop a method of printing using movable type (which was actually invented in China), together with movable molds and alloys, a special press and oil-based inks. It allowed, for the first time in history, the mass production of printed books and was far better than woodblock printing or handwritten manuscripts. Although his invention was created in 1440, with refinements and increased mechanisation, it remained the principal means of printing until the late 20th century. Gutenberg's best-known book was his 42-line Bible, often called the Gutenberg Bible (c 1455).

Guthriea (Achariaceae) Bolus
For Francis Guthrie (1831–1899), English lawyer, mathematician and botanist. He studied at London University, getting a BA degree (1850) and an LLB (1852) before coming to South Africa in 1861 to teach mathematics at Graaff-Reinet College. There he met Harry Bolus, and they became life-long friends. Guthrie moved to Cape Town, where he practised at the Bar and edited a newspaper before becoming professor of mathematics at the South African College (now the University of Cape Town) from 1876–1898. When Bolus wrote up the *Ericaceae* family for *Flora Capensis*, he invited Guthrie's collaboration. Guthrie's extensive collection of the Cape Peninsula flora is now housed at the Guthrie Herbarium at the University of Cape Town botany department.

Gymnachaena = ***Stoebe*** (Asteraceae) Rchb.
Gk. *gymnos* = naked; *khainein* = to gape. Meaning obscure.

Gymnadenia (Orchidaceae) R.Br.
Gk. *gymnos* = naked; *aden-* = gland; referring to the sticky disk of the pollinia, which are free on each side of the rosettelum, not in a bursicula (R.Brown, 1813).

Gymnanthemum (Asteraceae) Cass.
Gk. *gymnos* = naked, bare; *anthemon* = flower; referring to the bare stalks.

Gymnocarpium* (Cystoperidaceae) Newman.
Gk. *gymnos* = naked; *carpium* = fruit; referring to the lack of an indusium to cover the sporangia.

Gymnema (Apocynaceae) R.Br.
Gk. *gymnos* = naked; *nema* = a thread, filament. The stamen filaments lack corona lobes.

Gymnodiscus (Asteraceae) Less.
Gk. *gymnos* = naked; *diskos* = disk. The receptacle of the flowerhead is flat, naked (nude); the disk florets functionally male.

Gymnogramma (Adiantaceae) Desv.
Gk. *gymnos* = naked; *gramma* = a line. The sori on the veins are in lines, and naked.

Gymnonychium = ***Agathosma*** (Rutaceae) Bartl.
Gk. *gymnos* = naked; *onyx* = claw, nail; *-ium* = diminutive. Meaning unclear.

Gymnopentzia (Asteraceae) Benth.
Gk. *gymnos* = naked; *Pentzia* (q.v.).

Gymnopteris** (Pteridaceae) Bernh.
Gk. *gymnos* = naked; *pteris* = fern; referring to an exindusiate fern – the fructification is naked.

Gymnosporia (Celastraceae) (Wight & Arn.) Hook.f.
Gk. *gymnos* = naked; *spora* = seed; referring to the seeds that remain attached and exposed when the fruits open. The seeds are exposed (naked) when capsules split open. Technically this is not 100 per cent correct, because the seeds have arils that enclose them totally or partially or

are reduced to a rim at the base (*G. tenuispina*) (Marie Jordaan).

Gymnostephium (Asteraceae) Less.
Gk. *gymnos* = naked; *stephos* = wreath, crown; alluding to the lack of pappus in the radiate flowers.

Gymnostyles*** = **Soliva** (Asteraceae) Juss.
Gk. *gymnos* = naked; *stylos* = styles; referring to the long, permanent naked style.

Gymnothrix*** = **Pennisetum** (Poaceae) Spreng.
Gk. *gymnos* = naked; *thrix* = hair; referring to the bristles, being minute and rough, in the lower spikelets, hence naked, rather than feathery.

Gynandriris*** = **Moraea** (Iridaceae) Parl.
Gk. *gyne* = female, woman (pistil); *andros* = male, the stamens; *iris* = the rainbow; referring to the united pistil and stamens.

Gynandropsis*** = **Cleome** (Capparaceae) DC.
Gk. *gyne* = female, woman (pistil); *andros* = male, the stamens; *-opsis* = resembling; referring to the position of the stamens with respect to the ovary – they appear to be inserted on the top of the ovary.

Gypsophila* (Caryophyllaceae) L.
Gk. *gypsum* = chalk; *philos* = loving; referring to the type of soil in which this plant grows.

Gynura* (Asteraceae) Cass.
Gk. *gyne* = female; *oura* = a tail. The styles end in a long, rough stigma.

Gyrocarpus (Hernandiaceae) Jacq.
Gk. *gyros* = turning, revolving; *karpos* = fruit. The winged fruits gyrate like propellers when falling to the ground.

Gyrocarpus (TW)

Gyrophora*** (Serpulaceae) Pat.
Gk. *gyros* = turning, revolving; *phora* = bearing, carrying; probably referring to the revolving seed as it falls.

Gyrosorium = **Pyrrosia** (Polypodiaceae) C.Presl
Gk. *gyros* = round, circle; *soros* = spores on the underside of fern fronds. The fern has round sori.

H

David Gwynne-Evans (1978–) completed a diploma in horticulture in 1999, after which he pursued botany at the University of Cape Town. Visiting Oxford University in 2000, he conceived of a revolutionary crowd-sourced platform to use digital cameras to document flora and fauna. In 2004 he sought support for the development of this 'biodiversity engine', which the South African National Biodiversity Institute declared 'too futuristic' at the time. Disillusioned, he spent nine years completing a PhD. He collected plants throughout Southern Africa – including in the Mabu Forest, Mozambique – and revised the genus *Hermannia* (Malvaceae) and uncovered over 100 new species. His ambitions include conserving roadsides to create the world's longest and most species-rich nature reserve and pioneering electric car travel through Africa.

Habenaria (Orchidaceae) Willd.

(HS)

La. *habena* = a strap or thong, reins; *-aria* = possessing; alluding to the orchid's long spur, 'from the long, strap-like divisions of the petals and lip' (Willdenow, 1805).

*Habranthus** (Amaryllidaceae) Herb.
Gk. *(h)abros* = delicate, splendid, graceful; *anthos* = flower; referring to the features of the 'rain lily'.

Hackelochloa (Poaceae) Kuntze
For Eduard Hackel (1850–1926), Bohemian-born Austrian botanist and agrostologist, professor of natural history at the Polytechnical Institute at Vienna (1871–1900), and a world authority on the Poaceae. While he only made one collecting trip – to Spain and Portugal – he was charged with writing up collections of grasses mainly from Japan, Taiwan, New Guinea, Brazil and Argentina. Apart from systematics, Hackel also contributed to the morphology and histology of members of the grass family. He authored many papers and books, including *Monographia Festucarum Europaearum* (1882), a section on true grasses in *Die Natürlichen Pflanzenfamilien* (1890), *Enumeratio Graminum Japoniae* (1899) and *Plantae Ex Asia Media Fragmentum* (1906).

Haemanthus (Amaryllidaceae) (Images on following page)
Gk. *haima* = blood; *anthos* = flower. The colour of the (flower) perianth is red in many species.

*Haematobanche*** (Orobanchaceae) C.Presl
Gk. *haimato-* from *haima* = blood; *ancho* = to strangle; referring to the parsitic properties of this plant.

Haematoxylum (Fabaceae) L.
Gk. *haimato-* from *haima* = blood; *xylon* = wood; alluding to the red dye obtained from this wood.

Haenelia = *Ammellus* (Asteraceae) Walp.
For Eduard Gustav Haenel (Hänel) (1804–1856), German book publisher and type foundry owner, active in gardening societies. He learned the art of printing from his father, Jacob Haenel, and in 1824 inherited the latter's print shop and in 1828 brought the Congreve high speed printing press to Germany. In 1830 he established a font foundry and created a number of polytypes to cater to the burgeoning lithographic press. His printing office acquired extensive fame through their skillful preparation of securities. The foundry went to Haenert's colleague Wilhelm Gronau upon his death; Gronau retired in 1885.

Hagenia (Rosaceae) J.F.Gmel.
For Karl Gottfried Hagen (1749–1829), German chemist and professor at the University of Königsberg, Prussia (now Kaliningrad, Russia). While running his family-owned pharmacy (1772–1816), he became a doctor (1775), professor of medicine (1788), assessor (1789), medical advisor to the provincial health department (1800), and professor of chemistry, physics and natural history (1807). He founded the first German pharmaceutical (chemical) laboratory and introduced pharmacology as a university-taught discipline. He wrote *Textbook on the Apothecary Art* (1778) and *Outline of Experimental Chemistry* (1786), which Immanuel Kant declared 'a logical masterpiece'. In 1817, he founded the *Beiträge zur Kunde Preussens* aimed at the popularisation of local history and science.

*Hainardia** (Poaceae) Greuter
Presumably for Pierre Hainard (1936–), Swiss professor, geobotanist, phytogeographer and ecologist at the University of Lausanne's department of ecology and evolution and curator (1965–1981) of the Conservatoire et Jardin Botaniques of Geneva. He took part in the geobotanical survey of Switzerland in

TOP & BOTTOM LEFT: *Haemanthus* (CM)   *Haemanthus* – one of the few white species (MC)

1998 relating, *inter alia*, to the effect of climate change, and one of the publications used was *Effect of Reduced Flow of Rivers on the Flora and Vegetation*, a booklet he co-authored in 1987.

Hakea* (Proteaceae) Schrad.
For Baron Christian Ludwig von Hake (1745–1818), German patron of botany and councillor in Hanover. He was a ranked state minister in the Duchy of Bremen and the Principality of Verden (Bremen-Verden), two separate entities ruled in 'personal union', that is, governed by the same monarch although their boundaries, their laws and their interests remain distinct. He served as president of the Royal British and Electoral Brunswick-Lunenburgian Privy Council for the duchies of Bremen and Verden (effectively, of the government) for seven years under Hanoverian rule (1800–1807) and for three years under Westphalian rule (1808–1810).

Halenbergia* (Aizoaceae) Dinter
Named after Halenberg, near Luderitz, South-West Africa/Namibia, where it was discovered; a desert succulent.

Halesia (Styracaceae) Ellis ex L.
For Stephen Hales (1677–1761), English clergyman, scientist and philanthropist. He studied  at Cambridge University (MA 1703, BDiv 1711) and was ordained a priest (1709). He did pioneering research in plant physiology (transpiration and root pressure measurement), animal physiology (blood pressure measurements), air quality improvements in mines, ships and prisons through ventilators, distillation of fresh water from sea water, preservation of water

and meat on sea voyages, and the development of surgical forceps for the removal of bladder stones. He wrote *Vegetable Staticks* (1727) and *Haemastaticks* (1733). He was a Fellow of the Royal Society, and the Academy of Sciences in Paris and Bologna. He was also a philanthropist and outspoken against alcoholic intemperance and the gin trade.

Hallackia*** = **Huttonaea** (Orchidaceae) Harv.
For Russell Hallack (1824–1903), English businessman, amateur botanist and plant collector who came to South Africa in 1843 and settled in Port Elizabeth. He collected here and in Natal, to which he travelled on horseback in 1854. He sent his specimens to William Henry Harvey, and also to Peter McOwan. He was the father of collector Florence Mary Paterson.

Halleria (Scrophulariaceae) L.

Said to be one of the best garden trees for birds (EJM)

For Albrecht von Haller (1708–1777), Swiss botanist, physician, poet, experimental physiologist, professor of anatomy, surgery and botany at Göttingen (1736–1753) and founder of the Göttingen University herbarium. Haller studied under Herman Boerhaave at Leiden, gaining his MD in 1727. He wrote the poem *Die Alpen* while doing botanical research (1732), produced a major work on Swiss flora, *Enumeratio Methodica Stirpium Helveticarum* (*A Methodical Enumeration of Swiss Plants*) (1742) and wrote an eight-volume compendium of information on physiology, *Elementa Physiologiae Corporis Humani* (*Physiological Elements of the Human Body*) (1747–1766). His publications, numbering in the thousands, guided development in physiology for a century.

Hallia = Psoralea (Fabaceae) Thunb.
For Birger Mårten Hall (1741–1815), Swedish botanist and physician who studied medicine at

Uppsala University under Linnaeus, then spent four years abroad (1755–1759), mainly in Germany, to further his medical studies. He returned to Uppsala, where he obtained a PhD in 1762 for his work on the presence of honey glands in flowers.
From 1764 Hall worked at the university hospital in Uppsala (Nosoconium academicum), obtaining his surgical degree in 1768. He was appointed district medical officer in Västerås in 1773, a post he held for 20 years before retiring because of illness in 1793 aged 53. His legacy included a collection of medieval coins and banknotes, manuscripts, an extensive medicine and science library and a large insect collection and herbarium.

Hallianthus (Aizoaceae) H.E.K.Hartmann.
For Harry Hall (1906–1986), an English gardener and collector of succulent plants, trained at the

Cheshire Agricultural College (1925–1926), Reading University (1926–1927) and John Innes Institution (1928–1929). He worked at Kew Gardens (1930–1933), was curator of the Darrah cactus collection at Manchester (1933–1947) and horticulturist at the famed Kirstenbosch National Botanical Garden in Cape Town (1947–1968). He was a major explorer of euphorbias in South Africa and made three collecting visits to Zimbabawe. It is estimated he collected some 5 000 specimens and an equal number of living plants. He was awarded a Fellow of the Cactus and Succulent Society of America in 1981.

Halodule (Cymodoceaceae) Endl.
Gk. (*h*)*als, halos* = salt; *modulo* = slave. The plant lives exclusively in salt water.

Halopeplis** (Amaranthaceae) Bunge ex Unge & Sternb.
Gk. (*h*)*als, halos* = salt; *peplis* = a name for 'purslane'; referring to small, typically fleshy-leaved plants whose preferred habitat is marshland and wasteland; widespread in salty habitats throughout the countries of the Mediterranean.

Halophila (Hydrocharitaceae) Thouars
Gk. *halo-* = salt; *phila-* = loving. This is a salt-loving marine plant.

Halosarcia** (Amaranthaceae) Paul G.Wilson
Gk. *halo-* = salt; *sarkos* = flesh. A fleshy, succulent, salt-tolerant plant. Meaning unresolved, but some data suggests it is synonymous with *Tecticornia*.

Hammeria (Aizoaceae) Burgoyne
For Steven Allen Hammer (1951–), US pianist, plant collector and explorer, horticulturist and specialist on Mesembs especially *Conophytum*, lecturer and research fellow at the University of Cape Town, author of *The Genus Conophytum* (1993), *Lithops (Treasures of the Veld)* (1999), and a contributor to *Mesembs of the World* (1998) and many scientific papers. He is regarded internationally as one of the foremost authorities on Mesembs, manages a 'botanic garden disguised as a nursery' and has done extensive field work, especially in Namaqualand. He became a Fellow of the Cactus and Succulent Society in 1997 and is assistant journal editor of the US *Cactus and Succulent Journal*.

Haplocarpha (Asteraceae) Less.
Gk. *haplo-* = simple, single; *karphos* = chaff, any small dry body. The pappus is uniseriate. William Henry Harvey (*Flora Capensis*) states that the name is from *apolos* = soft; *karphos* = a scale, again referring to the pappus.

Haplocoelum (Sapindaceae) Radlk.
Gk. *haplo-* = single; *koilos* = hollow; probably referring to the single perianth whorl.

Hardwickia (Fabaceae) Roxb.
For Thomas Hardwicke (1755–1833), English military officer, ornithologist, naturalist and zoologist. He joined the British East India Company in 1777 and retired in 1823 as a major general in the Bengal Artillery. During his career he travelled extensively over the subcontinent, visiting the Cape and Mauritius and collecting mainly zoological specimens. He also amassed the largest collection of drawings of Indian animals ever formed by a single individual; these were drawn by Indian artists trained to make technical illustrations using water colours. His collection, some 4 500 illustrations, was bequeathed to the British Museum in 1835. He corresponded with the leading naturalists in England and was made a Fellow of both the Royal and Linnaean societies.

Harpagophytum (Pedaliaceae) DC. ex Meisn.
Gk. (*h*)*arpago* = grappling hook; *phyton* = plant; referring to the armed capsules.

Harpephyllum (Anacardiaceae) Bernh. ex Krauss.
Gk. *harpe* = sickle; *phyllon* = leaf. The leaflets are sickle shaped.

Harpochloa (Poaceae) Kunth
Gk. *harpe* = sickle; *chloa* = grass; referring to the sickle-shaped spikelets' 'flowers'.

Harrisia* (Cactaceae) Britton
For William Harris (1860–1920), Irish botanist, gardener and plant collector, trained in horticulture at Kew Gardens. He went to Jamaica in 1881, becoming superintendent of the Public Gardens (1908), and later government botanist (1917) and assistant to the director (1920). He made a study of the flora of Jamaica, and his extensive collections from 1890 onward led to the discovery in Jamaica of almost 400 species above those recorded for the island in Grisebach's *Flora* (1859–1864). He was elected a Fellow of the Linnaean Society in 1899. The genera *Harrisia* (*Cactaceae*) and *Harrisella* (*Orchidaceae*) are named after him.

Hartmanthus (Aizoaceae) S.A.Hammer
For Heidrun Elsbeth Klara Osterwald Hartmann (1942–), German botanist at the University of Hamburg, where she obtained her PhD in 1973. She was a specialist in the Aizoaceae with particular interest in, *inter alia*, taxonomy and phylogeny. She published more than 100 scientific papers and books – her top four scholarly works cited (by the University of Hamburg) being *Aizoaceae – a New Approach* (1988), co-authored with Bittrich; *Mesembryanthema Conrib* (1991); and two volumes of the *Illustrated Handbook of Succulent Plants* (2002), co-edited with Eggli. She made collecting tours to the Cape Province, especially Namaqualand, in 1969, 1974 and 1978 as well as the Transvaal and Namibia (1971) and also visited Argentina, Mexico and the southwestern United States.

Hartogia*** = **Agathosma** (Rutaceae) L.
For Johannes (Jan, Johan) Hartog (Hartogius, Hartogh, Hertog) (1663–1722), German/Dutch (born Aachen) plant collector and gardener, worked in Sri Lanka and Cape Town, master-gardener and first official botanist for the Dutch

East India Company's garden in Cape Town (f 1697) who collected indigenous seed while working under Simon van der Stel, first governor of the Cape. By orders from the directors of the Dutch East India Company, plant material was sent to the Amsterdam Hortus Botanicus, but as the Hortus did not wish to maintain a herbarium, the specimens came into the possession of Nicolaas Laurens Burman, author of *Flora Indica* (1768), which had a 33-page supplement on Cape flora.

Hartogiella = *Cassine* (Celastraceae) Codd
Hartogia (q.v.); Gk. *-iella* = diminutive.

*Hartwegia**** = *Chlorophytum*
(Asparagaceae) Nees
For Andreas Johann Hartweg (1777–1831), German garden inspector in Karlsruhe who bred the first dahlia there in 1808. His son, Karl Theodore Hartweg (1812–1871), also worked as a garden inspector in Karlsruhe before moving on to the Jardin des Plantes Paris, and later to the Royal Horticultural Society in London, which sent him to Columbia, Ecuador, Guatemala, Mexico, northern South America and Jamaica (1826–1843), where he discovered many new orchids and other species as described in *Plantae Hartwegianae* by George Bentham in 1839. He went on a second mission to Mexico and California from 1845–1848 finding many newer species. On his return he was appointed director of the Grand Ducal Gardens of Swetzingen in Baden, Germany.

Harveya (Orobanchaceae) Hook.
For William Henry Harvey (1811–1866), Irish-born botanist, algologist and pioneer of South African systematic botany, colonial treasurer general of the Cape Colony, keeper of the herbarium at Trinity College, Dublin, professor of botany for the Royal Dublin Society and at Trinity College, Dublin, and Fellow of the Linnaean and Royal societies. He came to the Cape in 1834, aged 23, and stayed about four years. He wrote *The Genera of South African Plants* (1838), *Manual of British Algae* (1841), *Phycologia Britannica* (1846–1851), *Phycologia Australica* (1858–1863) and was co-author, with Dr OW Sonder of Hamburg, of the first three volumes of *Flora Capensis* (1860–1865). He collected along the Atlantic coast of the United States and Australia and Tasmania in the South Seas.

Haumaniastrum (Lamiaceae) P.A.Duvig & Plancke
For Lucien Leon Hauman (1880–1965), Belgian-

Harveya is a root parasite. (CM)

born botanist who studied and collected plants in East Africa but whose main experience was with South America. He arrived in Argentina in 1804 and was a professor at the newly established faculty of agronomy and veterinary medicine of the University of Buenos Aires until 1925. He collected plants in Argentina, Paraguay, Uruguay and Chile and published many research papers. The Lucien Hauman Botanical Gardens of the University of Buenos Aires was named in his honour.

Haworthia (Asphodelaceae) Duval
For Adrian Hardy Haworth (1768–1833), English botanist, entomologist, carcinologist and an authority on succulents and lepidoptera. He did pioneering work in North America, Canada and Mexico focusing on cacti, and published *Synopsis Plantarum Succulentarum* (1819) with subsequent supplements. In England he collected and studied butterflies, publishing *Lepidoptera Britannica* (1803–1828). During his life he amassed a collection of over 40 000 insects. He was a Fellow of the Linnaean and Royal Horticultural societies and a friend of Sir Joseph Banks. In 1833 he lent support to the founding of what was to become the Royal Entomological Society of London.

Hebe (Plantaginaceae) Comm. ex Juss.
In Greek mythology, Hebe (ancient Greek) was the personification of youth, described as the daughter of Zeus and Hera. According to the *Iliad* she was a helper of the gods, filling their cups with nectar. An allusion to the beauty of this genus. (Despite the mention of 'youth', the ancestor of this genus has been traced back over 5 million years.)

Hebea*** = **Tritoniopsis** (Iridaceae) Bolus
Gk. *hebe* = the Greek goddess of youth; possibly symbolising strength and vigour.

Hebenstretia (Scrophulariaceae) L.
For Johann Christian Hebenstreit (1720–1791), German physician and botanist. He studied

medicine at the University of Leipzig from 1740–1748 and practised in Naumburg before becoming professor of botany and natural history at the Russian Academy of Sciences at St Petersburg. In 1751 he became a personal physician to Count Kyrylo Rosumowskyj, the president of the academy, for two years and was stationed in the Ukraine before returning to Leipzig. In 1755 he accepted the position of professor of botany and natural history in St Petersburg, but deteriorating health forced him to return to Leipzig in 1961. Little is known of his life thereafter.

Hedera (Araliaceae) L.
The classical name for ivy, possibly from Celtic *hedra* = a cord; alluding to the vining habit. The ivy leaf was also associated with Bacchus, the Greek and Roman god of wine.

Hedychium* (Zingiberaceae) J. König
Gk. (*h*)*edys* = sweet; *chion* = snow; referring to the fragrant white flowers of *Hedychium coronarium*.

Hedyotis (Rubiaceae) L.
Gk. (*h*)*edys* = sweet; *ous*, -*otis* = ear. The firm, soft oval leaves are compared to ears.

Hedypnois* (Asteraceae) Scop.
Gk. (*h*)*edys* = sweet; (*h*)*ypnos* = sleep or *pno, pnoi* = air, breathing. Pliny the Elder used the name for a kind of wild endive.

Hedysarum (Fabaceae) L.
Gk. (*h*)*edys* = sweet; *arum* = 'description of'; an ancient Greek name for the common European *Asarum* (ginger). The seeds of some species of *Hedysarum* are toxic.

Heeria (Anacardiaceae) Meisn.
For Oswald von Heer (1809–1883), Swiss paleobotanist, entomologist, phytogeographer,

traveller, theologian, plant and insect collector, director of the botanical gardens in Zurich, and professor of botany and entomology at the University of Zurich (1852–1883) and the Swiss Federal institute of Technology. He published a prodigious number of books and papers on, *inter alia,* the primaeval world of Switzerland, the fossil flora of Alaska, Arctica, Croatia, the Netherlands and Portugal including *Flora Tertiaria Helvetiae* (three volumes, 1855–1859), *Die Urwelt der Schweiz* (1865), and *Flora Fossilis Arctica* (seven volumes, 1868–1883) with Eduard Heinrich Graeff. He described over 1 600 new species in his three major paleobotanical works alone.

Heimia (Lythraceae) Link
For Ernst Ludwig Heim (1747–1834), German physician, amateur botanist and mycologist

(person who studies mosses). In 1772 he received his PhD from the University of Halle and thereafter served as medical officer in the rural districts of Spandau and Osthavelland before moving to Berlin in 1783, where he became the prototype of the faithful family physician ('Old Heim' is a symbolic figure for responsible medical activities). In 1822, on the 50th anniversary of earning his PhD, he was made an honorary citizen of Berlin in recognition of his decades of altruistic service, treating the poor free of charge. He is said to have introduced the Jennerian vaccination (a procedure using cowpox matter to prevent smallpox) to Berlin in 1798.

Heinsia (Rubiaceae) DC.
For Daniel Heinsius (1580–1655), a Dutch classical scholar, philologist and poet, one of the most famous scholars of

the Dutch Renaissance. He studied at the universities of Franeker and Leiden where he was appointed professor of Latin (1602) and Greek (1605) and librarian of the Leiden University library (1607). He published profusely, *inter alia,* many editions on classical figures, e.g. Hesiod, Aristotle, etc., three volumes of Latin poems, a collection of Dutch poems, Latin orations, a treatise on political philosophy and a tragedy. He spent all his working life at Leiden.

Helianthus* (Asteraceae) L.
Gk. (*h*)*elios* = sun; *anthos* = flower. The flower heads of most species in this genus follow the direction of the sun, 'sunflower'.

Helichrysopsis (Asteraceae) Kirp.
Helichrysum (q.v.); Gk. *-opsis* = resembling.

Helichrysum (Asteraceae) Mill.

(EJM)

(CM)

Gk. (*h*)*elios* = sun; *chrysos* = gold; referring to the bright yellow flowerheads of many of the flowers of species in this genus.

Heliconia* (Heliconiaceae*) L.
Gk. *helikonios, heliko* = spirally twisted; referring to the twist-turning stem and inflorescences.

Helictotrichon (Poaceae) Besser
Gk. *heliktos-* = spirally twisted; *trichoma-* = growth of hair, from *thrix* = hair; referring to the column of the lemma awn.

Helinus (Rhamnaceae) E.Mey. ex Endl.
Gk. *helinos* = a tendril. This plant is a shrub with climbing tendrils that can grow 6 metres or more right up to the ends of branches.

Heliophila (Brassicaceae) L.

(CM)

Gk. (*h*)*elios* = sun; *philein* = to love. The plant likes a sunny position.

Heliophytum = Heliotropium (Boraginaceae) Cham.
Gk. (*h*)*elios* = sun; *phyton* = plant; a plant that like sunshine.

Heliotropium (Boraginaceae) L.
Gk. (*h*)*elios* = sun; *tropein* = to turn. Many flowers in this genus turn their flower heads to follow the sun: 'heliotrope' – diurnal motion in response to the sun's movement.

Helipterum (Asteraceae) DC.
Gk. (*h*)*elios* = sun; *pteron* = wing, feather; referring to the plumed and feathery pappus.

Helixanthera (Loranthaceae) Lour.
Gk. *helix* = a spiral, winding around; *anthera* = anther; referring to its twisted anthers.

Helleborus (Ranunculaceae) L.
An ancient name for this genus. Gk. *ellos/hellos* = fawn; *bora* = food, hence food for a fawn or possibly *hele* = to take away (food) or (*h*)*elein* = to injure, or to cause death; alluding to the plant's poisonous nature.

Helixyra* = ***Moraea*** (Iridaceae) Salisb.
Gk. *helix* = a spiral, spirally twisting; *xyron* = razor; referring to the long lance-shaped leaves that twist slightly and the leaf edges that look 'razor sharp'.

Hellmuthia (Cyperaceae) Steud.
For Hellmuth Steudel (1816–1886), doctor of medicine and surgery at the hydropathy establishment in Esslingen near Kennenburg, son of German botanist and physician Ernest Gottlieb von Steudel (1783–1856). He maintained that health depended upon the proper use of water, fresh air, exercise, proper diet and washing for cleanliness. He advocated the use of cold water to reduce temperature and fever (*Quarterly Homeopathic Journal*, Volume 1, 1849). He wrote *Die Medizinische Praxis, Ihre Illusionen und ihr Streben zur Gewissheit* (*Medical Practice, Their Illusions and the Quest for Certainty*) (1853), *Praktik der Heilgymnastik* (*Practice of Physiotherapy*) (1860), and *Der Nihilismus, Das Einzig Wahre in Der Medizin* (*Nihilism, the One Truth in Medicine*) (1887) with Paul Niemeyer.

Helminthostachys (Ophioglossaceae) Kaulf.
Gk. *helminthos* = a worm; *stachys* = spike. The fertile segment of the frond is worm-shaped, referring to the shape of the sporophore.

Helophytum* = ***Crassula*** (Crassulaceae) Eck. & Zeyh.
Gk. (*h*)*elos* = marsh; *phyton* = plant; referring to the plant's habitat.

Helosciadium* = ***Apium*** (Apiaceae) W.D.J.Koch
Gk. (*h*)*elos* = marsh; *skiadeion* = parasol, umbrella or canopy; referring to the marsh flower's umbel.

Hemarthria (Poaceae) R.Br.
Gk. *hemi* = half; *arthron* = a joint; possibly referring to the resistance of the raceme joints to breaking.

Hemicarpha* = ***Lipocarpha*** (Cyperaceae) Nees & Arn.
Gk. *hemi-* = half; *karphos* = a dry stalk, chaff; referring to the perianth member.

Hemichlaena* = ***Ficinia*** (Cyperaceae) Schrad.
Gk. *hemi-* = half; *khlaina* = cloak, covering. Possibly referring to the gynophore attached to the base of the achene, which looks 'half covered' (having a 'ficinioid' appearance).

Hemigraphis* (Acanthaceae) Nees
Gk. *hemi-* = half; *graphis* = pencil. The flowers are spike-like and the fruit cylindrical (Robert Allen Dyer).

Hemimeris (Scrophulariaceae) L.f.

(CM)

Gk. *hemi-* = half; *meros* = a part or fragment; referring to the flower that is cut away on one side, that is, lacking a spur.

Hemionitis (Pteridaceae) L.
Gk. *hemionitis* = mule. A humourous way of explaining that these ferns were once regarded as sterile and were worn as a charm against pregnancy.

Hemitelia = ***Cyathea*** (Cyatheaceae) R.Br.
Gk. *hemi* = half; *telia* = cover. The indusia are free and later bent back, exposing the sporangia.

Hemizygia* = ***Syncolostemon*** (Lamiaceae) (Benth.) Briq.
Gk. *hemi* = half; *zygon* = joined. The name refers to the deeply two-lipped flowers.

Hereroa (Aizoaceae) (Schwantes) Dinter & Schwantes
After the Hereros, a native tribe of South-West Africa/Namibia, where the plants grow; referring to the plant's preferred habitat.

Hermannia (Malvaceae) L.
For Paul Hermann (1646–1695), German-born Dutch physician and botanist. He graduated in

medicine at the universities of Leiden and Padua, became a ship's medical officer (1672–1677) for the Dutch East India Company and went to Sri Lanka via the Cape, where he made the first known herbarium collection of local plants, now housed in the Sloane Herbarium, British Museum of Natural History and at Oxford. In 1679 he became professor of botany at the University of Leiden and director of the Hortus Botanicus in Leiden, Europe's finest botanical garden. His 1687 publication *Horti Academici Lugduno-Batavi Catalogus* includes 34 Cape plants, and his proposed *Prodomus Plantaerum Africanarum* was to contain 791 items, but untimely death intervened.

Hermas (Apiaceae) L.
The origin of the genus name Hermas is not known. Some sources state that Hermas, a freed slave who lived in Rome in the first or second century, was the seer of an apocalypse entitled *The Shepherd*, a work treated with great authority in ancient times and ranked with the Holy Bible. According to the Muratorian Canon and also stated in the *Liberian Catalogue*, he was the brother of Bishop Pius I, (c 145), who occupied the chair of the church of the city of Rome. However, there is no evidence that the genus name referred to him when Linnaeus created it in 1771.

Hermbstaedtia (Amaranthaceae) Rchb.
For Sigismund Friedrich Hermbstaedt (1760–1833), German botanist, Prussian court apothecary in Berlin, professor of technological chemistry, author of many treatises, textbooks and works on chemistry, technology and agriculture, and consultant. His *Grundriss der Technologie* (1814) was widely consulted by merchants, factory owners and officials and, as a member of the Technical Industrial and Trade Commission, he performed a valuable service to the Prussian industry.

Herminium (Orchidaceae) L.
An unexplained name. Gk. *herminium* = bedpost; Linnaeus also mentions *hermes* = mercury, *ermineus* – white like ermine; possibly referring to the stunted staminodia that stand on each side of the anther or referring to the resemblance of the inflorescence to a carved bedpost – the tubers represent the knobs of antique beds.

Herniaria (Caryophyllaceae) L.
La. *hernia* = rupture; *-aria* = possessing, a thing like; alluding to the plant's supposed property of curing hernia (i.e. rupture).

Herpestis = Bacopa (Plantaginaceae) C.F. Gaertn.
Gk. *herpes* from *herpein* = to creep; *tis* = life, existence; referring to a creeping herb whose habitat includes wetlands and muddy shores.

Herpolirion (Xanthorrhoeaceae) Hook.f.
Gk. *herpes,* from *herpein* = to creep; *lirion* = a lily; referring to its low-growing habit.

Herrea*** = Neohenricia (Aizoaceae) Schwantes
For Adolar Gottlieb Julius (Hans) Herre (1895–1979), German botanist, explorer, horticulturalist, curator of Stellenbosch University Botanical Garden, succulent plant specialist and author of *The Genera of the Aizoaceae* (1971). He collected over 300 species new to science. In May 1965, he received the Fellow Award, the highest honour the Cactus and Succulent Society of America can bestow.

Herreanthus** = Conophytum (Aizoaceae) Schwantes
Herrea (q.v.); *anthos* = flower.

Herschelia = Disa (Orchidaceae) Lindl.
For Sir John Frederick William Herschel (1792–1871), English astronomer, chemist, naturalist and social reformer, and son of the famous astronomer Sir William Herschel (1738–1822) who discovered the planet Uranus. Herschel spent four years (1834–1838) in Cape Town as astronomer royal, in order to complete the catalogue of the stars, nebulae and other objects of the Southern skies, started by his father, the results of which were published in 1847. While in South Africa, he took time for other interests, such as seeing and drawing the wildflowers and landscapes around the Cape Peninsula, Paarl and Stellenbosch, which led to a friendship with William Henry Harvey and Mary Pitt, who would later become his wife. He even cultivated orchids. Later, he recalled it was 'probably the happiest time in [his] life'.

Herschelianthe*** = Disa (Orchidaceae) Rauschert
Herschelia (q.v.); *anthos* = flower.

Hertia (Asteraceae) Less.
For Johann Casimir Hertius (1679–1748), German physician. He studied at the University of Halle and became a professor of botany, anatomy and surgery at Universität Giessen (1714–1748), being elected rector thrice and dean of the medical faculty on nine occasions. Although, seemingly, he did not leave behind a legacy of scientific literature, he is acclaimed for his teaching skills, and four of his students held a chair at the university. His doctoral dissertation, *De Crepitu Ossium* (*On the Clashing of the Bones*) (1704) certainly sounds interesting.

Hesperantha (Iridaceae) Ker Gawl.

(MC)

Gk. *hesperos* = evening; *anthos* = flower. Many flowers open late in the day, toward evening, Afrikaans *aandblom* = evening bloom/flower.

Hessea (Amaryllidaceae) Herb.
For Christian Henrich Friedrich Hesse (1772–1837), German Lutheran minister, scholar and naturalist who came to Cape Town in 1800 before returning to Germany in 1817. Hesse extended his hospitality to many naturalists visiting the Cape such as Martin Heinrich Karl Lichtenstein, William John Burchell, Dugald Carmichael, Carl Bergius and others and was friends with Peter Heinrich Poleman and an apothecary and active collector of plants who was a partner in the firm Pallas & Poleman, which sent consignments of seeds, bulbs and plants back to Europe. The website Amaryllidaceae.org says *Hessea* was named after Paul Hesse (1837), but no supportive evidence for this could be found.

Heteractis* = *Gymnostephium (Asteraceae) DC.
Gk. *heteros* = other, different, 'one or the other'; *aktis, aktinos* = ray; possibly relating to the uncertainty of placement of this genus. Now in *Gymnostephium*.

Heteranthera (Pontederiaceae) Ruiz & Pav.
Gk. *heteros* = variable, different; *anthera* = anther; referring to the variability of the anthers – some appear to have deformaties and look different to the others. '*Genus dictum ab antheris dfformibus*' (Ruiz and Pav.).

Heterocarpha* = *Drake-brochmania (Poaceae) Stapf & C.E.Hubb.
Gk. *hetero-* = different, unlike; *karphos* = chaff, dry stick. Possibly because the glumes differ morphologically.

Heterocentron* (Melastomataceae) Hook. & Arn.
Gk. *hetero-* = different, unlike; *kentron* = spur. Two of the anthers have bristly appendages, while two have basal swellings.

Heterolathus* = *Aspalathus*** (Fabaceae) C.Presl
Gk. *hetero-* = unlike, dissimilar; *lathos* = mistake, *lethas* or *lethe* = oblivion (Liddell and Scott). Meaning obscure.

Heterolepis (Asteraceae) Cass.
Gk. *hetero-* = different; *lepis* = a scale. *Heterolepis* has differing scales of strobili or spikelets – somewhat dissimilar to other genera in Asteraceae.

Heteromma (Asteraceae) Benth.
Gk. *hetero-* = different; *-omma* = an eye. The plant has different appearances and forms.

Heteromorpha (Apiaceae) Cham. & Schltdl.
Gk. *hetero-* = different; *morpha* = shape; referring to the variability in appearance between species from area to area (Palmer).

Heteropogon (Poaceae) Pers. (Image on opposite page)
Gk. *heteros* = variable, different, other; *pogon* = beard; alluding to the variability of the awns – some spikelets have beards, some do not, e.g. stalked spikelets without awns, sessile spikelets awned if bisexual.

Heteroptilis* = *Dasispermum (Apiaceae) E.Mey. ex Meisn.
Gk. *hetero-* = different; *ptilis* = feather, wing; referring to the wings, often heteromericarpic.

Heteropogon: The tangled awns form a roadside mass. (EJM)

Heteropyxis (Heteropyxidaceae) Harv.
Gk. *heteros* = different; *pyxis* = a jar with a lid; referring to the capsule (a fruit), sometimes called the pyxidium, because it looks like it has a lid.

Heterorhachis (Asteraceae) Sch.Bip. ex Walp.
Gk. *hetero-* = different, unlike; *rhachis* = axis, spine; possibly referring to the main axis of a stem that can be variable.

Heterosamara (Polygalaceae) Kuntze
Gk. *hetero-* = different; *amara* = bitter to the taste.

Heudelotia = Commiphora (Burseraceae) A.Rich.
For Jean P. Heudelot (1802–1837), French botanist and plant collector in tropical Africa, mainly Senegal and Gambia from 1828, and director of the royal cultivations in Senegal from 1835–1837. He worked as an associate and co-collector with François Mathias René Leprieur (1799–1869) and also collected fish, which were sent from West Africa to Paris.

Hewittia (Convolvulaceae) Wight & Arn.
Possibly for Hewett Cottrell Watson (1804–1881), English phrenologist, botanist and evolutionary theorist, who, among many achievements, edited *The London Catalogue of British Plants* from 1844–1874. Robert Wight (1796–1872) met Watson after returning from India in 1831 and spent three years recovering from illness before returning to india. In 1832 Wight met Hewett, and they went botanising together in the Scottish Highlands. Wight often made spelling mistakes when naming flowers, and it is possible that 'Hewittia' would

then be a misspelling of 'Hewett'. The genus was first described in the *Madras Journal of Literature and Science* in 1837.

Hexacyrtis (Colchicaceae) Dinter
Gk. *hexa-* = six; *kyrtos* = curved. The perianth segments carry enlarged incurved side-lobes.

Hexaglottis*** = Moraea (Iridaceae) Vent.
Gk. *hexa-* = six; *glottis* = tongue; alluding to the six spreading style branches.

Hexalobus (Annonaceae) A.DC.
Gk. *hexa-* = six; *lobos* = lobe; referring to the six basally fused petal lobes transversally folded in bud.

Hexastemon*** = Erica (Ericaceae) Klotzsch
Gk. *hexa-* = six; *stemon* = stamen; referring to the six stamens, much exserted, in some species.

Heywoodia (Phyllanthaceae) Sim
For Arthur William Heywood (1853–1918), a conservator of forests in South Africa from at least 1886–1906, author of the Cape woods and forests section in *Official Handbook: Indian and Colonial Exhibition* (London, 1886) and reports such as *Report on Forestry in Basutoland* [now Lesotho] *to Government Secretary, Maseru* (1908). He fought an uphill battle to conserve the Eastern Cape's indigenous natural forests, such as the Dwesa-Cwebe forested area located in the rugged Wild Coast of the former Transkei, from rapid destruction by local inhabitants who used merchantable timber as well as young trees as poles or 'wall' material in the construction of huts and kraals, leaving behind impoverished soil.

Hibbertia* (Dilleniaceae) Andrews
For George Hibbert (1757–1837), English amateur botanist, West India merchant, politician,  slave- and ship-owner, book collector, antiquarian and philanthropist. In 1781 he joined the family counting house Hibbert, Purrier and Horton, and became involved in the shipping and distribution of slave-produced goods, particularly sugar from Jamaica. He became chairman of the West India Dock Company, opposed William Wilberforce's proposal to abolish slavery on economic grounds, became an alderman for the City of London (1798–1802), a member of parliament for Seaford (1806–1812), and agent general for Jamaica (1812–1831). He sponsored James Niven to collect plants in the Cape region of South Africa

(1798–1803) and helped fund the National Lifeboat Institution (1824). He collected books, prints and art and was a member the Linnaean and Royal societies and the Society of Antiquaries.

Hibiscus (Malvaceae) L.

(MC)

Gk. *hibiskos* = name for the 'marsh-mallow' used by Virgil (Palmer), possibly derived from Gk. *ibis* = a stork that fed on some species of mallow.

Close-up of reproductive parts (CM)

Hieracium* (Asteraceae) L.
Gk. *hierax* = a hawk. Pliny the Elder named this plant for an eye ointment made from its juice. Hawks were believed to eat it in order to sharpen their eyesight. Hawkweed is the common name.

Hiernia (Orobanchaceae) S.Moore
For William Philip Hiern (1839–1925), British mathematician, botanist, bryologist and plant collector. He studied at Cambridge University (1857–1861) and later at Oxford (1886). He lived in Devon as a country squire with many public duties, and it is here that he developed an interest in botany, collecting 15 new species, and publishing 50 works, many in the *Journal of Botany*. His major works were *A Monograph on Ebenaceae* (1872) and the dicotyledon section he wrote for *Catalogue of the African Plants Collected by Dr Friedrich Welwitsch in 1853–61* (Volume 1). He was a Fellow of the Linnaean and Royal societies (1903). His herbarium specimens and manuscripts are preserved in the National History Museum, London and the Royal Botanical Kew Gardens.

Hilleria (Phytolaccaceae) Vell.
For Matthaeus Hiller (1646–1725), German professor of logic, metaphysics and the Hebrew

language at the University of Tübingen (1692), who was a 'competent expert in the Hebrew language with considerable grammatical and vocabulary [skills] which he used in his exploration of the Old Testament'. He became professor of theology in 1698 and domus master (headmaster) of the evangelical lodgings (1694–1700). In 1716 he became the abbot prelate, an ecclesiastical position of superior rank in Königsbronn. Among his many religious publications, he wrote *Hierophyticon, Sive Commentarius in Loca Scripturae Sacrae Quæ Plantarum Faciunt Mentionem Distinctus in Duas Partes* (1705), a commentary on the plants mentioned in the Bible.

Hilliardia (Asteraceae) B.Nord.
For Olive Mary Hilliard (1925–), South African botanist and plant collector. She obtained a PhD from Natal University and worked at the National Herbarium, Pretoria (1947–1948), as a lecturer of botany at Natal University, Pietermaritzburg (1954–1962), and as curator and research fellow at the Natal University Herbarium from 1963. She authored, *inter alia*, *Streptocarpus: An African Plant Study* (1971), *The Botany of the Southern Natal Drakensberg* (1987), and *Dierama: The Hairbells of Africa* (1991) with Brian Lawrence Burt (1913–2008), and collected some 8 000 specimens, some 5 000 with Burt, mostly from the Natal Drakensberg and Malawi. Her special interest was the taxonomy of *Streptocarpus,* the Compositae and Scrophulariaceae.

Hilliardiella (Asteraceae) H.Rob.
Hilliardia (q.v.); Gk. *-iella* = diminutive.

Hippia (Asteraceae) L.
Derivation unknown. Suggestions include after 'Hippia', a title given to the Roman goddess Minerva or possibly after Hippias of Elis, a Greek philosopher or Gk. *hippos* = horse.

Hippobromus (Sapindaceae) Eck. & Zeyh.
Gk. *hippos* = a horse; *bromos* = a kind of smelly oat; referring to unpleasant smelling aromatic leaves. The common name for this genus, in the US at least, is horsewood, but the more descriptive Afrikaans common name is *perdepis*, meaning horse urine.

Hippochaete = Equisetum (Equisetaceae) Milde
Gk. *hippos* = horse; *khaite* = flowing hair, mane, bristle; possibly referring to the bunched stems

that can grow erectly in tight clusters, giving the appearance of a tail.

Hippocratea* (Celastraceae) L.
For Hippocrates of Kos (Cos) (c. 460–370 BCE), Greek physician, contemporary of Herodotus,

Thucydides, Socrates and Plato, father of Western medicine, pharmaceutical botanist and founder of the Hippocratic School of Medicine and the Hippocratic Oath, although evidence suggests this was written after his death. Hippocrates is credited with being the first person to believe that diseases were caused naturally, the product of environmental factors, diet and living habits.

Hirpicium (Asteraceae) Cass.
Possibly La. *irpex* or *hirpex* = a harrow or rake with iron teeth; perhaps referring to the thin reflexed leaves that look somewhat like a rake (and certainly this is a fast-spreading invasive plant that needs to be raked out).

Hippuris (Plantaginaceae) L.
Gk. *hippuris* = horse's or mare's tail. This aquatic plant's stem and foliage rises above the water and looks somewhat like a horse's tail.

Hirschfeldia = Erucastrum (Brassicaceae) Moench
For Christian Caius (Cayus) Lorenz Hirschfeld (1742–1792), German philosopher, professor of aesthetics at Kiel University, best known for

his magnum opus, the five-volume *Theorie der Gartenkunst* (*Theory of Garden Art*) (1779–1785), comprising 1 300 pages and 250 illustrations, translated in French as *Théorie de l'Art des Jardins*. This work discusses all aspects of the garden – design, aesthetics, structure, functionality linked to locality, etc. and all garden types including public and cemetery gardens. He also wrote two other books on the design of landscape gardens, *Anmerkungen über die Landhäuser und die Gartenkunst* (*Notes on Country Houses and Garden Design*) (1773) and *Theorie der Gartenkunst* (*Theory of Garden Design*) (1775).

Histiopteris (Dennstaedtiaceae) (J. Agardh) J.Sm.
Gk. *histion, histos* = web, cloth; *pteris* = fern; referring to the conspicuous network of veins or to the sail-like shape of the pinnae in these ferns.

Hoarea* * (Geraniaceae) Sweet
For Sir Richard Colt Hoare (1758–1838), English baronet, antiquarian, archaeologist, artist and

traveller. He inherited a large estate from his grandfather Henry Hoare, which enabled him to pursue his interests in archaeological studies and travel. He visited Europe in 1785 and 1788, as described in his *Recollections Abroad* (1815) and *A Classical Tour through Italy and Sicily* (1819). He also visited Wales and Ireland. He worked on the first recorded excavations at Stonehenge (with William Cunnington) in 1798 and 1810 and excavated 379 burial sites on Salisbury Plain. His two-volume book *The Ancient History of Wiltshire* (1812, 1821) outlined his findings. He became a Fellow of the Royal Society (1792) and the Society of Antiquaries of London.

Hoffmannia = Psilotum (Rubiaceae) Willd.
For Georg Franz Hoffmann (1760–1826), German botanist and lichenologist. He studied at the

University of Erlangen (1786) where he became a professor of botany (1787–1792). Subsequently, he became head of the botany department and director of the Botanical Gardens at Göttingen University (1792–1803) and at the University of Moscow (from 1804). He published several important treatises on the taxonomy of lichens such as *Enumeratio Lichenum Iconibus et Descriptionibus Illustrata* (1784); the first German flora, *Deutschlands Flora* (1794); and a classic monograph on the Apiaceae family, *Genera Plantarum Umbelliferarum* (1814, 1816). He described the 3 528 species at the Moscow State University Botanical Garden in the 1808 *Enumeratio Plantarum et Seminum Hort Botanici Mosquensis*. He also collected a large herbarium for the university.

Hoffmannseggia (Fabaceae) Cav.
For Johann Centurius, Count Von Hoffmansegg (1766–1849), German botanist, entomologist and ornithologist. He studied at the universities of Leipzig and Göttingen before travelling through Hungary, Austria and Italy (1795–1796) and Portugal (1797–1801) where he met Johann Heinrich Friedrich Link (1767–1851). They

co-authored *Flore Portugaise* (*Flora of Portugal*), which remained a standard work for a long time.  He sent a large collection of plant and animal specimens to Johann Karl Wilhelm Illiger (1775–1813), a German entomologist and zoologist living in Brunswick. Hoffmansegg worked in Berlin from 1804–1816 where he founded the Zoological Museum in 1809. He became a member of the city's Academy of Science in 1815.

Holarrhena (Apocynaceae) R.Br.
Gk. *holos* = entire; *arrhen* = male; alluding to the included anthers (enclosed in the flower tube).

Holcus (Poaceae) L.
Gk. *holkos* = a kind of grain, barley, perhaps sorghum; Old Latin name for a barley, *Hordeum murinum*.

Holmskioldia (Verbenaceae) Retz.
For Johan Theodor Holm (Holmskjold) (1732–1794), Danish botanist, physician and writer. A graduate of the University of Copenhagen (1760),  he became a professor in medicine and natural history at Sorø Academy in 1762 before becoming, in 1767, director general of the Danish Postal Services in Copenhagen until he died. He also served as cabinet secretary for Queen Juliana Maria, consort of King Christian VII, first director in chief of the Royal Danish Porcelain Factory (all of which he embezzled!) and was co-director with Christen Friis Rottbøll of the Charlottenborg Botanical Garden. He authored a two-volume work on Danish fungi, *Beata Ruris Otia Fungis Danicis Impensa*. He became a member of the Royal Danish Academy of Sciences and Letters.

Holopetalum* = ***Oligomeris*** (Resedaceae) Turcz.
Gk. *holos,* entire; *petalum* = petals. The plant is completely petaled, some fused together.

Holophyllum* = ***Athanasia*** (Asteraceae) Less.
Gk. *holos,* entire; *phyllum* = leaf. The plant has a leafy involucre.

Holosteum* (Caryophyllaceae) L.
Greek name for the plant, *holosteon,* from *holos* = entire; *osteon* = bone. A humorous allusion to the frailty of the plant.

Holostylon = ***Plectranthus*** (Lamiaceae) Robyns & J.-P. Lebrun
Gk. *holos* = entire; *stylos, stylon* = pillar, tube, column; referring to the erect, virgate stems arising from a perennial base.

Holothrix (Orchidaceae) Rich. ex Lindl. (Image on folloing page)
Gk. *holos* = entire, whole; *thrix* (*thricos*) = hair; the plant is hairy all over.

Holubia (Pedaliaceae) Oliv.
For Dr Emil Holub (1847–1902), Czech author, physician, cartographer, naturalist and explorer in Southern Africa. In 1872 he went to South Africa,  where he practised as a surgeon in the Kimberley diamond fields. In 1879 he made several expeditions to the Northern Transvaal, Mashonaland (Southern Rhodesia, now Zimbabwe) and through Bechuanaland (Botswana) to Victoria Falls (which he mapped). In 1883 he attempted to traverse from the Cape to Cairo, but illness and hostile tribes in Zambia forced him to retreat. He wrote *Sieben Jahre in Süd-Afrika 1872–1879* (*Seven Years in Africa*) (1881) and *Reisen, 1883–87* (*Travels, 1883–87*) (1890). Some of the 13 000 items from his collections are housed in various museums.

Homalanthus* (Euphorbiaceae) A.Juss.
Gk. (h)*omalos* = flat; *anthos* = flower. The flowers appear flattened.

Homalium (Salicaceae) Jacq.
Gk. (h)*omalos* = even, regular. The stamens are evenly divided into bundles.

Homeria** = ***Moraea*** (Iridaceae) Vent.
Gk. *omereo* = to meet together; referring to the filaments being united around the style.

Homochaete** = ***Macowania*** (Asteraceae) Benth.
Gk. *homos* = one of the same, similar, equal; *khaite* = flowing hair, mane, bristle. Corolla of disk florets hairy immediately below the expanded upper part.

Homochroma** = ***Zyrphelis*** (Asteraceae) DC.
Gk. *homos* = one of the same, similar, equal;

chromos = colour; possibly referring to the plant petals, which are consistently the same colour.

Homoglossum = Gladiolus (Iridaceae) Salisb.
Gk. *homo-* = similar, equal; *glossa* = tongue; referring to the almost equal perianth segments.

Hoodia (Apocynaceae) Sweet ex Decne.

This species is claimed to be an appetite suppessent. (EJM)

Probably for Dr William Chamberlain Hood (1790–1879), British surgeon who lived in South Lambeth, London, and collected succulents. Many publications state that the genus *Hoodia* is named after a 'Mr van Hood', which would be an incorrectly spelled Dutch or Afrikaans version of this name. The author, Robert Sweet (1783–1835), who published the name in 1830, states the name was after a 'Mr Hood, a cultivator of succulent plants'. David Hollombe, a researcher, adds that, 'There are many references in journals of the time to "Mr Hood, Surgeon, South Lambeth, [London] who possesses a fine collection of rare succulent plants, which he cultivates with great success".' As South Lambeth and Chelsea, where Sweet lived, are both in central London, they may have met there, and they were approximately the same age.

Hoplophyllum (Asteraceae) DC.
Gk. *hoplo* = armed; *phyllon* = leaf. The leaves are spiny.

Holothrix – note the hairy stems (HS)

Hordeum (Poaceae) L.
Latin for an ancient Roman name *H. vulgare* = barley.

Hoslundia (Lamiaceae) Vahl
For Ole Haaslund-Smith (Schmidt) (d August, 1801, not 1802 as often stated), Danish botanist and naturalist, traveller and plant collector. In 1799 he went to Ghana with Peter Thonning (1775–1848), Danish physician and botanist. His contract was to do service there for three years, but he contracted a tropical illness and died.

Hottonia = ***Limnophila*** (Plantaginaceae) Burm.
For Pieter (Petrus) Hotton (Houttuyn) (1648–1709), Dutch botanist and physician, professor of botany at Leiden University (1678–1680) and supervisor of the university's botanical garden, for which he imported exotic plants from Sri Lanka. In 1681 he returned to his medical practice in Amsterdam, but when Jan Commelin (1629–1692) died, he took over his responsibilities as officer in charge of the Amsterdam Hortus Botanicus. Three years later, in 1695, Paul Herman, German professor of botany at Leiden, died, and Hotton took over his post. He did not publish much but had extensive correspondence with his contemporaries. In 1703 he wrote a revised edition of Ray's *Methodus Plantarum*. He was a member of the Royal Society of London.

Hoya* (Apocynaceae) R.Br.
For Thomas Hoy (c 1750–1822), English botanist and gardener to the Duke of Northumberland, a position he held for 40 years. He became a Fellow of the Linnaean Society in 1788. He often presented flowering plants to the Linnaean society. These included the Australian plants *Acacia suaveolens*, *Acacia myrtifolia* and *Goodenia ovata*. Robert Brown (1773–1858) named the plant genus in his honour, describing Hoy as someone 'whose merits as an intelligent and successful cultivator have been long known to the botanists of this country' (Wikipedia.org is the only source available on the internet).

Huernia (Apocynaceae) R.Br.
For Justus Heurnius (1587–1652), Dutch missionary, doctor and an early collector at the Cape, South Africa. His drawings constituted the iconotypes for *Stapelia*, which is what the first taxa of *Huernia* was described as. He was the author of *De Legatione Evangelica ad Indos capessenda admonitio* (1618) and discovered *Orbea variegata* at the Cape in April 1624, while on his way to Batavia (present-day Jakarta) as a missionary. In 1639 he returned to the Netherlands, where he became a minister at Wijk bij Duurstede and helped to translate the Bible into Malay. The genus name *Huernia* was misspelled by Robert Brown, who published it in 1810.

Huerniopsis = ***Piaranthus*** (Apocynaceae) N.E.Br.
Huernia (q.v.); Gk. *-opsis* = resembling.

Hugonia (Linaceae) L.
Probably named after Augustus Johannes Hugo (?–1753) German physician who studied at the University of Leiden, where he obtained a PhD in medicine. His disertation, *Dissertatio Botanicalinauguralis de Variis Plantarum Methodis* (1711), was published by Abraham Elzevier. His advisor on this dissertation was Herman Boerhaave. He travelled in Switzerland in 1732 with Swiss botanist and physician Albrecht von Haller (1708–1777). Linnaeus published the name *Hugonia* in the year Hugo died. Less likely would be Augustus Ludovicus de Hugo (1722–?), who may have been Augustus Johannes's son, author of *Dissertatio Anatomico-medica Inauguralis de Glandulis in Genere, et Speciatim de Thymo* (1746), which was presided over by Professor Albrecht von Haller at the University of Göttingen.

Humata = ***Davallia*** (Davalliaceae) Cav.
Gk. *humata* = of the earth; referring to the creeping rhizomes.

Humea = ***Calomeria*** (Asteraceae) Sm.
For Lady Amelia Hume (née Egerton) (1751–1809), English amateur botanist, wife

of Sir Abraham Hume, floriculturalist and Tory politician. In 1790 she took lessons in botany from Sir James Smith, president of the Linnaean Society. She developed their large garden at Wurmleybury, significantly incorporating an extensive range of stoves, hothouses and greenhouses, while Sir Abraham used his connections with the East India Company to obtain plants from South Africa, China and India. Some of the plants introduced were *Jasminum hirsutum* from India, *Magnolia coco* from China (1786) and *Camellia japonica* (1806) also known as 'Lady Hume's Blush', later called 'Incarnatata'. It was the last plant Lady Hume introduced before she died after a long and painful illness.

Huperzia (Lycopodiaceae) Bernh.
For Johann Peter Huperz (1771–1816), German botanist, physician and fern horticulturist, specialist on ferns of Australia, who wrote a book on fern propagation, *Specimen Inaugurale*

Medico-botanicum de Filicum Propagatione (1798).

Hura (Euphorbiaceae) L.
The vernacular name for a South American tree.

Huttonaea (Orchidaceae) Harv.
For Caroline Hutton (née Atherstone) (1826–1908), South African amateur botanist and plant collector, wife of the English civil servant Henry Hutton, who emigrated to South Africa in 1844, and sister of British-born botanist William Guybon Atherstone (1814–1898). Henry Hutton was a staff officer during the frontier war of 1846 and subsequently spent some 20 years in Grahamstown in the Eastern Cape. He and his wife made various plant collections in this area, as acknowledged by William Henry Harvey in *Thesaurus Capensis 1*: 4 (1859) and *Flora Capensis 1*: ix (1860). After Henry Hutton died in 1897, Caroline Hutton continued to collect specimens near Howick in Natal. She found the type species, *H. pulchra,* on the Katberg (Eastern Cape).

Hyacinthus (Hyacinthaceae) L.
Hyacinth, in Greek mythology a youth much beloved by Apollo as well as Zephyrus, the wind god. Apollo and Hyacinth took turns throwing the discus to each other and Hyacinth was accidentally struck by the discus as he ran to catch it, and fell to the ground and died. Apollo believed that the wind god, Zephyrus, a rival for Hyacinth's affections, purposely blew the discus off course, resulting in Hyacinth's death. In grief, Apollo caused the hyacinth flower to rise from the youth's blood.

Hyaenanche (Picrodendraceae) Lamb. & Vahl
La. *hyaena* = hyena or jackal; *-anche* = poison, from Gk. *anchein* = to strangle. The fruits were used as poisonous bait.

Hyalosepalum*** = *Tinospora*
(Menispermaceae) Troupin
Gk. *hualos* = of glass, transparent; *sepalum* = sepal; possibly referring to the near transparent sepals.

Hybanthus** (Violaceae) Jacq.
Gk. *hubos* = a hump; *anthos* = a flower; referring, variously and depending on the species, to the anterior pouched petal, the main petal, the spurred lower petal, and drooping flowers. The 'humpback flower'.

Hydnora (Hydnoraceae) Thunb.
Gk. (*h*)*udnon* = truffle ('fungus-looking'). The soft spines on the inner surface of the perianth lobes resemble those of the hymenium of a fungus of the genus *Hydnum*.

Hydrangea* (Hydrangeaceae) L.
Gk. *hudor* = water; *angos* = jar or vessel; referring to the hydrangea's need for plenty of water and its cup-shaped flower.

Hydrocotyle (Araliaceae) L.
Gk. *hudros* = water; *kotyle* = cup. The plant lives in damp places and has somewhat cup-shaped leaves.

Hydrodea*** = *Mesembryanthemum*
(Aizoaceae) R.Br.
Gr. *hudros* = water; *-odes* = of the nature of; referring to its preferred habitat.

Hydroidea (Hydrocharitaceae) P.O. Karis
Gk. *hudros* = water, water-loving; *-oidea* = of the nature of (a variant suffix). *Hydroidea elsiae* is the only known species native to South Africa.

Hydrolea (Hydroleaceae) L.
Gr. *hudros* = water; *eleia* = olive; referring to its water habitat and the resemblance of its leaves to the olive.

Hydrophilus (Restionaceae) H.P.Linder
Gk. *hudros* = water; *phila* = loving; alluding to its preferred habitat.

Hydrophylax (Rubiaceae) L.f.
Gk. *hudro* = water; *phylax* = keeper, guardian, sentine. The plant always grows by the seaside.

Hydrostachys (Hydrostachyaceae) Thouars
Gk. *hudros* = water; *stachys* = spike. This plant thrives on fast-moving white water, and its inflorescence can be seen sticking out of the water.

Hygrophila (Acanthaceae) R.Br. emend. Heine
Gk. *hugros* = damp, moist; *-phila* = loving; referring to the plant's 'love' of a moist habitat.

Hylocereus* (Cactaceae) (A. Berger) Britton & Rose
Gk. *hule-* = forest, woods, thicket; *cereus* = cactus; referring to the association of this genus with shady, wooded habitats.

Hymenasplenium = *Asplenium*
(Aspleniaceae) Hayata
Gk. *hymen-* = membrane, membranacious; *splenium* = spleen; referring to an enclosing membrane that covers the spleen-shaped sori on the backs of the fronds.

Hymenocardia (Phyllanthaceae) Wall. ex Lind.
Gk. *hymen* = membrane; *kardia* = heart; referring to the heart-shaped fruits, which have a transparent membrane cover.

Hymenocyclus = *Malephora* (Aizoaceae) Dinter & Schwantes
Gk. *hymen* = membrane; *cyclos* = circular, wheel; referring to the transparent membrane cover of the plant, the details of which could not be established.

Hymenodictyon (Rubiaceae) Wall.
Gk. *hymen* = membrane; *diktyon* = net; referring to the wing, a net-like membrane that surrounds the seed.

Hymenogyne (Aizoaceae) Haw.
Gk. *hymen* = membrane; *gyne* = female, woman; alluding to the pulling together of the styles into a membranous tube.

Hymenolepis (Asteraceae) Cass.
Gk. *hymen* = membrane; *lepis* = scale; referring to the membranous scales that cover the fruitification.

Hymenolobus = *Hornungia* (Brassicaceae) Nutt. in Torr. & A.Gray
Gk. *hymen* = membrane; *lobos* = pod; referring to the elongated dry dehiscent seed pods that at maturity have narrow partitions to which the seeds are attached.

Hymenophyllum (Hymenophyllaceae) Sm.

(JB)

Gk. *hymen* = membrane; *phyllon* = leaf; said to refer to the very thin translucent tissue of the fronds, thus the common name filmy fern.

Hymenosicyos = *Oreosyce* (Cucurbitaceae) Chiov.
Gk. *hymen* = membrane; *sikyos* = cucumber. The fruit, when sliced, has a membrane-like look.

Hymenosporum (Pittosporaceae) R.Br. ex F.Muell.
Gk. *hymen* = membrane; *spora* = seed; referring to the membranaceous wing on the seed.

Hymenostachys*** = *Trichomanes* (Hymemophyllaceae) Bory
Gk. *hymen* = membrane, *stachys* = a spike. This could refer to the spore body on the end of a filmy (membranous) frond. The plant is known as the 'filmy' fern.

Hyobanche (Orobanchaceae) L.

A root parasite (CM)

Gk. *hys* = swine; *anchein* = strangle. This parasitic plant eventually 'strangles' its prey.

Hyoseris (Asteraceae) L.
Gk. *hyos* = pig, swine; *seris* = a kind of chicory; referring to the use of this edible lettuce-like plant as fodder for pigs.

Hyparrhenia (Poaceae) E.Fourn.
Gk. *hypo* = under, below; *arren* (*arrhen*) = masculine, male; alluding to the male spikelets at the base of the racemes; the lowest part of the spikelet bears two male flowers.

Hyperacanthus (Rubiaceae) E.Mey. ex Bridson.
Gk. *hyper* = above; *akantha* = spine, thorn. The sharply pointed spines emerge just above the leaf axils.

Hypericophyllum (Asteraceae) Steetz
Gk. *phyllon* = leaf; like that of *Hypericum* (q.v.).

Hypericum (Hypericaceae) L.
Gk. *hyper* = above; *eikon* = a figure, icon, image. From the ancient practice of placing flowers above an image in the house to ward off evil spirits, celebrated at the midsummer festival of Walpurgisnacht, named after Saint Walpurga (c 710–777), which later became the feast of St John held in late June when they are in bloom, and thus took the name of St John's wort.

Hypertelis (Molluginaceae) E.Mey. ex Fenzl
Gk. *hyper-* = above; *telos* = end, consummation; referring to the fleshy leaves.

Hyperthelia (Poaceae) Clayton
Gk. *hyper-* = over, above; *thele* = nipple, female; referring to the fertile grass florets, which occur above the male homogamous spikelets at the base of the raceme.

Hyphaene (Arecaceae) Gaertn.

Fruits showing the bristles under the smooth, hard and brittle outer shell (EJM)

Gk. *hyphainein* = to entwine, weave; referring to the net-like fibres surrounding the seeds of these fruits (palms).

Hypobathrum (Rubiaceae) Blume
Gk. *hypo* = below; *bathrum* = seat, step, base, pedestal. The flowers are seated on flat axilliary receptacles (George Don); one seed is superimposed upon the other (David Gledhill).

Hypocalyptus (Fabaceae) Thunb.
Gk. *hypo-* = beneath; *kalypto* = covered, to veil. These species have large bracts under which the young flowers are hidden or veiled (William Henry Harvey, *Flora Capensis*).

Hypochaeris* (Asteraceae) L.
Gk. *hypo-* = beneath, below; *khaeris* = young pig. William Henry Harvey wrote that 'pigs eat the roots [of this plant] greedily' (*Flora Capensis*).

Hypodematium (Hypodematiaceae) Kunze
Gk. *hypo-* = beneath; *demation* = small slipper; referring to the shape of the indusium.

Hypodiscus (Restionaceae) Nees
Gk. *hypo-* = beneath; *diskos* = disk; referring to the toothed or lobed disk that crowns the ovary in some species.

Hypoestes (Acanthaceae) Sol. ex R.Br.
Gk. *hypo-* = beneath; *estia* = house; referring to the way the bracts cover the calyx (Plantzafrica.com).

Hypogynium = ***Andropogon*** (Poaceae) Nees
Gk. *hypo-* = below, beneath; *gyne* = woman; referring to the stalked members of the paired spikelets at the base of the racemes that are male and arise from below the hermaphroditic floret.

Hypolaena* (Restionaceae) R.Br.
Gk. *hypo-* = beneath; *chlaina* = a blanket; alluding to the fruit, much of which is covered by the perianth.

Hypolepis (Dennstaedtiaceae) Bernh.
Gk. *hypo-* = below; *lepis* = a scale; referring to the position of the sori under the revolute leaf margin.

Hypopeltis** = ***Polystichum***
(Dryopteridaceae) Michx.
Gk. *hypo-* = below; *peltis* = small, round shield. The sorus is beneath the shield-shaped indusium.

Hypoporum** = ***Scleria*** (Cyperaceae) Nees
Gk. *hypo-* = below; *poros* = pore, passage. The culm is triangular below, each side furnished with a row of very minute pores.

Hypoxis (Hypoxidaceae) L.
Gk. *hyp(o)-* = beneath, less than; *oxys* = sharp-pointed, sour; alluding to the leaves, which are acid; 'Quod folia sint acidula (*Hypoxis*)'.

Hypserpa = ***Tiliacora*** (Menispermaceae) Miers
Gk. *hypo* = under, beneath, below; *serpens* = snake; possibly referring to the climbing stems with young growing tips sometimes tendril-like, 'snake-like'.

Hyptis (Lamiaceae) Jacq.
Gk. *hyptios* = resupinate (upside down). 'The saccate lip is abruptly deflexed' (Robert Allen Dyer), i.e. the bag-shaped lip is bent abruptly downward.

I

Indigofera: A large genus of over 750 species of mainly herbaceous or small shrubs with occasional trees

Brian Schrire (1953–) was born in Johannesburg and took a BSc honours in ecology at the University of the Witwatersrand under Oliver Kerfoot and Brian Walker. He began focusing on the taxonomy of legumes in the 1980s when he completed his MSc at the then University of Durban-Westville with Esmé Hennessy. He later completed his PhD on the taxonomy of Southern African *Indigofera* (Leguminosae) at the then University of Natal under Charles Stirton and Peter Linder. Schrire was curator of the Natal Herbarium in Durban for five years before becoming the South African botanical liaison officer at the Royal Botanic Gardens in Kew in the United Kingdom. His time at Kew was interrupted by a stint at the Botanical Research Institute in Pretoria, South Africa, after which he returned to take up a position as legume researcher until his retirement in 2013. He remains an honorary research associate at Kew and his research in the tribe Indigofereae (Leguminosae) continues.

Ibbetsonia* (Fabaceae) Sims.
For Agnes Ibbetson (1757–1827), British vegetable physiologist who contributed more than 50 papers on the microscopic structure and physiology of plants to *Nicholson's Journal* and the *Philosophical Magazine* between 1800 and 1822 with papers such as *Dissection of the Petals of Various Flowers, Showing the Structure from which the Beauty of their Colour Arises, Dissections of Plants, Showing the Growth of the Bud, On the Interior of Plants, On the Perspiration of Plants,* and others. Her topics included such subjects as air-vessels, pollen, perspiration, sleep, winter-buds, grafting, impregnation, germination, and the Jussieuean method.

Iberis* (Brassicaceae) L.
Name used by Dioscorides for an Iberian plant. Several species came from Spain.

Ibicella (Pedaliaceae) Van Eselt.
La. *Ibis* = wading bird; *-ella* = diminutive. The woody capsule has a stout, curved beak like the ibis.

Iboza = ***Tetradenia*** (Lamiaceae) N.E.Br.
The isiZulu vernacular name of *Tetradenia riparia;* apparently referring to the aromatic qualities of its leaves.

Idiospermum* (Annonaceae) S.T. Blake
Gk. *idio-* = unusual; *spermum* = seed. The plants have adapted a unique poison, a chemical called Idiospermuline contained within the seed, to prevent animals eating them.

Idothea** = ***Drimia*** (Asparagaceae) Kunth
Latin form of Gk. *eidothea* = a prophetic sea-nymph, a daughter of the shape-shifting sea god Proteus. The 'knowing goddess'. Many of the species of this genus are coastline plants.

Ifloga (Asteraceae) Cass.
Anagram of the related genus *Filago*.

Ihlenfeldtia (Aizoaceae) H.E.K.Hartmann.
For Hans-Dieter Ihlenfeldt (1932–), German professor of botany at Hamburg specialising in morphology and taxonomy of the families Mesembryanthemaceae (now in Aizoaceae) and Pedaliaceae. He conducted succulent plant research in South Africa, mainly in Namaqualand and the Transvaal (now Gauteng), where he collected over 2 000 specimens. Among his many publications, he authored *Schumannia* 4/Biodiversity & Ecology 2: 1–250. He was the 8th recipient of the Cactus d'Or award by the International Organisation for Succulent Plant Study in 1994.

Ilex (Aquifoliaceae) L.
Ilex aquifolium refers to the common name 'holly'. This is often confused *Quercus ilex,* the classical Latin name for the 'holm oak' used by Virgil. The leaves of these two different genera are said to be somewhat similar.

Illecebrum* = ***Guilleminea*** (Amaranthaceae) L.
La. *illecebra* = attraction, charm, allurement; referring to the tiny white flowers that are borne in clusters along the stem like an attractive coral necklace.

Illigera* (Hernandiaceae) Blume
For Johann Karl Wilhelm Illiger (1775–1813), German entomologist and zoologist. He studied under the entomologist Johann Hellwig (who became his father-in-law) at the University of Helmstadt and later worked on the vast plant and animal collection of Johan Centurius Hoffmannsegg in Brunswick. Hoffmannsegg was the founder of the Zoological Museum of Berlin in 1809 and proposed Illiger for the position of curator of the Zoological Garden of Berlin as well as professor and director of the Zoological Museum from its formation until his death. Illiger was the author of *Prodromus Systematis Mammalium et Avium* (1811), which was an overhaul of the Linnaean system. It was a major influence on the adoption of the concept of the 'family'. He published and edited a large number of works, many of which appeared in *Magazin für Insektenkunde,* widely known as *Illiger's Magazine*.

Ilysanthes = ***Lindernia*** (Linderniaceae) Raf.
Gk. *ilys, ilyos* = mud; *anthos* = flower. Some species are aquatic or semi-aquatic.

Imantophyllum = ***Clivia*** (Amaryllidaceae) Benth. & Hook.f.
La. *imantophyllum* = strap-, ribbon-, or thong-like leaves. The long thin leaves are characteristic of this genus.

Imitaria* = ***Gibbaeum*** (Aizoaceae) N.E.Br.
La. *imita* = imitate, copy; *aria* = pertaining to. Brown called the genus *Imitaria*, as it imitates the shape of a *Conophytum*. However, the plants are now viewed as *Gibbaeums* and *Imitarias*.

Impatiens (Balsaminaceae) L.
La. *impatiens* = impatient; referring to the bursting

of the capsule and forceful scattering of seeds upon the slightest touch, which occurs in some species.

Imperata (Poaceae) Cirillo

A pan-tropic grass in sandy areas, commonly known as Blady-grass (EJM)

For Ferrante Imperato (1550–1625), Italian apothecary and author of *Dell'historia Naturale,* an illustrated catalogue divided into 28 books relating to mining, alchemy, animals and vegetable specimens collected in southern Italy. He also formed a museum ('*museo*') in Naples of natural history specimens, one of Europe's first, which was continued by his son Francesco Imperato who assisted him by writing up his observations. Ferrante Imperato was among the first naturalists to identify the processes through which fossils formed. He corresponded with many of the leading Italian scholars of the day, including botanists. He had a small garden, but botanical historians discount his interest in plants as 'curiosa'.

Indigastrum (Fabaceae) Jaub. & Spach.
La. *indicus,* Gk. *indikos,* referring to India; *astrum* = partially resembling, inferior; referring to some similarity to *Indigofera.*

Indigofera (Fabaceae) L. (Flowers on page 168)
Indigo is derived from the La. *indicus,* Gk. *indikos,* referring to India; La. *ferax* = bearing. Indigo is blue dye (cf *I. tinctoria*).

Inezia (Asteraceae) E.Phillips
For Inez Clare Verdoorn (1896–1989), South African botanist, taxonomist and plant collector.

After her working experience as a herbarium assistant in Pretoria (1917–1924) and at Kew Gardens (1925–1927), she was placed in charge of the National Herbarium, Pretoria, with the rank of senior professional officer (1944–1951). She is credited with more than 300 botanical and general publications, many major revisions of plant families, which appeared mainly in *Bothalia, Flowering Plants of Africa, Flora of Southern Africa, Kew Bulletin,* and the *Journal of South African Botany,* and she was a co-author of *Wildflowers of the Transvaal* (1962). She became president of the South African Biological Society in 1957 and, *inter alia,* received an honorary PhD from the University of Natal in 1957.

Inga = Entada (Fabaceae) Scop.
A Tupi (West Indian) name for these trees.

Ingenhoussia** (Fabaceae) E.Mey.
For Jan Ingenhousz or Ingen-Housz (1730–1799), Dutch physiologist, biologist and chemist. He

is best remembered for showing that light is essential to plant cellular respiration, a vital step in the discovery of photosynthesis. He was a physician to the Austrian Empress Maria Theresa. He carried out research in electricity, heat conduction and chemistry and met both Benjamin Franklin and Henry Cavendish. In 1785 he described the irregular movement of coal dust on the surface of alcohol, and therefore has a claim as discoverer of what came to be known as Brownian motion. Ingenhousz was elected a Fellow of the Royal Society of London in 1779.

Inhambanella (Sapotaceae) Engl. Dubard
From its place of origin, Inhambane in Mozambique, plus La. *-ella* = diminutive.

Intsia (Fabaceae) Thouars
From the Malagasy word *intsia*, meaning 'there it is'.

Inula (Asteraceae) L.
Gk. *inaein* = to clean (e.g. wounds); *inula*, an ancient name associated with the plant *elecampane* (now called *Inula helenium*). Inulin, a carbohydrate fibre, can be extracted from the rhizomatous roots of these plants, which have debatable medicinal value. Pliny the Elder wrote that *I. viscosa* strengthened teeth and healed wounds and stomach upsets.

Inulanthera (Asteraceae) Källersjö.
Inula (q.v.); *anthera* = anther.

Inuloides = *Osteospermum* (Asteraceae) B.Nord.
Inula (q.v.); Gk. *-oides* = resembling.

Inversodicraea = *Ledermanniella* (Podostemaceae) Engl.
La. *inversus* = inverse, turned over, bent downward; Gk. *dikraios* = cleft, forked. The flower buds of *Inversodicraea* have an inverted position inside the spathella.

Iocaste = *Phymaspermum* (Asteraceae) E.Mey.
Meaning obscure. Possibly named after Jocasta (Iocaste; in Homer, Epicaste). In Greek mythology, Oedipus, meaning 'swollen foot', was born to King Laius and Queen Jocasta. For reasons not given here, Laius wanted to kill his son. Thus, he fastened the infant's feet together with a large pin and left him to die on a mountainside. Some species of *Phymaspermum*, such as *Phymaspermum acerosum*, have root stems virtually 'pinned together'. Gk. *phyma* = an external nodule or swelling.

Iphigenia = *Wurmbea* (Colchicaceae) Kunth
Meaning obscure. Gk. *iphios* = strong, mighty; *genia* = born. In Greek myths, Iphigenia was the daughter of Agamemnon and Clytaemnestra.

Iphiona (Asteraceae) Cass.
Gk. *iphios* = strong, mighty. *Iphion* was a potherb of some kind.

Ipomoea (Convolvulaceae) L.
Gk. *ips, ipos* = worm or bindweed; *(h)omoios* = similar to, resembling; therefore, worm-like, referring to the plant's twining habit or possibly its coiled flower bud.

The common Morning glory (EJM)

*Ireon*** (Roridulaceae) Burm.f.
Meaning obscure. Speculatively derived from Gk. *ireos* = a hot spice. Called *cassia* in Greece, possibly made from *Cinnamomum cassia* (family Lauraceae), the inner bark of several species make the spice cinnamon. This genus has an unresolved name.

Iresine (Amaranthaceae) P.Browne
Gk. *eiresione* = a wreath or staff entwined with strips of wool; referring to the (hairy) trichome-covered flowers or possibly the woolly appearance of the seed.

Iris (Iridaceae) L.
Named after Iris, the Greek goddess of the rainbow = rainbow coloured, iridescent; referring to the wide variety of colours that this species has.

Ischaemum (Poaceae) L.
Gk. *ischaimon* = styptic, stopping the blood flow, from *ischo* = to restrain and *haima* = blood. The woolly seeds of the type species were reported as being used to control or stop bleeding.

*Ischnolepis** (Apocynaceae) Jum. & H.Perrier.
Gk. *ischnos* = thin, weak, withered; *lepis* = scale; probably referring to the thin and filiform corona segment of the species composing the genus.

Ischyrolepis (Restionaceae) Steud.
Gk. *iskhyros* = strong; *lepis* = scale. The restio is strongly scaled, referring to the sheaths of the plant or the rhizomes usually covered with imbricate scales.

Isoberlinia (Fabaceae) Craib & Stapf
Gk. *isos* = equal; *Berlinia* (q.v.). Flowers of this genus are described as being similar to those of *Berlinia* but with subequal petals about as long as the sepals.

Isoetes (Isoetaceae) L.
Gk. *isos* = the same, equal, identical; *etos* = the year (*etas* = green); referring to the 'evergreen' character of the plant; some of the species have the same look throughout the year.

Isoglossa (Acanthaceae) Oerst.
Gk. *isos* = equal; *glossa* = tongue; possibly referring to the tongue-like petals. The corolla is two-lipped.

Isolepis (Cyperaceae) R.Br.
Gk. *isos* = equal; *lepis* = scale; referring to the glumes of the flower.

Isolobus = **Lobelia** (Lobeliaceae) A.DC.
Gk. *isos* = equal; *lobus* = lobes. The five corolla lobes are identical in size and shape.

Isolona* (Annonaceae) Engl.
Gk. *isolona* = equal petals. The six petals of this genus are clearly fused into a tube, and the lobes are equal in size and shape.

Isopogon (Proteaceae) R.Br. ex Knight
Gk. *isos* = equal; *pogon* = beard, bearded; referring to the equal-length hairs on the fruit.

Itasina (Apiaceae) Raf.
La. *ita* = in this matter, thus; *sinus* = recess, rounded inward between two projecting lobes, bay. Meaning unclear.

Ixia (Iridaceae) L.

(CM)

Ancient Gk. *Ixia* = a Linnaeus-derived name for a plant noted for the variability of its flower colour or Gk. *ixos* = mistletoe (viscum), birdlime; referring to the viscous sap (WPU Jackson).

Ixianthes (Scrophulariaceae) Benth.
Gk. *Ixia* (q.v.); *anthos* = flower.

Ixora (Rubiaceae) L.
A Portuguese rendering of the Sanskrit *israra* (= Lord) referring to the Hindu god Siva (Hugh Glen).

J

Jamesbrittenia: a genus of well over 100 species mostly found in South Africa

John Manning (1962–) was born and grew up in the KwaZulu-Natal Midlands and read for his PhD at the then University of Natal (Pietermaritzburg) under the guidance of Olive Hilliard, Joyce Stewart and Hannes van Staden. Although primarily a plant taxonomist, his wide-ranging interests include pollination biology, anatomy, evolution and speciation, and popular botany. He has published several Southern African wildflower guides, mostly illustrated with his photographs. He works in close association with Peter Goldblatt of the Missouri Botanic Garden, and together they have described over 200 new species, mostly in Iridaceae but also in several other families. He works at the South African National Biodiversity Institute in Kirstenbosch, Cape Town.

Jacaranda * (Bignoniaceae) Juss.
Derived from the Brazilian vernacular name of this genus.

Jacksonago* = *Wiborgia (Fabaceae) Kuntze
For Benjamin Daydon Jackson (1846–1927), British botanist and taxonomist, who wrote the first volume of *Index Kewensis,* a reference book that appeared from 1893–1895, describing all the flowering plants, and which was accepted as the authority throughout the world for names of flowering plants. He authored nine books (botanical, botanic history and bibliographies) and many shorter publications. Aged 22 he became a Fellow of the Linnaean Society and then secretary from 1880–1902. Jackson was awarded an honorary PhD from Uppsala, created a knight of the Swedish Order of the Polar Star (1907), and in honour of his 80th birthday in 1926, the Linnaean Society appointed him curator of the Linnaean Collections. (La. *ago* = resemblance or connection).

Jacobaea (Asteraceae) Mill.
Possibly after St James (Jacobus), one of the 12 apostles, because the plant flowers around St Jacobs day, July 25. The phrase '*conditio Jacobaea*' means 'If the Lord wills it'. The author may have had in mind that this ragwort plant was only tolerated as part of God's will. St James is usually depicted as an older man with grey or greying hair and somewhat unkempt beard. The plant's leaves and stems are covered with long, thinly to thickly matted grey-white to white hairs.

Jacobaeastrum* = *Euryops (Asteraceae) Kuntze
Jaobaea (q.v.); *-astrum* = partially resembling, inferior, wild.

Jacobsenia (Aizoaceae) Bolus & Schwantes
For Hermann Johannes Heinrich Jacobsen (1898–1978), German botanist and algologist, a specialist in succulent plants who was curator of the Kiel Botanical Garden (1929–1963). Among other achievements, he built, with Gustav Schwantes, one of the largest collections of *Mesembryanthemum* (Aizoaceae) plants. He was a prolific author and wrote numerous horticultural and botanical books (some translated to English) as well as many articles. Two of his invaluable reference works are *A Handbook of Succulent Plants* (1960) and *Lexicon of Succulent Plants* (1974). He was a founding member of the International Organisation for Succulent Plant Study; vice-president of the African Plant Society and a member of the Cactus and Succulent Society of Great Britain, among others. He was elected Fellow of the Linnaean Society of London in 1974 and received a Cactus and Succulent Society of America Fellow Award in 1971.

Jacquemontia (Convolvulaceae) Choisy
For Venceslas Victor Jacquemont (1801–1832), French natural historian and botanist, explorer and plant collector for the Royal Museum of Paris. Jacquemont's travels took him from France to the United States, Haiti, and finally India in 1828. For over three years, he travelled extensively from Calcutta (now Kolkata) to Delhi and Lahore gathering plants, minerals, making animal sketches and conducting field surveys. He reached the Himalayas, where he made many early descriptions of (Alpine Himalayan) Indian flora. He also made an incursion into Tibet and China before returning to Delhi. He died of amoebic fever in a military hospital in Bombay in 1832 aged 31. The story of his travels in India was published by the French government under the title *Letters from India* in 1835.

Jacquesfelixia* = *Danthoniopsis (Poaceae) J.B.Phipps.
For Henri Jacques-Félix (1907–2008), French botanist, tropical agronomist, explorer and plant collector who worked in West Africa (Guinea, Ivory Coast, Cameroon) and was affiliated with the Musée National d'Histoire Naturelle, Paris, France. He authored La *Vie et La Mort du Lac Tchad* (*The Life and Death of Lake Chad*) (1947), *Géographie des Dénudations et Dégradations du Sol au Cameroun* (*Geography of Denudation and Degradation of Soil in Cameroon*) (1950) and *Les graminées* (Poaceae) *d'Afrique Tropicale* (*Grasses in Tropical Africa*) (1962), among many others. Note: The longest-living botanist we have come across.

Jamesbrittenia (Scrophulariaceae) Kuntze
(Flowers on page 173 and opposite)
For James Britten (1846–1924), who was born in London and lived there his entire life. He was educated privately with the intention of becoming a medical doctor but favoured botany and accepted a position as an assistant at the Kew Gardens herbarium from 1869–1871. He was

Jamesbrittenia (CM)

subsequently transferred to the botany department at the British Museum and worked there until his retirement in 1909. Britten published a number of dictionaries of British plants and botanists but was also an expert on Old English dialects and folklore and a devout Catholic who devoted time to social upliftment projects. He was evidently much admired by Otto Kuntze, who named *Jamesbrittenia* for him, as a strong upholder of the Principle of Priority in plant nomenclature and as a longtime editor of the *Journal of Botany*, a post he filled for 45 years.

Jamesonia (Pteridaceae) Hook & Grev.
For William Jameson (1796–1873), Scottish botanist, chemist, explorer and plant collector.

Jameson studied at Edinburgh's Royal College of Surgeons (c 1814–1818), then served as ship's surgeon from 1818–1826, travelling to Greenland, Baffin's Bay, Ecuador and other South American countries, sending specimens back to Britain. He described some Greenland flora in *Memoirs of the Wernerian Natural History Society* (Volume III: 416, 1821). In 1827 he was appointed professor of chemistry and botany at Universidad Central del Ecuador. He was also appointed an assayer to the mint in 1832 and director in 1861. He wrote a flora of Ecuador, *Synopsis Plantarum Aequatoriensium* (in Spanish), the first two volumes being published in 1865 and a partial third volume later. He went back to Edinburgh in 1869 as a result of a revolution at the university and returned to Quito in 1872 to see two of his sons (one of whom had died), and he died shortly thereafter.

Jasminum (Oleaceae) L.
Latinised from the Persian name *yasmin* = a fragrant shrub.

Jateorhiza (Menispermaceae) Miers
Gk. *iatros* = physician; *rhiza* = root; alluding to the plant's healing virtues for anorexia and anaemia.

Jatropha (Euphorbiaceae) L.
Gk. *iatros* = physician; *trophe* = food, nutrition. Hugh Glen comments that, 'some species have medicinal value, but most are poisonous.' (In 2005, Western Australia banned *J. gossypiifolia*, as it is invasive and highly toxic to humans and animals.)

Jaumea (Asteraceae) Pers.
For Jean Henri Jaume Saint-Hilaire (1772–1845), French botanist, naturalist, historian and artist who compiled, *inter alia*, a major work, *Exposition des Familles Naturelles et de la Germination des Plantes*, comprising some 2 337 genera and about 4 000 species with 112 plates drawn by himself (1805), and between 1808–1822 published *Plants of France Described and Painted from Nature* (10 volumes, 1808–1822), incorporating 1 000 engravings (prints) made by himself, and also contributed to *Dictionaire des Sciences Naturelle*, *Journal de Devaux* and *Annales de L'Agriculture Françoise*. During the 1820s he became interested in issues related to forests and later still in alien plants and those deemed harmful to livestock. He became a member of the Royal Agricultural Society in 1831.

Jensenobotrya** (Aizoaceae) A.G.J.Herre
For Emil Jensen (1889–1963), German naval officer, accountant and amateur botanist. During World War I, he joined the German Royal Navy and became a lieutenant. In 1936 he emigrated to Tsumeb, South-West Africa, where his interest in succulents commenced. From 1940 to 1943 he was held at the Andalusia internment camp in South Africa, then repatriated to Germany. In Andalusia he attended lectures by Hans Herre, custodian of Stellenbosch Herbarium, which considerably improved his botanical knowledge. Jensen had a special interest in Namib flora, especially *Welwitschia bainesii*, and for a number

of years made expeditions to the central and southern Namib Desert (and elsewhere), often with botanical colleagues such as Herre, Willy Giess and others. He published a few botanical papers and made a detailed study of the Nara plant (*Acanthosicyos horrida*). In 1955 he returned to South-West Africa, now Namibia, where, *inter alia*, he was responsible for developing the outstanding botany section of the Swakopmund museum. Gk. *botrys* = bunch of grapes; referring to the thick, rounded leaves.

Jordaaniella (Aizoaceae) H.E.K.Hartmann
For Pieter Gerhardus Jordaan (1913–1987), South African professor of botany at Stellenbosch

University and plant collector. He studied at Stellenbosch University (1931–1937), graduating with an MSc and PhD in 1944 with post-graduate research at Cambridge and Leiden universities in 1952. He was assistant director of the National Zoological Gardens, Pretoria (1937–1939), lecturer in botany at Stellenbosch University (1940–1946), senior lecturer (1947–1952) and professor and head of department from 1953–1978. He collected about 1 000 specimens from Bredasdorp, Caledon and Stellenbosch and worked mainly with Proteaceae and on biographies.

Jubaeopsis (Arecaceae) Becc.
For the genus *Jubaea,* named for King Juba of Numidia, ancient North African country; Gk. *-opsis* = resembling. This palm is closely related to *Jubaea.*

Juglans* (Juglandaceae) L.
Latin name for a walnut, from La. *jovis* = of Jupiter; *glans* = nut. Juglans comes from Roman mythology – the glans of Jupiter, which has sexual connotations. The drooping glans (seeds) on the tree are walnut sized.

Julbernardia (Fabaceae) Pellegr.
For Marie Joseph Jules Pierre Bernard (1876–?), French lieutenant-governor of Gabon from 1924–1931. He was awarded the Legion of Honour for reasons unknown to the authors.

Jumellea (Orchidaceae) Schltr.
For Henri Lucien Jumelle (1866–1935), French botanist, professor at the Marseilles Faculty of Sciences and Director of the Musée Colonial of Marseilles. With Joseph Marie Henry Alfred Perrier de la Bâthie (1873–1958), he wrote *Palms of Madagascar* (1913), though there is no

evidence he ever went there, and published, posthumously, *Flore de Madagascar et des Comores* (1945). A prolific author and researcher, he wrote *The History of Rubber Plants* (1898), *Physiological Research on the Development of Plants* (1889), *Cocoa: Its Culture and Operations in All Producing Countries* (1900), *Industrial and Medicinal plants* (1901), *The Agricultural and Forest Resources of the French Colonies* (1907) and *Colonial Cultures* (1916), just to mention a few in his early career.

Juncellus* = ***Cyperus*** (Cyperaceae) C.B. Clarke
Gk. *juncus* = rush; *-ellus* = diminutive.

Juncus (Juncaceae) L.

(CM)

Name for a rush, possibly from La. *jungere* = to tie together, bind; referring to the ancient practice of using rushes to make ropes.

Juniperus* (Cupressaceae) L.
La. *juniperus,* classic name for the genus, possibly derived from La. *junio (iunio)* = young; *parere* = to produce; hence, youth producing – some of the shrubs are evergreen.

Juncus islands in the Mgeni River mouth (GN)

Juncus: Bundles of 100s of sleeping mats made from the stems – awaiting transport to Johannesburg (EJM)

Justicia (Acanthaceae) L.
For James Justice (1698–1763), Scottish horticulturist and writer and owner of two estates, Justice Hall (Berwickshire) and Crichton (Midlothian), where most of his gardening experiments took place. He introduced many ornamental plants to Scotland and cultured new varieties, including the first pineapple in Scotland. He spent large sums in importing foreign seeds, roots and trees; collecting tulips was one of his passions. He authored a well-received book, *The Scots Gardener's Director* (1755), a work based on his practical experience of fruit gardening relating to the soil and climate of Scotland.
His passion for introducing new plants brought about his financial ruin, and he had to sell his house and garden. He was a Fellow of the Royal Society.

Juttadinteria (Aizoaceae) Schwantes
For Helena Jutta Dinter (née Schilde), wife of German botanist and explorer Moritz Kurt Dinter

(1868–1945). She met him in 1905 when he returned to Germany from South-West Africa (now Namibia) to hand over some 9 000 specimens of his herbarium to the Berlin-Dahlem Museum. She followed him to South-West Africa, and they married in 1906. Moritz Dinter, who was the botanist for the German-controlled territory, went on many expeditions between 1906–1914 accompanied by his wife, an energetic and enthusiastic collaborator. The couple went back to South-West Africa again from 1922–1925 to collect botanical material, after which they returned to Germany.

K

Kniphofia: a genus native to Africa

Himansu Baijnath (1943–) hails from Durban in KwaZulu-Natal. He completed his PhD on the African genus *Bulbine* at the University of Reading while based at the Jodrell Laboratory, Royal Botanic Garden, Kew. His research interests include the petaloid monocots (e.g. *Kniphofia*). On his retirement in 1997 as the first curator of the university's Ward Herbarium, he was appointed as an honorary research professor at the then University of Durban-Westville. He continues in an honorary capacity at the University of KwaZulu-Natal's Westville campus with research and postgraduate teaching.

Kaempferia* (Zingiberaceae) L.
For Engelbert Kaempfer (1651–1716), German physician, scholar, explorer, collector, musician, naturalist and linguist. He visited Russia, Persia, India, Southeast Asia and Japan between 1683 and 1693 and was the chief surgeon for the United East India Company in the Persian Gulf (1685–1693); in Japan, where he stayed for two years, he 'discovered' the *Ginkgo biloba* tree thought to be extinct by Western scientists. On his return to Europe in 1695 he became physician to the Count of Lippe. Among his works he authored the monumental *The History of Japan* – published posthumously, in 1727, in English – and *Amoenitatum Exoticarum* (1712), which contains the first extensive description of Japanese plants, the electric eel and acupuncture.

Kalaharia (Lamiaceae) Baill.
Tswana *Kgala* = the great thirst, or *Khalagari, Kgalagadi* or *Kalagare,* meaning 'a waterless place'. This is the name for a desert in southwestern Africa, mainly Botswana.

Kalanchoe (Crassulaceae) Adans.
The Chinese name for one of the species, *jiā lán cài* = temple vegetable = *K. ceratophyllum*.

Kalmia* (Ericaceae) L.
For Pehr (Pietsri) Kalm (1716–1779), Swedish-Finnish explorer, botanist, naturalist and agricultural economist. He studied at the universities of Åbo (now Helsinki) (1735–1739) and Uppsala (1740–1741), under Linnaeus. He did research in Sweden, Russia and the Ukraine (1742–1745) before becoming a senior lecturer (*docent*) (1746) then professor  of natural history and economics at the Academy of Turku (1747). In 1748 he was sent by the Academy of Sciences to North America and Canada to find seeds and plants that might be useful for agriculture or industry. He wrote a three-volume account with the English title *Travels into North America* (1770–1772), and visited the Niagara Falls, Montreal and Quebec. He returned to Finland in 1751 to his post as professor at the Turku Academy and taught there until his death.

Kalosanthes** (Crassulaceae) Haw.
Gk. *kalos* = beautiful; *anthos* = flower. The flowers are elegant.

Kamiesbergia = Hessea (Amaryllidaceae) Snijman
Named after a mountain range in the Northern Cape, where the genus was discovered.

Kanahia (Apocynaceae) R.Br.
Latinised form of *kanak*, the Arabic name for plants of this genus.

Kaokochloa (Poaceae) De Winter
Herero/Khoisan *kaoko* = referring to the 'left side' of the Kunene (Cunene) river in northwestern Namibia; Gk. *chloa* = grass; probably named after the Kaokoveld region in Namibia (South-West Africa) where the genus was first identified.

Kappia = Chlorocyathus (Apocynaceae) Venter, A.P.Dold & R.Verh.
Kappia is derived from the name of the reserve in which the plant was found. This plant was found near the confluence of the Fish and Kap rivers not far from Ndlambe (Port Alfred) in the Eastern Cape.

Karomia (Lamiaceae) Dop

A small tree found in semi-arid bushveld with persistent calyx lobes (EJM)

For Karom in Vietnam, where the first species of this genus was found.

Karroochloa = Rytidosperma (Poaceae) Conert & Türpe
Khoisan Karoo *garo, karo, !garo-b* of uncertain etymology = dry (desert); Gk. *chloa* = grass; grass from the Karoo.

Kaukenia*** = Manilkara, etc (Sapotaceae) Kuntze
Supposedly a Javan name, *kauken* (or *kauki*) *indorum,* for this or some similar plant.

Kaulfussia = Felicia (Asteraceae) Nees
For Georg Friedrich Kaulfuss (1786–1830), German botanist, bryologist and pteridologist, author and professor of botany and a professor

at the University of Halle from 1823. In 1824 he published *Enumeratio Filicium* (*A Detailed Listing of Ferns*), a text based upon the travels and collections of Ludolf Karl Adelbert von Chamisso, who served as botanist on the Russian voyage of exploration on the *Rurik* led by Otto von Kotzebue. Chamisso collected some 165 ferns, of which 77 were new to science. Kaulfuss honoured Chamisso by naming a small tree fern collected in Hawaii in 1821 after him.

Kedrostis (Cucurbitaceae) Medik.
Gk. *kedros* = cedar; *kedrostos* = made from cedar wood. Ancient Greek name for white bryony, another member of the *Cucurbitaceae* family.

Keetia (Rubiaceae) E.Phillips
For Johan Diederik Möhr Keet (1882–1976), South African forester and plant collector, director of forestry in South Africa 1935–1942. During his career he was controller of timber during World War II, a technical advisor to the division of soil conservation and extension at the department of agriculture (1945–1955), also acting in an advisory capacity in South-West Africa, and technical advisor for the development of the Westfalia Estate, northeastern Transvaal, and residential director of afforestation from 1956–1969. He collected fungi and other specimens around Stellenbosch, in the Knysna area, in the eastern and northeastern Transvaal, and in South-West Africa. Stellenbosch University awarded him an honorary PhD in 1964.

Kellaua = Euclea (Ebenaceae) A.DC.
A derivation of the names *Khella* (possibly Arabic) or *kellau* (possibly Ethiopian).

Kennedia* (Fabaceae) Vent.
For John Kennedy (1759–1842) English nurseryman, son of Lewis Kennedy (c 1721–1782) the original founder of the Kennedy and Lee nursery, together with his partner James Lee (1715–1795). Started in about 1745 and in operation for three generations, the nursery they ran in Hammersmith, west of London, was called The Vineyard. John Kennedy, who was raised in the business from a young age, succeeded his father on his death as did James Lee (the younger) (1754–1824). Among John Kennedy's clients was Empress Josephine, Napoleon I's first wife. He was a frequent contributor to the first five volumes of *The Botanist's Repository* (1799–1803), writing most of the notes, as well as providing the illustrations, and he also wrote *A Treatise Upon Planting, Gardening, and the Management of the Hot House* (1777).

Kensitia = Erepsia (Aizoaceae) Fedde
For Harriet Margaret Louisa Bolus (née Kensit) (1877–1970), South African botanist, daughter-in-law and grand-niece of Harry Bolus, and curator of the Bolus Herbarium from 1903 until her retirement in 1955. In her early career she did research in several overseas herbaria, including Kew Gardens (five visits), Carl Peter Thunberg's collection in Uppsala and Jacquin's types in Vienna. She studied *Ericaceae* and *Orchidaceae,* later *Iridaceae* and *Aizoaceae,* wrote many scholarly and popular publications, including the two-volume *Book of South African Flowers* (1925–1936). She was elected a Fellow of the Royal Society of South Africa in 1920 and was awarded an honorary DSc by Stellenbosch University in 1942.

Kentrophyllum (Asteraceae) Neck.
Gk. *kentron* = a spur; *phyllon* = a leaf; referring to its fleshy, rounded, spur-shaped leaves.

Kentrosiphon*** = Gladiolus (Iridaceae) N.E.Br.
Gk. *kentron* = a spur; *siphon* = tube; referring to the dimorphic perianth tube, slender below and abruptly widening into a cylindrical upper part.

Khadia (Aizoaceae) N.E.Br.
South Sotho *khadi* = starchy roots. The name of a beverage prepared from the roots of plants by the indigenous people of that area.

Khaya (Meliaceae) A.Juss.
From a West African (Senegambian) vernacular name for these trees, related to mahogany.

Kickxia* (Scrophulariaceae) Dumort.
For Jean Kickx Sr (1775–1831) – Belgian professor of botany, pharmacy and mineralogy at a medical school in Brussels, and author of *Flora bruxellensis* (1812) and *Tentamen Mineralogicum* (1820) – and his son Jean Kickx Jr (1803–1864), who succeeded him (having obtained PhDs in pharmacy and science from Leuven in 1825), and took charge of the Science Museum in 1831. Later, he became a professor at the University of Brussels (1834)

and Ghent (1835). He was a great mycologist, describing over 500 new taxa and authoring *Flore Cryptogamique Des Environs de Louvain* (*Cryptogamic Plants of Leuven*) (1835), which contains 754 species, and *Flore Cryptogamique des Flandres* (*Cryptogamic Plants of Flanders*), a work completed and published posthumously in 1867. A grandson, Jean Jacques Kickx (1842–1887), also became a professor of botany at the University of Ghent.

Kigelia (Bignoniaceae) DC.

A single fruit may weight 7–10 kg and be ±400 mm long. (EJM)

The large flowers hang inside the foliage and are bat pollinated. (EJM)

For the Mozambican name *kigeli-keia* for this tree.

Kiggelaria (Achariaceae) L.
For Francois (Franz) Kiggelaer (1648–1722), Dutch botanist, apothecary, traveller, plant collector and curator of Dutch plant collector Simon van Beaumont's garden in the Hague (some sources refer to Leiden), and author of *Horti Beaumontii Catalogus Plantarum Exoticarum* (1690), which listed Cape plants. He also collaborated with Frederik Ruysch (1638–1731), a Dutch anatomist and professor of botany at the Hortus Botanicus of Amsterdam, with the compilation of the first volume of Jan Commelin's (1629–1692) *Amstelodamensis Rariorum Horti Medici* published in 1697, dealing with mainly the plants of the West Indies.

Killickia (Lamiaceae) Bräuchler, Heubl & Doroszenko
For Donald Joseph Boomer Killick (1926–2008), South African botanist and plant collector. He

studied at the University of Natal, obtaining a PhD in 1962. He worked for the Botanical Research Institute from 1950, became liaison officer at Kew Gardens (1954–1957) and officer in charge of the Botanical Survey (1963–1966), then in charge of the flora research team. He returned to Kew as liaison officer (1969–1971) and was promoted to assistant director and then director of the Botanical Research Institute in 1973. He became a Fellow of the Linnaean Society in 1966 and was also president of the Biological Society. He published a number of papers on ecology and taxonomy and authored *A Field Guide to the Flora of the Natal Drakensberg* (1990). He is credited with having collected some 5 000 specimens, mainly in the Drakensberg mountains.

Kirkia (Kirkiaceae) Oliv.
For John Kirk (1832–1922), Scottish explorer, naturalist, physician and diplomat. He studied

medicine at the University of Edinburgh (1847–1854), served with civil medical staff during the Crimean war and joined David Livingstone's second expedition (1858) but returned to Britain in 1863 because of ill health. In 1866 he returned to Zanzibar and became consul general and acting surgeon in 1873. He played a major role in securing the abolition of slavery from Sultan Seyyid Bargash and negotiated a lease of territory that led to the founding of British East

Africa. He was very interested in the practical and pharmaceutical uses of plants, wrote a revision of economically important East African palms, and had a substantial collection of non-marine molluscs.

Kissenia (Loasaceae) R.Br. ex Endl.
For Kishin or Qishn, a locality on the south coast of Arabia, where the original specimen was collected. Originally, through a comedy of errors, it was thought that Kissen was a traveller in Arabia, but it has now been clearly established that this genus is named after a location, not a person.

Klattia (Iridaceae) Baker
For Friedrich Wilhelm Klatt (1825–1897), German botanist, a high school teacher in Hamburg, researcher and author. He obtained an honorary PhD from the University of Rostock for his revision of the Iridaceae family, contributed to many publications, including the multi-volume *Conspectus Florae Africae*, *Flora Brasiliensis*, *Flora of Central Brazil* and *The Botany of German East Africa,* and wrote extensively about the Compositae in Australia, Brazil, Columbia, Costa Rica, Guatemala, German East Africa and Madagascar. His name is on the Klatt Herbarium of Compositae at Harvard University.

Kleinia (Asteraceae) Mill.
For Dr Jacob Theodor Klein (1685–1759), German mathematician, lawyer and diplomat  employed by the Danziger administration and in the diplomatic service. He was a keen naturalist, and his countless diplomatic journeys enabled him to collect many specimens and have one of the largest private nature collections of the 18th century. He also founded the world, famous Danzig botanical garden. An author of many works, he questioned the Linnaean classification (unsuccessfully), and the taxonomic method in his book *Summa Dubiorum Circa Classes Quadrupedum et Amphibiorum in Celebris Domini Caroli Linnaei Systemate Naturae* (1743) is based entirely on externally visible characteristics. He was a member of the Royal Society of London.

Klenzea *** (Asteraceae) Sch.Bip.
Possibly for Leo von Klenze (Franz Karl Leopold von Klenze) (1784–1864), German architect, painter, draughtsman, art collector and writer. After completing his architectural studies and apprenticeship, he became court architect to Jérôme Bonaparte, King of Westphallia (1808–1813) and to Bavarian King Ludwig I, from 1816. He designed, *inter alia,* many neoclassical buildings – temples, museums, galleries – in Munich and Regensberg, submitted plans for the restoration of the Acropolis in Athens, Greece, and designed the New Hermitage public museum in St Petersburg, Russia. Klenze's collection of contemporary German painters is now housed in the Neue Pinakothek Museum in Munich.

Klingia = ***Gethyllis*** (Amaryllidaceae) Schonl.
For Erich Kling (1854–1892), German traveller, explorer, naturalist in West Africa, army officer and plant collector, possibly working for the German East Africa Company, in search of treaties of friendship and protection with local tribes as part of imperial Germany's colonial efforts. In 1889 Kling went to West Africa, specifically to Togo, bordered by Ghana to the west, Benin to the east and Burkina Faso to the north. He went to Dapaon in Togo, Borgou, one of the 12 'departments' of Benin, Sousou in Guinea, and Tchaoudjo in the central region of Togo, but was turned back because of the hostile population and African resistance. He fell ill several times during the expedition and died shortly after he returned to Germany.

Knightia **** (Proteaceae) R.Br.
For Thomas Andrew Knight (1759–1838), English vegetable physiologist and horticulturalist. His practical, large-scale research on 4 000 hectares of land he inherited involved  physiological experiments on plants aimed at improving fruit and vegetable qualities and preventing diseases among them. He raised new varieties of apples, cherries, strawberries, plums, nectarines, pears, potatoes, cabbages and peas, many of which bear his name. His major work was *Treatise on the Culture of the Apple and Pear: And on the Manufacture of Cider and Perry* (1797); he also wrote over 100 papers. He was president of the London (later 'Royal') Horticultural Society, founded in 1804, from 1811–1838, and elected a Fellow of the Royal Society (1805), which awarded him the Copley Medal in 1806, and became a Fellow of the Linnaean Society in 1807.

Kniphofia (Asphodelaceae) Moench
For Johann(es) Hieronymus Kniphof (1704–1763), German physician, lecturer, professor of medicine at the University of Erfurt (1737), becoming dean of the medical faculty (1747) and rector from 1661 until his death. Author of a folio of nature-printed illustrations of plants in 1733, followed by a significantly expanded second edition in

Just another beautiful Red-hot poker (EJM)

1747 and third edition in 1757–1759 containing 1 200 botanical images that were produced by a somewhat unique process, whereby dried plant specimens were coated with printer's ink and pressed on paper, resulting in a silhouette effect. His book *Botanica in Originali Seu Herbarium Vivum* (1757–1764) was the first significant work to follow Linnaeus's nomenclature.

Knowltonia = Anemone (Ranunculaceae) Salisb.
For Thomas Knowlton (1691–1781), English horticulturist, well known in his lifetime as a botanist

and gardener with a special interest in nature, wildflowers and hothouse exotics. He was the director of the once famous botanical garden at Eltham. His life story, *No Ordinary Gardener*, was written by Blanche Henrey (British museum, 1896).
He designed many gardens for the wealthy and collected and grew plants from around the world.

Kobresia = Schoenoxiphium (Cyperaceae) Willd.
For Paul von Kobres (Cobres) (1747–1823), Austrian botanist, plant collector, geologist, mineralogist, banker and patron of botany. He was the owner of a rich natural history collection and library, which was purchased in 1811 for the Bavarian Academy of Sciences.

Kochia (Amaranthaceae) Roth
For Wilhelm Daniel Joseph Koch (1771–1849), German botanist, state physician (1798–1823)

and professor of botany at Erlangen (1824–1849), where he was also director of its botanical gardens. He studied medicine at the universities of Jena and Marburg. Among his publications, he wrote *De Plantis labiates. Programma* (1833), a treatise on German and Swiss flora titled *Synopsis Florae Germanicae and Helveticae* (1835–1837), and *Catalogus Plantarum Florae Palatinae* (*Catalogue of Palatinate Flora*) (1814). In 1833, he was elected a foreign member of the Royal Swedish Academy of Sciences.

Koeleria (Poaceae) Persl.
For Georg Ludwig Koeler (1765–1807), German botanist, physician and pharmacologist, professor of botany at Mainz, and author of a work on the grasses of Germany and France, *Descriptio Graminum in Gallia et Germania* (1802).

Kogelbergia (Stilbaceae) Rourke
Named after the 100 000-hectare UNESCO-registered Kogelberg Biosphere Reserve, which lies 60 kilometres east of Cape Town.

Kohautia (Rubiaceae) Cham. & Schltdl.
For Francisci (Franz) Kohaut (?–1822), Czech plant collector and gardener from Neuhaus, Bohemia, who accompanied Franz Wilhelm Sieber, a Czech naturalist and explorer, to Crete, Egypt and Palestine (1816–1818) and afterward was also contracted to collect specimens in Martinique (1819–1821). Kohaut died in Senegal in 1822, while on an expedition.

Kohleria (Gesneriaceae) Regel
For Johann Michael Kohler (1810/1816–c 1884), Swiss teacher of natural history in Zurich, horticulturalist and author. He was deeply involved in the botanical gardens and horticulture and edited horticultural journals for a number of years, also publishing books associated with horticulture such as *Landwirtschaftliche Ortsbeschreibungen aus dem Kanton Zurich* (*Land Management of the Canton of Zurich*) (1853) and *Aufzählung und Beschreibung der wichtigsten Kern-Obstsorten* (*Enumeration and Description of the Main Core of Fruit in the Canton of Zurich*) (1869). Later in his

life his interest seems to have shifted toward wine making. He published books on new techniques in wine making such as *Der Weinstock und der Wein: mit Besonderer* Berücksichtigung *des Schweizerischen Weinbaus* (*The Vine and the Wine: With Special Reference to Swiss Viticulture*) (1869, 1874) and *Neueste Fortschritte in der Weinbereitung, als Ergänzung* (*Recent Advances in Wine Making*) (1871) and other publications.

Kohlrauschia** = Petrohagia
(Caryophyllaceae) Kunth
For Henriette Kohlrausch (née Eichmann) (1781–1841), a German botanist, educator and historian, wife of Dr Heinrich Kohlrausch.

Kolleria*** (Aizoaceae) C.Presl
For Franz von Koller (1767–1826), Austrian military officer from 1784 who worked his way up the ranks – colonel (1805), general (1809), field marshall (1813) – to become commander in chief of the Austrian army in 1820. He fought in France, Belgium, Germany and Italy, and took part in the final battle against Napoleon. He accompanied Napoleon to Elbe, where he completed the construction of a castle (1821–1826). He was honoured with aristocracy and made a baron and received numerous honours, including the Aspro Cross Maria Theresa Award for bravery. During his career he also served as governor of Bohemia and Austrian ambassador in London. Von Koller left a huge collection of antiques, especially large vases from Pompeii, now in the Berlin Museum. Carl Presl describes him as a lover and promoter of science. He died of typhus.

Kosteletzkya (Malvaceae) C.Presl
For Vincenz Franz Kosteletzky (1801–1887), Czech botanist and physician, professor of medicine and botany in Prague and director of the Botanical Garden Smíchov (Prague). In 1834 he published a monumental 2 237-page work on medicinal plants entitled *Allgemeine Medizinisch-Pharmazeutische Flora*. He was also the author of *Clavis Analytica in Florum Bohemiae Phanerogamicam* (1824) and *Index Plantarum Horti Caesarii Regii Botanici Pragensis* (1844).

Kotschya (Fabaceae) Endl.
For Georg Theodor Kotschy (1813–1866), Austrian botanist who collected plants in North Africa and western Asia and discovered hundreds of new species. It is estimated he collected over 300 000 specimens. Over the period 1835–1862 he took part in numerous expeditions to Egypt, Sudan, Cyprus, southern Iran and Zagros mountains, northern Tehran, southern Turkey, Palestine, Lebanon, Kurdistan, and northern Syria, during which time he became an assistant (1847) and then curator (1852) of the herbarium of the Vienna Natural History Museum. Among his works are *Abbildungen und Beschreibungen Neuer und Seltener Thiere und Pflanzen, in Syrien und im Westlichen Taurus Gesammelt* (*Illustrations and Descriptions of New and Rare Animals and Plants, in Syria and Western Turkey*) (1843), *Überblick der Vegetation Mexico* (*An Overview of Mexico's Vegetation*) (1852) and *Coniferen des Cilicischen Taurus* (*Conifers of the Cilician Taurus*) (1855) with Franz Antoine, and many others.

Krauseola** (Caryophyllaceae) Pax & K.Hoffm.
For Ernst Hans Ludwig Krause (1859–1942), German botanist, medical doctor, cryptogamist, prolific batologist (person who studies brambles) and plant collector. He was a professor of botany at Strasbourg (1904–1914) and Rostock (1919–1933), served in the Naval Medical Corps in Kiel and on the battleship Gneisenau and became a surgeon in Saarlouis and Rostock after World War I. He collected plants in France and Germany in Europe, in Liberia in Africa, the West Indies, and Grenadines and Virgin Islands in the Caribbean (c 1884–1930). He authored *Flora of Rostock and The Neighbourhood* (1879), *Mecklenburg Flora* (1893), *Rostock Moss Flora* (1921–1922) and others. His original general herbarium was deposited at Berlin-Dahlem (1929) as was his bryophyte herbarium of some 2 000 specimens (1935) and his fungal herbarium (1941). The remaining parts, those not destroyed during World War II, are housed in the Mecklenburg Herbarium at the University of Rostock.

Kraussia (Rubiaceae) Harv.
For Christian Ferdinand Friedrich von Krauss (1812–90), German naturalist, explorer and collector. In 1838, Krauss, an apothecary with a PhD in mineralogy, zoology and chemistry from Tübingen and Heidelberg (1836), sailed for the Cape. Here, he collected many specimens, especially molluscs and crustaceans, but also made a study of the geology, flora and fauna. From 1838–1839 he explored the areas between Cape Town and Port Elizabeth and in 1840

the bush and seashore around the Congella River, Pietermaritzburg. He left to join the Natural History Museum, Stuttgart, in 1940, becoming its director in 1956. He wrote *Die Südafrikanischen Crusaceen* (1843) and *Die Südafrikanischen Mollusken* (1848).

Krebsia = **Stenostelma** (Apocynaceae) Harv.
For Georg Ludwig Engelhard Krebs (1792–1844), German apothecary, naturalist and chemist,

botanist and botanical collector who came to the Cape in 1817 on a four-year contract to the apothecary firm Pallas & Poleman. From 1820 to 1840, Krebs collected, mainly in the Eastern Cape, for the Berlin Natural History Museum on behalf of the Royal Prussian Department of Education and Medicine. He sent 14 huge consignments to Berlin. His 12th consignment consisted of 7 245 dried plant specimens, 900 birds, 7 000 insects, a rhinoceros, an elephant, a quagga (now extinct), and a complete preserved 'Bushman' (Khoisan). In 1844 he was recommended for the King of Prussia's Order of the Red Eagle for meritorious contributions to science but died before it could be awarded.

Kumara = **Aloe** (Asphodelaceae) Medik.
An Indian name – *kumari* or *kumara* – for *Aloe vera*. Some sources state a Polynesian name *kumara* for sweet potato.

Kunzea* (Myrtaceae) Rchb.
For Gustav Kunze (1793–1851), German botanist and doctor. He studied medicine at the University

of Leipzig in 1813, joined the Wernerian Natural History Society in Edinburgh in 1817, qualified as a doctor in 1819 and was appointed as a zoology professor in 1822. As a result of his keen interest in botany and entomology, he was made an associate professor of botany at Leipzig in 1835, appointed director of the Botanical Gardens in Leipzig in 1837 and became a full professor in 1845. He collected in America, specifically in Louisiana in the early 1800s. His special interests were ferns, orchids and microscopic fungi. He was elected a foreign member of the Royal Swedish Academy of Sciences in 1851. He published a number of monographs on beetles, fungi and lichens. He published, with J.K. Schmidt, *Mykologische Hefte* (*Mycological Notebooks*) in two volumes (1817–1823).

Kyllinga (Cyperaceae) Rottb.
For Peder Lauridsen Kylling (c 1640–1696), Danish botanist. He studied theology at the University of Copenhagen and briefly served as a parish minister before withdrawing from this profession to study botany. His major work is *Viridarium Danicum* (*The Danish Garden*) (1688), which describes some 1 100 species that occurred in the crown lands of the Danish king, mainly in Zealand, Justland and Slesvig. It is alleged that he was working on an even larger edition, but this book was not published.

Kyllingiella = **Cyperus** (Cyperaceae)
R.W.Haines & Lye
Kyllinga (q.v.); Gk. *-iella* = diminutive.

Kyphocarpa (Amaranthaceae) (Fenzl) Lopr.
Gk. *kypho-* = bent; *karpos* = fruit. The ovary carries a 'distinct horn on one side below its apex' (Robert Allen Dyer).

L

Lampranthus is a genus of succulent plants indigenous to southern Africa.

Ernst van Jaarsveld (1953–) obtained his national diploma in horticulture in 1974 and started work in the Lowveld Botanical Gardens, Nelspruit (now Mbombela) and was transferred to Kirstenbosch in 1976 to take over the succulent section. In 1990 he graduated from what is now the University of KwaZulu-Natal with an MSc and in 2012 with his PhD from the University of Pretoria for his detailed study of the taxonomy and ecology of cliff dwelling plants (the first such detailed study globally of what are termed Cremnophytes). He is an acknowledged world authority on succulent plants and has published many scientific and popular articles and written many books. In 2015 he took early retirement to join the Babylonstoren horticultural team at the historic Cape Dutch farm Babylonstoren in the Western Cape.

Lablab (Fabaceae) Adans.
An ancient Arabian name for this single species genus, originally from Africa but now widespread. It has at least 34 common names.

Laburnum (Fabaceae) Fabr.
An old name given to this tree by Pliny the Elder, still retained. Speculatively, derived from La. *laborum* = labour. The tree has historically been used for cabinetmaking.

Labourdonnaisia (Sapotaceae) Bojer
For Count Bertrand François Mahé de La Bourdonnais (1699–1753), French naval officer and governor of Mauritius.  He joined the French East India Company and became a lieutenant in 1718 and captain in 1824 while taking part in the capture of Mahé off the Malabar Coast (southwestern India), then French governor of Mauritius and Réunion in 1735. In 1740 he was given command of the French fleet in Indian waters and captured Madras from the British in 1746. He quarrelled with Dupleix (in charge of India) over the surrender terms and also learned that Dupleix had been appointed governor of Mauritius. In anger he returned to France, where he was jailed for charge of gubernatorial wrongdoing – misusing his position as governor. After two years as a prisoner in the Bastille, despite being acquitted, his spirit was broken and he died of ill health.

Lachenalia (Hyacinthaceae) J.Jacq. ex Murray
For Werner de Lachenal (1736–1800), Swiss professor of botany and anatomy at the University  of Basel from 1776, eminent for his knowledge of European plants. He obtained his PhD in 1763. He was a pupil of Haller, who was one of his main correspondents, providing him with details of flora and their location around Basel, the Jura mountains, Alsat and Bruntrutain. He was a friend of Linnaeus. He authored several monographs in *Acta Helvetica*. While at the university he substantially improved its botanical garden, the oldest in Switzerland, that had fallen into disrepair. He continually strived to obtain funds to reconstruct and develop the garden and to pay for its gardener. He opened the garden to the public to cover expenditures.

Lachnaea (Thymelaeaceae) L.
Gk. *lachne* = woolly hair; alluding to the downy calyx.

Lachenalia: One of the many beautiful small Cape geophytes (GAN)

Lachnopylis = **Nuxia** (Stilbaceae) Hochst.
Gk. *lachne* = woolly hair; *pyle* = gate. The mouth of the corolla tube is woolly.

Lachnosiphonium = **Catunaregam** (Rubiaceae) Hochst.
Gk. *lachne* = woolly hair; *siphonium* = little tube. Meaning unclear.

Lachnospermum (Asteraceae) Willd.
Gk. *lachne* = woolly hair; *sperma* = seed. The seed is woolly.

Lachnostylis (Phyllanthaceae) Turcz.
Gk. *lachne* = woolly hair; *stylis* = style. The styles are woolly.

Lactuca* (Asteraceae) L.
Latin for lettuce, from *lactis* = milky; alluding to the milky latex exuding from the plant when pressed or crushed.

Lafoensia* (Lythraceae) Vand.
For João Carlos de Bragança (1719–1806), Portuguese Duke of Lafões (Lafoens), a nobleman, politician, traveller and founder of the Royal Academy of Science of Lisbon. When his elder brother died in 1761, King José I refused

him succession to the Duchy of Lafões. Such was the king's enmity that in 1757 De Bragança, aged 38, left Portugal and did not to return for 21 years. He visited England, becoming a member of Royal Society of London, and took part in the seven-year Austro-Prussian War. Between 1763–1777 he visited most Eastern and Western European countries and parts of Asia and North Africa. In 1778, he returned to Portugal aged 59, on the death of José I. The new Queen Maria I restored his title. With the help of the young scientist José Correia da Serra he founded the Academy of Sciences of Lisbon and later became Portugal's army marshal.

Lagarinthus = Schizoglossum (Apocynaceae) E.Mey.
Gk. *lagaros* = narrow; *inthos* = a pre-Greek ending, the Anatolian suffix meaning 'belonging to'. Meaning unclear.

Lagarosiphon (Hydrocharitaceae) Harv.
Gk. *lagaros* = narrow; *siphon* = tube; referring to the long thin tubes that allow the tiny white female flowers to reach above the water's surface.

Lagenaria (Cucurbitaceae) Ser.
La. *lagena* = flask; *-aria* = pertaining to. The fruits are bottle-shaped.

Lagenocarpus = Erica (Ericaceae) Klotzsch
La. *lagena* = flask; Gk. *karpos* = fruit. The small dry fruit (achenes) are flask- or bottle-shaped.

Lagenophora (Asteraceae) Cass.
Gk. *lagenos* = flask, flagon; *phoros, pherein* = bearing; referring to the achene, which has a pronounced neck.

Lagerstroemia* (Lythraceae) L.
For Carl Magnus von Lagerström (1691–1759), Swedish (-Pomeranian) director of the Swedish East India Company. He worked in various jobs before he became an accountant and treasurer (1731) for the Swedish East India Company, being promoted to the positions of secretary (1743) then junior director (1746). Between 1743 and 1745 he took a trip to China and brought back, for Linnaeus, plants that were hitherto unknown. Magnus had a good knowledge of science and worked for the Swedish

Academy of Sciences where he translated texts into Swedish. In 1748 and 1750 he gave his natural history collection to Linnaeus, who named *Lagerstroemia* in his honour upon his death.

Laggera (Asteraceae) Benth.
For Dr Franz Josef Lagger (1802–1870), Swiss physician and botanist. He studied medicine in Dijon and Freiburg (Fribourg) and practised in Freiburg in Switzerland. He collected specimens in Switzerland and Italy.

Lagunaria* (Malvaceae) (DC.) Rchb.
For Andrés Laguna de Segovia (1499–1559), Spanish physician, pharmacologist and botanist. He studied art and medicine at the universities of Salamanca and Paris and practised medicine in Spain, France, England, the Netherlands and Italy. Wherever he went he collected herbal remedies and not only verified all the prescriptions of Dioscorides – he was fluent in Greek and Latin – but wrote *Annotations on Dioscorides of Anazarbus* (Lyon, 1554), a much expanded version of Dioscorides's *Materia Medica* (c 65). He became Pope Julius III's (1487–1555) personal physician while in Italy (1545–1554), as well as doctor to Charles V (1500–1588), Holy Roman Emperor and King of Spain, and to his son Phillip II (1527–1598). In 1557 he returned to Spain, where he created the Botanical Garden of Aranjuez.

Lagurus* (Poaceae) L.
Gk. *lagos* = a hare; *oura* = tail; referring to the woolly heads at the end of the grass spikelets resembling 'hare's-tail' or 'bunnytail'.

Lagynias*** = Vangueria (Rubiaceae) E.Mey. ex
Origin unknown, but *lagynos* was a flask that Aristotle said contained 12 cupfuls.

Lamarckia* (Poaceae) Moench
For Jean-Baptiste Pierre Antoine de Monet Lamarck (1744–1829), French soldier, naturalist, zoologist, palaeontologist, conchologist and evolutionist. After a military injury in 1766 he resigned from the army and studied medicine, then botany, for 10 years. In 1778 he published a three-volume work, *Flora Française*, and became a royal botanist (1781), keeper of the Royal

Garden herbarium (1788), and curator of this garden and professor of invertebrate zoology at the Museum d'Histoire Naturelle in Paris (1793). While Lamarck's evolutionary theories (evolution through inheritance of acquired characters) have been discredited, his seven-volume *Histoire Naturelle des Animaux Sans Vertèbres* (1815–1822) is recognised as a lasting contribution to zoology.

Lambertia* (Proteaceae) Sm.
For Aylmer Bourke Lambert (1761–1842), British botanist. He studied at Oxford University for three  years, and thereafter he devoted himself to the study of natural history, which he could do as he had become financially independent when he inherited estates in Jamaica and Ireland. He developed one of the largest herbaria and libraries of its time. He published *A Description of the Genus Cinchona* (1797), *A Description of the Genus Pinus* (1803, 1824) and was a contributor to *Botanist's Guide through England and Wales* (1805) by Dawson Turner and Lewis Western Dillwyn, and *English Botany* (1790–1814) by James Sowerby and James Edward Smith. Thanks to JE Smith and Joseph Smith, he became one of the first fellows of the Linnaean Society (1788), was appointed vice-president of the society (1796–1842), and became a council member of the Royal Society in 1810.

Lamium* (Lamiaceae) L.
Gk. *lamion*, diminutive of *lamia* = a devouring monster, from *lamyros* = glutinous; referring to the galeate (helmet) shape of the flower, which has the general appearance of open jaws.

Lampranthus (Aizoaceae) N.E.Br. (Flowers on page 186)
Gk. *lampros* = bright, shining; *anthos* = flower; referring to the light reflecting off the glossy petals.

Lamprocaulis = Elegia (Restionaceae) Mast.
Gk. *lampros* = bright, shining; *caulis* = stem; possibly referring to the bright green stems.

Lamprocephalus (Asteraceae) B.Nord.
Gk. *lampros* = bright, shining; *kephale* = head; possibly referring to the inflorescence.

Lanaria (Lanariaceae) Aiton
Gk. *lana* = wool; *-aria* = connected with. *Lanaria lantana* is known as 'Cape Edelweiss' or 'lambtails', referring to the flower's woolly inflorescence.

Lanaria (CM)

Lancisia = Cotula (Asteraceae) Lam.
For Giovanni Maria Lancisi (1654–1720), Italian cardiologist, epidemiologist, professor of anatomy  at the Sapienza University of Rome and physician to Pope Innocent XI, Pope XII and Pope Clement XI. He studied at the University of Rome and qualified in medicine aged 18. Among his many epidemiological insights was that mosquito-infested swamps are the breeding ground for malaria. He recommended these be drained to prevent the illness and proposed that the most effective way to get rid of rinderpest in cattle was 'to kill all sick and suspect animals' instead of allowing the disease to spread while searching for a cure. He also published extensively on cardiology and the classification of heart disease. His landmark book, *De Motu Cordis et Aneurysmatibus* (*On the Motion of the Heart and on Aneurysms*) was published posthumously in 1728.

Landolphia (Apocynaceae) P.Beauv.
For Jean-François Landolphe (1747–1825), French naval officer and business owner. In 1769 he went to Benin in West Africa (the 'slave coast') and traded there between 1769 and 1792. He had a factory at a river mouth and tried to establish businesses like timber, sugar cultivation and trading of slaves. In 1972, he became involved in

a fight against the British, who wished to prevent progress in the colony, and was captured but later released. In 1795 he was raised to the rank of captain and participated in the capture of 63 ships in the Caribbean. During the Napoleonic Wars (1799), he commanded a French cruising squadron that was defeated by the British Navy. He was captured but released in 1802 with the signing of the Peace Treaty of Amiens and returned to France, retired and published his memoirs in 1823.

Landtia = Haplocarpha (Asteraceae) Less.
For Jørgen Landt (c 1753–1804), Danish pastor, botanist and researcher of the Faroe Islands. From 1777–1791, he was a private teacher with a special interest in botany, in Copenhagen and Frederiksborg. In 1791, he was appointed pastor of the Faeroe Islands, where he worked for the Nordstreymoy municipality. On behalf of the Natural History Society of Denmark, he gathered details about the natural history of the Faroe Islands. After seven years' work there, he left the islands in 1798 as a result of a knee injury. He wrote *Forsøg til en Beskrivelse over Færøerne* (1800) and (as George Landt) *A Description of the Faroe Islands* (1810).

Langebergia (Asteraceae) Anderb.
Presumably after the Langeberg mountains in the Western Cape.

Lannea (Anacardiaceae) A.Rich.
This name is derived from *lanné* = native name of the plant in Senegal and/or Gambia. Palmer suggests La. *lana* = wool; referring to the dense woolly hairs that cover the young parts of the tree or the roots of some species.

Lantana (Verbenaceae) L.
The name is derived from the medieval Latin name for a species of *Viburnum lantana,* whose foliage it resembles slightly in terms of its flexible branches, leaf shape and black fruits.

Lapeirousia (Iridaceae) Pourr.
For Philippe Isidore Picot, former Lord of  Lapeyrouse (Lapeirouse), Baron de Bazus (1744–1818), French botanist, geologist, and mineralogist. He was professor of natural history at Toulouse University, first Dean of the Faculty of Science, Mayor of Toulouse (1800–1807) under Napoleon, Secretary of the Academy of Sciences of Toulouse (from 1811), and creator of the Natural History Museum. He explored the entire Pyrenees twice, the second

These small irises are widespread in the Cape in spring. (CM)

time with Déodat de Dolomieu (1750–1801). He was the author of papers on public education, geology, the mines and agriculture, and of the work *Histoire abrégée des plantes des Pyrénées et itinéraire des botanistes dans ces montagnes* (*Brief history of the plants from the Pyrenees and botanists' routes in the mountains*) (1818).

Lapidaria (Aizoaceae) (Dinter & Schwantes) N.E.Br.
La. *lapis* = a stone; *-aria* (cf *Lithops*); alluding to the plant looking like a group of stones or growing in a stony area.

Laportea (Urticaceae) Gaudich.
For François Louis Nompar de Caumat de Laporte, Count of Castelnau (1810–1880), English-born  French diplomat, naturalist, traveller, entomologist and plant collector. He led an expedition to Canada (1837–1841) and another in South America (1843–1847) exploring the Amazon basin for the Duc d'Orléans and Musêe National d'Histoire Naturelle. During 1856–1857 he came to the Cape travelling eastward to Port Elizabeth (Algoa Bay) and subsequently wrote a treatise on South African fish (1861) and a number of papers, although he lost many of his notes when his deranged servant used them to light a fire. He served as French consul in Bahia, Brazil (1848), Siam, Thailand (1848–1862) and Australia (1864–1877).

Lappula (Boraginaceae) Moench
La. *lappa* = a burr; *-ula* = small, little (diminutive); referring to the seeds.

Lapsana* (Asteraceae) L.
Latin name for wild mustard or 'charlock', Gk. *lapsane* = a weed alleged to have healing properties, common name 'nipplewort'. Each of the lower leaves of *communis* typically has a large terminal lobe and two small side lobes, the latter resembling a pair of elongated 'nipples'.

Larochea** (Crassulaceae) Persl.
For Daniel Delaroche (de la Roche) (1743–1813), Swiss botanist and physician. He received a medical degree with a botanical thesis from the University of Leiden in 1766 and later studied at Edinburgh, and then practised as a physician in Geneva for 10 years. In 1782 he moved to Paris for political reasons. He was appointed physician to the Swiss Guards (protecting the king). In 1792, during the Revolution he left Paris for London and then Lausanne, Switzerland. In 1798 he returned to Paris where he continued pursuing his interest in botany and aided the young Swiss botanist Augustin Pyramus de Candolle. From 1802 he was a physician at the Maision de Sante. He contributed to the eight-volumed *Bibliothèque Germanique Medical Chirurgicale* (1798–1802) and wrote *Encyclopédie Méthodique: Chirurgie* (1790–1792).

Larryleachia** (Apocynaceae) Plowes
For Leslie Charles (Larry) Leach (1909–1996), British-born amateur taxonomic botanist, electrical  engineer and plant collector in Angola, Botswana, Namibia, Mozambique, South Africa and Zimbabwe. He emigrated to Rhodesia (now Zimbabwe) in 1938. From 1956, he devoted himself to a study of succulent plants, particularly Stapeliae, Euphorbieae and the genus *Aloe*. He was honorary botanist at the National Herbarium of Zimbabwe (1972–1981), then employed at the National Botanical Garden, Worcester, South Africa (1982–1989) and from 1990 at the department of botany, University of Pietersburg (now University of Limpopo). He authored a number of works on succulents and was a Fellow of the Cactus and Succulent Society of America (1983). He received, *inter alia*, the Harry Bolus Medal (1968).

Lasiagrostis* = *Stipa (Poaceae) Link
Gk. *lasios* = hairy, shaggy; *Agrostis* (q.v.). The genus resembles *Agrostis*, except that the lemmas and awns are hairier.

Lasiochloa* = *Tribolium (Poaceae) Kunth
Gk. *lasios* = hairy, shaggy; *chloa* = grass; referring to the lemmas.

Lasiocoma* = *Euryops (Asteraceae) Bolus
Gk. *lasios* = hairy, shaggy; *coma* = tuft of hair or leaves. Possibly because the foliage on some species is woolly looking.

Lasiocorys* = *Leucas (Lamiaceae) Benth.
Gk. *lasios* = hairy, shaggy; *corys* = helmet. The upper lip of the corolla is fringed.

Lasiodiscus (Rhamnaceae) Hook.f.
Gk. *lasios* = hairy, shaggy; *diskos* = disk. The disk is thick, the fruit velvety.

Lasiopogon (Asteraceae) Cass.
Gk. *lasios* = hairy, shaggy; *pogon* = beard; referring to the feathered pappus.

Lasiopus* = *Gerbera (Asteraceae) Cass.
Gk. *lasios* = hairy, shaggy; *-pus* = foot (*pous*, a foot) alluding to the woolly (foot)stalks reaching up to the flower heads.

Lasiosiphon* = *Gnidia (Thymelaeaceae) Fresen.

(NH)

Gk. *lasios* = hairy, shaggy; *siphon* = tube. This group of plants (now included in *Gnidia*, q.v.) is characterised by woolly flower tubes.

Lasiospermum (Asteraceae) Lag.
Gk. *lasios* = hairy, shaggy; *sperma* = seed; referring to the woolly seeds.

Lasiostelma** = *Brachystelma*** (Apocynaceae) Benth.
Gk. *lasios* = hairy, shaggy; *stelma* = crown; referring to the appearance of the corona.

Lastrea = ***Thelypteris*** (Thelypteridaceae) Bory
For Charles Jean Louis Delastre (1792–1859), French lawyer and botanist, mayor of St Benedict, sub-prefect of Loudun of Gien and Melle, and author of *Flore Analytique et Descriptive du Département de la Vienne* (1842). He corresponded a great deal with the South African-born mycologist Christiaan Hendrik Persoon.

Lastreopsis* (Dryopteridaceae) Ching
Lastrea (q.v.); Gk. *-opsis* = resembling; a related genus.

Lathriogyna* (Fabaceae) Eck. & Zeyh.
Gk. *lathraios* = hidden; *gyne* = woman (female parts). William Henry Harvey states that the reference is to the ovary being largely hidden in the extremely hairy calyx (*Flora Capensis*).

Lathyrus* (Fabaceae) L.
Gk. *lathyros* = very passionate; referring to the ancient belief in the plant's supposed aphrodisiacal powers.

Launaea (Asteraceae) Cass.
For Jean Claude Mien Mordant de Launay (c 1750–1816), French lawyer, naturalist and lover of the arts. He became an assistant librarian at Museum d'Histoire Naturelle in Paris from 1794 and later became librarian of the Natural History Museum of Havre. From 1798–1794 he was in charge of the menagerie at the Jardin des Plantes in Paris. He authored a number of works that include (in translation) *The Good Gardener*, an almanac that he edited every year from 1804 and the *General Herbarium [for the] Amateur* (1811–1812).

Laurembergia (Haloragaceae) P.J.Bergius
For Peter Lauremberg (1585–1639), German botanist, school teacher and professor of medicine. He studied medicine and astronomy at Rostock (1606) and medicine at Leiden (1608) and became a doctor of medicine in Paris (1611). That year he lectured at the University of Montauban. He returned to Germany in 1614, becoming professor of mathematics and physics at the Hamburg Academic Gymnasium. From 1624–1639, he was professor of poetry, mathematics and medicine at the University of Rostock, becoming rector in 1635. He wrote many textbooks, including *Introduction to Anatomy* (1619), *Mathematics Textbook* (1621), *Introduction to Medicine* (1630), *Horticulture Textbook* (1632) and an encyclopaedic school book.

Laurentia = ***Solenopsis*** (Lobeliaceae) Mich. ex Adans.
For Laurent Garcin (1683–1751) French physician and botanical explorer. He studied medicine and surgery in the Netherlands from 1695–1705 (university unknown) but may not have completed his dissertation. From 1705–1720 he served in a Dutch military regiment visiting Spain, Portugal and Flanders. Between 1720 and 1729 he made three voyages to the East Indies for the Dutch East India Company, visiting Java, Arabia, Persia, India, Ceylon and Malaysia as ship's surgeon and botanical explorer, having been briefed by Herman Boerhaave (1668–1738), as regards his botanical explorations and collections, the results of which were so satisfactory that Linnaeus named a tree, *Garcinia moluccas*, after him. On his return he obtained his PhD in medicine and thereafter practised medicine, but seemingly during the period 1734–1549 also wrote articles for the *Mercury Swiss* and *Swiss Journal* of Neuchâtel. He was made a Fellow of the Royal Society of London in 1730.

Lauridia (Celastraceae) Eck. & Zeyh.
La. *laurus* = laurel; *-idia* = diminutive; indicating a fairly distant resemblance to a laurel tree.

Laurophyllus (Anacardiaceae) Thunb.
La. *laurus* = laurel; Gk. *phyllon* = a leaf. The tree has leaves like the laurel family.

Laurus* (Lauraceae) L.
La. *laurus* = laurel. The leaves of this genus are often associated with the wreath of bay laurels given at the Pythian games and as a symbol of success, hence 'resting on one's laurels'.

Lavatera* (Malvaceae) L.
For the Lavater brothers: Johann Heinrich (1611–1691), Swiss physician, professor of medicine and natural history at the Collegium Carolinum, Zurich; and Johann Jacob (1594–1636), also a physician and naturalist, about whom little is known. They were the sons of Heinrich Lavater (1560–1623). Johann Heinrich obtained his PhD in Basel in 1647, became town physician in Bern in 1653 and later worked in Zurich, where he drew up the Zurich Ordinance, relating to the plague, in 1668. The Lavater family is so extensive that details are clouded. A genealogy search of Lavater shows that Johann Heinrich was also called by six other name variations.

Lavandula* (Lamiaceae) L.
Possibly Old French *lavandre,* Latinised as *lavare* = to wash. The early Egyptians, Greek and Romans used the essential oils of this plant for

many purposes, e.g. its healing and antiseptic qualities, its usefulness in deterring insects, but also as a cleanser and for its fragrance.

Lavrania** (Apocynaceae) Plowes
For John Jacob Lavranos (1926–), Greek-born South African insurance broker, botanist, botanical explorer and plant collector of over 4 500 specimens. He obtained a BSc in economics from Athens University, came to South Africa in 1953 and obtained a BSc and MSc from the University of the Witwatersrand in natural science. He undertook over 14 journeys to little-known areas in southern Arabia, Somalia, Socotra, Kenya and Tanzania concentrating on succulent flora and discovering about 50 new species. He published a number of articles in the US *Cactus and Succulent Journal*, dealing chiefly with succulent Asclepiadaceae, *Aloe* and other succulent groups. Volume 61 of *Flowering Plants of Africa* is dedicated to him.

Leachia*** = **Larryleachia** (Apocynaceae) Plowes
For Leslie Charles (Larry) Leach (1909–1996), British-born amateur taxonomic botanist (q.v. *Larryleachia*).

Leachiella = **Lavrania** (Apocynaceae) Plowes
Leachia (q.v.); Gk. *-iella* = diminutive.

Lebeckia (Fabaceae) Thunb.
For Heinrich Julius Lebeck (1772–1800), Dutch botanist born in Ceylon (now Sri Lanka), whose family made the acquaintance of Carl Peter Thunberg, the renowned botanist, ichthyologist and disciple of Linnaeus, who was on his way to Japan. Subsequently, Lebeck studied natural history at Uppsala, Sweden, under Thunberg. In 1795 he made a voyage to the Cape, Ceylon, India and Java. While at the Cape he visited the Caledon area and published a paper in *Der Naturforscher* (1802) on his experience. He sent his herbaria – plants, insects, minerals, shells, snakes, birds, etc. – to Thunberg, who named the South African plant genus *Lebeckia* after him.

Lecomtedoxa (Sapotaceae) (Pierre ex Engl.) Dubard
For Paul Henri Lecomte (1856–1934), French botanist and professor at Lycée Saint-Louis in Paris who worked voluntarily in the botany laboratory at the Museum d'Histoire Naturelle in Paris. In 1906, having worked as a volunteer for 20 years in the museum, he became paid

director of the Laboratoire de Phanerogamie (spermatophyte division) and an author of some 15 books, including works on the trees of Indochina and the trees and flowers of Madagascar. He took part in many scientific expeditions to North Africa, Egypt, the Antilles, French Guiana and French Indo-China. He was elected a member of the French Academy of Sciences. Gk. *doxa* = glory.

Ledebouria (Hyacinthaceae) Roth
For Carl Friedrich von Ledebour (Ledebur) (1785–1851), German-Estonian botanist and professor of botany at the University of Tartu, Estonia (1811–1836), traveller and plant collector and previously a teacher in natural history and director of the botanical gardens in Greifswalde (Prussian Pomerania). His most significant works were *Flora Altaica*, the first flora of the Altay Mountains (1833) and the four-volume *Flora Rossica* (1841–1853), the first complete flora of the Russian empire. One of the new species he discovered was *Malus sieversii* (described as *Pyrus sieversii*), which is considered to be the sole ancestor of the cultivated apple.

Ledermanniella (Podostemaceae) Engl.
For Carl Ludwig Ledermann (1875–1958), Swiss-born German horticulturist, explorer and prolific plant collector in Europe, the Malesian region, Papua New Guinea, Cameroon, Congo, Polynesia and Micronesia (Caroline Islands) for the Berlin-Dahlem Botanical Museum, Royal Botanical Gardens and other institutions. He studied at the University of Heidelberg, went to the tropics in 1904 and collected some 6 500 plants in Victoria, German Cameroon (now Limbe, Cameroon). In 1912–1913 he botanised in German New Guinea, collecting more than 6 600 plants and, in 1913–1914, he travelled alone to the Caroline Islands, where he made 1 400 collections, interrupted by World War I. He wrote *A Botanical Walk to German Adamawa* (1912).

Leersia (Poaceae) Sw.
For Johann (Georg) Daniel Leers (1727–1774), German botanist, mycologist, bryologist, algologist

and apothecary. He trained in Nuremberg as a chemist and became a pharmacy manager and finally in 1756 a university pharmacist in Herborn. His main activity was collecting plants around Herborn, which he identified according to the Linnaean classification system. Much of his time was spent drawing and engraving grasses – he etched 16 copper plates with 104 grass varieties. His 422-page book, *Flora Herbornensis,* was only published after his death in 1775. The advertisement for his book said, 'To be carried in the pocket during walks', a precursor of the paperback.

Lefebvrea (Apiaceae) A.Rich.
For Charlemagne Théophile Lefebvre (1811–1860), French explorer and writer who participated in a voyage from 1838 to 1844 to study the natural history, geography, anthropology, linguistics, archaeology and customs of Abyssinia. He published *Voyage en Abyssinie Exécutés Pendant les Années 1839, 1840, 1841, 1842, 1843 ...* in 1845 with many co-authors including Achille Richard, one of the leading botanists of his day. In 1842 Stephan Friedrich Endlicher apparently changed the genus name to *Lefeburea,* with no explanation, leading to confusion, resulting in WPU Jackson's mistaken attribution of the genus to Louis-François Henri Lefébure (1754–1839), a French botanist.

Leidesia (Euphorbiaceae) Müll. Arg.
Anagram of *Seidelia* (q.v.).

Leiocarya = ***Trichodesma*** (Boraginaceae) Hochst.
Gk. *leios* = smooth; *karyon* = nut.

Leiothylax (Podostemaceae) Warm.
Gk. *leios* = smooth; *thylakos* = a bag, sack, pouch. Meaning obscure.

Leipoldtia* (Aizoaceae) L.Bolus
For Christian Frederik Louis Leipoldt (1880–1947), South African physician, poet, author, epicure, journalist, traveller and plant collector of over 1 000 specimens, especially of aloes and succulents. He started his career as a journalist (1898–1902 and 1922–1925), his first volume of poems being published in 1911. He studied medicine at Guy's Hospital, London (1903–1907), specialised in children's diseases (1910–1911), became chief medical inspector of schools (1914–1919) and part-time lecturer on children's diseases at UCT, also being secretary of the medical council and editor of the *South*

African Medical Journal until 1944. He received an honorary DLitt from the University of the Witwatersrand in recognition of his literary output.

Lellingeria* (Polypodiaceae) A.R.Sm. & R.C.Moran
For David Bruce Lellinger (1937–), US botanist specialising in ferns. He obtained a BA from the University of Urbana and an MSc (1960) and PhD (1965) from the University of Michigan. He joined the staff of the Smithsonian Institute in 1963 and became curator of pteridophytes at the institute, which houses the largest and most diverse collection of ferns in the Western Hemisphere, in excess of 20 000 species. Among his many publications, he authored *A Field Manual of the Ferns & Fern-Allies of the United States & Canada* (1985), *The Ferns and Fern-Allies of Costa Rica, Panama, and the Choco* (1989), where he conducted field work between 1967 and 1987, and *A Modern Multilingual Glossary for Taxonomic Pteridology* (2002) among others.

Lemna (Araceae) L.
An ancient Greek name for a water plant used by Theophrastus, possibly derived from *limne,* referring to a lake or swampy area. This aquatic plant's common name is 'duckweed' – perhaps related to the Aegean island of Lemnos.

Leobordea** = ***Lotononis*** (Fabaceae) Delile.
For Léon Emmanuel de Laborde (1807–1869), French politician, art historian and explorer of the Middle East. In 1824, when he was just 17, he travelled from France to Cairo with his father, who turned back there. De Laborde explored further with a friend, Linant de Bellefonds, as described in *A Flora of Arabia* and *Voyage dans L'Arabie-Pétrée* (1830). He later joined the diplomatic corps, becoming private secretary to Charles Maurice de Talleyrand at the French Embassy, London. With his deep interest in art and historical research, he became *conservateur des antiques* at the Louvre, founder of the Archive Museum and was elected to the Academy Française. In 1868, he 'inherited' his father's parliamentary seat and was elected to the senate.

Leonotis (Lamiaceae) (Pers.) R.Br. (Image on opposite page)
Gk. *leon* = lion; *ous, otis* = ear; alluding to the resemblance of the corolla to a lion's ear.

Leontodon (Asteraceae) L.
Gk. *leon* = lion; *odon* = tooth. This dandelion-like

Leonotis: These small shrubby plants can be very showy in autumn. (EJM)

plant has deeply lobed leaves somewhat similar to those of *Taraxacum*, which look somewhat like lion's teeth. 'Dandelion' is anglicised from the French 'dents de lion'.

Leontonyx = *Helichrysum* (Asteraceae) Cass.
Gk. *leon* = lion; *onyx* = claw; alluding to the hooked involucral scales.

Leonurus (Lamiaceae) L.
Gk. *leon* = lion; *oura, -urus* = tail. The flowers look vaguely like a lion's tail.

Lepicystis = *Pleopeltis* (Polypodiaceae) J.Sm.
Gk. *lepis* = scale; *kystis* = bladder. The sorus is immersed in scales that form a calyciform indusium.

Lepidagathis (Acanthaceae) Willd.
Gk. *lepis* = scale; *agathis* = a ball of thread, clump. Tangled clumps of yellow or yellow green threads are found at the end of branches.

Lepidanthus = *Hypodiscus* (Restionaceae) Nees
Gk. *lepis* = scale; *anthos* = flower; referring to the particularly scaly flowers.

Lepidium (Brassicaceae) L.
Gk. *lepidion,* from *lepis* = scale; *-idium*, diminutive; referring to the shape of the seed pods; 'cress'.

Lepidobolus (Restionaceae) Nees
Gk. *lepidotos* = scaly; *bolis, bolos* = casting, throwing away; referring to the quickly deciduous stem-bracts.

Lepidoneuron = *Nephrolepis* (Nephrolepidaceae) Fée
Gk. *lepidotos* = scaly; *neuron* = nerve, vein. The scale-like indusium is located on a veinlet.

Lepidosperma (Cyperaceae) Labill.
Gk. *lepidotos* = scaly; *sperma* = seed; referring to the hypogynous scales associated with the fruit.

Lepidostephium (Asteraceae) Oliv.
Gk. *lepidotos* = scaly; *stephos* (poetical form of *stephanos*) = crown, wreath. Possibly because the corona has small scurfy scales.

Lepidotis = *Lycopodium* (Lycopodiaceae) Mirb.
Gk. *lepidotos* = scaly; referring to the scale-like leaves.

Lepidoturus = *Alchornea* (Euphorbiaceae) Baill.
Gk. *lepis* = a scale; *ouros* = a tail. The bracts and form of the male inflorescence give the appearance of a scaly tail.

Lepigonum*** = *Spergularia* (Caryophyllaceae) Wahlb.
Gk. *lepis* = a scale; *gonu* = knee, node; referring to the scale-like stipules at the node.

Lepipogon = *Catunaregam* (Rubiaceae) Bertol.
Gk. *lepis* = scale; *pogon* = bearded; referring to the densely bearded scales.

Lepisia = *Tetraria* (Cyperaceae) C.Presl
Gk. *lepis* = scale; referring to the sedges' spikelets consisting of florets and their scales, ovate in shape, glabrous, and sharply folded along their keels.

Lepisiphon = *Chrysantemoides* (Asteraceae) Turcz.
Gk. *lepis* = scale; *siphon* = tube. The plant has scaly tubes.

Lepisorus (Polypodiaceae) (J.Sm.) Ching
Gk. *lepis* = scale; sorus = *heap*; referring to the large scales arising within or from the edge of the sorus.

Leptactina (Rubiaceae) Hook.f.
Gk. *leptos* = slender, lightweight; *aktis* = a ray. The corolla lobes form a ring of narrow 'rays'.

Leptaloe = *Aloe* (Asphodelaceae) Stapf
Gk. *leptos* = fine, delicate, slender; *Aloe* (q.v.). A grass aloe having thin, long, dull green leaves in solitary, upright rosettes (Davesgarden.com).

Lepterica* = *Scyphogyne (Ericaceae) N.E.Br.
Gk. *leptos* = fine, thin, slender; *Erica* (q.v.); referring to the numerous branchlets that are long and very slender.

Leptobryum (Meesiaceae) (Schimp.) Wilson
Gk. *leptos* = slender, thin; *bryon* = lichen, moss; referring to the delicate, long and narrow, yellow-green leaves.

Leptocarpus (Restionaceae) R.Br.
Gk. *leptos* = slender; *karpos* = fruit; referring to the thin, slender fruit.

Leptocarydion (Poaceae) Stapf
Gk. *leptos* = slender; *karyon* = nut (*karydion* = small nut); referring to the *caryopsis* or grain (dry fruit) that resembles a small nut.

Leptochilus (Polypodiaceae) Kaulf.
Gk. *leptos* = slender, thin; *cheilos, chilos* = lip; referring to the shape of the fertile frond that resembles a pair of narrow lips or the blade that ends in a slender curled tip.

Leptochloa (Poaceae) P.Beauv.
Gk. *leptos* = slender, thin; *chloa* = grass; referring to its slender panicle branches.

Leptocodon* = *Treichelia* (Campanulaceae) Sond.
Gk. *leptos* = slender, thin; *kodon* = a bell; referring to the slender stalks and bell-shaped flowers.

Leptogramma* = *Cyclosorus (Thelypteridaceae) J.Sm.
Gk. *leptos* = slender, thin; *gramma* = a line; referring to the short linear sori.

Leptopteris* (Osmundaceae) C.Presl
Gk. *leptos* = slender, thin; *pteris* = fern; referring to the thin texture, the membranous pinnae, of these ferns.

Leptospermum* (Myrtaceae) J.R. & G.Forst.
Gk. *leptos* = slender, thin; *sperma* = seed. The seeds are relatively long and narrow.

Leptostachya* = *Justicia (Acanthaceae) Nees
Gk. *leptos* = slender, thin; *stachys* = spike; alluding to the slender flower-spikes or stems.

Leptothamnus* = *Nolletia* (Asteraceae) DC.
Gk. *leptos* = slender, thin; *thamnos* = bush; slender bush.

Lepturus (Poaceae) R.Br.
Gk. *leptos* = slender, thin; *oura, -urus* = tail; alluding to the spikelet axis being thin, like a pointed tail.

Lespedeza* (Fabaceae) Michx.
For Vicente Manuel de Céspedes y Velasco (c 1721–1794). Spanish military man, obtaining the rank of colonel and field marshal while serving in the Spanish royal army. He became Spanish governor of Santiago de Cuba (1781–1782) and East Florida (1784–1790) and was replaced by Juan Nepomuceno de Quesada y Barnuevo. The genus name *Lespedeza* is the result of a spelling error.

Lessertia (Fabaceae) DC.
For Jules Paul Benjamin Delessert (1773–1847), French banker, industrialist, philanthropist and amateur botanist. After serving with the Paris National Guard from 1790–1793, where he became an artillery officer, he joined his father's bank. A gifted and energetic entrepreneur, he started many commercial enterprises – a cotton factory in 1801 and a beet-sugar factory in 1802. He became regent of the Bank of France (1802) and introduced the idea of a savings bank in France (with Jean-Conrad Hottingerces) in 1818. He was an ardent botanist and conchologist with a notable herbarium and a botanical library of 30 000 volumes. He wrote several books and financed several exquisitely illustrated shell books. He was made a baron by Napoleon.

Letestuella (Podostemaceae) G.Taylor
For Georges Marie Patrice Charles Le Testu (1877–1967), French colonial administrator and specialist in West African flora. He worked in Dahomey for an agricultural company and collected plant specimens there (1900–1902) and in Mozambique (1904–1906), Gabon and Umbangi-Shari (1907–1934), now called the Central African Republic, and thereafter returned to France and worked at the botanical garden at Caen, France from 1935 onward. He authored *Notes sur les Coutumes Bapounou Dans la*

Circonscription de la Nyanga (*Notes on the Customs of Bapounou in the District of Nyanga*) (1918).

Letrouitia*** (Letrouitiaceae) Hafellner & Bellem.
For Marie-Agnès Letrouit-Galinou (1931–), French lichenologist, director of research at

the Centre National de la Recherche Scientifique. She studied lichen structures and development at Université Pierre et Marie Curie in Paris, obtaining her PhD under Marius Chadefaud (1900–1984). She had a deep interest in lichen systematics, ecology and the effects of air pollution on lichens. She made significant contributions to lichenology over a period of 50 years. She helped establish the Association Française de Lichénologie in 1976 and was its first vice-president and second president (1978–80). On her retirement in 1999, she donated her library to the Musée Nationale d'Histoire Naturelle. She was honoured with an Acharius Medal in 2004.

Leucadendron (Proteaceae) R.Br.
Gk. *leukos* = white; *dendron* = tree; referring to commonly called 'witteboom' or 'silver tree'.

Leucaena* (Fabaceae) Benth.
Gk. *leukos* = white; referring to the white flowers.

Leucanthemum* (Asteraceae) Mill.
Gk. *leukos* = white; *anthemon* = flower; referring to the white flowers.

Leucas (Lamiaceae) Burm. ex R.Br.
Gk. *leukos* = white; referring to the white flowers.

Leucojum* (Amaryllidaceae)
Gk. *leukos* = white; *ion* = violet; referring to the colour of the flowers. *Leucoion* was the name used by Dioscorides and Theophrastus for a white-flowered bulbous plant.

Leucophrys*** = **Brachiaria** (Poaceae) Rendle
Gk. *leukos* = white; *ophrys* = eyebrow, fringe; possibly referring to the lemma, which has a continuous fringe of hairs on either side, eyebrow-like.

Leucoptera (Asteraceae) B.Nord.
Gk. *leukos* = white; *pteron* = wing; referring to the unusual white wings on the fruits (cypselas).

Leucosidea (Rosaceae) Eck. & Zeyh.
Gk. *leukos* = white; *idea* = appearance; referring

Leucadendron (EJM)

Leucadendron: Female fruits. This species is restricted to the far Western Cape. (EJM)

Leucadendron: Male inflorescence (EJM)

to the overall white hairiness or pubescence of the shrub.

Leucospermum (Proteaceae) R.Br.

The seeds are large and ant dispersed. (MC)

Gk. *leukos* = white; *sperma* = seed. The tree has white seeds.

Leucosphaera (Amaranthaceae) Gilg
Gk. *leukos* = white; *sphaira* = sphere; referring to the hairy, globose heads.

Leysera (Asteraceae) L.
For Friedrich Wilhelm von Leysser (1731–1815), German botanist, mineralogist and Prussian civil servant – advisor to the crown on military and domestic issues. From 1758–1765 he studied botany at the University of Halle, becoming a lecturer there. His special interests were algae, bryophytes (mosses), fungi (mushrooms), lichens and spermatophytes (seed plants). He also made a name for himself as a mineralogist collecting some 6 000 specimens on behalf of Karoline Luise of Hesse-Darmstadt. His major work was *Flora Halensis* (*Flora of Halle*) (1761), in which he used Linnaean nomenclature. He became first president of the Halle Naturforschende Gesellschaft zu Halle (Naturalist Society in Halle) and founded the Etymological Society of Halle. The genus name was incorrectly spelled.

Lichtensteinia (Apiaceae) Cham. & Schltdl.
For Martin Heinrich Karl Lichtenstein (1780–1857), German physician, naturalist and botanical

explorer. After qualifying as an MD in 1802 at Jena and Helmsted, he travelled to the Cape where he became the governor's family doctor and his son's tutor. He went on three major expeditions to the northern and eastern boundaries of the colony and into Bechuana (now Botswana) making natural history collections. He returned to Europe in 1806 and wrote his two-volume *Travels in Southern Africa* (1810). He was appointed a lecturer in zoology, later professor at Berlin University (1911), with an honorary PhD, and became Director of the Zoological Museum in 1813 and Berlin Zoological Garden in 1844. He was elected a member of the Academy of Sciences in 1817.

Lichtensteinia = *Ornithoglossum*
(Colchicaceae) Willd.
Lichtensteinia (q.v.).

Lichtensteinia = *Tapinanthus*
(Loranthaceae) J.C.Wendl.
Lichtensteinia (q.v.).

Lidbeckia (Asteraceae) P.J.Bergius
For Professor Erik Gustavus Lidbeck (1724–1803), Swedish naturalist and botanist. He studied

under Linnaeus at Uppsala University, who appointed him his personal secretary (1746). Lidbeck was Linnaeus's only companion and secretary on his travels in Västergötland (a Swedish province). Subsequently, he was appointed associate professor of economics and natural history at Uppsala (1748), a lecturer in medicine at Lund University (1750), where he also became professor of natural history (1756), acting director of the botanical gardens (1772) (where he planted 50 000 mulberry trees produced silk) and professor of economics (1786). He was made a member of the Swedish Academy of Sciences in 1755.

Lightfootia = *Wahlenbergia*
(Campanulaceae) L'Hér.
For John Lightfoot (1735–1788), British botanist, conchologist, lichenologist and clergyman. A graduate of Oxford University in 1760, he was a meticulous organiser, researcher and recorder of information and chaplain and librarian of Margaret Bentinck, Duchess of Portland; also, a close friend of Joseph Banks. He is best known for his two-volume *Flora Scotica* (1777), which contains

hundreds of plant species and cryptogams and was by far the greatest contribution to Scottish mycology until Thomas Hopkirk published his *Flora Glottiana* some 36 years later. Lightfoot also published *An Account of Some Minute British Shells* (1786). He was elected a Fellow of the Royal Society and became one of the original fellows of the Linnaean Society in 1785.

Ligustrum* (Oleaceae) L.
A classical Latin name for privet.

Lilium* (Liliaceae*) L.
Latin name (used by Virgil) for lilies, adapted from the Gk. *leirion* and Persian *laleh* = lily; also possibly from the Celtic *li* = whiteness, hence Middle English *lilie*.

Limeum* (Molluginaceae) L.
Ancient classical name for a poisonous plant, Gk. *loimos* = a plague, pestilence. 'These small weeds are acrid poisons' (Bonder, *Flora Capensis*).

Limnophila (Plantaginaceae) R.Br.
Gk. *limne* = marsh; *philos* = lover; referring to the plant's preferred habitat.

Limnophyton (Alismataceae) Miq.
Gk. *limne* = marsh or lake; *phyton* = plant; alluding to the habitat of this aquatic species.

Limodorum (Orchidaceae) Boehm
Gk. *leimon* = meadow; *dōron* = gift; an ancient name used by Theophrastus. Some manuscripts apparently have the name *haimodoron* in its place. The meadow provides the gift of a beautiful lily.

Limonium (Plumbaginaceae) Mill.
Gk. *leimōnion*, the 'sea-lavender', from *leimon* = meadow (not marsh, which is *limné*). Many species flourish in saline soils and are therefore common near coasts and in salt marshes.

Limosella (Scrophulariaceae) L.
La. *limosus* = muddy; *-ella* = diminutive; referring to the habitat of these plants.

Linaria* (Plantaginaceae) Mill.
Gk. *linum* = flax; *-aria* = belonging to, connected with; a name given to this plant by Linnaeus from its likeness to a flax plant before flowering.

Linconia (Bruniaceae) L.
Named by Linnaeus in 1771 after an unidentified person or place.

Lindernia (Linderniaceae) All.
For Franz Balthazar von Lindern (1682–1755),

German physician, author and botanist. He studied from 1700 in Strasbourg, then in Halle, Leipzig, Wittenberg, Erfurt, and Jena, obtaining a PhD in 1708. Thereafter he taught botany, chemistry and pharmacology at the University of Strasbourg medical faculty. He authored a work on venereal disease entitled *Venus Krankheiten* (*Venus Diseases*) (1728) in German (rather than in Latin, to ensure it could reach less-educated citizens) and the two-volume *Medicinisch Passe-partou* (*Master Key to Common Diseases*) (1739/1741). He was also an avid botanist and director of the botanical garden in Strasbourg, publishing a book on the flora of Alsace called *Tournefortius Alsaticus* (1728) containing 920 plants, renamed *Hortus Alsaticus* in 1747.

Limonium: The dried flowers make excellent vase specimens. (MC)

Lindneria = Pseudogaltonia (Hyacinthaceae) T.Durand & Lubbers
After Otto Lindner (1852–1915), German-born technician who adopted Belgian nationality in 1888. He was sent by the L'État independent du Congo (now the Democratic Republic of the Congo) to Damaraland in northeastern Namibia in order to recruit labour to work in the Congo. King Leopold II of Belgium exercised a *de facto* sovereignty over the Congo from 1885 to 1908. During his stay from February to July 1886, Lindner collected plants that he sent to the director of the Jardin Botanique de L'Etat. Among

these was a live specimen of a *Hyacinth* genus that was recognised by Théophile Durand and Louis Lubbers as new. They named it in Lindner's honour (see *Un Nouveau Genre de Liliacées*, 1889, in the *Bulletin of the Botanical Society of France*).

Lindsaea (Lindsaeaceae) Dryand. ex Sm.
For John Lindsay (c 1750–1803), British surgeon, discoverer of how ferns reproduce. While working in Jamaica in 1794, he observed that ferns produce seed in the form of tiny dust-like round or bean-shaped bodies (spores). He grew full-sized ferns from a mote of fern dust to prove it. When Sir Joseph Banks, president of the Royal Society, asked Lindsay to collect Jamaican plants, especially ferns, for the Kew Gardens, he replied that he would send seeds along with instructions for their sowing. Banks was astonished, as the mechanism by which ferns reproduced was hitherto unknown. James Edward Smith, a leading pteridologist, commemorated Lindsay for his discovery by naming a genus of tropical ferns after him.

Linociera = Chionanthus (Oleaceae) Sw. ex Schreb.
For Geoffrey Linocier, 16th century French physician and botanist who, in 1584, translated Antoine du Pinet's (c 1510–1584) *Historia Plantarum* (Lyon, 1561) from Latin to French. He renamed the two parts *L'Histoire des Plantes* and *L'Histoire des Plantes Aromatiques,* an account of the aromatic plants of the East and West Indies. He also wrote *Mythologia Musarum* (1605).

Lintonia (Poaceae) Stapf
For Andrew Linton of Gilmanscleugh, Selkirk (died 1951), Scottish agriculturalist. He worked at the School of Agriculture in Cairo and was director of agriculture at government farms in Nairobi and Naivasha in the East Africa Protectorate during the early 1900s, when he collected plants. He also researched, wrote and corresponded about veterinary medicine and animal disease.

Linum (Linaceae) L.
Gk. *linon* = flax, name used by Theophrastus, successor to Aristotle in the Paripatetic School in Athens (372–287 BCE). Possibly Celtic *lin* = thread, used from making fabric.

Linzia (Asteraceae) Sch.Bip. ex Walp.
For Johann Michael Linz (Johannis Michaelis Linzius) (1771–1855), German entomologist and botanist of Speyer, Germany. Linz was misspelled as Lniz in the original title, *Genus Dicavi Memoriae Nestoris Botanicorum Palatinorum Joannis Michaelis Lniz, Praefecti Horti Botanici Spirensis* (*Genus Dedicated to the Memory of the Botanist John Michael Lniz, a Platine and Prefect of the Botanic Garden of Spires*).

Liparia (Fabaceae) L.
Gk. *liparos* = oily, shiny; referring to the shiny, hairless leaves of the genus.

Liparis (Orchidaceae) Rich.

Fresh flowers (HS)

Older flowers (HS)

Gk. *liparos* = fat, greasy, shiny; referring to the smooth, shining, oily feel of the surface of many species.

Lipocarpha (Cyperaceae) R.Br.
Gk. *leipo* = to fall; *karphos* = chaff, dry stick; referring to the deciduous transparent inner secondary scale of the spikelet in many species (Gordon Tucker).

Lipozygis* = **Lotononis** (Fabaceae) E.Mey.
Gk. *lipo, leipo-* = fatty; *zygon* = yoke; *lipo* = to abandon, fail; *zygis* = wild thyme. Meaning obscure.

Lippia (Verbenaceae) L.
For Augustin Lippi (1678–1705), French-born Italian naturalist, botanist, physician and traveller. He studied medicine at the University of Paris. In 1703 Louis XIV, King of France, sent a delegation to Abyssinia (now Ethiopia) to establish trade and missionary relationships in Egypt. Lippi went along as a botanical collector. It was well known that the mission was dangerous, as there was a lot of hostility toward the missionaries. The group was delayed time and again and were attacked and murdered at Sennar on the Blue Nile on 10 November 1705. Only one of the party survived. Lippi was just 27 when he died.

Liriodendron* (Magnoliaceae) Juss.
Gk. *leirion* = lily; *dendron* = tree; perhaps named as such because the flowers generally resemble lilies.

Liriothamnus = **Trachyandra** (Asphodelaceae) Schltr.
Gk. *leirion* = lily; *thamnus* = shrub, bush. As above.

Lissochilus = **Eulophia** (Orchidaceae) R.Br.
Gk. *lissos* = smooth; *cheilos* = lip; referring to the uncrested lip of many of the species.

Listia = **Lotononis** (Fabaceae) E.Mey.
For Friedrich Ludwig List, flourishing around 1837, German botanist and high school teacher at Tilsit, author of *Spicilegium Botanicum Continens Stirpes Nuperrime in Lithuana Detectas et Observationes Criticas ad C Hagenii Chloridem Prussicam* (1828) and *Plantae Lithuanae, Quae Chloridi Borussicae cl Hagenii Inserendae Sunt. Salicum, Quae Prope Tilsam Sponte Crescunt Adumbrationes* (1837).

Listrostachys = **Cyrtorchis** (Orchidaceae) Rchb.f.
Gk. *listron* = spade; *stachys* = spike; alluding, probably, to the compact, many-flowered, distichous inflorescence.

Litanthus = **Drimia** (Hyacinthaceae) Harv.
Gk. *lithos* = stone; *anthos* = flower, or *litos* = smooth, plain, simple; referring to the stony soil in which it grows.

Lithops (Aizoaceae) N.E.Br.
Gk. *lithos* = stone, rock; *opsis* = appearance, resemblance; referring to the stone-like appearance of the plant.

Lithops (MC)

Lithospermum (Boraginaceae) L.
Gk. *lithos* = stone; *sperma* = seed. The white nutlets are bony. Species are known generally as gromwells or stone-seeds.

Litobrochia = **Pteris** (Pteridaceae) C.Presl
La. *litos* = small; *brochos* = loop; referring to the fine meshes of the vein reticulation.

Litogyne (Asteraceae) Harv.
Gk. *litos* = simple, plain, smooth; *gyne* = woman, female (gynaecium); possibly referring to the simple, much exerted style. Meaning unclear.

Litsea* (Lauraceae) Lam.
French name *litsé*, from the Chinese *leitsai* = small plum, cherry; derived from the Japanese name of these trees.

Littonia* = **Gloriosa** (Colchicaceae) Hook.
For Samuel Litton (1781–1847), Irish physician, professor of botany in Dublin (Glasnevin), librarian of the Royal Dublin Society and honorary fellow of the College of Physicians. He was a collector of Irish flowering plants and ferns, which he obtained from the west of Ireland (unknown date) and Dublin and Wicklow counties (both in 1842). This collection is housed in the herbarium of the National Botanical Gardens, Glasnevin, and Dublin.

Lloydia* (Liliaceae) Salisb. ex Rchb.
For Edward Lhuyd (Lhwyd, Eduardus Luidius) (1660 –1709), Welsh naturalist, botanist, linguist, geographer and fossil collector.  He studied at Oxford University in 1682 before being appointed in 1684 as assistant to Robert Plot, the keeper of the Ashmolean Museum at Oxford. Lhuyd took over this post in 1690 and held it until his death. In 1688 he visited Snowdonia and recorded the flora in that region, which John Ray used in his *Synopsis Methodica Stirpium Britannicorum*. Among Lhuyd's many accomplishments, he wrote *Lithophylacii Britannici Ichnographia* (1699), the first illustrated catalogue of fossils to be published in England, and published the first volume of *Archaeologia Britannica* (1707), an account of the languages, histories and customs of Great Britain, from travels through Wales, Cornwall, Bas-Bretagne, Ireland and Scotland. He also made a detailed study of Celtic language families. Lhuyd was given an honorary MA by the University of Oxford (1701) and was elected Fellow of the Royal Society (1708).

Lobelia (Lobeliaceae) L.

These are tough annulas that can make a local show of hazy-blue. (CM)

For Mathias de L'Obel (Lobel, Lobelius) (1538–1616), Flemish botanist, traveller, plant collector. He studied medicine in Leuven and Montpellier and practised medicine from 1571–1581 in Antwerp and Delft, where he was physician to William, Prince of Orange. In 1584 he left the Netherlands for England to escape the civil war and never returned. He became physician to King James I of England and also the king's botanist. His major work, written in collaboration with Pierre Pena, was *Stirpium Adversaria Nova* (1571), which describes some 1 500 species in the vicinity of Montpellier, also of Tyrol, Switzerland and the Netherlands. A second volume, *Plantarum Historia Stirpium*, was published in 1576 with more than 2 000 illustrations, and a further work, *Icones Stirpium, seu, Plantarum Tam Exoticarum* in 1591.

Lobophyllum = ***Coldenia*** (Boraginaceae) F.Muell.
Gk. *lobos* = lobe, pod; *phyllon* = a leaf; presumably referring to the lobed leaves.

Lobostemon (Boraginaceae) Lehm. (Images on opposite page)
Gk. *lobos* = lobe; *stemon* = thread, stamen; referring to the filaments being opposite the corolla lobes.

Lobularia* (Brassicaceae) Desv.
La. *lobulus* = little lobe (Gk. *lobos*); -*aria* = connected with, hence a 'small lobe'; alluding to the small silicles (the dry, dehiscent fruit).

Locellaria** (Fabaceae) Welw.
La. *locellus* = small compartment (from La. *locus* = a place); -*aria* = belonging to, connected with; referring to the chamber formed by the false partition in the unilocular ovary.

Lochnera****** = ***Catharanthus (Apocynaceae) Rchb.
For Michael Friedrich Lochner von Hummelstein (1662–1720), German botanist, polymath, physician, historian, artist and prolific writer.  He studied at the gymnasium at Nuremberg and thereafter theology at Rostock, Copenhagen and Kiel, during which time he travelled extensively in Central Europe. In 1680 his interest switched to medicine, which he studied at Altdorf,

Lobostemon shrubs can be rather spectacular. (CM)

Lobostemon (CM)

qualifying in 1684. He joined the College of Nuremberg physicians in 1686, served as dean for three terms and in 1711 became director, then president of the Imperial Academy of Natural Philosophers. He was a prolific contributor to the academy's periodical and corresponded with many of the illustrious men and historians of his era.

Loddigesia* = *Hypocalyptus* (Fabaceae) Sims.
For Joachim Conrad Loddiges (1738–1826), German-born gardener and nurseryman. His

father, Casper Lochlie, served as gardener to George Augustus (1683–1760), Duke of Brunswick-Lünburg, Hanover, who later became King George II of England in 1727. Loddiges emigrated to Britain in 1757, aged 19, during the Seven Years' War where he worked for a Dr Silvester of Hackney, an influential medical practitioner. In his forties, he had accumulated sufficient knowledge and savings to expand a small seed and nursery business, started by fellow German émigré John Busch, by writing to people all over the world, urging them to send him packets of seeds collected from their localities. He introduced many new US species to Great Britain, such as rhubarb and rhododendron.

Loeseneriella (Celastraceae) A.C.Sm.
For Ludwig Eduard Theodor Loesener (1865–1941), German botanist. He studied in Lausanne then Berlin, obtaining a PhD in 1890. In 1896 he became an assistant to the Berlin-Dahlem Museum as a botanist and then curator in 1904, and in 1912 received the title of professor. In 1920 he went into retirement for political reasons but worked without pay at the botanical institute. His activities included systematic work on phanerogams and the processing of Eduard Seler's collections from Mexico and Guatemala. He also studied the flora in the German colony of Kiaochow in China (published 1920) and areas in Europe. He wrote many papers and a number of obituaries for German botanists, such as Seler, Gustav Lindau and Otto Eugen Schultz.

Loethainia** = *Wiborgia* (Fabaceae) Heynh.
For Rudolf Benno von Römer (Roemer) of Neumark and Löthain (1803–1870), German farmer, botanist and wealthy Saxon estate owner who maintained an extensive botanical library, some 2 600 volumes of plants, many of which were 'valuable prints', which he bequeathed to the Leipzig University library.

Lolium * (Poaceae) L.
Latin name for darnel, *Lolium ternulentum,* ryegrass.

Lomandra * (Alliaceae) Labill.
Gk. *loma* = edge or fringe; *andros* = male; referring to the circular margin of the anthers.

Lomaria* = *Blechnum (Blechnaceae) Willd.
Gk. *loma* = edge or fringe; *-aria* = belonging to, connected with; referring to the sori on the edge of the fronds – bordering the midrib of the narrow pinnae.

Lomariobotrys *** = *Stenochlaena* (Blechnaceae) Fée
Gk. *loma* = edge or fringe; *botrys* = bunch of grapes; referring to the clustered sori.

Lomariopsis * (Lomariopsidaceae) Fée
Lomaria (q.v.); Gk. *-opsis* = resembling, appearance.

Lonchitis * (Lonchitidaceae*) L.
An ancient fern name used by Dioscorides. Gk. *lonchits* = spear-shaped; referring to the frond shape.

Lonchitis-aspera* = *Blechnum (Blechnaceae) Hill ex Farw.
Gk. *lonchits* = spear-shaped; *aspera* = rough; referring to the leaves that are almost doubly winged – spear-shaped. The fern is like *Lonchitis,* but rougher.

Lonchocarpus (Fabaceae) Kunth
Gk. *lonche* = lance; *karpos* = fruit; tree with lance-shaped fruits (Hugh Glen); the pod is linear-oblong, flat, membranous or leathery (Robert Allen Dyer).

Lonchostoma* = *Brunia (Bruniaceae) Wikstr.
Gk. *lonche* = lance; *stoma* = mouth; alluding to the lance-shaped sepals and petals of some species, e.g. *acutiflorum* (now *myrtoides*).

Lonicera * (Caprifoliaceae*) L.
For Adam Lonitzer (Lonicer, Lonicerus) (1528–1586), German professor of mathematics at Nuremberg, who later obtained a medical degree from the university in Mainz and became city physician in Frankfurt. Lonitzer married Magdalena Egenolph, daughter of a Frankfurt printer who specialised in 'herbals' – books containing the names and descriptions of plants extolling their medicinal, culinary, aromatic and other virtues. When his father-in-law died (1533), Lonitzer inherited a substantial share of the business and ran it with his brothers-in-law. His interest in natural history grew and in 1551 he published his most famous work, *Natvralis Historia,* in Latin and in 1557 in German, *Kreuterbüch,* which went through several subsequent editions, being published as late as 1783.

Lophacme (Poaceae) Stapf
Gk. *lophos* = crest, tuft; *akme* = highest point, point. The plant grows on ridge tops.

Lophiocarpus (Lophiocarpaceae) Turcz.
Gk. *lophlon* = small crest, from *lophos,* a crest; *karpos* = fruit. The fruit is enclosed in a persistent perianth.

Lophochloa *** = *Rostraria* (Poaceae) Rchb.
Gk. *lophos* = crest, tuft; *chloa* = grass; referring to a tufted annual with grass blades erect.

Lopholaena (Asteraceae) DC.
Gk. *lophos* = crest, tuft; *chlaina* = cloak; referring to the involucre – the whorl (ring) of bracts around each flower.

Loranthus * (Loranthaceae) Jacq.
Gk. *loron* = a thong; *anthos* = a flower, hence 'strap-flower'. This group of parasitic plants has been split into many genera but could be recognised in the field by its strap-like flowers.

Lotononis (Fabaceae) (DC.) Eck. & Zeyh.
Combination of the two generic names *Lotus* and *Ononis,* both of which are legumes.

Lotus (Fabaceae) L.
Gk. *lotos,* but of semitic origin, a name used by Theophrastus for the legendary fruit eaten by the lotophagi, which could produce a state of intoxicated relaxation, euphoria, dreaminess and forgetfulness.

Loudetia* = *Tristachya (Poaceae) Hochst. ex Steud.
Possibly for Dr Eduard Loudet (1811–1867), a German dentist and surgeon at Karlsruhe. No more details could be found.

Lovoa (Meliaceae) Harms
From the Lovoi River, Democratic Republic of the Congo.

Loxocarya * (Restionaceae) R.Br.
Gk. *loxos* = slanting, oblique; *carya* = nut. The nut is egg shaped (ovoid) with the narrower end at the base (obovid).

Loxogramme (Polypodiaceae) (Blume) C.Presl
Gk. *loxos* = slanting, oblique; *gramme* = line. The elongated sori are oblique to the costa (on each side of the midrib).

Loxoscaphe*** = **Asplenium** (Aspleniaceae) T.Moore
Gk. *loxos* = slanting, oblique; *skaphis* = vessel, boat, bow; referring to the indusium, forming a compressed, suborbicular or cup-shaped sac, oblique boat-shaped, broader than long, open at the top.

Loxostylis (Anacardiaceae) A. Spreng. ex Rchb.

Because both flowers and fruits are showy, this is a good small garden tree. (EJM)

Gk. *loxos* = slanting, oblique; *stylis* = style; refers to the lateral attachment of the style to the ovary.

Luckhoffia = X Hoodiapelia (Apocynaceae) A.C.White & B.Sloane.
For Dr James Lückhoff, medical practitioner,  educated at the University of Cape Town (1937) and in the Netherlands, a prominent doctor who practised in Cape Town and did studies on vitamin deficiencies in the African population. Also for his son Carl August Lückhoff (1914–1960), South African botanical artist, photographer, medical practitioner, naturalist, author of *Table Mountain* (1951) and *Stapelieae of Southern Africa* (1952). He also described several species of this group in semi-popular journals. Both father and son were friends of Rudolf Marloth and Harry Bolus and collected mainly living plants in the Cape Peninsula.

Luculia* (Rubiaceae) Sweet
La. *Luculia* is the Latinised form of the Nepalese name *Luculi Swa*.

Lüderitzia*** (Malvaceae) K. Schum.
For August Lüderitz (Luderitz) (1838–1922), German merchant and collector who came to  the Cape in 1884, younger brother of Franz Adolf Eduard Lüderitz (Luderitz). In his younger days he farmed in the United States (1860–1863), travelled in Central and South America (1866–1867), lived in Columbia (1870–1878) and from 1882–1883 worked for his brother's trading business in Lagos, Nigeria. In 1884 he moved to Walvis Bay and tried to obtain trading concessions from Herero chief Kamaherero, but without success. He returned to Lübeck, Germany. It is thought that he collected plant specimens in Hereroland for his older brother, who sent them to the Berlin-Dahlem Museum and was credited for them.

Ludwigia (Onagraceae) L.
For Christian Gottlieb Ludwig (1709–1773), German botanist and physician and professor  of medicine at Leipzig. He studied medicine and botany from 1728 but was forced to discontinue as a result of a lack of funds and took a job as a botanist on an expedition to Africa under Johann Ernst Hebenstreit (1703–1757). He resumed his studies in 1733 at the University of Leipzig, obtaining his PhD in 1737, becoming an associate professor (1740), and full professor of medicine (1747), pathology (1755) and therapy (1758). Ludwig is remembered for his correspondence with Linnaeus, often critical, particularly regarding discussions of the latter's classification system. He published works on both plants and mineralogy.

Luffa* (Cucurbitaceae) Mill.
Arabic *louff* (or similar) for the plant *cylindrica*, 'loofah'; usually refers to the fruit of the two

species *aegyptiaca* and *acutangula*, which are eaten as vegetables.

Lumnitzera (Combretaceae) Willd.
For István (Stefani, Stefan) Lumnitzer (1750–1806), Hungarian botanist and physician who pioneered the systematic description of Central European plants. He was the author of *De Rerum naturalium Adfinitatibus* (1777) and *Flora Posoniensis Exhibens Plantas Circa Posonium Sponte Crescentes Secundum Systema Sexuale Linnaeanum Digestas* (1791).

Lunathyrium = Deparia (Athyriaceae) Koidz.
La. *luna* = moon (lunate = crescentic); *Athyrium* = a genus of ferns (q.v.); referring to the shape of the sori.

Lunularia = Botrychium (Ophioglossaceae) Adans.
La. *lunula* = little moon; *-aria* = pertaining to; alluding to the crescentic gemma cups on the dorsal surface.

Lupinus* (Fabaceae) L.
Gk. *lupnus, lupinus* = wolf; referring to an old belief that this plant devours the good parts (nutrients) of the soil.

Luzula (Juncaceae) DC.
Related to Old Italian *luzziola* (mod. *lucciola*) = glow-worm, fire-fly or adder's tongue; referring to its shiny capsules. Paxton's Botanical Dictionary says that 'luzulai' was German for 'glow-worm grass'.

Lychnidea** (Scrophulariaceae) Burm.
La. *lychnis* = lamp; *idea* = like; alluding to its resemblance to lychnis, meaning a lamp or lamp-wick.

Lychnis* = Silene (Caryophyllaceae) L.
La. *lychnis* = a red flower, from Gk. *lukhnis;* akin to *lukhnos, lychnos* = lamp; *idea* = like; alluding to the brilliantly coloured flowers; also refers to the fact that the felt-like leaves were formerly used for lamp wicks.

Lycium (Solanaceae) L.
Gk. *lykion* = name of a thorny bush/tree from Lycia in Asia Minor (Turkey) (Lycien, near Xanthos, ancient Anatolia), where the plant grows.

Lycopersicon (Solanaceae) Mill.
Gk. *lukos* = a wolf; *persikon* = a peach. Originally the name of an Egyptian plant, the word was later re-used for the tomato. Georg Christian Wittstein suggests that the significance is that a 'wolf's peach' is a conspicuous fruit that looks better than it tastes.

Lycium (EJM)

Lycopodiella (Lycopodiaceae) Holub
Gk. *Lycopodium* (q.v.) *-iella* = diminutive.

Lycopodioides = Selaginella (Selaginellaceae) Kuntze
Gk. *lukos* = wolf; *pous, podos* = foot; *odes* = resembling; bearing the characteristics of the club moss *Lycopodium* (q.v.)

Lycopodium (Lycopodiaceae) L. (Image on opposite page)
Gk. *lykos* = wolf; *-podion* = diminutive of *pous,* foot, stalk or pedice. The branch tips resemble a wolf's paw; other sources suggest this refers to the leafy branches.

Lycopsis (Boraginaceae) L.
Gk. *lycos* = wolf; *ops* = eye; *-opsis* = resembling; referring to the blue rounded flowers having a resemblance to the eyes of a wolf, which are grey.

Lydenburgia (Celastraceae) N.Robson
Named after the South African town of Lydenburg, Mpumalanga.

Lygodium (Lygodiaceae) Sw.
Gk. *lygodes* = flexible; referring to the lengthy and flexible rachis of the frond that twines around supports, thus enabling this climbing fern to grow.

Lycopodium (JB)

Lyperia (Scrophulariaceae) Benth .
Gk. *lyperos* = mournful, grief; referring to the fact that one species turns black after flowering.

Lysichlamys*** = **Euryops** (Asteraceae) Compton
Gk. *lysis-* = loose; *khlamus* = military cloak, mantle. Possibly because as the fruit ripens, the spathe is removed from the spadix. For the rest the meaning is unclear.

Lysimachia (Primulaceae) L.
A Linnaean name after a Greek physician, Lycimachus (fourth or fifth century) or after King Lacimachos (c 360–281 BCE) of Thracia. Their names mean *lysi-* = loosen, dissolution; *mache* = strife. The plant is alleged to have medicinal, calming properties; 'loosestrife'.

Lythrum* (Lythraceae) L.
Gk. *lythron* = black blood; possibly referring to the flower colour of some species.

Lyperia: fresh flowers (CM)

Lyperia: older flowers (CM)

M

Marsilea **ferns (NC)**

Mohria **ferns (JB)**

John Eric Burrows (1950–) was born in Salisbury (now Harare), Zimbabwe and married Sandra Margaret Schultz in 1980. He graduated with a national diploma in horticulture in 1979 and with an MSc (plant taxonomy) *cum laude* from what is now the University of Kwazulu-Natal in 1992 with a thesis entitled *The Taxonomy of the Genus Ophioglossum (Ophioglossaceae) in Southern Africa*. He has worked as a research technician and horticulturalist and, from 1988, has been the managing trustee of the Buffelskloof Private Nature Reserve in South Africa's Mpumalanga Province. In 1990 he started the Buffelskloof Herbarium, which now has more than 25 000 specimens (upward of 15 000 are his own). In 1994 he started the local plant specialist group that has attracted both local and international interest. He has specialised in the taxonomy of ferns and their allies, written a book with Sandra on the figs of Southern and Central Africa, and they have worked and are still working on books including *Trees and Shrubs of Mpumalanga and the Kruger Park* (second edition) and on the 1 750 species of trees and shrubs of Mozambique.

Maba* = ***Diospyros*** (Ebenaceae) J.R. & G.Forst.
Named after a Tongan vernacular word for plants of this genus.

Macadamia* (Proteaceae) F.Muell.
For John Macadam (1827–1865), Scottish-born Australian analytical chemist, medical teacher

and politician. He studied chemistry at the Andersonian and Edinburgh universities and medicine at Glasgow University (MD 1854). In 1855, he went to Melbourne, where he lectured in chemistry and natural science at Scotch College. In 1858 he became a Victorian government analytical chemist. In 1859 he was elected to the legislative assembly in the Victorian parliament and lost his seat in 1861 but regained it in 1862. In 1862 he was appointed a lecturer in chemistry at the University of Melbourne's school of medicine. Between 1857 and 1862 he served as honorary secretary of what became the Royal Society of Victoria, becoming its president in 1863. He died at sea after a shipboard accident.

Macaranga (Euphorbiaceae) Thouars

(EJM)

The Malagasy name for the first species described, which was *M. tanarius*. Common name 'David's heart'. The plant has large heart-shaped leaves.

Macfadyena* = ***Bignonia*** (Bignoniaceae) A. DC.
For James Macfadyen (1799–1850), Scottish botanist, doctor and classicist. He qualified as a doctor in 1822. On the recommendation

of Sir William Hooker, he went to Jamaica, then a colony of Britain, and from 1826–1828, Macfadyen collected specimens for plant collector Gilbert Hibert. He also established a botanical garden. He was the first person to describe the grapefruit scientifically. He authored *A Sketch of a Short Botanical Journey in Jamaica* (1931), which appeared in Hooker's *Miscellany, Flora of Jamaica* Volume 1 (1837) and *Description of Nelumbium Jamaicense* (1847). In addition to botany, he was actively involved in many social organisations. From 1837 onward he practised medicine and died while treating patients during a cholera outbreak, catching the disease himself. He was elected a Fellow of the Linnaean Society of London in 1838 and the Geological Society of London in 1850 (posthumously).

Machairophyllum (Aizoaceae) Schwantes
Gk. *machaira* = dagger, sabre, sword; *phyllon* = leaf; referring to the dagger-shaped leaves.

Mackaya (Acanthaceae) Harv.
For James Townsend Mackay (1775–1862), Scottish horticulturalist and curator of the Durban University's botanical gardens from 1804 until his death. He came to Trinity College, Dublin, as assistant to Dr Robert Scott, then professor of botany. He was a keen field botanist and author of *A Systematic Catalogue of the Rare Plants Found in Ireland* (1806–1807), in which he noted that up to this time botanists only recorded plants considered to be rare or of medicinal value, a practice that probably resulted in the loss of many records and locations of then common plants. He also authored *Catalogue of the Indigenous Plants of Ireland* (1825) and was the principal author of *Flora Hibernica* (*Flora of Ireland*) (1836), which included all wild species known in the area at the time. Dublin University honoured him with an LLD and PhD in 1850 for his service to botany.

Macledium (Asteraceae) Cass.
Probably for William Sharp Macleay (1792–1865), English diplomat, commissioner, judge and

natural historian, son of the entomologist Alexander Macleay (1767–1848). He was well known to Georges Cuvier (1769–1832), editor of *Dictionnaire des Sciences Naturelles,* in which the genus *Macledium* is published. He studied at Trinity College, Cambridge, then became an attaché to the British Embassy in Paris (1818), then secretary to the board for liquidating British claims on the French government. He moved to Cuba in 1825, becoming British commissioner in Havana for the abolition of the slave trade, then commissary judge (1830) and then judge (1833). He retired in 1836 and emigrated to Australia in 1839. Throughout his life, he followed his father's interest in entomology, authoring the two-part *Horae Entomologicae or, Essays on the Annulose Animals* (1819–1821) and *Annulosa Javanica or an Attempt to Illustrate the Natural Affinities and Analogies of the Insects Collected in Java by T. Horsfield, no. 1* (1925). He was a Fellow of the Royal Society, a member of both the Linnaean Society and the Zoological Society of London and was elected president of the natural history section of the British Association for the Advancement of Science.

Maclura (Moraceae) Nutt.
For William Maclure (1763–1840), Scottish-born North American geologist, educational reformer, agriculturist, traveller and president of the Academy of Natural Sciences of Philadelphia (1817–1839). Through his partnership with a London firm of US merchants, he owned businesses in England and travelled extensively in Europe. In 1800 he joined the American Philosophical Society and thereafter devoted his life to science and philanthropy. This included representing US citizens on a United States Commission in Paris (1803–1807) concerning their losses resulting from the French Revolution, surveying and making the first widely available geographical chart of the United States (1809), and introducing agricultural colleges in the latter country and Spain.

Macowania (Asteraceae) Oliv.
For Dr Peter MacOwan (1830–1909), academic and plant collector and professor of chemistry at Huddersfield, England. He moved to South Africa for health reasons, was principal of Shaw College, Grahamstown (1862), science master at Gill College, Somerset East (1869), director of the Cape Town Botanical Gardens and curator of the Cape Government Herbarium (1881) to which he contributed many specimens. He was one of the first professors of botany at the South African College, Cape Town (now the University of Cape Town) and in 1892 became a Cape government botanist doing advisory work for the farming community and writing nearly 1 200 scholarly articles for the *Cape Agricultural Journal*. After his retirement (1905), he worked at the Albany Museum herbarium, where many of his early specimens were housed.

Macrochaetium = Tetraria (Cyperaceae) Steud.
Gk. *makros* = large or long; *khaite* = bristle; referring to the conspicuous hairy beak on the fruit.

Macrolinum*** = Erica (Ericaceae) Klotzsch
Gk. *makros* = large; *linum* = flax, linen, cloth, line; implication obscure.

Macropetalum*** = Brachystelma (Apocynaceae) Burch. ex Decne.
Gk. *makros* = large; *petalum* = petals; referring to the large, long, yellow-green wiry and curving petals.

Macrorungia*** = Anisotes (Acanthaceae) C.B. Clarke
Gk. *makros* = large, long; *Rungia* (q.v.); perhaps looking like *Rungia,* but larger.

Macrosiphon*** = Rhamphicarpa (Orobanchaceae) Hochst.
Gk. *makros* = large; *siphon* = tube; referring to the long corolla tube.

Macrostylis** (Rutaceae) Bart & H.Wendl.
Gk. *makros* = large; *stylos* = pillar style; referring to the plant's long, exerted style.

Macrothelypteris (Thelypteridaceae) (H. Itô) Ching (Image on opposite page)
Gk. *makros* = large; *Thelypteris* (q.v.). The large fern is related to *Thelypteris.*

Macrotyloma (Fabaceae) (Wight & Arn.) Verdc.
Gk. *makros* = large; *tylos* = lump, swelling; *loma* = edge or fringe. Meaning unclear.

Maerua (Capparaceae) Forssk.
Probably from the Arabic *meru,* meaning high, according to the 11th-century Arabic encyclopaedia *The Canon of Medicine*. Some species grow in excess of nine metres.

Maesa (Maesaceae) Forssk.
Arabic vernacular name *maas*, meaning unknown.

Mafekingia = Raphionacme (Apocynaceae) Baill.
Named after the South African town of Mafeking, now renamed Mahikeng.

Macrothelypteris: A tough fern preferring stream bank habitats (NC)

Magnolia* (Magnoliaceae) L.
For Pierre Magnol (1638–1715), French botanist. Magnol obtained a PhD from the University of Montpellier in 1659 and then did some field work in France. Facing religious discrimination as a Calvinist, he was not accepted for many jobs that he applied for early in his career. In 1685, he converted to Catholicism, the official state religion, and subsequently became a demonstrator of plants at the botanical garden of Montpellier (1687), professor of medicine at Montpellier (1694) and director of the botanical garden (1696). His major contribution to science, as described in his *Prodromus Historiae Generalis Plantarum, in quo Familiae Plantarum per Tabulas Disponuntur* (1689), was the discovery that plants could be placed in plant families, a natural classification, based on combinations of morphological characters.

Mahea = Manilkara (Sapotaceae) Pierre
Named after Mahé, which is the largest island of the Seychelles.

Maianthemum*** (Ruscaceae) F.H.Wigg.
Named after Maia, the goddess of growth of plants or from Old French *mai* from La. *maius mensis* implying the fifth month; Gk. *anthemum* = flowers; alluding to the time when the flowers bloom (May).

Mahernia*** = Hermannia (Malvaceae) Schum.
Near anagram of *Hermannia*.

Mairia (Asteraceae) Nees
For Louis Maire (f 1815–1833), Prussian plant collector. He served in the Napoleonic Wars with Johannes Ludwig Leopold Mund (*Mundia* q.v.) and Peter Jonas Bergius (*Bergia* q.v.). After training as a gardener, he was sent to the Cape with Mund by the Berlin Museum to collect plants and natural history specimens. At least two large consignments were sent to Europe, but their productivity slackened, and Berlin recalled them. They ignored the order. Mund became a surveyor, and Maire apparently set up as a doctor. In 1833 the French Protestant missionary Eugene Casalis met with a 'Dr Lemaire', possibly the same man, at Graaff-Reinet, who said he had been a surgeon in the Prussian army when they entered Paris in 1815 and that afterward the King of Prussia had dispatched him on a botanical expedition to South Africa.

Malephora (Aizoaceae) N.E.Br.
Gk. *male* = armhole (La. equiv. *ala*); *phorein* = to bear (*pherein,* to bear) (Hans Herre); referring to the seed pockets of the fruit.

Mallotus (Euphorbiaceae) Lour.
Gk. *mallotus* = lined in wool, fleecy; referring to the woolly hairs on the leaves of some of the species.

Maltebrunia (Poaceae) Kunth
For Konrad Bruun (Conrad Malte-Brun) (1755–1826), Danish-French geographer. In his youth, activism for press freedom and criticism of Crown Prince Frederick's censorship laws led to his indictment. Fearing an unfair trial, he left for Sweden (1779), then Hamburg and eventually France (1799). He authored the six-volume *Géographie Mathématique, Physique et Politique de Toutes les Parties du Monde* (1803–1807) with Edme Mentelle, *Les Annales des Voyages, de la Géographie et de l'Histoire* (1819), and a treatise on Polish geography. He founded *Les Annales des Voyages* (in 1807) and was the first general secretary of the newly founded Société de Géographie (1822–1824). He died while drafting his major work, *Précis de Géographie Universelle*.

Malus* (Rosaceae) Mill.
Classical Latin name for an apple. The name *Malus* evolved from the Greek *melon,* a word used for tree fruits with a fleshy exterior.

Malva* (Malvaceae) L.
Latin for a mallow, used by Cicero; Gk. *malakos* = soft, soothing; referring to the emollient properties of the seeds.

Malvastrum* (Malvaceae) A.Gray
Malva (q.v.); *-astrum* = partially resembling, inferior, wild.

Malvaviscus (Malvaceae) Fabr.
Malva (q.v.); *viscus* = sticky; referring to the thick, gluey sap produced by many members of the genus.

Malveopsis** = ***Anisodontea*** (Malvaceae) C.Presl
Malva (q.v.); Gk. *-opsis* = resembling.

Mandevilla* (Apocynaceae) Lindl.
For Henry John Mandeville (1773–1861), English diplomat, minister plenipotentiary to the Argentine Republic and keen gardener. He entered the navy when still a boy and later held a commission in the first dragoon regiment. From 1801–1803 he took jobs as an unpaid attaché. He became a paid attaché in Paris (1824), secretary to the embassy at Lisbon (1828), minister to Constantinople (1831–1833), and spent a decade as minister in Buenos Aires (1835–1845), where he mediated in the rivalry between the British merchants there and in Montevideo and managed relations with the Argentinian dictator Juan Manuel de Rosas, whom he saw as essential for maintaining stability in the country. He retired in 1845. He was responsible for introducing to Europe *Ipomoea indica*, commonly called 'morning glory'.

Mangifera* (Anacardiaceae) L.
Mango is the Malayalam name for the fruit. La. *-fer* = bearer, hence 'The tree that bears mangoes.' The Malayalam name was recorded by Hendrik Adriaan van Rheede in the 17th century and is still used in Kerala, South India.

Manihot* (Euphorbiaceae) Mill.
Derived from the Brazilian (Tupian) vernacular name *manioc*. The botanical name for manioc is *Manihot esculenta,* a species well known as a food plant, as its starchy tuberous root is used to make cassava and tapioca.

Manilkara (Sapotaceae*) Adans.
Derived from *manil-kara*, a vernacular name for *M. kauki* in Malayalam, the language most widely spoken in Kerala, India. The word also refers to the Malabar region in the state of Kerala.

Manochlamys (Amaranthaceae) Aellen
Gk. *manos* = loose, flaccid (or sparse); *khlamys* = a cloak. The fleshy bracts encircle the fruit.

Mantisalca* (Asteraceae) Cass.
Anagram of *Salamantica,* from Salamanca in Spain, where the genus occurs.

Manulea (Scrophulariaceae) L.

(NH)

La. *manus* = a hand, plus diminutive; referring to the corolla's finger-like divisions – the appearance of the five spreading (upright) corolla lobes.

Manuleopsis (Scrophulariaceae) Thell.
Manuela (q.v.)*;* Gk. *-opsis* = resembling.

Maprounea (Euphorbiaceae) Aubl.
A Guyanese vernacular name for these trees.

Marasmodes (Asteraceae) DC.
Gk. *marasmos* = a wasting away (cf *marasmus,* med.); *-odes* = resembling. The 'plants look lean or ill-fed' (William Henry Harvey, *Flora Capensis*).

Marattia (Marattiaceae) Sw.
For Giovanni Francesco Maratti (1704–1777), Italian Benedictine monk and botanist with a special interest in the flora of Rome and Lasio (central Italy). He entered the abbey of Vallombrozskom in 1721, taking his vows the following year, and completed his studies at the monastery of San Michele di Passignano. He studied under Bruno Tozzi (1656–1743), a highly respected botanist and mycologist. In 1739 Maratti became abbot of the sanctuary of the Virgin Mary Gallorskoy and in 1747 professor of botany and medicine at the University of Rome and director of the botanical gardens of Rome, which, at that

stage, were badly neglected. During the next 30 years he enriched the garden with new species he discovered during his explorations in the vicinity of Rome and through plant exchanges with his many contacts in other cities. He wrote papers on a variety of subjects, and his work *Flora Roma* (*Flora of Rome*) was published posthumously (1822).

Marcellia**** = *Marcelliopsis
(Amaranthaceae) Baill.
Possibly for Claudia Marcella, daughter of Roman senator and consul Gaius Claudius Marcellus Minor, sister of Emperor Augustus (Octavian) and wife of Marcus Vipsanius Agrippa, Augustus's close friend and lieutenant (WPU Jackson). Author Louis Antoine Francois Baillon did not explain the name (David Hollombe).

Marcelliopsis* (Amaranthaceae) Schinz
Marcellia (q.v.); Gk. *-opsis* = resembling.

Margaritaria (Phyllanthaceae) f.
Latin name for a collection of pearls; probably referring to the unusually coloured metallic blue fruit (coccus).

Margelliantha* = *Diaphananthe
(Orchidaceae) P.J.Cribb.
La. *margarita* = pearl; *anthos* = flower; for the pearly white flowers of the type species.

Marginaria* = *Pleopeltis (Polypodiaceae) Bory
La. *marginata* = margined; *-aria* = pertain to. The sporangia are along the margins of the frond.

Mariscus* = *Cyperus (Cyperaceae) Vahl
Old name used by Pliny the Elder for a kind of rush. (*Paxton's Botanical Dictionary* gives Celtic *mar* = marsh.)

Markhamia (Bignoniaceae) Seem. ex Baill.
For Clements Robert Markham (1830–1916), English geographer, explorer and author of  over two dozen books on history, biography and travel (excluding his Spanish-English translations). He served in the Royal Navy (1844–1851), during which he went to the Arctic in search of Sir John Franklin, visited Peru twice where he collected *Cinchona* trees, a source of quinine, introducing these to India (1859–1861) and served as army surveyor and naturalist in Abyssinia (1867–1868). He went on a second Arctic voyage (1875–1876) and served as honorary secretary of the Royal Geographical Society (1863–1888) and president (1893–1905). He was a Fellow of the Linnaean and Royal societies, was made a companion of the Order of Bath and received many other honours.

Trumpet-shaped flowers (TW)

Marlothia* = *Helinus (Rhamnaceae) Engl.
For Hermann Wilhelm Rudolf Marloth (1855–1931), German-born South African botanist, pharmacist, explorer and plant collector. He  studied pharmacy and chemistry at the universities of Berlin and Rostocka and obtained his PhD in 1883. He worked in various capacities in South Africa including professor of chemistry at (now) Stellenbosch University and botanised widely in this country and South-West Africa (now Namibia), discovering many new and rare species. He wrote many papers on phytogeography, ecology, economic botany, ethnobotany and education, his major work being the six-volume *Flora of South Africa* (1913–1932) and its supplement *Dictionary of the Common Names of Plants* (1917). He was chairman of the Mountain Club of South Africa (1901–1906).

Marlothiella** (Apiaceae) H.Wolff.
Marlothia (q.v.) plus Gk. *-iella,* diminutive.

Marlothistella (Aizoaceae) Schwantes
Marlothia (q.v.). La. *stella* = star.

Maronea*** (Fuscideaceae) A.Massal.
For Nicolai Marogna or Nicholas Maronae, a 17th-century Veronese doctor of philosophy

and medicine. He was the author of the book *Commentarius in Tractatus Dioscoridis et Plinii de Amomo* (1608) about Pedanius Dioscorides (c 40–90), a physician, pharmacologist, botanist and author of a five-volume encyclopaedia about herbal medicine, *De Materia Medica* and Pliny the Elder, a Roman naturalist and naturalist philosopher and author of the encyclopedic *Naturalis Historia* (*Natural History*).

Marrubium* (Lamiaceae) L.
From the Hebrew *marrob* = bitter juice; referring to the leaves, which are said to have a wide range of beneficial uses such as for coughs, colds and as an antidote to snake bite poison. Recent research (2012) shows it contains 'antidiabetic, anti-atherogenic and anti-inflammatory properties' (Wikipedia.org).

Marsdenia (Apocynaceae) R.Br.
For William Marsden (1754–1836), Irish-born British traveller, Orientalist, numismatist and antiquarian who worked for the British East India Company in Sumatra for eight years. Later, he became first secretary of the Admiralty. He was one of the founders of the Royal Asiatic Society and the first Englishman to study and learn Malayan, publishing a grammar and a dictionary of the Malay language in 1812. He also translated the *Travels of Marco Polo* (1818). His *History of Sumatra* (1784) was the first serious English-language attempt to fully describe that great island just off the coast of the Malayan peninsula. As an antiquarian, he had a matchless collection of printed Bibles from all over the world. He was a Fellow of the Royal Society in 1783.

Marsilea (Marsileaceae) L.
For Luigi Ferdinando Marsigli (Marsili) (1658–1730), Italian botanist, naturalist, soldier, military engineer, surveyor and author of over 20 publications. He collected a vast amount of scientific information, specimens, antiquities, fossils, instruments, artefacts of flora and fauna etc., which he gave to the senate of Bologna, and founded the Institute of Sciences and Arts (1712). He travelled extensively in Europe, mapping the 850-kilometres-long Hapsburg-Ottoman border, did scientific research on the Danube River written up in *Danubius Pannonico-mysicus* (1726) and studied the nature of the seas from Marseilles,

France, publishing *Histoire Physique de la Mer* (1725). He was a foreign associate of the Paris Academy of Sciences and a Fellow of the Royal Society of London and of Montpellier.

Martynia (Martyniaceae) L.
For John Martyn (1699–1768), British botanist who also practised as a physician although he never qualified as such. He studied as an apothecaries' apprentice at Chelsea's Physic Garden and became secretary of a botanical society that existed for a few years. As a result of his writings and lectures he was appointed professor of botany at Cambridge University in 1732. He produced a flora of the Cambridge area in 1727, but his reputation chiefly rests upon his *Historia Plantarum Rariorum* (1728–1737) and his translations of *Eclogues* (1749) and *Georgics* (1741). He resigned in 1762 in favour of his son Thomas (1735–1825), author of *Flora Rustica* (*Rural Flora*) (1792) and gave the university his botanical specimens and books. He was elected a Fellow of the Royal Society in 1727.

Mascarenhasia* (Apocynaceae) A.DC.
For Don Pedro Mascarenhas (1470–1555), Portuguese fleet commander, explorer and colonial administrator, who was the first European to discover the island of Diego Garcia in the Indian Ocean, in 1512, and may also have seen Mauritius. His helmsman on that trip to Goa was Diogo Rodrigues (c 1490/1501–1577). Years later, in 1528, when making his way back to Portugal across the Indian Ocean from Goa, Rodrigues navigated via the islands of Réunion, Mauritius and Rodrigues (their current names) naming this entire archipelago, including several small islands nearby, the Mascarene Island, Mascarene, or Mascarenhas Islands, after his countryman and commander Don Pedro Mascarenhas.

Massonia (Hyacinthaceae) Thunb. ex Houtt.
For Francis Masson (1741–1805), British gardener and plant collector for Kew Gardens. He was sent by Sir Joseph Banks to collect plants in South Africa and sailed with Captain James Cook to the Cape, where he stayed from 1772–1775. Two of his three expeditions were made jointly with Carl Peter Thunberg, who named this genus for him. From 1786–1795, he visited Madeira, the

Canary Islands and Azores, West Indies, North America and North Africa. He collected more than 500 specimens including, now household names, the bird-of-paradise flower *Strelitzia reginae* and the arum lily *Zantedeschia aethiopica* among others such as *Gladioli, Lobelia, Geranium, Pelargonium, Protea* and *Mesembryanthemum*. He authored *Stapeliae Novae* on new South African succulents he discovered (1796).

Mastersiella (Restionaceae) Gilg-Ben.
For Maxwell Tylden Masters (1833–1907), British physician and taxonomist. He studied medicine  at King's College Hospital, London, became a member of the Royal College of Surgeons (1856), and obtained an MD from the University of St Andrews (1862). After a few years' practice he was appointed a lecturer on botany at St George's Hospital. He was the editor of the *Gardeners' Chronicle* from 1865 and a contributor to Martius's massive *Flora Brasiliensis* (1840–1906) and Daniel Oliver's *Flora of Tropical Africa*. His major work was *Vegetable Teratology* (1869), and he wrote many other books such as *Botany for Beginners* (1872), *On the Conifers of Japan* (1881), and *Plant Life on the Farm* (1885). He was elected a Fellow of the Royal and Linnaean societies.

Mataxa* = *Lasiospermum (Asteraceae) Spreng.
Gk. *mataxa* = a silk worm cocoon; referring to the pubescent synsepala.

Matricaria* (Asteraceae) L.
La. *mater* = mother or *matrix* = womb; referring to the herbs of old that were used in midwifery matters. The plant's reputed value was in treating menstrual disorders; 'chamomile'.

Matthiola (Brassicaceae) R.Br.
For Pietro Andrea Gregorio Mattioli (1501–1577), Italian physician and botanist. He qualified as an MD from the University of Padua in 1523 and became personal physician to Maximilian II, Holy Roman Emperor in Vienna, and to Ferdinand I, Archduke of Austria. He published many works and described over 100 plants, and his *Di Pedacio Dioscoride Anazarbeo Libri Cinque* (1544) is a celebrated translation of the herbal of Dioscorides, complete with a commentary

in which Mattioli aimed to identify the plant species described by Dioscorides. He also described species not of medical value, marking a transition to the study of plants for non-medical interest. It is thought that Mattioli was the first physician to describe a case of feline allergy.

Maughaniella* = *Diplosoma (Aizoaceae)
Bolus
For Dr Herbert Maugham-Brown (1883–1940), South African physician and plant collector. He qualified with an MBChB in 1905, MD in 1907 from the University of Edinburgh, and took his DPH, RCPS Eng, in 1911, devoting himself to public health work in England. In 1919, he applied for the position of medical inspector of schools in the Cape Province and was appointed, with Dr Elsie Chubb, as joint medical inspector of schools. He lectured in school hygiene at the University of Cape Town and was an examiner in that subject for the department of education. He was an excellent field botanist and discovered several new plant species mainly in the south- and northwestern districts of the country.

Mauhlia* = *Agapanthus*** (Agapanthaceae)
Dahl
For Johannis Mauhle, a Swedish promoter of natural science, who tried to save Linnaeus's collection of natural history, his herbarium, letters and manuscripts from falling into foreign hands. When Linnaeus's son, Carl Linnaeus the Younger, died in 1783, his wife, desperate for funds, offered the entire collection to Joseph Banks, who declined. Dr James Edward Smith (1759–1828), a British naturalist, took up the offer and bought the collection for 1 000 guineas. Mauhle, at that time in China on business for the Swedish East India Company, instructed a Dr Dahl, a pupil of Linnaeus, to put in a counter-offer of up to 2 000 guineas. Unfortunately for Sweden this failed as Dr Smith had first option, and by the time an appeal was made to the king to stop the transaction, it was too late and the collection had already been shipped out of the country. As to who this Mr Mauhle was, little is known.

Maurocenia (Celastraceae) Mill.
For Giovanni Francesco Morosini (1658–1739), Italian Venetian senator, patrician and patron of science. He was elected a senator of the Great Council in 1690 and re-elected more than 30 times until 1738. He became ambassador to the papal state in Rome from 1702–1706 under Pope Clement XI (1649–1721) and from 1709–1711

Like many woody plants, the new leaves are colourful. (EJM)

became ambassador extraordinary to the Holy Roman Emperor Joseph I (1678–1711). He was elected reformer at the University of Padua on six occasions between 1719–1737 and made significant reforms to the university, especially in regards to press law and the teachings of the University of Padua. He also developed a magnificent botanical garden in Padua, which became famed throughout Europe.

Maytenus (Celastraceae) Molina
From the Chilean vernacular name *maitén* or *mayten,* derived from the Mapuche Indian name *mantun,* the local name for the species. The Mapuche are one of three surviving tribes of the ancient Araucanian linguistic group of Indians who lived in Chile and Argentina before the Incas or the Spaniards arrived.

Mecodium = Hymenophyllum (Hymenophyllaceae) (C.Presl ex Copel.) Copel.
Gk. *mecodios* = seen by the way. As regards this name, the American botanist Edwin Bingham Copeland scathingly remarks, '[Carl Borivo] Presl certainly did not adequately describe his genus *Mecodium*. His description would be inaccurate, even if what he wrote were correct, which it is not!'

Medicago* (Fabaceae) L.
Gk. *medike* = a grass, lucerne, a kind of clover; also the Greek name for alfalfa, which came to Greece from Medea.

Megaclinium = Bulbophyllum (Orchidaceae) Lindl.
Gk. *mega* = wide, large; *klinion* = little bed; referring to the flattened inflorescence.

Megalastrum (Dryopteridaceae) Holttum
Gk. *mega* = large; *Lastrea* (q.v.); referring to the fern genus *Lastrea;* a large *Lastrea.*

Megalochlamys (Acanthaceae) Lindau
Gk. *megalo-* = large; *khlamys* = cloak; possibly referring to the widely spread overlapping leaves in some species.

Megaloprotachne (Poaceae) C.E.Hubb.
Gk. *megalo-* = large; *proto-* = first, chief; *achne* = glume, scale; referring to the upper glume, which is considerably bigger than the lower glume.

Megastachya (Poaceae) P.Beauv.
Gk. *megas* = large; *stachys* = ear of corn, spike. The genus's spikelets have more florets than other related grasses.

Melaleuca* (Myrtaceae) L.
Gk. *melas* = black; *leucos* = white; referring to the black trunk and white barked branches of some of the Asian species, possibly due to fire charring.

Melampyrum (Orobanchaeae) L.
Gk. *melas* = black; *pryos* = wheat; referring to the black-coloured bread that results from adding the seeds to flour made from other grains.

Melancranis* = Ficinia** (Cyperaceae) Vahl
Gk. *melas* = black; Gk. *kranion* = head, skull; referring to a female rush that has black seeds toward the head of the plant.

Melandrium = Vaccaria (Caryophyllaceae) Röhl.
Gk. *melas* = black*; andros* = man; *ium* = diminutive; referring to the dark stamen.

Melanospermum (Scrophulariaceae) Hilliard
Gk. *melanos* = black, dark coloured; *spermum* = seed; referring to the seed colour.

Melanosticta = Hoffmannseggia (Fabaceae) DC.
Gk. *melas, melano-* = black; *stiktos* = dotted, dappled. 'All parts of the plant except the petals and stamens thickly sprinkled with black, hemispherical resinous dots' (William Henry Harvey, *Flora Capensis*).

Melanthera* (Asteraceae) Rohr
Gk. *melas* = black; *anthera* = anther; referring to the plant's black anthers (stamens).

Melanthium (Melanthiaceae*) L.
Gk. *melas, melano-* = black; *anthos* = flower. The perianth darkens with age.

Melasma (Orobanchaceae) P.J.Bergius
Gk. *melas* = black. The plants turn black on drying.

Melasphaerula (Iridaceae) Ker Gawl.
Gk. *melas* = black; *sphaerula* = little ball; referring to the plant's small black bulbs (corms and cormlets).

Melhania (Malvaceae) Forssk.
From Mount Melhan in Yemen, where the type species was found.

Melia* (Meliaceae) L.
Gk. *melia*, manna-ash from *meli* = honey. Named by Linnaeus in 1753 because of the ash-like leaves.

Melianthus (Melianthaceae) L.
Gk. *meli,* La. *mel* = honey; *anthos* = flower. The honey flowers 'contain abundant nectar, but, judging by their stink, I would not expect edible honey! 'Kruidjie-roer-my-nie' (WPU Jackson).

Melica (Poaceae) L.
La. *meli* = honey, sweetness, pleasant thing; *-ica* = belonging to. Ancient Italian *melica, meliga* = sorghum; refers to the sweetness of the stem. The derivation is uncertain, and other possibilities have been suggested.

Melilotus* (Fabaceae) Mill.
Gk. *meli,* La. *mel* = honey; *lotus* = clover; referring to the fragrant foliage enjoyed by bees, 'the flowers resembling those of the *Lotus* and smelling of honey' (Terence Macleane Salter).

Melinis (Poaceae) P.Beauv.
Gk. *meline* – name for grain (millet), derivation uncertain; possibly Gk. *melinis* = quince-yellow (Deon Kesting).

Melochia (Malvaceae) L.
Related to Arabic *melukhiyah,* Hebrew *maluach,* Gk. *malachē* = mallow.

Melolobium (Fabaceae) Eck. & Zeyh.
Gk. *melos* = a joint; *lobion* = diminutive of *lobos* = lobe, pod, capsule; referring to the joint-like constrictions between seeds.

Melothria (Cucurbitaceae) L.
Gk. *melon* = apple, but also the generic name for all fruits; *thrion* = a certain food. In ancient times this classical name was for a different plant, probably *Bryonia*. Georg Christian Wittstein states that this cucurbit is very similar to *B. cretica*.

Melpomene (Polypodiaceae) A.R.Sm. & R.C.Moran
For Melpomene, mythological songstress and the muse of tragedy. She was the daughter of Zeus, and her sisters included Calliope (muse of poetry), Clio (muse of history), Euterpe (muse of flute playing), Terpsichore (muse of dancing), Erato (muse of erotic poetry), Thalia (muse of comedy), Polyhymnia (muse of hymns), and Urania (muse of astronomy).

Memecylon (Memecylaceae) L.
Gk. *memaikulon,* a corruption of *mimaikulon* = name used by Dioscorides and by Pliny the Elder, fruit of *Arbutus unedo*. Linnaean name used by Dioscorides and Pliny the Elder for the fruit of the strawberry tree, 'perhaps because the fruits of some species are edible' (WPU Jackson).

Meniscium* = **Cyclosorus**
(Thelypteridaceae) Schreb.
Gk. *meniskos,* diminutive of *mene* = moon; referring to the crescent moon's shape, which is similar to the form of the sorus.

Menispermum (Menispermaceae) L.
Gk. *mene* = moon; *sperma* = seed. The seed is curved; 'moonseed'.

Menodora (Oleaceae) Humb. & Bonpl.
Gk. *menos* = vigour, force; *doron* = gift. The author states (in French), 'This plant gives strength [the gift] to the animals. This name has already been used by ancient botanists.'

Mentha (Lamiaceae) L.
Latin name for mint, from the nymph Minthe, mistress of Pluto, daughter of Cocytus, who was turned into mint by the jealous Proserpine. Old English *minte*.

Mentocalyx = Gibbaeum (Aizoaceae) N.E.Br.
La. *mentum* = chin; *calyx*; alluding to the chin-shaped bumps on the calyx.

Menyanthes* (Menyanthaceae) L.
Gk. *menyein* = disclosing, revealing; *anthos* = flower; referring to the sequential opening of the flowers on the inflorescence.

Merciera (Campanulaceae) A.DC.
For Marie Philippe Mercier (1781–1831), French botanist, plant collector and traveller. Born on the island of Martinique, he worked briefly for the military and police before making lengthy trips to the United States, Mexico, West Indies, Brazil and

One of the species in the genus (CM)

Chile as a trader collecting plants that he sent to Augustin Pyramus de Candolle and Stephano Moricand in Geneva. In 1822 Mercier moved to Geneva, where he studied botany under the direction of De Candolle. He died before his *Choix des Plantes Exotiques Rares ou Nouvelles* was completed, although an extract was published by Nicolas Seringe in his bulletin. His large herbarium of some 300 000 items containing many neotropical plants was purchased by the British naturalist Philip Barker Webb.

Mercurialis* (Euphorbiaceae) L.
Probably for Hieronymus Mercurialis (real name Girolamo Mercuriale (Mercuriali)) (1530–1606),

Italian physician and philologist. He studied at the universities of Bologna and Padua, obtaining a medical PhD in 1555, then went to Rome where he studied the classical and medical literature of the Greeks and Romans and how these factors could prevent or cure disease. This resulted in his best known work, *De Arte Gymnastica* (1569), in which he emphasised the importance of diet, exercise and hygiene for healthy living. He became professor of practical medicine at Padua (1569–1586) and during his tenure was called to Vienna in 1573, by Holy Roman Emperor Maximilian II who – pleased with Mercurialis's treatment – made him Count Palatine. From 1587–1592 he was a professor at the University of Bologna before being 'bought' with an exorbitant salary in 1593 by the University of Pisa, where he stayed until retiring, aged 75. His publications included, *inter alia*, *De Morbis Cutaneis* (*On Skin Disease*) (1572), *De Morbis Mulieribus* (*On the Diseases of Women*) (1582), *De Morbis Puerorum* (*On the diseases of children*) (1583) and *De Oculorum et Aurium Affectibus* (*On Ears, Eyes and Emotions*) (1583).

Meringium = Hymenophyllum
(Hymenophyllaceae) C.Presl
Gk. *merinx* = bristle. The sporangia are borne on a stiff, bristle-like axis that projects from the indusium.

Merremia (Convolvulaceae) Dennst.
For Basius Merrem (1761–1824), German naturalist and botanist, mathematician and

professor of political economy at Marburg. He studied at the University of Göttingen and developed an interest in zoology, particularly ornithology. Best known for his work *Tentamen Systematis Naturalis Avium*, published in Berlin in 1816, dealing with bird classification, in which he proposed the division of birds into ratites (running birds, with a flat sternum) and carinatus (flying birds, with a keeled sternum). He was a prolific author of many books in German, Latin and English on a wide range of subjects mainly related to birds, amphibians, Linnaeus's global classification system but also about economic issues such as castles and the household economy.

Mertensia = Sticherus (Gleicheniaceae) Willd.
For Franz Karl (Carl) Mertens (1764–1831), German botanist and algologist, professor

of botany at Bremen Polytechnic College from 1795. In his early years he studied theology and languages at the University of Halle and was a preacher but studied botany in his spare time. He met the German botanist Albrecht Wilhelm Roth (1757–1834), with whom he went on botanical explorations in Europe. Arising from this trip Mertens described a number of species of algae and also drew illustrations for Volume 3 of Roth's *Catalecta Botanica*. He was instrumental (with Erlangen professor Wilhelm Daniel Joseph Koch, 1771–1849) in updating the third edition of Johann Christoph Röhling's *Deutschland's Flora*. He had a large private herbarium.

Merwilla (Hyacinthaceae) Speta.
For Frederick Ziervogel van der Merwe (1894–1968), South African botanist and medical doctor.

After obtaining a BA, he went to Trinity College, Dublin, obtaining an MBBCh, BAO (1921) followed by a DTM from Liverpool University and a DPH from the University of the Witwatersrand (1930). As medical inspector of schools he travelled widely in the then Transvaal and Natal. He had an interest in botany, especially the *Aloe* and *Scilla*. His other interests were collecting sheet music of Africana value, and he authored a music bibliography, *Suid-Afrikaanse Musiekbibliografie* (1958), as well as a glossary of Afrikaans medical terms (with JD Louw), *Mediese Woordeboek* (1935). He described two new genera, *Schizocarphus* and *Resnova*.

Merxmuellera (Poaceae) Conert
For Hermann Merxmüller (1920–1988), German botanist, professor of botany at the University

of Munich, director of National Botanical Archives (Botanische Staatssammlung), director of the Munich Botanical Gardens (1969–1985), and director of Institut für Systematische Botanik (Institute of Systematic Botany) at the University of Munich (1958). Merxmüller made six visits to Southern Africa and discovered nearly 100 species new to science; he also edited/published the *Prodromus of the Flora of SWA* (now Namibia) from 1966–1972. Of the 32 000 specimens he collected in Central Europe, Iran, Egypt, Morocco, Botswana, Namibia, South Africa, Canada, America, Venezuela, Brazil and Chile, about 6 000 were from Southern Africa.

Mesembryanthemum (Aizoaceae) L.
Gk. *mesos* = in the middle, between; *embryon* = fruit (embryo); *anthemon* = flower. The flowers need mid-day sunshine to open.

Mesembryanthemum (EJM)

Mesogramma* (Asteraceae) DC.
Gk. *mesos* = in the middle, between; *gramma* = a line, character; alluding to the strongly marked medial nerve of the corolla.

Mesothema = Blechnum (Blechnaceae) C.Presl
Gk. *mesos* = in the middle, between; *thema* = a theme; referring to the frond, the sori of which are midway between the midrib and the margin.

Mestoklema (Aizoaceae) N.E.Br. ex Glen.
Gk. *mestos* = full; *klema* = twig, small branch; alluding to its dense branching.

Metalasia (Asteraceae) R.Br. (Images on following page)
Gk. *meta-* = meaning reverse; *lasios* = shaggy, woolly. The leaves are twisted, rolled upward, to present the woolly side of the leaf from the top to the bottom.

Metaporana = Bonamia (Convolvulaceae) N.E.Br.
Gk. *meta* = after, beyond; *Porana* (q.v.). A genus related to *Porana*.

Metarungia (Acanthaceae) Baden
Gk. *meta* = after, beyond; *Rungia* (q.v.). This genus was originally called *Macrorungia* but had to be renamed when that name was found to be illegitimate. So this new name was 'after' *Rungia* (Acanthaceae).

Metastelma = Parapodium (Apocynaceae) R.Br.
Gk. *meta* = after, beyond; *stelma* = crown. The segments of the crown not hooded, narrow, inserted at the base of the column, surpassing the stigma.

Metalasia (EJM)   *Metalasia* (EJM)

Methyscophyllum = **Catha** (Celastraceae) Eck. & Zeyh.
Gk. *methysko* = to intoxicate; *phyllon* = leaf. The leaves of *M. glaucum* (= *Catha edulis*) contain an intoxicating and narcotic alkaloid.

Metrosideros (Myrtaceae) Banks ex Gaertn.
Gk. *metra* = heartwood; *sideros* = iron; referring to the tough, hard wood.

Metzgeria (Metzgeriaceae) Raddi
For Johann Baptiste Metzger (1771–1844), German copper engraver and art restorer  who studied under the great Florentine engraver Raphael Sanzio Morghen (1753–1833). Metzer was a friend of the Italian botanist Giuseppe Raddi (1770–1829), who named the genus after him. Metzger became an important art agent in Florence and around 1828 acquired artworks by David Ghirlandaio now owned by the Metropolitan Museum of Modern Art in New York. Koperski (1991) wrongly aligned the name with Johann Christian Metzger (1789–1852), landscape architect and director of the botanical gardens in Heidelberg.

Meyerophytum (Aizoaceae) Schwantes
For Louis Gottlieb Meyer (1867–1958), German missionary and naturalist. He was sent by the Rhenish mission to Kommagass, Namaqualand, in 1894 and later served in Steinkopf, which also included the Richtersveld. An agriculturalist by training, he had a keen interest in plant and insect collecting. When Rudolf Marloth paid visits to his area, they collected together. Meyer sent Marloth many of the plants he discovered in his area as well as to Hans Herre (1895–1979), curator of the University of Stellenbosch Botanical Garden and a specialist in *Mesembryanthemaceae* (now *Aizoaceae*). He also sent insects to Hans KC Andrae (?–1966), Marloth's assistant who had a special interest in this field.

Mezleria*** = **Lobelia** (Lobeliaceae) C.Presl
Derivation uncertain. Probably for Johann Georg Metzler (1761–1833) who, for personal reasons,  took the pseudonym of Karl Ludwig Giesecke. Giesecke studied law and mineralogy at the University of Göttingen (1781–1784). From 1784–1800 he became an actor, stage manager, writer, poet and librettist for more than 15 operas. From 1801 he studied mineralogy, surveying and mining and became a mineral dealer, collector and tutor in Copenhagen. He obtained permission from Danish King Christian VII to explore the geology of the Faroe Islands (in 1805) and Greenland (1806–1807). As a result of the Napoleonic Wars he had to stay in Greenland until 1813. During this time he collected minerals and botanical specimens, mainly bryophytes, which were sent to the botanist Lorenz Chrysanth von Vest, who named *Campanula gieseckiana* in honour of Giesecke. In 1813 Giesecke obtained the position of professorship of mineralogy in the Royal Dublin Society, which he held until his death. (**Note**: WPU Jackson suggests that 'Mezler' and 'Metzler' were interchangeable or that the genus name was typography.)

Micranthus (Iridaceae) Eckl. (Image on opposite page)
Gk. *mikros* = very small; *anthos* = flower. A tiny but lovely flower. (Seen above, a close-up view)

Microcharis (Fabaceae) Benth.
Gk. *mikros* = very small; *charis* = grace(ful); referring to the delicate appearance of the plant.

Microchloa (Poaceae) R.Br.
Gk. *mikros* = very small, tiny; *chloa* = grass; referring to its size.

Micrococca (Euphorbiaceae) Benth.
Gk. *mikros* = small; *kokkos* = a berry. The fruit breaks up into small cocci when ripe.

Micranthus: close-up view (CM)

Microcodon (Campanulaceae) A.DC.
Gk. *mikros* = very small; *kodon* = a bell; referring to the plant having bell-shaped (campanulate) flowers.

Microcoelia (Orchidaceae) Lindl.
Gk. *mikros* = very small; *koilia* = abdomen; alluding probably to the small lip that is ventricosely calcarate.

Microdon (Scrophulariaceae) Choisy
Gk. *mikros* = very small; *-odon* = toothed. With small teeth – the calyx is five-toothed, the two anterior teeth sometimes more deeply incised than the other three.

Microglossa (Asteraceae) DC.
Gk. *mikros* = small; *glossa* = tongue. These daisies have short 'tongues' (ray-florets).

Microgonium = Didymoglossum (Hymenophyllaceae) C.Presl
Gk. *mikros* = small; *gonos, gonium* = seed. The seeds are small.

Microgramma (Polypodiaceae) C.Presl
Gk. *mikros* = very small; *gramma* = line; referring to the two lines of relatively small sori.

Microlaena = Ehrharta (Poaceae) R.Br.
Gk. *mikros* = very small; *laina* = cloak; alluding to the two minute glumes.

Microlepia (Dennstaedtiaceae) C.Presl
Gk. *mikros* = very small; *lepis* = a scale; referring to the small scale-like indusia protecting the sori.

Microloma (Apocynaceae) R.Br.
Gk. *mikros* = very small; *loma* = edge or fringe; referring to the hairy corolla tube. The hairs are minute.

Microlonchus (Asteraceae) Cass.
Gk. *mikros* = very small; *lonche* = lance, spine. The involucral bracts are tipped with sharp spines.

Micromeria (Lamiaceae) Benth.
Gk. *mikros* = very small; *meris* = a part; referring to the tiny leaves and flowers.

Micropterum = Cleretum (Aizoaceae) Schwantes
Gk. *mikros* = very small; *pteron* = wing, feather; referring to the fruit capsule that has large expanding keels and broad valve wings.

Micropteryx = Erythrina (Fabaceae) Walp.
Gk. *mikros* = very small; *pteryx* = wing; small-winged. Possibly referring to the wing-like flowers.

Microrhynchus*** = launaea (Asteraceae) Sch.Bip.
Gk. *mikros* = very small; *rhynchos* = beak, bill, snout; referring to the tiny, paired, beak-like fruit.

Microschizaea = Schizaeaceae (Schizaeaceae) C.F. Reed
Gk. *mikros* = very small; *schizein* = to split; referring to the incised, dichotomously divided frond segments.

Microsorium (Polypodiaceae) Link
Gk. *mikros* = very small; *sorus* = sori; referring to the small sori (spores on the underside of fern fronds).

Microstegium (Poaceae) Nees
Gk. *mikros* = very small; *stege* = cover; referring to the usually very minute lemma of the fertile floret.

Microtea = **Lophiocarpus** (Lophiocarpaceae) Sw.
Gk. *mikros* = very small; *ous, ot-* = ear; referring to the tiny flower.

Mikania (Asteraceae) Willd.
For Josef (Joseph) Gottfried Mikan (1743–1814), Austrian-Czech botanist and medical professor at the University of Prague, and father of the naturalist and zoologist Johann C Mikan (1769–1844). He studied at Dresden, Prague and Vienna and started his career as a doctor before becoming associate professor (1773), professor of botany and chemistry (1775) and then rector of the University of Prague (1798). He was the author of *Catalogus Plantarum Omnium* (1776), which was dedicated to the Botanical Garden of Prague.

Mikaniopsis (Asteraceae) Milne-Redh.
Mikania (q.v.); *iopsis* = resembling.

Milicia* (Moraceae) Sim
For a Senhor Milicia, sometime administrator of Maganja da Costa, Mozambique (Hugh Glen).

Millettia (Fabaceae) Wight & Arn.
For Charles Millet (1792–1873), a plant collector engaged by the British East India Company in the 1830s, stationed in Canton, China (now Guangzhou), where he was a member of an organisation known as the 'Canton Factory' established by Joseph Banks, a group of naturalists and collectors. He collected botanical specimens and corresponded with William Hooker, director of Kew Gardens, and John Henslow, professor of botany at Cambridge University. He collected plants around Macau and also in Ceylon, Malabar and Java. He returned to London in 1834. There is a suggestion that this Charles Millet may have been the French naturalist whose major study was freshwater fauna, but this is not the case.

Mimetes (Proteaceae) Salisb.
Gk. *mimetes* = imitator, mimic. Possibly given this name because some of its features, like the toothed leaves, bear a close resemblance to other family members, like *Leucospermum*, better known as pincushion. The genus itself is distinctive.

Mimosa (Fabaceae) L.
Gk. *mimos* = a mimic. The plant is sensitive to light, heat and movement. When touched, this 'sensitive plant' will collapse its long, thin leaves in unison as if wilted. Each leaf 'mimics' the other.

Mimulus (Phrymaceae) L.
La. *mimulus* = little mask, derived from *mimus* =

Mimetes (EGHO)

an actor, mime, mimic; *ulus* = diminutive. Some flowers are said to look a little like a monkey's face or the grinning mask used by actors.

Mimusops (Sapotaceae) L.
Gk. *mimus* = an actor, mime, mimic; *ops* = face; sometimes interpreted as *mimo* = an ape; *ops* = face, hence like a monkey; referring to the corolla and the shape of the flowers, which allegedly resemble a monkey face. 'The resemblance is at best obscure' (Hugh Glen).

Minurothamnus = **Herolepis** (Asteraceae) DC.
La. *minurus* = small, tiny; *thamnos* = bush. William Henry Harvey in *Flora Capensis* says *minuros* = slender.

Mirabilis* (Nyctaginaceae) L.
La. *mirabilis* = wonderful, remarkable; perhaps referring to *M. jalapa* for three reasons: flowers of different colours appear on the plant simultaneously; an individual flower may be touched with different colours; flowers can change colour – e.g. yellow to dark pink, white to light violet – as they mature.

Miraglossum (Apocynaceae) Kupicha
La. *mirus* = wondrous; Gk. *glossa* = tongue; referring to its bright yellow tongue-like 'head'.

Mirasolia* (Asteraceae) Sch.Bip.
La. *mirus* = wondrous; *solium* = seat, throne; referring to the wonderful showy flowers.

Miscanthidium* = **Miscanthus** (Poaceae) Stapf
Miscanthus (q.v.); *idium* = diminutive.

Miscanthus (Poaceae) Andersson
Gk. *miskos* = stalk, pedicel; *anthos* = flower; referring to both spikelets of the pair being pedicellate.

Misopates* (Plantaginaceae) Raf.
Gk. *misos* = hate; *patein* = step on; inferring that stepping on this beautiful plant would spoil its beauty.

Mitracarpum* = **Spermacoce** (Rubiaceae) Zucc
Gk. *mitra* = headband, turban, girdle; *karpos*, fruit; alluding to the capsule being 'circumcised round the middle' (Sonder in *Flora Capensis*).

Mitriostigma (Rubiaceae) Hochst.
Gk. *mitrion* = diminutive of a girdle, turban or bishop's mitre; *stigma* = stigma; referring to the stigma, which is thick and split in two.

Mitrophyllum (Aizoaceae) Schwantes
Gk. *mitra* = bishop's cap, headband, turban; *phyllon* = a leaf; the second pair of leaves are united for most of their length into a conical body resembling a bishop's mitre.

Mniothamnea* = *Brunia (Bruniaceae) Nied.
Gk. *mnion* = moss, seaweed; *thamnos* = bush; referring to a moss-like clump, often like matted bush.

Modecca* = **Adenia** (Passifloraceae) Lam.
The native Indian name for one species.

Modiola* (Malvaceae) Moench
La. *modiolus* = a little measure; referring to the fruit, which is in the form of a wheel-shaped schizocarp.

Moenchia* = *Paspalum (Caryophyllaceae) Ehrh.
For Conrad (Konrad) Moench (Mönch) (1744–1805), German professor of botany at the Collegium Medicum Carolinianum (1781) and at Marburg University (1785) until his death. He was the founder of the Marburg Botanical Garden. An outspoken critic of Linnaeus's sexual system (together with other German botanists such as Heister, Haller, Gleditsch and Medikus) and strong opponent of Linnaean classification, generic concepts and nomenclature. He wrote *Methodus Plantas Horti Botanici et Agri Marburgensis* (1794), describing 674 species in the garden and surrounding areas, restoring the names and genera of Tournefort wherever possible, as did his *Supplementum ad Methodum Plantas* (1802), which added 634 more flowering plants.

Moesslera* = *Tittmannia (Bruniaceae) Rchb.
For Johann Christoph Mössler (Moessler) (1770–1840), German botanist with expertise in the classification of plants. He authored a number of books, including *Taschenbuch der Botanik* (1805) (*Handbook of Botany*), which explained botanical language and examined 23 classes using the Linnaean system, and *Handbuch der Gewachskunde* (1827) (*A Manual of Living Plants*), which was written with Heinrich Gottlieb Ludwig Reichenbach (1793–1879), a German botanist, zoologist, naturalist and ornithologist who classified 2 900 new species. Their handbook examined the flora of Germany, including the major foreign crop plants.

Mohria (Anemiaceae) Sw.
For Daniel Matthias Heinrich Mohr (1780–1808), German professor of zoology and botany, plant collector and author. In 1799, he studied zoology under Johann Christian Fabricius, and from 1801 botany under Heinrich Adolph Schrader (1767–1836) at the University of Göttingen. He became associate professor of philosophy at the Christian Albrecht University of Kiel in 1805 and two years later associate professor of zoology and botany. He had a special interest in cryptogams (non-flowering plants), algae and mosses and published a considerable amount including *Contributions to Natural History* (1805) with professor Friedrick Weber and *Botanical Observations* (1803). He died at an early age.

Mollugo (Molluginaceae) L.
Derived from La. *mollis* = soft, pliant; -*ugo*, the standardised Latin suffix of plant family names. 'The name comes from *Galium mollugo*, or Indian chickweed, supposed to be the *Mollugo* of the ancients' (Sonder in *Flora Capensis*). Probably because of the similarity of their whorled leaves.

Momordica (Cucurbitaceae) L.
Originally an Arabic word for a tropical Old World vine; *mordere* = to bite; referring to the seeds that look as if they have been bitten.

Monactinocephalus* = *Inula (Asteraceae) Klatt
Gk. *mono-* = one; *aktino-* = rayed; *kephale* = head; possibly referring to the solitary rayed flower head.

Monadenia *** = ***Disa*** (Orchidaceae) Lindl.
Gk. *mono-* = one, only; *aden* = gland; referring to the solitary gland.

Monadenium = ***Euphorbia*** (Euphorbiaceae) Pax
Gk. *monos* = one, only; *aden* = a gland. The genus is distinguished by the single, horseshoe-shaped involucral gland surrounding the flowers.

Monanthotaxis (Annonaceae) Baill.
Gk. *monos* = one, solitary; *anthos* = flower; *taxis* = arrangement. The flowers in some but not all species are solitary (others cymose).

Mondia (Apocynaceae) Skeels
Latinised from *uMondi*, the isiZulu name of plants of this genus.

Monechma (Acanthaceae) Hochst.
Gk. *monos* = one; *echma* = that which holds fast, (obstacle, support); hence possibly referring to the root.

Monelytrum (Poaceae) Hack. ex Schinz
Gk. *monos* = one, single; *elutron* = a sheath, cover; the spikelet lacks a lower glume.

Monerma = ***Lepturus*** (Poaceae) P.Beauv.
Gk. *monos-* = one; *(h)nerma* = support; referring to the single spike or one glume (*Hortus Britannicus,* 1830).

Monilaria (Aizoaceae) Schwantes
La. *monile* = necklace (worn by boys, according to Cicero); *-aria* = connected with; referring to the succulent stems – 'The thick stems are shortly jointed, resembling a bead necklace' (Court).

Monniera (Scrophulariaceae) Kunze
For Louis Guillaume le Monnier (1717–1799), French natural scientist. He studied physics,  geology, medicine and botany and worked in all these areas as a researcher of electrical phenomena, documenting mines, working as a physician at the Saint Germain En Laye hospital and being appointed professor of botany at the Jardin du Roi (later the Jardin des Plantes) in 1759. He was also one of the original organisers of Louis XV's botanic collection at Petit Trianon, a small château on the grounds of the Palace of Versailles and also became the king's physician in 1761. He wrote several entries for Diderot's *Encyclopédie* and scientific papers on electricity. He was elected a member of the Académie des Sciences, Royal Society and Prussian Academy of Sciences.

Monochilus = ***Zeuxine*** (Orchidaceae) Wall. ex Lindl.
Gk. *mono-* = one, single; *cheilos, chilos* = lip; referring to the large and conspicuous lip that dominates the other floral segments.

Monochlaena = ***Didymochlaena*** (Hypodematiaceae) Cass.
Gk. *mono-* = one; *khlaina* = cloak. The two sori have a common indusium.

Monochoria (Pontederiaceae) C.Presl
Gk. *mono-* = one; *chori* = separate, apart; referring to one stamen being larger than the other five in some species.

Monoculus = ***Osteospermum*** (Asteraceae) B.Nord.
Gk. *mono-* = one; *oculus* = eye; referring to the fenestrated window on the fruits.

Monocymbium (Poaceae) Stapf

(GN) A bristly spikelet (GG)

Gk. *mono-* = one; *cymbium* = boat. Each raceme is enclosed by a single boat-shaped spathe.

Monodora (Annonaceae) Dunal
Gk. *mono-* = one, solitary; *doron* = gift. The flowers on these trees are solitary.

Monogramma (Pteridaceae) Schkuhr
Gk. m*ono-* = single; *gramma* = line; possibly referring to the frond's single costal vein or to the sori, which are in a single line in each frond.

Monographidium = ***Cliffortia*** (Rosaceae) C.Presl
Gk. *mono-* = single; *graphe* = drawing, writing; *-idion* = diminutive; referring to a little monograph (possibly describing the plant).

Monopsis (Lobeliaceae) Salisb.
Gk. *mono-* = one, single; *-opsis* = resembling; referring to the appearance of the corolla for most species. Every flower looks the same.

Monotes (Dipterocarpaceae) A.DC.
Gk. *monotes* = solitary. Trees of the family *Dipterocarpaceae* are 'scattered through the tropical forests' and occur 'sparingly' (Richards).

Monotris = ***Holothrix*** (Orchidaceae) Lindl.
Gk. *mono-* = only; *tris* = three times. This genus was separated by Lindley from the known species of *Holothrix* because its lip has only three (not five or seven) lobes.

Monsonia (Geraniaceae) L.
For Lady Ann Monson (*née* Vane) (1714–1776), English naturalist and great-granddaughter of Charles II. In 1774, aged 60, she came to the Cape on her way to India. Carl Peter Thunberg (1743–1728), who had arrived two years earlier to collect plant specimens and learn Dutch, took Lady Ann to a number of farms adjacent to Cape Town. She seemingly had more interest in the animal kingdom than the floral one. Lady Ann corresponded with Linnaeus, who seemed besotted with her, and he named the genus in her honour, writing: 'Nature has never produced a woman who is your equal – you are a phoenix among women.'

Montanoa* (Asteraceae) La Llave & Lex.
For Luis José Montaña Carrascó (1755–1820), Mexican doctor and naturalist. He studied humanities and philosophy at Palafoxiano seminary, theology at St Ignatius seminary and obtained his medical PhD at the Real y Pontificia Universidad de México (1793), where he became a professor in the medical faculty. He played a significant role in modernising medical science in Mexico through his emphasis on using the latest research to improve medical practices and healing, introducing the monitoring of hospital boards and promoting herbal medication through research into the medicinal properties of plants. He also encouraged medical practitioners to adhere to the great aphorisms, to which end he translated Anuce Foë's (1528–1595) Latin version of the great aphorisms of Hippocrates in his book *Praelectiones et Concertationes Medicae pro Hippocratis Magni Aphorismis* (1817).

Montbretia*** = ***Tritonia*** (Iridaceae) DC.
For Antoine François Ernest Coquebert de Montbret (1781–1801), French botanist and plant collector. He was part of Napoleon's military and scientific expedition to Egypt in 1798 and studied the flora of Egypt. He was the first librarian of the Institute of Egypt in Cairo and died of the plague there. He was co-author with JC Fabricius (1745–1808), HT Stainton (1822–1892) and JF Stephens (1792–1853) of *Illustratio Iconographica Insectorum quæ in Musæis Parisinis* (1799).

Montbretiopsis = ***Tritonia*** (Iridaceae) L.Bolus
Montbretia (q.v.); *iopsis* = resembling.

Montinia (Montiniaceae) Thunb.

The dried capsules are charateristic. (EJM)

For Lars Jonasson Montin (1723–1785), Swedish botanist and physician, whose specimens form the basis of the Swedish Academy of Sciences herbarium in Stockholm. Aged 21, he studied mining engineering at the University of Lund for two years before moving to the University of Uppsala, where he was inspired by Linnaeus. He graduated as a medical doctor in 1751 and became the district medical officer for the County of Halland on the west coast of Sweden in 1756. Here he met Pehr Osbeck (1723–1805), and together they made an inventory of flora of that area. While Osbeck went to China at Linnaeus's behest, Montin did botanical, zoological and ornithological research in Sweden. He discovered many new species and reported on the wild herbs in Halland. He was elected a member of the Royal Swedish Academy of Sciences in 1771 and given the title of assessor in 1782.

Moquinia = ***Moquiniella*** (Loranthaceae) Spreng.f.
For Christian Horace Bénédict Alfred Moquin-Tandon (1804–1863), French botanist and doctor. He was professor at Marsailles (1829–1833) and director of the Botanical Garden of Toulouse 1834–1853. He studied the flora of Corsica for the French government (1850–1852), became professor of botany at the Faculté de Médecine at Paris (1853) and later director of the Jardin des Plantes and the Académie des Sciences and was one of the founders of the Société Botanique de France. A specialist in seed plants and a prolific author, he wrote the three-volume *Histoire Naturelle des Mollusques Terrestres et Fluviatiles de France* (1855) and *Natural History of the Canary Islands* (Paris, 1836–1844) with Philip Barker Webb (1793–1854) and Sabin Berthelot (1794–1880).

Moquiniella (Loranthaceae) Balle
Moquinia (q.v.); Gk. *-iella* = diminutive.

Moraea (Iridaceae) Mill. (Images on opposite page)
For Sara Elisabeth Moraea (1716–1806), daughter of Dr Johan Moraeus, the town physician of Falun. When Linnaeus proposed to Sara in 1735, Dr Moraeus agreed but insisted that before any marriage could take place, Linnaeus should qualify. Four years later, in 1739, when they married, Linnaeus had obtained a doctorate in medicine cum laude from Hardewijk University and published *Systema Naturae,* his major work, and *Fundamenta Botanica,* and so gained a reputation. The genus *Moraea* was first described as *Morea* in 1758 by Philip Miller in honour of the English natural historian Robert More (1703–1780) but renamed in 1762 by Linnaeus to *Moraea.*

Morella (Myricaceae) Lour.
La. *morus* = a mulberry; *-ella* = diminutive. The fruit looks like a mulberry. (Other derivations for this name have been suggested.)

Moringa (Moringaceae) Adans.
A Sinhalese (Sri Lanka) vernacular name, *murunga.*

Morphixia* = ***Ixia*** (Iridaceae) Ker Gawl.
Gk. *morphe* = shape; *Ixia* (q.v.). *Morphixia* is a form of *Ixia.*

Morus (Moraceae) L.
The Latin name for the black mulberry, *M. nigra.* (Celtic *mor* = black); referring to the colour of the fruit.

Morysia* = ***Athanasia*** (Asteraceae) Cass.
For Charles Bourgevin of Vialart, Count of Saint Morys (1772–1817), French brigadier, councillor of the parliament of Paris serving King Louis XVIII who became mayor of Houdainville. He was made a knight of St Louis and an officer of the Royal Order of the Legion of Honour. He wrote many books, including *Picturesque Journey in Scandinavia* (1802) and *European Politics and the Interior Administration of France* (1815), in which he advocated abolishing the slave trade. He was killed by a Colonel Barbier in a duel in 1817 in an argument over land rights. Cassine, the plant author, writes: '[H]e was preparing to write a monograph on his many observations [relating to the cultivation of willows taken from different parts of Europe]. This appeared as a "memorial on the means of planting uncultivated plants".'

Moschosma* = ***Basilicum*** (Lamiaceae) Rchb.
La. *moschus* = musk; Gk. *osme* = fragrance odour; referring to the musky odour of the flower.

Mosdenia (Poaceae) Stent
For Mosdene, a farm near Naboomspruit to which Ernest Edward Galpin (1858–1941) retired in 1917 and where he found this genus. *Galpinia* (q.v.).

Mossia (Aizoaceae) N.E.Br.
For Charles Edward Moss (1870–1930), British botanist, teacher, lecturer and curator of the Cambridge Herbarium from 1908–1916, who played an important role in the formation of the British Ecological Society. He came to South Africa in 1917 and took up the newly created post of professor of botany at the South African School

Moraea (CM – All above)   Commonly called 'tulps' by farmers, they increase in density with over-grazing.

of Mines and Technology. This later became the University of the Witwatersrand. He built up the botany department from scratch, and his collection of flora in the Transvaal and neighbouring regions and his own herbarium laid the foundation for the university herbarium, which later was named the CE Moss Herbarium after him. He was elected a Fellow of the Linnaean Society in 1912.

Mucuna (Fabaceae) Adans.
A Portuguese word – the Brazilian name for these vines, the pods of which often bare irritating hairs.

Muiria*** = *Gibbaeum* (Aizoaceae) N.E.Br.
For Dr John Muir (1874–1947), Scottish physician, naturalist and cultural historian and  plant collector. An Edinburgh University graduate, he came to the Cape in 1896 and finally settled in Riversdale. He contributed greatly to our knowledge of the plants of this area, discovering many new species. He had a special interest in sea shells, ocean-borne fruits, seed-drift and folklore. He authored *Seed-drift in South Africa* (1937), *Gewone Plantname in Riversdal* (1929) and many articles on botany, medicine, folklore, and others of a socio-historical and genealogical nature. He donated his collection of seashells to the South Africa Museum, his drift-seed collection to Stellenbosch University and his personal herbarium to the National Herbarium, Pretoria.

Mukia = *Cucumis* (Cucurbitaceae) Arn.
A Malayalam name *mucci-pirri: mucci* = three, *piri* = spring, possibly referring to the curled tendrils (Umberto Quattrocchi).

Mundia*** = *Nylandtia* (Polygalaceae) Kunth
For Johannes Ludwig Leopold Mund (1791–1831), Prussian pharmacist, botanist, land surveyor and plant collector. Both he and Louis Maire (f 1815–1833) arrived in the Cape in 1816 as official collectors of plants and animals for the Prussian government. Mund collected around Cape Town and as far east as Uitenhage. In 1821 both men were recalled by the Prussian government for failing to deliver but ignored this instruction and were dismissed. Despite Dr Thom in a letter to Hooker stating that both men spent much of their time 'in sloth and gaiety in Cape Town', Mund set himself up as a land surveyor but continued to collect. His name appears on various taxa by botanists such as Johann Klotsch, Carl Meisner, Christian Nees von Esenbeck and Karl Ludwig, which suggests his contribution was not insignificant.

Mundulea (Fabaceae) (DC.) Benth.

The crushed leaves can be used as a fish poison. (EJM)

La. *mundulus* = neat, trim or possibly a diminutive of *Mundia* (q.v.).

Munichia**** (Asteraceae) Cass.
Gk. *mounichion* = spring or possibly after Artemis Mounichia, a Greek goddess of that name.

Muraltia (Polygalaceae) DC.

Leaves are spiny tipped and flowers bi-coloured. (CM) (CM)

For Johannes von Muralt (1645–1733), Swiss anatomist and naturalist who studied medicine, surgery and obstetrics in Basel, Leiden, London, Oxford, Paris, Montpellier and Lyon. He became

professor of physics and mathematics at the Zurich Collegium Carolinum in 1688 and contributed substantially to the foundations of teaching anatomy and medicine there. He was a prolific writer on surgery, anatomy, obstetrics, biology, pathology, philosophy, zoology, botany and general medicine and wrote several medical books. He was superstitious and believed that the devil played a large part in the ills of mankind. He was elected to the Academia Naturae Curiosorum in 1685.

Murdannia (Commelinaceae) Royle
For Munshi Murdan Ali, plant collector and keeper of the herbarium at the Saharanpur Botanical Gardens in India in the 1840s, an expert on Himalayan flora. According to David Arnold of London's School of Oriental and African Studies, beside *Plurality and Transition of Knowledge Systems in Nineteenth-century India* (2003), Ali was said to have written a Hindustani treatise on botany.

Muriea*** = **Manilkara** (Sapotaceae) Hartog
For Dr James Murie (1832–1925), Scottish pathologist, surgeon and medical officer. He obtained his MD from the University of Glasgow in 1857, specialising in zoology, then worked at the Royal Infirmary, Glasgow, and later did comparative anatomy on acquatic mammals at the Royal College of Surgeons' Museum. After two years at the museum he travelled through Europe and made several voyages as ship's surgeon and naturalist, including the expedition to the White Nile under John Petherick to meet JH Speke and JA Grant (1861–1863). He gathered much valuable zoological material, and the fish collection was declared by Dr Gunther to be the finest ever received from that part of Africa. For much of his life he practised in London and made a comprehensive study of the ecology in the Thames estuary still admired today. He was a member of the Royal Geographic Society and sometime assistant secretary of the Linnaean Society.

Murraya* (Rutaceae) L.
For Johan Andreas (Anders) Murray (1740–1791) Swedish physician, pharmacologist and botanist of German parentage. He studied at the universities of Uppsala (1756–1759) – under Linnaeus – and at Göttingen (1763), qualifying as a doctor of medicine, and was appointed a professor and director of the botanical garden (1769). He led investigations into the properties of medicinal plants and into the ways in which

 plant-derived medicines could be prepared and administered. His major work was the six-volume *Apparatus Medicaminum Tame Simplicium Quam Præparatorum et Compositorum* (1776–1792), a comprehensive compilation of vegetable drugs. He also wrote a German translation of Linnaeus's *Systema Naturae* (13th edition) renamed *Systema Vegetabilium* (*System of the Vegetable Kingdom*), with a new introduction called *Regnum Vegetabile* (*The Vegetable Kingdom*).

Musa*** = **Heliconia (Musaceae) L.
Arabic *mauz* meaning *musa* or banana-tree.

Muscari* (Hyacinthaceae) Mill.
Gk. *muschos* = musk; referring to the scent.

Mussaenda* (Rubiaceae) L.
Latinised form of Sinhalese vernacular name *mussenda* meaning 'red-leaved' (more correctly 'red-sepaled'); referring to the stunning blood-red sepals of *M. erythrophylla*, but other species have a wide range of colours.

Myagrum** = **Rapistrum (Brassicaceae) L.
Gk. *muagron* or *myagros,* name for species of mustard by Dioscorides and Pliny the Elder, but alluding to a wider range of crops. Gk. *mys* = mouse; *agra* = to ensnare, hence a mouse-trap, or *agros* = field, hence mice field. Some species are invasive and compete with broadleaf crops such as chickpeas, canola, lupins, etc. – all areas infested with mice.

Myaris** (Rutaceae) C.Presl
Anagram of *Amyris,* a genus of flowering plants in the citrus family, *Rutaceae.*

Mycelis* (Asteraceae) Cass.
An ancient name, derivation obscure. Possibly, *mykes* = fungus; *-elis* = diminutive; possibly referring to *Mycelis muralis* (Linnaeus) = a little fungus that grows on walls or buildings.

Myoporum (Scrophulariaceae) G.Forst.
Gk. *Myein, myo* = to close or to be shut; *poros* = an opening, pore; referring to the plant's ability to close the pores on the leaves – and save water going into the atmosphere – and so cope well in drier areas.

Myosotis (Boraginaceae) L.
La. and Gk., from *myos,* gen. of *mys* = mouse; *-ous, -otis,* = ear; referring to the shape of the leaves.

Myosurus* (Ranunculaceae) L.
Gk. *mys* = mouse; *oura* = tail; translated from the herbalists' *cauda murina,* referring to the apiculate achenes.

Myrica** = **Morella (Myricaceae) L.
Gk. *myrike* = fragrance; also *murikē* (*myrike*) meaning a Tamarisk (= a family of mainly Old World desert shrubs and trees); referring to the plant's aromatic leaves in many species.

Myriophyllum* (Haloragaceae) L.
Gk. *myrios* = countless, myriad; *phyllon* = leaf; alluding to the many divisions of the submerged leaves of these aquatic plants; water milfoil.

Myriotheca***** = **Marratia (Marattiaceae) Comm. ex Juss.
Gk. *myrios* = countless, myriad; *theka* = a case, capsule. The sporangium produces large quantities of small spores.

Myrobalanus***** = **Terminalia (Combretaceae) Gaertn.
Gk. *myro, myron* = ointment, perfume, uguent; *balanos* = nut, often acorn. The nuts are used for making perfume.

Myrothamnus (Myrothamnaceae) Welw.
(Images on following page)
Gk. *myron* = perfume; *thamnos* = shrub, bush; referring to the plant's fragrance.

Myrsine (Myrsinaceae) L.
Gk. name used by Dioscorides for the common myrtle, *Myrtus communis,* which it resembles.

Myrsiphyllum** = **Asparagus (Asparagaceae) Willd.
Myrsine (q.v.); *phyllon* = a leaf.

Myrtillocactus (Cactaceae) Console
La. *myrtillo* = small, myrtle-like; *cactus* = cactus. Blueberry cactus; referring to the myrtle-like fruits.

Myrtus* (Myrtaceae) L.
The Greek name for myrtle. Georg Christian Wittstein derives the word from Gk. *myron* = myrrh, in allusion to the aroma of the leaves and fruit.

Mystacidium (Orchidaceae) Lindl.
Gk. *mystax* = moustache (hence *mystacine,* fringed or striped in a moustache-like manner); *-idion* = diminutive; probably from the papillose-barbate lateral segments of the rostellum of the type species.

Mystropetalon (Balanophoraceae) Harv.
Gk. *mystron* = a spoon; *petalon* = peta. The perianth segments are spoon-shaped (William Henry Harvey).

Mystroxylon (Celastraceae) Eck. & Zeyh.
Gk. *mystron* = spoon; *xylon* = wood. A direct translation of the vernacular *lepelhout* or spoonwood.

Myxopappus (Asteraceae) Källersjö.
Gk. *myxa* = mucus, slime; *pappus* = fluff, pappus. An ancient name given by Pliny for a kind of plum tree, *Cordia myxa* (Umberto Quattrocchi).

Myrothamnus: Take the dried plant, emerse in water for a minute then 24 hours later the leaves are green and fully expanded. (EJM)

Myrothamnus: Dessicated form. (EJM)

N

Nymphaea (EJM)

Nymphoides (CM)

Nymphaea (CM)

Nymphoides (CM)

Rene Glen (*née Henrici*) (1946–) grew up on a farm near Villiersdorp, in the Western Cape in South Africa. She has always been interested in wetlands and their inhabitants. Her first ecological work in this field was an undergraduate study on the algae of Princess Vlei, on the Cape Peninsula. She then did her MSc on the coccolithophorids (tiny single-celled algae) of Swartkops River Estuary, Port Elizabeth. Moving on to larger, if not higher, things, she has produced a checklist of aquatic and wetland plants and, in conjunction with Carina Cilliers and others, a guide to wetland grasses, sedges and similar plants of Southern Africa. Although retired, she is working on various projects with colleagues in Durban, Cape Town and the UK.

Nachtigalia** = Phaeoptilum (Nyctaginaceae) Schinz
For Gustav Hermann Nachtigal (1834–1885), German physician, botanist, explorer and  consul-general in Tunis and commissioner for West Africa. He studied medicine at Halle, Würzburg and Greifswald, and practised in Cologne from 1859–1863. He then moved to Algiers and Tunisia for health reasons. He undertook a hazardous expedition across the Sahara as described in his three-volume book *Sahara and Sudan* (1879–1881) visiting places not known to Europeans. In 1884 Chancellor Otto von Bismarck appointed him commissioner for West Africa, and he was instrumental in safeguarding German business interests in Africa by annexing Togoland and the Cameroon, which subsequently became colonies of Germany. He died of malaria on his return voyage to Germany.

Nageia (Podocarpaceae) Gaertn.
A Japanese vernacular name; origin and meaning unknown.

Nagelocarpus = Erica (Ericaceae) Bullock
Anagram of *Lagenocarpus*, Gk. *lagenos* = flask; *karpos* = fruit; referring to the flask-shaped fruit.

Najas (Najadaceae) L.
Linnaean name for a genus of aquatic plants, perhaps derived from Naiad, a water nymph, or the Gk. *nāias* = floating, swimming, that is, in the water; alluding to its habitat.

Namacodon (Campanulaceae) Thulin
Nama = ethnic name, from Namaqualand region; *kodon* = a bell; alluding to the bell-shaped flowers and where the plant is found.

Namaquanthus** (Aizoaceae) Bolus
Gk. *anthos* = flower; (from) Namaqualand.

Namaquanula (Amaryllidaceae) U.& D.Mül-Doblies
Namaqualand; *-ula* = diminutive. This genus is found in Namibia and South Africa.

Namibia (Aizoaceae) (Schwantes) Dinter & Schwantes
From the Namib desert, Namibia.

Namophila (Hyacinthaceae) U. & D.Mül-Doblies
Namibia; *phila* = loving. There is only one species of this genus, *Namophila urotepaia*, which is found in Namibia.

Nananthus (Aizoaceae) N.E.Br.
Gk. *nanos* = dwarf; *anthos* = flower; referring to the small size of the flowers.

Nandina* (Berberidaceae) Thunb.
The derivation of this scientific name is a Latinised version of the word *nan-ten*, a Japanese name for the plant, meaning southern sky.

Nanobubon (Apiaceae) Magee.
Gk. *nano* = dwarf; *bubon* = swollen gland; alludes to the similarity to *Notobubon* and the difference in habit, hence the Greek prefix *nano* (AR Magee).

Nanolirion*** = Caesia (Hemerocalliaceae) Benth.
Gk. *nanos* = dwarf; *lirion* = a white lily; referring to its size.

Nanozostera*** (Zosteraceae)
Gk. *nanos* = dwarf; *zoster* = belt, girdle; referring to the ribbon-shaped leaves.

Narcissus* (Amaryllidaceae) L.
La. *narcissus,* possibly derived from Gk. *nárkissos* (*narkē* = numbness, sleep); referring to the narcotic properties attributed to species of the plant.

Narthecium* (Melianthaceae) Huds.
Gk. *narthex* = a hollow stem or cane; referring to the stem.

Nassella* (Poaceae) (Trin.) E.Desv.
La. *nassa* = wicker basket with a narrow neck, fish basket; *-ella* = diminutive; alluding to the resemblance of the spikelet to such a basket in lateral view.

Nasturtium* (Brassicaceae) R.Br.
La. *nasus* = nose; *tortus* = tormented, twisted; from *torquere* = to twist. The acrid smell of *N. officinalis* (watercress) causes the nose to wrinkle. (No relation to the gardener's 'nasturtium', which is a *Tropaeolum*.)

Nastus = Cenchrus (Poaceae) Juss.
Classical name used by Dioscorides for *Cenchrus frutescens* from Gk. *nastos* = close-pressed, firm; referring to the tree-like stem.

Natalia*** = Bersama (Melianthaceae) Hochst.
The Portuguese navigator Vasco de Gama sighted the coast along what is now Durban on Christmas Day in 1497 and named the country Terra Natalis, a name which embraces 'the earth' and 'Christmas Day', which the Portuguese call '*Dia de Natal*', Latinised as *Natale Domini*, hence La. *natalia*. (Natal's full name is now KwaZulu-Natal.)

Nathusia = **Schrebera** (Oleaceae) Hochst.
For Hermann Engelhard von Nathusius (1809–1879), German zoologist and agriculturalist and animal breeder who collected a huge amount of information on cattle breeding. He studied under Johannes Müller from 1827–1829. He was a member of the Prussian Land Economic Council, director of the province of Saxony's Central Agricultural Council, and president of the German Agricultural Society. He wrote a number of books on botany and zoology, including *Über die Rassens des Schweines* (1860), but also books about breeding of sheep (1856), animals (1860), pigs (1860), shorthorn cattle (1861), and husbandry and breed knowledge (1872–1880).

Nauclea (Rubiaceae) L.
Perhaps derived from La. *naucula* = a small ship; referring to the valves of the fruit.

Nautochilus = **Orthosiphon** (Lamiaceae) Bremek.
Gk. *naus* = boat; *cheilos* = lip. The lower corolla lip is boat-shaped in some species.

Nebelia = **Brunia** (Bruniaceae) Neck. ex Sweet
For Daniel Nebel (1663–1733) who studied medicine at the University of Marburg, obtaining his PhD in 1686 and becoming a full professor in the medical faculty in 1707. From 1708–1728 he worked as a doctor and apothecary at the Heidelberg hospital and Sapienzkolleg and from 1728 exclusive personal physician of Elector Carl Philip at his court in Mannheim. Also for Nebel's son William Bernhard Nebel (1699–1748), physician and botanist, professor of mathematics, physics and medicine at Heidelberg. Daniel Nebel, in particular, was a prolific author both in the fields of botany and medicine, and a considerable amount of his work was published posthumously.

Nectaropetalum (Erythroxylaceae) Engl.
La. *nectar* = nectar, drink of the gods; *petalum* = petals; referring to the nectar glands on the petals.

Nectouxia (Solanaceae) Kunth
For Hyppolyte Nectoux (1759–1836), French botanist and horticulturalist. In 1788 he was appointed a director of the botanical garden of Saint-Dominique in the Carribean, where he introduced breadfruit. Between 1798 and 1801 he was a member of the Commission des Sciences et des Arts during the French military expedition in Egypt. During this campaign he travelled across Egypt as far as Nubia, as recorded in *Voyage Dans la Haute Égypte, au-dessus des Cataractes: Avec des Observations sur …* (1808) where he discovered *Senna alexandrina* (Caesalpiniaceae). He also introduced the potato to Egypt. In Egypt, he lost vision in his right eye and returned to France where he briefly became head gardener at Fontainebleau (1802). He was asked by Napoleon in 1808 to redesign the public gardens of Rome, which he did in part but could not complete them as a result of Napoleon's fall (1814). For his work, Pope Pius VII conferred on him the Order of the Golden Spur.

Neesenbeckia (Cyperaceae) Levyns
For Christian Gottfried Daniel Nees von Esenbeck (1776–1858), German botanist, plant collector, physician and professor of botany at Erlangen (1817), Bonn (1818–1830) and Breslau (1831). He was a prolific author in numerous fields and described about 7 000 plant species, almost as many as Linnaeus. His special interest was fungi. In later life he became involved in radical politics and became president of the workers' union in conflict with the government, resulting in him losing his professorship and pension. His resulting poverty forced him to sell his library and herbarium of some 80 000 sheets to pay off debts. He was president of the German Academy of Sciences Leopoldina, and his last act was to admit Charles Darwin as a member.

Negria = **Lintonia** (Poaceae) F.Muell.
For Cristoforo Negri (1809–1896), Italian geographer, politician and professor of political science. He studied at universities in Pavia, Graz, Vienna and Prague. In 1843 he became professor of political science at the University of Padua. During the 1848 war of independence he sided with the Italians, but after the fall of Vicenza, in June 1848, was compelled to leave for political reasons. He moved to Turin where he became university rector. He was the founder in 1866 and first president of the Italian Geographic Society, which he directed until 1872, and was re-elected president again in 1880. During his career he became involved in diplomatic missions to many European countries. His last official activity was the management of the general consulate in Hamburg, from 1873–1874. He became a senator in 1890. After that he retired to Turin.

Nelia (Aizoaceae) Schwantes
For Gert Cornelius Nel (1885–1950), South African botanist, plant collector and cactus specialist, and professor of botany at Stellenbosch University, 1921–1950. He studied in Stellenbosch before moving to Germany, where he studied chemistry and obtained a PhD from Berlin University under Adolf Engler in 1914. He had a special interest in succulent plants, especially *Stapelieae, Euphorbias* and *Mesems*. He collected extensively in the drier parts of South Africa and took responsibility for the university garden. He produced an illustrated volume on *Lithops* (undated) and one on *Gibbaeum* (1953), edited by PG Jordaan

Nelsia* (Amaranthaceae) Schinz
Probably for Louis Nels (1855–1910), German government official and sometime plant collector. Nels was born in France as his father fled the 1848 Prussian revolts. He studied in France, trained as a lawyer, then joined the German civil service in 1885 as secretary to Acting Reichskommisar Heinrich Ernst Göring (1839–1913) in Otjimbingwe, colonial headquarters, German South-West Africa (now Namibia). When Göring left the country in 1890, Nels became acting reichskommissar. About this time he collected plants near Windhoek, according to the Harvard University Herbaria index of botanists. In 1897 he became German consul in Johannesburg, South Africa. After the South African War (1902), he returned to Berlin for health reasons. From 1904–1910, when he retired, he did consular work in Rotterdam in the Netherlands.

Nelsonia* (Acanthaceae) R.Br.
For David Nelson (?–1789), English gardener-botanist to James Cook on his third voyage (1776–1780) and to William Bligh on his voyage to Tahiti to obtain breadfruit trees (1787–1789). Nothing is known of Nelson's early life. He received some botanical training from Joseph Banks and William Aiton prior to his voyage with Cook. After the ship's return, he worked for Kew Gardens for seven years before accepting an appointment as botanist to Bligh on the HMS Bounty. When Bligh's crew mutinied, Nelson, together with 19 men loyal to Bligh, were cast adrift without arms in a small boat. Nelson survived the famous 3 800-mile voyage to Timor. A few days after arriving, he spent a day botanising in the mountains, caught a cold and died.

Nematanthus = Willdenowia (Restionaceae) Nees
Gk. *nemata* = thread; *anthos* = flower; referring to the thread-like flower stalks.

Nemesia (Scrophulariaceae) Vent.

(CM)

Gk. *nemesion, nemeseion* from *nemo* = to distribute, to enjoy, to pasture, to feed; or *nemos* = wooded pasture, glade, a grove; name used by Dioscorides for a similar plant, referring to their habitat.

Nemia*** = Manulea (Scrophulariaceae) P.J.Bergius
Gk. *aneimon* = naked, unclad, so possibly *nemia, neimon* = fully covered. Meaning not explained by Bergius.

Nenax (Rubiaceae) Gaertn.
Gaertner does not explain this name.

Neobakeria = Daubenya (Hyacinthaceae) Schltr.
La. *neo-* = new. For John Gilbert Baker (1834–1920), British botanist, botanical collector at the herbarium of Kew Gardens for 33 years, during the last nine years of which he was keeper of the herbarium, Fellow of the Royal and Linnaean societies and a prolific writer. He wrote handbooks on many plant groups, including *Amaryllidaceae, Bromeliaceae, Iridaceae, Liliaceae* and ferns. His published works include *Flora of Mauritius and the Seychelles* (1877) and *Handbook of the Irideae* (1892). He was the father of the botanist Edmund Gilbert Baker.

Neoboivinella*** = *Englerophytum*
(Sapotaceae) Aubrév. & Pellegr.
La. *neo-* = new. For Louis Hyacinthe Boivin (1808–1852), French botanist, plant collector for the French National Museum of Natural History in Paris. He visited the islands of the Indian Ocean (Comoros, Seychelles and Réunion) as botanist on the Oise expedition (1846–1852) and Mauritius in 1847–1849 as well as the coasts of Africa, the Canary Islands and Madagascar. He gathered a vast and important collection of specimens, which were deposited in the Museum of Paris for identification and classification of the new species. Unfortunately, shortly after he returned from his mission, exhausted and beset by malaria, he died at a hospital in Brest.

Neobolusia (Orchidaceae) Schltr.

Another of our many genera of ground orchids (HS)

La., Gk. *neo-* = new. For Harry Bolus (1834–1911), English-born South African botanist, botanical artist, stockbroker and plant collector who arrived in South Africa in 1850. He started collecting in 1865 and organised six major collecting expeditions within South Africa, Mozambique and Swaziland during 1883–1904. He had a passion for orchids, writing, *inter alia,* the three-volume work *Icones Orchidearum Austro-Africanum Extra-tropicarum* (1893–1913), the third volume published posthumously. He founded the Bolus Herbarium and bequeathed his library to the (now) University of Cape Town and left sufficient funds to develop a chair of botany. He was a Fellow of the Linnaean Society and member and president of the South African Philosophical Society (later the Royal Society of South Africa).

Neodregea = *Wurmbea* (Colchicaceae)
C.H.Wright.
La. *neo-* = new. For Isaac Louis Drège (1853–1921), an apothecary and plant collector like his father, Carl Friedrich Drège (1791–1867), who collected plants in the Albany, Uitenhage and Port Elizabeth areas. The original monotype discovered by James Glass in (1896) was in poor condition, and the immature fruiting material sent to Kew Gardens resulted in some errors in the original generic description. In 1909, Isaac Louis wrote about finding a number of plants in flowering condition 'in damp ground at Cradock Place' and 'at Baakens River' both near Port Elizabeth. Selmar Schönland of the Albany Museum in Grahamstown sent these to Kew Gardens, hence the new species *Neodregea glassii* came into being. Drège published 'A Preliminary List of Flowering Plants, Ferns and Fern Allies in the Port Elizabeth District' in the *South African Journal of Science* (1913).

Neoglaziovia (Bromeliaceae) Mez.
La. *neo-* = new; for Dr Auguste François Marie Glaziou (1828–1906), French landscape gardener and botanist who, in 1858, at the request of Emperor Dom Pedro II, relocated to Brazil to become director of parks and gardens in Rio de Janeiro.  During his many years in Brazil he explored widely and described over 700 species. Where possible, he distributed at least 10 sets of plants or seeds found to herbariums in Berlin, Paris, Brussels, Kew, St Petersburg, Copenhagen, Stockholm, Geneva, Montpellier and Rio de Janeiro. He returned to France in 1897, where he worked on his personal herbarium, which contained over 12 000 species. He was the author of a two-volume work, *Plantae Brasiliae Centralis a Glaziou Lectae,* which described the species in his collection. He died of pulmonary disease shortly after the publication of the second volume. More than 200 species have been named in his honour.

Neohenricia (Aizoaceae) Bolus

La. *neo-* = new. For Dr Marguerite Gertrude Anna Henrici (1892–1971), Swiss plant physiologist and plant collector who, after obtaining a PhD from the University of Basel, spent much of her life in the Orange Free State (now Free State). She obtained a DSc from the University of South Africa for work on the content of Karoo shrubs and grasses, and the division of plant industry built a well equipped laboratory for her in Fauresmith to study problems connected mainly with Karoo vegetation. She collected some 6 000–7 000 specimens mainly from the western Orange Free State and Ermelo.

Neohyptis = *Plectranthus* (Lamiaceae) J.K.Morton.
La. *neo-* = new; *Hyptis* (q.v.).

Neoluederitzia** (Zygophyllaceae) Schinz
La. *neo-* = new. For Franz Adolph Eduard Lüderitz (1834–1886), merchant and colonial pioneer, brother of botanical collector August Lüderitz. In 1878 he inherited his father's trading business and established a trading station at Lagos, later turning his attention to South-West Africa (now Namibia). In 1883 he acquired land extending from the Orange River mouth to latitude 26° south and 32 kilometres inland, which became known as Lüderitzland, and in 1884 concluded treaties with Nama chiefs to acquire the whole Nama area to the Kunene River. He obtained support from Bismarck and the German government for protection of this territory, which established German interests in South-West Africa. He died at sea off the Orange River mouth in unknown circumstances.

Neomarica* (Iridaceae) Sprague.
La. *neo-* = new; *Marica* = the name of a water nymph; renamed the new *Marica*.

Neomuellera*** = ***Plectranthus*** (Lamiaceae) Briq.
La. *neo-* = new; *Muellera* (q.v.).

Neonotonia (Fabaceae) J.A.Lackey
La. *neo-* = new; *Notonia* (q.v.).

Neopatersonia*** = ***Ornithogalum*** (Hyacinthaceae) Schönland
La. *neo-* = new. For Florence Mary Paterson (*née* Hallack) (1869–1936), botanical collector. She collected in the area around Port Elizabeth and Uitenhage, and was the wife of TV Paterson of South Africa. She made a comprehensive collection of specimens, some of which had not been seen since Carl (Karl) Ludwig Philipp Zeyher (1799–1858) collected in that area in 1838. She sent her specimens mainly to Selmar Schönland (1860–1940), who was professor of botany at Rhodes University and curator (later director) of the Albany Museum, who specifically thanked her in Volume 1 of his *Memoirs of the Botanical Survey of South Africa* (1919) as did Harry Bolus in Volume 3 of *Orchids of South Africa* (1913).

Neorautanenia (Fabaceae) Schinz.
La. *neo-* = new. For Martti Rautanen (1845–1926), Russian-born Lutheran church missionary and pioneer of the Finnish Mission Society, who in 1868 went to South-West Africa (specifically Ovamboland in present-day Namibia), where he served more than 50 years and was the director of the missionary station. His most important work was the translation of the Bible into Oshindonga, a Herero language. He greatly assisted the botanist Dr Hans Schinz during his expedition through South-West Africa (1884–1887), and the two became firm friends. He amassed a significant collection of ethnography materials, which are now housed at the National Museum of Finland, and sent plants that he collected to the University of Zurich.

Neorhine = ***Rhinephyllum*** (Aizoaceae) Schwantes
La. *neo-* = new; a synonym of *Rhinephyllum*.

Neorosea*** = ***Sericanthe*** (Rubiaceae) Hallé.
La. *neo* = new. For Valentin Rose the Younger (1762–1807), German apothecary and pharmacologist, son of the discoverer of the low-melting alloy Rose's metal, Valentin Rose the Elder (1736–1771). After Rose the Elder died, his sons were educated by Martin Heinrich Klaproth (1743–1817), who also looked after Rose the Elder's business for nine years. Klaproth made many brilliant contributions to analytical and mineralogical chemistry and became the first professor of chemistry at the University of Berlin. Rose the Younger collaborated with him in his research and verified

all his analyses before publication. Rose the Younger was the first person to demonstrate the presence of chromium in serpentine soils.

Neostenanthera* (Annonaceae) Exell.
La. *neo-* = new; *steno-, sten* = narrow, contracted, short; *anthera* = anther; possibly referring to the anther's shape.

Nepeta (Lamiaceae) L.
Latin for an ancient city in Etruria, modern Tuscany, possibly Nepi or Nepet, that may have supplied herbs to Rome, or invented some use for or from the plant.

Nephrodium = Dryopteris (Dryopteridaceae) Michx.
Gk. *nephros* = kidney; *-odes, odium* = resembling; alluding to the shape of the spores.

Nephrolepis (Nephrolepidaceae) Schott
Gk. *nephros* = kidney; *lepis* = a scale; referring to the kidney-shaped indusia in the type species.

Nephroma**** (Nephromataceae) Ach.
Gk. *nephros* = kidney; referring to the shape of the apothecia.

Nephrotheca = Osteospermum (Asteraceae) B.Nord. & Källersjö.
Gk. *nephros*= kidney; *theka* = a case, capsule; referring to the shape of the seed capsule.

Neptunia (Fabaceae) Lour.
La. *Neptunus,* God of the sea and rivers; referring to one of the few aquatic legumes in which some species are aquatic, floating or prostrate near water's edge.

Nerine (Amaryllidaceae) Herb.

(HC)

For Nerine, in Greek mythology a sea-nymph or nereid, daughter of Doris and Nereus, and granddaughter of Oceanus and Tethys. The Nereids were meant to protect sailors and their ships. Common name 'Guernsey lily'. In 1820, William Herbert named this indigenous South African plant *Nerine* (previously *Imhofia*), when a ship carrying boxes of the bulbs of this species was shipwrecked on Guernsey. The boxes were washed ashore, and flowers grew around the coast, hence the common name.

Nerium* (Apocynaceae) L.
Classical Gk. *nerion* = name for Oleander.

Nervilia (Orchidaceae) Comm. ex Gaudich.

(HS)

La. *nervus* = vein; referring to the distinctly fine veins in the leaves.

Nesaea (Lythraceae) Kunth
For *Nesaea* or *Nesaie,* in Greek mythology a sea nymph, one of the Neriads; referring to an aquatic plant.

Nestlera (Asteraceae) Spreng.
For Chrétien Géofroy (also Christian Gottfried) Nestler (1778–1832), Alsatian botanist, professor of botany and pharmacy, faculty of medicine, chief pharmacist of the hospitals of Strasbourg and director of the botanical garden of Strasbourg in 1816. He wrote a thesis on *Potentilla*. He studied in Paris under Louis Claude Marie Richard, a French botanist. With JB Mougeot, he made a fabulous collection of plants of Alsace and the Vosges. In 1830 he discovered, on the banks of the Rhine, the moss *Trichostomum viridulum* from the family Pottiaceae. He was a prolific author of

both botanical and pharmacologicial works, which included *Index Plantarum quae in Horto Academ. Argentinensi* (1818).

Neumannia *** = ***Aphloia*** (Aphloiaceae) A.Rich.
For Joseph Henri François Neumann (1800–1858), French head gardener of the greenhouses of the Imperial Museum of Natural History, Paris. According to *Popular Science Monthly* (Volume 15, Sept. 1879): 'Neumann, head gardener at the Paris Museum of Natural History, was the first to obtain … in 1838 … (from) a single stock … over two hundred vanilla fruits of excellent quality.' In the wild, *Vanilla planifolia*, a species of vanilla orchid native to Mexico, has less than a 1 per cent chance of being pollinated. Neumann discovered a simple method to transfer the pollen from the anther to the stigma. He wrote *The Art of Building and Governing the Serres* (= Greenhouses) (1846).

Neuracanthus (Acanthaceae) Nees
Gk. *neuron* = nerve, vein; *akanthos* = thorn; genus *Acanthus* (q.v.); possibly referring to the compact inflorescences with overlapping strongly-veined bracts.

Neurada *** (Neuradaceae)
Gk. *neuron* = nerve, vein, nerve fibre; referring to the typical venation of the leaves.

Neuradopsis (Neuradaceae) Bremek. & Oberm.
Neurada (q.v.); *opsis* = resembling.

Neurolobium = ***Diplorhynchus***
(Apocynaceae) Baill.
Gk. *neuron* = nerve, vein; *lobion* = diminutive of *lobos* = lobe, pod, capsule. The leaves are distinctly veined, and the anthers have twice the normal number of lobes.

Neurotheca (Gentianaceae) Salisb. ex Benth.
Gk. *neuron* = nerve, vein; *theka* = a case, capsule; referring to the prominent veins on the calyx.

Nevillea (Restionaceae) Esterh. & H.P.Linder
For Neville Stuart Pillans (1884–1964), South African botanist who assisted Professor Henry Harold Welch Pearson (1870–1960) in selecting the Kirstenbosch site for the future National Botanical Garden, son of Charles Eustace Pillans (1850–1919). He spent two years at Cambridge University studying agriculture but had to give it up because of ill health. After various jobs he joined the Bolus Herbarium, where he remained until his retirement and even worked there afterward. As a schoolboy he grew indigenous plants, especially succulents, and was 'the most eminent collector of *Stapeliads* in the eventful history of the tribe' (White & Sloane). He devoted himself to the taxonomy of Restionaceae, Bruniaceae, Phyllica, Agathosma and Metalasia.

Newtonia (Fabaceae) Baill.
For Isaac Newton (1642–1726), English mathematician, physicist, astronomer and natural philosopher, who held the Lucasian Chair of Mathematics at Cambridge. One of the world's foremost scientists, he wrote *Philosophiae Naturalis Principia Mathematica* (1687), which defines the laws of gravity and planetary motion (thus reinforcing Kepler's laws), co-founded the field of integral and differential calculus (with his arch-rival Gottfried Leibniz), wrote *Opticks* (1704) explaining the laws of light and colour, became the first person to build a practical refractive telescope, formulated a law of empirical cooling, and studied the speed of sound, etc. In 1705 he became the first English scientist to be knighted when this honour was conferred on him by Queen Anne.

Newtonia = ***Distephanus*** (Asteraceae) O.Hoffm.
For Francisco Xavier Oakley de Aguiar Newton (1864–1909), British-born Portuguese plant collector. His father, the British botanist Isaac Newton (1840–1906), worked as a curator in the Botanical Garden of Porto, Portugal. For 17 years, Francisco collected for his father, initially in West Africa in the 1880s – Dahomey (1882–1883), Angola (1885–1886) – but later also in Portugal, Java, Cape Verde, Benin, Equatorial Guinea, São Tomé and Príncipe.

Nicandra * (Solanaceae) Adans.
For Nikander of Colophon (Nikandros Kolophonios) (c 100–150), Greek botanist, physician, poet, grammarian and medical writer, who wrote about plants. He authored *Alexipharmaca*, a hexameter poem relating to the treatment of poisons and their antidotes, and *Theriaca,* a hexameter poem on the nature of venomous animals and the wounds they inflict. His medical knowledge is attributed to the physician Apollodorus (possibly of Alexandria). Most of his poems and prose work, such as the

mythical epic *Heteroeumena*, have been lost or only fragments survive, such as *Georgica*, which was perhaps imitated by Virgil. The works of Nikander were praised by Cicero, imitated by Ovid, Lucan and perhaps Virgil and frequently quoted by Pliny the Elder and other writers.

Nicolasia (Asteraceae) S.Moore
For Nicholas Edward Brown (1849–1934), English plant taxonomist and authority on succulents. He worked for Kew Gardens, becoming assistant keeper in 1909 and held the post until 1914. Although he never visited South Africa, he made a considerable contribution toward the taxonomy of South African plants, particularly Asclepiadaceae, Aizoaceae, Labiatae, Mesembryanaceae and other Cape plants, with publications in the *Kew Bulletin* and *Flora Capensis*. He revised the genus *Mesembryanthemeum* in 1931. He was awarded the Captain Scott Memorial Medal by the South African Biological Society in recognition of his work on South African flora (1921) and an honorary DSc from the University of the Witwatersrand (1932).

Nicolsonia = Desmodium (Fabaceae) Span.
For Jean Barthélemy Maximilien Nicolson (1734–1773), a Dominican priest and keen naturalist from Paris and superior in Haiti. He served as a dominican superior in Léogane from 1769 until his death. On the royal command of Louis XV, Nicolson's family published his *Essai sur l'Histoire Naturelle de St Dominique* (1776) posthumously. This large work (376 pages, 10 plates) describes the colony to a degree and has a chapter on archaeology, but the bulk of the book focuses on botany and the fauna and flora in the colony. Nicolson's main 'complaint' was that the colonialists didn't even have a name for most plants.

Nicotiana* (Solanaceae) L.
For Jean Nicot (1530–1600), French diplomat and ambassador in Lisbon, Portugal, from 1559–1561. He was sent to Portugal to negotiate the marriage of six-year-old Princess Marguerite de Valois (who became queen of France in 1589) to five-year-old King Sebastian of Portugal, but unsuccessfully. When Nicot returned to France he brought with him both tobacco plants and snuff (made from crushed tobacco leaves) to the French Court. This proved to be an instant success, although cigarette smoking only emerged in Europe in the 1830s. Nicot became something of a celebrity. He also compiled one of the first French dictionaries, *Thresor de la Langue Françoyse Tante Ancienne que Modern* (published posthumously in 1606).

Nidorella (Asteraceae) Cass.
La. *nidor* = vapour, smell; *-ella* = diminutive; alluding to the strong smell these plants have.

Niebuhria*** = Bachmannia (Capparaceae) Oliv.
For Carsten Niebuhr (1733–1815), German-born Danish botanist, explorer, surveyor and sole survivor of Pehr Forsskål's six-man 1760 expedition to Arabia, the others dying by 1764, probably from malaria. Niebuhr continued the expedition's work by himself before returning to Copenhagen in 1767. He measured landscapes, described everything he saw in notebooks, copied thousands of cuneiform characters from the ruins of Persepolis (which enabled language experts to decipher the writing) and made detailed drawings and maps that were used for over 100 years. He authored *Description of Arabia* (1772) and *Travels through Arabia* (1774–1778). He was a member of the Royal Society of Göttingen and the Royal Swedish Academy of Sciences.

Niemeyera (Sapotaceae) F.Muell.
For Felix von Niemeyer (1820–1871), German physician and pathologist. After completing his medical studies at the University of Halle in 1843, he practised as a physician in Magdeburg. In 1855, he became a professor of 'internist' medicine at the University of Greifswald and then professor at the University of Tübingen from 1860. His main textbooks were *Lehrbuch der Speziellen Pathologie und Therapie ...* (*A Textbook of Special Pathology and Therapy ...*) (1858–1861) published in seven languages, and *A Textbook of Practical Medicine ...* (1869). He became a consulting physician to King Charles I of Württemberg in 1865, was elected a foreign member of the Royal Swedish Academy of Sciences in 1870 and served as medical consultant during the Franco-Prussian War (1870–1871). He died two months later.

Nierembergia* (Solanaceae) Ruíz & Pav.
For Juan Eusebio Nieremberg (1595–1658), Spanish Jesuit born of German parents, author, mystic, and first professor of natural history at Madrid. He studied the classics at the Royal Court, science at the University of Alcalá, and canon law at the University of Salamanca. He became a Jesuit in 1614 and a lecturer on scripture at the Jesuit seminary until his death. He authored a number of books that, at the time, were published in many languages but are now largely forgotten. His treatise *De la Hermosura de Dios y su Amabilidad* (*In God's Beauty and Kindness*) (1649) is regarded as a classical manifestation of mysticism but cloyingly 'sugary sweet' when compared with, say, the enraptured vision of St Theresa.

Nigrina*** = Melasma (Orobanchaceae) L.
La. *niger* = black; -*ina* = diminutive; referring to the stems that are often dark crimson to purple-red but black when dry.

Niphobolus*** = Pyrrosia (Polypodiaceae) Kaulf.
Gk. *niphos* = snow, snowflake; *bolos* = clump. The fronds are covered with a somewhat white starry pubescence; when examined with a magnifying glass, the fronds are observed to be scattered over with snow-like crystals.

Nivenia (Iridaceae) Vent. (Image on opposite page)
For James Niven (1776–1827), Scottish gardener at the Royal Botanical Garden of Edinburgh and at Syon House, Middlesex. He collected plants in South Africa from 1798–1803 for his patron, George Hibbert, in Clapham, London. Three months after his return to England he went back to the Cape as botanical collector for Empress Josephine of France and James Lee and John Kennedy of the Vineyard Nursery, Hammersmith, near London. He spent a further nine years at the Cape collecting herbarium specimens, seeds and bulbs but also visiting areas such as Grahamstown in the Eastern Cape to Clanwilliam northwest of Cape Town, returning to England in 1812 and setting up his own business, unrelated to botany.

Nivenia = Paranomus (Proteaceae) R.Br.
Nivenia (q.v.) above.

Nolletia (Asteraceae) Cass.
For Jean Antoine Nollet (1700–1770), French clergyman and physicist. He was the first professor of experimental physics at the University of Paris. His main research area was the new field of electricity. He constructed one of the first electrometers, invented the electroscope (a device to detect the presence of an electrical charge), developed a theory of electrical attraction and repulsion that supposed the existence of a continuous flow of electrical matter between charged bodies and is credited with the discovery of osmosis in natural membranes. He was a contemporary of Benjamin Franklin (1706–1790) and wrote many letters to him trying to discredit his experimental results. Nollet became a member of the Royal Society of London in 1734.

Noltea (Rhamnaceae) Rchb.f.
For Ernst Ferdinand Nolte (1791–1875), German botanist and physician, professor of botany and director of the botanical garden at Kiel from 1826. He studied medicine at the University of Göttingen from 1813 and later at Charité, a university hospital in Berlin. From 1815 he explored floristically, and from 1821–1823 had the financial support of the Danish government and the Elbe Duchies of Lauenburg to study the flora of the Schleswig-Holstein mainland and coastal islands, and he made many contributions to the *Flora Danica,* a massive work initially designed to cover all plant species in the crown lands of the Danish King. In 1825 he wrote *Botanical Observations on Stratiotes and Sagittaria*, which received acclaim, probably leading to his professorship. His most famous student was Ferdinand von Mueller, who explored the flora of Australia.

Nomaphila** (Acanthaceae) Blume
Gk. *nomos* = a pasture; *philos* = loving; referring to the plant's favoured habitat.

Norlindhia = Osteospermum (Asteraceae) B.Nord.
For Nils Tycho Norlindh (1906–1995), Swedish botanist and plant collector. After graduating with a DSc (1943), he became an associate then acting professor at Lund (1943–1949) and then curator, Botanical Museum of Lund (1950–1961). He moved to Stockholm, accepting the position of acting professor of botany and director of the Swedish Museum of Natural History, Stockholm, from 1961, and professor from 1965–1972. His interests were taxonomy, morphology, cytology, phytogeography and

The genus *Nivenia* has many species that are now rare. (CM)

ecology. He collected 6 000 specimens in South Africa from 1930–1931 (including the eastern side of Southern Rhodesia = Zimbabwe) and in 1963. He collected in Mongolia in 1970 and 1972, authoring *Flora of the Mongolian Steppe and Desert Areas.* He was an honorary member of the South African Association of Botanists.

Nortenia* = **Torenia** (Linderaceae) Thouars
Near anagram of *Torenia* (q.v.).

Notechidnopsis* (Apocynaceae) Lavranos & Bleck
Gk. *notos* = the south; *Echidnopsis* (q.v.); alluding to its distribution in the Southern Hemisphere.

Nothofagus* (Nothofagaceae*) Blume
Gk. *nothos* = false; *fagus* = beech. Genetic tests by the Angiosperm Phylogeny Group revealed these 'beech' trees, originally grouped with the Fagaceae family, to be genetically distinct and now in a family of their own, the Nothofagaceae 'Southern beeches'.

Notholaena* = **Cheilanthes** (Pteridaceae) R.Br.
Gk. *notho-* = false; *laina* = cloak; possibly referring to the whitish or yellowish farina (powdery wax) that coats the surface of the leaves to prevent the leaves from drying out.

Nothoperanema = **Dryopteris** (Dryopteridaceae) (Tagawa) Ching

Gk. *notho-* = false; *Peramena* = a fern genus. Peramena, the name given to a flagellate by Félix Dujardin in 1842, but this name had also been claimed earlier for a fern genus first collected in Nepal.

Nothosaerva (Amaranthaceae) Wight
Gk. *notho-* = false; *Aerva* (q.v.).

Nothoscordum* (Alliaceae) Kunth
Gk. *notho-* = false; *skordon* = garlic. This genus is closely related to *Allium*, which includes the 'real' garlic.

Nothria* (Frankeniaceae) P.J.Bergius
Gk. *nothros* = sluggish. The inference is obscure.

Notiophrys* = **Platylepis** (Orchidaceae) Lindl.
Gk. *notis* = southern; *Ophrys* = a genus of orchids; alluding to its distribution in the Southern Hemisphere.

Notobubon (Apiaceae) B.-E. van Wyk
Gk. *noto-* = of Southern (African) origin; *Bubon* (q.v.). This genus was once part of the *Bubon* genus.

Notobuxus (Buxaceae) Oliv. (Image on following page)
Gk. *notos* = south; *Buxus* (q.v.). A Southern Hemisphere genus closely related to *Buxus*.

Notobuxus: A forest under-shrub (EJM)

Notochloe (Poaceae) Domin
Gk. *notos* = south; *chloe* = young green grass; referring to its occurrence in the Southern Hemisphere.

Notonia (Asteraceae) DC.
For Benjamin Noton (1784–1869), English assay master of the East India Company's Mint at Bombay and plant collector, largely in South India, especially in the Nilgiri Hills. Little is known about him. Robert Wight and Walker-Arnott in *Prodromus Florae Peninsulae Indiae Orientalis* (1834), write: 'In 1928, Dr Wallich [...] superintendent of the Botanical Gardens, arrived in England with an enormous number of specimens of plants [...] collected by himself (and many others including) Noton in the (O.E. Neelgherries = Nilgiri Hills).' These specimens were distributed to various authorities. 'Professor De Candolle [...] named for us the greater part of the *Compositae*.' Noton also published the earliest record of tidal observations made at Bombay (1832), which appeared in *Rushton's Gazetteer* (1842).

Notoniopsis*** = ***Kleinia*** (Asteraceae)
B.Nord.
Notonia (q.v.); Gk. *-opsis* = resembling.

Notosceptrum*** = ***Kniphofia***
(Asphodelaceae) Benth.
Gk. *notos* = south; *skeptron* = regal staff, staff or stick. Some species have erect stems at the top of which are elongated inflorescence with stalkless flowers. With imagination this could be a 'royal sceptre or staff'.

Nurmonia = ***Turraea*** (Meliaceae) Harms
An anagram of *Munronia*, after William Munro (1818–1880), English military officer and plant collector. He rose through the ranks from being an ensign in 1834 to becoming a general in 1878, serving in India, Gibraltar, the Crimea, Canada, the Caribbean, Bermuda and the Windward and Leeward islands. From an early age he had an interest in collecting plants. His nine publications include *Hortus Bangalorensis* (1837), *Hortus Agrensis* (1844), *On Antidotes to Snake Bites* (1848), *Report of the Timber Trees of Bengal* (1849) and *A Monograph of the Bambusaceae* (1868), describing the 219 bamboo species then known. He became perhaps the greatest authority on grasses of his day, identifying grasses from around the world for many of the great botanists. About a dozen or so were new to science.

Nuxia (Stilbaceae) Comm. ex Lam.
For Jean Baptiste François de Lanux (la Nux) (1702–1772), a French amateur naturalist. He became chief clerk (1725), then commander of Saint-Denis, the administrative capital on Réunion Island (1736). He tried to develop silkworm farming on the island and was interested in all aspects of natural history. He became a correspondent of the Royal Academy of Sciences (1762). Both Mathurin Jacques Brisson (1723–1806), French zoologist (*Le Règne Animal,* 1756; *Ornithologie,* 1760) and Georges-Louis Leclerc Comte de Buffon (1707–1788), French naturalist (*Histoire Naturelle des Oiseaux,* 1771–1786), relied on his detailed descriptions and specimens. He described the now extinct lesser Mascarene flying fox, *Pteropus subniger*.

Nycteranthus*** = ***Mesembryanthemum***
(Aizoaceae) Neck.
Gk. *nykteris* = a bat; *nykteinos* = by night; *anthos* = flower; therefore, night-flowering, possibly bat-pollinated.

Nycterinia = ***Zaluzianskya*** (Scrophulariaceae)
D.Don
Gk. *nykteris* = a bat; *nykteinos* = by night. As above.

Nylandtia*** = ***Muraltia*** (Polygalaceae)
Dumort.
For Petrus (Peter, Pierre) Nylandt, (c 1635–1710) Dutch botanist, pharmacist and physician. A prolific author, he published some 50 works on a wide range of topics such a bee-trafficking, garden

design, what to do with garden produce, etc., mainly in Dutch between 1670 and 1710, many of the publications with co-authors. His major works are considered to be *Herbarius Belgicus* (1670), which was the first record of the flora of the Low Countries – mainly Belgium, the Netherlands and Luxembourg – and also of *De Nederlandtse Herbarius of Kruydt-boeck* (1682), which not only describes all medicinal plants and herbs growing wild or cultivated in the Netherlands, but also the imported dried herbs found at chemists.

Nymania (Meliaceae) Lindb.
For Carl Fredrik Nyman (1820–1893), Swedish botanist and curator of the Swedish Museum of Natural History in Stockholm for some 40 years. In 1838 he studied at Uppsala University with Karl Greg Theodor Kotschy (1794–1865), an Austrian botanist, and Heinrich Wilhelm Schott (1794–1865). They produced *Analecta Botanica* (1854), of which he was an editor. His major works include *Conspectus Florae Europaeae* in four volumes (1878–1885) and *Sylloge Florae Europaeae* (1854–1855).

Nymphaea (Nymphaeaceae) L.
La. *nymphaea* = water-lily; Gk. *nymphe* = goddess of springs, general name for delightful, mythological, semi-divine water-maidens (Virgil), also bride; (*Nymphaios* = sacred to the nymphs); referring to the water lilies, of which there are many in this genus.

Nymphoides (Menyanthaceae) Ség.
Gk. *nymphe* = goddess of springs; *-oides* = resembling. The genus name refers to their resemblance to the water lily *Nymphaea*.

Nymania: The common name for this plant is Chinese lanterns because of the inflated fruits. (EJM)

O

The *Olea* genus has various forms, ranging from our Olive trees to Black ironwood.

Hugh Glen (1950–) grew up in Johannesburg, South Africa but is now living in Durban. His first botanical interest was in succulents, but finding that field overcrowded, he moved on to the study of cultivated plants. He has served on various international committees concerned with both plant nomenclature generally and the nomenclature of cultivated plants in particular. His publications include a forerunner of this book. He claims to be one of the few surviving botanists with any fluency in Latin but admits his command of the language resembles Chaucer's Nun's French (which she spoke '*ful fyne and fetisly, after the scole of Stratford atte Bowe, for Frenssh of Parys was to hir unknowe*'). He is retired and authored *Guide to Trees Introduced into Southern Africa* (Struik Nature, 2016).

Oberonia (Orchidaceae) Lindl.
For Oberon, the mythological fairy king and husband of Titania. Commonly known as 'fairy orchids'.

Obesia *** = ***Stapelia*** (Apocynaceae) Haw.
Gk. *obesa* = fat; alluding to these succulent gouty plants.

Obetia * (Urticaceae) Gaudich.
Probably for Louis Jean Marie Obet (1777–1856), naval surgeon, second chief physician of the navy in 1812. He was a professor of anatomy at the Saint Bernard Hospital and did much to improve the standards of nutrition and hygiene. He was elected to the Royal Academy of Medicine in 1822. He published *Essai sur la Nutrition* (1806). He was honoured with Chevalier of the Legion d'Honneur in 1821 and became an officer of the Legion d'Honneur in 1833.

Ochna (Ochnaceae) L.

Best know for the fruits which some say resemble monkey faces (EJM)

Gk. *ochne* = wild pear; referring to the leaves that resemble those of the pear, or at least Linnaeus thought so.

Ocimum (Lamiaceae) L.
Gk. *okimon* = smell, fragrance. The genus contains aromatic herbs such as Basil (*O. basilicum*).

Ocotea (Lauraceae) Aubl.
Probably from a vernacular name for stinkwood in French Guiana.

Octogonia *** = ***Erica*** (Ericaceae) Klotzsch
La. *octos* = eight; *gonia* = angle; octagonal. Referring to the leaf-shaped bracts, equal, oblong, underneath the calyx reaching upward longitudinally and octagonally (octopus-like).

Octopoma (Aizoaceae) N.E.Br.
La. *octos* = eight; *poma* = cover, lid; alluding to the capsule's structure.

Odina *** = ***Lannea*** (Anacardiaceae) Engl.
An Indian name for a species of this genus.

Odontella = ***Oncocalyx*** (Loranthaceae) Tiegh.
Gk. *odontos-* = toothed; *-ella* = diminutive; small-toothed; possibly referring to the lateral appendages or teeth present 1–2 millimetres below anthers.

Odontelytrum (Poaceae) Hack.
Gk. *odontos-* = toothed; *elektryon* = a sheath. The involucre has an involucre of fused bristles.

Odontophorus (Aizoaceae) N.E.Br.
Gk. *odontos-* = teeth; *phoros* = bearing, carrying; bearing teeth. Probably referring to tomato hornworms whose big white teeth are really just harmless suction cups.

Odontosoria * (Lindsaeaceae) Fée
Gk. *odontos-* = toothed; sorus. The sori are at the tips of toothed segments.

Odyssea (Poaceae) Stapf
For Odysseus, Homer's hero of the *Odyssey*. Author Otto Stapf writes in the *Journal of Botany*: '[t]he grass described here […] has become connected with no fewer than nine mostly widely different genera – a veritable odyssey, hence the name – but it may be hoped that it has at last reached a safe port.'

Oeceoclades (Orchidaceae) Lindl.
Gk. *oikeios* = private; *-klados* = branch, shoot; referring, possibly, to James Lindley's separation of certain species from *Angraecum* to form a distinct tribe or 'private branch'.

Oedera (Asteraceae) L.
For Georg Christian Edler von Oldenburg Oeder (1728–1791), German botanist, physician  and economist. He studied medicine at the University of Göttingen under Albrecht von Haller who, in 1751, persuaded King Frederick V of Denmark to appoint Oeder as *professor botanices regius* (royal professor). He developed a botanical garden and in 1753 he became the founding author of *Flora Danica*, a massive work initially designed to cover all plant species in the crown lands of the Danish King, which was only completed 153 years later. Oeder served on many commissions and was involved

in agrarian and social reforms. In 1771 Oeder lost his professorship as a result of a financial crisis in Denmark and was given a lesser post as a bailiff in Oldenberg, then under Danish rule. Two years before his death he was ennobled by Joseph 11, Holy Roman Emperor of the Hapsburg lands (present-day Austria).

Oenanthe (Apiaceae) L.
Gk. *oenos* = wine; *anthe* = flowery. The type species has a wine-like fragrance.

Oenostachys = Gladiolus (Iridaceae) Bullock
Gk. *oenos* = wine; *stachys* = spike; referring to the brightly coloured bracts that hide the flowers almost completely and form a wine-coloured spike (Eurekamag.com).

Oenothera* (Onagraceae) L.
Gk. *Oenothera,* derived from Gk. *oenos* = wine; *thera* = to catch, hunt, pursue, seek, imbibe. La. *Oenothera* = a plant whose juices may cause sleep. The roots of *O. biennis* were supposed to, variously, be edible, smell of wine, be used to flavour wine, give one a thirst for wine or, when the juices were drunk with wine, cause sleep.

Oetosis = Neurodium (Polypodiaceae) Kuntze
Possibly, La. *oetos* = resting-place. Etymology unknown.

Oftia (Scrophulariaceae) Adans.
Derivation uncertain. Michel Adanson (1727–1806) may have named the genus for a friend, M Oft, but the name was unexplained.

Ohlendorffia*** = Aptosimum
(Scrophulariaceae) Lehm.
For Johann Heinrich Ohlendorff (1788–1857), founder of the nursery Ohlendorff JH & Sons and inspector of the Hamburg Botanical Garden. This garden was founded in 1820 by the German botanist Johan Georg Christian Lehmann (1792–1860) for whom Ohlendorff worked, and it was Lehmann who gave the name to the genus *Ohlendorffia*. Ohlendorff came from a family of landscape gardeners and nurserymen. Perhaps better known is his famous son Baron Jacob Heinrich Bernard von Ohlendorff (1836–1928), who became immensely rich and a land baron through being involved in the guano fertiliser business and owning both a bank and a publishing and printing business.

Olax (Olacaceae) Doric.
Gk. *olax* = a furrow. The bark is furrowed.

Oldenburgia (Asteraceae) Less.
For Franz (Frantz) Pehr Oldenburg (1740–1773),

Restricted to quartz ridges round Grahamstown (EJM)

Swedish soldier in the Dutch East India Company. He was an amateur botanist and plant collector who arrived at the Cape in 1765 and collected specimens for Joseph Banks in London and Peter Jonas Bergius in Stockholm. He went on collecting expeditions with Carl Peter Thunberg (1742–1828) around Cape Town (1772) and with Francis Masson (1741–1805) as far as the Breede River for two months in 1772 and 1773. Later that year, Governor Joachim van Plettenberg asked Thunberg to go to Madagascar as ship's surgeon and itinerant botanist. He declined but recommended Oldenburg, who collected briefly in Madagascar and the Comoros, before he died at sea of a fever, probably malaria, in 1773, and was buried on the Barren Islands near Madagascar.

Oldenlandia (Rubiaceae) L.
For Heinrich (Hendrik) Bernhard Oldenland (Oldeland) (c 1663–1697), German-born South African botanist and plant collector. He studied medicine at Leiden University under Paul Hermann (1646–1695) before coming to the Cape in 1688. He was the compiler of the first plant list at the Cape. He went on an expedition to near Aberdeen, Eastern Cape (1689), at the time the most easterly point explored. He served the government in various capacities: master-gardener of the Company's Garden, land surveyor and the equivalent of town-engineer (superintendent of streets, roads, bridges, buildings). He compiled the 13-volume *Herbarus Vivus* consisting of 380 indigenous plants with a second list of exotic plants. He was working on a *kruidboek* of dried and mounted plants when he died.

Olea (Oleaceae) L.
Gk. *elaia,* La. *olea* = classical Latin name for the olive.

Oleandra (Oleandraceae) Cav.
Gk. *elaia,* La. *olea* = olive; *andros* = male or stamen; possibly because the fronds resemble the leaves of *Nerium oleander.*

Olfersia* (Dryopteridaceae) Raddi
For Ignaz Franz Werner Maria von Olfers (1793–1871), German naturalist, historian and diplomat.

He studied medicine, science and linguistics at the University of Göttingen (1812–1815) before joining the diplomatic corps. He served Brazil (1816), Lisbon and Naples (1826–1828), Switzerland – as chargé d'affaires (1831–1834) – before becoming a privy councillor of the court (1835). He became general director of the Royal Museums (1839–1869), which he redeveloped, and increased their collections significantly, introducing many innovations, expanded the medieval art and Renaissance sculptures sections, created an antiquarium and developed the library. Von Olfers described a number of new mammal species in Wilhelm Ludwig von Eschwege's *Journal von Brasilien* (1818).

Oligocarpha = Brachylaena (Asteraceae) Cass.
Gk. *oligos-* = few; *karphos* = a dry stick, chaff; possibly referring to its appearance.

Oligocarpus = Osteospermum (Asteraceae) Less.
Gk. *oligos-* = few; *karpos* = fruit. The plant has only a few fruit.

Oligodora = Hymnolepis (Asteraceae) DC.
Gk. *oligo-* =little; *dora* = gift; alluding to the tiny bright yellow flowerheads which, when massed together in a tight compound inflorescence, give off a showy glow worth seeing – the gift!

Oligodorella = Marasmodes (Asteraceae) Turcz.
Gk. *oligos-* = few; *dora* = gift; *-ella* = diminutive.

Oligomeris (Resedaceae) Cambess.
Gk. *oligos-* = few; *meris* = part; possibly referring to the two whitish petals that may be fused together or even missing, hence 'few parts'.

Oligothrix (Asteraceae) DC.
Gk. *oligos-* = few; *thrix* = hair; referring to the few delicate bristles that comprise the pappus.

Olinia (Oliniaceae) Thunb.
For Johan Henrik Olin (1769–1824), Swedish botanist, student of Carl Peter Thunberg, and author. He studied theology in 1789 and obtained a BA degree in 1793, became an assistant at the botanical garden in Uppsala, during which he obtained further degrees and a licentiate in medicine in 1797. He was the acting medical officer in Växjö in 1800, the district medical officer in Eksjö in 1802, district medical officer in Växjö in 1815, and the nation's curator in Växjö from 1794–1796. He translated from German to Swedish Paul Erdman Isert's *Journey to Guinea and the Caribbean Islands in Colombia* (1788). He was also the author of *Plantae Svecanae* (1797) and *Dissertatio Arnica* (1799).

Oliverella (Loranthaceae) Tiegh.
For Daniel Oliver (1830–1916), British botanist. He was librarian of the Herbarium at Kew Gardens from 1860–1890 and keeper there from 1864–1890, as well as being professor of botany at University College, London from 1861–1890. He authored *Lessons in Elementary Biology* (1864) based upon material left in manuscript by John Stevens Henslow (1796–1861), a clergyman, botanist and mentor to Charles Darwin. He was also author of *First Book of Indian Botany* (1885), *Flora of Tropical Africa* (1868), *Illustrations of the Principal Natural Orders of the Vegetable Kingdom* (1874) (with WH Fitch), and other works. He wrote monographs on new genera *Hillebrandia* (1866) and *Begoniella* (1873), of the family Begoniaceae. Oliver was a Fellow of the Royal and Linnaean societies.

Olyra* (Poaceae) L.
Old Greek word used by Theophrastus and Dioscorides for a kind of grain, not identified, but probably rye.

Omentaria*** = Tulbaghia (Alliaceae) Salisb.
La. *oment* = fat, fat skin, adipose tissue; *-aria* = pertaining to, possession. Meaning obscure. Perhaps referring to their fat, tuberous roots or somewhat fleshy leaves.

Ommatodium = Pterygodium (Orchidaceae) Lindl.
Gk. *omma* = an eye; *-odes* = resembling; referring to the 'eye-like spots on the labellum' (Lindley, 1838).

Omphalobium** (Fabaceae) Jacq.
Gk. *omphalos* = navel, umbilicus; *lobion, lobos* = lobe, pod, small fruit; referring to the shape of the fruits.

Omphalocaryon *** = Erica** (Ericaceae) Klotzsch
Gk. *omphalos* = navel, hub, central part of the flower; *caryon* = nut.

Onagra **= Oenothera** (Onagraceae) Mill.
Gk. *onagros* = the wild ass, from *onos* = ass, *agrios, agra* = wild; referring to the shape of the calyx.

Oncinema (Apocynaceae) Arn.
Gk. *onkos* = tumour, tubercle, or from *onkinos* = hooked; *nema* = a thread, filament. The anthers carry ovate appendages.

Oncinotis (Apocynaceae) Benth.
Gk. *onkinos* = hooked. The anthers and appendages are curiously hooked (Hugh Glen).

Oncoba (Salicaceae) Forssk.
From *onkob*, the Arabic name for this genus.

Oncocalyx (Loranthaceae) Tiegh.
Gk. *onkos* = bulb, swollen mass; *calyx* = calyx; possibly referring to the corolla, which is swollen at the base.

Oncosiphon (Asteraceae) Källersjö.
Gk. *onkos* = bulb, swollen mass; *siphon* = tube. The base of the corolla tube is tumorous (swollen).

Ondetia (Asteraceae) Benth.
The type specimen was collected at 'Elephantskloof' in Damaraland and called 'Ondetu' by the Damara (Benth., author).

Onixotis *** = Wurmbea** (Colchicaceae) Raf.
Gk. *onyx* = claws (hence clawed base of petals); *otos* = ear; alluding to the auriculate petal claws.

Onoclea * (Onocleaceae*) L.
Gk. *onos* = vessel, one; *kleio* = to close. The 'closed vessel' refers to the rolled up bead-like segments that enclose the sori.

Ononis (Fabaceae) L.
Gk. name *onōnis;* from *onos* = an ass and perhaps *onis* = ass's dung, or *oneris* = delight; alluding to the asses enjoying this particular fodder; rest-harrow.

Onychium (Adantiaceae) Kaulf.
Gk. *onyion* = little nail or claws; resembling the shape of the fertile segments of the fronds.

Oophytum (Aizoaceae) N.E.Br.
Gk. *ōion, oon* = egg; *phyton* = plant. The leaves are egg-shaped.

Operculina (Convolvulaceae) Silva Manso
La. *operculum* = lid, cover; *-ina* = diminutive. The seed pod has a lid-like top.

Ophioglossum (Ophioglossaceae) L.
Gk. *ophis* = serpent, snake(-like); *glossa* = tongue; referring to the snake-like tongue, the bifid apex above the fertile spike. Commonly called adder's tongue, the folk name in many countries.

Ophionella (Apocynaceae) Bruyns
Gk. *ophis* = serpent, snake-like; *-ella* = diminutive; referring to the snake-like stems.

Ophiopogon * (Ruscaceae) Ker Gawl.
Gk. *ophis* = snake(-like); *pogon* = beard; most probably referring to its beard-like leaves and tufted growth.

Ophrestia (Fabaceae) H.M.Forbes.
Anagram of *Tephrosia* (q.v.).

Ophrys (Orchidaceae) L.
Gk. *ophrys* = eyebrow. The original name may have referred to a two-leaved plant (possibly *Listera ovata*), called *Ophrys* by Pliny the Elder, who reported that it was used to blacken eyebrows or hair. Also perhaps alluding to the arched calyx of some species, resembling the eyebrow, or referring to the furry edges of the lips of several species.

Ophthalmophyllum **= Conophytum** (Aizoaceae) Dinter & Schwantes
Gk. *ophthalmos* = eye; *phyllon* = leaf. The upper epidermis is lens-like, letting in light for photosynthesis.

Opilia * (Opiliaceae*) Roxb.
Latinised from an Indian vernacular name for a species of this genus.

Oplismenus (Poaceae) P.Beauv.
Gk. *oplismenus* is derived from *hoplismos* = equipment for war; referring to the awns, which resemble spears.

Opophytum **= Mesembryanthemum** (Aizoaceae) N.E.Br.
Gk. *opo-* = juice (sap); *phytum* = plant. The plant exudes a milky juice, which is said to help asthma and catarrh.

Opuntia * (Cactaceae) Mill.
La. *opuntia* = a plant from Opus, ancient town of Locris, Greece; prickly pear.

Orbea (Apocynaceae) Haw.
La. *orbis* = ring, circle, disk; probably a reference

to the well-developed annulus (ring-shaped structure).

Orbeanthus = Orbea (Apocynaceae) L.C.Leach
Orbea (q.v.); *anthos* = flower. *Orbea*-flowered.

Orbeopsis = Orbea (Apocynaceae) L.C.Leach
Orbea (q.v.); Gk. *-opsis* = resembling.

Orbivestus (Asteraceae) H.Rob.
Orbea (q.v.); La. *vestus* = to dress or clothe; possibly alluding to its resemblance to *Orbea*.

Orchis (Orchidaceae) L.
Gk. *orchis* = testicle; referring to the resemblance to the twin pseudo-bulbs (tubers) of some genera. Dioscorides used this name for a plant with tuberoids in the shape of testicles.

Oreoleysera (Asteraceae) K.Bremer.
Gk. *oreo-* = mountain; *Leysera* (q.v.) (*Leyssera*); referring to its habitat.

Oreoselinum*** (Apiaceae) Hill
Gk. *oreo-* = mountain; *Selinum* = old genus name for the carrot family; also known as 'mountain parsley', referring to its habitat.

Oreosyce = Cucumis (Cucurbitaceae) Hook.f.
Gk. *oreo-* = mountain; *syce, skyon* = fig. Possibly a 'mountain fig'.

Oresbia (Asteraceae) Cron & B.Nord.
Gk. *oresbios* = living in the mountains; referring to its habitat.

Oricia (Rutaceae) Pierre
Origin obscure. Perhaps after Oricus, a town of Epirus, on the Ionian Sea. Some sources state after Gk. *oros* = mountain, but its habitat is mainly subtropical or tropical moist lowland forests.

Ormiscus = Heliophila (Brassicaceae) DC.
Gk. *ormiskos* = a small necklace; referring to the slender pods that are contracted between the seeds, giving them the appearance of a necklace (George Don).

Ormocarpum (Fabaceae) P.Beauv.
Gk. *(h)ormos* = necklace; *karpos* = fruit. The fruits are constricted between the seeds.

Ornithogalum (Hyacinthaceae) L.
Gk. *ornithos* = bird; *gala* = milk, presumably referring to the colostrum-like, high fat secretions produced by the *Colombidae* ('pigeon's milk') and stored in the crop for feeding the young. Maybe this somewhat resembles the gooey sap that

(CM)

(CM)

exudes from the cut stems. Some authors suggest that the name merely refers to the milky whiteness of some flowers, while 'bird's milk' to the ancient Greeks was a colloquial expression for something wonderful.

Ornithoglossum (Colchicaceae) Salisb.

(CM)

Gk. *ornithos* = bird; *glossa* = tongue; referring to the narrow tepals.

Ornithopus* (Fabaceae) L.
Gk. *ornithos* = bird; *-pus* = foot (*pous,* a foot); referring to the shape of the leaves.

Orobanche (Orobanchaceae) L.
Gk. *orobanche* = broomrape; from *orobus* = vetch; *ancho, anchein* = to choke or strangle; referring to the parasitic properties of this plant – some species may kill their host plant.

Oropetium (Poaceae) Trin.
Possibly Gk. *oros* = mountain; *petium* = to inhabit; referring to its preferred habitat.

Orothamnus (Proteaceae) Pappe ex Hook.
Gk. *oros* = mountain; *thamnos* = bush; referring to its habitat.

Orphium (Gentianaceae) E.Mey.
After Orpheus in Greek mythology, a poet and miraculous lute player, one of the Argonauts.

Orthanthera (Apocynaceae) Wight
Gk. *ortho* = straight, upright; *anthera* = anther; referring to the anther's appearance.

Orthochilus = Eulophia (Orchidaceae) Hochst. ex A.Rich.
Gk. *ortho* = straight, upright; *cheilos, chilos* = lip; referring possibly to the long claw that is adnate to the column, giving the whole lip a straight or rigid appearance.

Orthopenthea = Disa (Orchidaceae) Rolfe
Gk. *ortho* = straight, upright; *Penthea* (q.v.); alluding to a similarity with *Penthea* (q.v.).

Orthopterum (Aizoaceae) Bolus
Gk. *ortho* = straight, upright; *pteron* = wing; on account of the erect wings on the cell lid of the capsule (Hans Herre).

Orthosiphon = Ocimum (Lamiaceae) Benth.
Gk. *ortho* = straight, erect; *siphon* = tube. The corolla tubes are more or less straight.

Orthrosanthus* (Iridaceae) Sweet
Gk. *orthros* = morning; *anthos* = flower; referring to the flowers' habit of blooming in the morning and quickly fading.

Orygia* = Corbichonia** (Molluginaceae) Forssk.
Possibly from the classical Gk. *oryza* = rice. Orygia is a Greek island.

Oryza (Poaceae) L.
Gk. *oryza* = rice, from Arabic *eruz.*

Oryzidium (Poaceae) C.E.Hubb. & Schweick.
Oryza (q.v.); *-idium* (name-forming suffix) = resembling.

Oryzopsis (Poaceae) Michx.
Oryza (q.v.); Gk. *-opsis* = resembling.

Oscularia (Aizoaceae) Schwantes
La. *osculum* = small mouth; *aria* = possessing; from the fanciful likeness of the toothed leaves.

Osmanthus* (Oleaceae) Lour.
Gk. *osma* = fragrance; *anthos* = flower. The flowers have a strong fragrance.

Osmites** (Asteraceae) L.
Gk. *osme* = scent. The plants have a balsamic odour according to William Henry Harvey (*Flora Capensis*).

Osmitopsis (Asteraceae) Cass. emend. K.Bremer.
Osmites (q.v.); Gk. *-opsis* = resembling.

Osmunda (Osmundaceae) L.
Linnaean name of uncertain derivation. Possibly from La. *os* = bone; *munda* = cure, because the root was a remedy for rickets; perhaps from *mundae* = to clean, as it was used medically to clean bones; or even Osmunder, the Saxon name for Thor, the god of war, or Osmundus (c 1025), Scandinavian writer of runes, or Osmund, Bishop of Salisbury, who died in 1099. Origin uncertain.

Oxalis (On following page)

Osteospermum (Asteraceae) L.

Many *Osteospermum* species have yellow flowers. (CM)

Gk. *osteon* = bone; *sperma* = seed. The achenes are bone-hard.

Ostryoderris = Xeroderris (Fabaceae) Dunn
Gk. *ostryon* = oyster, related to *ostrakon* = hard shell as of a tortoise; the genus *Derris* (q.v.). The name refers to the woody fruit, which opens like an oyster.

Osyridicarpos (Santalaceae) A.DC.
Osyris (q.v.); Gk. *karpos* = a fruit. The fruits resemble those of *Osyris* (q.v.).

Osyris (Santalaceae) L.
Gk. *ozos* = a branch, branched; referring to the branching habit of this fast-growing bush. *Osyris* was used as a plant name by Dioscorides and Pliny the Elder.

Otholobium (Fabaceae) C.H.Stirt.
Gk. *otheo* = to burst forth; *lobion* = a small pod. The fruits of *O. caffrum* suddenly seem to be pushed out from the calyx.

Othonna (Asteraceae) L.
Ancient Gk. name, probably from *othonne* = linen, cloth; referring to the soft texture of the leaves of some species.

Otiophora (Rubiaceae) Zucc
Gk. *otion* = ear (from *ous* = ear); *phoros* = bearing, producing. Referring to the plant's brilliant 'ear' lobes.

Otochlamys = Cotula (Asteraceae) DC.
Gk. *ous, otos* = ear; *chlamys* = cloak. The ear-like appendage to the corolla wraps round the achene like a cloak (William Henry Harvey, *Flora Capensis*).

Otoptera (Fabaceae) DC.
Gk. *ous, otos* = ear; *pteron* = wing. The wing petals have an auricle (small ear-shaped appendage) on the unquis (claw-like base of some petals).

Ottelia (Hydrocharitaceae) Pers.
Latinised form of *ottel-ambel*, the Malabar name for this plant.

Ottosonderia (Aizoaceae) Bolus
For Otto Wilhelm Sonder (1812–1881), German botanist and pharmacist, practising in Hamburg.

He accumulated an enormous private herbarium in excess of 250 000 specimens, from some of the leading botanists and collectors of his day. He had a special interest in algae and wrote an algal supplement to Mueller's *Fragmenta Phytographiae Australiae* and a major paper on Australian tropical algae. Although he never actually visited the Cape, he became co-author with William Henry Harvey of the first three volumes of the seven-volume *Flora Capensis*. He also wrote a local flora, *Flora Hamburgensis*, and was editor and author of several families of *Plantae Muellerianae* in the journal *Linnaea*.

Oxalis (Oxalidaceae) L. (Images on previous page)
Gk. *oxys* = sharp, sour or acid; *(h)als* = salt. The plant's sour taste is a result of the oxalic acid content in its leaves and stem.

Oxyanthus (Rubiaceae) DC.
Gk. *oxys* = sharp-pointed; *anthos* = flower. The lobes of both calyx and corolla have sharp teeth (Hugh Glen).

Oxybaphus = Mirabilis (Nyctaginaceae) Vahl
Gk. *oxys* = sharp; *baphe* = a dye. Meaning obscure. Possibly because the leaves were used to make a dye.

Oxycaryum* = *Cyperus (Cyperaceae) Nees
Gk. *oxys* = sharp, pungent; *karyon* = nut; referring to the sharp-pointed achene (nuts).

Oxygonum (Polygonaceae) Burch. ex Campd.
Gk. *oxys* = sharp; *gonia* = an angle. The seeds are three-angled, and in some cases there is also a sharp point at each angle.

Oxylaena (Asteraceae) Benth. ex Anderb.
Gk. *oxys* = sharp, pungent; *chlaena, laina* = cloak; referring to the pungent smell that emerges from the long, deflexed, glandular, hairy, tightly packed leaves.

Oxyrhachis (Poaceae) Pilg.
Gk. *oxys* = sharp; *rhachis* = axis, spine. The inflorescence axis that extends above the upper spikelet is narrow and spear-shaped.

Oxyspermum* = *Galopina (Rubiaceae) Eckl.
Gk. *oxys* = sharp, pointed; *sperma* = seed. The seed is sharp-pointed.

Oxytenanthera (Poaceae) Munro
Gk. *oxytenes* = sharp-pointed; *anthera* = bristle; referring to the anthers that each have a conspicuous, sharp-pointed appendage at the upper end.

Oziroë* (*Oziroe*) (Hyacinthaceae) Raf.
Origin unknown – possibly a contrived Rafinesque name.

Ozoroa (Anacardiaceae) Delile.
Probably after an Arabian name, or an Ethiopian word for 'queen'. Could also be an Arabic name. Unlikely to be Gk. *ozein* = to smell (Jackson).

P

Protea – possibly the best-known flower in South Africa

John Rourke (1942–) was the curator of the Compton Herbarium at Kirstenbosch for over 30 years. A graduate of the University of Cape Town, he has published extensively on the Proteaceae and the Cape endemic family Stilbaceae. He has authored and co-authored several books on early botanical explorers at the Cape as well as publications on South African botanical art. He is the only South African botanist to have been elected a foreign member of the Linnaean Society of London, a distinction limited to only 50 persons internationally.

Pachidendron = Aloe (Asphodelaceae) Haw.
Gk. *pachys* = thick; *dendron* = tree; referring to the thick, dense leaves.

Pachites (Orchidaceae) Lindl.
Gk. *pachytes* = thickness; alluding to the column. 'The rostellum [is] so thick and large as completely to cut off the anther from the stigmatic process or arms' (Lindley, 1835).

Pachycalyx*** = Erica (Ericaceae) Klotzsch
Gk. *pachys* = thick; calyx; referring to the thick 'gum nut' or capsule.

Pachycarpus (Apocynaceae) E.Mey.
Gk. *pachys* = thick; *karpos* = fruit; referring to the large thick fruit of some species.

Pachycymbium = Orbea (Apocynaceae) C.Leach
Gk. *pachys* = thick; *kumbivon* = wine cup or vessel; referring to the cup-shaped corona.

Pachypodium = Sisymbrium (Apocynaceae) Lindl.
Gk. *pachys* = thick, fat; *podion* = foot (of a vase); referring to the thick stem or trunk of this plant.

Pachypodanthium* (Annonaceae) Engl. & Diels
Gk. *pachys* = thick, fat; *pous, pod* = foot; *anthos* = flower; *-ium* = diminutive; alluding to the thick-footed flowers (the crowded stalkless carpels).

Pachyraphea*** = Aspalathus (Fabaceae) C.Presl
Gk. *pachys* = thick; *raphe* (*rapha*) = seam, suture, raphe; referring to the commonly visible line or ridge on the seed coat.

Pachystela = Synsepalum (Sapotaceae) Pierre ex Radlk.
Gk. *pachys* = thick; *stele* = a pillar. The thick style is noted as a distinguishing character of this genus.

Pachystigma*** = Vangueria (Rubiaceae) Hochst.
Gk. *pachys* = thick, fat; *stigma* = stigma. The stigma is thick and cylindrical.

Paederia (Rubiaceae) L.
Gk. *paedor* = a stench, offensive smell. Some species exude a strong, fetid smell when the plant's leaves or stem are crushed or bruised.

Paederota (Plantaginaceae) L.
Gk. *Paederia* (q.v.); *-ota* = resembling.

Pachypodium (EJM)   *Pachypodium* (EJM)

Paepalanthus = Syngonanthus (Eriocaulaceae) Mart.
Gk. *paipale* = finest flour or meal; *anthos* = flower. This genus is related to the Bromeliaceae family – and the name possibly derives from the fact that the plant makes edible 'flour'.

Pagella = Crassula (Crassulaceae) Schönland
For Mary Maud Page (1867–1925), English botanical artist. She studied art at the School of Art (Caldrons) but stopped as a result of poor

eyesight, yet still painted. She emigrated to South Africa in 1911. While at the Cape in 1915, she showed some of her flower paintings to Louisa Bolus (1877–1970), curator of Bolus Herbarium, who encouraged her to learn 'to paint from a botanical point of view'. Page contributed over 200 painting to the Bolus Herbarium, with which she was associated from 1917 until she died. Many of her paintings appeared in the botanical magazine *Flowering Plants of Africa* and in Hans Herre's *The Genera of the Aizoaceae*. She wrote a handbook on culinary herbs, published by the Royal Horticultural Society.

Pahudia = **Afzelia** (Fabaceae) Miq.
For Charles Ferdinand Pahud de Montanges (Mortanges) (1803–1873). He went to India (then Batavia) in 1823 as a tax and import/export official and held a number of positions before becoming inspector of finances in 1841. He later became director of the government and civil products warehouse (1844–1847), minister for the colonies (1849–1855) and governor-general of the Dutch East Indies (1856–1861). While governor-general he put through legislation abolishing slavery (1860) among many other regulations and worked at building infrastructure, such as telegraph lines and the first high school. He also went on expeditions to Sumatra, Borneo, Timor and elsewhere. While in Java he received many gifts, some of historical value, including ceremonial swords, which he donated to the Museum of the Batavian Society in 1860.

Palhinhaea = **Lycopodiella** (Lycopodiaceae) Vasc. & Franco
For Ruy Telles (Teles) Palhinha (1871–1957), Portuguese (Azores-born) botanist, director of both the botanical garden of Lisbon and also a professor at the Botanical Institute of the University of Lisbon. He made a systematic exploration of the Azorean flora in 1934, 1937 and 1938 and published a number of articles as a result of various expeditions. Before he could finish working on his catalogue of Azores spermatophytes, a work that would conclude his studies on the flora of the Azores, he died as a result of injuries sustained in a car crash. His work *Catalogue of the Vascular Plants of the Azores*, was published posthumously with the revised text prepared for publication by AR Pinto da Silva.

Palmstruckia = **Chaenostoma** (Scrophulariaceae) Retz.
For Wilhelm Palmstruch (1770–1811), Swedish army officer and botanical artist, editor of the 11-volume *Svensk Botanik*, now in the Academy of Sciences. Palmstruch's 455 drawings of plants in the first six volumes have been characterised as the best pictures of plants ever produced in Sweden. This magnificent work was used as models for Georg Oeder's *Flora Danica* and James Sowerby's *English Botany*. The text was written by some of the most distinguished botanists of the period: Conrad Quensel, Olof Swartz, Göran Wahlenberg, CW Venus and Gustav Billberg.

Palmstruckia = **Helophila** (Brassicaceae) Sond.
Palmstruckia (q.v.).

Pancovia (Sapindaceae) Willd.
For Thomas Panckow (1622–1665), physician to King Wilhelm von Brandenburg and author of *Herbarium Portatilea* (1656), a very early field guide to plants.

Pancratium (Amaryllidaceae) L.
Gk. *pankratos* = strength; from La. *pan* = all; *katrys* = potent; referring to the supposed medicinal values.

Pandiaka (Amaranthaceae) (Moq.) Hook.f.
From the locality of the type species, *P. involucrata*, near Pandiaki, on the river Niger.

Pandorea (Bignoniaceae) Spach.
For Pandora's box. According to Greek mythology Pandora was the first mortal woman sent to earth. The botanist was reminded of it when he saw the seed pod.

Panicum (Poaceae) L.
The Latin name for wild millet, *Setaria italica*, from La. *panis* = bread. Used in bread-making.

Papaver (Papaveraceae) L.
La. *poppy*; perhaps from 'pap', etymologically known as a 'child's lip word', imitative of a chewing sound, as when munching poppy seeds; *papaver* (*pappa*) = food, milk. William Henry Harvey (in *Flora Capensis*) suggests the name was given because opium 'is administered to children with

pap (papa in Celtic) to induce sleep'. For infants the seeds were boiled before being added to pap(a). Batten says the name has been used for more than 4 000 years.

Pappea (Sapindaceae) Eck. & Zeyh.

A widespread bushveld and karoo shrub or small tree (EJM)

For Carl (Karl) Wilhelm Ludwig Pappe (1803–1862), German physician, economic botanist and plant collector. He studied medicine and botany at Leipzig before moving to Cape Town in 1831, initially practising as a doctor. He was the first colonial botanist and South Africa's first professor of botany at the South African College in Cape Town (later the University of Cape Town), 1858, and started South Africa's oldest herbarium, which forms the basis of the South African Museum collection. He was an international government advisor on botanical issues and friend of Baron von Ludwig and Carl Zeyher.

Paralepistemon (Convolvulaceae) Lejoly & Lisowski
Gk. *para* = next, near to; the genus *Lepistemon* (*lepis* = scale; *stemon* = stamen); referring to the scale at the base of each stamen.

Paranomus (Proteaceae) Salisb.
Gk. *para* = beyond, irregular, contrary to; *nomos* = custom or law; referring to the leaves that in adult plants vary from dissected to entire, and many species have both at the same time; the veins fork repeatedly and don't meet up or unite; the leaves don't lie flat and so on (Plantzafrica.com).

Parapetalifera*** (Rutaceae) H.L.Wendl.
Gk. *para* = near, beside; *petalon* = petal; *-fera* = bearing. Meaning obscure.

Parapholis (Poaceae) C.E.Hubb.
Gk. *para* = near, beside; *pholis* = horny scale, scale; alluding to the resemblance of the glumes to the scales of a snake.

Parapodium (Apocynaceae) E.Mey.
Gk. *para* = near, beside; *pous, pod-* = foot. These herbs have a thickened rootstock.

Paraserianthes* (Fabaceae) I.C.Nielsen
Gk. *para* = near to; *seri* = silk; *anthos* = flower; presumably meaning somewhat resembling the genus *Serianthes*.

Paraspalathus*** = Aspalathus (Fabaceae) C.Presl
Gk. *para* = near; *Aspalathus* (q.v.); presumably meaning somewhat resembling *Aspalathus*.

Parastranthus*** = Lobelia (Lobeliaceae) G. Don
Gk. *parastrepho* = to twist aside; *anthos* = flower; alluding to the corolla being 'upside down' compared to that of *Lobelia*.

Paratheria (Poaceae) Griseb.
Gk. *para* = near, beside; *theria* = beasts, animals. Possibly alluding to the fact that the grass is found where there are animals.

Parietaria (Urticaceae) L.
La. *parietarius* = of a wall, pertaining to a wall; from *paries* = wall; referring to the habit of *P. officinalis* of growing on the walls of houses.

Parinari (Chrysobalanaceae) Aubl.
The Guyanese vernacular name for *P. montanum*. Parinarium is an obsolete variant spelling.

Paris* (Melanthiaceae*) L.
The name Paris has many meanings, none of which seem to relate to the genus. Significance unknown.

Paritium = Hibiscus (Malvaceae) A. St.-Hil.
From the Malay *paritti* = cotton tree.

Parkinsonia (Fabaceae) L.
For John Parkinson (1567–1650), famous English apothecary to King James I, herbalist, botanist, gardener and author of two monumental works: *Paradisi in Sole Paradisus Terrestris* (Park-in-Sun's Terrestrial Paradise) (1629), which generally describes the proper cultivation of plants and illustrates some 800 plants in 108 full-page plates;

and *Theatrum Botanicum* (*The Botanical Theatre* or *Theatre of Plants*) (1640), the beautifully presented and most complete English treatise of its time, featuring some 3 800 plants. In his later years, Charles I conferred on Parkinson the title *botanicus regis primarius* (royal botanist of the first rank).

Paronychia* (Caryophyllaceae) Mill.
Gk. *Paronuchia*; *pars* = near to; *onux, onuchi-* = nail. The plant was reputed to cure paronychia, a kind of whitlow, an infection around the nail.

Paropsia (Passifloraceae) Noronha ex Thouars
Gk. *paropsis* = a dainty side-dish used by Xenophon, Greek historian, essayist, and soldier (c 430–354 BC). The aril (fleshy thickening of the seed coat) is in some species both conspicuous and tasty.

Parthenium* (Asteraceae) L.
Gk. *parthenos* = virgin or *parthenion* = maidenly, an ancient name for the flower. Significance unclear.

Paspalidium (Poaceae) Stapf
Paspalum (q.v.); Gk. *idium* = resembling.

Paspalum (Poaceae) L.
Gk. *paspalos, paspalum* = a kind of millet.

Passerina (Thymelaeaceae) L.
La. *passer* = a sparrow. The black seeds resemble a sparrow's beak.

Passiflora* (Passifloraceae) L.

(EJM)

La. *passio* = passion; *flos* = flower. Spanish missionaries in South America read a complete allegory of Christ's crucifixion into the flower. In South America *Passiflora edulis* or granadilla is often called passion fruit.

Pastinaca* (Apiaceae) L.
Latin name for parsnip possibly derived from *pastinium* = a dibble; that is, a pointed instrument for making holes in the ground for bulbs (such as parsnip). Possibly from La. *pastus* = food, referring to the edible qualities of the parsnip's root (Davesgarden.com).

Patersonia* (Iridaceae) R.Br.
For William Paterson (1755–1810), Scottish soldier, explorer, lieutenant-governor and

naturalist. After studying horticulture at Syon, London, he went to the Cape Colony (1777–1780) to collect plants for the Countess of Strathmore, as recorded in his *Narrative of Four Journeys into the Country of the Hottentots and Caffraria* (1789) dedicated to Sir Joseph Banks. Later he joined the military, served in India and Australia, and reached the rank of lieutenant-colonel in 1794. From 1791–1793 he collected botanical, geological and insect specimens on Norfolk Island and sent them to Banks and seed to James Lee and John Kennedy. He became lieutenant-governor of New South Wales in 1794. He led expeditions up the Hunter and Paterson rivers, New South Wales (1801) and Tamar and North Esk rivers in Tasmania (1804). He was elected a Fellow of the Royal Society (1798).

Pauleti*** = *Bauhinia* (Fabaceae) Cav.
For Jean Jacques Paulet (1740–1526), French physician, mycologist and author. Among

Paulet's important works were *The Secret of Medicine or Preservative Against Smallpox* (Paris, 1768), *Smallpox Destroyed, or New Facts and Observations …* (1776), *Complete Treatise on Fungi* (1775), and *Traite des Champignons* (1793), a major field guide on collecting fungi, which gave detailed advice on how to distinguish between good and bad fungi, such as mushrooms. He fiercely criticised Franz Anton Mesmer's (1734–1815) belief in 'animal magnetism' or 'mesmerism' as occultist and lacking in curative properties, a view supported by the Royal Society of Medicine and Academy of Sciences, neither of which approved Mesmer's doctrines.

Paullinia (Sapindaceae) L.
For Simon Paulli (1603–1680), professor of anatomy, botany and surgery at Copenhagen, physician to King Christian V of Denmark, author of *Flora Danica* (1648), describing herbal plants of medical interest. This is not to be confused with the great work of botany *Flora Danica*, first proposed by Georg Oeder in 1753 and published over a 123-year period from 1761–1883.

Paulownia (Paulowniaceae) Paulownia Siebold & Zucc
For Anna Pavlovna (Anna Paulowna) (1795–1865), Russian queen consort of the Netherlands.

She was the sixth daughter of Paul I of Russia (1754–1801) and Empress Maria Feodorovna (1759–1828). She received private education. Although Napoleon Bonaparte asked her to marry him when she was 14, this did not materialise, and when she was 21 she married Willem Frederik George Lodewijk (1792–1849) who would become King William II of the Netherlands. They had a stormy marriage, and from 1829–1843 she lived separately from him and founded 50 orphanages during this time. When her husband became king in 1840 she became queen consort. Shortly thereafter she left the royal palace and retired from court life.

Pauridia (Hypoxidaceae) Harv.
Gk. *pauros* = small; *-idia* = diminutive; referring to the minute size of some species.

Pavetta (Rubiaceae) L.
From the Sinhalese (Sri Lankan) vernacular name for this genus.

Pavonia (Malvaceae) Cav.
For José Antonio Pavón y Jiménez (José Antonio Pavón) (1754–1840), Spanish pharmacologist, botanist, and explorer who, in 1777, went on an expedition led by Hipólito Ruiz López to research the flora of Peru and Chile together with French physician Joseph Dombey and two botanical illustrators, Joseph Bonete and Isidro Gálvez. They explored Peru and Chile for 10 years (1778–1788), collected 3 000 specimens, made 2 500 life-sized botanical illustrations and brought back many living plants. Ruiz and Pavón published an illustrated ten-volume work, *Flora Peruviana et Chilensis*. They discovered 150 new genera, 70 of which were named by Pavón, and 500 new species.

Pearsonia (Fabaceae) Dummer
For Henry Harold Welch Pearson (1870–1916), British-born South African botanist and the first director of the former National Botanical Institute of Southern Africa.

He worked at the Cambridge herbarium, was a professor of botany at the South African College, Cape Town (now the University of Cape Town), a plant collector and botanical explorer and founder and honorary director of the Kirstenbosch National Botanical Garden in Cape Town. He was a Fellow of the Linnaean and Royal societies and made several expeditions to South-West Africa (now Namibia) to study the monotypic *Welwitschia*.

Pechuel-Loeschea (Asteraceae) O.Hoffm.
For Moritz Eduard Pechuël-Loesche (1840–1913), German naturalist, geographer and traveller. After finishing school, he joined the merchant navy and travelled extensively for many years in both the Pacific and Atlantic oceans as well as visiting both poles. He then studied geography and natural history at Leipzig University, obtaining his PhD in 1872. Thereafter he collected plants in the Cape, Hereroland and the Congo (Loango expedition 1873–1876), became Henry Morton Stanley's agent (1882–1883), visiting the Congo again and German South-West Africa (now Namibia) (1884–1885). In 1886 he became a lecturer at Leipzig before becoming professor of geography in Jena and Erlangen (1895–1912) and publisher of the third edition of *Brehm's Animal Life* (1890–1893).

Pectinaria = ***Etiserica*** (Apocynaceae) Haw.
La. *pectin* = a comb (*pectinate* = comb-like); referring to the comb-like outer corona edge in the type species *P. articulata*, a 'feature not shared by later found species' (Court).

Pedalium = ***Rogeria*** (Pedaliaceae) L.
Gk. *pedal ion* = rudder, from *peon* = oar-blade; referring to the leaves at the end of long petioles equally spaced along a branch, giving the impression of oars.

Peddiea (Thymelaeaceae) Harv.
For John Peddie (?–1840), lieutenant-colonel and plant collector. He joined the army in 1805, fought in Portugal (1808) and Spain (1812) where he lost his right arm at Salamanca. He came up through the ranks and was in command of the 72nd Foot Regiment in the Eastern Cape in the 1830s. It was from here that he sent plant specimens to William Henry Harvey, then in Dublin. It is possible that he did not collect them directly, but through Thomas Williamson (1807–?), a soldier in the 72nd Foot

Regiment, who had been employed by Harvey earlier to collect plants. Such collecting could only have been done with Peddie's consent. Peddie was transferred to Ceylon in 1840 and died shortly thereafter. Both men were commemorated: Peddie in the genus name *Peddiea* and Williamson in *Amphithalea williamsonii.*

Pedicularis (Orobanchaceae) L.
La. *pediculus* = louse, from = *pes, pedis* = foot or louse; *-ulus* = little, hence 'little feet'. So-called because there was a belief that the plant gave lice to people and cattle. Or, according to some sources, the plant was thought to cure people or cattle of lice.

Pedilanthus = **Euphorbia** (Euphorbiaceae) A.Poit.
Gk. *pedilon* = sandal, shoe, boot; *anthos* = flower; possibly referring to the foot-shaped flower or possibly the need to wear shoes as some species are highly toxic.

Pedistylis (Loranthaceae) Wiens
Gk. *pedion* = flat; *stylos* = pillar, style; possibly referring to the filaments arising at the base of corolla lobes, curving out at about 90 degees.

Peersia = **Rhinephyllum** (Aizoaceae) L.Bolus
For Victor Stanley Peers (1874–1940), Australian  botanist and amateur archaeologist. He came to South Africa in 1899 during the South African War. Later he joined the South African Railways where he worked for 30 years. He lived in Fish Hoek with his family, where his son Bertie was born in 1903. Together father and son, keen amateur naturalists, archaeologists and palaeontologists, explored various caves and rock shelters in the area. In one of these caves, since named Peers Cave, they came across the fossil remains of 'the Fish Hoek Man'. Further researchers found other skeletons and artefacts going back 200 000 years. Peers's discovery created great interest at the time but no really significant findings were made. Peers died in Cape Town.

Peganum (Nitrariaceae) L.
Gk. *peganon* = rue, referring to a drought-resistant perennial evergreen shrub. *Peganum harmala* (originally from eastern Iran to western India) is remarkable in that the plant's seed have been traced back thousands of years and used in the rites of many cultures, folk medicine and spiritual practices.

Peglera* = **Nectaropetalum** (Erythroxylaceae) Bolus
For Alice Marguerite Pegler (1861–1929), teacher, painter and East Cape collector around the area  of Kentani where she lived. She corresponded with the leading botanists of South Africa including MacOwan, Bolus, Pearson, Selmar Schönland, Pole Evans, Kolbe and others. She collected over 2 000 specimens, most of which were from an area with a radius of 8 kilometres from the village of Kentani where she had settled. Her observations 'On the flora of Kentani' were published in *Annals of the Bolus Herbarium* 2:1–32 (1918), as they changed month to month throughout the year. When her eyesight began to fail she turned her attention to algae and fungi but had to give this up completely as her health deteriorated. She donated her personal herbarium to the Pretoria Museum.

Pegolettia (Asteraceae) Cass.
Origin obscure. Georg Christian Wittstein writes 'perhaps after FB Pegoletti, a 14th century Italian.' Possibly refers to Francesco Balducci Pegolotti (f 1310–1347), a prominent Florentine merchant and politician who wrote *Libro di divisamenti di paesi e di misuri di mercatanzie e daltre cose bisognevoli di sapere a mercatanti* but we cannot confirm this. The author does not explain the name.

Pelargonium (Geraniaceae) L'Hér ex Aiton

(CM)

Gk. *pelargos* = a stork; referring to the beak of the fruit which resembles a stork's bill (cf *Geranium, Erodium*).

Peliostomum (Scrophulariaceae) E.Mey. ex Benth.
Gk. *pelion* = livid, purple; *stoma* = mouth; referring to the colour of the flowers.

Pellaea (Pteridaceae) Link

Note the dark-brown to black 'stems' and the diagnostic sori clustered round the frond margin below. (EJM)

Gk. *pellos* = dusky; referring to the dark or black colour of the stipe in most species.

Pellaeopsis* = Pellaea** (Pteridaceae) J.Sm.
Gk. *Pellaea* (q.v.); *-opsis* = resembling.

Peltaria (Brassicaceae) Jacq.
La. *pelta*, Gk. *pelte* = a light, often crescent-shaped shield; *-aria* = like, resembling; referring to the leaf shape.

Peltaria = Wiborgia (Fabaceae) DC.
La. *pelta* = small shield; Gk. *pelte*; *-aria* = pertaining to; referring to the round and flat seed-vessel, like a shield.

Peltophorum (Fabaceae) (Vogel) Benth.
Gk. *pelte* (La. *pelta*) = small shield; *-phorum* = carrier; referring to the shape of the stigma.

Penaea (Penaeaceae) L.
For Pierre Pena (1535–1605), French physician to Henri III of France, botanist, and ecclesiastical assistant to Matthias de L'Obel (1538–1616) (see *Lobelia*). In collaboration with Mattias de L'Obel, he published *Stirpium adversary nova* (1571)

and *Seu Plantarum stirpum historia* (1576) which was later translated into Flemish. In *Stirpium* they described some 1 300 species, including the localities where they were collected, mainly around Montpellier, France, and in Switzerland and the Netherlands. The two works had a huge number of wood engravings. These early botanical books describing plants followed pre-Linnaean principles with the emphasis on leaf analysis.

Pennisetum (Poaceae) Rich.
La. *penna* = a feather; *setum* = a bristle. The spikelets are surrounded by feathery bristles.

Pentacoilanthus** = Mesembryanthemum** (Aizoaceae) Rappa & Camarrone
Gk. *penta-* = five; *koilos* = a hollow; *anthos* = flower. Meaning obscure.

Pentameris (Poaceae) P.Beauv.
Gk. *penta-* = five; *meros* = a part; referring to the five-awned lemma.

Pentanema* (Asteraceae) Cass.
Gk. *penta-* = five; *nema* = a thread, filament; referring to the shape of the pappus, consisting of five filiform bristles in some species.

Pentanisia (Rubiaceae) Harv.
Gk. *penta-* = five; *anisos, aniso-* = unequal; referring to the calyx lobes; the longest is up to 3.5 mm, the rest minute.

Pentarrhinum (Apocynaceae) E.Mey.
Gk. *penta* = five; *rhis, rhinos* = nose; probably referring to the shape of the corona.

Pentas (Rubiaceae) Benth.
Gk. *pentas* = a group of five. The flower parts (petals, stamens, etc.) are in fives, whereas related genera have them in fours.

Pentaschistis (Poaceae) (Nees) Spach.
Gk. *penta-* = five; *schistos* = a cut, split, divide. The cleft lemma has five divisions.

Pentatrichia (Asteraceae) Klatt
Gk. *penta* = five; *thrix, trichos-* = hair; probably referring to the cypselas which are 'sparsely hairy with elongated twin hairs and glandular hairs' (*Gateway to African Plants*).

Pentatropis (Apocynaceae) R.Br.
Gk. *pente* = five; *tropis, tropidos* = a keel; referring to the corona lobes, the flowers have five keels.

Penthea* = Disa** (Orchidaceae) Lindl.
Gk. *pentheia* = grief; referring, possibly, to the

drab, dark colour of the flowers in the type species.

Pentheriella = ***Heromma*** (Asteraceae) O. Hoffm.
For Arnold Penther (1865–1931), Austrian zoologist born in Italy who collected in South Africa and Southern Rhodesia (now Zimbabwe) from 1894–1895. He worked in the zoology department at the Vienna Natural History Museum from 1898–1924. From an analysis of the specimens he collected it seems he first collected from Cape Town northward to Clanwilliam where he met up with Rudolf Schlechter (1872–1975). A month after they returned to Cape Town the two men went eastward, eventually to Durban; being joined by a P Krook, about whom little is known. Penther's third major excursion was from Durban to Southern Rhodesia. His plants were written up under the title 'Plantae Pentherianae' in *Annalen des Naturhistorischen Museum in Wien*.

Pentodon (Rubiaceae) Hochst.
Gk. *pente* = five; *-odon* = toothed; referring to the calyx of five triangular lobes.

Pentopetia = ***Ischnolepis*** (Apocynaceae) Decne.
Gk. *penta* = five; La. *petia* = rock. This genus grows on almost bare granitic, gneissic or sandstone rocks, but also in savannah and grassland habitats.

Pentzia (Asteraceae) Thunb.
For Carolus Johannes Pentz, 18th-century Swedish botanist, author of *De Diosma* (*Dissertatio Botanica de Diosma*) (1797), and a student of Carl Peter Thunberg who published the genus in 1800.

Peperomia (Piperaceae) Ruíz & Pav.
Gk. *peperi* = pepper; (*h*)*omoios* = resembling. A small fleshy-leaved tropical plant of the pepper family. Many are grown as houseplants, chiefly for their decorative foliage (Wikipedia.org). This genus is related to and closely resembles the true peppers, *Piper* spp.

Peponia = ***Peponium*** (Cucurbitaceae) Naudin
Gk. *pepon*, La. *pepo* = a pumpkin (originally), later melon, subsequently the word was used for 'large fruit'. Pliny the Elder says distinctly that *cucumeres* (a form of cucumber), when of excessive size, are called *pepones*.

Peponium (Cucurbitaceae) Engl.
Gk. *pepon* or La. *pepo* = a melon. The name is derived from the characteristic fruit of the family Cucurbitaceae.

Perapentacoilanthus**** = ***Mesembryanthemum*** (Aizoaceae) Rappa & Camarrone
Gk. *pera* = a pouch; *Pentacoilanthus* (q.v.).

Peratetracoilanthus**** = ***Mesembryanthemum*** (Aizoaceae) Rappa & Camarrone
Gk. *pera* = a pouch; *tetra* = four; *koilos* = a hollow; *anthos* = flower.

Perdicium (Asteraceae) L.
La. *perdere* = to lose (cf perdition); a name given by Pliny the Elder to a plant (now not recognised) of which the partridge, *perdix,* was very fond.

Pereskia* (Cactaceae) Mill.
For Nicholas-Claude Fabri (Fabry) de Peiresc (1580–1637), French antiquarian, numismatist, patron of botany, naturalist, historian and archaeologist. He was regarded as one of Europe's leading scholars and was in contact with many of the leading figures of his day and a noted politician in his home region, Provence. His interests included astronomy; he discovered the Orion Nebula, studied the moons of Jupiter and began a map of the surface of the moon. Despite his network of scholarly friends, Peiresc published nothing in his lifetime, although there are many finished essays and countless drafts among his vast collection of papers, including over 10 000 letters, which have survived. His *Abridged History of Provence* was published posthumously.

Pergularia (Apocynaceae) L.
La. *pergula* = arbour, projecting roof. The plants are suitable for climbing up a trellis (*Paxton's Botanical Dictionary*).

Periballia* (Poaceae) Trin.
Gk. *peri* = around, about; *ballo* = to dance; referring to the awns which move, sometimes swelling, shrinking or changing position due to changes in moisture content.

Periboea*** = ***Lachenalia*** (Hyacinthaceae) Kunth
Gk. *peri* = round, about; *bou* = cow. The name *Periboia*, 'surrounding the cattle', might suggest the plant's favoured habitat is pastureland or a river stream.

Periglossum* = *Cordylogyne (Apocynaceae) Decne.
Gk. *peri-* = around; *glossa* = tongue; possibly referring to the tongue-like petals surrounding the centre of the flower.

Periphanes* = *Hessea*** (Asparagaceae) Salisb.
Gk. *peri-* = around; *phanos* = lamp, torch. The original genus '*Periphanes*' was placed in the Amaryllidaceae family. '*Periphanes*' also referred to moths of the Noctuidae family. Moths cluster around light, hence Salisbury might have been referring to nocturnal pollination.

Periploca* = *Secamone (Apocynaceae) L.
Gk. *peri* = around; *plokein* = to twine. These plants usually climb by twining around a support.

Peristrophe (Acanthaceae) Nees
Gk. *peri-* = around; *strophes* = a twisted band, belt; alluding to the bracts surrounding and enclosing the calyx.

Peristylus (Orchidaceae) Blume
Gk. *peri-* = around; *stylos* = column, style; alluding to the prominent glands borne on each side of the gymnostemium.

Perlebia* = *Bauhinia (Fabaceae) Spix & Mart.
For Karl (Carl) Julius Perleb (1794–1845), German botanist and physician, associate professor (1821), then full professor of natural history at the University of Freiburg (1823). In 1826 he became director of the Freiburg Botanical Garden and in 1838 he became the director of Freiburg University. His main work seems to have been a diagnostic analysis of the natural plant system classes, families, genera, etc., *Clavis, Classium, Ordinum et Familiarum Atque Index Generum Regni Vegetabilis* (*Diagnostische Uebersichtstafeln des Natürlichen Pflanzensystems: Nebst Vollständigem Gattungsregister*) (1838).

Perotis (Poaceae) Aiton
Gk. *peros* = mutilated, maimed; *otos* = ear. The lemma is awnless.

Perotriche* = *Stoebe (Asteraceae) Cass.
Gk. *piros-* = wanting; *-triche* = hair. The plant lacks a pappus.

Persea* (Lauraceae) Mill.
Derived from the Greek Περσέα which Phillip Miller (author) Latinised as Persea. This Greek name was applied by Theophrastus and Hippocrates to an uncertain Egyptian tree, possibly *Cordia myxa* or a *Mimusops* species.

Persea, alia Perseus the son of Zeus and Danae, a mythological hero celebrated for many achievements.

Persica* = *Prunus*** (Rosaceae) Mill.
La. *persicum* = peach.

Persicaria (Polygonaceae) Mill.
La. *persicum* = peach; *-aria* = pertaining to; referring to the leaves, which resemble *Persica* (q.v.).

Petalacte (Asteraceae) D.Don

(CM)

Gk. *petalon* = leaf, petal; *aktis* = ray; referring to the prominent bracts looking like radiate florets.

Petalactella* = *Trichogyne (Asteraceae) N.E.Br.
Petalacte (q.v.) plus Gk. *-ella* = diminutive.

Petalidium (Acanthaceae) Nees
Gk. *petalon* = leaf, petal; *idium* = diminutive; referring to the small leaves.

Petamenes* = *Gladiolus*** (Iridaceae) Salisb.
Gk. *petamenes* (Gk. plural participle of *petannumi*) = to spread out, open wide; probably a reference to the wide open petals.

Petopentia (Apocynaceae) Bullock
Anagram of *Pentopetia* (q.v.).

Petrina* = ***Danthoniopsis*** (Poaceae)
J.B.Phipps.
La. *petra* = rock; *-ina* = belonging to; referring to its habitat – it grows in rocky places.

Petrogeton* = ***Crassula*** (Crassulaceae)
Eck. & Zeyh.
Gk. *petros* = rock; *geton* = a neighbour; referring to the habitat in which these plants occur.

Petrorhagia* (Caryophyllaceae) (Ser.) Link
Gk. *petros* = rock; *rhagas* = tear, rent or crack; referring to the plants liking to grow in rock crevices.

Petroselinum (Apiaceae) Hill
Gk. *petros* = rock; *selinon* = parsley. Victors at the Isthmian and Nemean games were crowned with chaplets made from this plant's leaves.

Petunia* (Solanaceae) Juss.
Brazilian *petyn, petyma,* from the Tupi-Guaranti language; Middle French *petun* = tobacco. Petunia flowers are only indirectly linked to tobacco in that they are part of the large Solanaceae family – a family that includes tomatoes, potatoes, chilli peppers, cape gooseberries, etc.

Peucedanum = ***Notobubon*** (Apiaceae) L.
Gk. *peukedanon* = Greek name for *Ferula,* the 'hogs-fennel'. The name is said to derive from Gk. *peuke* = fir, pitch-tree (which leaves it resembles) and *-edanon* = dry or burning; alluding to the very pungent quality of the root and resinous-smelling seeds.

Peyrousea = ***Schistostephium*** (Asteraceae) DC.
For Jean François de Galaup, Comte de la Pèrouse (1741–1788), French navigator, explorer  and naturalist. He fought against the British off North America in the Seven Years' War and was promoted to the rank of commodore. In 1785 he led an expedition to the Pacific which included 10 scientists, an astronomer, a botanist, a physicist and three naturalists. He went to Easter Island, Hawaii, Alaska, California, the Philippines, Korea, the Kurile Islands, Russia, Japan, the South Seas, and Australia, but then he and all his men disappeared and were never seen again. Thirty-seven years later it was determined that both his ships had been wrecked on reefs and sunk.

Phacelia* (Boraginaceae) Juss.
Gk. *phakelos* = a bundle, cluster; referring to the inflorescence.

Phacelurus (Poaceae) Griseb.
Gk. *phakelos* = a bundle; *urus* = a tail; possibly referring to the tightly compacted tail-like shape of the inflorescence.

Phacocapnos = ***Cysticapnos*** (Papaveraceae) Bernh.
Gk. *phakos* = a lentil or lentil-shaped object; *kapnos* = smoke. The seeds are black and lenticular.

Phaenocoma (Asteraceae) D.Don

(CM)

Gk. *phainos* = to shine; *coma* = hair, head of hair; referring to the bright pink shiny bracts and the tiny bristle (hence hair-like) pappus.

Phaenohoffmannia = ***Pearsonia*** (Fabaceae) Kuntze
For Heinrich Karl Hermann Hoffmann (1819–1891), German botanist, mycologist, plant geographer, professor of botany and director of the botanical gardens at Giessen. He studied medicine at the University of Gliessen and physiology at the University of Berlin under Johannes Peter Müller (1801–1858). Hoffman was a pioneer in botanical phenology, especially in relation to climatic conditions and geography – plant distribution and migration – and wrote many other works related to plant physiology and mycology. He conducted research into the biological aspects of fungi, especially in relation to fermentation, putrefaction and disease, and also performed early investigations in the field of bacteriology.

Phaeocephalus = ***Hymenolepis*** (Asteraceae) S.Moore
Gk. *phaios* = dusky, brownish grey; *kephale* = head; referring to the colour of the flowerhead. [Authors' comment: This applies to *Juncus phaeocephalus* but questionable for *Phaeocephaus gnidioides*.]

Phaeoptilum (Nyctaginaceae) Radlk.
Gk. *phaios* = dusky, brownish grey; *ptilon* = feather. The mature fruits are brown and wrinkled (Hugh Glen).

Phaianthes = ***Moraea*** (Iridaceae) Raf.
Gk. *phaios* = dusky, brownish grey; *anthos* = flower; referring to the flower which has considerable colour variation.

Phalaris* (Poaceae) L.
Gk. ancient name, perhaps from *phaleros* = shining; or *phalos* = the ridge of a helmet; referring to the shining seeds or possibly to the crest-like seed head.

Phaneroglossa (Asteraceae) B.Nord.
Gk. *phaneros* = visible, conspicuous; *glossa* = tongue; referring to the conspicuous rays.

Phanerophlebia (Dryopteridaceae) C.Presl
Gk. *phaneros* = evident, obvious, visible; *phlebia* = vein; alluding to the conspicuous vein in the pinnae of the first species included in this genus.

Pharnaceum (Molluginaceae) L.
For Pharnaces II (63–47 BCE), son of Mithradates VI Emperor, King of Pontus, in North East Anatalia, on the Black Sea. He was defeated by Julius Caesar (100–140 BCE) at Sinopa, the actual occasion on which Caeser gave the extraordinarily concise message 'veni, vidi, vici' ('I came, I saw, I conquered') which he dispatched to Rome.

Phaseolus (Fabaceae) L.
Gk. *phaselos* = a bean or light boat (skiff, canoe); alluding to the appearance of the pod including 'runner, kidney, French, Lima and haricot beans'.

Phaulopsis (Acanthaceae) Willd.
Gk. *phaulos* = slight, trivial; *-opsis* = appearance. Perhaps referring to the flowers, which are 'trivial' compared with the leaves, which are fairly large and well-veined.

Phegopteris = ***Thelypteris*** (Thelypteridaceae) Fée
Gk. *phegos* = beech; *pteris* = fern, from *pteron* = wing; originally called the 'Beech fern'.

Phelypaea (Orobanchaceae) Tourn. ex L.
For Louis Phélypeaux, Comte de Pontchartrain (1643–1727), French nobleman and politician.

He served France in various senior positions – head of the Brittany parliament, controller-general of finances, navy secretary, the king's secretary of State (1690) and chancellor of France (1699–1714). He lived during a period of considerable volatility. A capable administrator, he was held partly responsible for the French navy's defeat in 1692 at the battles of Barfleur and La Houge. King Louis IV planned to land an army in England and restore James II to the throne. The 44 French ships, some ill-equipped, came up against an 82-ship Anglo-Dutch force and were defeated. Phélypeaux retired when he came into disagreement with Louis XIV over a religious matter.

Pherolobus* = ***Cleretum*** (Aizoaceae) N.E.Br.
Gk. *pherein* = to bring forth; *lobos* = lobe; referring to the five fleshy processes carried by the ovary.

Phialodiscus = ***Blighia*** (Sapindaceae) Radlk.
Gk. *fiale, phiale* = a shell; *diskos* = a disk. The disk of the flower is somewhat shell-like in shape.

Phiambolia (Aizoaceae) Klak
Anagram of *Amphibola* (q.v.). Most of the species were originally classified as belonging to the genus *Amphibola*.

Philenoptera (Fabaceae) Fenzl ex A.Rich.
Gk. *philenos* = tractable; *pteron* = a wing. The pod wing assists the dispersal of seeds.

Philippia = ***Erica*** (Ericaceae) Klotzsch
For Rodolfo Amando (Rudolph Amandus) Philippi (1808–1904), German botanist, traveller,

botanical explorer and plant collector, professor of natural sciences at the Polytechnic School in Kassel (1834–1850), professor of natural history at the University of Chile, and director of the Museo Nacional de Chile (1853), professor of botany and zoology at Santiago where he directed the National Museum and created the Santiago Botanical Gardens. He published some 453 articles in the fields of palaeontology, entomology, ornithology, marine mammalogy, anthropology and mineralogy and described 3 720 new species from Chile.

Phragmites (EJM)

Philyrophyllum (Asteraceae) O.Hoffm.
Gk. *philyra* = lime tree; *phyllon* = leaf. The plant's leaves are similar to those of a lime tree.

Phlebodium* (Polypodiaceae) (R.Br.) J.Sm.
Gk. *phlebos* = veins; referring to the numerous veins on the fronds.

Phlomis (Lamiaceae) L.
Gk. *phlomos* = plant name for mullein (*Verbascum* spp.) or similar woolly plant; perhaps from *phlogmos* = a blaze. The down of this plant is used to make lamp-wicks.

Phlyctidocarpa** (Apiaceae) Cannon & W. Theob.
Gk. *phlyktaina* = a blister, from *phlyein* = to boil over; *karpos* = fruit. The mericarps are covered with cone-shaped papillae.

Phoberos = Scolopia (Salicaceae) Lour.
Gk. *phoberos* = formidable; referring to the thorns.

Phoenix (Arecaceae) L.
The Greek name used by Theophrastus for the date palm.

Porphyrostemma (Asteraceae) Benth ex Oliv.
Gk. *porpyrius* = clad in purple; *stemma* = wreath, garland; referring to the florets, which are mauve or purple; the styles are also mauve in some species.

Phragmanthera (Loranthaceae) Tiegh.
Gk. *phragma* = a hedge, fence; *anthera* = anther. The anthers form a barrier to pollinators probing for nectar.

Phragmites (Poaceae) Adans.
Gk. *phragmites* = growing in hedges, from *phragma* = a fence, hedge, from *phrassein* = to enclose. Dioscorides alluded to this name, mentioning that the tall stems were used for making hedges.

Phygelius (Scrophulariaceae) E.Mey. ex Benth.
Gk. *phugo* = to shun; *(h)elios* = the sun. Plants of this genus prefer shade, not sunlight.

Phyla* (Verbenaceae) Lour.
Gk. *phyla* = referring to leaves, or possibly relating to a new taxonomic division. Some of the species of this genus are very densely leaved.

Phylica (Rhamnaceae) L.
Gk. *phyllikos* = leafy; referring to the plentiful foliage.

Phyllanthus (Phyllanthaceae) L.
Gk. *phyllon* = leaf; *anthos* = a flower. In some species the flowers are borne on leaf-like branches (cladodes).

Phyllobolus = Mesembryanthemum (Aizoaceae) N.E.Br.
Gk. *phyllon* = leaf; *bolos* = throwing; referring to the deciduous shedding of leaves by some species.

Phyllocomos = Anthochortus (Restionaceae) Mast.
Gk. *phyllon* = leaf; *comé* = hair (as of head); referring to the plant's tufted appearance.

Phyllogeiton = **Berchemia** (Rhamnaceae)
Weberb.
Gk. *phyllon* = a leaf; *geiton* = a neighbour. The flowers are borne next to the leaves.

Phyllopodium (Scrophulariaceae) Benth.
Gk. *phyllon* = leaf; *pons, pod-* = foot; *-ium* = diminutive; referring to the stem or perhaps the shape of the leaves.

Phyllosma (Rutaceae) Bolus ex Schltr.
Gk. *phyllon* = leaf; *osme* = scent; referring to the fragrant leaves.

Phylohydrax (Rubiaceae) Puff
Gk. *phyl-* = referring to leaves; *hydrax* = water. The preferred habitat of this genus is the coastline.

Phymaspermum (Asteraceae) Less. emend. Källersjö.
Gk. *phyma* = swelling; *sperma* = seed. The achenes are papillated.

Phymatodes = ***Phymatosorus***
(Polypodiaceae) C.Presl
Gk. *phyma* = tubercle, swelling; *odes* = resembling; referring to the impressed sori having the appearance of tubercles on the upper side of the frond.

Phymatosorus (Polypodiaceae) Pic. Serm.
Gk. *phymatos* = a swelling; *sorus* = a cluster of sporangia on the back of the fronds of ferns; alluding to the pustulate swelling of upper surface of the lamina above the sorus in most of the species in this genus.

Physalis* (Solanaceae) L.
Gk. *physalis* = genus name referring to a bladder. The inflated calyx surrounding the fruits resembles a bladder.

Phytolacca (Phytolaccaceae) L.
Gk. *phyton* = plant; Hindi *lakh* = dye. The fruits stain badly. The juice of the fruit is crimson red because the plant contains betalain – red nitrogenous pigments found in certain plants, such as beetroot.

Piaranthus (Apocynaceae) R.Br.
Gk. *piaros* = fat; *anthos* = flower; referring to the fleshy corolla.

Picris* (Asteraceae) L.
Gk. *pikris* = sharp or bitter. Although many sources state 'allusion unclear', these Old World yellow-flowered herbs usually contain a bitter-tasting substance, hence 'bitter-weed', and some species have hooked bristles, hence common names such as 'prickly ox-tongue' and 'hawkweed ox-tongue'.

Pilea (Urticaceae) Lindl.
Gk. *pilos* = pileus (the name for the Roman felt cap); alluding to the appearance of calyx covering the achene.

Piliostigma (Fabaceae) Hochst.
La. *pileus* = a felt cap; *stigma*; referring to the cap-shaped stigma.

Pillansia (Iridaceae) Bolus
For Neville Stuart Pillans (1884–1964), South African botanist, son of Charles Eustace Pillans (1850–1919). He studied agriculture at Cambridge University for two years but had to stop due to ill health. He joined the staff of the Bolus Herbarium in Cape Town in 1918 and became its assistant curator for many years. He had a special interest in succulents and was an authority on  Stapeliads and collected plants near Clanwilliam. He assisted HHW Pearson in selecting the Kirstenbosch site for the future National Botanical Garden. He spent much time on the taxonomy of South African flora, including *Restionaceae*, *Bruniaceae*, *Phylica*, *Agathosma* and *Metalasia*. He was revising *Hermannia* but gave this up on medical advice two years before he died.

Piloselloides** = **Gerbera** (Asteraceae)
C.Jeffrey
Gk. *pilosella* = hair, small hair; *-oides* = resembling. Many species of this genus have basal pilose (small, hairy) leaves, as do the stems toward the apex.

Pimpinella (Apiaceae) L.
A medicinal plant in Latin possibly being a typographic corruption of *bipinnata* = two-winged or feathered; referring to the leaves of this genus which are pinnate.

Pinguicula (Lentibulariaceae) L.
La. *pinguis* = fat or grease; referring to the oily texture of the trapping surface of the leaf.

Pinus* (Pinaceae) L.
La. *pinus* = pine, probably from *picnus* = pitch or pit resin. Classical Latin name for a pine.

Piper (Piperaceae) L. (Image on opposite page)
Gk. *peperi, piper* = pepper, probably from Sanskrit, *pippali*. An old name for pepper, in this case referring to *P. longum*.

Piper: Leaves showing diagnostic venation (EJM)

Piptatherum (Poaceae) P.Beauv.
La. *pipto-* = to fall; *ather* = barb or spine (spike as of an ear of wheat); alluding to the awns dropping readily from the lemmas.

Piptochaetium (Poaceae) J. Presl
La. *pipto* = to fall; *chaite* = bristle; alluding to deciduous awns.

Piptostigma* (Annonaceae) Oliv.
La. *pipto* = to fall; *stigma* = stigma. The stigma falls off after flowering.

Pircunia** = ***Phytolacca*** (Phytolaccaceae) Bertero
Mapuché name – *piricún*. The Mapuché are a group of indigenous inhabitants of south-central Chile and southwestern Argentina.

Piriqueta* (Passifloraceae) Aubl.
La. *pirus* = pear, hence piriform (pyriform); *queta* = views. Not explained by the author Jean Baptiste Aublet.

Pisonia (Nyctaginaceae) L.
For Willem Piso (Willem Pies) (c 1611–1678), Dutch physician, pharmacologist, botanist, pioneer of tropical medicine, and author. He studied medicine at the University of Leiden (1633), practised medicine in Amsterdam, and in 1637 was sent by the Dutch West Indian Company to be the physician of Johan Maurits of Nassau-Siegen, governor of the Dutch colony in northeastern corner of Brazil. He also served as a surgeon to the Dutch troops. During his six-years stay, he wrote, in conjunction with Markgraf (Macgrave), *Historia Naturalis Brasiliae* (1648); a compendium

of tropical medicine, pharmacology and natural history. He returned to the Netherlands in 1644, where he established a successful practice and also became dean of the Amsterdam Medical College.

Pisosperma = ***Kedrostis*** (Cucurbitaceae) Sond.
Gk. *pison* = pea; *sperma* = seed. Pea-like seed.

Pistia* (Araceae) L.

(JB)

Gk. *pistos* = aquatic, water; referring to the acquatic nature of the plants.

Pistorinia (Crassulaceae) DC.
For Santiago Pistorini (?–1776) Spanish physician of Italian descent. In 1766 he was appointed a royal physician to the Spanish King Charles III (1716–1788) and in 1770 became assistant royal physician and assistant Chief Medical Officer to Dr Asceso. He went on sick leave in 1774, requested retirement in 1775 and died in 1776.

Pitcairnia* (Bromeliaceae*) L'Her.
For William Pitcairn (1712– 1791), British physician and amateur botanist. He studied at the universities of Leiden under Boerhaave and at Rheims, where he was awarded his MD, and obtained a further degree in medicine from Oxford University in 1749. He was a physician at St Batholemew's Hospital, London and a Fellow of the Royal College

of Physicians, becoming its president from 1775–1784. He was elected a Fellow of the Royal Society in 1770, 'distinguished by his application to Botany and success in rearing scarce and foreign plants'.

Pittosporum (Pittosporaceae) Banks ex Sol.
Gk. *pitta* = pitch; *spora* = seed. The seeds are covered with a dark, sticky resin.

Pituranthos = Deverra (Apiaceae) Viv.
Gk. *pituron* = chaff, bran; *anthos* = flower; referring to the chaffy-flowers.

Pityrogramma (Pteridaceae) Link
Gk. *pityron* = bran, chaff; *gramma* = line; referring to powdery indumentum and naked sporangia on the undersurface of the frond, which leaves an outline of the frond when pressed on to a surface.

Plagiochloa = Tribolium (Poaceae) Adamson.
Gk. *plagios* = oblique (angle), slanting; *chloa* = grass; referring to the positioning of the spikelets to the central axis.

Plagiostemon = Erica** (Ericaceae) Klotzsch
Gk. *plagios* = oblique, sideward; *stemon* = stamen.

Plagiostigma* (Fabaceae) C.Presl
Gk. *plagios* = oblique (angle), slanting; *stigma* = stigma; alluding to the typically angled stigma.

Plananthus = Huperezia (Lycopodiaceae) P.Beauv. ex Mirb.
La. *planus* = flat, level; *anthos* = flower. The upright and lateral shoots are round or flat in cross section.

Planea (Asteraceae) P.O.Karis
La. *planus* = flat, even level, smooth; referring to the keeled (carinate) leaves, curling inward (slightly involute) but with flat margins.

Plantago (Plantaginaceae) L.
La. *plantagin-* = foot-sole-like; referring to the broad, flat leaves lying close on the ground.

Platanthera (Orchidaceae) Rich.
Gk. *platys* = broad, flat; *anthera* = anther; referring to the unusual width of the anther.

Platostoma (Lamiaceae) P.Beauv.
Gk. *platys* = flat; *stoma* = an opening, mouth; alluding to the wide mouth of the flower.

Platycalyx = Erica (Ericaceae) N.E.Br.
Gk. *platys* = broad, flat; calyx. The four-lobed calyx is 'almost flat and square in outline' (Robert Allen Dyer).

Platycarpha (Asteraceae) Less.
Gk. *platys* = broad, flat; *karphos* = a dry stalk, scale; referring to the flat scales (paleae) of the pappus.

Platycarphella (Asteraceae) V.A.Funk & H.Rob.
Gk. *platys* = broad, flat; *karphos* = a dry stalk, chip of straw or wood, a scale.; *-ella* = diminutive; referring to its stalk and flat leaves. 'Leaves prostate in rosettes radiating from a woolly root crown' (VA Funk & H Rob.).

Platycaulos (Restionaceae) H.P.Linder
Gk. *platys* = broad, flat; *caulos* = stem; referring to the large round culms typical of this species (Plantzafrica.com).

Platycerium* (Polypodiaceae) Desv.
Gk. *platys* = broad, flat; *keras* = horn, hence *platycerium* = flat horn; 'stag's horn fern'; referring to the shape of the broad, flat fronds, divided in segments like stags' horns.

Platycoryne (Orchidaceae) Rchb.f.
Gk. *platys* = broad, flat; *koryne* = a club; referring to the broad and thickened rostellum.

Platylepis = Ascolepis (Cyperaceae) Kunth
Gk. *platys* = broad, flat; *lepis* = scale; referring to the prominent floral bracts.

Platylepis (Orchidaceae) A.Rich.
Gk. *platys* = broad; *lepis* = scale. Descriptive, probably, of the very large, ovate, floral bracts.

Platyloma = Pellaea** (Pteridaceae) J.Sm.
Gk. *platys* = broad, flat; *loma* = edge or fringe; referring to the broad sori close to the margin.

Platylophus (Cunoniaceae) D.Don
Gk. *platys* = broad; *lophos* = crest. The fruit is crested with persistent styles (Hugh Glen).

Platythyra = Mesembryanthemum (Aizoaceae) N.E.Br.
Gk. *platys* = broad, flat; *thyra* = door, entrance, access. Meaning unclear.

Plecostachys (Asteraceae) Hilliard & B.Burtt
Gk. *pleko, plekein* = to twist or plait; *stachys* = spike; alluding to its tangled growth habit and the shape of its inflorescence.

Plectranthus (Lamiaceae) L'Hér. (Image on opposite page)
Gk. *plektron* = a spur; *anthos* = a flower. A genus with conspicuously spurred flowers.

Plectroniella (Rubiaceae)
Gk. *plektron* = a spur; *-iella* = diminutive. This genus has only one species, characterised by its large spines.

Plectranthus (EJM)

Pleiospilos (Aizoaceae) N.E.Br.
Gk. *pleios* = full of; *spilos* = a stain, spot; referring to the leaves speckled with small spots.

Pleiospora*** = Pearsonia (Fabaceae) Harv.
Gk. *pleios* = full of, more; *spora* = seeds; referring the numerous that which distinguish it from every *Psoralea*.

Pleiotaxis (Asteraceae) Steetz
Gk. *pleios* = full of, more, multiple; *taxis* = arrangement, order; possibly referring to its numerous taxonomic characteristics.

Pleopeltis (Polypodiaceae) Humb. & Bonpl. ex Willd.

(CM)

Gk. *pleios* = many, full of; *peltis* = small, round shield: referring to the peltate scales covering the immature sori.

Pleurachne*** = Ficinia (Cyperaceae) Schrad.
Gk. *pleuron* = side, rib; *achne* = chaff, husks. Possibly referring to the plant's glumes 'glumes [husks] boat-shaped, ovate, very thick, stiff, the sides nearly occupied by the close thick white ribs (*Flora Capensis*).'

Pleurosorus*** = Asplenium (Aspleniaceae) Fée
Gk. *pleuron* = side, rib; *sorus* = the spores; referring to the shape of the sorus.

Pleurostylia** (Celastraceae) Wight & Arn.
Gk. *pleuron* = side, rib; *stylis* = style. The styles are lateral, referring to the position of the style in the mature ovary.

Plexipus = Chascanum (Verbenaceae) Raf.
Gk. *plexis* = weaving, plaiting; *pous* = foot. The synonym of *Plexipus* is *Chascanum cunefolius*, and with imagination this plant's 'foot-shaped' petals moving around (the weave) would suit the the brief description.

Plicosepalus (Loranthaceae) Tiegh.
La. *plicatus* = folded, pleated; sepal; referring to the marked folds on the inner surfaces of the interlocking lower part of the sepals (petals).

Plinthanthesis (Poaceae) Steud.
Gk. *plinthos* = plinth; *anthos* = flower; referring to the raceme-like inflorescence.

Plinthus (Aizoaceae) Fenzl
Gk. *plinthos* = plinth, tile, brick, squared stone. Meaning unclear. Possibly referring to some structural aspect of the plant, the plinth being a firm base for some columnar aspect.

Pluchea (Asteraceae) Cass.
For Noël-Antoine Pluche (1688–1761), French abbot, seminary teacher and naturalist, lecturer  (professor) in humanities, then rhetoric at Rheims before taking holy orders. He was a man of strong principles, which led him to clash with the Catholic Church over various issues. He is best remembered for his writings, which combined natural history and religion. His major work, *Spectacle de la nature*, was an eight-volume study of life and creation that was translated into virtually all European languages, still appearing

in abridged editions in the early 19th century. His other works include *Histoire du ciel* (1739), *La Méchanique des langues* (1751), and *Concorde de la Géographie des différents âges* (1765), as well as works on the Bible and French royal coronation ceremonies.

Plukenetia (Euphorbiaceae) L.
For Leonard Plukenet (1641–1706), British physician, royal professor of botany and gardener

to Queen Mary II of England and superintendent of the Hampton Court garden. He published *Phytographia* (London, 1691–1692) in four parts in which he described and illustrated rare exotic plants. It is a copiously illustrated work of more than 2 700 figures and is frequently cited in books and papers from the 17th century to the present. He collaborated with John Ray in the second volume of *Historia Plantarum* (London, 1686–1704). He was also the first author to mention a Cape species of *Erica* in his book *Almagesti Mantisssa Botanici*, published in 1700.

Plumbago (Plumbaginaceae) L.

Now widespread in horticulture (EJM)

La. *plumbum* = lead; *-ago* = resemblance or connection. The roots contain a fatty, lead-coloured dyestuff. Pliny the Elder mentions a plant of this name (probably *P. europaea*) said to cure an eye disease called plumbum, probably cataracts.

Plumeria* (Apocynaceae) L.
For Charles Plumier (1646–1704), French botanist and explorer. He entered the religious order of the Minims in 1662 where he studied, *inter alia*, art and later botany under Paolo Boccone (1633–1704)

and Joseph Pitton de Tournefort (1656 –1708). Apart from local explorations, he journeyed to the West Indies in 1689, 1693 and 1695, recording what he observed, mainly plants, animals and fish. After his first journey he wrote *Description des Plantes d'Amérique* (1693), and was appointed royal botanist. He undertook two further expeditions on the command of Louis XIV, the result of which was a massive work, *Nova Plantarum Americanarum Genera* (1703–1704), and other publications. He died of pleurisy before he could start his fourth journey, leaving behind 31 manuscript volumes containing notes, descriptions and 6 000 illustrations.

Pneumatopteris =
Cyclosorus (Thelypteridaceae) Nakai
Gk. *pneuma* = air, breathe; *pteris* = a fern; referring to the breathing pores (aerophores) at the base of the pinnae.

Poa (Poaceae) L.
Gk. *poa* = grass.

Poagrostis (Poaceae) Stapf
Gk. *Poa* (q.v.); *Agrostis* (q.v.). The genus has characteristics of both these genera.

Podalyria (Fabaceae) Willd.
For Podalirius (La.), from Podaleirios (Gk.), son of Asklepios, god of healing. He and his brother, Machaon were physicians to the Greek army during the Trojan wars, as described in the *Iliad*. The brothers' great feat was the healing of the festering foot of Philoctetes, who was badly needed for his arrows, but whose fetid stench sorely disturbed the warriors. The flowers of this genus are strongly fragrant but not unpleasantly so, rather sweet-smelling.

Podanthes = ***Orbea*** (Apocynaceae) Haw.
Gk. *podos, pous* = a foot; *anthos* = flower; alluding to the flowers being on long pedicels.

Podocarpus (Podocarpaceae) L'Hér. ex Pers.
(Image on opposite page)
Gk. *podos* = foot; *karpos* = fruit; referring to the fleshy 'foot', the receptacle, on which the fruit of some species develops (Plantzafrica.com).

Podophyllum* = ***Jeffersonia***
(Berberidaceae*)
Gk. *podo* = foot; *phyllum* = leaves (originally *anapodophyllum,* from Latin *anas* = a duck); referring to the palmately-lobed leaves in some species, which look a bit foot-like.

Podocarpus: Note the reddish foot below the seed (EJM)

Podranea (Bignoniaceae) Sprague.
Anagram of *Pandorea,* the name of a genus of *Bignoniaceae* to which this one is related. Pandora (meaning 'all-gifted') was the first woman mentioned in Greek mythology. She was given a box by Prometheus's brother that contained all of humanity's ills; when she opened the box, they all flew out.

Poecilolepis (Asteraceae) Grau
Gk. *poecila* = variable, many-coloured; *lepis* = a scale. Meaning obscure.

Poellnitzia* = Astroloba** (Asphodelaceae) Uitewaal
For Karl Joseph Leopoldt Arndt von Poellnitz (1896–1945), German botanist, agriculturist

and specialist in succulent plant systematics. He studied agronomy and natural science at Leipzig University and owned an estate at Oberlödla, near Altenburg, Thüringen. He became a keen collector and observer of *Haworthias* and described some 25 'new' *Gasteria* species between 1929 and 1940 and published, as author or co-author, 215 new names, 175 of them were new taxa and 40 were new combinations. During the last days of World War II he died when a bomb destroyed his castle. Most of the *Haworthias* he described had been sent to the botanical museum of Berlin-Dahlem, photographed and preserved, and some survived despite the war.

Pogonarthria (Poaceae) Stapf
Gk. *pogonatos* = bearded; *arthron* = joint; alluding to the hairy tips of the rachilla joints.

Pogonia (Orchidaceae) Juss.
Gk. *pogonias* = bearded; referring to the fringed lip (labellum).

Poinciana = Caesalpinia (Fabaceae) (Tourn. Ex L.) L.
For Phillippe de Longvilliers de Poincy (1583–1660), French nobleman, bailiff Grand Cross of the Knights of Malta, patron of botany and French governor on St Kitts in the West Indies. He was sent to the Caribbean in 1638 and expanded his control to include St Croix, St Bartholomew and St Martin. The official website of the island of Martinique indicates he governed from 1639–1645 and from 1647–1660. In 1642 he started building an elaborate building, the Chateau de Montagne on his estate called La Fontaine, at the time one of the grandest ever constructed in the Americas. On the grounds he planted exotic tropical plants, including a specifically imported flamboyant plant from Madagascar which was named (or misnamed) Royal Poinciana in his honour.

Poinsettia = Euphorbia (Euphorbiaceae) Graham
For Joel Roberts Poinsett (1775–1851), US statesman, member of the House of

Representatives, secretary of war under president Martin van Buren, amateur botanist, gardener, traveller and diplomat. He travelled extensively through Europe 1802–1804, through Russia 1806–1807 (only three of the eight in the party survived), South America (Chile and Argentina, 1810–1814, where he served as the US consul general), and Mexico 1822–1830 as America's special envoy, later minister to Mexico. It was here he discovered what in Mexico is called '*Flor de Noche Buena*' (Christmas Eve flower). He sent samples of the plant back to the US. By 1836 the plant was widely known as 'poinsettia'.

Poivrea* = Combretum** (Combretaceae) Comm.
For Pierre Poivre (1719–1786), French horticulturalist, plant collector, traveller and author of *Voyages d'un Philosophe.* He visited

China, Indochina, the Philippines, and the Île de France (Madagascar, Mauritius, and Seychelles), first as a Catholic missionary, later for French colonial agriculture. In the 1760s, Poivre became administrator of Île de France. For six years he worked to develop the islands, introducing spice plants, such as clove, nutmeg, pepper and cinnamon and other Asian species. He also smuggled spices to France to break the Dutch monopoly of the spice trade. In Mauritius he created the now world famous botanical garden of Pamplemousses. His name translated from the French means 'pepper'.

Polanisia = Cleome (Cleomaceae) Raf.
Gk. *poly-* = many; *anisos, aniso-* = unequal, different; referring to the numerous stamens with filaments of different lengths.

Polemannia (Apiaceae) Eck. & Zeyh.
For Peter Heinrich Poleman (Polemann, Pohlmann) (c 1780–1839), German chemist, apothecary and keen naturalist. He came to the Cape in 1802 and made friends with botanical explorers and collectors such as William John Burchell, William Henry Harvey, Georg Ludwig Krebs and Martin Lichtenstein, who also came to the Cape in 1802 for three years and with whom Poleman collected on Sunday walks. Poleman joined the pharmaceutical firm of which he was a partner, Pallas & Poleman. They employed several botanists and collectors such as Krebs, Peter Jonas Bergius, CF Drège and Ecklon and shipped consignments of wildflowers and animal specimens to Europe, and particularly Germany. He died in Cape Town.

Polemanniopsis (Apiaceae) B. Burtt
Polemannia (q.v.); Gk. *-opsis* = resembling, likeness. The only species in the genus until 2010, when another was found in Namibia. Found in the Richtersveld (Northern Cape) and the Clanwilliam district (Western Cape).

Polemonium (Polemoniaceae*) (Tourn.) L.
Gk. *Polemonium,* La. *polemonia* = ancient name of a plant, for a species of *Hypericum,* perhaps for the Greek valerian, otherwise called philetaeria and philetaeris (Plinius), Greek philetairion. Greek valerian or Jacob's ladder are common names for *Polemonium caeruleum,* a hardy perennial flowering plant.

Polevansia (Poaceae) De Winter
For Illtyd Buller Pole Evans (1879–1968), Welsh-

born South African botanist, mycologist, plant pathologist and collector. A graduate of the University College of South Wales and Cambridge, he joined the Transvaal Department of Agriculture and took charge of the Division of Botany and Plant Pathology. His botanical contribution was immense including his superb *Aloe* collection, research of the major vegetation types in Southern Africa, travels throughout South Africa and Southern Africa and extensive collecting, building up of the national herbarium, and his many publications. He was the first editor of *Flowering Plants of South Africa* and *Bothalia.* He was a Fellow of the Linnaean Society and director of the Botanical Survey of South Africa 1918–1939.

Polhillia (Fabaceae) C.H.Stirt.
For Dr Roger Marcus Polhill (1937–), botanist at Kew Gardens, authority on legumes, plant collector in the Congo, Kenya, Tanzania, Botswana and Malawi, and editor of *Flora of Tropical East Africa* 1966–1997.

Pollichia (Caryophyllaceae) Aiton
For Johann Adam Pollich (1741–1780), German doctor, botanist, entomologist, naturalist and author. After completing his medical studies in Strasbourg, he practised medicine for a while before turning to natural sciences in 1764. He published numerous works relating to physiology and body structure (1763), the history of the plants in the Palatinate, an area of southwestern Germany, titled *Historia plantarum in Palatinatu* (1776–1777) with CF Schwan (1733–1815), descriptions of some insects (1779), and posthumously two more works on insects (1781, 1783).

Polpoda* (Molluginaceae) C.Presl
Gk. *polypodes* = a fringe. Both the stipules and the perianth segments are fringed.

Polyalthia* (Annonaceae) Blume
Gk. *poly* = many; *althia* from *àltheo* = to cure, or *althainein* = *to heal*; possibly, referring to medicinal properties.

Polyarrhena (Asteraceae) Cass.
Gk. *poly-* = many; *arrhen* = male. Many, but not all, of the disk florets are functionally male.

Polybotrya (Dryopteridaceae) Humb. & Bonpl. ex Willd.
Gk. *polys* = many; *botrya* = bunch; referring to the clusters of sporangia on the fronds.

Polycarena (Scrophulariaceae) Benth.
Gk. *poly-* = many; *karenon* = head. The inflorescence consists of numerous small heads of flowers.

Polycarpaea (Caryophyllaceae) Lam.
Gk. *poly-* = many; *karpos* = fruit; alluding to the numerous capsules.

Polycarpon (Caryophyllaceae) L.
Gk. *poly-* = many; *karpos* = fruit. Much fruited or bearing much fruit.

Polycenia = **Hebenstretia**
(Scrophulariaceae) Choisy
Gk. *poly-* = many, much; *kenos* = hollow, empty; possibly referring to the many flowers with slender hollow tubes opening to four separate lobes.

Polyceratocarpus* (Annonaceae) Engl.& Diels.
Gk. *poly-* = many, much; *keras* = horn; *carpus* = fruit; referring to the horned fruit.

Polychaetia = **Relhania** (Asteraceae) Less.
Gk. *poly-* = many, much; *chaete* = flowing hair, mane, bristle. The plant has very many short, stiff hairs or barbs.

Polydora* = **Veronia** (Asteraceae) Fenzl
Gk. *poly* = many; *doros* = gifts. Perhaps named after one of the Polydoras in mythology.

Polygala (Polygalaceae) L.

The brush-like stigma is characteristic. (CM)

Gk. *poly* = much; *gala* = milk; so-called from the belief that cattle grazing in fields with this plant produced more milk. (San Marcos growers in the United States claim that Polygala virgata 'Portola' has this property but the authors could find no scientific evidence).

Polygonatum (Ruscaceae) Mill.
Gk. *poly* = many; *gonatum* = knees or joints; referring to the multiple jointed rhizomes.

Polygonum (Polygonaceae) L.
Gk. *poly-* = many; *gonu* = knee, node, joint; referring to the many-jointed stems with many nodes.

Polymita* (Aizoaceae) N.E.Br.
Gk. *poly-* = many; *mita* = small petal, according to Hans Herre, but may mean threads. Alluding to the structure of the flower.

Polymnia (Asteraceae) L.
Possibly, after Polyhymnia, the muse of song and oratory in Greek mythology or *mnion* = moss, seaweed. Meaning obscure.

Polyphlebium (Hymenophyllaceae) Copel.
Gk. *poly-* = many; *phlebium* = veins; referring to the frond's easily seen, bright green, translucent venation.

Polypodium (Polypodiaceae) L.
Gk. *polys* = many; *podion* = foot. The branching rhizome and its roots appear many-footed.

Polypogon (Poaceae) Desf.
Gk. *polys* = many; *pogon* = beard; referring to the hairy glumes, alternatively the many awns on the inflorescence.

Polystachya (Orchidaceae) Hook.

(HS – Both)

Gk. *polys* = many; *stachys* = ear of grain or spike; referring to the many branchlets which make up the inflorescence in some species and which may resemble spikes of wheat.

Polystichum (Dryopteridaceae) Roth
Gk. *polys* = many; *stichos* = row; alluding to the many rows of sori (spore masses) in these ferns.

Polyxena *** = **Lachenalia** (Hyacinthaceae) Kunth
For Polyxena, the name of a daughter of Priam, the last king of Troy and his queen, Hecuba.

Pomaria = **Cercidium** (Fabaceae) Cavaco.
Named after a town in Roman Africa (Algeria) and its nearby apple orchards. La. *pomum* = an apple or fruit shaped like an apple.

Pontederia* (Pontederiaceae) L.
For Giulio (Julius) Pontedera (1688–1757), Italian botanist, physician, and plant collector. He studied

at the University of Padua, qualifying as a doctor in 1715 and becoming professor of botany at Padua in 1719, as well as praefectus of the botanical garden of Padua. He authored *Compendium tabularum botanicarum*, which described 252 plant species growing in Italy, and other books. He rejected Linnaeus's 'biological' taxonomy system, which led the angered Linnaeus to use the derogatory name of the family *Narcissoides Pontederia*, after him. (The word *Narcissus* Gk. *Narkissos* is the name of a youth who fell in love with his reflection in water, hence self-worship.)

Popowia* (Annonaceae) Endl.
For Johannes Siegmund Valentin Popowitsch (1705–1774), Austrian/Slovenian linguist and

naturalist, professor of linguistics in Vienna (1753–1766) where he taught Austrian German. He also had a great interest in botany, especially sponges and fungi. He travelled extensively and in the process built up an extensive herbarium. His two main publications were in the field of linguistics.

Populus* (Salicaceae) L.
La. *populus* = people. Name for the poplar, used by Virgil. One species was much planted in Roman cities, hence 'arbor populi', tree of the people.

Porana* (Convolvulaceae) Burm.
Derivation uncertain. Probably from the Indian Marathi (Bhauri) name for Porana paniculata; (Umberto Quadrocchi); other sources suggest from the Greek *poreno* = to travel, carry over. Referring to the extensive twining stems.

Porcellites *** = **Hypochaeris** (Asteraceae) Less.
Speculatively, this word could be associated with the La. *porcus* = a pig or sow. The Italian word *sow* = scatter, as in the context of 'scattering of seeds'. The new name for this genus *Hypochaeris* has the common name 'false dandelion' and is a weed, as is the common dandelion.

Portulaca (Portulacaceae) L.
La. *portare* = to carry; *lac* = milk; referring to the milky sap. Another source suggests *portula* = little door, diminutive of *porta* = gate; referring to the lid of its capsule that opens like a gate.

Portulacaria (Didieriaceae) Jacq.
Portulaca (q.v.); *aria* = pertaining to; somewhat like a *Portulaca*.

Potamogeton (Potamogetonaceae) L.
Gk. *potomos* = river; *geiton* = neighbour; referring to the habitat in which these plants occur.

Potamophila (Poaceae) R.Br.
Gk. *potomos* = river; *philos* = loving; alluding to the habitat of this river-loving species.

Potentilla (Rosaceae) L.
La. *potentia*, diminutive of *potens* = potent, powerful; referring to its potent medicinal properties.

Poterium* = **Sanguisorba** (Rosaceae) L.
Gk. *Poterion*, La. *poterium* = a vessel, drinking cup. The leaves of the *Poterium sanguisorba* were used in the preparation of drink. Parkinson relates in 1629 'the greatest use that burnett (*P. sanguisorba*) is put unto, is to put a few leaves into a claret cup of wine which ... giveth a pleasant taste there unto, very pleasant unto the palate, and is accounted a help to make the heart merrie.' Jack Saub (*75 Exceptional Herbs For Your Garden*).

Pouteria (Sapotaceae) Aubl.
Derived from the Galibi (Guyana) vernacular name *pourama-pouteri* for a species of this genus.

Pouzolzia (Urticaceae) Gaudich.
For Pierre Marie Casimir de Pouzolz (1785–1858), French soldier, botanist and writer on flora of France, member of the Linnaean Society of Paris and the Académie du Gard. He served in the military from 1805–1830, retiring early with the rank of captain. He authored *Catalogue des plantes qui croissent naturellement dans le Gard* (1842) and began work on volume one of *Flore du département du Gard* (1854), a work which Professor Philip Courciere, professor of natural history at a school in Nîmes, completed after his death as *Graminees et Cryptogames Vasculaires de la Flore du Gard* (1862). He had a herbarium containing more than 20 000 specimens.

Premna (Lamiaceae) L.
Gk. *premnon* = a tree stump; referring to the dwarf size of the type species. Some species make excellent bonsai trees.

Prenanthes (Asteraceae) L.
Gk. *prenes* = prone, prostrate; *anthos* = flower; alluding to the flowers' 'drooping heads'.

Prenia** = ***Mesembryanthemum*** (Aizoaceae) N.E.Br.
Gk. *prenes* = bent forward, downward, prone, prostrate; referring to the creeping habit of this plant.

Prepodesma (Aizoaceae) N.E.Br.
Gk. *prepo* = to be conspicuous; *desma* = bundle, cable or band; referring to the stamens in a noticeable bundle in the middle of the flower.

Pretrea** = ***Dicerocaryum*** (Pedaliaceae) J. Gay ex Meisn.
Gk. *petraios* = rocky; referring to its habitat.

Priestleya (Fabaceae) DC.

For Joseph Priestley (1733–1804), English chemist, clergyman, philosopher, political theorist, and scientist, best known for his discovery of oxygen, possibly the inventor of soda-water, and he had an interest in electricity which he shared with Benjamin Franklin. A multi-talented man, he studied French, Italian, and German and published over 150 works ranging from political philosophy, education, theology, natural philosophy, history and a seminal work on English grammar. He was a member of almost every major scientific society, including being a Fellow of the Royal Society. An advocate of religious toleration and equal rights, he helped found Unitarianism in England. His views were controversial and he was forced to live his final 10 years in the United States.

Primula (Primulaceae) L.
La. *primus* = prime, first, *-ula* = diminutive; referring to the early flowers of spring, which come into bloom earlier than many others.

Printzia (Asteraceae) Cass.
For Jacob Gabriel Printz (1740–1779), Swedish physician and botanist, pupil of Linnaeus, co-author with Linnaeus of *Plantae rariores africanae* (1760), based upon Printz's PhD dissertation. This publication describes 100 South African plants, based on a collection sent from the Cape of Good Hope. Jacob Printz himself never visited the Cape.

Prionachne** = ***Prionanthium*** (Poaceae) Nees
Gk. *prion* = a saw; *achne* = chaff, husks; referring to the serrated keels of the glumes.

Prionanthium (Poaceae) Desv.
Gk. *prion* = a saw; *anthos* = flower; referring to the serrated lemma keels formed by the glands.

Prionium (Juncaceae) E.Mey.

Plants grow in dense clumps in swampy land. (EJM)

Pronium: saw-teeth leaf edge (EJM)

Gk. *prion* = a saw; referring to the leaf blades, which have serrated edges.

Prionodon (Prionodontaceae) Müll.Hal.
Gk. *prion* = a saw; *-odon* = toothed; referring to the leaves, which have serrated edges.

Prionostemma** (Celastraceae) Miers
Gk. *prion* = a saw; *stemma* = a crown. The corona of the flower is composed of five saw-edged lobes.

Prismatocarpus (Campanulaceae) L'Hér.
Gk. *prisma* = a thin saw, hence angular; *karpos* = fruit. The slender inferior ovary becomes greatly elongated; the fruits are shaped like prisms.

Pristimera (Celastraceae) Miers
Gk. *pristis* = serrated; *meris, meros* = part, portion; referring to the parts of the flower (Umberto Qauttrocchi).

Pristocarpha* = *Athanasia (Asteraceae) E.Mey. ex DC.
Gk. *pristis* = serrated; *carphos* = any small dry body, twigs, etc.; referring to the serrated leaves and the dry involucral bracts around the flowers which remain on the plant long after it has finished flowering.

Priva (Verbenaceae) Adans.
Unclear, possibly an Indian vernacular name, or from Latin *privus* = without, individual, single, deprived of. Meaning unclear.

Proboscidea* (Martyniaceae) Schmidel
Gk. *proboskis* = snout, proboscis, from *pro* = in front and *boskein* = to feed; referring to the shape of the fruit, the so-called 'unicorn plant'.

Prosopis* (Fabaceae) L.
Gk. *prosopon* = countenance, face. A classical Greek name for a very different plants originally.

Prosphytochloa (Poaceae) Schweick.
Gk. *prosphytuo-* = grow upon; *chloa* = grass. This genus is a climbing forest grass which 'grow upon' other plants.

Protasparagus* = *Asparagus*** (Asparagaceae) Oberm.
Gk. *protos* = first, earliest, foremost; *Asparagus* (q.v.). Many species, particularly from Africa, were once included in this genus as they seemed different to the asparagus genus found elsewhere having formidable hooked spines, etc. These species have now been incorporated back into *Asparagus*.

Protea (Proteaceae) L.
Gk. After Proteus, a mythological sea-god, who could change his form at will, taking new shapes. Seemingly Linnaeus was so over-awed by the variety of plants sent to him from the Cape that he named the genus *Protea*. The authors could not confirm this.

Protium* = *Commiphora (Burseraceae) sensu Wight & Arn., non Burm.f.
Gk. *protos-* = first, foremost, earliest. Possibly a Javanese name, or from Proteus, alluding to its rate of growth (Burman, *Flora Indica*). Meaning unclear.

Protorhus (Anacardiaceae) Engl.
Gk. *protos* = first; *rhous* (*rhus*) = red. There are

(EJM)

nearly always a few red leaves on these trees, so, 'first to turn red [in autumn]'.

Prototulbaghia (Alliaceae) Vosa.
Gk. *protos* = first; *Tulbaghia* (q.v.). referring to a 'first' finding of a new genus, a kind of transitional plant, intermediate between the *Allium* and *Tulbaghia* genera.

Prunella* (Lamiaceae) L.
Linnaeus's misspelling of *brunella*, now corrected as *Brunella* (q.v.), from the German *braune* = quinsy; Gk. *-ella* = diminutive. This genus was supposed to have curative medical properties for this condition. Quinsey is a tonsillar abscess at the back of the throat caused by streptococcus.

Prunus (Rosaceae) L.
La. *prunus* = the classical Latin name for a plum tree. This large genus, around 430 species, contains a wide range of fruit in addition to plums such as peaches, apricots, cherries, nectarines, and almonds (Wikipedia.org).

Psammophora (Aizoaceae) Dinter & Schwantes
Gk. *psammos* = sand; *phorein, phorus* = to bear or carry. In its natural habitat the leaves

are protected from excessive transpiration by a coating of sand.

Psammotropha (Molluginaceae) Eck. & Zeyh.
Gk. *psammos* = sand; *trophos* = a nurse, feeder; referring to the environment which nurtures and cares for the flowering plants.

Psathurochaeta = Melanthera (Asteraceae) DC.
Gk. *psatharos* = friable, scratch, rub; *chaete* = mane, bristle; possibly referring to the barbed pappus bristles immediately surrounding the corolla.

Psednotrichia (Asteraceae) Hiern
Gk. *psednos* = scanty, spare, thin; *trichion* = small hair; referring to the scant pubescence of the plants, which are glabrous except for the distinct or tiny tufts of fine hairs in the leaf-axils (Bertil Nordenstam).

Pseudalthenia = Zannichellia (Potamogetonaceae) Nakai
Gk. *pseudo* = false; *Althenia* (q.v.).

Pseudarthria (Fabaceae) Wight & Arn.
La. *pseudo,* Gk. *pseudes* = false; *arthron* = joint. The pods are only partially articulated between the seeds.

Pseudechinolaena (Poaceae) Stapf
Gk. *pseudein* = false, or resembling; *Echinolaena* (q.v.).

Pseuderanthemum (Acanthaceae) Radlk.
Gk. *pseudes* = false; *er* = spring; *anthemon* = flower. Some plants of this genus were at first placed in the genus *Eranthemum,* a related genus and confusingly similar.

Pseudobaeckea = Brunia (Bruniaceae) Nied.
For Dr Abraham Baeck (Baekea, Bäck) (1713–1795), Swedish physician, botanist and writer on natural history. He studied medicine at Uppsala University and after he qualified in 1740 travelled for five years 1740–1745, then practised in Stockholm. He became assessor of the Collegium Medicum and president of the Royal College of Medicine. He became physician-in-ordinary to King Frederick I, his successor Adolf Frederick and later Gustav III. He was a keen botanist and claimed that his herbarium was second in size in Sweden only to that of Linnaeus of whom he was a good friend. *Pseudo-* = false,

or resembling *Baeckea* (or *Baekea*) a genus in *Myrtaceae.*

Pseudobarleria = Petalidium (Acanthaceae) T. Anderson.
Gk. *pseudo-* = false, resembling but not the real *Barleria* (q.v.).

Pseudoberlinia (Fabaceae) P.A. Duvign.
Gk. *pseudo-* = false, resembling but not the real *Berlina* (q.v.).

Pseudobersama (Meliaceae) Verdc.
Gk. *pseudo-* = false, resembling but not the real *Bersame.* (q.v.).

Pseudobrachiaria = Brachiaria (Poaceae) Launert
Gk. *pseudo-* = false, resembling but not the real *Brachiaria* (q.v.).

Pseudobromus*** = Festuca (Poaceae) K. Schum.
Gk. *pseudo-* = false, resembling but not the real *Bromus* (q.v.).

Pseudobrownanthus = Mesembryanthemum (Aizoaceae) Ihlenf. & Bittrich
Gk. *pseudo-* = false, resembling but not the real *Brownanthus* (q.v.).

Pseudocadia*** = Xanthocercis (Fabaceae) Harms
Gk. *pseudo-* = false, resembling but not the real *Cadia* (q.v.) = Arabic qadhy.

Pseudocassine (Celastraceae) Bredell
Gk. *pseudo-* = false, resembling but not the real *Cassine* (q.v.).

Pseudocimum*** = Endostemon (Lamiaceae) Briq.
Gk. *pseudo-* = false, resembling but not the real *Ocimum* (q.v.).

Pseudoconyza (Asteraceae) Cuatrec.
Gk. *pseudo-* = false, resembling but not the real *Conzya* (q.v.).

Pseudocyclosorus = Cyclosorus (Thelypteridaceae) Ching
Gk. *pseudo-* = false, resembling but not the real *Clyosurus* (q.v.).

Pseudogaltonia (Hyacinthaceae) (Kuntze) Engl.
Gk. *pseudo-* = false, resembling but not the real *Galtonia* (q.v.).

Pseudognaphalium (Asteraceae) Kirp.
Gk. *pseudo-* = false, resembling but not the real *Gnaphalium* (q.v.). Similar to a plant whose soft white leaves are used as cushion stuffing.

Pseudognidia = Gnidia (Thymelaeaceae) E.Phillips
Gk. *pseudo-* = false, resembling but not the real *Gnidia* (q.v.).

Pseudolachnostylis (Phyllanthaceae) Pax
Gk. *pseudo-* = false, resembling but not the real genus *Lachnostylis* (q.v.). Rock Coalwood.

Pseudolycopodiella = Lycopodiella (Lycopodiaceae) Holub
Gk. *pseudo-* = false, resembling but not the real *Lycopodia* (q.v.); *-ella* = diminutive.

Pseudomariscus = Courtoisina (Cyperaceae) Rauschert
Gk. *pseudo-* = false, resembling but not the real *Mariscus* (q.v.).

Pseudopentameris (Poaceae) Conert
Gk. *pseudo-* = false, resembling but not the real *Pentameris* (q.v.).

Pseudophegopteris (Thelypteridaceae) Ching
Gk. *pseudo-* = false, resembling but not the real *Phegopteris* (q.v.).

Pseudoprospero (Hyacinthaceae) Speta
Gk. *pseudo-* = false or resembling; La. *prosperus* = thrive, flourish, succeed; referring to the genus *Prospero*.

Pseudosalacia (Celastraceae) Codd
Gk. *pseudo-* = false, resembling but not the real *Salacia* (q.v.).

Pseudoschoenus (Cyperaceae) (C.B. Clarke) Oteng-Yeb.
Gk. *pseudo-* = false, resembling but not the real *Schoenus* (q.v.).

Pseudoscolopia (Salicaceae) Gilg
Gk. *pseudo-* = false, resembling but not the real *Scolopia* (q.v.). Somewhat like *Scolopia* except the leaves are opposite and the fruit is a capsule and not a berry.

Pseudoselago (Scrophulariaceae) Hilliard
Gk. *pseudo-* = false, resembling but not the real *Selago* (q.v.). Looks like *Selago* whose seeds are tapered on both ends compared with *Pseudoselago* only at one end; also *Pseudoselago* has a yellow-orange patch inside the flower which is not present in *Selago*.

Pseudotragia* = Plukenetia** (Euphorbiaceae) Pax.
Gk. *pseudo-* = false, resembling but not the real *Tragia* (q.v.).

Pseudowolffia* = Wolffiella** (Araceae) Hartog & Plas
Gk. *pseudo-* = false, resembling but not the real *Wolffia* (q.v.).

Psiadia (Asteraceae) Jacq.
Gk. *psias* = a (dew) drop. The leaves exude drops of a sticky substance.

Psidium* (Myrtaceae) L.
Gk. *psidion* = originally, it is claimed, the Greek name for the pomegranate, now given to the guava. Ancient Greeks mainly used *rhoa* for both the pomegranate tree and its fruit. Modern Greek for 'pomegranate' is *rhodi*.

Psilocarya* = Rhynchospora** (Cyperaceae) Torr.
Gk. *psilos* = bare, bald, smooth; *karyon* = nut, hence naked nuts. The achenium lacks bristles, hence is 'naked'.

Psilocaulon = Mesembryanthemum (Aizoaceae) N.E.Br.
Gk. *psilos* = bare; *kaulos* = stalk, stem; referring to the leafless stems.

Psilochloa = Panicum (Poaceae) Launert
Gk. *psilos* = naked; *chloa* = grass. The upper glume and lemmas lack apical appendages.

Psilolepus = Aspalathus** (Fabaceae) C.Presl
Gk. *psilos* = bare; *lepus* = hare. Meaning obscure.

Psilothamnus = Euryops** (Asteraceae) DC.
Gk. *psilos* = bare; *thamnos* = shrub. Many species have few mid branches and the stems are bare toward the base.

Psilothonna = Steirodiscus (Asteraceae) E.Mey.
Gk. *psilos* = bare; *Othonna* (q.v.).

Psilotrichum (Amaranthaceae) Blume
Gk. *psilos* = bare; *trichos* = three-fold. The plants are trichotomously branched.

Psilotum (Psilotaceae) Sw. (Image on opposite page)
Gk. *psilotes* = nakedness; alluding to this fern's leafless aerial stems and naked, scattered fruit (synangia) not normally found in other ferns.

Psilotum (NC)

Psoralea (Fabaceae) L.
Gk. *psoraleos* = scabby. The plants are covered with rough warty-looking glandular dots.

Psychotria (Rubiaceae) L.
Gk. *psychotria* = vivifying; from *psyche* = life; referring to the several supposed healing properties of some species.

Psydrax (Rubiaceae) Gaertn.
Gk. *psydrax* = a blister on the tongue; possibly because the leaves have domatia (small depressions partly enclosed by leaf tissue or hairs).

Ptaeroxylon (Rutaceae) Eck. & Zeyh.
Gk. *ptairein* = to sneeze; *xylon* = wood. Common name 'sneezewood'. The freshly-cut wood contains an irritant, giving rise to both scientific and vernacular names.

Pteleopsis (Combretaceae) Engl.
Gk. *ptelea* = an elm tree; *-opsis* = resembling. The two-winged fruits of *P. myrtifolia* resemble those of the common European elm in shape and colour, but not in size (Plantzafrica.com).

Pteridella = Pellaea (Pteridaceae) Mett. ex Kuhn.
Gk. *pteris* = fern; *-ella* = diminutive; referring to a small fern named pteris in antiquity.

Pteridium (Dennstaedtiaceae) Gled. ex Scop.
Gk. *pteridion* = a small fern; referring to the fronds, small, wing-like.

Pteris (Pteridaceae) L.
Gk. *pteris* = fern, from *pteron* = wing, feather; referring to the shape of the pinnae: the symmetrical fronds resemble wings.

Pterocarpus (Fabaceae) Jacq.
Gk. *pteron* = wing; *karpon* = fruit. The pod is conspicuously winged; winged fruit.

Pterocelastrus (Celastraceae) Meisn.
Gk. *pteron* = wing; *kelastros* = holly. The fruit carries horns or wing-like outgrowths.

Pterococcus*** = Plukenetia (Euphorbiaceae) Hassk.
Gk. *pteron* = a wing; *kokkos* = berry. The capsule breaks into winged cocci; the seeds are also winged in some species.

Pterodiscus (Pedaliaceae) Hook.
Gk. *pteron* = a wing; *diskos* = disk. The fruit is winged, the disk 'inconspicuous' (Robert Allen Dyer).

Pterolepis = Schoenoplectus (Cyperaceae) Schrad.
Gk. *pteron* = a wing; *lepis* = scale. Scaly wings.

Pterolobium (Fabaceae) R.Br. ex Wight & Arn.
Gk. *pteron* = wing; *lobion* = diminutive of *lobos* = lobe, pod, capsule. The little pods are winged.

Pteronia (Asteraceae) L.
Gk. *pteron* = a wing; probably referring to seeds which are wind-dispersed.

Pteropsis = Pyrrosia (Polypodiaceae) Desv.
Gk. *pteron* = wing; *-opsis* = resembling. Probably referring to the fronds which wave in the wind, birdlike.

Pterospermum (Malvaceae) Schreb.
Gk. *pteron* = wing; *sperma* = seed. The seeds are winged.

Pterothrix*** = Amphiglossa (Asteraceae) DC.
Gk. *pteron* = wing; *thrix* = hair. The pappus bristles are plumose (feathery, with fine hairs or bristles).

Pterygodium (Orchidaceae) Sw.
Gk. *pterygodium* = diminutive of *pteryx* = wing. Descriptive probably of the dorsal sepal and petals, which are united into an erect, hood-shaped segment, and of the lateral sepals, similar in shape and size to the dorsal, giving a wing-like appearance to the flower.

Ptisana (Marattiaceae) Murdock
La. *ptisane* = pearl barley or barley-gruel; alluding to the shape of the synangia.

Ptycholobium (Fabaceae) Harms
Gk. *ptychos* = a fold, a wrinkle; *lobion* = diminutive of *lobos* = lobe, pod, capsule. The pod is compressed and membranous, sometimes contorted.

Ptychotis** (Apiaceae) W.D.J.Koch
Gk. *ptychos* = fold; *otis* = ear; referring to the shape of the petals folded to look like an ear.

Puccinellia (Poaceae) Parl.
For Benedetto Luigi Puccinelli (1808–1850), Italian botanist, professor of botany at the University of Lucca and director of the botanical garden of Lucca from 1830–1850. He wrote a book about the flora of the Lucca, the two-volume *Synopsis Plantarum in agro Lucensi sponte nascentium* (*The History of Plants Born in the Territory of Lucca*) (1841).

Pueraria* (Fabaceae) DC.
For Marc Nicolas Puerari (1766–1845), Swiss botanist, pupil of Danish-Norwegian botanist and zoologist Martin Vahl Hendrickson in Copenhagen; a political refugee from Geneva. He came to Copenhagen in 1794 where he became tutor to the merchant Fr. de Coninck and later professor of French at Copenhagen University. In 1820 he returned to Geneva, and left his herbarium to a younger compatriot, Alphonse Louis Pierre Pyrame de Candolle (1806–1893), but retained links with Denmark and in 1826–1828 was advisor and French teacher for Prince Frederik VIII.

Pulicaria (Asteraceae) Gaertn.
La. *pulex* = a flea; *aria* = pertaining to; alluding to the use of the strong smelling plant as a flea repellent. Hence 'fleabane'.

Pulsatilla* = *Anenome (Ranunculaceae) Mill.
La. *pulsatilla* = pulsate a little. The plant grows in windy areas, often on mountains.

Punica (Lythraceae) L.
La. *punica* = the Pomegranate; '*punic-*' also refers to a reddish-purple colour.

Punctillaria* = *Pleiospilos (Aizoaceae) N.E.Br.
La. *punctum* = small spot or dot; *-illaria* = having the property of dots; referring to the leaf surface, which is patterned with many tiny dark dots.

Pupalia (Amaranthaceae) Juss.
A vernacular, possibly eastern name, *Pupali*. Meaning unknown.

Purgosea** (Crassulaceae) Haw.
Gk. *purgos* = tower; referring to its vertical growth and appearance.

Pusaetha* = *Entada (Fabaceae) ex Kuntze
Meaning unknown to authors. Possibly a Sri Lankan name, although the plant is also found in Africa, Asia and South America.

Putterlickia (Celastraceae) Endl.
For Aloys Putterlick (1810–1845), Austrian botanist and physician, botanical artist, bryologist. He obtained a PhD in medicine in 1839 based on his thesis *Pittosporum*, authored *Synopsis Pittosporearum* (1839) and was a contributor to five other books. He became curator at the Imperial Natural History Museum of Vienna from 1849–1845. While some sources say 'Dutch', he was actually born in Jihlava, which is now in the Czech Republic.

Pycnodoria***** = *Pteris (Pteridaceae) C. Presl
Gk. *pyknos* = close, thick, dense; *-odora* = scent. These plants are noticeably scented.

Pycnosphaera (Gentianaceae) Gilg
Gk. *pyknos* = dense; *sphalra* = ball, sphere; referring to the dense 'ball-like' shape of the inflorescence corymbosely arranged at the end of stems and branches.

Pycnostachys (Lamiaceae) Hook.
Gk. *pyknos* = dense; *stachys* = a spike. These plants have flowers in very dense spikes.

Pycreus (Cyperaceae) P.Beauv.
Anagram of *Cyperus*. The inflorescences of these two genera are similar.

Pygeum* = *Prunus (Rosaceae) Gaertn.
Gk. *pyge* = rump. The stone of the fruit resembles a pair of buttocks.

Pygmaeothamnus (Rubiaceae)
Gk. *pugmalos* (La. *pygmaeus*) = dwarf; *thamnos* = bush; referring to the very small flowers and leaves.

Pyracantha * (Rosaceae) M. Roem.
Gk. *pyr* = fire; *akanthos* = thorn; alluding to the brilliant fiery colour of the berries and, to some extent, the spines. 'Firethorn'.

Pyrenacantha (Icacinaceae) Wight
Gk. *pyren* = a fruit-stone, kernel; *akanthos* = thorn; referring to 'peg-like' (thorny) protuberances impacting the inner walls of the fruit.

Pyrethrum (Asteraceae) Zinn
Gk. *pyrethron*, from *pyr* = fire. The roots of some species have a hot taste, hence 'fire'. The petals of some species when crushed release pyrethrum, which has mosquito and insecticide usage.

Pyrostria (Rubiaceae) Comm. ex Juss.
La. *pyrum* = a pear; *stria* = a stripe. The fruit is pear-shaped, with eight stripes.

Pyrrosia (Polypodiaceae) Mirb.
Gk. *pyrros* = flame-coloured; *pyrro-* = red, tawny from *pyr* = fire; referring to the coat of red stellate hairs or reddish lamina scales of some species or the burnt appearance of the fertile portions of the fronds.

Pyrus * (Rosaceae) L.
Gk. *pyrus* = pear; classical Latin name for a pear tree.

Richard James Poynton (1925–2013) was one of the last forest scientists whose meticulous life's work underpinned the development of what is South Africa's most successful and productive forestry industry. He graduated from the University of Stellenbosch's forestry department in 1948 and obtained a PhD from Wits University in 1979. He spent all of his working life working for government. He was a founder member of the SA Institute of Forestry and received the Distinguished Forestry Award in 2000. Poynton wrote books on pines (1975) and gums (1979) and is probably best know for his three-volume encyclopaedic work *Tree Planting in Southern Africa* and for his *Silvicultural Map of South Africa and Neighbouring States* that flowed from his pioneering assessment of the suitability of various introduced species to South African conditions.

Quaqua (Apocynaceae) N.E.Br.
Nama word *Kam-qua-qua*. La. *quaqua* = in every direction; everywhere. The plant is indigenous to southern Namibia and western South Africa.

Quartinia* = *Rotala (Fabaceae) A.Rich.
For Léon Richard Quartin-Dillon (1814–1841), Portuguese botanist, physician, explorer and plant collector. He studied medicine under the physician and botanist Achille Richard (1794–1852) at the Faculty of Medicine at the Université de Paris, receiving his doctorate in 1839. His dissertation was titled *Questions sur Diverses Branches des Sciences Médicales* (1839). After graduating, he joined the Lefebvre expedition (1839–1843) to explore Ethiopia with botanist colleague Antoine Petit. Quartin-Dillon died of illness in 1841, and Petit drowned in the Tacazze River in 1843.

Quercus* (Fagaceae*) L. (Images on opposite page)
Latin name for an oak.

Quisqualis (Combretaceae) L.
From the Malay name *udani*, which Rumphius (1627–1702) translated into Dutch, in a play on words, as *hoedanig* (Dutch = how, what), and then translated that into Latin as *quis* = who? *qualis* = what? This was possibly a reference to the variability of features in this genus, which made it difficult to classify.

Quisqualis: This woody climber uses hooked lateral branches to climb. (RB)

Quisqualis: Without flowers this plant can be confused with Combretum spp. (RB)

R

Western Cape mountain view with a monospecific restio sward in the foreground (PL)

Restio inflorescences female (PL)

Restio inflorescences male (PL)

Peter Linder (1954–) from Piketberg in the Western Cape, South Africa, studied taxonomy under Ted Schelpe of the Bolus Herbarium, completed his PhD on *Disa* and then wrote an account of the Southern African orchids. Thereafter he started working on the Restionaceae and the danthonioid grasses, investigating their taxonomy, ecology, evolutionary history and biogeographical patterns. He works at the universities of Cape Town and Zurich.

Rabdosia **= *Isodon*** (Lamiaceae) Hassk.
Gk. *rhabdos* = rod. The plant has rod-like branches.

Rabdosiella **= *Plectranthus*** (Lamiaceae) Codd
Gk. *Rabdosia* (q.v.); *-iella* = diminutive; referring to its shape.

Rabiea (Aizoaceae) N.E.Br.
Probably for William (Bill) Abbot Rabie (1869–1936), South African farmer and plant collector.

His family farm, Groenkloof, was just outside Fauresmith, the second-oldest town in southern Free State. He was the second of four brothers who all went to Grey College in Bloemfontein, the third-oldest school in South Africa. After leaving school, he worked on the farm for 30 years. In 1899, the South African War broke out, and after joining the republican forces, Rabie was taken captive and sent to the Amritsar camp for prisoners of war in Punjab State, India, close to Lahore. After the war in 1902, he took up farming again at the family farm. He had to rebuild the house and lived there until his death in 1936. He served on numerous committees and was a popular public speaker in the local community, serving as an active deacon in the church. As a farmer, he must have had an interest in nature, and it is known that other members of his family did. In 1926, the plant *Salsola rabieana* (Amaranthaceae) was named after him, and later, in 1930, the genus *Rabiea* was also named after him.

Radinosiphon (Iridaceae) N.E.Br.
Gk. *radinos* = delicate, slender, supple; *siphon* = tube; referring to this slender plant's small, long-tubed flowers.

Radyera (Malvaceae) Bullock
For Robert Allen Dyer (1900–1987), South African botanist, taxonomist and author of 450 publications, mainly taxonomic.

After completing his studies at Natal University (1923), he worked as assistant curator, then curator of the Albany Museum Herbarium. From 1931 to 1934, he held the post of liaison officer at Kew Gardens before moving to the National Herbarium in Pretoria, where he became director of the Botanical Research Institute (1944–1963). During this time, he founded the Pretoria National Botanical Garden. He was, *inter alia,* president of the Pretoria Horticultural Society (1961–1972), and during his career received many local and overseas awards, including the first Gold Medal from the South African Association of Botanists (1973) and an honorary DSc from the University of the Witwatersrand (1976).

Rafnia (Fabaceae) Thunb.
For Carl (Karl) Gottlob Rafn (1769–1808), Danish civil servant, botanist and science writer. He studied medicine and botany at the University of Copenhagen in 1788, and later veterinary science, but did not take the exams. He had a range of jobs such as an agriculture assessor and director of a distillery, but his main interests were natural history and science. He authored or co-authored a range of publications, including the *Flora of Denmarks and Holstein*, a book on plant physiology (1798), a paper on animal hibernation with JD Herholdt, and a book on life-saving measures for drowning persons. He became a member of the Royal Danish Academy of Sciences in 1798.

Randia * (Rubiaceae) L.
For Isaac Rand (1674–1743), English botanist, apothecary, and gardener at the Society of Apothecaries' Physic Garden, Chelsea, one of the most esteemed botanical gardens in Europe during the 18th century. He became an official demonstrator of the plants to the Society of Apothecaries in 1718, and its first director in 1724. During his tenure, he produced two catalogues of the garden and, in 1739, *Horti medici Chelseiani Index Compendiarius,* a 214-page alphabetical list of plants in the garden. Each year, new plants were grown at the garden, 50 of which were given to the Royal Society, of which he became a Fellow in 1739. Linnaeus met Rand when he visited the garden in 1736, and 10 years after Rand's death, named a genus after him.

Rangaeris (Orchidaceae) (Schltr.) Summerh.
An anagram of *Aerangis,* genus name of a related orchid.

Ranunculus (Ranunculaceae) L.
La. *rana* = frog; *unculus* = diminutive, hence 'a little frog or tadpole'; referring to the wetland, marshes or running streams that many frogs inhabit. Many species of *Ranunculus* can be found near water.

Rapanea (Myrsinaceae) Aubl.
Perhaps derived from the native Guinean name. Meaning unknown.

Raphanocarpus *** = *Momordica***
(Cucurbitaceae) Hook.
The genus *Raphanus* (q.v.); *karpos* = fruit.

Raphanus* (Brassicaceae) L.
Gk. *raphanos* = quick appearing; referring to the rapid germination and growth of the genus. Radish.

Raphia (Arecaceae) P. Beauv.
Gk. *rhaphis, rafis, raphia* = a needle; almost certainly referring the sharply pointed fruits, but could also refer to the long, pointed fronds of the palm, or perhaps from the Malagasy word *rafia,* which describes the fibre derived from the leaf stalks of *R. ruffia*.

Raphidophyllum* = *Striga*
(Orobanchaceae) Hochst.
Gk. *rhaphis* = a needle; *phyllon* = leaf. The leaves are as sharp as needles.

Raphionacme (Apocynaceae) Harv.
Gk. *rhaphys* = needle or beetroot; *akme* = sharpness; referring either to the leaves on short stalks that end in a sharp point, or to the bitter taste of the tuberous roots.

Raphistemma (Apocynaceae) Wall.
Gk. *rhaphis* = needle; *stemma* = garland, wreath; referring to the linear scales of the crown.

Rapistrum* (Brassicaceae) Crantz
Gk. *rhapis,* La. *rapia* = rape-flower or turnip; *astrum* = of inferior resemblance, appearance; implies inferiority of wild plants compared to cultivated plants.

Rapuntium* = *Lobelia* (Lobeliaceae) Mill.
La. *rapuntium,* from La. *rapum, rapa* = turnip; referring to the shape of the root.

Raspalia = *Brunia* (Bruniaceae) Brongn.
For François Vincent Raspail (1794–1878), French botanist, politician, physician and naturalist. A republican, with strong political and moral beliefs, he campaigned for social justice, freedom of the press, prison reform, and better working conditions in factories. He was wounded in revolutionary battle, jailed and banished from France for 10 years to Belgium. He was one of the originators of cell theory in biology and an early proponent of the use of the microscope for the study of plants. Among his publications were *Essai de Chimie Microscopique* (*Testing the Chemistry Microscope*) (1830), Nouveau *Système de Chimie Organique* (*New System for Organic Chemistry*) (1833), and *Manuel Annuaire de la Santé* (*Manual Directory of Health*) (1845), which promoted the use of hygiene and antiseptics by the general public.

Rauvolfia (Apocynaceae) L.

A Swamp Forest tree with glossy leaves (EJM)

For Leonhard Rauwolf (Rauwolff) (1535–1596), German physician and botanist, born in Augsburg. He studied at Montpellier and at Valence where he received his MD degree. He became city physician at Augsburg in 1570. In 1573, he went on a 33-month field trip to Marseilles and the Near East, visiting Syria, Palestine, Armenia, Mesopotamia, Assyria and Chaldea. His aim was to find new medicinal plants and drugs that could be traded profitably by his firm. He was the first European to describe the preparation and drinking of coffee in Aleppo. The alkaloids in *Rauvolfia serpentina* (Plum.) are still in medical use today. In 1596, he joined the imperial troops fighting the Turks in Hungary, where he died.

Ravenala* (Strelitziaceae) Adans.
Malagasy name for the plant *Ravinala,* meaning leaves of the forest.

Rawsonia (Achariaceae) Harv. & Sond.
For Rawson William Rawson (1812–1899), British colonial administrator, pteridologist, conchologist, fern collector, traveller, and Cape Colonial secretary (1854–1864). He served on the board of trustees for the South African Museum, rendered assistance to the initiation of *Flora Capensis*, and persuaded the Cape government to buy Pappe's herbarium on the latter's death. His personal

A forest under-shrub with tough leaves (EJM)

herbarium of 2 000 ferns collected in Mauritius, the Cape and West Indies was purchased by the British Museum in 1900. Rawson became governor of the Bahamas in 1864, Jamaica in 1865, and the Windward Islands between 1869 and 1875.

Rehmannia (Orobanchaceae) Libosch. ex Fisch. & C.D.Mey.
For Matthew Augustine Joseph Rehmann (1779–1831), German physician who studied medicine

in Vienna (1794–1801). He became a practitioner of Count Rasumovskij, the Russian ambassador in Vienna, and followed him to St Petersburg. From 1805 to 1806, he was the doctor on the embassy tour of Count Yuri Aleksandrovich Golovkin (1762–1846) to China, but only went as far as the Mongolian-Chinese border. After his return, he became a general practitioner in Moscow before being appointed in 1810 as a professor at the University of Moscow. In 1812, he moved to St Petersburg where he was appointed councillor, and two years later he became personal physician to Emperor Alexander I (1777–1825). In 1821, he became head of the Russian Civil Medical System and full professor of pharmacology at the Medico-Surgical Academy in Moscow (1821–1831). He died from cholera in 1831.

Reichardia *** = ***Caesalpinia*** (Fabaceae) Roth
For Johann Jakob (Jacob) Reichard (1743–1782), German physician and botanist. From 1768, he was a civil doctor at the hospital in Frankfurt am Main, and was also responsible for Frankfurt's first botanical garden, created in 1773–1774 by the Senckenberg Foundation (founded by Johann Christian Senckenberg, 1707–1772). This garden served as a *hortus medicus* for the cultivation of medicinal herbs for the foundation's public hospital and medical institute. Reichard authored *Flora Moeno-Frankfurtana* (*The Flora of Frankfurt*) (1772–1778) and *Syolloge, Opusculorum Botanicorum* (1782); edited a medical weekly paper (1780–1791); and also edited and co-authored several of Linnaeus's works, including *Genera Plantarum* (1778) and *Sytema Planetarium* (1787).

Reissantia (Celastraceae) N.Hallé.
For a rocky island, Reissant, off the coast of South Finistère, France. The author compares the branching of the plant to the debris of a shipwrecked wooden fishing vessel.

Relhania (Asteraceae) L'Hér. emend K.Bremer.
For Reverend Richard Relhan (1754–1823), Irish botanist, bryologist, lichenologist and plant collector. He was born in Dublin, educated at Trinity College, Cambridge, where he obtained an MA degree (1779), and became chaplin of Kings College, Cambridge. While there, he studied and collected plants in the neighbourhood, publishing a book *Flora Cantabrigiensis* (1785). He became a Fellow of the Royal Society (1787) and one of the founders of the Linnaean Society (1787). Later, he became rector in Lincolnshire and devoted himself to a scholarly study of the works of Tacitus.

Rendlia = ***Microchloa*** (Poaceae) Chiov.
For Alfred Barton Rendle (1865–1938), British botanist, assistant keeper (1888–1906), then

keeper (1906–1930) in the department of botany at the British Museum. A graduate of St John's College, Cambridge (MA 1891) and London University, (DSc 1898), he lectured in botany at Birkbeck College (1894–1906). He was botanical editor of *Encyclopaedia Britannica* (1911), editor of the *Journal of Botany* (1924–1938) and wrote *The Classification of Flowering Plants* (Volume 1, 1904, Volume 2, 1925), also co-authoring five other books. He served on the editorial committee dealing with botanical nomenclature (1905–1935). He became

a Fellow of the Royal Society (1909), president of the botanical section of the British Association (1916), and president of the Linnaean Society (1923–1927).

Renealmia* (Zingiberaceae) L.f.
For Paul Reneaulme (Reneaume, Reneaulm) (1560–1624), French botanist and eminent

physician of Blois who botanised mainly in Switzerland, southern France and Italy, as well as in Bois and Paris. He had an interest in the medical use of plants and, in 1606, wrote a book on pharmacopoeia, which he had to withdraw as a result of a lawsuit forbidding the publication of the recipes in his book. This publication also described various plants indigenous to America, including tobacco. His major work was *Historia Specimen Plantarum* (1611), a book of 150 pages of text, in Latin and some Greek, featuring 108 plants, with 24 superbly illustrated pages made from woodcuts etched by Reneaulme himself.

Rennera (Asteraceae) Merxm. emend P.P.J.Herman
For Otto Renner (1883–1960), German botanist, geneticist and bacteriologist,

professor of plant physiology and pharmacognosy (1913) in Munich and professor of botany at Jena (1920–1948), where he was also director of the Botanical Garden of Jena. He undertook research in Algeria (1914) and Java, Bali, Sumatra and Ceylon (1930–1931). In addition to his numerous publications, he was the editor of *Flora* from 1933–1944. His laboratory was totally destroyed during the war. Thereafter, he edited *Plant* from 1947–1956 and *Fortschritte der Botanik* (Botanical Progress) from 1949–1955. Following the work of Erwin Baur (1875–1933), Renner established the theory of maternal plastid inheritance as a widely accepted genetic theory.

Requienia (Fabaceae) DC.
For Esprit Requien (1788–1851), French botanist, palaeontologist, malacologist, traveller, botanical explorer, director of the Botanical Garden of Avignon, and historical conservationist. Early in his life, he explored the Mont Ventoux, southern France and Corsica, and formed large collections of plants, minerals, shells and fossils. He developed a herbarium of international importance

and discovered many taxa. He was a director of the Calvet Museum in Avignon for 34 years, where he founded the department of natural history. In 1840, he bequeathed his natural history collections and library reference collection to the Calvet Museum.

Reseda (Resedaceae) L.
La. *resedo* = heal, calm, appease. Name given by Pliny the Elder to a species of mignonette that was believed to heal external bruises.

Resedella* = **Oligomeris** (Resedaceae) Webb & Berthel.
La. *resedo* = heal; *-ella* = diminutive.

Resnova = **Ledebouria** (Hyacinthaceae) Van der Merwe
La. *resnova* = something new; referring to a new genus.

Restio (Restionaceae) Rottb.
La. *restis* = rope, cord-like; referring to a common use for the plant. Restios is known to have been used for rope.

Retzia (Stilbaceae) Thunb.
For Anders Jahan Retzius (1742–1821), Swedish botanist, lichenologist, bryologist and

entomologist. He studied to become an apothecary, and became an associate professor of chemistry (1766), professor of natural history at the University of Lund (1767), and professor of natural science (1795). He also did work in the fields of zoology, mineralogy and palaeontology. During his career, he described many new species of insects and did fundamental work on their classification. Retzius was elected a member of the Royal Swedish Academy of Sciences 1782.

Reyemia (Scrophulariaceae) Hilliard
For Heinrich Meyer (fl 1861–1886), German medical practitioner and amateur botanist. He graduated as an MD from the University of Würzburg (1861), then emigrated from Germany to the Cape and practiced medicine in Calvinia in the 1860s, later practising medicine at Cape Town in the 1880s. He collected in the Hantam Mountains, which are in the Namaqualand region of the Northern Cape Province where he found several new species which were sent to Berlin.

Reynoutria **=** ***Fallopia*** (Polygonaceae) Houtt. For Karel van Sint Omaars (Omaers) (c 1532 or 1533–1569), Flemish nobleman, botanist, and botanical illustrator, otherwise known as Charles de Saint Omer van Dranouter, Renouteren, Reynoutre van Reynoutre in West Flanders. He studied at the University of Leuven, and entered the military before becoming incapacitated by disease. He spent the rest of his life in the castle of Moerkerke, and was devoted to botany and botanical painting. Carolus Clusius commissioned him to illustrate his book *Centuriae Plantarum Rariorum,* but civil war and Omaars's early death prevented full completion. Most of the paintings that make up a collection called *Libri Picturati,* created between 1564 and 1569, and now housed in Krakow, were commissioned by Karel van Sint Omaars, painted for the greater part by Jacques van Corenhuyse, annotated by Clusius, and afterwards rearranged and supplemented by Karel van Arenberg.

Rhabdostigma ***** =** ***Kraussia*** (Rubiaceae) Hook.f.
Gk. *rhabdos* = rod, rod-like, columnar; stigma. The stigmas are described in the protologue as being fusiform-fluted, which would give them the appearance of tiny Greek pillars.

Rhadamanthus ***** =** ***Drimia*** (Hyacinthaceae) Salisb.
Gk. *rhada* = slender; *anthos* = flower; also the name of a mythological god.

Rhagodia **=** ***Einadia*** (Amaranthaceae) R.Br.
Gk. *rhagos* = a berry; *rhagodes* = grape-like; *-odes* = resembling; referring to the fruit berries.

Rhamnus (Rhamnaceae) L.
Celtic *ram,* later Gk. *rhamnos* = tuft of branches; referring to about 150 species of shrubs and trees generally known as Blackthorn.

Rhamphicarpa (Orobanchaceae) Benth. emend Engl.
Gk. *ramphos* = a curved beak; *karpos* = fruit. The fruit is usually a beaked capsule.

Raphionachme (Apocynaceae) Harv.
Gk. *rhaphis* = beetroot; *akme* = sharpness; referring to the edible tuberous root of this species.

Rheome ***** =** ***Moraea*** (Iridaceae) Goldblatt
An old name, semi-anagram of *Homeria.*

Rhigiophyllum (Campanulaceae) Hochst.
Gk. *rhigios* = stiff (Eng. rigid); *phyllon* = leaf; referring to the leaf structure.

Rhigozum (Bignoniaceae) Burch.

(EJM)

(EJM)

Gk. *rhigos* = cold, hence rigid, stiff; *ozos* = a branch; alluding to the rigid branches held stiffly out from the stem, a characteristic of this species.

Rhinacanthus (Acanthaceae) Nees
Gk. *rhinos* = nose, snout; *akanthos* = thorn; genus *Acanthus.* Presumably snouted acanthus.

Rhinanthus (Orobanchaceae) L.
Gk. *rhinos* = nose, snout; *anthos* = flower; alluding to the flower's upper labellum, which forms a hooked nose.

Rhinephyllum (Aizoaceae) N.E. Br.
Gk. *rhine* = file, rasp; *phyllon* = leaf; referring to the succulent leaves' rough surface.

Rhinolobium* = **Aspidoglossum**
(Apocynaceae) Arn.
Gk. *rhinos* = nose, snout; *lobion* = a little lobe, pod, capsule.

Rhipidodendron** (Asphodelaceae) Willd.
Gk. *rhipis* = bellows; *rhipidi* = fan, bellows, fan-like structure; *dendron* = tree; possibly referring to the often circular-shaped, fan-like flowers.

Rhipidoglossum (Orchidaceae) Schltr.
Gk. *rhipidi* = fan, fan-like structure; *glossum* = tongue; referring to the very rough, papillose, rasp-like roots.

Rhipsalis (Cactaceae) Gaertn.

(EJM)

Gk. *rhipsalis* = wickerwork-like. Many species of this genus have slender, pliant stems resembling flexible wickerwork.

Rhizoglossum = **Ophioglossum**
(Ophioglossaceae) C.Presl
Gk. *rhiza* = root; *glossum* = tongue; alluding to the shape of the roots (or fronds), which are tongue-like.

Rhizophora (Rhizophoraceae) L.
Gk. *rhiza* = a root; *phora* = bearer; referring to the mangrove trees that are borne up on stilt-like roots.

Rhodocoma (Restionaceae) Nees
Gk. *rhodon* = rose-red; *kome* = hair, tuft of hair; refers to the terminal clusters of the reddish inflorescence.

Rhizophora (EJM)

Rhododendron (Ericaceae) L.
Gk. *rhodon* = rose-red; *dendron* = tree; referring to a genus with more than 1 000 species, many with showy flowers, not all pink in colour.

Rhodohypoxis (Hypoxidaceae) Nel.
Gk. *rhodon* = rose; *Hypoxis* (q.v.); referring to its pink and white flowers and anthers borne at two levels.

Rhoiacarpos* (Santalaceae) A.DC.
Gk. *rhoia* = pomegranate; *karpos* = fruit. The fruit is a red drupe topped with a persistent perianth.

Rhoicissus (Vitaceae) Planch.
Gk. *rhoia* = pomegranate; *kissos* = ivy; *kissus* = ivy. Most plants of this genus climb and have tendrils, but the reference to pomegranate is obscure.

Rhombonema* = **Parapodium**
(Apocynaceae) Schltr.
Gk. *rhombos* = a rhombus, equilateral parallelogram; *nema* = a thread (filament); possibly referring to the shape of the filament.

Rhombophyllum (Aizoaceae) Schwantes
Gk. *rhombos* = a rhombus, equilateral parallelogram; *phyllon* = leaf; referring to the rhomboid- (diamond-)shaped leaves.

Rhopalocyclus = *Leipoldtia* (Aizoaceae) Schwantes
Gk. *rhopalon* = a club, stick; *cyclos* = circular; referring to the circular mass of succulent club-shaped leaves.

Rhopalota = *Crassula* (Crassulaceae) N.E.Br.
Gk. *rhopalon* = a club. The stem and branches are club-shaped.

Rhus* = *Searsia* (Anacardiaceae) L.
Gk. *rhous*, from *rhodos* = red (the word can be traced back for centuries); referring to the fruits or autumn leaves of some species.

Rhynchelytrum* = *Melinis* (Poaceae) Nees
Gk. *rhynchos* = snout; *elytron* = cover; referring to beaked glumes and lemmas.

Rhynchocalyx (Rhynchocalycaceae) Oliv.
Gk. *rhynchos* = beak, snout; *calyx* = a shallow bow. The calyx is beaked-shaped in bud.

Rhynchocarpa* = *Kedrostis* (Cucurbitaceae) Schrad. ex Endl.
Gk. *rhynchos* = beak, snout; *karpos* = fruit; referring to the shape of the fruit of this species.

Rhynchopsidium (Asteraceae) DC.
Gk. *rhynchos* = beak, snout; *-opsis* = resembling; *-ium* = little; alluding to the shape of the fruit.

Rhynchosia (Fabaceae) Lour.
Gk. *rhynchos* = beak, snout, horn; referring to the keel of the flowers.

Rhynchospora (Cyperaceae) Vahl
Gk. *rhynchos* = a beak; *spora* = seed; alluding to the beaked achenes.

Rhynea = *Tenrhynea* (Asteraceae) DC.
For Willem ten Rhyne (Wilhelmi ten Rhijne) (1647–1700), Dutch physician with the Dutch East India Company. He visited the Cape on his way to Deshima, Japan, wrote *An Account of the Cape of Good Hope and the Hottentotes*, describing the lives of the Khoisan (then 'Hottentots') during the early days of Dutch settlement, visited Japan, and was allegedly consulted by the Japanese emperor, who was very ill, but recovered. Ten Rhyne spent most of his life engaged in botanical and medical studies. He contributed to the great work *Hortus Malabaricus* (*Flora of Kerala* – a state in southwestern India). He was the first European to give an account of the Japanese practices of acupuncture and moxibustion, as well as a study of tea.

Rhyssolobium (Apocynaceae) E.Mey.
Gk. *rhyssos* = wrinkled, shrivelled; *lobion* = small lobe, pod, fruit; referring to the follicles.

Rhytachne (Poaceae) Desv.
Gk. *rhytos* (equiv. to *rhyssos*) = wrinkled; *achne* = glume, scale; referring to the wrinkled appearance of the glumes.

Rhyticarpus (Apiaceae) Sond.
Gk. *rhytos* (equiv. to *rhyssos*) = wrinkled; *karpos* = fruit; referring to the fruit's appearance.

Rhytiglossa* = *Isoglossa* (Acanthaceae) Nees
Gk. *rhytos* = wrinkled; *glossa* = tongue. The palate of the lower lip is wrinkled.

Ribes (Hamamelidaceae) R.Br.
Arabic or Persian (Syrian or Kurdish) *ribas* = acid tasting; referring to the sharpness in the taste of some fruit.

Richardia (Rubiaceae) L.
For Richard Richardson (1663–1741), English botanist, physician, county magistrate, and collector of valuable botanical and historical books. He studied at Oxford, seemingly without graduating, then for three years at the Dutch University of Leiden in the 1680s, where he trained as a physician while lodging with Paul Hermann, professor of botany. From 1687, he practised as a physician but spent more time on botany. His special interest was bryophytes (mosses and lichens). His garden was one of the best in England, and included one of the first hothouses in the country. He was elected a Fellow of the Royal Society in 1712.

Richardia* = *Zantedeschia* (Araceae) Kunth
For Louis Claude Marie Richard (1754–1821), French botanist and author, father of noted botanist Achille Richard (1794–1852). Between 1781 and 1789, he collected botanical specimens in Central America and the West Indies. He became a professor at the École de Médecine in Paris. He authored *Demonstrations Botaniques* (1808), *De Orchideis Europaeis* (1817), *Commentatio Botanica de Conifereis et Cycadeis* (1826), and *De Musaceis Commentatio Botanica* (1831), and

is noted especially for his work on orchids, coining the terms 'pollinium' and 'gynostemium'.

Richardsonia* = *Richardia* (Rubiaceae) Kunth
Richardia (q.v.). In 1818, Carl Sigismund Kunth announced that the rubiaceous plants *Richardia* should be called *Richardsonia*. This has since been rejected.

Richtersveldia (Apocynaceae) Meve & Liede.
For the Richtersveld, Northern Cape, where it occurs.

Ricinocarpos (Euphorbiaceae) Desf.
La. *ricinus* = a tick; *carpos* = fruit. These trees look superficially like the castor oil plant (*Ricinus communis*), but are much larger. Also, the seeds resemble ticks.

Ricinodendron (Euphorbiaceae) Müll.Arg.
La. *ricinus* = a tick; Gk. *dendron* = a tree; The tree's seeds are thought to look like ticks.

Ricinus* (Euphorbiaceae) L.

(EJM)

La. *ricinus* = a tick; referring to the seeds of this genus, which resemble a Mediterranean sheep tick of the same name.

Rikliella* = *Lipocarpha* (Cyperaceae) J.Raynal
For Martin Albert Rikli (1868–1951), Swiss botanist, phytogeographer, traveller and explorer. In 1908, he did polar exploration with H Bachmann on the west coast of Greenland and, in 1912, explored the Caucasus and many of its summits. He became assistant professor of plant geography from 1899 to 1930, and adjunct professor (1909) at the Swiss Federal Institute of Technology (Eldgenössische Technische Hochschule), Zurich, which involved also having the responsibility of being curator of the institute's botanical museum.

He authored of more than 20 publications relating to botany, travel and higher education.

Rimaria* = *Vanheerdea* (Aizoaceae) N.E.Br.
La. *rima* = cleft, crack; referring to the nature of the upper surface, which has a shrivelled appearance during winter as it nourishes new growth when the surface splits, revealing the new growth in spring.

Rinorea (Violaceae) Aubl.
Gk. *rhinos* = nose; *oros* = mountain. With imagination, the flower has a nose-like shape with an exerted anther and grows in tropical, hilly rainforests. Author Jean Baptiste Aublet does not explain the meaning of this name.

Riocreuxia (Apocynaceae) Decne.
Alfred Riocreux (1820–1912), French artist and botanical illustrator. He was trained by his artist father, and showed precocious talent. His drawing and paintings as a 13-year-old were so good that the State Porcelain Factory at Sevres, where his father worked, considered conserving them. The botanist Adolphe Théodore Brongniart (1801–1876) saw his sketches and was probably responsible for bringing him to the Paris Muséum d'Histoire Naturelle. TG Hill, author of *The Essentials of Illustration* (London, 1915), has described Riocreux's work on seaweeds for a work by Gustave Thuret as 'the finest plates ever published in a botanical work', while the distinguished Dutch botanist FA Stafleu, writing in 1966, commented: 'Riocreux was one of the great botanical artists of all time'.

Ritchiea (Capparaceae) R.Br. ex G.Don
For Joseph Ritchie (1792–1819), English surgeon, explorer and naturalist. In 1818, he and George Francis Lyon (1795–1832), a naval officer who had explored in both the Arctic and Africa, were sent by Sir John Barrow (1764–1848), second secretary of the admiralty, to find the course of the Niger River and the location of Timbuktu. Barrow decided they should depart from Tripoli and cross the Sahara as part of their journey. The expedition was underfunded, poorly supplied, and burdened with ill-selected merchandise that proved unsalable. After long delays, due to officialdom, the expedition reached Tripoli in March 1819, and 39 days later, Murzuk in Libya bordering on the Murzuk Desert. There they both fell ill, and Ritchie died.

Rivina* (Phytolaccaceae) L.
For Augustus Quirinus Rivinus (1652–1723), born August Bachmann, German botanist and physician, professor of botany and physiology at the University of Leipzig (1691), and also curator

of the university's garden of medicinal plants. In 1701, he became professor of pathology and, in 1719, professor of therapeutics and permanent dean of the faculty of medicine. That same year, he became a Fellow of the Royal Society. He authored *Introductio Generalis in rem Herbariam* (Leipzig, 1690), a classification of plants based on the structure of flowers, mainly the shape of the corolla, as well as three other books on plant orders.

Robinia* (Fabaceae) L.
For Jean Robin (Ioannes Robinus) (1550–1629), French botanist, royal gardener and herbalist

to King Henry III, Henry IV and Louis XIII of France. He worked at the Jardin des Plantes in Paris. He published a catalogue of 1 300 species grown in 1601 under the title *Catalogus Stirpium tam Indigenarum Exoticarum Quam*. The Paris Faculty of Medicine appointed Jean Robin in 1597 to supervise the construction of a botanical garden. Trees planted early in the 17th century by Robin, one at Jardin des Plantes and the other off the north facade of the church of Saint Julien-le-Pauvre near the Notre Dame, still stand.

Robsonodendron (Celastraceae) R.H.Archer
For Norman Keith Bonner Robson (1928–), English botanist and author who worked for the British Museum (1962–1988). He made important botanical expeditions to Morocco, Zambia and Malawi between 1958 and 1959. He authored several works: *New Research in Plant Anatomy* (1970) with Norman Keith Bonner, DF Cutler and M Gregory; *Ochnaceae* (1973) with JG Garcia; *Botanical Prints* (1989); *Celastraceae: Flora of Tropical East Africa* (1994) with N Halle, B Mathew and R Blakelock; and others. Between 1977 and 2012, he described all 490 species of *hypericum*. He was made a Fellow of the Linnaean Society in 1960. La. *dendron* = tree.

Rochea = Crassula (Crassulaceae) DC.
For Daniel de la Roche (Delaroche) (1743–1813), Swiss botanist and physician. De la Roche studied in Geneva, Leiden and Edinburgh, where he stayed until 1771, then practised in Geneva until 1881. He moved to Paris in 1782 as physician to the Swiss guards and consulting physician to the Duke of Orleans. When the 1792 revolution broke out, he emigrated to London, then Lausanne, returning to Paris in 1798 where he worked in various hospitals and on vaccines for smallpox. He gave botanical help to Augustin Pyramus de Candolle (1778–1841), who named the genus *Rochea* after him in 1802. He died from typhus brought to Paris by Napoleon's troops returning from Russia, as did his doctor son, François Etienne (1780–1813).

Rochelia* (Boraginaceae) Rchb.
For Anton Rochel (1770–1847), Austrian surgeon and botanist. He studied medicine at the Medical Faculty in Vienna and qualified as a surgeon in 1792. He served as surgical assistant in the Austrian army from 1788–1798. For the next 22 years, he worked in Moravia in various capacities such as being personal physician for Baron Alojza Mednyánsky and Count Aspremonta, where he became not only interested but an expert in the flowers in what is now Slovakia. He also had an interest in entomology. From 1821 to 1840, he was curator of the University Botanical Gardens in Budapest, which he improved substantially. Rochel authored *Flora of Banat* (1828), an eastern province of (old) Hungary, now part of Romania, and *Naturhistorische Miscellen über den Nordwestlichen Karpath in Ober-Ungarn* (1821) (loosely translated as *'A Natural History Journey into the Nothwestern Parts of Hungary'*). He was made a member of the Botanical Societies of Regensburg and Marktbreit in Germany.

Roella (Campanulaceae) L.

The bell-shaped flowers are characteristic of the family. (CM)

For Willem (Wilhelm) Röell (1700–1775), Dutch professor of anatomy in Amsterdam and horticulturist. He commenced his medical studies in 1718 and graduated at the University of Leiden in 1725. In 1731, he succeeded Frederick Ruysch as *praelector anatomiae*. This proved to be a poor choice as he neglected his teaching duties and the quality of training was poor, thus

earning the ire of the Surgeons Guilds in Delft and Amsterdam. In 1754, he became a governor of the West Indian Company and owned two plantations in British Guiana. In 1755, he was made professor of anatomy and surgery at the *Amsterdam Athaeum*, but resigned the same year because of ill health. He owned the mansion De Keukenhof at Lisse, where he kept a botanical collection, and donated specimens, *Lithophyta* (stone plants) and African seed to the Clifford collection.

Rogeria (Pedaliaceae) J.Gay ex Delile.
For Baron Jacques François Roger du Loiret (1787–1849), a lawyer and French colonial administrator who became governor of Senegal from 1821 to 1827. He studied Senegal's agricultural possibilities and tried to set up plantations, unsuccessfully, both because the Senegalese trading communities were reluctant to take up agriculture and because such attempts were always subject to petty harassment by neighbouring African states. In 1826, he explored much of the country with naturalist Georges Samuel Perrottet (1793–1870) and collected many plants. In 1828, he published *Fables Sénégalaises,* a collection of 43 Wolof (Senegalise language) folktales, with notes about Senegambia, its climate, main products, civilization and customs.

Roggeveldia* = *Moraea (Iridaceae) Goldblatt
Afrik. *roggeveld* = rye veld, an area that extends from Calvinia southward along the Roggeveld Escarpment to the foothills at Matjiesfontein in the Karoo. Referring to the location where it is found.

Rohria* = *Cullumia (Asteraceae) Thunb.
For Julius Philip Benjamin von Röhr (1737–1793), Prussian-born botanist, plant collector, medical doctor and watercolour painter. He took refuge in Denmark during the Seven Years' War (1756), and the following year was appointed surveyor in the Danish West Indies, with instruction to make collections in natural history. He sent many plants to Europe from South America and the West Indies. In 1784, he made an intensive study of cotton botany and agriculture under commission for King Christian VII. He published a treatise on this, which was also translated into French. During his career, he received the title Royal Naturalist. In the 1790s, Denmark was considering abolishing the slave trade. Von Röhr sailed from New York to Guinea with the idea of setting up a plantation colony, staffed by (free) blacks in the vicinity of the old Danish slaving forts, but died shortly after landing as a result of a malignant fever.

Romulea (Iridaceae) Maratti.

This species often occurs in marshy places. (CM)

For the legendary Romulus, founder and first king of Rome.

Rondeletia (Rubiaceae) L.
For Guillaume Rondelet (1507–1566), French physician, zoologist and ichthyologist, professor of medicine at the University of Montpellier (1545), who became chancellor of the Medical Faculty (1556). He studied Latin and philosophy in Paris (1525), and medicine at the University of Montpellier (1529). He learned Greek in Paris while living with Viscount Turenne as preceptor to his sons. He obtained his MD in 1537 and became physician to Cardinal Tournon, whom he accompanied on his journeys as ambassador to Italy and other countries. Rondelet's two main interests were medicine, particularly anatomy, and fish. He authored *Methodus Materia Medicinali et Compositionism Medicamentorum* (1556) and *De Piscibus Marinis* (1554 and 1558). He dissected his infant son in an attempt to discover his cause of death.

Roodebergia (Asteraceae) B.Nord.
For Roodeberg, Hex River Mountains; referring to the location where the type of species was found.

Roodia* = *Argyroderma (Aizoaceae) N.E.Br.
For Petrusa Benjamina Rood (1861–1946), South African plant collector. She attended school in Clanwilliam, and one of her friends was Marian van Wyk, who later became the wife of Rudolf Marloth (1855–1931), the noted pharmacist, analytical chemist and botanist. Rood became a keen collector of seeds, succulents and bulbous plants, which she sent to Kew Gardens and various South African collectors. She also sent grasses to General

Smuts. Many of her plants were illustrated and appeared in the early volumes of *The Flowering Plants of South Africa*.

Roridula (Roridulaceae) L.

(CM – Both)

La. *roridus* = dewy; *-ula* = diminutive. Probably referring to this carnivorous plant's possession of insect-trapping sticky hairs, which look 'dewy'.

Rorippa (Brassicaceae) Scop.
Latinised form of *rorippen,* a Saxon vernacular name for watercress used by Euricius Cordus.

Rosa* (Rosaceae) L.
Uncertain origin, denoting colour, Celtic *rhod* = red; Gk. *rhodo-* = rosy red; referring either to the rose genus or to the colour.

Rosenia (Asteraceae) Thunb. emend K.Bremer.
For the two Rosén brothers. Eberhard Rosén (Rosenblad, 1714–1796), Swedish physician, botanist, professor of practical medicine, and skilled humanist. He was a founder of Lund Hospital, and senior curator of the local mineral water well. He became rector of Lund University in 1754 and was knighted in 1770 when he took the name Rosenblad.

Nils Rosén (1706–1773) was a Swedish professor of medicine and anatomy at Lund University. He was physician to various kings of Sweden and a member of the Royal Swedish Academy of Sciences. He was made a knight of the Polar Star, and changed his name to Nils Rosén von Rosten. He is often considered the founder of paediatrics, and authored *The Diseases of Children, and their Remedy* (1764).

Rotala (Lythraceae) L.
La. *rotalis* = wheeled, wheel-like; referring to the whorled leaves.

Rotheca (Lamiaceae) Raf.
Originally named *Tsjerou-Theka* (the Malayalam name, *cheriya* = small; *thekku* = teak) as stated in Rheede's *Hortus Malabaricus,* later Latinised by Rafinesque as *Rotheca*.

Rothia (Fabaceae) Pers.
For Albrecht Wilhelm Roth (1757–1834), German botanist and physician. He studied medicine at

the universities of Halle and Erlangen, qualified in 1778, practised for a while, and then worked at the Botanical Institute, University of Jena. His major publications are regarded as his treatise of German flora *Tenetamen Florae Germanica* (1788–1800) and a book on Indian flora, *Novae Plantarum Species Praesertim Indiae Oriental* (1821), based largely on botanical specimens collected by the Moravian missionary Benjamin Heyne (1770–1819). He also wrote the third volume of *Catalecta Botanica* (1806) with Franz Carl Mertens (1764–1831), an algae plant collector who illustrated all the algae and described a number of algal species. In this book, Roth provided the first definition of algae.

Rothmannia (Rubiaceae) Thunb. (Image on following page)
For Jöran (Georg) Johansson Rothman (1739–1778), Swedish botanist, physician and Swedish translator. He was a student of Linnaeus who, in turn, had been a student of Jöran's father, Dr Rothman, during Linnaeus's senior school time in Växjö. Jöran Rothman obtained a BA (1757), MA (1761) and doctorate (1763) from the University of Uppsala. He was also a friend of Carl Peter Thunberg. He worked in Tunisia and Libya

Rothmannia: Like many of the species in the family the flowers are rather fragrant (gardenias). (EJM)

from 1773–1776, funded by the Swedish Royal Academy of Science to work. A lack of sufficient funds drove him to bankruptcy and forced his return to Stockholm, where he died shortly thereafter. His unpublished notes and diary, *Resa till år Tripoli from 1773 to 1776,* are housed in the Bergianska collections, Kungliga Vetenskaps Academies Stockholm. (He translated both Voltaire and Alexander Pope into Swedish.)

Rottboellia (Poaceae) L.
For Christen Friis Rottbøll (Rottboell, 1727–1797), Danish botanist and physician, traveller, professor of medicine from 1776 at the University of Copenhagen, and director of the Copenhagen Botanical Garden. As a student, he studied theology, then medicine at the University of Copenhagen, obtaining his doctorate in 1755, and completed further studies in chemistry, botany and medicine at Uppsala University under Linnaeus. Rottbøll made a significant contribution to the way smallpox was prevented or treated in Copenhagen, and as a botanist, he drew up the first comprehensive *Flora of Greenland.*

Roubieva*** = Dysphania (Amaranthaceae) Moq.
For Guillaume Joseph Roubieu (1757–1834), a French physician (anatomist) and professor of botany at Montpellier. He studied at Montpellier, obtaining his PhD in 1829. He authored a number of publications, including *Competitio ad Aggregationem, De Utilitate Chemiae in Medicina* (1829) and *Opuscules d'Anatomie d'Histoire Naturelle* (1816). The genus name was wrongly spelled.

Rourea = Cnestidium (Connaraceae) Aubl.
Named after the Aroura Parish, French Guiana (Hugh Glen). The plant is native to the woods of New Guiana.

Royena = Diospyros (Ebenaceae) L.
For Adriaan van Royen (1704–1779), Dutch botanist and physician, professor of botany (1732–1755) and later of medicine (1755–1775) at Leiden University, and director of the botanicalgarden at Leiden (1730–1754). He studied medicine and botany at the University of Leiden under Herman Boerhaave, and graduated as a medical doctor in 1728. He was a friend and close associate of Linnaeus (who published the genus in 1753), and a colleague of George Clifford, a wealthy English merchant and governor of the Dutch East India Company. Van Royen's major work was *Flora of Southeast Asia.*

Rubia (Rubiaceae) L.
La. *ruber* = name for madder (red); referring to the reddish dye obtained from the roots of this plant.

Rubus (Rosaceae) L.
Old name; also from *ruber* = red; referring to the colour of the fruit of some species. Latin name for brambles.

Ruckeria*** = Euryops (Asteraceae) DC.
For Johann Friedrich Rücker (17th century), German pharmacist and plant collector at the Cape. He was employed by the Dutch East India Company for five years prior to 16 5, and brought back seeds and bulbs for Christ al Mentzel (1622–1701), counsellor and physician to the elector of Brandenburg. Mentzel wrote a multilingual compilation of common names entitled *Index Nominum Plantarum Universalis Multilinguis* (1696), and mentions Rücker in this publication.

Ruellia (Acanthaceae) L.
For Jean Ruel (Jean de la Ruelle) (1474–1537), a French physician, who obtained his doctorate in medicine at the University of Paris in 1502 and became dean of the medical faculty from 1508 to 1510. He was called to be personal physician to François I of France, but tactfully declined, saying that it would interfere with his studies. In about 1512, after the death of his wife, he was

free to become a priest and was ordained canon of Notre-Dame de Paris. For the next 20 years, he dedicated his life to translating, commenting on and restoring the real text of the ancient Greek medical authors such as Dioscorides, Hippocrates, Galen, Euclid, Celsus and Pliny the Elder. He also wrote a treatise on botany *De Natura Stirpium Libri Tres* (1536), a massive work of some 666 pages.

Ruelliopsis (Acanthaceae) C.B.Clarke
Ruellia (q.v.); *-opsis* = resembling.

Rumex (Polygonaceae) L.
The Latin name for culinary sorrel, as used by Pliny the Elder, probably derived from *rumo* = to suck or I suck; alluding to the practice among Romans of sucking the leaves to allay thirst.

Rumohra (Dryopteridaceae) Raddi
For Karl Friedrich Felix von Rumohr (1785–1843), German art historian, art expert and writer,  collector of antiquities, author and poet. He studied foreign languages and the history of art at university in Göttingen, and later studied art and painting in Hamburg. From 1806 to 1821, he made a comprehensive study of Italian art. His major work was his three-volume *Italienischen Forschungen* (*Italian Investigations*) (1826–1831). He also wrote a book on culinary art, *Der Geist der Kochkunst,* a history of the Royal Danish collection of engravings, a four-volume novel (1832), stories, *Novellen* (1833–1835), and a satirical poem (1835). Towards the end of his life, he increasingly devoted time to agriculture and cooking. He was a member of the Berlin School of Art History.

Rungia** (Acanthaceae) Nees
For Friedlieb Ferdinand Runge (1795–1867), German analytical chemist. He obtained a  medical degree from Jena (1819) and a doctorate in chemistry from Berlin (1822). He taught chemistry shortly at the University of Breslau (now Wrocław, Poland.) From 1831–1852, he worked as a chemist in a chemical factory at Oranienburg. During his career, he analysed coffee and identified caffeine, examined the dilation effects of belladonna (deadly nightshade) on the eye, determined the various components of coal-tar oil, did research on synthetic dyes, developed a process for obtaining sugar from beet, and pioneered the concept of paper chromatography, about which he wrote two books.

Ruppia (Ruppiaceae) L.
For Heinrich Bernhard Rupp (Ruppius, Ruppee) (1688–1719), German physician and botanist. He studied medicine at the Universities of Giessen (1704), Jena (1711), Leiden (1712), and again in Jena (1713). Rupp wrote *Flora Jenensis* (1718), describing the flora around Jena and a large part of Thuringia (one of the federal states of Germany), with a second edition being published after his death (1726). A further expanded edition was completed in 1745 by Albrecht von Haller (1708–1777).

Ruschia (Aizoaceae) Schwantes

Many of the Aizoaceae have similar looking flowers. (CM)

For Ernst Julius Rusch (1867–1957), Namibian farmer, businessman and plant collector. He came to South-West Africa (Namibia) in 1890, where he grew succulent plants and later established a nursery at Lichtenstein, near Windhoek, Namibia. He and his son Ernst Franz Rusch Jnr (1897–1964) made many collecting trips together. He was one of the founders of Windhoek, and was given freedom of the city on his 60th birthday.

Ruschianthemum = **Stoeberia** (Aizoaceae) Friedr.
For Ernst Franz Theodor Rusch (1897–1964), farmer and businessman, son of Ernst Julius Rusch, who, like his father, collected succulent plants. Gk. *anthemon* = flower.

Ruschianthus ** (Aizoaceae) Bolus
Ruschianthemum (q.v.); *anthos* = flower.

Ruschiella (Aizoaceae) Klak
Ruschianthemum (q.v.); La. *-iella* = diminutive.

Ruscus* (Ruscaceae) L.
Derived from Anglo-Saxon = holly or box; referring to cladodes that have the appearance of stiff, spine-tipped leaves. Native to western and southern Europe (north to southern England), Macronesia, northwest Africa, and southwestern Asia, east to the Caucasus.

Ruspolia (Acanthaceae) Lindau
For Prince Eugenio Ruspoli (1866–1893), Italian explorer, ethnologist, naturalist, botanical and zoological collector, hunter, and son of the mayor of Rome. He authored *In the Land of Unexplored Africa and Mirra*, in Italian in 1892. He was sent by the Italian government to Ethiopia (1891–1893) as it wished to secure the country's colonial pretensions that Ethiopia would become a protectorate. In 1893, while leading an expedition to the Amara mountains in southwestern Ethiopia, a large elephant appeared at the foot of these mountains in an open plain. Ruspoli walked out to try to shoot the animal, which suddenly charged him and trampled him to death. He was 28.

Russelia = ***Vahlia*** (Vahliaceae) L.f.
For Alexander Russell (c 1715–1768), British physician and traveller. He qualified as a doctor at Edinburgh University (1732–1734) and worked in Aleppo, Syria, from 1740–1754 as appointed physician to the English factory, and soon became the principal practitioner in the city. He witnessed the outbreaks of the plague between 1742 and 1744, and the resultant collapse of the Syrian economy in the 1750s. He recorded the plague's progress in his diary, and an account of this is given in his *Natural History of Aleppo* (1756). In 1759, a vacancy occurred in St Thomas's Hospital, London, and Russell was elected physician to that institution, an office he held until his death. He was a Fellow of the Royal Society and a member of the Medical Society of London, to which he contributed several papers.

Ruta* (Rutaceae) L.
Gk. *rutos* = flowing (La. *rute*), perhaps from *ruomai* = to preserve; referring to the plant's medicinal value in maintaining health. English 'rue' (*R. graveolens* is 'herb of grace').

Ruthea** (Apiaceae) Bolle.
For Johann Friedrich Ruthe (Ruthé or von Ruthe) (1788–1859), German teacher, botanist and entomologist with a special interest in *Hymenoptera* and *Diptera*. He studied botany at the University of Berlin in 1811 and became a schoolteacher in Berlin and Frankfurt, but retired early in 1842 because of ill health. Among his publications was a textbook on zoology, *Handbuch der Zoologie*, with co-author Arend Friedrich Wiegmann (1802–1841), and *Flora der Mark Brandenburg und der Niederlausitz Pflanzen* (1827, 1834).

Ruttya (Acanthaceae) Harv.
For John Rutty (1697–1775), Irish physician, naturalist, entomologist, lichenologist, Quaker, and author. He studied medicine at Leiden University and practised as a physician in Dublin all his life. He took part in the activities of the Physico-Historical Society of Dublin, and directed the later activities of its botanical collector, Isaac Butler (f 1744). A prolific author, he wrote, *inter alia*, *A History of the Rise and Progress of the People called Quakers in Ireland, from 1653–1751* (1751); an *Essay towards a Natural History of the County of Dublin* (in two volumes) (1772), which included its flora, fauna, geology and meteorology; and a treatise on drugs in Latin, *Materia Medico, Antigua and Nova* (1775), on which he worked for 40 years.

Rytigynia (Rubiaceae) Blume
Gk. *rhytis* = wrinkle, fold; *gynaia* = little woman. Plants of this genus have wrinkled ovaries.

Rytidosperma (Poaceae) Steud.
Gk. *rhytis* = wrinkle, fold; *sperma* = seed; referring to the back of a larvae mistaken for the seed.

S

Searsia: These trees, previously call *Rhus*, occur throughout South Africa.

Rodney Moffett (1937–) retired in 2000 as professor of botany at the Qwaqwa campus of the University of the North. He is currently an honorary research associate of the department of plant sciences at the University of the Free State, where he is involved in research on the medicinal uses of plants by the Basotho and on biographies of Free State and Lesotho naturalists. Details of the acceptance of *Searsia* may be found in *Name Changes in the Old World* Rhus *and recognition of* Searsia (*Anacardiaceae*) in *Bothalia* 37(2): 165–175 (2007).

Saccidium = Holothrix (Orchidaceae) Lindl.
Gk. *sakkos* = bag; *sakkion* = small bag, pouch; referring to the saccate lateral sepals.

Sacciolepis (Poaceae) Nash
Gk. *sakkion* = small bag, pouch; *lepis* = scale; referring to the inflated shape of the upper glume.

Saccolabium (Orchidaceae) Blume
La. *saccus* = bag, pouch; *labium* = lip; alluding to the bag-like shape of the lip.

Sagenia = Tectaria (Tectariaceae) C. Presl
Gk. *sagēnē*, La. *sagena* = seine, fishing net; referring to the network of veins (veinlet anastomosis).

Sagina* (Caryophyllaceae) L.
La. *sagina* = fattening, nourishment; a word for fattening fodder. Some species of *Sagina* (e.g. *S. spergula* = 'spurrey') were thought to fatten sheep.

Saintpaulia*** (Gesneriaceae) H.Wendl.
For Baron Adalbert Emil. Walter Redcliffe Le Tanneaux von Saint Paul-Illaire (1860–1940),

German official in East Africa who in 1892 discovered this Tanzanian (then German East Africa) plant in the wooded Usambara mountains located southeast of Lake Victoria, near the border of Kenya. He sent seeds of the so-called 'African violet' (which resemble violets) to his father in Germany, Baron Ulrich von Saint Paul, who was an avid amateur horticulturalist. He not only sowed the tiny seeds in his greenhouse at his hometown of Fischbach and cultivated the plants until they bloomed, but took seeds and plant samples to Hermann Wendlend, Master Gardener of the Herrenhaus Greenhouses who described the plant in the journal *Gartenflora*, Berlin, and in *Möllers Deutsche Gärtner-Zeitung*, Erfurt, in 1893 (Edition 16, Volume 8).

Salacia (Celastraceae) L.
For Salacia, in Roman mythology the goddess of saltwater who presided over the depths of the ocean, wife and queen of Neptune. Relationship to the genus is unclear.

Salaxis*** = Erica (Ericaceae) Salisb.
La. *salax, salacis* = lustful or Gk. *salos* = unsteady; axis. The flowers are axillary, with stamens connate only on one side, giving the flowering branch an irregularly lopsided appearance.

Salicornia (Amaranthaceae) L.
Catalan *salicorn*, from La. *sali* = salt; *cornu* = horn; referring to the saline habitat and the horn-shaped stems.

Salix (Salicaceae) L.

(EJM)

La. *salia* = willow, implying to spring or leap. Willow branches are highly flexible and, when bent and released, spring forward.

Salpichroa (Solanaceae) Miers
Gk. *salpinx* = trumpet or tube; *khroia* = skin or complexion; referring to the form and texture of the flower.

Salpinctium*** = Asystasia (Acanthaceae) T.J.Edwards.
Gk. *salpinx* = trumpet; *salpinctēs* = trumpeter; referring to the flared mouth of the flowers.

Salsola (Amaranthaceae) L.
La. *salsola* = diminutive, of *sal* = salty. Plants mostly grow on coastlines and salt-impregnated soils. The name *salsola* is also Italian.

Saltera (Penaeaceae) Bullock
For Terence Macleane Salter (1883–1969), English paymaster-captain in the Royal Navy and amateur

botanist. He retired to the Cape in 1931, and for the next 30 years devoted most of his time to botany. He worked at the Bolus Herbarium, then in Kirstenbosch, and with Professor RS Adamson was co-editor of the *Flora of the Cape Peninsula* (1950), to which he contributed a major part. He was an authority on *Oxalis* and contributed many taxonomic articles to the *Journal of South African Botany* and *The Flowering Plants of South Africa*. Among various distinguished awards, he received an honorary DSc degree from the University of Cape Town in 1962.

Salvadora (Salvadoraceae) L.
For Juan (Joan) Salvador y Bosca (1598–1681), Spanish apothecary, botanist, and plant collector from Barcelona. As a young naturalist, Salvador collected animal and flora specimens in the Mediterranean wildlife areas such as Iberia, Catalan and the Balearic islands. Several generations of his family carried on his tradition. The herbarium Salvador at the Botanical Institute of Barcelona houses some 4 025 specimen sheets, one of the oldest, most well-documented and preserved collections in existence. Salvador's correspondence with the foremost botanists of his era, such as Herman Boerhaave, Sébastien Vaillant and the like, is also preserved.

Salvia (Lamiaceae) L.
La. *salvia* = the sage plant; a name used by Pliny the Elder, from La. *salvere* = to heal; referring to the medicinal properties claimed for some species.

Salvinia* (Salviniaceae) Ség.
For Anton Maria Salvini (1653–1729), Italian scholar, polyglot and translator. He studied law at the University of Padua to accommodate his father's wishes but thereafter studied Greek and Latin under the Italian naturalist and linguist Francesco Redi (1626–1697). He also learned French, English, Hebrew and Spanish. In 1677, Salvini was appointed as the chair of Greek at the Studio Fiorentino, and translated many Greek literary works and poems, as well as a few Latin, French and English works. His writing and poetry did not achieve the excellence of his translations. As a result of his erudition and linguistic skills, he became a member of the Accademia della Cruscia, the most prestigious language institution in Italy, and was called upon to comment on many translations.

Samolus (Theophrastaceae) L.
The ancient Latin name for this water-loving plant, possibly of Celtic origin.

Sandersonia = Littonia (Colchicaceae) Hook.
For John Sanderson (1820–1881), Scottish journalist, trader and plant collector. He worked for the Scottish press before coming to South Africa in 1850. From November 1851 to April 1852, he made an extensive inland trip as described in the *Journal of the Geographical Society* 30 (1860). He became editor of the *Natal Colonist,* office bearer of the Natural History Association of Natal, Durban Horticultural Society, and Natal Agricultural and Horticultural Society, of which he was president for 14 years. He was an active plant collector, and corresponded with Hooker at Kew Gardens and William Henry Harvey in Dublin, sending them herbarium material, seeds and bulbs. He wrote *Rough Notes of the Flora of Natal*, an appendix to James Chapman's *Travels* (1868).

Sanguisorba* (Rosaceae) L.
La. *sanguis* = blood; *sorbere* = to absorb; referring to its reputed ability to stop bleeding by causing coagulation. 'Burnet'.

Sanicula (Apiaceae) L.
La. probably from *sanus* = healthy; *-ula* = diminutive; alluding to its healing properties.

Saniella = Paurida (Hypoxidaceae) Hilliard & B.Burtt
Named after the 'Sani Pass' located in the west of KwaZulu-Natal, South Africa, a road between Underberg, KwaZulu-Natal and Mokhotlong, Lesotho. The 'Pass' itself is a challenging drive usually by 4x4 vehicles especially in bad weather. It rises 1332 metres from bottom to top.

Sansevieria (Ruscaceae) Thunb.
Named in error by Carl Peter Thunberg for Raimondo di Sangro (Sansgrio), prince of Sanservol (Sanseveria, Sanseviera) (1710–1771), Italian nobleman, inventor, soldier, writer and scientist. He studied at a Jesuit college in Rome, and spoke many European languages, including Hebrew and Arabic. He was a prolific inventor of, *inter alia*, a lightweight cannon that could outshoot standard ones, a hydraulic device that could pump water to any height, a printing press that could print many colours in a single impression, and a wooden carriage that could travel on both land and water, among others. The discoverer Vincenzo Petanga of Naples actually wanted the plant to be named after Pietro Antonio Sanseverino (1724–1771), Duke of Chiaromonte. He established a garden of rare and exotic plants in the south of Italy. No further details are available.

Santolina (Asteraceae) L.
La. *sanctus* = holy; *linum* = flax; referring to the herb's reputed medicinal properties. There are other derivations. Unfortunately, Linnaeus does not explain the name.

Saphesia (Aizoaceae) N.E.Br.
Gk. *saphos* = distinct. The plant is distinct from every other known species in the family (NE Brown).

Sapindus* (Sapindaceae) L.
La. *sapo* = soap; *indicus* = Indian. The fruit pulp of *S. saponaria* contains saponin, which was used by Native Americans to make soap.

Sapium (Euphorbiaceae) P.Brown
La. *sapinus* = smelling like a fir tree or pines. The word is derived from an ancient Latin name for a resinous pine whose sap lathers like soap. This genus, which is not related to resinous pines, has been historically cultivated as a seed-oil crop for soap production and other tallow products, hence its name.

Saponaria* (Caryophyllaceae) L.
La. *sapo* = soap; *aria* = pertaining to sap. The bruised leaves of *S. officinalis* lathers like soap. 'Soapwort'.

Sarcocalyx*** = Aspalathus (Fabaceae) Walp.
Gk. *sarx, sarkos* (*sarco*) = flesh; referring to the appearance of the calyx.

Sarcocaulon = Monsonia (Geraniaceae) (DC.) Sweet
Gk. *sarx, sarkos* (*sarco*) = flesh; *kaulon* = stem; referring to the fleshy appearance of the stem.

Sarcocolla*** = Saltera (Penaeaceae) Kunth
Gk. *sarx, sarkos* (*sarco*) = flesh, fleshy; *kolla* = glue. The fleshy stem just below the inflorescence is sticky.

Sarcocornia (Amaranthaceae) A.J.Scott
Gk. *sarx, sarkos* (*sarco*) = flesh, fleshy; La. *cornis* = horned; referring to the appearance of fleshy lobes.

Sarcocyphula = Cynanchum (Apocynaceae) Harv.
Gk. *sarx, sarkos* (*sarco*) = flesh, fleshy; *skyphos* = drinking bowl; *-ula* = diminutive; referring to the cup-shaped fruiting bodies (*apothecia*).

Sarcolipes** (Crassulaceae) Eck. & Zeyh.
Gk. *sarkolipes* = with a fleshy base or a fleshy foot-stalk, 'forsaken by flesh'. The entire plant is fleshy.

Sarcophagophilus = Quaqua (Apocynaceae) Dinter
Gk. *sarx, sarkos* (*sarco*) = flesh, fleshy; *phagein* = to eat; *philos* = to love, loving; referring to its parasitic love of flesh.

Sarcophyllus = Aspalathus (Fabaceae) Thunb.
Gk. *sarx, sarkos* (*sarco*) = flesh, fleshy; *phyllon* = leaf; referring to the thick, fleshy leaves.

Sarcophyte (Balanophoraceae) Sparrm.
Gk. *sarx* = flesh; *phyte, phyton* = plant; referring to the carnivorous nature of this parasitic plant.

Sarcostemma (Apocynaceae) R.Br.

(EJM)

Gk. *sarx, sarkos* (*sarco*) = flesh, fleshy; *stemma* = a wreath; alluding to the fleshy inner corona lobes of the flower and not from the twining, fleshy stems.

Sarothamnus = Cytisus (Fabaceae) St.Lag.
Gk. *saron* = brush, broom; *thammos* = bush; referring to the dense slender stems and small leaves. 'Broom'.

Sartidia (Poaceae) DeWinter
Anagram of *Aristida*.

Satureja (Lamiaceae) L.
The Latin name for savoury; from the Arabic *satta*, applied to several different labiates. 'Catmint' = savoury. The etymology of this word is unclear.

Satyridium = Satyrium (Orchidaceae) Lindl.
Gk. *Satyrium* (q.v.); *-idion* = diminutive.

Satyrium (Orchidaceae) Sw. (Image on opposite page)
Referring to the two-horned satyr, a demigod in Greek mythology, half man, half goat; possibly from *satyrion*, a name used by Dioscorides and Pliny the Elder for an orchid, *Aceras anthropophorum*, from the presumed aphrodisiacal properties possessed by the plant. The satyrs were closely associated with Dionysius. The allusion is to the two-spurred lip.

Savia (Euphorbiaceae) Willd.
For Gaetano Savi (1769–1844), Italian physician and botanist. He studied at the University of

Pisa. He was appointed curator of the Pisa botanical garden and museum (1791), became professor of physics and botany at the University of Pisa (1810), and director of the Third Botanical Garden of Pisa (1814–1843). He authored *Flora Pisana* (1798), *Trattato Degli Alberi della Toscana* (*Treatise on Trees of Tuscany*) (1801), *Botanicum Etruscum* (1808) and *Flora Italiana* (1818). In 1816, he was elected to the Royal Swedish Academy of Sciences. His son Paolo (1798–1871) became professor of natural history in 1823, also at Pisa, and contributed greatly to the study of the geological formation of Tuscany, as well as to ornithology.

Satyrium (HS)

Scabiosa (Dipsacaceae) L.
La. *scabios* = rough, scaly, from La. *scapies* = roughness, scurf, itch, referring to leprosy; alluding to the plant's supposed ability to cure cutaneous diseases and as remedies for relief from 'the itch' (scabies).

Scadoxus (Amaryllidaceae) Raf.
Gk. *scias* = that which you know; *doxus* = glorious, splendid; referring to a plant with magnificent bright orange-red flowers that look a bit like a shaving brush. Rafinesque, comments 'umb. Glor.', glorious umbel.

Scadoxus (MC)

Scaevola (Goodeniaceae) L.
For Gaius Mucius Scaevola, a 6th-century BCE Roman hero (known as 'Scaevola' meaning 'left-handed' from La. *scaevus* = left) who attempted to assassinate the Etruscan King, Lars Porsenna, who had Rome under siege, by sneaking out of Rome and infiltrating the enemy camp. By error, he killed the wrong man, was apprehended and dragged before Porsenna, where he admitted to his assassination attempt, saying 300 more Romans had sworn to kill the king. To demonstrate his resolution and indifference to physical suffering, Mucius thrust his right hand into a flame burning on an altar next to which he was standing. Impressed by such bravery, Porsenna decided to allow Mucius to return home to Rome. Subsequently, he sent envoys to Rome suing for peace.

Scaraboides (Apiaceae) Magee & B.E.van Wyk
Gk. *scarabeus* = beetle; -*oides* = resembling; referring to its beetle-like appearance in parts, such as involute marginal wings.

Sceletium = *Mesembryanthemum*
(Aizoaceae) N.E.Br.
Gk. *skeletos* = dried up, withered; referring to

the foliage that dies back after flowering to leave skeletonised remains.

Schefflera (Araliaceae) J.R. & G.Forst.
For Johann Peter Ernst von Scheffler (1739–c 1808) German physician and botanist of Gdańsk, and later of Warsaw, who contributed plants to Gottfried Reyger (1704–1788) for his book *Tentamen Florae Gedanensis Methodo Sexuali Adcommodatae* (1764), the first book in Germany that tried to describe local flora according to Linnaeus's 'sexual system' of classification. Linnaeus became convinced that all organisms reproduce sexually. He reasoned that each plant possessed male and female sexual organs (stamens and pistils) or 'husbands and wives', as he also put it. On this basis, he designed a simple system of distinctive characteristics to classify each plant.

Schinus* (Anacardiaceae) L.
Gk. *schinos* = a Greek word which seems to have an undefined meaning, used originally for *Pistacia lenticus,* the mastic tree.

Schinziophyton (Euphorbiaceae) Hutch. ex Radc-Sm.
For Hans Schinz (1858–1941), Swiss botanist and botanical collector. He studied natural history at Zürich Polytechnic, obtaining his PhD in 1883. He accepted a job as botanist to explore South-West Africa, now Namibia (1884–1886), going as far north as the Kunene River bordering Angola, where he also collected plants (1885). On his return to Zürich, he became director of the botanical garden in Zurich (1891), extraordinary professor (1892), and professor of systematic botany from 1895–1928. He published many articles on African plants, two parts of *Florae Africae* (with TH Durand) Volume 1, Part 2 (1898) and Volume 5 (1895), and edited *Vierteljahrschrift* for 26 years. He made Zurich one of the leading centres for the study of African plants.

Schismus (Poaceae) P.Beauv.
Gk. *schisma* = cleft, division; referring to the lemma tip, which is cleft in two, or alluding to the outer palea.

Schistanthe = Alansoa (Scrophulariaceae) Kunze
Gk. *schistos* = split, cut, divided; *anthos* = flower; perhaps referring to the serrated leaves.

Schistostephium (Asteraceae) Less.
Gk. *schistos* = split, cut divided; *stephos* = a crown; alluding to the deeply toothed ray florets.

Schizachyrium (Poaceae) Nees
Gk. *schizen* = to split; *achuron* = chaff; referring to the bilobed lemma of the sessile spikelet.

Schizaea (Schizaeaceae) Sm.

(CM – Both)

Gk. s*chizein* (*skhizein*) = to split; referring to the distichous or two-ranked spikes of sporangia.

Schizobasis = Drimia (Hyacinthaceae) Baker
Gk. *schizein* = to split; *basis* = step or base; perhaps referring to its web-like structure and firm base.

Schizocarphus = Scilla (Hyacinthaceae) Van der Merwe
Gk. *schizein* = to split; *carpus* = fruit; bearing a dry pericarp of two or more united carpels.

Schizochilus (Orchidaceae) Sond.
Gr. *schizein* = to split; *cheilos* = lip; alluding to the trifid apex or divided condition of the lip; relating to the incised lip.

Schizodium*** = Disa (Orchidaceae) Lindl.
Gr. *schizein* = to split*; odium* = resemblance;

referring to either the bipartite column, which has the stigma and anthers separated, or to the bilobed petals.

Schizoglossum (Apocynaceae) E.Mey.
Gk. *schizo* = split; *glossa* = a tongue. The corolla is lobed almost to its base in some species.

Schizolegnia = Lindsaea (Lindsaeaceae)
Alston.
Gk. *schizo* = split; *legnon* = border; possibly alluding to the fringed borders.

Schizoloma = Lindsaea (Lindsaeacea)
Gaudich.
Gk. *schizo* = I cut, split; *loma* = edge, fringe or margin. The sori is in the form of a slit on the margin of the frond.

Schizostephanus = Cynanchum (Apocynaceae) Hochst.
Gk. *schizo* = split; *stephanus* = garland, wreath. Meaning unclear.

Schizostylis*** = Hesperantha (Iridaceae)
Backh. & Harv.
Gk. *schizo* = I cut; *stylis* = style; referring to the split style.

Schkuhria (Asteraceae) Roth
For Christian Schkuhr (1741–1811), German gardener and physical scientist at the University

of Wittenberg. He conducted botanical studies throughout his life, and was an advocate of the Linnaean system. He taught himself to draw, engrave and use a microscope (using self-made instruments), and had a keen interest in taxonomy. He published his *Botanisches Handbuch* (*Handbook of Botany*) covering flora in the vicinity of Wittenberg, providing names of the flowers in Latin and German; describing their characteristics, habitat, locations and abundance; and the usefulness of the species, such as their economic value or agricultural use.

Schlechteranthus (Aizoaceae) Schwantes

For Max (Maximilian) Schlechter (1874–1960), amateur botanist and businessman who settled in Port Nolloth, Namaqualand, who helped his renowned brother, Friedrich Richard Rudolf Schlecter (1872–1925) (*Schlecteria* q.v.) collect plants in South Africa, mainly in Namaqualand, between 1896 and 1897. He built up a good collection of succulents plants, especially mesembs. On his death, these were distributed to various botanical institutes, including the Bolus Herbarium, the department of botany at Stellenbosch University, and the Division of Botany and Plant Pathology in Pretoria.

Schlechteria = Heliophila (Brassicaceae)
Bolus
For Friedrich Richard Rudolf Schlecter (1872–1925), German botanist, botanical explorer and

plant collector for 19 years in South Africa (1891–1897), where he worked for Harry Bolus as a herbarium and library assistant, and in Mozambique, West Africa, Indonesia, and New Guinea between 1898 and 1910. In 1913, he was appointed to the staff, later curator, of the Berlin-Dahlem Botanical Museum, where, except for during World War I, he worked until his death. A world authority on orchids, he described more than 1 000 species of orchids and published more than 300 scientific papers, as well as other major publications, including those relating to New and Old worlds, and tropical US and Chinese-Japanese regions.

Schlechterina (Passifloraceae) Harms
Schlecteria (q.v.); *ina* = belonging to a Latin language; 'ina' added to a short name. The same Schlecter as above.

Schlumbergera (Cactaceae) Lem.
For Frédéric Schlumberger (1823–1893), French gardener and cacti collector who lived in Les Château de la Haye des Authieux in Authieux-sur-le-Port-Saint-Ouen on the outskirts of Rouen, France. Charles Lemaire named this cactus for him in 1858 in recognition of a number of submissions Schlumberger made to *Revue Horticole* between 1854 and 1857. Other writers have suggested that the attribution should be made for Frederick Schlumberger (1804–1865), but an internet article, *Origin of the Generic Name Schlumbergera*, makes a strong case in favour of the first-mentioned Schlumberger (*www.pxc.me.uk/misc/schlumbergera.html*).

Schmidelia = Allophylus (Sapindaceae) L.
For Casimir Christoph Schmidel (1718–1792), German physician, naturalist and professor. He studied medicine at the universities of Jena and Halle, and also botany and geology in Saxony, Holland and Switzerland. In 1742, he

was appointed professor of pharmacology at Freidrichs-Akademie in Bayreuth, where he also opened a private practice. When the university was moved from Bayreuth to Erlangen in 1744, he became professor of medicine and later dean of the medical faculty. His interest in geology resulted in him publishing *Erz Stuffen und Berg Arten* (1753), which was a comprehensive, well-illustrated book to help miners and prospectors recognise different ore minerals. Because he did not get on well with colleagues, he resigned from the university in 1763 and thereafter served as physician ordinary to Carl Alexander, margrave of Brandenburg-Ansbach, who made him a privy councillor and president of the board of health. He was awarded an honorary MD at the University of Erlangen in 1783.

Schmidtia (Poaceae) Steud. ex J.A.Schmidt
For Johann Anton Schmidt (1823–1905), German botanist and plant collector. He obtained a PhD from the University of Göttingen and lectured at the University of Heidelberg from 1852 to 1863, after which he lived as a private scholar in Hamburg. He collected Bryophyta for the Herbarium Hamburgense between 1846 and 1905 in Bavaria, Lower Saxony, North Rhine-Westphalia, Austria, France and Poland. He went on an expedition to the Cape Verde Islands in 1851. He authored *Beiträge zur Flora der Cap Verdischen Inseln* (*Contribution to the Flora of the Cape Verde Islands*) (1852), *Flora von Heidelberg* (1857), and *Anleitung zur Kenntniss der natürlichen Familien der Phanerogamen* (*Instructions to the Knowledge of the Nature Family Phanerogams*) (1866, 1870).

Schoenefeldia (Poaceae) Kunth
For Wladimir de Schoenefeld (1816–1875), German-born French botanist and pteridologist of Russian parentage. Seemingly, when only one year old, he became an orphan and went to live with the botanist Carl Sigismund Kunth (1788–1850) in Berlin until he was 24, and from whom he developed an interest in botany. In 1840, he returned to Paris, where he took botanical lessons from Adrien-Henri de Jussieu, the French physician and botanist. In Paris, he networked with a number of other botanists and was one of the 15 members who founded the *Société Botanique de France* in 1854, of which he became the secretary general from 1862 to 1875.

Scleria (Cyperaceae) P.J.Bergius
Gk. *scleros* = hard; perhaps referring to the hypogynium.

Schoenoplectiella (Cyperaceae) Lye
Schoenoplectus (q.v.); Gk. -*ella* = diminutive.

Schoenoplectus (Cyperaceae) (Rchb.) Palla
Gk. *schoinos* = a rush, reed; *plektos* = twisted; a plaited or twisted rush.

Schoenoxiphium (Cyperaceae) Nees
Gk. *schoinos* = a rush, reed; *xiphias* = swordfish, from Gk. *xiphos* = sword (c f *xiphisternum*); alluding to the sword-shaped leaves.

Schoenus (Cyperaceae) L.
Gk. *schoinos* = a rush, reed.

Schonlandia = *Corpuscularia* (Aizoaceae) L.Bolus
For Selmar Schönland (1860–1940), German botanist. He studied at Berlin and Kriel universities, obtaining a PhD in 1883. In 1886, he was appointed curator of the Fielding Herbarium, Oxford, also obtaining an MA degree. He came to the Eastern Cape in 1889 to take up an appointment as curator, and later director, of the Albany Museum. His main interest was the herbarium, founded by WG Atherstone in 1860, which became the second-largest herbarium in South Africa. When Rhodes University was founded in 1904, he became a member of the council and the first professor of botany (1905–1926). He played a leading role in the Botanical Survey of South Africa (1919), which was initiated by Illtyd Buller Pole Evans.

Schotia (Fabaceae) Jacq. (Image on opposite page)
For Richard van der Schot (c 1730–1790), Dutch gardener who studied in Leiden. He became head gardener of the Imperial Gardens at Schönbrunn, Vienna, Austria, having been appointed by the French botanist Nicolaus (Nicolaas) Joseph von Jacquin (1727–1817). In 1755, Jacquin was asked by the Emperor Franz Ferdinand (Emperor Francis I) to lead an expedition to the Caribbean to collect tropical plants and other 'curiosities' for the palace's natural history collections. Van der Schot joined Jacquin on this four-year journey, visiting Grenada, Saint Vincent, Aruba, Cuba and Curacao, and returned to Vienna in 1759. Although some sources state Van der Schot visited South Africa between 1785 and 1788, just two years before his death, he did not, both according to our research and confirmed in a 1970 Taxon paper *Jacquin Names, Some Notes on Their Typification,* by WG D'Arcy.

Schrebera (Oleaceae) Roxb.
For Johann Christian Daniel von Schreber (1739–1810), German botanist and zoologist,

and correspondent of Linnaeus. He was the author of a multivolume set of books illustrating the mammals of the world, *Die Säugethiere in Abbildungen nach der Natur mit Beschreibungen* (*Images of Mammals from Nature with Descriptions*), in which Linnaeus's binomial system was used for the first time for mammals; as well as an entomology on newly discovered insects, *Schreberi Novae Species Insectorvm*. Schreber was elected a member of the Royal Swedish Academy of Sciences (1787), and president of the German Academy of Sciences Leopoldina (1791–1810). His herbarium collection has been preserved in the Botanische Staatssammlung München since 1813.

Schotia: These generally large bushveld trees with glorious red flowers are loved by sunbirds. (EJM)

*Schwabea**** = *Monechma* (Acanthaceae)
Endl.
For Samuel Heinrich Schwabe (1789–1875), German botanist, pharmacist and astronomer,  best remembered for his work on sunspots. In 1826, he started trying to find a new planet, tentatively called Vulcan, inside the orbit of Mercury, by looking at dark spots when it passed in front of the sun. For 17 years, from 1826–1843, Schwabe recorded sunspots trying to detect the new planet among them. He noticed regular variation in the number of sunspots over time with the number of spots reaching a maximum every 10 years. This 'periodicity of sunspots' is now fully recognised as one of the most important discoveries in astronomy. In 1857, Schwabe was awarded the gold medal of the Royal Astronomical Society. He was a member of the Royal Society of London.

*Schwantesia**** = *Mitrophyllum* (Aizoaceae)
Dinter
For Martin Heinrich Gustav (Georg) Schwantes (1881–1960), German botanist and archaeologist.  A teacher from 1903 to 1923, he studied ethnology, geology and botany at Hamburg University, graduating in 1923. He subsequently became curator of the Museum of Ethnology and Prehistory of Hamburg (1926), professor of history at the University of Hamburg (1928), director of the National Archaeological Museum of Kiel (1929), and professor of prehistory and early history in Kiel (1937). He wrote several books on the prehistory of northern Germany, a contribution on *Aizoaceae* for *Parleys Blumengärtnerei* (1957), and two botanical books, *The Cultivation of the Aizoaceae* (1953) and *Flowering Stones and Mid-Day Flowers* (1957).

Schweiggera = *Tritoniopsis* (Iridaceae)
E.Mey. ex Baker
For August Friedrich Schweigger (1783–1821), German naturalist, professor of botany and  medicine at the University of Königsberg, and his younger brother, Johann Salomo Christoph Schweigger (1779–1857), chemist, physicist and professor of mathematics, who developed the first sensitive galvanometer. AF Schweigger studied medicine, zoology and botany at Erlangen, and after his graduation, spent some time in Berlin (from 1804) and in Paris (1806) before becoming a professor (1809). Throughout his life, he maintained a keen interest in botany, writing *Specimen Flora Erlangensis* (1805) and *De Plantarum Classificatione Naturalis* (*The Natural Classification of Plants*) (1821); but he wrote on his other interests as well. In 1815, he was elected a member of the Royal Swedish Academy of Sciences. On a research trip to Sicily, he was murdered near Agrigento.

Schwenckia (Solanaceae) L.
For Martin Wilhelm Schwenke (1707–1785),

surgeon of the Hague, city doctor and pharmacy inspector of the city, who was appointed in 1750 as professor of botany, physician, and botanist of the Hague.

Scilla * (Hyacinthaceae) L.
Gk. *skilla* = squill or sea leek. Ancient Gk. and Latin name for *Urginea maritima*. Gk. *skyllo* = I injure; referring to the poisonous bulbs. The medicinal 'squill', containing active cardiac glycosides, was obtained from another species of *Urginea*.

Scirpoides (Cyperaceae) Ség.
La. *Scirpus* (q.v.); Gk. *-oides* = resembling.

Scirpus (Cyperaceae) L.
The Latin name for rush or bullrush.

Scleranthus * (Caryophyllaceae) L.
Gk. *skleros* = hard; *anthos* = flower; referring to the hard, dry perianth.

Scleria (Cyperaceae) P.J.Bergius
Gk. *skleros* = hard, rough; referring to the stony nutlet.

Sclerocarpus (Asteraceae) Jacq.
Gk. *skleros* = hard, rough; *karpos* = fruit; alluding to hardened paleae enfolding disk cypselae.

Sclerocarya (Anacardiaceae) Hochst.
Gk. *skleros* = hard, rough; *karya* = a walnut. The kernel of the drupe is hard and indestructible, but the seeds are tasty.

Sclerochaetium *** = ***Tetraria*** (Cyperaceae) Nees
Gk. *skleros* = hard, rough; *khaite* = bristle; *-ium* = diminutive.

Sclerochiton (Acanthaceae) Harv.
Gk. *skleros* = hard, rough; *chiton* = a covering or outer garment. The capsules are woody.

Sclerochloa (Poaceae) P.Beauv.
Gk. *skleros* = hard, rough; *chloe* = the ancient name for grass or fodder. The grass is hard and tough.

Sclerocroton = ***Shirakiopsis*** (Euphorbiaceae) Hochst.
GK. *skleros* = hard, rough; *kroton* = a tick. The seeds are hard, relatively large, and resemble ticks.

Scleropoa = ***Catapodium*** (Poaceae) Griseb.
Gk. *skleros* = hard; *Poa* (q.v.), *poa* = grass. The spikelets resemble those of *Poa,* but the glumes and lemmas are much more leathery.

Scolopia (Salicaceae) Schreb.
Possibly Gk. *skolops* = a stake or other pointed object; referring to the long, sharp spines or thorns on some species.

Scoparia * (Plantaginaceae) L.
Gk. *skopa* = a broom; *aria* = pertaining to. Some species are like a 'broom', such as *Cytisus scoparius*.

Scopelogena (Aizoaceae) L.Bolus
Gk. *skopelos* = rock; *gena* = genus. This genus, for instance *S. verruculata*, grows among rocks in the mountains.

Scopularia *** = ***Holothrix*** (Orchidaceae) Lindl.
La. *scopulae* = little broom. The plant was often used for making small hand brooms.

Scorzonera * (Asteraceae) L.
Italian, from *scorzone* = viper or adder. The plant root was seen by Celtic and Germanic peoples as an antidote against venomous snakes until the 16th century (Wikipedia.org). There are other possible derivations.

Scurrula (Loranthaceae) L.
La. *scurrula* = a little clown, buffoon; alluding to the appearance of the flower.

Scutellaria * (Lamiaceae) L.
La. *scutella* = a drinking bowl or flat dish, from *scutum* = a shield; alluding to the shape of the persistent calyx or sepals in fruit.

Scutia (Rhamnaceae) (Comm. ex DC.) Brongn.
La. *scutum* = a shield. The calyx surrounds the fruit like a shield.

Scyphofilix *** = ***Microlepia*** (Dennstaedtiaceae) Thouars
Gk. *skuphos* = cup*; filix* = fern. The indusium is cup-shaped.

Scyphogyne = ***Erica*** (Ericaceae) Brongn.
Gk. *skuphos* = cup; *gune* = woman; referring to the shape of the stigma.

Scytalis *** = ***Vigna*** (Fabaceae) E. Mey.
La. *scytale* = staff, rod, baton; referring to a use for the reeds – the Spartans sent secret messages on a parchment wound around a rod.

Scytophyllum *** = ***Maytenus*** (Celastraceae) Eck. & Zeyh.
Gk. *skytos* = skin, leather; *phyllon* = leaf; alluding to the texture of the leaves.

Searsia (Anacardiaceae) F.A.Barkley (Images on page 301)
For Paul Bigelow Sears (1891–1990), US plant ecologist, professor of botany at Oberlin College

(1938–1950), chair of the conservation programme and Yale University (1950–1960), and author of many books, including *Deserts on the March* (1935), his most popular book explaining ecological principles to the general public. During the 1920s and 1930s, he pioneered 'palynology', the study of fossil pollen as a cue to past vegetation and climate. He was president of the Ecological Society of America (1948), the American Association of Science (1956), and the American Society of Naturalists (1959), among others, and named an eminent ecologist by the Ecological Society of America (1965).

Sebaea (Gentianaceae) Sol. ex R.Br.

Seemingly always with bright yellow flowers (CM)

For Albertus Seba (1665–1736), Dutch pharmacist, zoologist and naturalist. In 1700, he opened an 'apothecary shop' in Amsterdam and collected exotic plants and animal products from sailors and ship surgeons from which he could make 'medicines'. In 1716, he sold his first collection (as well as the Dutch botanist Frederik Ruysch's collection) to the Russian Tsar, Peter the Great, on his visit to the Netherlands. Seba immediately set about building an even larger collection. In 1734, he published his magnificently illustrated four-volume *Thesaurus* (1734, 1735), with 446 plates (2 volumes published posthumously), which displays marine animals, insects and reptiles. Linnaeus must have seen this collection when he visited Seba twice in 1735. Seba became a Fellow of Royal Society in 1728.

Secale (Poaceae) L.
Latin name for rye (*Secale cereale*).

Secamone (Apocynaceae) R.Br.
Latinised from the Arabic name *Secamone,* which refers to *S. aegyptiaca.*

Securidaca (Polygalaceae) L.

(BW)

La. *securis* = a hatchet; the wing of the fruit is hatchet-shaped.

Securinega (Phyllanthaceae) Comm. ex Juss.
La. *securis* = a hatchet, axe; *nego* = to deny. The wood is so hard that it is difficult to cut.

Seddera (Convolvulaceae) Hochst.
The species, *S. virgata*, was collected at Mount Sedder in Arabia.

Sedum* (Crassulaceae) Tourn. ex L.
Gk. *sedum* = sedentary, from *sedere* = to sit. La. word for houseleek; referring to the manner in which some plants attach themselves to stones or walls and 'sit' there.

Seemannaralia (Araliaceae) R.Vig.
For Berthold Carl Seemann (1825–1871), German botanist, naturalist, explorer and botanical collector who studied at the Royal Botanical Garden, Kew, and was associated with a number of British missions. He accompanied Captain H Kellet on HMS *Herald* (1846–1851) during a Pacific expedition, visiting the Cape on his return to England. He authored *The Botany of the Voyage of HMS Herald* (1845–1851). He travelled to Fiji and published a catalogue of the flora of the islands (1859), visited South and Central America (Venezuela, Nicaragua and Panama) in the 1860s, and died shortly thereafter. He was the editor of *Bonplandia* (1853–1862), the *Journal of Botany* (1863–1869), and the *Journal of Botany, British and Foreign* (1863–1871).

Seetzenia (Zygophyllaceae) R.Br. ex Decne.
For Ulrich Jasper Seetzen (Iospar Sentzen) (1767–1811), German explorer and naturalist. He obtained his doctorate in medicine and natural sciences at the University of Göttingen (1785 to 1789). From 1802–1810, he made extensive sponsored travels through Asia Minor and the Middle East. He became fluent in Arabic, often travelled disguised as a pilgrim/beggar, and acquired art objects, mummies, eastern manuscripts, plants and minerals for his promoter, the duke of Gotha. His collections and diaries/letters recording his observations of Eastern society form an important base for Orient research. He identified the physical locations of several sites from the Judaeo-Christian Bible, and was the first European to travel completely around the Dead Sea. He died on his way to Muscat, seemingly poisoned by his guides on orders from the imam of Sana'a for reasons unknown.

Sehima (Poaceae) Forssk.
Derived from the Arabic *sehim, saehim* (Wittsein); the common name for the type species in Yemen and Egypt.

Seidelia (Euphorbiaceae) Baill.
Possibly for Christopher Friedrich Seidel (f 1869), German botanist and horticulturalist in Dresden. (Not explained by author.) He imported a camellia from Japan in 1893 and changed its name from *usu-otome* to *frau minna siedel*.

Selaginastrum *** = ***Antherothamnus***
(Scrophulariaceae) Schinz & Thell.

Selago (q.v.); *astrum* = partially resembling, inferior, wild.

Selaginella (Selaginellaceae) P.Beauv.
Gk. *Salaginella* is the name given by Pliny the Elder for an ancient plant name that was formerly called *Selago*. The name *Salaginella* means 'little Selago', but it is only distantly related to *Selago* (q.v.). *Selaginella* is an evergreen, low-growing, flowerless, moss-like plant, and relates to *Lycopodium slago* and *Huperzia* selago. In contrast, the genus *Selago* belongs to a family of flowering plants called Scrophulariaceae.

Selago (Scrophulariaceae) L.
Celtic. *sel* = sight; *jach* = salutary, beneficial; referring to its supposed medicinal properties, especially for diseases of the eyes. Another source suggests the derivation is Greek, *selagh* = flashing. (Allusion unknown.)

Selinum (Apiaceae) L.
Gk. An old genus name for the carrot family; the word used for various fragrant, leafy herbs, most commonly parsley.

Selliguea * (Polypodiaceae) Bory
For Alexandre François Gilles, called Selligue (Paris, 1784–1845), French engineer and a pioneer in 'fracking' who made a contribution towards the development of the achromatic microscope. At the end of 1823, Vincent and Charles Chevalier, opticians in Paris, began, on the recommendation of M Selligue, to construct an achromatic microscope by using the concept of combining achromatic object-glasses, which was an innovative breakthrough. His early pioneering efforts were described in *Rapport Sur le Microscope Achromatique de M Selligue* (1824) for the Académie Royale des Sciences. More substantially, from 1834–1836, he pioneered and patented a way to make 'water gas', distilling oils made from industrial shale, later used for oil for cars, machinery, lighting and candles, and gas engines for irrigation.

Semele * (Ruscaceae) Kunth
In Greek mythology, daughter of the Boeotian hero Cadmus and Harmonia. Reason for selection unknown.

Semnanthe = ***Erepsia*** (Aizoaceae) N.E.Br.
Gk. *semnos* = holy, august, sacred, inspiring respect; *anthos* = flower. The bright-coloured flowers command attention.

Semonvillea = ***Limeum*** (Molluginaceae) Heimerl.
For Charles-Louis Huguet of Montaran, Earl

and Marquis de Semonville (1759–1839), French politician, diplomat and son of Charles de Montaran Huguet, advisor to King Louis XV, secretary of the Royal Finances. Huguet lived a life of political intrigue and positioning. Highly intelligent, he became an advisor to the Parliament of Paris (1777) at age 18, held several diplomatic posts, was imprisoned in Austria for 29 months, and became Napoleon Bonaparte's state councillor and minister plenipotentiary. He was made Marquis (1817) and awarded the Grand Cross of the Legion of Honour (1823). His connection with botany is that, in 1811, he gave Jacques Étienne Gay (1786–1864), the author of some 478 plant species, a job in his office, working for the senate. He worked there until 1848. Gay named the genus Semonvillea after him in 1829.

Senebiera = Coronopus (Brassicaceae) DC.
For Jean Senebier (1742–1809), Swiss botanist, bibliographer, linguist, clergyman, physiologist,

librarian of the city of Genève, and voluminous writer on plant physiology, best remembered for his account of the influence of light on vegetation. He conducted and published experiments in the 1780s and 1790s, summarised in *Physiologie Végétale* (1800), showing that plants 'fix' carbon dioxide. He largely explained how photosynthesis works, and in particular showed that its efficacy is dependent on the intensity and duration of sunlight, and is confined to the above-ground green parts. He was an active member of the Society of Arts, and the Society of Physics and Natural Sciences.

Senecio (Asteraceae) L.
La. *senex* = an old man. The white, hairy pappus of the seeds is reminiscent of an old man's beard.

Senegalia (Fabaceae) Raf.
Fr. *le Sénégal*; referring to a plant presumably found in Senegal, West Africa.

Senna (Fabaceae) Mill.
La. *senna* = Latin form of an Arabic word for a thorny bush. *Sana* is the Arabic name for a species of this genus.

Septas*** (Crassulaceae) L.
La. *septem* = seven. 'The calyx of this plant (*Septas capensis*) consists of seven parts: there are seven petals, seven germs, and seven capsules.' *Elements of Botany* (1827), Benjamin Smith Barton.

Senecio flowers are often yellow. (MC)

Septulina (Loranthaceae) Tiegh.
From *septulum* = a small septum; La. *septum* = an enclosure, fence (septum); possibly referring to the size of the partitions of the locules of an ovary.

Serapias (Orchidaceae) L.
Possibly named after the Egyptian god Serapis, in whose temple pilgrims engaged in licentious living. The name was used by the Greeks for an orchid (probably *Orchis morio*), which was a reputed aphrodisiac.

Sericanthe (Rubiaceae) Robbr.
Gk. *serikos* = silky; *anthos* = flower; referring to the silky hair-coat of the flower parts.

Sericocoma (Amaranthaceae) Fenzl
Gk. *serikos* = silky; *kome* = hair of the head. The ovary is woolly or tomentose.

Sericocomopsis*** **= Leucosphaera** (Amaranthaceae) Schinz
Sericocoma (q.v.); *iopsis* = resembling, likeness.

Sericorema (Amaranthaceae) (Hook.f.) Lopr.
Gk. *serikos* = silky; *korema* = a broom; referring to the bracteoles of the sterile flowers. The tufts of woolly hair almost enclose the flowers.

Seriphium (Asteraceae) L.
Seriph is a stroke or line of a letter, from La.

scribere = to write. The pattern of the closely knit, wiry branches form straight 'lines' at right angles to the stem.

Serpicula = Laurembergia (Haloragaceae) L.
La. *serp* = creep, like a (small) snake; *-ula* = diminutive; referring to its creeping habit.

Serratula = Saussurea (Asteraceae) L.
La. *serratus* = serrate, from *serra* = a saw; *-ula* = diminutive; referring to the serrated leaf margins.

Serruria (Proteaceae) Salisb.
For Joseph (Josephus) Serrurier (1663–1742),

Dutch physician, philosopher, mathematician, botanist and physicist. He studied at the University of Utrecht, obtaining an MA degree in 1690 and qualifying as a doctor of medicine on the same day. During his career at Utrecht, he became professor of philosophy and mathematics (1705), professor of botany and medicine (1716), and the university's rector on three occasions. Among his works, he authored an extensive treatise on experimental physics, *Physicae Experimentis Innixae Compendiosa Tractatio* (1700), a commemorative volume on Adrianus Reland, orientalist, linguist and physicist; and created a new *hortus botanicus*, which he modelled on the University of Leiden's widely acclaimed botanical garden designed by Herman Boerhaave.

Sesamothamnus (Pedaliaceae) Welw.
Sesamum (q.v.); Gk. *thamnos* = shrub. This genus differs from sesame in being woody.

Sesamum (Pedaliaceae) L.
Gk. and La. *Sesamum* = an oily plant, name for the plant *S. indicum* = sesame. An ancient name used by Hippocrates and Ali Baba.

Sesbania (Fabaceae) Scop.
From the Arabic vernacular name *saisaban* = Persian *sisaban*.

Seseli** (Apiaceae) L.
Ancient Greek name for an umbelliferous plant name for certain members of the family, used by Hippocrates.

Sessilistigma = Moraea (Iridaceae) Goldblatt
La. *sessilis* = low, sitting, fixed, attached; Gk. *stigma* = stigma; referring to the positioning of the stigma (sessile).

Sesamothamnus: A woody, spiny small tree in arid areas (EJM)

Sesuvium (Aizoaceae) L.
La. *Sesuvium*, named by Linnaeus after the country of the Sesuvii and a Gallic tribe mentioned by Caesar. The reason for this genus name is unknown.

Setaria (Poaceae) P. Beauv.
La. *seta* = a bristle; *-aria* = pertaining to; referring to the bristly awns in the involucrum.

Shantzia = Azanza (Malvaceae) Lewton
For Homer Leroy Shantz (1876–1958), US plant physiologist and ecologist. He studied botany at

the University of Nebraska obtaining a PhD in 1905. He had a varied career, teaching in four universities (Nebraska, Missouri, Illinois and Arizona). He travelled expensively in the United States and Africa, recording geographic and climatic conditions. In 1918 and 1919 he toured Africa, photographing the vegetation and studying the natural plant resources and crop production possibilities. In 1924, he traversed the continent from Cairo to the Cape, and repeated this in 1953. His publications include *Agricultural Regions of Africa* (1940–1943), *The Vegetation and Soils of Africa* (1923), and *Photographic Documentation of Vegetational Changes in Africa over a Third of*

a Century (1958). He received many meritorious awards, including the Association of American Geographers' Outstanding Achievement Award (1954).

Sheilanthera (Rutaceae) I.Williams
Named for Sheila Williams by her husband, Ion James Muirhead Williams (1912–2001), civil engineer and botanist. Sheila accompanied Ion on many field trips while he was conducting extensive field research into the genus *Leucadendron* for which he obtained a PhD in 1972. The couple were heavily involved with nature conservation in the Hermanus area in the Western Cape. Sheila's collection of coastal plants is now preserved in the herbarium named after her in the Fernkloof Nature Reserve in the area.

Sherardia* (Rubiaceae) L.
For William Sherard (1659–1727), British botanist, plant collector, patron of botanists. He studied at St John's College, Oxford (1677–1683) and in Paris (1686–1688) at the Jardin des Plantes under Joseph Pitton de Tournefort and Herman Boerhaave. He bequeathed 3 000 pounds to Oxford University for a chair of botany, on the condition that the initial chair went to Johan Jacob Dillenius (1687–1747); he also donated his library and herbarium. He contributed to John Ray's *Stirpium* (1694), was editor of Paul Hermann's *Paradisus Batavus* (1698) and was instrumental in the publication of Sebastien Valliant's *Botanicon Parisiense* (1727). He became British consul to Smyrna (1703–1716), during which time he amassed a fortune. He was a Fellow of the Royal Society.

Shirakiopsis (Euphorbiaceae) Esser
The genus *Shirakia*; *-opsis* = resembles.

Shutereia = Hewittia (Convolvulaceae) Choisy
For Dr James Shuter (1775–1826), Irish assistant surgeon and government naturalist and botanist at Madras. In 1800, he obtained an MD from Edinburgh University. In 1809, he went to Madeira, and to Madras, India, and in 1819 began work in the service of the East India Company (effectively the government). While there, he received a written request for Indian plant specimens, and even though he was ill, he was entrusted with a valuable collection of *Materia Medica* of India, many of the seeds and roots of which were in fine condition. These were presented to professor of botany at Edinburgh, Robert Graham (1786–1784), by Sir Thomas Munro, governor of Madras. Some of these specimens were probably sent to William J Hooker, thus increasing the already fine collection of tropical plants in the Royal Botanical Gardens at Kew. Shuter died in Dorset, England, in 1826.

Sickmannia = Ficinia (Cyperaceae) Nees
For Johann Rudolph Sickmann (1779–1849), German, soldier, botanist, plant collector and faculty member of the University of Hamburg from 1844 to 1849, although he was not a professor. He collected plants in Germany, France and Switzerland, and also ran an antiquarian bookshop. At times Sickmann stood in for Professor Johann Georg Christian Lehmann (1792–1860), when the professor was temporarily on leave. Lehman's interests were physics, natural history and botany. He founded the University of Hamburg's botanical garden. During these times, Sickmann, his assistant, took charge of this botanical garden. He was the author of *Enumeratio Stirpium Phanerogamicarum circa Hamburgum Sponte Crescentium* (1836).

Sicyos* (Cucurbitaceae) L.
Gk. *sisyos* = cucumber. This is a 'burr cucumber', not the cucumber we eat, which is *Cucumis sativus*.

Sida (Malvaceae) L.
Gk. *sidē* = name used by Thecophrastus for a water plant, perhaps a water lily.

Sideroxylon (Sapotaceae) L.
Gk. *sideros* = iron; *xylon* = wood; referring to the hard wood (though the common 'ironwoods' are in the entirely different oleaceous group).

Sigesbeckia* (Asteraceae) L.
For Johann Georg Siegesbeck (1686–1755), Prussian physician, botanist and director of the botanical garden at St Petersburg (1735–1747). He was a fierce critic of Linnaeus's system for the classification of plants, the so-called sexual system, on 'moral grounds' rather than scientific merit, asking, *inter alia*, whether God really would allow that 20 men or more (i.e. the stamens) have one wife in common (i.e. the pistil). Such was their enmity that Linnaeus named a stinking weed *Siegesbeckia,* and in a publication on contemporary botanists, ranked Siegesbeck as the last of 33 botanists. So outraged was Siegesbeck at these and other incidents that he stopped Russian and Siberian plant trade with Uppsala, Sweden. The genus name was misspelled by Linnaeus.

Sigmatosiphon = Sesamothamnus
(Pedaliaceae) Engl.

Gk. letter *sigma* (*sigmate* = S-shaped); *siphon* = tube. The corolla tube is curved.

Silene (Caryophyllaceae) L.

White petals topping a groved tube is characteristic. (CM)

Gk. *Silenos* = a Greek woodland deity, half man, half horse or goat, companion of Dionysius, always portrayed as old, bald and bearded, and usually covered with foam or slaver; referring to the sticky secretion of the stems that entrap small insects, thus foiling predators. 'Catchfly.'

Silicularia = Heliophila (Brassicaceae) Compton
La. *silicula* = a little pod or husk; *-aria* = pertaining to; possibly referring to the silicle – the dry dehiscent fruit.

Silybum* (Asteraceae) Adans.
Gk. *silybos* = (Dioscorides used *silybon*) an ancient name for some edible thistles.

Simocheilus = Erica (Ericaceae) Klotzsch
Gk. *simos* = flat-nosed, steep or concave; *chellos* = lip. Meaning obscure.

Sinapis* (Brassicaceae) L.
Gk. and La. *sinapi* = mustard; referring to the flavour of the seeds.

Sineoperculum = Cleretum (Aizoaceae) Van Jaarsv.
La. *sine* = without; *operculum* = lid, cover. The capsule does not have an operculum.

Siphocodon (Campanulaceae) Turcz.
Gk. *siphon* = tube; *kodon* = a bell; referring to the shape of the corolla tube.

Siphonochilus (Zingiberaceae) J.M.Wood & Franks

This ginger is almost extinct in the wild from over harvesting for medicinal purposes. (GN)

Gk. *siphono-* = tube; *chilos* = lip; referring to the shape of the flower.

Siphonoglossa = Justicia. (Acanthaceae) Oerst.
Gk. *siphon* = tube; *glossa* = tongue. The corolla is bilabiate, with a cylindrical tube.

Sison = Trachyspermum (Apiaceae) L.
Gk. *sinon, sison* = stone parsley. A name used by Dioscorides.

Sisymbrium (Brassicaceae) L.
Gk. *sisymbrion* = watercress; used by Theophrastus. Name for a fragrant herb, perhaps mint.

Sisyndite (Zygophyllaceae) E.Mey. ex Sond.
Name derivation uncertain. Sonder (*Flora Capensis*) suggests the Gk. *syndeo* = to link together; possibly referring to the tightly forked branches.

Sisyranthus (Apocynaceae) E.Mey.
Gk. sus(sys) = pig; anthos = flower; possibly referring to the shape (*Paxton's Botanical Dictionary*) of the flower head.

Sisyrinchium* (Iridaceae) L.
Gk. *sisyrinchion*, possibly from *sys* = pig; *rhynchos* = snout; possibly alluding to swine (pigs) grubbing the plant roots for food or even eating this plant. Hollombe suggests Gk. *sisyra* = goat-hair cloak; or *sisyrna* = garment of skin and relating to corm tunics.

Sium = Falcaria (Apiaceae) L.
Gk. *sion* = Greek name for a marsh herb.

Skiatophytum (Aizoaceae) L.Bolus
Gk. *skias* = shade; *phyton* = plant, hence 'shade plant'; referring to its preferred habitat.

Smelophyllum (Sapindaceae) Radlk.
For Timofei Andreevich Smielowski (Timotheus Smielovsky) (1769–1815), a Russian pharmacist and botanist. He studied medicine at the Imperial Academy of Sciences in St Petersburg. In 1802, he was hired by the university as an 'adjunct' professor to teach but was not a full member of the faculty. He authored *Hortus Petropolitanus seu Descriptiones et Icones Plantarum ... Part 1* (1806). In 1815, he was promoted to full professor of pharmacy at the Medico-Surgical Academy in St Petersburg. Gk. *phyllon* = leaf; presumably meaning 'leaves like those of *Smelowskia*, a Northern Hemisphere genus also with pinnate leaves' (Palmer).

Smicrostigma (Aizoaceae) N.E.Br.
Gk. *smikro-* = small (equivalent to *mikro-*); *stigma* = stigma.

Smilacina* = Maianthemum (Ruscaceae) Desf.
Gk. *Smilax* (q.v.); *-ina* = diminutive.

Smilax (Smilacaceae) L.

(GN)

Gk. *smilax*, from *smilakos* = twining. This genus, related to the bindweed and prickly ivy, is a vigorously climbing evergreen with hooked thorns allowing it to hang on to and over branches of its host tree and overwhelm it.

Smithia (Fabaceae) Aiton
For Edward Smith (1759–1828), British botanist and physician who, in 1784, after Linnaeus's  (1707–1778) death purchased the latter's entire collection of books, manuscripts and specimens for 900 guineas. He founded the Linnaean Society of London (1788) with Samuel Goodenough (1743–1827) and Thomas Marsham (1748–1819), and was its life president. In his last 30 years, he devoted himself to natural history and wrote, *inter alia*, *Flora Britannica* and *The English Flora* (four volumes), 3 348 botanical articles for *Rees's Cyclopaedia* (1808–1819), the earliest US book on *Lepidopterous Insects of Georgia* (1791), and made substantial contributions to *Flora Graeca*. He was a Fellow of the Royal Society (1786) and the Royal Swedish Academy of Sciences (1792).

Smodingium (Anacardiaceae) E.Mey. ex Sond.
Gk. *smodix* = a weal caused by a blow. Jackson refers to the calloused fruit, but it is probably the irritant leaves that can raise nasty wheals and blisters.

Solanecio (Asteraceae) (Sch.Bip.) Walp.
A combination name derived from *Solanum* (q.v.) and *Senecio* (q.v.); referring to a daisy-like plant with tuberous roots.

Solanum (Solanaceae) L.
The Latin name for nightshade, perhaps from *solamen* = soothing; referring to the narcotic properties of some species.

Solenangis (Orchidaceae) Schltr.
Gk. *solen* = pipe, tube (or can be a canal, channel); *angis* = vessel; referring to the funnel-like lip or alluding to the infundibuliform lip, which, lacking a disk, consists entirely of a spur.

Solenostemon = Plectranthus (Lamiaceae) Thonn.
Gk. *solen* = a tube; *stemon* = a stamen. The stamens are united at the base to form a tube.

Solidago (Asteraceae) L.
La. *solido* = to make whole or strengthen; *-ago* to make; probably alluding to its healing properties.

Soliva* (Asteraceae) Ruíz & Pav.
For Salvador Soliva (1750–1793), Spanish botanist and medical practitioner. He obtained his PhD from the University of Madrid in 1771, became physician to the Spanish Court and the royal family of Spain, and director of the Madrid Botanical Garden. In 1785, he and Joaquin Rodriguez (1839–1905) began studying the medicinal properties of some plants. He co-authored, with Rodriguez, the three-volume work *Observaciones de las Eficaces Virtudes Nuevame*nte *Descubiertas ó Comprobadas en Varias Plantas* (*Observations of Newly Discovered Virtues Effective or Proven in Several Plants*) (1787–1890).

Sonchus* (Asteraceae) L.
Gk. *sonchos* = an ancient name for a kind of thistle. 'Sow thistle', sometimes incorrectly called 'milk thistle', which belong to the genus *Silybum*. This genus exudes a milky latex when the plant is cut or damaged. As a result, this thistle was fed to lactating pigs (sows) in the belief that milk production would increase.

Sonderina = Dasispermum (Apiaceae) H. Wolff
For Otto Wilhelm Sonder (1812–1881), German botanist and pharmacist, practising in Hamburg.

He accumulated an enormous private herbarium in excess of 250 000 specimens from some of the leading botanists and collectors of his day. He had a special interest in algae, and wrote an algal supplement to Mueller's *Fragmenta Phytographiae Australiae* and a major paper on Australian tropical algae. Although he never actually visited the Cape, he co-authored with William Henry Harvey the first three volumes of the seven-volume *Flora Capensis*. He also wrote *Flora Hamburgensis,* and was editor and author of several families of *Plantae Muellerianae* in the journal *Linnaea.*

Sonderothamnus (Penaeaceae) R.Dahlgren
Sonderina (q.v.); Gk. *thamnos* = bush, shrub.

Sophora (Fabaceae) L.
Arabic name *sophera* or *sufayra* for a tree with pea-like flowers.

Sopubia (Orobanchaceae) Buch.Ham. ex D.Don
The Indian vernacular name.

Sorghastrum (Poaceae) Nash
New Latin, from Italian *sorgo* = a tall cereal grass; *-astrum* = somewhat resembling or inferior to. Literally, a 'poor imitation of *Sorghum*', referring to its similarity to *Sorghum,* another wild grass genus.

Sorghum (Poaceae) Moench
Origin uncertain. Possibly derived from the Indian *Sorghi*, or Latin *surgum*, *suricum*; referring to *Syricum gramen* = Syrian grain.

Sorocephalus (Proteaceae) R.Br.
Gk. *soros* = a heap; *kephale* = head. The flowers are in clustered heads.

Soroveta (Restionaceae) H.P.Linder & C.R.Hardy
La. *soror* = sister; *vetus* = ancient; referring to the topological position as the sister to the rest of the *Restionaceae.*

Sowerbaea* (Alliaceae) Sm.
For James Sowerby (1757–1822), English naturalist and illustrator. He studied art at the

Royal Academy and took an apprenticeship under Richard Wright. His worked initially with an English botanist and entomologist, William Curtis (1746–1799), who recognised his outstanding ability. They created the early volumes of the first British botany journal, *The Botanical Magazine,* containing 70 of his works. He also illustrated *Flora Londinensis*, a venture with Curtis. His major projects were a 36-volume work on the botany of England known as *Sowerby's Botany* (1790–1814) featuring 2 592 hand-coloured illustrations. Besides the renowned botanical works, Sowerby produced extensive volumes on mycology, conchology and mineralogy, and a seminal work on his colour system.

Sparaxis (Iridaceae) Ker Gawl. (Images on opposite page)
Gk. *sparasso* = to rend or tear; relating to the bracts that are lacerated or cut into segments.

Sparrmannia (Malvaceae) L.f.
For Anders Erikson Sparrman (1748–1820), Swedish botanist and physician, and pupil of

Linnaeus. He visited Canton (1765–1767), completed his medical studies (1770), and visited the Cape in 1772, where he botanised around Cape Town with Carl Peter Thunberg. That year, Captain Cook's expedition to circumnavigate the world reached the Cape. Sparrman was invited to join them. They returned in 1775, and Sparrman

Sparaxis (CM)   *Sparaxis*: The torn bracts are not easily visible. (MC)   (MC)

spent from July 1775 to April 1776 exploring and collecting in the Eastern Cape. He recorded these experiences in his *A Journey to the Cape of Good Hope ...* (1777). He was elected to the Royal Swedish Academy of Sciences in 1777 and was its president from 1778–1785. From 1790–1803, he was professor of natural history and pharmacy at the Royal Collegium Medicum in Stockholm.

Spartina (Poaceae) Schreb.
Gk. *spartine* = a cord made from *Spartium junceum* (Spanish broom); possibly named as such because of its tough fibrous leaves or stems, which were used for making cord.

Spartium* (Fabaceae) L.
Gk. *sparton* = a kind of grass used for weaving and rope-making. Although *Spartium junceum* is known as 'Spanish broom' or 'weaver's broom', the stiff plant fibres were mainly used to make cordage, nets, bags and even sails.

Spatalla (Proteaceae) Salisb.
Gk. *spatalos* = wanton, riotous. A sexual allusion to the unusually large pollen presenters on so small a flower. Commonly known as 'spoons'.

Spatallopsis*** = Spatalla (Proteaceae) E.Phillips
Spatalla (q.v.); Gk. *-iopsis* = resembling, likeness.

Spathodea* (Bignoniaceae) P.Beauv.
Gk. *spathe* = a large bract enclosing a flower cluster; *-odes* = of the nature of, having the form of; referring to the spathe-like calyx.

Spergula* (Caryophyllaceae) L.
La. *spergere* = to scatter, strew; *-ula* = diminutive; probably referring to the seed dispersal – 'the species are widely dispersed' (Sonder in *Flora Capensis*).

Spergularia (Caryophyllaceae) (Pers.) J. & C.Presl
Spergula (q.v.); *-aria* = resembling.

Spermacoce (Rubiaceae) Gaertn.
Gk. *sperma* = a seed; *akoke* = a point; alluding to the capsule being crowned by calycina teeth.

Spetaea (Hyacinthaceae) Wetschnig & Pfosser
For Franz Speta (1941–), Austrian botanist. He studied at botany and zoology at the University  of Vienna, graduating in 1972. From 1970, he worked as a research assistant at the Upper Austrian State Museum, first as head of the department of botany and invertebrates, then, in 1985, as deputy director of the museum and, in 1990 and 1991, as its interim director. From 1993–2003, he headed up the biology centre of the National Museum. Speta's research focus was bulbous plants, especially *Hyacinthaceae*, with a special interest in *Scilla* and *Ornithogalum*. He published about 100 scientific papers and 50 biographical works (mainly of botanists). He was awarded the *venia lengendi* for systematic botany by the University of Salzburg and, in 1994, was appointed 'real councillor'.

Sphaeralcea (Malvaceae) A.St.Hil.
Gk. *sphair-* = spherical, globe-shaped; *alkaia* = mallow; referring to the seed capsule's spherical appearance.

Sphaeranthus (Asteraceae) L.
Gk. *sphair-* = spherical, globe-shaped; *anthos* = flower. William Henry Harvey says the 'heads are crowded into a globose glomerule'; 'in globular clusters' (Don Perrin).

Sphaeritis = Crassula (Crassulaceae) Eck. & Zeyh.
La. *sphaer-, sphaero-* = spherical, globe-shaped; referring to the shape of the inflorescence.

Sphaerocionium = Hymenophyllum (Hymenophyllaceae) C.Presl
La. *sphaer-, sphaero-* = spherical, globe-shaped; *kionium* = small column. The sporangia are borne on a small pedicel.

Sphaeroclinium = Cotula (Asteraceae) Sch. Bip.
La. *sphaer-, sphaero-* = spherical, globe-shaped; *klinion*, diminutive of *kline* = couch, receptacle; referring to the flower-head.

Sphaerocodon (Apocynaceae) Benth.
La. *sphaer-, sphaero-* = spherical, globe-shaped; *codon* = bel. The tube is campanulate.

Sphaeroma = Anisodontea (Malvaceae) DC.
Gk. *sphairoma* = something made round; referring to the inflorescence.

Sphaeropteris* (Cyatheaceae) Bernh.
Gk. *sphair-* = spherical, globe-shaped; *pteris* = fern, from *pteron* = wing; alluding to the globose sori.

Sphaerospora* = **Gladiolus*** (Iridaceae) Klatt
La. *sphaer-, sphaero-* = spherical, globe-shaped; *spora* = seed. The seeds are ovate to round.

Sphaerostylis (Euphorbiaceae) Baill.
La. *sphaer-, sphaero-* = spherical, globe-shaped; Gk. *stylis* = style (column, pillar); referring to the globe-shaped styles.

Sphaerothylax (Podostemaceae) Bisch.
La. *sphaer-, sphaero-* = spherical, globe-shaped; *thylakos* = (leather) bag, sack, pouch, scrotum.

Sphagneticola* (Asteraceae) O.Hoffm.
Gk. *sphagnos* = peat moss; La. *etum* = place (where something grows); La. *cola* = to inhabit; an ancient name used by Pliny the Elder for a lichen or moss.

Sphalmanthus = Mesembryanthemum (Aizoaceae) N.E.Br.
Gk. *sphalma* = a stumble, false step, error; *anthos* = flower; apparently referring to its rooting from surface runners. This genus does not have deep roots.

Sphedamnocarpus (Malpighiaceae) Planch. ex Benth. & Hook.
Gk. *sphe* = wasp; *damnos* = a curse, 'damn it'; *karpos* = fruit. Some species provide pollinators, such as wasps, with pollen or nectar for their young.

Sphenandra* = **Sutera*** (Scrophulariaceae) Benth.
Gk. *spheno-* = wedge-shaped, sphenoid; *-andros* = male; referring to the wedge-shaped anthers.

Sphenoclea (Sphenocleaceae) Gaertn.
Gk. *spheno-* = wedge-shaped, sphenoid; *-kleio* = to enclose; alluding to the wedge-shaped capsules.

Sphenogyne = Ursinia (Asteraceae) R.Br.
Gk. *spheno-* = wedge-shaped; *gyne* = woman; referring to the wedge-shaped stigmas.

Sphenomeris* (Lindsaeaceae) Maxon.
Gk. *spheno-* = wedge-shaped; *meris* = a part, portion; referring to the wedge-shaped fronds or pinnules.

Sphenopus* (Poaceae) Trin.
Gk. *spheno-* = wedge-shaped; *-pus, pous* = foot; referring to the distally thickened stalks or pedicels.

Sphenostylis (Fabaceae) E.Mey.
Gk. *spheno-* = wedge-shaped; *stylis* = column, pillar, style. The style is wedge-shaped.

Spicanta = Blechnum (Blechnaceae) C.Presl
La. *spica* (ear of corn, spike) = spiked or tufted; referring to the central fertile fronds.

Spielmannia (Scrophulariaceae) Willd.
For Jakob Reinhold Spielmann (1722–1783), German physician and pharmacist. He qualified  as a pharmacist in 1743 and worked in his family pharmacy until 1755, when he became professor of medicine at the University of Strassbourg. In 1756, he became professor of poetry and, in 1759, professor of medicine, chemistry and pharmacognosy. A highly respected European scholar, he was interested in both the theory and practice of science, especially chemistry. He wrote many books and textbooks that were translated into foreign languages based upon his lectures. He was one of the last chemists of note who supported the phlogiston theory. He was also Goethe's chemical teacher. For 34 years, he was director of the botanical garden at Krutenau.

Spilanthes (Asteraceae) Jacq.
Gk. *spilos* = a spot; *anthos* = flower. The type species has spot-like disk florets.

Spiloxene* = *Pauridia (Hypoxidaceae) Salisb.

(CM)

Gk. *spilos* = a spot, stain; *xenos* = host or stranger; referring to the spotted base of the tepals in some species.

Spiralepis** (Asteraceae) D.Don
Gk. *speiros* = spiralled, twisted; *lepis* = scale; referring to the twisted scales.

Spirodela (Lemnaceae) Schleid.
Gk. *speiros* = spiralled, twisted; *delos* = visible. The spiral vessels are visible through transparent tissues.

Spirostachys (Euphorbiaceae) Sond.
Gk. *speiros* = spiralled, twisted; *stachys* = spike. The flowers are in a tight spiral on the spike.

Sponia** = ***Trema*** (Urticaceae) Comm. ex Lam.
For Jacob (Jacques) Spon (1647–1685), French physician, archaeologist, antiquarian and traveller. He studied medicine at Strasbourg and Montpellier, graduating in 1668 and practiced in Lyon focusing on wealthy clientele. He developed an interest in classical antiquity and his *Recherche des Antiquités et Curiosités de la Ville de Lyon* (1668) launched him into a wider circle of erudite scholars, philosophers and theologians. Between 1675 and 1676, he travelled to Italy, Greece, (modern) Turkey, and the Levant with Sir George Wheeler (1650–1723), collecting coins, inscriptions and manuscripts and the like, as described in *Voyage d'Italie, de Dalmatie, de Grèce et du Levant* (1678). Before his early death from tuberculosis, he authored many papers, such as *Récherches Curieuses d'Antiquité* (1683).

Sporledera* = *Ceratotheca (Pedaliaceae) Bernh.
For Friedrich Wilhelm Sporleder (1787–1875), German botanist and government administrator, and a member of the Scientific Association of Wernigerode. He collected a herbarium of plants from Austria, South Tirol (Monte Baldo), North America, South Africa, Java and Greenland, all of which was destroyed during World War II. His most significant and outstanding scientific contribution is without doubt the recording of the flora of the county of Wernigerode. He published his results in the 'list of Wernigerode in the county and the immediate neighbourhood wild Phanerogams and vascular Kryptogamnen', listing 1 300 species. By 1950, 22% of them could not be found and many were endangered species and on the 'red list' of Saxony-Anhalt.

Sporobolus (Poaceae) R.Br.
Gk. *sporos* = seed; *bolos* = casting (*bollein* = to throw); referring to the ease with which the ripe seed is released.

Sprekelia* (Amaryllidaceae) Heist.
For Johann Heinrich von Spreckelsen (1691–1764), German lawyer and collector of natural history specimens, and secretary of the Hamburg city council. He was a personal friend of Linnaeus (1707–1778), who admired his fine garden and the many exotics and orange trees. Linnaeus mentions that he had 45 aloes and 50 kinds of mesembryanthemums. He also had 'a big collection of fossils, I've never seen larger'. In addition, he had a splendid library from which his guest could borrow Patrick Blair's *Botanck Essays*. In return, Linnaeus displayed his collection of insects.

Staavia (Bruniaceae) Dahl
For Martin Staaf (1731–1788), Swedish chief financial officer and director of the East India Company of Gothenburg and correspondent of Linnaeus in 1772. As a businessman, it is known that he had an interest in materials such as porcelain obtained from China and, in 1784, was present at a meeting of the Académie Royale des Sciences to discuss Chinese pharmacopoeia – the use of drugs and their preparation.

Staberoha (Restionaceae) Kunth

Male inflorescence (Left) Female inflorescence (Right) Because these are so different the plants were originally thought to be a separate species. (CM – Both)

For Johann Heinrich Julius Staberoh (1785–1857), German pharmacist, medical assessor and co-owner with Georg Friedrich Albrecht Hempel (died 1836) of a chemical factory in Berlin. He was a member of the German Pharmaeutical Society, and a member of the board of examiners and co-workers at the Pharmacopoea Borussica. He published many articles in pharmaceutical journals, and was a co-author with Heinrich Friedrich Link (1767–1851) of *Pharmacopoeia Borussica* (1829), originally published in Latin. Carl Sigismund Kunth (1788–1850), professor of botany at the University of Berlin, author of the genus name, could have known Staberoh personally. Staberoh seems to have visited Norway and Scotland in 1838 in connection with a typhus epidemic. The *Gazette Médical de Paris* (1839) reports, '*Le docteur Staberoh, de Berlin, qui a observé la dernière épidémie de fièvre typhoïde à Glasçow ...*' ('Dr Staberoh, from Berlin, who observed the last epidemic of typhoid fever in Glasgow.'), but perhaps this was another Staberoh.

Stachygynandrum*** = Selaginella
(Selaginellaceae) P.Beauv. ex Mirb.
Gk. *stachys* = spike; *gyne* = female, woman (pistil); *andrum* = male, the stamens; possibly referring to the column bearing stamens and pistils.

Stachys (Lamiaceae) L.
Gk. for a *spike* (originally ear of wheat); referring to the inflorescence that is often a spike.

Stachytarpheta* (Verbenaceae) Vahl
Gk. *stachus* (*stachys*) = spike; probably from *tarpheia* = fem. of *tarphys* = thick, dense; referring to the flowers growing in dense spikes.

Stadmannia (Sapindaceae) Lam.
For Jean Frédéric Stadtmann (1762–1807), German physician, botanist and painter. He

obtained his medical qualifications from the medical school of the University of Strasbourg and, shortly after, in 1786, emigrated to the Île-de-France (Mauritius). He practised there as a doctor, studied botany, and became an excellent botanical artist.

He collected in Madagascar, Mauritius and South Africa, with a keen interest in mainly algae and spermatophytes. He was asked to contribute paintings for the five-volume *Encyclopédie Méthodique: Botanique* by Jean-Baptiste Lamarck and Jean Louis Poiret on the Île-de-France (Mauritius).

Staehelina (Asteraceae) L.
For Benedict Staehelin (Staehelin) (1695–1750), a Swiss botanist and professor of physics in Basel. He studied medicine in Basel and Paris, obtaining his doctorate in 1716, and joined the medical faculty. He distinguished himself by his research in cryptogamous plants, and published several works. He was a good friend of Albrecht Haller, Swiss anatomist, physiologist and naturalist. In 1727, he became professor of physics. From his mid-40s, he experienced increasing mental illness, and his position was taken over in 1747 by Daniel Bernoulli (1700–1782), the Dutch-Swiss mathematician and experimental physicist. In 1750, Staehelin wanted to become a vicar but died that year.

Stangeria (Zamiaceae) T.Moore
For William Stanger (1811–1854), English physician, geologist and first surveyor general

of Natal. He survived the ill-fated Niger expedition under Captain H Trotter (1841), came to the Cape in 1843, moved to Natal in 1845, and produced the first accurate map of Natal in 1850. When he went to England in 1851, he gave a Dr NB Ward a

live fern-like plant that later produced a cone. The English botanist T Moore described this new cycad *Stangeria paradoxa*, a reference to the confusion with ferns. The German botanist Otto Kuntze had previously described this plant as a 'fern', *Lomaria eriopus,* in 1839. This was inadmissible under the rules of botanical nomenclature, and the French botanist Henri Baillon renamed it *Stangeria eriopus* in 1892.

Stapelia (Apocynaceae) L.

Stapelia flowers are fly pollinated and smell of carrion. (CM)

For Johannes Bodaeus van Stapel (1602–1636), Dutch physician and botanist. He received a medical degree in 1625 from Leiden University and studied botany under Adolphus Vorstius. His life's ambition was to publish an annotated edition of the botanical works of Theophrastus (370–287 BCE), but he died before the book was finished. The content was edited and published by his father as *Theophrasti Eresii de Historia Plantarum* in 1644. One of the plants in the book, drawn by Justus Heurnius (1587–1653) from his brief stay at the Cape in 1624, was *Fritillaria crassa* (*Stapelia variegata*), now known as *Orbea variegata*. The genus was named *Stapelia* in 1753 by Linnaeus in his *Species Plantarum*.

Stapelianthus (Apocynaceae) Choux
Stapelia (q.v.); *anthos* = flower; Stapelia-flowered.

Stapeliopsis (Apocynaceae) Pillans
Stapelia (q.v.); Gk. *-opsis* = resembling.

Staphylea (Staphyleaceae) L.
Gk. *staphyle* = a cluster; referring to the inflorescence.

Stathmostelma = Ascelepis (Apocynaceae) K.Schum.
Gk. *stathmos* = plumb line; *stelma* = crown, garland, wreath; referring to the straight appendages of the inner corona segments.

Statice*** = Limonium (Plumbaginaceae) Boiss.
Gk. *statike, statikos* for an astringent herb, mentioned by Dioscorides; from *statikos* = causing to stand still; referring to the efficacy of this flower to stop diarrhoea.

Staudtia* (Myristicaceae) R.Br.
For Alois Staudt (?–1897), German gardener and plant collector who collected specimens in Togo and Cameroon from 1893–1895 with Georg August Zenker (1855–1922) at Yaounde in Cameroon. Zenker left Yaounde in 1895 and moved to Bipinde, also in Cameroon, but Staudt stayed at Yaounde until his death.

Stayneria (Aizoaceae) Bolus
For Frank James Stayner (1907–1981), South African horticulturist, specialist on succulent  plants, trained with the Port Elizabeth parks department from 1928 to 1932 and at the Royal Botanical Gardens in Kew from 1933–1934. He was assistant superintendent of parks in the Port Elizabeth parks department from 1935–1946, farmed from 1947–1948, and worked as a horticulturalist for Ford Motor Company from 1949–1954. He farmed again in 1955, was with the parks department again from 1956–1959, and was curator of the (newly named) Karoo Desert National Botanical Garden at Worcester from 1959–1969.

Steganotaenia (Apiaceae) Hochst.
Gk. *steganos* = covered; *tainia, taenia* = a band. The meaning is obscure.

Stegnogramma = Cyclosorus (Thelypteridaceae) Blume
Gk. *steganos* = covered; *gramma* = a line. The elongated sori is covered with lines.

Steinchisma* (Poaceae) Raf.
Gk. *steinos* = narrow; *khasma* (*chasma*) = yawning; alluding to the gaping lower floret. Possibly referring to the gaping glumes and somewhat narrow spikelet when compared to *Panicum*.

Steirodiscus (Asteraceae) Less.
Gk. *steiros* = sterile; *diskos* = disk. The disk florets are sterile.

Stellaria* (Caryophyllaceae) L.

The leaves can be eaten as a 'lettuce'. (EJM)

La. *stella* = a star; *-aria* = like; the flowers are star-like in shape.

Stemodia* (Plantaginaceae) L.
Gk. *stemon* = thread, stamen; *di* = two; referring to the double anthers.

Stemodiacra*** = *Stemodia* (Scrophulariaceae) Kuntze
Gk. *stemon* = stamen; *di* = two; *akros* = end, tip; referring to the stamens with two tips.

Stemodiopsis (Scrophulariaceae) Engl.
Stemodia (q.v.); Gk. *-opsis* = resembling. The four stamens are didynamous (occurring in pairs).

Stenochlaena (Blechnaceae) J.Sm.
Gk. *steno-* = narrow; *chlaina* = cloak, blanket; referring to the narrow, involute margins of the sterile pinnae.

Stenocline = *Helichrysum* (Asteraceae) DC.
Gk. *steno-* = narrow; *kline* = a bed, couch. The leaves and flowers of some species were used as bedding.

Stenoglottis (Orchidaceae) Lindl.
Gk. *steno-* = narrow; *glottis* = tongue; referring to the lobes of the tongue-like lip.

Stenolobus*** = *Davallia* (Davalliaceae) C.Presl
Gk. *steno-* = narrow; *lobus* = lobe; referring to the plant's narrower calyx lobes and bracts.

Stenosemis (Apiaceae) E.Mey. ex Sond.
Gk. *stenos* = narrow; *semis* = 'half of anything'. The petals of some species of this genus, such as *Stenosemis caffra*, are curiously shaped as though sliced in places (emarginate).

Stenostelma (Apocynaceae) Schltr.
Gk. *stenos* = narrow; *stelma* = crown, garland, wreath (equivalent to *stemma*); referring to the shape of the corona.

Stenotaphrum (Poaceae) Trin.
Gk. *steno-* = narrow; *taphros* = a trench or ditch; alluding to the small cavities (pits) on the surface of the rachis.

Stephania (Menispermaceae) Lour.
Gk. *stephanē* = crown; alluding to the anthers being united and forming a crown.

Stephanocoma*** = *Berkheya* (Asteraceae) Less.
Gk. *stephanos* = crown; *coma* = head of hair; alluding to the crown-like pappus (William Henry Harvey, *Flora Capensis*).

Sterculia (Malvaceae) L.
Stercullus was the god of cultivation and manuring. La. *stercus* = dung; referring to the smell of the flowers and leaves of some species.

Stereochlaena (Poaceae) Hack.
Gk. *stereo-* = stiff, solid; *chlaena* = cloak, covering. The fertile lemma is rigid.

Steudelia*** = *Adenogramma* (Molluginaceae) C.Presl
For Ernst Gottlieb von Steudel (1783–1856), German botanist and chief state physician for the Kingdom of Württemberg. He studied at Tübingen University, and received his DM in 1805. With Christian Ferdinand Hochstetter (1787–1860), he founded Württembergischer Botanische Reiseverein, aimed at encouraging young botanists to discover and collect plants throughout the world. Steudel's herbarium contained more than 20 000 species. Steudel and Hochstetter co-authored *Enumeratio Plantarum Germaniae* (1826), *Synopsis Planterum Glumacearum* (two volumes, 1853–1855), while Steudel wrote *Nomenclator Botanicus* (two volumes, 1821–1824), listing of more than 3 300 genera and nearly 40 000 species. After 50 years as a doctor, he was made a knight of the Royal Order and received an honorary doctorate from the University of Tübingen.

Stiburus*** = *Eragrostis* (Poaceae) Stapf
Gk. *stib-* = anatomy; *-urus* = tailed. The grass spikelets looks like a (fox's) tail.

Sticherus = *Gleichenia* (Gleicheniaceae) C.Presl
Possibly derived from the Gk. *stichos* = row or

Stipagrostis: Namibian desert with grass (EJM)   *Stipagrostis*: Note the 'feathery' awns. (EJM)

line; referring to the double rows of sori on the pinnules.

Stictocardia (Convolvulaceae) Hallier.
Gk. *stiktos* = spotted; *kardia* = a heart; the leaves are heart-shaped and spotted below.

Stigmatocarpum = Cleretum (Aizoaceae) L.Bolus
Gk. *stigmatos* = tattoo mark, spot; *karpos* = fruit; possibly referring to the spotted fruit.

Stigmatorhynchus (Apocynaceae) Schltr.
Gk. *stigmatos* = tattoo mark, spot; *rhynchos* = snout, beak; referring to the paired beak-like fruit.

Stilbe (Stilbaceae) P.J.Bergius
Gk. *stilbo* = to shine; also a nymph in Greek mythology, daughter of the river god Peneus and the Naiad Creusa, but 'the allusion is not clear' (WPU Jackson). The inflorescence of *S. vestita* often shines like lighted candles at the top of a candelabra (Hugh Clarke).

Stilpnogyne (Asteraceae) DC.
Gk. *stilpnos* = glittering; *gyne* = female, woman (pistil); referring to some aspect of the gynoecium.

Stilpnophyton*** = Athanasia (Asteraceae) Less.
Gk. *stilpnos* = glittering; *phyton* = plant; referring to the glossy involucre.

Stipa (Poaceae) L.
Gk. s*type* = fibre, such as hemp or plant fibre; referring to the feathery awns in some species.

Stipagrostis (Poaceae) Nees
Stipa (q.v.); *Agrostis* (q.v.). The two 'conjoined' genera have common characteristics, such as feathery awns.

Stirtonanthus (Fabaceae) B.E.van Wyk & A.Schutte.
Stirtonia (q.v.); *anthos* = flower.

Stirtonia = Stirtonanthus (Fabaceae) B.E.van Wyk & A.Schutte.
For Charles Howard Stirton (1946–), South African botanist, university professor at Natal and Wales,

plant collector and author. He has an MSc from the University of Natal (now KwaZulu-Natal) (1974) and a PhD from the University of Cape Town (1989). He was chief professional officer at the Botanic Research Institute (South Africa), working mainly at the National Herbarium in Pretoria. He was deputy director of the Royal Botanic Gardens in Kew (1992–1996) and founding director of the National Botanical Garden of Wales (1996–2002) before setting up a consultancy, Calle Contextua. Between 1981 and 1992, he collected some 10 000 specimens, mainly in KwaZulu-Natal and Mpumalanga. He is the author of four books and 125 scientific publications, and has received many academic awards. He is a Fellow of the Linnaean Society.

Stiza*** (Fabaceae) E.Mey.
Gk. *stizo* = to sting; referring to the rigid, spiny, nearly leafless bush.

Stizolobium = Mucuna (Fabaceae) Pers.
Gk. *stizein* = to tattoo, *stizo* = to sting; *lobion* = dim. of *lobos* = lobe, pod, capsule; referring to the stinging hairs with which the pods are clothed.

Stobaea (Asteraceae) Thunb.
For Kilian Stobaeus (1690–1742), Swedish physician and naturalist. He studied at Lund University, the first individual in Sweden to receive a doctorate of medicine (1721). Later he became acting professor of medicine, then of natural philosophy and experimental physics (1728) and of history (1732). While at Lund University (1727), Linnaeus regularly attended his lectures on molluscs. Stobaeus befriended him, let him live and eat in his house, attend his lectures without payment, and showed him his collection of plants dried and glued to paper, something new to Linnaeus. Later, Linnaeus created his own herbarium and showed his gratitude by naming a plant after Stobaeus. In 1735, Stobaeus donated his collection to Lund University, which became Museum Stobaeanum.

Stoebe (Asteraceae) L.
Gk. *steibein* = to tread; *stoibe* = a stuffing (Jackson); a classical name for a plant used as packing material and for stuffing cushions.

Stoeberia (Aizoaceae) Dinter & Schwantes
For Ernst Stöber, (1889–?), German teacher and botanical explorer in (then) South-West Africa who taught at the Lüderitz Primary School, Namibia, during between 1923 and 1924. He would have been acquainted with Moritz Kurt Dinter (1868–1945), who was in Namibia at that time.

Stoibrax (Apiaceae) Raf.
Derivation unknown.

Stokoeanthus = Erica (Ericaceae) E.G.H.Oliv.
For Thomas Pearson Stokoe (1868–1959), Yorkshire-born legendary mountaineer and important fynbos collector. He came to South Africa in 1911, joined the Mountain Club in 1913, and climbed extensively the mountains of the southwestern Cape. He discovered nearly 150 plants previously unknown to science, and

'rediscovered' many plants only known from the early collectors at the Cape. Thirty-one Cape flowers and one beetle are named after him. He is perhaps best known for his discovery of *Mimetes hottentoticus* in the Kogelberg Reserve, named after the Hottentots Holland Mountains. He died shortly after his 91st birthday after becoming ill while climbing with Rycroft and never fully recovering.

Stoebe: The classic 'stuffing'
Stoebe (EJM – Both)

Stoebe (CM)

Stomatanthes (Asteraceae) R.M.King & H.Rob.
Gk. *stoma* = an opening, mouth; *anthos* = flower; referring to the flowers appearance.

Stomatium (Aizoaceae) Schwantes
Gk. *stoma* = an opening, mouth; *-ium* = characteristic of; referring to the pairs of toothed leaves that resemble an open mouth.

Stomatostemma (Apocynaceae) N.E.Br.
Gk. *stoma* = mouth; *stemma* = wreath, garland; referring to the corona at the mouth of the corolla tube.

Streblochaete (Poaceae) Pilg.
Gk. *streblos* = twisted; *chaete* = a bristle; referring to the lemma that have long, twisted awns.

Strelitzia (Strelitziaceae) Aiton

These 'crane-flowers' are now used globally in flower arrangements. (GAN)

For Charlotte of Mecklenburg-Strelitz (1744–1818), who married King George III of England in 1761 after being selected unseen from a list of German princesses. Remarkably, the marriage was a success and King George was devoted to her. She cared for her husband during his long slide into insanity, though terrified by his occasional outbursts of violence. She was a patroness of the arts and an amateur botanist who helped expand the Kew Gardens. She predeceased him and was buried in 1818, in St George's Chapel, Windsor. *Strelitzia reginae* arrived in England in 1773 being grown at the Royal Botanic Gardens in Kew. It was so named to commemorate Charlotte of Mecklenburg-Strelitz.

Streptanthera* = ***Sparaxis*** (Iridaceae) Sweet
Gk. *streptos* = bent, twisted; *anthera* = anther. The anthers are twisted.

Streptocarpus (Gesneriaceae) Lindl.
Gk. *streptos* = twisted; *karpos* = fruit. A characteristic of this genus is its spirally twisted fruit.

Streptopetalum (Turneraceae) Hochst.
Gk. *streptos* = twisted; *petalum* = petals.

Striga (Orobanchaceae) Lour.
La. *strigo,* from *stringo* = to grasp and hold fast. This word is associated with *strigis* = furrow, channel; and *strix* = screech owl, hag and witch. This parasitic plant, known as 'witchweed', is a vicious invasive species that seriously damages crop cereals by colonising the underground, and in so doing wipes out crops. The word *striga* can be used in the sense of rendering victims prematurely aged and weak.

Strobilopsis (Scrophulariaceae) Hilliard & B.Burtt
Gk. *strobilos* = round ball, spinning top, pine (*strobilus* in N.La. = a cone); *-opsis* = resembling; referring to lobes of the corolla: the petals have long, often twisted appendages.

Strophanthus (Apocynaceae) DC.
Gk. *strophos* = cord, twisted band; *anthos* = flower. The petals have long, often twisted appendages.

Strumaria (Amaryllidaceae) Jacq. ex Willd.
N.La. *struma* = cushion-shaped swelling of an organ from La. *stroma* = a tuberculosis or other swelling; *-aria* = relating to, possessing; referring to the conspicuously swollen base of all the flowers of this genus (Plantzafrica.com).

Struthiola (Thymelaeaceae) L.
Gk. *strouthos* = a starling. The seeds are pointed like a starling's beak (Hugh Glen). Alternately, Gk. *strouthion* = a small bird, sparrow (dim. of *strouthos* = ostrich). Perhaps the seed resembles a sparrow's beak (WPU Jackson).

Struthiolopsis** = ***Gnidia*** (Thymelaeaceae) E.Phillips
Struthiola (q.v.); Gk. *-opsis* = resembling.

Struthiopteris = ***Osmunda*** (Osmundaceae) Bernh.
Gk. *stroutheios* = ostrich; *pteris* = fern, from *pteron* = wing; alluding to the large, feathery fern with its distinctive, fertile frond.

Strychnos (Loganiaceae) L.
Classical Greek name for several poisonous plants meaning 'deadly'; referring to the poisonous alkaloids contained in the seed integuments.

Stultitia *** = ***Orbea*** (Apocynaceae) E.Phillips
La. *stulti-* = stupid, folly, adopted. The reason for this name was not given.

Sturmia *** = ***Liparis*** (Orchidaceae) Rchb.f.
For Jacob Sturm (1771–1848), German engraver, naturalist and botanical artist. He

learned the art of drawing and copperplate engraving from his father, Johan Georg Sturm. When aged 16, he met botanist Johann Christian Daniel Schreber (1739–1810) and the entomologist Georg Wolfgang Franz Panzer (1755–1829), and worked for them. In 1791–1792 he published *Insekten-Cabinet nach der Natur Gezeichnet und Gestochen*, and with Panzer, undertook a 20-year project, *Faunae Insectorum Germanicae Initia* (1792–1813), featuring 2 600 insects with Panzer's short textual descriptions. Sturm also produced his own insect catalogue, which went through four editions. He was a founding member of Naturhistorische Gesellschaft zu Nürnberg in 1801.

Stylapterus (Penaeaceae) A.Juss.
Gk. *stylos* = pillar, column; *-a-* = without; *pteron* = wing, as opposed to the winged style of Penaea (Umberto Quattrocchi).

Stylochaeton (Araceae) Lepr.
Gk. *stylos* = pillar, column; *-chiton* = covering, tunic; referring to a connate perigone surrounding the gynoecium.

Stylocoryne **** = ***Coptosperma*** (Rubiaceae) Wight & Arn.
Gk. *stylos* = style; *koryne* = a club; referring to the shape of the style.

Stylosanthes (Fabaceae) Sw.
Gk. *stylos* = style; *anthos* = flower. The flower has long styles.

Styppeiochloa (Poaceae) De Winter
Gk. *stuppeion* = coarse fibre; *chloa* = grass. The plant has sclerophyllous tufts, with hard, fibrous basal sheaths forming thick, dense, fire-resistant mats.

Suaeda (Amaranthaceae) Forssk. ex J.F.Gmel.
From the Arabic vernacular name, *suwed* = silty, possibly *suwaida* = blackish. This genus usually grows in dry, silty places.

Succisa (Dipsacaceae) Haller
Gk. *succissus,* past participle of *succidere* = to cut from below; alluding to its vertical rootstock, which looks like it has been bitten from below.

Suessenguthiella (Molluginaceae) Friedr.
For Karl Suessenguth (1893–1955), German

botanist, professor of botany at the University of München (1927–1955). His research centred on botanical systematics and the chemical physiology of plants. In the 1930s, he succeeded Gustav Hegi as editor and also contributed to the *Illustrated Flora of Central Europe*. A prolific author, he wrote *Ziele der Neue Botanik* (1939), *Nachruf* (1955) with Hermann Merxmüller, and *A Contribution to the Flora of the Marandellas District, Southern Rhodesia*, also with Merxmüller. From 1946 until he died, he was director of the Botanical State Collection in Munich. Under his aegis, the state library began publishing its own annual journal. In 1949, he was appointed honorary professor.

Suregada (Euphorbiaceae) Roxb. ex Rottler.
Said to be based on the Indian Telinga name *soora gade*. Meaning obscure. Possibly Latinised from the Gk. *surigglas* = a hollow reed.

Susanna (Asteraceae) E.Phillips
Named by Edwin Percy Phillips (1884–1967) botanist, taxonomist and plant collector. He

studied at the University of Cape Town, obtaining a DSc in 1915. Thereafter, he became curator of the South African Museum (1911) and curator of the National Herbarium in Pretoria (1918). He is noted for his monumental work, *The Genera of South African Flowering Plants* (1926), *South African Grasses* (1931), and *Weeds of South Africa* (1939). He also published more than 200 papers. He held many important and influential positions such as secretary of the South African Biological Society (1919–1944), and scientific liaison officer for the Council for Scientific and Industrial Research in Washington (1946–1948). He also represented South Africa at international conferences. He was a Fellow of the Linnaean Society and the Royal Society of South Africa.

Sutera (Scrophulariaceae) Roth
For Johann Rudolf Suter (1766–1827), Swiss physician, botanist, politician and professor. He studied classical philology and natural sciences at the University of Göttingen, PhD (1787), and

philosophy and medicine at Mainz, MD (1794). He practised as a doctor in Zofingen until 1798. He became politically involved in the Jacobin Republic of Mainz, and later became a subprefect of the district of Zofingen, a member of the Canton of Aargau, and of the Switzerland's Grand Council in the Helvetic government (1798–1800), belonging to the Reform Party. In 1802, he published his book on Swiss flora, *Flora Helvetica*, with the help of Johannes Hegetschweiler (1789–1839). In 1819 he became professor of philosophy, Greek literature and history at the Academy of Berne.

Sutherlandia (Fabaceae) R.Br. ex Aiton f.
For James Sutherland (1639–1719), Scottish botanist, king's botanist for Scotland, first superintendent of the Royal Botanical Gardens in Edinburgh, and professor of botany at the University of Edinburgh. One report (unconfirmed) states he started his career as a gardener without a college education to care for the university's Physic Garden. Through self-education, he mastered the subject of botany to such an extent that he published a famous treatise, *Hortus Medicus Edinburgensis* (1683), which led to his doctorate and professorship. The Physic Garden became the Royal Botanical Gardens in Edinburgh in the early 1700s.

Swartzia (Fabaceae) Schreb.
For Olof Peter Swartz (1760–1818), Swedish botanist. He commenced his studies under  Linnaeus and Carl Peter Thunberg, and received his doctorate in 1781. He then went on a number of plant-collecting voyages and visited Lapland (1780), North America and the West Indies, mainly Jamaica and the Caribbean island of Hispaniola (1783), collecting 6 000 botanical specimens now held by the Swedish Museum of Natural History. He visited England in 1788, returning in 1789, and became Bergian botanical professor at the Academy of Sciences at Stockholm (1791). He was the author of many systematic works, and is best known for taxonomic work on pteridophytes, and also for his critical review of orchid literature.

Swertia (Gentianaceae) L.
For Emanuel Sweert (rarely spelled Swert) (1552–1612), Dutch gardener, prefect of gardens for Kaiser Rudolf II (1552–1612), merchant and artist. He worked for Rudolf II as a florist, herbalist and bulb-grower in Haarlem, Holland, selling plants to other gardens. He also acted as a middleman,  buying merchandise from ships arriving at Amsterdam (not only plants from botanists, but novelties such as crystals, stuffed birds, and sea shells) and selling from his shop in Frankfurt. He was a highly talented and accurate illustrator, who was one of the first artists to show the bulb distinct from the plant. His major work was an important *Florilegium*, published in Frankfurt in 1612, which provides a good indication of the flowering plants grown in European gardens in the 1600s.

Sycomorus*** = **Ficus** (Moraceae) Gasp.
Gk. *sykomoros* = fig-mulberry.

Sylitra*** = **Ptycholobium** (Fabaceae) E.Mey.
Greek name used by Dioscorides for 'some *Glycyrhiza* and by Medicus, formerly given to the genus now called *Lessertia*' (William Henry Harvey, *Flora Capensis*). But now it has been renamed again.

Sympieza*** = **Erica** (Ericaceae) Licht. ex Roem. & Schult.
Gk. *sympiezo* = to squeeze together; referring to the flowers being matted together.

Synadenium = **Euphorbia** (Euphorbiaceae) Boiss.
Gk. *syn-* = together; *aden* = a gland. The glands in the inflorescence are united.

Synaptolepis (Thymelaeaceae) Oliv.
Gk. *synaptos* = joined; *lepis* = a scale. The disk of the flower is composed of joined scales.

Synaptophyllum = **Mesembryanthemum** (Aizoaceae) N.E.Br.
Gk. *synaptos* = joined; *phyllon* = leaf; referring to the leaves that are joined at the base.

Syncarpha (Asteraceae) DC.
Gk. *syn-* = together; *karphos* = a dry stalk, scale; possibly referring to the dry bracts that are united into a cone-like structure.

Syncarpia* (Myrtaceae) Ten.
Gk. *syn-* = together; *karpon* = fruit. The capsules in an inflorescence fuse as they ripen.

Synclisia = **Albertisia** (Menispermaceae) Benth.
Gk. *syn-* = together, united; *klisia* = place for lying down. The sepals are somewhat united at base into a very short tube (Bentham). 'Probably implying lying down together' (WPU Jackson).

Syncolostemon (Lamiaceae) E.Mey. ex Benth.
Gk. *syn-* = united; *kolos* = curtailed, stunted; *stemon* = pillar, stamen. At least two stamens are joined both to each other and to the corolla tube (Hugh Glen). The lower pair of filaments are connate and adnate to the corolla tube (Jackson).

Syndesmanthus*** = Erica (Ericaceae) Klotzsch
Gk. *Syn* = together; *desme* = a bundle; hence *syndesmos* = that which binds together. The flowers are closely bunched together.

Syngonanthus (Eriocaulaceae) Ruhland.
Gk. *syngonus* = joined together (reproductive parts); *anthos* = flower; referring to the connate petals of pistillate flowers.

Synnema = Hygrophila (Acanthaceae) Benth.
Gk. *syn-* = united; *nema* = a thread (filament); referring to the united stamens.

Synnotia = Sparaxis (Iridaceae) Sweet
For Walter Synnot (1773–1851), Irish captain of the 89th Regiment who led a group of settlers to South Africa in 1820, landing in Saldana Bay. Most of the settlers moved to Eastern Cape, but Synnot stayed in Clanwilliam and was appointed deputy landdrost in 1821. He sent many bulbs to England and the London nurseryman Robert Sweet, who credits him with the introduction of 'more rare and new bulbs from the Cape of Good Hope at one time than was done by any other individual'. Synnot returned to Ireland in 1825, taking numerous bulbs and seed samples with him and, in 1836, emigrated to Australia and settled in Tasmania.

Sypharissa*** = Drimia (Hyacinthaceae) Salisb.
Gk. *syphar* = wrinkled skin; alluding to the scarious stipules barred with transverse partitions. Resembles shed snake skin.

Syringa (Oleaceae) L.
Gk. *syrinx* = a hollow tube or pipe; referring to the broad pith in the shoots in some species, and easily hollowed.

Syringodea (Iridaceae) Hook.f.
Gk. *syrinx* = tube, pipe; *-odes* = resembling; alludes to the long perianth tube.

Syzygium (Myrtaceae) Gaertn.

(EJM)

Gk. *syn-* = together; *zygon* = a yoke, hence *syzygos* = joined; referring to the paired branches and leaves.

T

A collage of *Tritonia* flowers

Peter Goldblatt (1943–), born in Johannesburg, South Africa, obtained a BSc from the University of the Witwatersrand and PhD from University of Cape Town studying cytology and morphology of the Iridaceae under Ted Schelpe of the Bolus Herbarium. After lecturing at the University of Cape Town for four years he took up a position at the Missouri Botanical Garden in St Louis in the US, becoming curator of African botany and subsequently studying the taxonomy, evolution and pollination biology of African Iridaceae. He continues to work on the family, one of the largest in Southern Africa, completing monographic accounts of *Babiana*, *Ixia*, *Geissorrhiza*, *Gladiolus*, *Hesperantha*, *Moraea*, *Sparaxis* and other genera, working alone or in close collaboration with John Manning at the Compton Herbarium, Kirstenbosch.

Tabernaemontana (Apocynaceae) L.

The 'warty' fruits give this tree the common name 'Toad-tree'. (EJM)

For Jakob Theodor von Bergzaben (Jacobus Theodorus) (1522–1590), also known as 'Tabernaemontanus', German physician, pharmacist and botanist. He developed a passion for herbs and their effects, and studied under Hieronymus Bocks (1545); became a private physician to Count Phillipp III of Nassau-Saarbrücken-Weilburg (1549); studied medicine further at the University of Heidelberg (1562); and later became personal physician to Bishop Marquard and Elector John Casimir in Heidelberg. Throughout his life, he sought a universal cure for the plague (typhus). He is best known for his life's work, an illustrated book of herbals (medical plants) *Neuwe Kreuterbuch*, published in 1588, with more than 2 300 woodcut illustrations of herbals, which was reprinted up until the 18th century.

Tacazzea (Apocynaceae) Decne.
From the Tacazze River, Tigray, Ethiopia (not traced in modern gazetteers).

Tacca* (Dioscoreaceae) J.R.Forst. & G.Forst.
For Pietro Tacca (1577–1640), Italian sculptor and outstanding pupil of Giambologna (1529–1608) (born Jean Boulogne), who was famed for his marble and bronze statues. Tacca began working for Giambologna in 1592. When Giambologna died, Tacca finished off many of his incomplete projects and succeeded Giambologna almost immediately as court sculptor to the Medici Grand Dukes of Tuscany. Tacca's bronze castings were widely acclaimed, and are to be found not only in Italy, but also Spain and France. His last work was a huge equestrian bronze of Philip IV of Spain, which took six years (1634–1640) to complete and deliver. This massive statue places the entire weight of a rearing horse on its two back legs (and, discreetly, its tail). This was a unique first. Such was the uncertainly about its stability that Galileo Galilei was called in to make the necessary calculations.

Tagetes* (Asteraceae) L.
Thought to be named after Tages, an ancient Etruscan deity.

Talbotia (Velloziaceae) Balf.
For William Henry Fox Talbot (1800–1877), British inventor, member of parliament, and

author. He studied classics at Cambridge University and sent many papers to the Royal Society. In 1834, he began photographic experiments and invented the calotype process, precursor to most photographic processes of the 19th and 20th centuries, for which he received the Rumsford Medal from the Royal Society. He served as a member of parliament for Chippenham (1832–1835) and was high sheriff of Wiltshire (1840). He published many papers to the Royal Society and, *inter alia*, wrote books on Hermes or classical and antiquarian research topics (1838), and English etymology (1846).

Talinum (Anacamoserotaceae) Adans.
Derivation obscure. Webster dictionary provides an African vernacular name of one species in Senegal. Possibly from the Gk. *thalia* = a green branch.

Tamarindus* (Fabaceae) L.
Arabic name *tamr al-Hindi* = Indian date, (*tamr* = dried date; *al-Hindi* = of India); referring to the pulp of the pods.

Tamarix (Tamaricaceae) L.
La. *Tamaris*, from the Spanish *Támaris*, a river in the Pyrenees where it was first discovered. The classical Latin name for these trees.

Tanacetum* (Asteraceae) L.
Generally supposed to be a corruption of Athanasia, Gk. *athanatos* = without death, long-lasting; referring to the long-lived flowers; possibly derived from the medieval Latin name *tanazeta*, *tanasia*. 'Tansy'.

Tanaosolen* = **Tritoniopsis** (Iridaceae) N.E.Br.
Gk. *tanaos* = long; *solen* = tube; alluding to the long tube of the flower.

Tanquana (Aizoaceae) H.E.K.Hartmann & Liede.
From the Tanqua Karoo.

Tapinanthus (Loranthaceae) (Blume) Rchb.
Gk. *tapeinos* = low, humble; *anthos* = a flower. The flowers have narrow, linear petals, and are hidden between bracts.

Tapiphyllum* = **Vangueria** (Rubiaceae)
Gk. *tapez* = a rug, made of wool; *phyllon* = a leaf. The leaves are often woolly or hairy (Hugh Glen). The leaves are tomentose and often villous within (Jackson).

Tapura (Dichapetalaceae) Aubl.
From a Guyanese vernacular name for these trees.

Tara* (Fabaceae) Molina
The native Amerindian (Quechua) name.

Tarachia = **Asplenium** (Aspleniaceae) C.Presl
Tara = a word with many different meanings in different cultures. Most Greek names starting with *tara* imply having medical properties; *chia* is a suffix. The *Tarachia*'s synonym genus name is *Asplenium*, meaning 'spleenwort'. An old belief dating back to the times of Pedanius Dioscorides and Claudius Galen was that plant parts could be used to treat human body parts they resemble. The fern has a spleen-shaped sori on the backs of the fronds, so it was thought that it could be used to treat illnesses of the spleen.

Taraxacum* (Asteraceae) F.H.Wigg.
The name derivation of this plant has been traced back to the Persian scientist Al-Razi in about 900 CE who wrote 'the *tarashaquq* is like chicory' (Wikipedia.org). More recent names are Gk. *taraktokos* = cathartic or *tarassein* = to confuse, alter; referring to the culinary or medicinal properties of the leaves, which are said to have purgative or diuretic properties. 'Dandelion.'

Taraxis* (Restionaceae) B.G.Briggs & A.S.Johnson
Gk. *taraxis* = to stir up; it can also mean confusion; alluding to its medical properties. The root contains taraxacin, which is a 'diuretic, aperient, and tonic' (Charles M Skinner).

Tarchonanthus (Asteraceae) L.

The woolly seeds and crushed leaves smell of camphor. (EJM)

Gk. *tarchos* = funeral rite; *anthos* = flower; presumably from the camphorous odour of the leaves (Jackson) as used in incense sticks in places of worship.

Tardavel* = **Spermacoce** (Rubiaceae) Adans.
Malayalam name for *Spermacoce hispida*.

Tarenna (Rubiaceae) Gaertn.
From the Sinhalese name *tarana* for a species of this genus.

Tarigidia (Poaceae) Stent
Anagram of *Digitaria* (q.v.).

Tavaresia (Apocynaceae) Welw.
For Joaquim da Silva Tavares (1866–1931), Portuguese naturalist, clergyman and

entomologist. He entered the Jesuit order (1880), studied humanities and philosophy (1882–1888), taught Portuguese and Latin before teaching science, studied theology in Spain and France, and completed his training in Vienna, Austria (1898). He was a co-founder of the natural sciences journal *Brotéria* (1902), which published several of his scientific articles, became dean of the [Jesuit] College of St Faithful (1908), co-founded both the Portuguese and Iberian societies of Natural Sciences, and went to Argentina and Brazil (1910) before returning to Portugal (1928). He was regarded as a leading cecidiologist (the study of galls produced on trees and plants by fungi, insects or mites).

Taxillus = Septulina (Loranthaceae) Tiegh.
La. *taxillus* = small dice; alluding to the shape of a truncated cone or block.

Taxodium* (Cupressaceae) L.
La. *taxus* = yew; Gk. *eidos* = appearance. The leaves of swamp-cypress and yew are similar in shape.

Taxus* (Taxaceae) L.
Gk. *taxus* = ancient Latin name for a yew tree.

Teclea (Rutaceae) Delile.
Possibly for Mara Takla (Tekle) Haymanot (c 1137), Emperor of Ethiopia and founder of  the Zagwe dynasty, which dominated large parts of the Abyssinia (Ethiopia) for about 133 years until 1270. The author of this genus, Allire Raffeneau Delile (1778–1850), in his *Annales des Sciences Naturelles* (1843) states, in Latin, that Teclea is '[n]amed after an ancient respected Abyssinian emperor' (Nomen á Tecla Haimanout antiquo imperatore Abyssinorum venerato). Why Mara Takla Haymanot should be so honoured is unknown. The only other emperors of the same name were Tekle Haymanot I (1684–1708) 'the Cursed', who was allegedly murdered, aged 24, by two of his former courtiers; and his son Tekle Haymanot II (1754–1777), who succeeded him aged 15. After seven years of precarious rule, battles with enemies and brutal murders, he repudiated his throne, became a monk, lived like a hermit and died a few months later.

Tecoma (Bignoniaceae) Juss.
From the Mexican vernacular name *tecomaxochitl* = a certain earthenware vessel; a general referrence by indigenous Mexicans to plants with tubular flowers.

Tecomaria = Tecoma (Bignoniaceae) (End) Spach.
Tecoma (q.v.); *-aria* = resembling; resembling *Tecoma*, of which genus this is now considered to be a synonym.

Tectaria (Tectariaceae) Cav.
La. *tectus* = roof, covered, concealed; *-aria* = like; referring to the 'roof-like' indusium of some species that covers the sori.

Tecoma: Typical flowers are scarlet but red and yellow varieties are also available. (EJM)

Tedingea* = Strumaria** (Amaryllidaceae) D. & U.Mül-Doblies
For Edward 'Ted' George Hudson Oliver (1938–), South African botanist and world authority on the  subfamily *Ericoideae*, author of some 110 papers and books, and his wife Inge Oliver (born Nitzsche) (1947–2003), South African botanist and botanical artist. Oliver obtained an MSc and PhD from the University of Cape Town and during his career was curator of the government herbarium in Stellenbosch (1964–1966, 1984–1996), South African liaison botanist at Kew Gardens (1967–1969) and research taxonomist in Stellenbosch (1970–1975, 1982–1983) and at the Compton Herbarium, Kirstenbosch. Together, the couple wrote the *Field Guide to the Ericas of the Cape Peninsula* (2010).

Teedia (Scrophulariaceae) Rudolphi.
For Johann Georg Teede, German naturalist and plant collector living in the late 1700s, mentioned by his friend Heinrich Adolph Schrader (1767–1836), German botanist and mycologist, who in his journal, *Für die Botanik* (1799), states that Teede 'cultivated' (studied?) in Germany for a couple of years before he went to collect in Portugal and to Surinam, where he died shortly after his arrival. Schrader states, loosely translated: 'Had he lived longer, he would have perhaps become a famous botanist, worthy of a monument.' His surname was misspelled 'Taede' in the British Museum herbarium, and 'Tiede' in the Berlin herbarium.

Telina* = **Lotononis** (Fabaceae) E.Mey.
La. *tela* = web; *-ina* = resembling; Romanian *telina* = celery. Meaning obscure.

Telosma (Apocynaceae) Coville
Gk. *tele* = distant, far off; *osme* = scent; referring to the plant's fragrance, which can be smelled from a distance.

Tenagocharis* = **Butomopsis** (Alismataceae) Hochst.
Gk. *tenagos* = lagoon; *charis* = delight; referring to the plant's habitat.

Tenaris = **Brachystelma** (Apocynaceae) E.Mey.
Gk. *teino* = to stretch; referring to the lobes of the corolla, which are extended erectly.

Tenaxia (Poaceae) N.P.Barker & H.P.Linder
La. *tenax* = strong, tenacious, holding firmly together; referring to the densely tufted grass.

Tenicroa* = **Drimia** (Hyacinthaceae) Raf.
Gk. *taenia* = tape, ribbon; *croa* = unknown. An apparently meaningless word, which was Rafinesque's penchant.

Tenrhynea (Asteraceae) Hilliard & B.Burtt
For Willem ten Rhyne (Wilhelmi ten Rhijne) (1647–1700), Dutch physician *Rhynea* (q.v.).

Tephrocactus = **Opuntia** (Cactaceae) Lem.
Gk. *tephros* = ash-coloured; *kaktos* = cactus; referring to the cactus's appearance.

Tephrosia (Fabaceae) Pers.
Gk. *tephros* = ash-coloured; referring to the leaves of these plants.

Teramnus (Fabaceae) P.Browne
Gk. *teramnos* = soft; referring to the pods and leaves.

Terminalia (Combretaceae) L.

(TW)

La. *terminus* = end. The leaves are at the ends of the shoots (Hugh Glen). The leaves are clustered at the ends of branches (Jackson).

Terpsichore* (Polypodiaceae) A.R.Sm.
Named after the Greek muse and god of dancing and song, who brought rejoicing to the heart. Reason for this name unknown, possibly weddings.

Testudinaria* = **Dioscorea** (Dioscoreaceae) Salisb.
La. *testudo* = tortoise; *-aria* = relating to; referring to the aerial rootstock resembling the carapace of a tortoise.

Tetrachne (Poaceae) Nees
Gk. *tetra* = four; *achne* = chaff; referring to this genus's four glumes.

Tetradenia* (Lamiaceae) Benth.
Gk. *tetra* = four; *aden* = a gland. The inner structure of the flowers is hidden under four equal glands.

Tetragonia (Aizoaceae) L.

Here are the flowers of this succulent plant. (CM)

Gk. *tetra* = four; *gonia* = an angle; referring to the angular shape of the fruit.

Tetraphyle = **Crassula** (Crassulaceae) Eck. & Zeyh.
Gk. *tetra* = four; *phylē* = rank; alluding to the arrangement of leaves.

Tetrapogon (Poaceae) Desf.
Gk. *tetra* = four; *pogon* = beard; referring to the tufts of hairs on the plant.

Tetraria (Cyperaceae) P.Beauv.

LEFT: A flowering plant (CM) RIGHT: Showing the characteristic closed leaf-sheath (EJM)

Gk. *tetra* = four; *-aria* = concerning. The first described species had its floral parts in fours.

Tetraselago (Scrophulariaceae) Junell
Gk. *tetra* = four; *Selago* (q.v.); referring to its four stamens.

Tetrateleia* = **Cleome** (Capparaceae) Sond.
Gk. *tetra* = four; *teleia* = perfect; referring to the four perfect stamens (eight stamens: four sterile and short; four fertile and elongated).

Teucrium (Lamiaceae) L.
Gk. *teukrion* for germander, from *Teukros* (*Teucer*), first king of Troy, son of Scamandar and father-in-law of Dardanus. He was said to have introduced one species into medicine.

Thalassodendron (Cymodoceaceae) Hartog
Gk. *thalasso-* = sea; *dendron* = tree. A submerged marine herb; referring to its habitat.

Thalia* (Marantaceae) L.
For Johannes Thal (1542–1583), German physician and botanist. He studied medicine in Jena, worked a short while as a doctor before joining the public health department in Stolberg (Hartz) (1572). In 1581, he became Nordhausen town physician but died two years later in a horse-and-cart accident on his way to see a patient. Thal's major publication was *Sylva Hercynia* ...

(1577), a description of the resin highlands plants mainly in the forested mountain range in northern Germany. In five years, Thal described all the plants he came across, not just flowers and herbs, but new species never before described, including a plant he called *Pilosella siliquosa*, which was renamed in 1753 by Linnaeus as *Arabis thaliana* in honour of Thal. The genus *Thalia* named by Linnaeus in *Genera Plantarum* was previously named *Cortusa* by Plumier. The name *Cortusa* applies to a range of very different genera.

Thalictrum (Ranunculaceae) L.
Gk. *thaliktron* = name for some sort of meadow rue; from *thallos* = a new green shoot. Georg Christian Wittstein suggests Gk. *thallein* = to make green, on account of the [fresh] bright green colour of the young leaves.

Thaminophyllum (Asteraceae) Harv.
Gk. *thaminos* (= *thameios*) = crowded, dense; *phyllon* = a leaf; referring to the plant's dense foliage.

Thamnea (Bruniaceae) Sol. ex Brongn.
Gk. *thamnos* = a bush, shrub.

Thamnium* = *Erica (Ericaceae) Klotzsch
Gk. *thamnion* = a small bush, diminutive of *thamnos* = bush; alluding to the rather untidy, bushy appearance of the plants.

Thamnocalamus (Poaceae) Munro
Gk. *thamnos* = bush; *kalamus* = reed, hence 'bushy reed'; referring to a species of clump-forming bamboo found in the grass family.

Thamnochortus (Restionaceae) P.J.Bergius

LEFT: Female inflorescence RIGHT: Male inflorescence (EJM – Both)

Gk. *thamnos* = a shrub; *khortus* = green herbage; referring to the overall colour of this plant. Peter Jonas Bergius states '*Khortus* = a grass, alluding to the hard shrubby habit and natural affinity of the plant' (*Cyclopaedia Abraham Rees*, 1819).

Thamnosma (Rutaceae) Torr. & Frém.
Gk. *thamnos* = a bush; *osme* = scent. The flowers have glands filled with fragrant oils.

Thamnus = Erica (Ericaceae) Klotzsch
Gk. *thamnos* = a bush, shrub. Jackson writes: 'Some taxonomists have no imagination!'

Theilera (Campanulaceae) E.Phillips
For Sir Arnold Theiler (1867–1936), Swiss-born English veterinarian and botanist who worked in  South Africa, is considered the 'father of veterinary science in South Africa'. He developed vaccines against smallpox and rinderpest. 'Theiler was the first director of the Onderstepoort Veterinary Research Institute outside Pretoria. This institute under his leadership carried out research on African horse sickness, sleeping sickness, malaria, East Coast fever (*Theileria parva*) and tick-borne diseases such as redwater, heartwater and biliary. A faculty of veterinary science was established here in 1920, which enabled vets to train locally for the first time. Theiler became the first dean of this faculty' (Wikipedia.org).

Thelechitonia* = Sphagneticola**
(Asteraceae) Cuatrec.
Gk. *thele* = wart, nipple; *chiton* = covering or outer blanket. Meaning obscure.

Thelepogon (Poaceae) Roth ex Roem. & Schult.
Gk. *thele* = wart; *pogon* = beard; alluding to the rugose surface of the lower glume and the awned spikelets.

Thelypteris (Thelypteridaceae) Schmidel
Gk. *thelys* = female; *pteris* = fern; alluding to this plant's common name of 'maiden fern'.

Themeda (Poaceae) Forssk.
Arabic *thamada* = a depression filled with water after rain, Latinised as *thaemed*. Possibly referring to water storage cells on the upper surface of the leaves or to the habitat of the type specimen in Yemen.

Theodora** (Fabaceae) Medik.
For Charles Theodore (Karl Theodor), (1724–  1799), Prince-elector and Duke of Bavaria from 1777–1799. He had several mistresses and many illegitimate children, and was not a popular ruler among his subjects in Bavaria. He made many dubious political decisions, such as exchanging Bavaria for the Austrian Netherlands in 1784, which failed. To his credit, Theodore founded, in 1763, the Academia Theodoro Palatina of Mannheim and an associated botanical garden at Medikus's instigation. He was also at least partially responsible for the creation of the Englischer Garten (English Garden) in the centre of Munich, which was opened to the public in 1780. It is one of the world's largest urban public parks.

Thereianthus (Iridaceae) G.J.Lewis.

(JM)

Gk. *thereios* = summer; *anthos* = flower; summer-flowering.

Thesidium (Santalaceae) Sond.
Thesium (q.v.); Gk. *ium* = diminutive.

Thesium (Santalaceae) L.
Derivation uncertain. Gk. *thes* = a hired labourer. An ancient name for a species of *Linaria*, toad flax, used by Pliny the Elder. Georg Christian Wittstein traces this to the legendary hero Theseus, who slew the Minotaur and to whom Ariadne gave a wreath in which this plant was woven.

Thesmophora (Stilbaceae) Rourke
Gk. *thesmophoros* = law-giving, an epithet of the goddess Ceres. The plant was found near Ceres.

Thespesia (Malvaceae) Sol. ex Corrêa
Gk. *thespesios* = divine. *T. populnea* was formerly held sacred in Tahiti.

Thilachium (Capparaceae) Lour.
Gk. *thylax* = a bag or pouch; referring to the characteristic bag-like remains of the top of the calyx attached to the open flower (Georg Christian Wittstein), possibly also to the form of the fruit.

Thinopyrum *** = *Elymus* (Poaceae) A.Löve
Gk. *thino* = sand; *pyros* = wheat. The plant grows on beach dunes.

Thlaspeocarpa *** = *Heliophila* (Brassicaceae) C.A. Sm.
Gk. *thlaspis* = cress, from *thlao* = to compress, flatten and *aspis* = shield; *karpos* = fruit. The siliculae are compressed and discoid. (Jackson states 'I do not know where the *-peo-* comes from'.)

Thlaspi* (Brassicaceae) L.
La. and Gk. name for a kind of cress, as described above.

Thodaya *** = *Euryops* (Asteraceae) Compton
For David Thoday (1883–1964), English Cambridge-trained botanist, Harry Bolus professor at the University of Cape Town (1918–1922), professor of botany at the University College of North Wales, now the University of Wales, Bangor (1923–1949). He authored *Botany: A Textbook for Senior Students* (1915: 5 editions), many scientific papers on succulent and parasitic

plants, and a revision of the genus *Passerina,* etc., but his main interest was on plant physiology – he developed the Thoday potometer (measures the rate of water uptake of a leafy shoot) and Thoday respirometer (measures the rate of respiration of a living organism by measuring its rate of exchange of oxygen and/or carbon dioxide). His presidential lecture to the British Association, in 1939, on 'The interpretation of plant structures', was seminal. He was elected a Fellow of the Royal Society in 1942, and received an honorary DSc from the University of Wales in 1960.

Thoracosperma = *Erica* (Ericaceae) Klotzsch
Gk. *thoraco, thorax* = a breastplate, the thorax; *sperma* = seed; referring to the well-protected seed.

Thorncroftia (Lamiaceae) N.E.Br.
For George Thorncroft (1857–1934), English merchant with an interest in horticulture, who emigrated to South Africa in 1882 and settled in Barberton, eastern Transvaal. He started collecting seeds and bulbs in about 1889 and sent them to Medley Wood, a South African botanist in Durban, together with herbarium specimens.

He also sent specimens to England and the United States, and collected botanical specimens for Durban Herbarium in his later years, the National Herbarium in Pretoria, and for Archdeacon Rogers, a prolific collector. He is credited with having collected more than 4 000 specimens. He wrote occasional articles of a botanical nature under the pen name 'Kof Kof'.

Thryocephalon *** = *Kyllinga* (Cyperaceae) J.R. & G.Forst.
Gk. *thryon* = reed, rush; *kephale* = head.

Thuja (Cupressaceae) L.
Gk. *thuia, thyia* = a kind of juniper, probably a corruption of the thya of Theophrastus; alluding to the use of cedar wood as incense (Meagher), or for a resinous, fragrant-wooded tree.

Thunbergia (Acanthaceae) Retz.
For Carl Pehr (Peter) Thunberg (1743–1828), Swedish botanist, physician, student of Linnaeus, professor of botany and medicine at Uppsala University (1784–1828). He visited the Cape (1772–1775) to study Dutch and the Cape's flora, natural history and social history, and did extensive botanical exploration in the southern Cape where no botanical collecting had been done, amassing some 3 100

species. In 1775 he went to Japan, Java and Sri Lanka for 15 months. Thunberg published *Flora Japonica* (1784), his travel diaries (1788–1796), and *Flora Capensis* (in parts between 1807–1820). He presented his herbarium of 23 510 specimens and 25 000 insects to Uppsala University. He was made a knight of the Royal Order and received many honours.

Thunbergiella = *Itasina* (Apiaceae) H.Wolff
Thunbergia (q.v.); Gk. *-iella* = diminutive.

Thuranthos *** = *Drimia* (Hyacinthaceae) C.H.Wright.
La. *thur*, Gk. *thyos* = incense; *anthos* = flower; referring to its fragrance.

Thymus* = *Clinopodium (Lamiaceae) L.
Gk. *thymos* = thyme; from *thuos* = smelling like incense; referring to its fragrance.

Tieghemia* = *Oncocalyx*** (Loranthaceae) Balle
For Philippe Édouard Léon van Tieghem (1839–1914), French botanist, professor of botany.

After qualifying with a PhD in physics (1864) and PhD in natural history (1866), he taught at the École Centrale des arts et Manufactures (1873–1886), was professor at the Muséum National d'Histoire Naturelle (1878–1914), and an instructor at the Institut Agronomique (1899–1914) in Paris. Among his achievements, he created the 'Van Tieghem cell', a device mounted on a microscope slide that allows for observing the development of a fungus's mycelium; wrote many scientific papers and translated a botanical work by the German botanist Julius von Sach from German into French; and in 1876 was the first person to describe blastomycosis, a fungal infection that is also known as 'Gilchrist disease'. In the same year, he was elected a member of the Academy of Sciences.

Tigridia* (Iridaceae) Jussieu
Gk. *tigris, tigridis* = tiger-like; alludes to the coloration and spotting of the flowers of the type species *Tigridia pavonia*.

Tiliacora (Menispermaceae) Colebr.
Latinised from the Bengali vernacular name *tiliakora*.

Tillaea* = *Crassula (Crassulaceae) L.
For Michelangelo Tilli (Michele Angelo Tilli) (1655–1740), Italian physician and botanist. He

graduated in medicine and surgery at the University of Pisa (1677). As a naval surgeon in 1861, he travelled to the Balearic Islands and in 1863 to Constantinople on a 'mercy mission', together with the florentine surgeon Pier Francesco Pasquali, to tend to son of the Mehmed IV, Sultan of the Ottoman Empire (1648–1687) who had had a serious fall from his horse. From there they visited Albania. Tilli went on to Tunis to study the remains of Carthage and collect plants. He became professor of botany at the University of Pisa in 1685, and director of the botanical gardens, introducing plants from Asia and Africa.

He pioneered heated rooms (greenhouses) for growing exotic plants, many from Africa and Asia, and was able to cultivate pineapples and coffee in Italy. He authored *Catalogus Plantarum Horti Pisani* (1723). He became a member of the Royal Society in 1708.

Tillandsia* (Bromeliaceae*) L.
For Elias Tillandz (1640–1693), later Tillander, Swedish physician and botanist who lived in Finland. He studied medicine at the Academy of Turku, later renamed the University of Helsinki (1659–1663), and then moved to the universities of Uppsala and Leiden, obtaining his doctorate in 1670. He became professor of medicine at the Academy of Turku and, in the same year, 1690, its rector, a position he held twice. He wrote the first botanical work on Finnish flora, the *Catalogus Plantarum* (1673), featuring some 500 plants with Finnish and Latin names, with an expanded second edition in 1683. He performed the first official autopsy in the university's main hall in 1686, and was a pioneer in searching for new ways to treat illness and healing methods, such as leprosy. He used his extensive knowledge of plants to prepare medicines for his patients.

Tinnea (Lamiaceae) Kotschy & Peyr.
For Henrietta M Tinne, her daughter Alexandria Tinne (1835–1869) and her sister Adrienne van

Capellen, all from Holland, who undertook a scientific expedition to explore the Nile. They left Europe in 1861 and explored part of the 'White Nile' during 1862. In 1863, they were joined by two botanists, Theor von Heuglin and Hermann Steudner. The party set out from Khartum to explore how far westward the Nile basin extended. Once they reached the limits of navigation, they went overland and suffered from fever (malaria?). Steudner died in April, Henrietta Tinne died in June, and Adrienne van Capellen died on returning to Khartum. Alexandria subsequently tried to cross the Sahara in 1869. Between Murzuk and Ghat, Libya, she was murdered. It was an ill-fated group.

Tinospora (Menispermaceae) Miers
Gk. *teino* = to stretch; *spora* = a seed. The seeds are very long and narrow.

Titanopsis (Aizoaceae) Schwantes
Ancient Gk. *titanos* = limestone; *-opsis* = resembling. The succulent's appearance is somewhat like limestone.

Tithonia * (Asteraceae) Desf. ex Juss.
For Tithonus, son of Laomedon and consort of Aurora, the (Latin) goddess of dawn; symbolic of old age, referring to the grey to white covering of fine hairs on the leaves of some species.

Tittmannia *** = ***Audouinia*** (Bruniaceae) Brongn.
For Dr Johann August Tittman (1774–1840), German physician, agronomist and botanist. He was lecturer in pharmaceutical botany at Dresden (1804–1813) and later practised medicine in Dresden, working at the Ambulance College from 1814. He wrote both medical and botanical books such as *Lehrbuch der Chirurgie zu Vorlesungen*, a textbook of surgery to lectures (three volumess, Leipzig, 1800–1802, 1810–1811); *Chirurgische Verbandlehre* (*Surgical Association Teaching*) (1812); *Von den Topischen Arzneymitteln gegen Augenkrankheiten* (*Of the Topical Arzneymitteln against Eye Diseases*) (1804); *Über das Stadium der Botanik* (*On the Study of Botany*) (1802); and *Über den Embryo des Saamenkorns und Seine Entwicklung zur Pflanze* (*On the Embryo of the Grain of Seed and its Development to the Plant*) (1817).

Tmesipteris * (Psilotaceae) Bernh.
Gk. *tmesis* = cutting; *pteris* = fern (cut fern); alluding to the forked appendages on fertile fronds; another source states referring to the truncated leaf tips.

Toddalia (Rutaceae) Juss.
Latinized from the Malabar vernacular name *Kaki-Toddali* for *T. asiatica* (Jackson states for *T. aculeata*).

Toddaliopsis (Rutaceae) Engl.
Toddalia (q.v.); *-iopsis* = resembling.

Todea (Osmundaceae) Willd. ex Bernh.
For Heinrich Julius Tode (1733–1797), German Lutheran theologian, mycologist, cantata lyricist, poet, song writer and architect. He studied theology in Göttingen from 1757–1761, worked as a minister in Mecklenburg, and was appointed to several administrative positions in the council of the church and in the administration of the city. From 1765 to 1768, he was the draftsman and architect responsible for building St Trinity Church in Warlitz, wrote many lieder and cantatas, and the text for Rosetti's oratorio *Jesus in Gethsemane*, formulated new principles for mycological classification, and published *Fungi Mecklenburgenses Selecti* (*Selected Fungi from Mecklenburg*) (1790–1791) with copper engraved illustrations of excellent quality.

Tolpis (Asteraceae) Adans.
Origin unknown, possibly from the Greek *tolype* = a ball of wool, lump; referring to the fruiting capitula (the small flowering heads) (Umberto Quattrocchi); alternately, the long outer bracts.

Toona * (Meliaceae) (Endl.) M.Roem.
From the Indian vernacular name for *Toona ciliata*.

Torenia (Linderaceae) L.
For Reverend Olof Torén (1718–1753), Swedish clergyman, traveller, botanist and plant collector, and ship's chaplain with the Swedish East India Company. He obtained a job in 1747 as ship priest and travelled to Canton, China, twice: first on the East Indiaman *Hoppet* (Hope), and then, in 1750, on the *Götha Leijon* (Gothic Lion), returning in 1752. On his second trip, Torén collected many rare plants, but his health deteriorated and he died 13 months later in Näsinge, Bohuslän, Sweden. Before he died, he wrote seven letters to Linnaeus about his travels. Although he collected plants for Linnaeus, he failed to bring back China's famous tea bush as the plant withered and died. Ten years later, a Captain Ekeberg succeeded in this task.

Torilis (Apiaceae) Adans.
Gk. *torilis* from *torus* = a borer, *-ilis* = passive adjective); *toreo* = to bore through, pierce; referring to the prickled fruit.

Tounatea = ***Swartzia*** (Fabaceae) Aubl.
A name, somewhat arbitrarily constructed by autor Jean Baptiste Aublet, from the Guiana (Guyana) name Tounou, presumably the vernacular name.

Tournefortia (Boraginaceae) L.
For Joseph Pitton de Tournefort (1656–1708), French botanist and physician, naturalist, professor of medicine and botany. He studied medicine at Montpellier and was appointed professor of botany at the Jardin des Plantes in Paris in 1683. His principal work, *Eléments de Botanique, ou Méthode pour Reconnaître les Plantes* (1684) attempted to classify some 7 000 plants into class, section, genus and species. His system remained in practice until it was superseded by that of Linnaeus who used five ranks: class, order, genus, species and variety. Between 1700 and 1702, Tournefort travelled extensively in western Europe and eastern Anatolia (Armenia, Georgia, Turkey) collecting plants. He died in an accident when hit by a carriage in Paris.

Toxicodendrum* = *Hyaenanche*
(Picrodendraceae) Thunb.
Gk. *toxikon* = poison; *dendron* = tree. Some plants in this genus cause itching, irritation and painful rashes caused by the oozing sap, and even death when the 'poison ivy' is burnt and the smoke inhaled.

Toxicophlaea* = *Acokanthera*
(Apocynaceae) Harv.
Gk. *toxicon* = poison; *phloios* = bark. The shrub is used as a source of arrow poison.

Trachelium* (Campanulaceae) L.
La. *trachea* = windpipe; Gk. *trachelium* = rough throat. This flower was used for treating throat ailments in ancient times.

Trachyandra (Asphodelaceae) Kunth

The tiny hairs on the filaments are difficult to see without a strong magnifier. (CM)

Gk. *trachys* = rough; *andros* = male. The thick filaments are usually hairy.

Trachycalymma = *Asclepia* (Apocynaceae) Bullock
Gk. *trachys* = rough; *kalymma* = hood. The inside of the hoods are covered with papillae.

Trachydium (Apiaceae) Lindl.
Gk. *trachys* = rough; *-idion* = diminutive; referring to the umbel.

Trachypogon (Poaceae) Nees
Gk. *trachy-* = rough; *pogon* = beard; alluding to the plumose awn of the fertile spikelet.

Trachysciadium = *Ezosciadium* (Apiaceae) H.Wolff
Gk. *trachy-* = rough; *skiadion* = parasol, canopy; possibly referring to the plant's umbel.

Trachyspermum (Apiaceae) Link
Gk. *trachy-* = rough; *sperma* = seed; referring to the raw fruit pods ('seeds') often used as a spice to flavour a dish.

Tradescantia* (Commelinaceae) L.
For John Tradescant (aka 'Treadeskant') (c 1570–1638), English gardener of Dutch descent

and plant collector, and his son, John Tradescant (1608–1662). Tradescant (snr) worked, *inter alia*, for the Earl of Salisbury (1609), the first Duke of Buckingham, George Villiers (1623), and King Charles I (1630). He travelled to many countries collecting new plants such as Archangel, Russia (larch trees), Algiers (apricot trees), France (poppies), etc. His son worked with him and introduced many species to England. John (jnr) visited Virginia, the United States, three times (in 1637, 1642 and 1654) and brought back the Virginia creeper bush, the yucca plant, the scarlet runner bean, and many other species. He took over as head gardener for Charles I upon the death of his father in 1638.

Tragia (Euphorbiaceae) L.
For Hieronymus Tragus (Hieronymus Jerome Bock) (1498–1554), German botanist, physician, teacher,

Lutheran minister, and herbalist author. He possibly worked as caretaker in the gardens of Count Palatine Ludwig, who funded his theology and medicine studies at university. He authored *New Kreuterbuck*, published by Rihel in 1539, a botanical herbal featuring some 700 German plants, written in vernacular German, in which the plants were arranged to associate 'such plants as nature seems to have linked together by similarity of form', the first known attempt at a natural classification of plants. It was reprinted in 1545 with more than 500 illustrations by David Handel.

Tragiella (Euphorbiaceae) Pax & K.Hoffm.
Tragia (q.v.); La. *-iella* = diminutive; a small *Tragia*, or a plant that looks like *Tragia*.

Tragiopsis* = *Microstachys* (Apiaceae) Pomel
Gk. *tragia* = a goat; *-opsis* = resembling. Meaning unclear.

Tragopogon* (Asteraceae) L.
Gk. *tragos* = a he-goat; *pogon* = a beard; alluding to the hairy pappus that crowns each seed.

Tragus (Poaceae) Haller
Gk. *tragus* = goat or hairy part of the ear. Also possibly Hieronymus Tragus (Hieronymus Jerome Bock) (1498–1554); see *Tragia*; alluding to the similarity of the leaf hairs and spikelet bristles to a goat's beard.

Trapa (Lythraceae) L.
Originally said to be La. *calcitrapa*, Gk. *kalkitrapa* = a four-spiked iron ball thrown to maim horses in battle, now La. *trapa* = thistle; alluding to the four-horned fruits.

Treichelia (Campanulaceae) Vatke
For Alexander Johann August Treichel (1837–1901), Prussian scientist, folklorist and historian who studied law and economics in Berlin. In 1876, he took charge of his large family estate in Hochpaleschken. Treichel had an extensive interest in the natural sciences, zoology, pre- and early history, folklore and culture. He was a very active member of many scientific teams, especially the Rudolf Virchow, led by Berlin Gesellschaft Society for Anthropology, Ethnology and Prehistory. His research over 20 years resulted in the publication of books relating to botany, archaeology, ethnology, house construction, local legends, songs and anecdotes, and rural and religious folklore and customs.

Trema (Ulmaceae) Lour.
Gk. *trema* = aperture, hole, opening. The kernel of the fruit is pitted.

Trianoptiles (Cyperaceae) Fenzl
Gk. *triana* = trident; *ptilon* = a plume. The three perianth scales are usually terminated by three stiff bristles.

Trianthema (Aizoaceae) L.
Gk. *tri-* = three; *anthos, anthemon* = flower. The flowers are generally grouped in threes.

Triaspis (Malpighiaceae) Burch.
Gk. *tri-* = three; *aspis* = a round shield. The fruits have three roundish wings.

Tribolium (Poaceae) Desv.
Gk. *tribolo* = three-pronged; *tribulus* = three-pointed; possibly referring to the three florets encased in the very bristly glumes (Hubbard, 2001).

Tribulocarpus (Aizoaceae) S.Moore
Gk. *tri-* = three; *bolos* = point; *karpos* = fruit. See *Tribulus* (q.v.) for the similarly spiny fruits.

Tribulus (Zygophyllaceae) L.
La. *tribulus* = three-pointed, a caltrop, originally, a

(EJM)

Roman weapon used against cavalry, containing three prongs and a fourth projecting upward. The fruit has four or five hard, single-seeded nutlets, each of which bear two or three sharp spines.

Tricalysia (Rubiaceae) A.Rich. ex DC.
Gk. *tris* = three; *kylix* = a shallow bowl. The flowers have a true calyx and two epicalyces.

Trichelostylis = Fimbristylis (Cyperaceae) T.Lestib.
Gk. *thrichos* = hair; *stylis* = style. The styles are often hairy.

Trichilia (Meliaceae) P.Browne
Gk. *tricheilos* = three-lipped; *tricha* = in threes. The fruits are three-lobed and occasionally the leaflets are in threes, or possibly three-lipped.

Trichinium* = Ptilotus** (Amaranthaceae) Moq.
Gk. *trikhinos* = relating to hair, from *thrix* = a hair; *trichinos* = of hair. The plant is hairy.

Trichocaulon = Hoodia (Apocynaceae) N.E.Br.
Gk. *trichos* = hair, hairy; *kaulos* = stem; referring to the hairy stem.

Trichocephalus (Rhamnaceae) Brongn.
Gk. *trichos* = hair, hairy; *kephale* = head; referring to the heads of flowers that are surrounded by hairs.

Trichocereus * = ***Echinopsis*** (Cactaceae)
(A.Berger) Riccob.
Gk. *trichos* = hair, hairy; *cereus* = cactus; referring to the hairy floral tube.

Trichocladus (Hamamelidaceae) Pers.
Gk. *trichos* = hair, hairy; *klados* = branch; referring to the young branches (twigs) that are hairy.

Trichocyclus (Woodsiaceae) N.E.Br. (Lilj) Dulac
Gk. *trichos* = hair, hairy; *cyclos* = circular. The sori are round and surrounded by hairs.

Trichodesma (Boraginaceae) R.Br.
Gk. *trichos* = hair, hairy; *desme* = band, cable; referring to the long, spiralling hairs or awns.

Trichodiadema (Aizoaceae) Schwantes
Gk. *trichos* = hair, hairy; *diadema* = crown or band as around a turban; referring to the hair-like bristles that form a tufted crown at the leaf tips.

Trichogyne = ***Ifloga*** (Asteraceae) Less.
Gk. *tricho-* = hair; *gyne* = female, woman (pistil); referring to the slender, hair-like female flowers (William Henry Harvey, *Flora Capensis*).

Tricholaena (Poaceae) Schrad.
Gk. *tricho-* = hair; *chlaena* = a cloak; referring to glumes that cover the dairy spikelets.

Trichomanes = ***Asplenium***
(Hymenophyllaceae) L.
Gk. *thrix, trichos* = hair; *manos* = soft; referring to the delicate nature of the fronds.

Trichonema* = ***Romulea*** (Iridaceae) Ker Gawl.
Gk. *tricho-* = hair; *nema* = a thread, filament; referring to the hairiness of that part of the plant.

Trichoneura (Poaceae) Andersson
Gk. *tricho-* = hairy; *neurus* = nerved, veined; referring to the hairy lemma veins.

Trichopteryx (Poaceae) Nees
Gk. *tricho-* = hairy; *pteryx* = wing; referring to the lemma whose margins bear tufts of hair.

Trichosanthes (Cucurbitaceae) L.
Gk. *tricho-* = hair; *anthos* = flower. The flowers are fringed.

Tricliceras (Turneraceae) Thonn. ex DC.
Gk. *tri-* = three; *cliceras* = chambered. Meaning unclear.

Triclissa = ***Kniphofia*** (Asphodelaceae) Salisb.
Gk. *tri-* = three; *clissa* (old Latin name) = key. Perhaps referring to 'the three outer petals united into a straight cylindrical tube considerably beyond their middle, while its three inner petals are quite separate and loose both from the outer and from each other' (*The Genera of Plants,* 75, 1866, Richard Anthony Salisbury).

Tridactyle (Orchidaceae) Schltr.

(HS – Both)

Gk. *tri-* = three, thrice; *daktylos* = finger; alluding to the three-lobed lip.

Tridax* (Asteraceae) L.
Gk. *tridaknos* = eaten in three bites. Referring to the three-toothed ray florets.

Tridentea (Apocynaceae) Haw.
Gk. or *tri-* = three; *dens* = tooth; referring to the three-toothed interstaminal segments of the corona.

Trieenea (Scrophulariaceae) Hilliard
For Elizabeth ('Elsie') Esterhuysen (1912–2006), South African botanist and plant collector (q.v. *Elsiea*).

Trifolium (Fabaceae) L.
La. *tri-* = three; *folium* = leaf; three-leaved, as in clover.

Triglochin (Juncaginaceae) L.
Gk. *treis* = three; *glochin* = a point; referring to the pointed follicles of *T. palustris* or its fruit; or to the projections on the carpel; or to the three valves of the capsule, which separate from the base in maturity, resembling a three-barbed arrowhead.

Trigonella (Fabaceae) L.
Gk. *trigon-* = triangular; *gonia* = angle, cornered; *-ella* (diminutive). 'Little triangle' – obscure Linnaean appellation. From the appearance of the corolla. 'Fenugreek'.

Trigonocapnos (Papaveraceae) Schltr.
Gk. *trigon-* = triangular; *kapnos* = smoke (see *Cysticapnos*). The fruit is obtusely triangular in cross-section.

Trigonotheca *** = **Melanthera*** (Asteraceae) Sch.Bip.
Gk. *trigono-* = three-angled; *theka* = case (fruit); referring to the shape of the fruit.

Trilepisium (Moraceae) Thouars
Gk. *tris* = three; *lepis* = scale. The tube between the stamens and the stigma ends in three tiny lobes.

Trillidium* (Melanthiaceae*) Kunth
La. *trillium* = in three; *idium* = a little. The flower has three petals.

Trimeria (Salicaceae) Harv.
Gk. *tris* = three; *meris* = a part. The flower parts are in threes, which is unusual in dicotyledons.

Trineuria *** = **Aspalathus*** (Fabaceae) C.Presl
Gk. *tri-* = three; *neuron* = nerve, vein.

Triplocephalum (Asteraceae) O.Hoffm.
Gk. *triplo-* = threefold; *kephale* = head; referring to the flower heads.

Triplophyllum* (Tectariaceae) Holttum
Gk. *triplo-* = threefold; *phyllon* = leaf; possibly referring to the three-branched fronds growing from a creeping caudex.

Tripogon (Poaceae) Roem. & Schult.
Gk. *tri-* = three; *pogon* = beard; referring to the three-nerved lemma with the mid-nerve protruding to a short awn.

Tripterachaenium *** = **Tripteris***
(Asteraceae) Kuntze
Gk. *tri-* = three; *pteron* = wing; *chainein* = to gap.

Tripteris = ***Osteospermum*** (Asteraceae) Less.
Gk. *tri-* = three; *pteron* = wing; referring to the fruits that are three-winged, so are easily dispersed by wind.

Triraphis (Poaceae) R.Br.
Gk. *tri* = three; *raphis* = needle; referring to the three main nerves of the lemma.

Trisetaria (Poaceae) Forssk.
Gk. *tri-* = three; *seta* = bristle; *-aria* = pertaining to. The lemma is three-awned.

Trisetum = ***Pentaschistis*** (Poaceae) Pers.
Gk. *tri-* = three; *seta* = bristle; referring to the three-bristled appearance of the lemma, which is three-awned, but there are other interpretations.

Tristachya (Poaceae) Nees
Gk. *tri-* = three; *stachys* = spike. The spikelets are arranged in groups of three.

Tristemon *** = **Erica*** (Ericaceae) Klotzsch
Gk. *tri-* = three; *stemon* = stamens. The plant has three stamens.

Tristicha (Podostemaceae) Thouars
Gk. *tri-* = three; *stichos* = row. The leaves are distinctly arranged in three rows – one row being dorsal and of short leaves, and the other two lateral rows having longer leaves.

Triteleia* (Alliaceae) Douglas ex Lindl.
Gk. *tri-* = three; *teleios* = perfect; referring to the floral parts, all of which are in threes.

Tritoma *** = **Kniphofia*** (Asphodelaceae) Ker Gawl.
Gk. *tritomos* = thrice cut, from *tri-* = three; *temnein* = to cut; referring to the sharp edges of the leaves.

Tritomanthe *** = **Kniphofia*** (Asphodelaceae) Link
Gk. *Tritoma* (q.v.); *anthos* = flower; referring to the sharp edges of the leaves.

Tritomium = ***Kniphofia*** (Asphodelaceae) Link
Another variant of *Tritoma* (q.v.).

Tritonia (Iridaceae) Ker Gawl.
La. *triton* = weathercock; alluding to the variable positions of the stamens. (Triton was also a Greek sea demigod, often represented as half man, half fish.)

Tritoniopsis (Iridaceae) L.Bolus (Images on opposite page)
Tritonia (q.v.); Gk. *-iopsis* = resembling.

Tritonixia *** = **Tritonia*** (Iridaceae) Klatt
Tritonia (q.v.); *Ixia* (q.v.). *Tritonia* is a form of *Ixia*.

Triumfetta (Malvaceae) L.
For Giovanni Battista Trionfetti (1656–1708), Italian botanist and physician who became a professor of medicine at the University of La Sapienza in Rome in 1680, and director of the Royal Botanical Garden from 1676–1706. Under his administration, the botanical garden became one of the most famous in Europe, and the number of plants, both indigenous and exotic, was considerable. He collected some large herbaria,

Tritoniopsis (CM)

Tritoniopsis (CM)

one of which comprised 13 folio volumes. His book *Observationes de Ortu ac Vegetatione Plantarum* (1685), published by Dominicus Antonius Hercules, bitterly attacked the view held by Marcello Malpighi (1628–1694), an Italian physician, anatomist and physiologist, that the whole plant is actually in the seed.

Trixago*** = **Bartsia** (Orobanchaceae) Stev.
Gk. *trixis* = triple; *ago* = like, resembling; referring to the three-angled fruit.

Trochomeria (Cucurbitaceae) Hook.f.
Gk. *trochos* = a wheel; *meris* = portion, part. The connection to the plant is rather obscure – perhaps because the long, narrow corolla lobes look like the spokes of a wheel.

Troglophyton (Asteraceae) Hilliard & B.Burtt
Gk. *troglo-* = cave-dwelling; *phyton* = plant; referring to its habitat.

Trollius* (Ranunculaceae) L.
From German *troll, trollen* = globular, hence a 'globe flower'; referring to the form or shape of the flowers, globe flowers such as *T. europaeus* and *T.* x *cultorum* have petals that are curved over the top of the flower, forming a globe.

Tromotriche (Apocynaceae) Haw.
Gk. *tromos* = trembling; *trichos* = hair, hairy; referring to the vibratile corolla hairs.

Tyrbastes* (Restionaceae) B.G.Briggs & A.S.Johnson
La. *tyrbastes* = a troublemaker or agitator; referring to the difficulty in determining the affinities of the genus.

Tryphia*** = **Holothrix** (Orchidaceae) Lindl.
Gk. *tryphe* = softness, delicate; alluding probably to the membranaceous leaves.

Tryphostemma*** = **Basananthe** (Passifloraceae) Harv.
Gk. *tryphos* = fragment; *stemma* = a crown; alluding to the depauperate condition of the corona rays.

Tulbaghia (Alliaceae) L.
For Ryk Tulbagh (Rijk Tulbagh) (1699–1771), Dutch governor of the Cape Colony from 1751 to 1771. When only 16, he emigrated to the Cape as a Dutch East India Company employee on a five-year contract to be used as needed. The governor, Maurice Pasques Chavonnes, recognised the young man's ability and gave him an administrative post as assistant clerk of the secretary of the political council, the start of a career that ended in his being made governor of the Cape. He was a responsible governor who, *inter alia*, codified the slave laws of the country with set rules for slave management. He corresponded with Linnaeus in 1763 and sent him

a collection of 200 plant specimens, 50 bulbs and seeds, and several birds. The town of Tulbagh is named after him.

Tunica*** = **Petrorhagia** (Caryophyllaceae) Scop.
La. *tunica* = coat, tunic; referring to the outer petals (sepals) at the base of the flower, the calyx, which gives the flower a protective covering.

Turbina (Convolvulaceae) Raf.
La. *turbo,* rarely *turben* = a spinning-top. The fruits bear a faint resemblance to a spinning-top.

Turia*** = **Luffa** (Cucurbitaceae) Forssk. ex J.F.Gmel.
Georg Christian Wittstein suggests this is an Arabic name. *Turia* is also the name of an Indian vegetable, *badam beerakaaya* (almond turia). The synonym genus of *Turia* is called *Luffa*. At least two species of this tropical vine provide fruit, *luffa* (gourds), which are eaten as a vegetable, while Vietnamese *luffa* or Chinese *okra* (also spelled *loofah*) seem very similar.

Turnera (Turneraceae) L.
For William Turner (c 1508–1568), English botanist, naturalist, physician, clergyman, traveller and author. He studied physics, philosophy and medicine (which in the 16th century focused mostly on plants) at Cambridge University. He took Holy Orders and became a dean, but was forced into exile, twice, as a result of his outspoken religious views. He visited France and Germany, and later studied medicine at Bologna, Italy. He wrote *Libellus de re Herbaria Novus* (1538) to bring 238 native plant names into line with continental nomenclature, and *A New Herball*, in three parts, between 1551 and 1568, correctly identifying and describing more than 300 species of plants in English with reliable plant descriptions, habitats and common names where known.

Turraea (Meliaceae) L.
For Giorgio della Torre (Turre, Turra) (1607–1688), professor of botany at the University of Padua and prefect of its gardens. He was the author of *Catalogus Plantarum Horti Patavini Novo Incremento Locupletior* (*The Growth of the New Rich, Patavini Garden Plants*) (1660) and *Dryadum, Amadryadum Cloridisque Triumphus, Ubi Plantarum Universa Natura Spectatur, Affectiones Expenduntur, Facultates Explicantur* (*Dryadum, the Triumph of Amadryadum Cloridisque, Where the Whole of the Nature of Plants … Are Explained*) (1685).

Turritis* (Brassicaceae) L.
La. *turris* = a tower; *turritus* = furnished with towers. The plant has a pyramidal appearance.

Tylecodon (Crassulaceae) Toelken
Anagram of *Cotyledon*, in which genus the species were previously placed.

Tylophora (Apocynaceae) R.Br.
Gk. *tylos* = lump, knob; *phora* = bearing, carrying. The corona lobes are thick and fleshy (Hugh Glen); stamens bear pollen masses like weals (Don Perrin).

Tylosema (Fabaceae) (Schweinf.) Torre & Hillc.
Gk. *tylos* = lump; *sema* = a sign. The seeds of this genus bear a U-shaped mark on a swelling at the hilum.

Typha (Typhaceae) L.
Gk. *typhos* = marsh; referring to the plant's favoured habitat being wetlands.

Tysonia*** = **Afrotysonia** (Boraginaceae) Bolus
For William Tyson (1851–1920), Jamaican-born South African botanist, plant collector, schoolteacher. He taught in various South African schools (1874–1887), joined the forestry department as secretary to the superintendent (1888–1892), and became librarian and sub-editor of the *Agricultural Journal* for the agricultural department  (1893–1904). He was a main contributor to the Herbarium Normale Austro-Africanum organised by Bolus and MacOwan. On his retirement, he disposed of his phanerogam herbarium to the Cape government, and his personal set of marine algae to the Bolus Herbarium. In 1919, he was contracted to collect 10 sets of phanerogams for the Botanical Society, but took ill on the first collecting expedition to Coffee Bay and died shortly thereafter. He was a Fellow of the Linnaean Society.

U

Umtiza is a monotypic genus in the legume family containing the single species *Umtiza listeriana*. This tree is endemic to South Africa.

Braam van Wyk (1952–) holds a master's degree in botany and a diploma in higher education from the former Potchefstoom University for Christian Higher Education and a doctorate from the University of Pretoria. Since 1977 he has been a member of the lecturing staff of the University of Pretoria. Van Wyk is a plant taxonomist and has authored or co-authored numerous publications, including several books, on the botany of Southern Africa.

Ugena* = **Lygodium** (Lygodiaceae) Cav.
For Manuel Munõz de Ugena (Manuel Munõz y Matarranz) (1747–1807), Spanish botanical artist and court painter to Charles III, who ruled Spain and the Spanish Indies from 1759–1788. Munõz was a good friend of Casimiro Gómez de Ortega (1741–1818), Spanish physician and botanist and first professor of the Royal Botanical Garden of Madrid. Gómez published extensively on the new plant species collected during Spanish explorations of South America. Together, they authored the two-volume work on Spanish flora *Florae Hispanicae Delectus Florae, sive Plantarum per Hispaniensis Imperium Insigniorum Nascentium Icones Sponte, et Description* (1791–1792). Munõz de Ugena did woodcuts for this volume.

Ulex* (Fabaceae) L.
The Latin name, apparently for a different shrub, resembling rosemary (*Rosmarinus officinalis*).

Ulmus* (Ulmaceae) L.
The Latin name for the elm; possibly from the Celtic *ulm*.

Umtiza (Fabaceae) Sim
IsiXhosa name for this tree, *umThiza*.

Uncaria** = **Harpagophytum (Pedaliaceae) Burch.
La. *uncus* = hook; *aria* = like; referring to the hooks from old peduncles.

Uncinia (Cyperaceae) Pers.
La. *uncinus* = little hook (on seeds); referring to the bur-like fruiting heads.

Unigenes (Lobeliaceae) E. Wimm.
La. *uni-* = single; *-genes* = born, hence 'unparalled'. The Plant List put this genus in Campanulaceae.

Unona** = **Xylopia (Annonaceae) L.
Meaningless name modified from *Annona* (q.v.), previously *Anona* (Jackson). African species formerly included in this genus are now distributed among several others.

Urachne* = **Oryzopsis** (Poaceae) Trin.
Gk. *ur-* = tail or tail-like; *achne* = scale, glume; referring to its long-tailed appearance.

Uragoga* = **Psychotria** (Rubiaceae) Baill.
Gk. *ouron* = urine; *agogos* = a leader. Plants so named are now regarded as species of *Psychotria*. Their action is emetic (causing vomiting) rather than diuretic (causing urine release).

Urelytrum (Poaceae) Hack.
Gk. *oura* = tail; *elytron* = cover, scale, husk; alluding to the lower glume, which is drawn out into a long awn.

Urena* (Malvaceae) L.
From the Malaysian name *uren*.

Urera (Urticaceae) Gaudich.

(EJM)

La. *urere* = to burn; *urens* = stinging. These relatives of stinging nettles are armed with some of the most painful stinging hairs known on their leaves and stem (Jackson).

Urginea* = **Drimia** (Hyacinthaceae) Steinh.
From Beni urgen, a tribe in Algeria living where the type species was found.

Urochlaena (Poaceae) Nees
Gk. *oura* = tail; *chlaena* = cloak; referring to the lemma, which has a long awn.

Urochloa (Poaceae) P. Beauv.
Gk. *oura* = tail; *chloa* = grass; referring to the mucronate lemma in the upper floret.

Urospermum* (Asteraceae) Scop.
Gk. *oura* = tail; *sperma* = seed; alluding to the beaks of the cypselae.

Urostachys** = **Huperizia (Lycopodiaceae) Herter
Gk. *oura* = tail; *stachys* = spike; referring to the appearance of drooping spikes.

Urostigma** = **Ficus (Moraceae) Gasp.
Gk. *oura, ur* = tail; *stigma* = stigma. Meaning obscure.

Ursinia (EJM)

Ursinia (Asteraceae) Gaertn.

These daisies have papery bracts surrounding the inflorescense base. (CM)

For Johann Heinrich Ursinus (1608–1666), German Lutheran theologian, school principal (superintendent) in Regensburg, notary and prolific author – some 153 works in three languages – mainly concerning religious matters or related thereto. One of his main works was *Arboretum Biblicum* (1699), which describes the plants in the Bible (trees, fruits, herbs and spices). In 1664, he took a stance against Justinian Von Weltz (1621–1668) and Johann Georg Gichtel (1638–1710), who opposed the leadership of the Lutheran Church. They criticised it for failing to share the gospel with heathens and wrongfully spending money on amusement, food and dress. Ursinus (1608–1667) responded to Von Weltz's vehement attacks by calling him 'a heretic and fanatic', and published *A Sincere, Faithful and Earnest Admonition to Justinian* anonymously.

Ursiniopsis = Ursinia (Asteraceae) E.Phillips
Ursinia (q.v.); *-opsis* = resembling.

Urtica (Urticaceae) L.
La. *urtica* = probably from La. *urere* = to burn; referring to the stinging nettle.

Utricularia (Lentibulariaceae) L.
La. *utriculus* = small bag, bladder-like; *aria* = pertaining to; referring to the insectivorous leaf sacs. The small insect-trapping sacs are attached to the underground leaves. 'Bladderwort'.

Uvaria (Annonaceae) L.
La. *uva* = a bunch of grapes; *aria* = pertaining to. The fruits look like grapes.

V

Vernonia: a genus of a great many species of shrubs and herbaceous flowering plants, many of whose names are still unresolved or synonyms

Tony Dold (1965–) grew up in Ngcobo in the former Transkei, where he became interested in plants at a young age. He obtained a national diploma in forestry at Saasveld and later an *ad eundem gradum* MSc in botany at Rhodes University in Grahamstown, where he still works today. He has worked at the Selmar Schönland Herbarium since 1992, and assumed the curatorship in 2003. His interest in useful plants has contributed to scientific, popular and educational works on the subject. He is a co-founder of Inkcubeko Nendalo, a schools environmental education programme in Grahamstown.

Vaccaria* (Caryophyllaceae) Medik.
La. *vacca* = cow; *-aria* = pertaining to, from Med. La. *vaccaria* = for cow-pasture. Presumably the plant makes good fodder for cows.

Vaccinium (Ericaceae) L.
Derivation obscure. This Latin name, used by Virgil and Ovid, is possibly a corruption of *baccinum* = little berry; referring to bilberry. There are various common name forms, depending upon the species.

Vachellia = Acacia (Fabaceae) Wight & Arn.

(EJM)

For Rev. George Harvey Vachell (1798–1839), British priest, plant collector. He graduated from Cambridge University in 1821. He went to China in 1828 and stayed in Macau until 1836, except for a leave of absence in 1832. He became the chaplain to the British East India Company's factory in Macau (Macao) from 1825–1836. In his spare time, he and collected plants in China when he was in Canton, and also sailed around the islands of Macao in his small boat. He discovered several new taxa and gave his collections to Reverend Professor JS Henslow of Cambridge, who passed them on to WJ Hooker and John Lindley.

Vahlia (Vahliaceae) Thunb.
For Martin Henrichsen Vahl (1749–1804),

Norwegian-born Danish botanist, zoologist and author. He studied botany under Linnaeus at Uppsala University. He edited *Flora Danica*, published in booklets XVI–XXI (1787–1803), *Eclogæ Americanæ* I–IV (1796–1807), and *Enumeratio Plantarum* (1804), Volume 1, in the year he died – the first of 15 volumes he had worked on for 26 years. He lectured at the University of Copenhagen Botanical Garden (1779–1782), made research trips to Europe and North Africa (1783–1788), became professor at the Society of Natural History in Copenhagen (1786) and of Botany at Uppsala University (1801–1804). Vahl's library, about 3 000 volumes, was sold to the government on his death. He was a member of the Royal Swedish Academy of Sciences.

Valeriana (Valerianaceae) L.
Presumably La. *valere* = to be healthy; alluding to the plant's medicinal use. *Valerianus* = of Valeria, a Roman province. Publius Valerius Publicola, who helped drive out the Roman kings, is said to have initiated its use in medicine.

Vallisneria (Hydrocharitaceae) L.
For Antonio Vallisnieri (Vallisneri) (1661–1730), Italian physician, botanist, naturalist, biologist and

professor of medicine at the University of Padua. Author of many publications on insects, including (in translation) *The Curious Origin of Many Insects* (1696–1670), *Common Worms of the Human Body* (1710), *The Origin, Development and Habits of Various Insects* (1713), *Origin of Fountains* (1715), *History of the Generation of Man and Animals* (1721), and *Marine Bodies on Mountains – Their Origin* (1728), as well as writing about the unusual, such as ostriches and chameleons. Vallisnieri was a 'contrarian', and abandoned reliance on medieval theories in favour of experimentation, conducting trials, observations and deductive reasoning, and he wrote his treatises in Italian rather than Latin. He was a member of the Royal Society of London.

Vallota* = **Cyrtanthus** (Amaryllidaceae) Herb.
For Antoine Vallot (1594–1671), French botanist, physician, count of Archiâtres. He studied at the

University of Montpellier and became physician to Anne of Austria (1601–1666), the Queen Consort of France, and to her son, Louis XIV of France. Vallot kept a diary of the king's health, *Journal de la Santé du Roi Louis XIV de l'Année 1647 à l'Année 1711*. (He was a great believer in emetics as a cure for sickness.) He was made director of the king's botanical garden, Jardin du Roi, later to become the Jardin des Plantes. When Vallot took over,

the garden was in a state of decline, but with the help of Pierre Magnol (1658–1715), professor of botany at Montpellier, and other botanists (Denis Joncquet and Guy-Crescent Fagon), he created this world-famous garden.

Vandenboschia* = Trichomanes (Hymenophyllaceae) Copel.
For Benjamin Roelof van den Bosch (1810–1862), Dutch physician, botanist and specialist in ferns and mosses. He studied medicine at Leiden University (1828–1837), where he obtained his doctorate. Thereafter he established a practice in Goes, but was also highly in demand because of his botanical knowledge. On the death of the plant collectors JH Molkenboer (1854) and his co-author and co-collector F Dozy (1856), Van den Bosch was appointed to the (now) National Herbarium of the Netherlands. Together, he and CM van der Sande Lacoste of the National Herbarium studied Molkenboer and Dozy's moss collection, and published a book, *Bryologica Javanica* (1861), based on their uncompleted work. This book describes many Indonesian mosses already known, and includes more than 300 new mosses. Van den Bosch was also a founding member of the Royal Dutch Botanical Society in 1845.

Vangueria (Rubiaceae) Comm. ex Juss.
From the Malagasy name *voa vanguer* for *Vangueria edulis*.

Vangueriopsis (Rubiaceae)
Vangueria (q.v.); Gk. *-opsis* = resembling. The genus looks like *Vangueria*.

Vanheerdea (Aizoaceae) L. Bolus ex H.E.K.Hartmann.
For Pieter van Heerde (1893–1979), South African teacher and school principal, active field collector of succulents. He obtained an MA in chemistry and a teaching diploma from the University of Cape Town. After working in a number of lecturing and teaching posts, he accepted the position of headmaster of Springbok School, where he remained until he retired (1926–1952).

His special interest was in the succulent plants of Namaqualand, which he kept in his private garden, and later at the Hester Malan Nature Reserve, where he worked briefly from 1966–1967 until his health failed. He sent specimens of local discoveries to contemporary botanists such as Dr Louisa Bolus (1877–1970), who honoured him with the genus *Vanheerdea*.

Vanilla* (Orchidaceae) Mill.
Spanish *vainilla* = a small pod, from *vaina* = a sheath, plus- *illa* = diminutive; referring to the shape of the long, thin, cylindrical seed pods.

Vanwykia (Loranthaceae) Wiens
For Piet van Wyk (1931–2006), South African botanist, ecologist, photographer and author. A graduate of Potchefstroom University, he worked as a research biologist at the Kruger National Park prior to his retirement in 1991. Piet played a significant role in promoting an interest in trees among the general public. Among his published works, inter alia, were *Field Guide to the Trees of the Kruger National Park* (Struik, 1984), *Photographic Guide to Trees of Southern Africa* (Struik, 1993), *Field Guide to Trees of Southern Africa* (Struik, 1997), and *How to Identify Trees in Southern Africa* (Struik, 2007). He was awarded a chancellor's medal by the University of Pretoria and an honorary doctorate from Unisa in recognition of his contributions to botany, nature conservation, and environmental education in Southern Africa.

Vanzijlia (Aizoaceae) L.Bolus
For Dorothy Constantia van Zijl (1886–1938), South African plant collector and keen gardener, wife of the late Judge-President HS van Zijl. She collected in the Cape, and lived for many years in Clanwilliam.

Vascoa*** = Rafnia (Fabaceae) DC.
For Vasco da Gama (1460 or 1469–1524), Portuguese explorer who pioneered the first sailing route from Europe to India between 1498 and 1499. While this trip was commercially successful, more than half his men died, mainly of scurvy. He also failed to secure a commercial treaty with the King of Calicut. Subsequently war broke out between Portugal and Calicut. In 1502, Da Gama led a fleet of 15 ships and 800 men to India, during which he inflicted barbarous acts of cruelty upon competing traders and local inhabitants. (For instance, he looted a ship with more than 400 Muslim pilgrims, including 50 women, locked them in the ship, and burnt them to death.) In 1524, he returned to Portuguese India as governor and viceroy, but died shortly after from malaria.

Vauanthes*** = Crassula (Crassulaceae) Haw.
Gk. *Vau* = the name for a Greek symbol (the early

Greek digamma), shaped like two capital gammas one on top of the other, approximating in sound the English W. It fell into disuse; *anthos* = flower. The allusion is to the markings on the flower.

Vellereophyton (Asteraceae) Hilliard & B.Burtt

(CM)

La. *vellus, vellere* = fleece, wool, down; Gk. *phyton* = plant. The plant is silver and woolly all over.

Vellozia (Velloziaceae) Vand.
For Joaquim Velloso de Miranda (1733–1815), Portuguese-Brazilian Franciscan priest, naturalist and plant collector. He graduated in canonic law; studied at the University of Coimbra, Portugal (1772) under the Italian naturalist Dominicus Vandelli (1735–1816), obtaining a PhD (1778), then became a lecturer at the university and leading researcher. He was elected a member of the Royal Academy of Science of Lisbon. Years later, he returned to Brazil and collected plants in the Rio de Janeiro/Minas provinces, which he sent to Vandelli for the Royal Museum. In 1799, he established the botanical garden for Vila Rica following the orders of the Prince Regent.

Veltheimia = Kniphofia (Hyacinthaceae) Gled.
For August Ferdinand von Veltheim (1741–1801), German mineralogist, with interests in geology, archaeology, and civic matters related to his estates. He was inspector of the mines in the mining district of the Hartz Mountains (1766) but declined an offer as general inspector of mines and salt works in Russia's western regions. Among other books he  authored a geology book, *Etwas über die Bildung des Basalts* (1787), the first person to correctly attribute the origins of granite to volcanic action. He developed and beautified a botanical garden on the grounds of Castle Uarbke at Harbke, and was a patron of botany. He received an honorary PhD from the University of Helmstedt, was elected an honorary member of the Prussian Academy of Arts, and appointed a count by the king of Prussia.

Venidium = Arctotis (Asteraceae) Less.
Origin obscure; perhaps from La. *vena* = a vein; *idium* = little; referring to the ribbed fruits.

Vepris (Rutaceae) Comm. ex A. Juss.
La. *vepres* = a bramble or thorny shrub. While this characteristic may apply to the type species, this is not applicable to *Vepris lanceolata* (white ironwood) in Southern Africa.

Veratrum* (Melanthiaceae*) L.
La. *veratrum* = true black, from *vere* = true; *ater* = black; referring to highly poisonous black rhizomes, or possibly to the seed pods, which turn black as they ripen.

Verbascum* (Scrophulariaceae) L.
La. *verbascum* = a plant commonly called mullein. Probably derived from a corruption of the Latin word *barbascum* or *barba*, referring to a beard; possibly alluding to the hairy leaves in some species, or possibly referring to the stamens.

Verbena* (Verbenaceae) L.
Probably singular of La. *verbenae* = sacred boughs of olive, myrtle laurel; referring to a leafy branch or twig from aromatic trees or shrubs often used in herbal remedies.

Verbesina* (Asteraceae) L.
Probably from the Italian dialect *forbesina* = *Verbesina*. The name *Verbesina* refers to the similarity of the foliage to that of the (unrelated) *verbena* (Wikipedia.org).

Verea = Kalanchoe (Crassulaceae) Willd.
For James Vere (1739–1803) of Kensington-Gore, a wealthy silk merchant who had a large collection of plants, some very rare, and also collected botanical art that was displayed in Curtis's *Botanical Magazine* and elsewhere. The gentleman's magazine and historical chronicle, Volume 49, indicates he had 'literary talent'. Seemingly, he was a founder of the Royal Horticultural Society, but he was not at the founders meeting because, although initial discussions commenced in 1800, the founding meeting was in 1804, by which time James Vere had died.

Vernicia* (Euphorbiaceae) Lour.
La. *vernix* = varnish; referring to tung oil, used in quick-drying varnish, which is derived from the type species of the genus.

Vernonanthura (Asteraceae) H.Rob.
Veronica (q.v.); *anth-* = anther; *oura* = a tail; referring to the inflorescence, which looks like a tail.

Vernonia (Asteraceae) Schreb.
For William Vernon (1666–1711), English botanist and bryologist, fellow of St Peter's College, Cambridge, graduating with a BA (1688) and MA (1692); a Fellow of the Royal Society, who collected plants in Maryland, Virginia, North America, in 1696 with the English Reverend Hugh Jones (1671–1702) (a replacement for John Banister, who was accidently shot dead while collecting plants in 1692) and Dr David Krieg (1669–1710), German surgeon and botanist, a correspondent of Petiver. All together they collected more than 650 plants from Maryland.

Veronica (Linderaceae) L.
Probably after St Veronica, the maiden who handed her handkerchief to Jesus on his way to Calvary. He wiped his brow and returned the handkerchief, which now bore his likeness and so was called *vera-ikon* = true likeness, and the maiden became St Veronica (Brewer). The allusion is obscure.

Vernonella** (Asteraceae) Sond.
Vernonia (q.v.); Gk. *-ella* = diminutive; restored as a distinct genus from synonymy under *Vernonia* and *Centrapalus*.

Vetiveria = Chrysopogon (Poaceae) Bory
From the Tamil *vetti* = its popular or vernacular name, Khus khus or Khas; *ver* = root; referring to its thick, fibrous, adventitious roots, which are aromatic and highly valued.

Vexatorella (Proteaceae) Rourke
La. *vexator* = one who disturbs, vexes; *-ella* = diminutive; alluding to the taxonomic problems created by the new genus.

Viburnum* (Adoxaceae) L.
La. *Viburnum* = a kind of shrub. The generic name referred to originally *V. lantana*.

Vicia* (Fabaceae) L.
Old Latin name for a vetch, used by Ovid and Virgil; possibly related to *vincio* = I bind; referring to some species' clasping tendrils.

Vicoa = Pentanema (Asteraceae) Cass.
Possibly after Giovanni Battista (Giambattista)

Vico or Vigo (1688–1744), Italian philosopher and historian, professor in rhetoric at the University of Naples (1699), later royal historiographer to Charles III, king of Naples (1734). His major work was *Scienza Nuova* (*The New Science*) (1725), revised 1730. He describes 'history' as the account of the birth and development of human societies and their institutions, not of individual biographies. Each era in history goes through phases that have distinct characteristics that recur throughout history, always in the same sequence, but do not exactly replicate themselves, as the social and political character of each age is subject to the modifications and circumstances of each new era. Vico is regarded by many as the first modern historian.

Vieusseuxia*** = Moraea (Iridaceae) D. Delaroche
For Gaspard Vieusseux (1746–1814), Swiss botanist and physician. After qualifying in Leiden in 1766, he completed his studies in Vienna, Strasbourg, Paris, London and Edinburgh, and opened a practice in Geneva. In 1771, he opened an office in Geneva. His first major publication, *Traité de la Méthode Nouvelle d'Inoculer la Vérole Pepite* earned him honours not only in Switzerland, but also in France, England and Sweden. He was the first physician to describe, in 1805, during an outbreak in Geneva, cerebrospinal fever or brain fever, subsequently called meningitis (meningococcal disease). He also gave an early account of Wallenberg's Syndrome in 1810.

Vigna (Fabaceae) Savi
For Domenico Vigni, usually Latinised Domenicus Vigna (1577–1647), Italian botanist, professor of botany at Pisa for 38 years and ninth curator/director of the Orto Botanico di Pisa (Botanical Garden of Pisa), also known as the Orto Botanico dell'Universita di Pisa, the oldest in Europe, founded in 1543, a botanical garden where plants were studied for medical use, operated by the University of Pisa. Vigni was also a commentator on Theophrastus. No further details could be found.

Villarsia (Menyanthaceae) Vent.
For Dominique Villars (1745–1814), French physician and botanist, doctor of the Military Hospital of Grenoble (1782), professor of botany

Villarsia is a small semi-aquatic herb. (CM)

in Grenoble (1795–1803), and later professor of botany and dean of the faculty of medicine at the University of Strasbourg. His main work is *Histoire des Plantes du Dauphiné*, published in three volumes between 1786 and 1789, in which about 2 700 species (particularly alpine plants) are described after more than 20 years of observation. His herbarium and botanical manuscripts are preserved at the Muséum d'Histoire Naturelle de Grenoble. He was a correspondent of the Royal Society of Paris, the Royal Society of Sciences of Turin, and the Linnaean Society of London, among others.

Vinca* (Apocynaceae) L.
The name used by Pliny the Elder, probably from *vincio* = I bind; derived from *vincire* = to bind or wind around; referring to the plant's long, tough, smothering runners.

Vincetoxicum (Apocynaceae) Ruppius
La. *vincere* = to conquer; *toxicum* = poison; referring to the belief that this plant could counter poison.

Viola* (Violaceae) L.
La. *viola* = the name for a violet used by Virgil; one of several scented flowers not of Indo-European origin.

Virgilia (Fabaceae) Poir.
For Publius Vergilius Maro (70–19 BCE), known as Virgil or Vergilus, regarded as one of the greatest Roman poets. He studied mathematics, medicine and rhetoric in Rome and Naples, and thereafter entered literary circles, writing *Eclogues* (or *Bucloics*), the *Georgics* and the *Aeneid*, considered Virgil's finest

(CM)

work and one of the most important poems in the history of Western literature. This work, commissioned by Augustus Caesar, depicted the glories of Rome and the Roman Empire. It took 11 years to write (29–19 BCE). According to the tradition, Virgil travelled to Greece in about 19 BCE in order to revise the Aeneid. While there, he caught a fever, returned to Italy by ship, but weakened with disease and died on reaching Brundisium harbour.

Viola (CM)

Viscum (Santalaceae) L.
Latin name for birdlime, a sticky substance made from the plant's berries and smeared on branches to catch birds. Classical Latin name for mistletoe, derived ultimately from an Indo-European root meaning 'sticky'; hence the English word viscid.

Vitellariopsis (Sapotaceae) Baill. ex Dubard
La. *vitellus* = yolk of an egg; *-aria* = like; Gk.

-iopsis = resembling, likeness. *Vitellaria* is the name of a genus previously in the same family.

Vitex (Lamiaceae) L.
La. *Vitex*, a name used by Pliny the Elder for *Vitex agnus-castus*. It is derived from the La. *vieo* = to weave or to tie up, Gk. *vico* = to bind; alluding to the flexible stems of some species. *Vitex agnus-castus* is used in basketry.

Vitis = **Cissus** (Vitaceae) L.
The Latin name for vine, derived ultimately from the same root as Sanskrit *vja* and German *winden* = to wind, twine, bind; referring to the plant's growth characteristics.

Vittaria (Pteridaceae) Sm.
La. *vitta* = a narrow ceremonial headband, or ribbon; probably referring to the narrow string or ribbon-like leaves on the fronds.

Vleisia = **Pseudalthenia** (Zannichelliaceae) Tom & Pos.
Dutch/Afrikaans *vlei* = a shallow seasonal or intermittent lake, original meaning 'pond'; alluding to the preferred habitat of the genus.

Vlokia (Aizoaceae) S.A.Hammer
For Johannes Hendrik Jacobus Vlok (1957–), South African nature conservationist. He has a diploma in forestry (1982) and an MSc (*cum laude*) from the University of Natal (1997). He was a research forester at Saasveld (1982–1900), nature conservation officer (1990–1997), and since 1998 has acted in a consultant or advisory capacity to the World Wildlife Federation South Africa, Cape Nature, South African National Parks, and national, provincial and private institutions in the fields of land conservation, impact assessment and environmental control. He has authored more than 30 scientific and popular articles, and has made many presentations at scientific forums on ecological and floristic studies. He is the recipient of the Leslie Hill (2003) and Gold (2006) awards for his contribution to conservation.

Voacanga (Apocynaceae) Thouars
From the Malagasy name for a species of this genus: *voa* = fruit; *acanga* = spotted.

Vogelia*** = **Dyerophytum** (Plumbaginaceae) Lam.
Probably for Benedict Christian Vogel (1745–1825), Bavarian botanist and physician. He became a professor of botany at the University of Altdorf in 1767 or 1768, and worked in the botanical garden of the University of Erlangen-Nuremberg from 1769–1809. He was the author of a number of books, including *Supplementum Plantarum Selectarum*, a work written with Georg Dionysius Ehret, Johann Elias Haid and Christoph Jacob Trew, which Linnaeus described as one of '[t]he miracles of our century in the natural sciences', referring especially to Ehret's illustrations. One of Vogel's interests was nosology (the study of disease and how this can be classified in order to come to a better understanding of it).

Volkiella (Cyperaceae) Merxm. & Czech
For Otto Heinrich Volk (1903–2000), German professor of botany. He studied natural science at Munich, Vienna and Heidelberg, obtained a doctorate, and in 1930 started work at the University of Würzburg. In 1937, he went to Namibia to carry out research and was interned during World War II, during which he taught botany to many interns, including Johan Wilhelm Heinrich Giess. After the war, he developed an interest in medicinal plants and collected in Spain, before moving to Kabul, Afghanistan, as a visiting lecturer (1950–1952), where he collected some 1 600 specimens. In 1953, he returned to Würzburg and became director of the Institute of Pharmacognosy until he retired in 1972. During his career, he botanised in Tanzania, South African, Namibia (seven visits between 1956 and 1985, gathering 6 000 specimens), Ecuador, Peru and Argentina. His publications centred on plant geography.

Vossia (Poaceae) (Roxb.) Griff.
For Johann Heinrich Voss (1751–1826), German poet and translator who studied philology at  the University of Göttingen in 1772. He was editor of *Musenalmanach* (1775), rector of the Latin school at Otterndorf (1778), rector of the gymnasium at Eutin (1782–1802), professor at the University of Heidelberg (1805–1826), and author of two volumes of poems, including his idyllic poem *Luise*, and papers on mythology and religious freedom. A classical scholar and linguist, his translations into German of the works of Homer's epics (*Illiad* and *Odessy*), Virgil, Horace, Tibullus, Ovid, Theocritus, Propertius and others have received acclaim. He also translated Shakespeare's plays, which he completed with the help of his sons, Heinrich and Abraham.

Vulpia* (Poaceae) C.C.Gmel.
For Johann Samuel Vulpius (1760–1846), German chemist, physicist, pharmacist and amateur botanist who investigated the Flora of Baden. No further details could be found.

W

Welwitschia mirabilis: The only plant of its kind in the world

Ernst van Jaarsveld (1953–) obtained his national diploma in horticulture in 1974 and started work in the Lowveld Botanical Gardens, Nelspruit (now Mbombela) and was transferred to Kirstenbosch in 1976 to take over the succulent section. In 1990 he graduated from what is now the University of KwaZulu-Natal with an MSc and in 2012 with his PhD from the University of Pretoria for his detailed study of the taxonomy and ecology of cliff dwelling plants (the first such detailed study globally of what are termed Cremnophytes). He is an acknowledged world authority on succulent plants and has published many scientific and popular articles and written many books. In 2015 he took early retirement to join the team at the historic Cape Dutch farm Babylonstoren in the Western Cape.

Uschi Pond (1961–) from Namibia obtained her national diploma in conservation biology and nature conservation and has published numerous papers. She has worked with environmental consultants and qualified in architectural garden design. She has an interest in graphic design and digital art. She has designed and published two books, the latter on *Welwitschia mirabilis*. She is planning on establishing a nature reserve on her family farm in Namibia.

Wachendorfia (Haemodoraceae) Burm.

Typical flower (CM)

One can easily understand why the family is a bloodroot. (EJM)

For Evert Jacob van Wachendorff (1702–1758), Dutch physician and botanist. He studied medicine in Leiden and Utrecht, becoming a doctor of medicine in 1724, and became municipal physician of Utrecht (1726–1758). During this period, he also became a lecturer in chemistry at Utrecht University (1726–1743), then professor of medicine, chemistry and botany (1743–1758). He also became a director of the Botanical Gardens of Utrecht. Among his publications was *Horti Ultraiectini Index* (1747) and *Oratio Botanico-Medica de Plantis Immensitatis Intellectus Divini Testibus* (1743). He was a good friend of Clifford and Linnaeus.

Wahlenbergia (Campanulaceae) Schrad. ex Roth

For Georg Göran Wahlenberg (1780–1851),  Swedish naturalist, geographer and doctor, who became a demonstrator in botany (1815–1828) and professor of botany at the University of Uppsala (1828–1851), succeeding Carl Peter Thunberg. Wahlenberg made his main work in the field of plant geography, and published, among other things the *Flora Lapponica* (1812), a considerably extended version of the work of his compatriot Linnaeus, who wrote a publication of the same name (1737). His other works were based on his trips to Norway, Finland and the plant world of northernmost Sweden. Wahlenberg was elected a member of the Royal Swedish Academy of Sciences in 1808.

Walafrida = Selago (Scrophulariaceae) E.Mey.

For Walahfrid Strabo (or Walafrid Strabus) (c 809–849), German monk, poet, politician and  theologian, educated at the monastery of the Reichenau on Lake Constance. He became an Abbott and tutored the future King Charles the Bald of the West Franks, moved in high circles, and met many of the leading intellectuals of his age. He wrote many theological works and visionary poems, also a book on gardening called *Hortulus,* full name *Liber de Cultura Hortorum*. This medieval work, possibly based on a treatise on agriculture, *De re Rustica* (*Country Matters*), by the 1st-century writer Lucius Junius Columella (c 4–70) examines various plants with reference to their mythological, medicinal or Christian value or implications.

Walleria (Tecophilaeaceae) Kirk

For Horace Waller (1833–1896), British botanist, protestant pastor and plant collector. He  abandoned his career in his father's stock broking business to become a lay missionary to Central Africa (1860–1864), and took part in David Livingstone's failed Zambezi expedition (1858–1864), during which he discovered two species of this genus and a new gazelle, which was named after him (*Litocranius walleri*). On his return to England, he became a pastor of St John's Church, Chatham (1867), a member of the Society for the Abolition of Slavery (1870), vicar of Leytonstone (1874), then rector of Twywell (1874–1895), during which time he wrote many publications about both his African experiences and the need to liberate slaves.

Wallinia = **Lophiocarpus** (Lophiocarpaceae) Moq.
For Georg/Jöran Wallin the Younger (1686–1760), Swedish theologian, historian and antiquarian.

He obtained an arts degree from Uppsala University (1707), and later a doctorate of theology from Wittenberg. During his career, he became an associate professor of theology at the University of Härnösand (1710), court chaplin to King Frederick I (1724), Uppsala university librarian (1727), professor of theology at Uppsala (1732), superintendent of Gotland (1735), and bishop of Göteborg (Gothenburg) (1744). In the 1740s, he compiled a manuscript collection of historical, folkloristic and antiquarian matters, the *Analecta Othlandensia*, which is now kept in the Royal Library, Stockholm.

Waltheria (Malvaceae) L.
For Augustin Friedrich Walther (1688–1746), German anatomist, botanist, physician and

academic. He obtained a philosophy degree from the University of Wittenburg and a medical doctorate from the University of Leipzig, where he became professor of anatomy (1728), pathology (1732), therapy (1737), and in the same year university rector. He was director of the Leipzig Botanical Gardens and owned his own private botanical garden. In his book, *Designatio Plantarum quas Hortus AF Waltheri Complectitur* (1735), he describes thousands of species from his garden. As a physician, he made many contributions in the fields of mycology and angiology (the study of the circulatory and lymphatic systems), and has several medical and anatomical terms named after him.

Warburgia (Canellaceae) Engl.
For Otto Warburg (1859–1938), German botanist and industrial agriculture expert. He graduated from Strasbourg University (1883), travelled in

south and Southeast Asia from 1885–1889, and collected plant specimens, which he later donated to the Royal Botanical Museum in Berlin. He published his findings in *Die Pflanzenwelt* (three volumes). He founded and edited a journal, *Der Tropenpflanzer*, for 24 years. He became interested in economic botany, specialising in tropical agriculture, and started several companies, working in the colonies of the Deutsches Reich. In his later years, he became an active member of the World Zionist Organisation, and became its president between 1911 and 1920. He founded the botanical garden of the Hebrew University on Mount Scopus.

Wardia (Wardiaceae) Harv. & Hook. ex Hook.
For Nathaniel Bagshaw Ward (1791–1868), English surgeon and gardening enthusiast who, in 1829, accidently discovered, as described in his book *On the Growth of Plants in Closely Glazed Cases* (1852), what became known as the 'Wardian case' or terrarium. This is a sealed glass container for growing ferns and other plants without exposure to ambient air. They were used extensively for transport of plants back to England, also for rearing butterflies. Ward also had an interest in microscopy, and the Microscopical Society (1840) arose out of meetings he held. He was a member of the board of the Chelsea Physic Garden and helped develop it. He became a Fellow of the Linnaean Society in 1817, and was elected a Fellow of the Royal Society in 1852.

Warneckea (Melastomataceae) Gilg
For Otto Warnecke (f 1899–1908), German botanist and plant collector. Details are sketchy about his life, but it is believed he collected plants from the coastal area Lomé, Togo, on the west coast of Africa from 1899–1901, then worked as head gardener for the Biological-Agricultural Institute at the Amani Nature Reserve in Tanzania from 1903–1908, where he collected the type specimen of the fern *W. amaniensis*. During this period, many exotic trees and crops were introduced to Amani by Germans, who were interested in economic development, especially the export of agricultural goods to Germany. His original collections were deposited at Berlin but destroyed (1943) in a bombing during World War II.

Watsonia (Iridaceae) Mill. (Images on following page)
For William Watson (1715–1787), English physician, apothecary, botanist and naturalist. He introduced the work of Linnaeus and his

Watsonias are another iris that has become a much sought-after horticultural and floral species. (CM – All)

botanical classification system to Britain. He was the first scientist to observe the flash of light from the discharge of a Leyden jar and to show that electricity could pass through a vacuum and that it had a positive and negative charge; he coined the word 'circuit'. His articles, entitled *Experiments on the Nature of Electricity*, appeared from 1745 onward in the Philosophical Transactions of the Royal Society, of which he became a member (1741) and vice president (1772). Both he and Benjamin Franklin discovered some of the same characteristics of electricity at the same time, but independently. The two men became friends.

Webbia = *Conyza* (Asteraceae) Sch.Bip.
For Philip Barker Webb (1793–1854), English botanist and geologist who collected in the Mediterranean area and was the first person to collect in the Tetuan Mountains of Morocco. He also collected extensively on the Canary Islands from 1828 to 1830, and co-authored with Sabin Berthelot *L'Histoire Naturelle des Iles Canaries* in nine volumes, the text of which took 20 years to complete. Webb amassed a huge herbarium of his own and other collections, and a library, which is now housed in the Botanical Museum of Florence. The angiosperms alone consist of more than 300 000 specimens. Webb coined the biogeological term 'Macaronesia' to cover the five archipelagos of the Azores, Madeira, the Salvage Islands, the Canary Islands, and Cape Verde Islands.

Webera = *Tarenna* (Rubiaceae) Schreb.
After Georg Heinrich Weber (1752–1828), German physician and botanist. He obtained his medical doctorate in 1777 and in 1780 became professor of medicine and botany at the University of Kiel (1780). He started a polyclinic, which became the Institute of Christian-Albrechts-University (1802), developed the university's botanical garden and, in 1810, was appointed director of the Schleswig-Holstein Medical College in charge of the combined hospitals and the botanical garden. He published extensively on botany, entomology and medicine, including lichens as a life form, the healing-scientific use of mosses and ferns, the Flora of Göttingen, African 'medicine', etc. For his research, publications and for his humanitarian achievements he was awarded an honorary PhD by the Göttingen Philosophical Faculty in 1824.

Websteria (Cyperaceae) S.H.Wright.
For George W Webster (1833–1914), US botanist in Florida and farmer. No further details found.

Wedelia* (Asteraceae) Jacq.
For Georg Wolfgang Wedel (1645–1731), German physician and botanist. He qualified as a doctor at the University of Jena (1669), served as a district physician until 1672, and then was appointed professor of anatomy, surgery and botany at the University of Jena (1673), a position he held until 1719. During this period, he was voted rector on 10 occasions. From 1719–1721, he held the chair of theoretical medicine. He was also personal physician to the duke of Weimar (1679) and prince of Saxony (1685). His medical publications leaned heavily in the direction of pharmacological and pharmaceutical chemistry. He was also a defender of both of alchemy and astrology. He translated a new and accurate edition of the Greek Bible into German.

Weihea = *Cassipourea* (Rhizophoraceae) Spreng.
For Carl Ernst August Weihe (1779–1834), German botanist, physician and batologist (a person who studies brambles). He obtained his doctorate (medicine and botany) from the University of Halle in 1802, worked briefly as a doctor, established a small botanical garden in Mennighüffen, his birthplace, and described some 160 new plant species. He discovered that the bramble, genus *Rubus*, consists of many similar species, 18 of which he discovered near his home. Between 1822 and 1827, he and Christian Gottfried Nees von Esenbeck, a taxonomic botanist, published in 10 parts an elaborate folio monograph *Rubi Germanici Descripti et Illustrati*, describing 49 species of the genus.

Weinmannia = *Platylophus* (Cunoniaceae) L.
For Johann Wilhelm Weinmann (1683–1741), pharmacist and botanist. By age 40, he had become wealthy enough to follow his passion – botany. He established a botanical garden in Regensberg. His major work, *Phytanthoza Iconographia* (1737–1745), featured more than 1 000 hand-coloured engravings of several thousand

plants. Many illustrations were drawn by Georg Dionysius Ehret (1708–1770), later to become one of the foremost floral illustrators of the 18th century, but other illustrators were mediocre. The text was the work of the Regensburg physician Dr Johann Georg Nicolaus Dieterichs. Because of some 'untrue, even faked images', the volume is seen as impressive in size rather than quality.

Wellstedia (Boraginaceae) Balf.
Probably for James Raymond Wellsted (1805–1842), surveyor and traveller. He was appointed second lieutenant of the East India Company's surveying ship *Palinurus*, which made a survey of the Gulf of Aqaba, and the northern part of the Red Sea (1830) and southern coast of Arabia (1833). He spent two months exploring the island of Socotra and twice visited Oman, in 1835 and 1837, before retiring from service in 1839 for health reasons. He authored the two-volume *Travels in Arabia* (1838) and *Travels to the City of the Caliphs Along the Shores of the Persian Gulf and the Mediterranean* (1840). He was elected a Fellow of the Royal Geographical Society (1837) and Royal Astronomical Society in London.

Welwitschia (Welwitschiaceae) Hook.
For Friedrich Martin Joseph Welwitsch (1806–1872), Austrian naturalist, explorer and medical

doctor. He studied medicine at the University of Vienna and practised for a short while, but in 1839 gave up medicine and became director of the botanical gardens in Portugal. He collected plants in the Canary Islands, Madeira and Angola (from 1853), then a Portuguese colony. He 'discovered' *Welwitschia mirabilis* in 1859 in the Namib Desert of southern Angola. In 1861, he returned to Portugal and shortly after to London, where, from 1863, he categorised and catalogued his huge Angolan collection at the Natural History Museum and Kew Gardens. His publication, *Sertum Angolense*, describes 12 new categories and 48 new species. On his death, his collection was split between the Portuguese government (first set of duplicates) and Natural History Museum (second set).

Whiteheadia*** = Massonia (Hyacinthaceae) Harv.
For Henry Whitehead (1817–1884), English-Anglican missionary and plant collector. He came to South Africa in 1855 and spent a year at a mission station in Namaqualand, near Springbok. He collected in that area, also in the Clanwilliam district, sending specimens to William Henry Harvey in Dublin, who acknowledged his collection in *Flora Capensis* (1862). As a result of the failure of copper mines, Whitehead was transferred to Tulbagh before being sent to St Helena in 1861, where he stayed until he died. He sent collected ferns to Kew Gardens, also to Carl Wilhelm Ludwig Pappe (1803–1862) and William Rawson (1812–1899) in Cape Town, authors of *Synopsis Filicum Africae Australis* (1858).

Wiborgia (Fabaceae) Thunb.
For Eric Nissen Viborg (1759–1822), Danish veterinarian and botanist. He studied under Peter

Christian Abildgaard (1780), became professor of botany at the University of Copenhagen (1797–1801), and director of the botanical garden. In 1801, he became professor and rector of the Royal Veterinary and Agricultural School in Copenhagen. He wrote important works on animal medicine, horse breeding, sheep farming and the treatment of infectious diseases in pigs. His botanical publications include a paper on the use of sand plants to stabilise the coast of Jutland (1789), and he did much work on systematising Danish names for indigenous plants (1793). He was a member of the Academy of Sciences and other societies. His name was misspelled by Carl Peter Thunberg.

Wiborgiella (Fabaceae) Boatwr. & B.E.van Wyk
Wiborgia (q.v.); Gk. *-iella* (diminutive).

Widdringtonia (Cupressaceae) Endl.

A cluster of female cones (EJM)

For Samuel Edward Widdrington (formerly Cook) (1787–1856), a Royal Navy captain, author of two books on Spain: *Sketches in Spain* (1834)

and *Spain and the Spaniards* (1844). While living in Spain, he took a great interest in Spain's coniferous forests – hence a genus being named after him – which he scientifically detailed in his 1844 book. In 1840, he took the name Widdrington, when his mother became heiress of some estates of that family. He served as the high sheriff of Northumberland in 1854. He was an honorary knight commander of the Military Order of the Tower and Sword, an honour bestowed upon him by Dom João VI of Portugal, formerly prince regent. He was elected a fellow of the Royal Society (1842) as well as the Royal Geographical Society.

Wiesneria (Alismataceae) Micheli
For Julius Ritter von Wiesner (1838–1916), Austrian botanist, professor of plant anatomy

and physiology. He studied in Vienna and Jena, where he graduated in 1860, became associate professor at the Polytechnic Institute in Vienna in 1868, and at the Forest Academy in Mariabrunn in 1870. From 1873 to 1909, he was professor of anatomy and physiology at the University of Vienna and director of the Institute of Plant Physiology at the university, the first of its kind. He made scientific voyages to Egypt, India, the Dutch East Indies, the Arctic, and North America. A prolific author of many books, his first publication *Die Rohstoffe des Pflanzenreiches* (*The Raw Material in Plants*) (1873) was still being republished in 1962.

Wigandia* (Boraginaceae) (Humb.) Bonpl. & Kunth
For Johann Wigand (1523–1587), German Lutheran theologian and bishop, professor

of theology and bishop of Pomerania. He attended the University of Wittenberg, and after a short spell as a teacher, obtained his MA degree (1546) and became a pastor. In 1553, he started working on *Historia Ecclesiastica*, an analysis of church history from a Protestant viewpoint. This work was to consume a major portion of his life. During his career, he became professor of theology at Jena, where he studied botany, as well as accepting a position at the University of Koningsberg. Wigand played an important role in many of the 16th-century theological controversies that erupted in the Protestant church after the death of Luther.

Willdenowia (Restionaceae) Thunb.

A male inflorescence shielded by a characteristic bract (CM)

For Karl Ludwig Willdenow (1765–1812), German botanist and physician. He obtained an MD from the University of Halle (1789), became professor of natural history at the Berlin Medical-Surgical College (1798), director of the Berlin Botanical Garden (1801), professor of botany at the University of Berlin (1810), and author of *Florae Berolinensis Prodromus* (1787) and *Principles of Botany* (1805), originally in German (1792). He is

regarded as one of the early phytogeographers and noted that plant distribution patterns change over time – climate influences the numbers, new species arise and existing ones go extinct. His extensive plant collection was purchased by the Berlin Botanical Garden on his death. He was elected a member of the Berlin Academy of Sciences in 1794.

Willkommia (Poaceae) Hack.
For Heinrich Moriz Willkomm (1821–1895), German botanist, explorer, traveller, naturalist.

He studied medicine at the University of Leipzig, obtaining his PhD in 1850, during which time he also collected plants in Portugal and Spain (1844–1845 and 1850–1851). He became professor of botany at the University of Leipzig (1852–1855); professor of natural history at the Forest and Agricultural Academy at Tharandt (1855–1868); professor of forest botany at the University of Tartu, Estonia (1868–1874); and professor of botany at the University of Prague (1874–1892). He was an authority on the Iberian Peninsula flora; the forest flora of Austria, Germany, the Baltic provinces and Russia; and wrote extensively on forest botany, dendrology, pathology and entomology. With Robert Hartig, he is considered co-founder of forest phytopathology. He contributed to *The Vegetation of the Earth* by Adolf Engler and Oscar (Volume 1) dealing with the Iberian Peninsula.

Wimmerella (Lobeliaceae) L.Serra, M.B.Crespo & Lammers
For Elfrieda Franz Wimmer (1881–1961), an Austrian botanist, naturalist, teacher and Roman Catholic priest. From his early youth, he had an interest in insects and plants. He studied in Vienna and Graz, and taught at St George's College in Constantinople. In 1907, he was ordained and became chaplain in Vienna. Later, he became director of the Elisabeth Hospital Sisters of Mercy in Vienna until his retirement in 1958. In his spare time, he travelled to Asia Minor. From 1943–1953, Wimmer contributed to the book *Das Pflanzenreich* (*The Empire of Plants*) by Adolf Engler, in particular for the *Campanulaceae-Lobelioideae*. In 1944, he was appointed correspondent of the Natural History Museum in Vienna.

Wissadula (Malvaceae) Medik.
Derivation uncertain. Diminutive of Gk. *Wissasda* a genus name meaning 'always'. Possibly from Sinhalese *wisaduli*, from *wisa* = poison (as in the pain and inflammation of the bite of a cobra) and *duli* = a very fine powder.

Wisteria* (Fabaceae) Nutt.
For Caspar Wistar (1761–1818), US physician and anatomist. He studied medicine at the universities of Pennsylvania (1782) and Edinburgh (MD, 1786). From 1789–1792, he was professor of chemistry at the institutes of medicine at the College of Philadelphia, becoming adjunct professor of anatomy, midwifery, and surgery at the medical department of the University of Pennsylvania (1793–1808), then chair of the anatomy department (1808–1818). He published *A System of Anatomy* in two volumes (1811–1814), and was an acclaimed teacher of anatomy. Among his achievements, he was the chosen physician to the Pennsylvania Hospital, a promoter of vaccinations, and a friend of Thomas Jefferson, serving as president for three years after Jefferson died in 1815. He was a member of the American College of Physicians, the American Philosophy Society, and president of the Society for the Abolition of Slavery.

Withania (Solanaceae) Pauquy
For Henry Thomas Maire Witham (1779–1844) (misspelling of Withan), English palaeobotanist, geologist and philanthropist, who applied to plants James Nicol's (1810–1879) method of examining rocks by slicing thin sections of them, making full use of the microscope. In the 1830s, he published two works of similar content. These publications were *Observations on Fossil Vegetables* [fossilised wood] *Accompanied by Representations of their Internal Structure as Seen through the Microscope* (1831) and *The Internal Structure of Fossil Vegetables Found in the Carboniferous and Oolitic Deposits of Great Britain* (1833). In the same year, he discovered palaeozoic spores in thin petrological sections of coals from Lancashire. He was a Fellow of the Geological Society of London.

Witsenia (Iridaceae) Thunb.
For Nicolaas (Nicolaes) Witsen (1641–1717), Dutch patron of botany. He studied law, philosophy and languages at Leiden University, became a proficient cartographer, wrote a standard work on ship-building in 1671, and made the first

map of Siberia in 1690. He became mayor of Amsterdam and director of the Dutch East India Company. A prolific collector, his South African collection, *Codex Witsenii*, of early Cape botanical and zoological paintings and manuscripts relating to the fauna, flora and landscapes was estimated, by Herman Boerhaave (1668–1738), professor of botany at Leiden, to contain more than 1 500 paintings of plants. Witsen was a supporter of Carl Peter Thunberg, who visited South Africa and collected more than 3 000 previously unknown species.

Wolffia (Lemnaceae) Horkel ex Schleid.
For Johann Friedrich Wolff (1778–1806), German physician, botanist and entomologist, author of the monograph *Commentatio de Lemna* published by Altdorfii et Norimbergae (1801), and other papers. His most important work, published in five volumes (1800–1811), was done in two languages: the Latin version entitled *Icones Cimicum Descriptionibus Illustratae* (*An Illustrated Description of Bugs*) and the German version *Abbildungen mit der Wanzen Beschreibungen.* The work was completed by Johann Friedrich's father, John Philip Wolf, after his son's early death. Of interest, the smallest flower in the world is *Wolffia globosa*, at only 0.4 to 0.9 millimetres.

Wolffiella (Lemnaceae) (Hegelm.) Hegelm.
Wolffia (q.v.) plus Gk. *-iella* (diminutive).

Wolffiopsis = Wolffiella (Lemnaceae) Hartog & Plas
Wolffia (q.v.) plus Gk. *-iopsis* = resembling, likeness.

Woodia (Apocynaceae) Schltr.
For John Medley Wood (1827–1915), South African botanist, curator of the Natal Botanical Garden (1882–1903) and founder and director of the Natal Herbarium (1903–1913). He contributed greatly to the knowledge of Natal ferns writing publications such as *A Popular Description of Natal Ferns* (1877) and *The Classification of Ferns* (1877), as well as works about plants in general, including a *Preliminary Catalogue of Indigenous Plants* (1894), *A Handbook of the Flora of Natal* (1907), and he was preparing the seventh volume of his *Natal Plants* at the time of his death. He is generally credited with the establishment of Uba

sugar cane (*Saccharum sinense*) in Natal. He was awarded an honorary DSc from the University of Cape Town in 1913.

Woodsia (Woodsiaceae) R.Br.
For Joseph Woods (1776–1864), English architect and botanist, author and Fellow of the Linnaean Society. In 1806, Woods founded the London Architectural Society. After the Napoleonic Wars, he travelled extensively in France, Switzerland, and Italy; studying their architecture and botany. His publications include a paper on the genus *Rosa* in the *Transactions of the Linnaean Society* (1818), and many papers in the *Companion to the Botanical* Magazine (1835–1836 and onward) and *The Phytologist* (1843 onward) based on his travel notes to the continent. In 1850, he published *The Tourist's Flora*, a descriptive catalogue of the flowering plants and ferns of the British islands, France, Germany, Switzerland, Italy and the Italian islands.

Woodwardia (Blechnaceae) Sm.
For Thomas Jenkinson Woodward (c 1745–1820), English botanist, magistrate and lieutenant-colonel of the Diss Volunteers. He studied at Eton and Cambridge, graduating with an LB in 1769. He was regarded by Sir James Edward Smith, who named the fern plant *Woodwardia* after him, as 'one of the best English botanists, whose skill and accuracy are only equalled by his liberality and zeal in the service of the science'. He wrote *Observations on the British Fuci* (1797) with Samuel Goodenough, and contributed many papers to the *Philosophical Transactions* and *Transactions of the London Society* (1784–1794), as well as other works. He was elected a Fellow of the Linnaean Society in 1789.

Wooleya (Aizoaceae) Bolus
For Charles Hugh Frederick Wooley (1894–1969), English major in the Royal Marines, self-taught naturalist, and a citrus farmer at Addo, South Africa, later moving to Knysna. For many years, he contributed succulents to Kirstenbosch from various parts of South Africa. The species he found, *Wooleya farinosa*, was on the western coast of Namaqualand in the western part of South Africa.

Wormskioldia = Tricliceras (Turneraceae) Thonn.

For Morten Wormskjold (1783–1845), Danish botanist, explorer and plant collector. He studied botany at the University of Copenhagen (1806), and made a botanical trip to Greenland (1812–1813), where he collected molluscs, flora, and vascular plants – clubmosses, ferns, conifers – many of which were new to science. In 1815, he sailed with botanist Adelbert von Chamisso and zoologist JF Eschscholtz on Captain Otto Kotzebue's ship to circumnavigate the world. He left the expedition at Kamchatka (northeastern Russia) in 1816 before it reached North America, and collected there for two years. After his return, for reasons not clear, he gave up natural history. Most of his specimens, journals and notes were destroyed in a fire in 1842, although a few were preserved at the University of Oslo.

Wrightia (Apocynaceae) R.Br.
For William Wright (1735–1819), Scottish botanist and physician. He studied medicine at the

universities of Edinburgh and St Andrews, Scotland, taught surgery at Falkirk in Scotland, became a ship's surgeon in 1760, and an assistant to Dr Gray in Jamaica (1764–1777). He joined the navy in 1779 and was taken prisoner by the French. After his release in 1782, he returned to Jamaica and the following year became doctor of the colony. In 1785, he settled in Edinburgh. He participated in the Caribbean exploration expedition (1796–1798) led by Sir Ralph Abercromby (1734–1801). He wrote numerous articles on medicine and described more than 750 species of plants. His most important plant collection is housed in Hamaïque, Jamaica. He became a member of the Royal Society in 1788.

Wulfhorstia* = ***Entandrophragma***
(Meliaceae) C.DC.
For August Wulfhorst (1861–1936), German-Rhenish missionary and plant collector who was ordained in 1890 and sent to South-West Africa (now Namibia) to establish a Rhenish mission station. In 1891, he established a mission station in Ondjiva (today Angola), together with Rhenish Missionary Meisenholl, and a second mission station at Omupanda in 1892, where he stayed until 1917 when hostilities obliged them to close it. He returned to Germany. Wulfhorst returned to Namibia again and was stationed at Karibib until 1927. He collected a great quantity of ethnological material, probably at the instigation of Martti Rautanen (1845–1926), and he also sent plant specimens to Hans Schinz (1858–1941) at Zurich.

Wurmbea (Colchicaceae) Thunb.

(CM)

For Christoph Carl Friedrich von Wurmb (1742–1782), Saxony-born German naturalist and Dutch colonial administrator, who worked in Indonesia (Java) as a merchant in the service of the United East India Company. Later, in 1778, he moved to Batavia, where he became the first secretary and director of the Bataviaasch Genootschap van Kunsten en Wetenschappen (Batavian Society of Arts and Sciences) in charge of its library and small botanical garden, donated by a member. A keen naturalist – he had a special interest in palm trees – Wurmb was the first traveller to publish accurate observations on the Bornean orangutan in its adult state (it had never before been seen at that time and initially thought to be a new species). He called this animal 'Pongo', named after the Mpongwe nation.

X

Ted Woods (1946–) is an investment banker with a passion for botany. His particular fascination is trees of the South African Lowveld. *Xanthocercis zambesiaca*, or the nyala tree, shown on this page, is one of his favourites. Photography is another of Woods's great interests, and he is building a photographic database of 365 trees of Kruger National Park, including diagnostic features of each. Woods and Eugene Moll are also working on a related project designed to deepen the knowledge and appreciation of indigenous trees among all who live in, or visit, the Lowveld.

X Astroworthia (Asphodelaceae) G.D.Rowley
A hybrid between *Astroloba aspera* and *Haworthia pulima* that occurs naturally in the wild in South Africa.

X Duvaliaranthus (Apocynaceae) Bruyns
Duvalia (q.v.).

X Herscheliodisa = Disa (Orchidaceae) H.P.Linder
A hybrid between *Herschelianthe lugens* (blue form) and *Disa longicornu*, common name Darling Blue, was found on Table Mountain by Nicolas and Wilfred Duckett.

X Hoodiapelia* (Apocynaceae) G.D.Rowley
A hybrid between *Hoodia gordonii* and *Stapelia arenosa*. X Hoodiapelia beukmanii (= *Luckhoffia beukmanii*).

X Hoodiopsis (Apocynaceae) C.A.Lückh.
A hybrid between *Hoodia gordonii* and *Orbea lutea ssp. vaga*.

X Pleopodium = Pleopeltis (Polypodiaceae) Schelpe & N.C.Anthony
A putative intergeneric fern hybrid from Africa (e.g. *Pleopeltis polylepis* and *Polypodium polypodioides*).

X Ruttyruspolia (Acanthaceae) A.Meeuse & De Wet
A natural intergeneric hybrid between *Ruttya* (q.v.) and *Ruspolia* (q.v.).

Xanthium* (Asteraceae) L.
Gk. name *xanthion*, from *xanthes* = yellow. The plant does not have a yellow flower, but is used for dyeing hair yellow.

Xanthocercis (Fabaceae) Baill.
Gk. *xanthos* = yellow; *kerkis* = a weaver's shuttle; referring to the yellow fruit and the shape of the pod.

Xanthophyllum (Woodsiaceae) Roxb.
Gk. *xanthos* = yellow; *phylla* = leaves; referring to the yellowish colour of the dried leaf.

Xenismia = Osteospermum (Asteraceae) DC.
Gk. *xenismos* = foreigner. The name alludes to the habitation of the plants in Africa, at the time thought to be unique, one of a kind.

Xenoscapa (Iridaceae) (Goldblatt) Goldblatt & J.C.Manning
Gk. *xenos* = strange; *scapa* = flowering stem; referring to the genus bearing solitary flowers on the main and lateral branches instead of spikes of multiple flowers.

Xanthocercis: Ripe fruits turn nutty-brown. (EJM)

Xenostegia (Convolvulaceae) D.F.Austin & Staples
Gk. *xenos* = strange; *stegos* = roof, covering. The plant has pollen characteristics like *Calystegia*, but otherwise is not close to that genus.

Xeranthemum* (Asteraceae) L.
Gk. *xeros* = dry; *anthemon* = flower. An 'everlasting' flower.

Xerocarpus* = Rothia** (Fabaceae) Guil & Perr.
Gk. *xeros* = dry; *karpos* = fruit; referring to the dry fruit.

Xerocladia (Fabaceae) Harv.
Gk. *xeros* = dry; *klados* = branch; referring to its seemingly dry branches.

Xeroderris (Fabaceae) Roberty (Image on opposite page)
Gk. *xeros* = dry; genus name *Derris* (q.v.). The first species described was from a dryish part of the western Sahel.

Xeromphis* = Catunaregam** (Rubiaceae) Raf.
Gk. *xeros* = dry; *omphalos* = navel; referring to the dry umbillicate fruit.

Xeroderris: The pes-like flowers are dirty-white. (EJM)

Xerophyta (Velloziaceae) Juss.
Gk. *xeros* = dry; *phyton* = plant. Some species, particularly the larger ones, are 'resurrection plants', which can revive after severe drying out and survive in dry, arid places.

Xeroplana*** = *Stilbe* (Stilbaceae) Briq.
Gk. *xeros* = dry; *planos* = wandering. The genus does not have a precise locality.

Xerothamnus = *Gibbaria* (Asteraceae) DC.
Gk. *xeros* = dry; *thamnos* = a bush or shrub; referring to its habitat or appearance.

Ximenesia*** = *Verbenesia* (Asteraceae) Cav.
For José (Joseph) Salvador Ximénez (Ximenes) Peset, (1713–1803), Spanish apothecary, botanist and artist who compiled a *Flora of Castellon de la Plana*, in four volumes, in which he portrayed or described more than 700 plants, keeping a record of where they grew, when and whether they had any medicinal properties. He also recorded the butterflies and birds found in Castellon de la Plana on the east coast of Spain, about 71km northeast of Valencia. When the author, Antonio José Cavanilles met him, as recorded in his *Observations on the Natural History of the Kingdom of Valencia,* he was astonished to find Ximénez had no botanical training or books, had not seen gardens, and was extremely poor with barely enough to eat. Ximénez is said to have had an interest in the characteristics of flowers and the form of the floret, and an interest in the question of 'sex' in flowers.

Ximenia (Olacaceae) L.

(TW)

Probably for Francisco Ximénez (1666–1729), Spanish Dominican monk who went to Guatemala in 1688, where he learned Kaqchikel, or Kaqchiquel, a Mayan language, and is accredited with the conservation and survival of the *Popol Vuh*, a manuscript dealing with Meso-American mythologies. From 1693, he was a doctrinero (teacher of Christian doctrine) in various parishes before being appointed vicar, then predicador general of Rabinal (1704–1714) and its surrounding districts, where he founded a hospital. Among his writings is *Historia Natural del Reino de Guatemala* (*Natural History of the Kingdom of Guatemala*), concerning the flora and fauna of Guatemala, their medicinal and industrial use, and indigenous beliefs about the properties of drugs.

Xiphopteris (Grammitidaceae) Kaulf.
Gk. *xiphos* = sword; *pteris* = fern; referring to the shape of the fronds – 'sword fern'.

Xiphotheca*** (Fabaceae) Eck. & Zeyh.
Gk. *xiphos* = sword; *theka* = case, capsule; referring to the shape of the pod.

Xylia (Fabaceae) Benth.
Gk. *xylon* = wood. The tree has very hard wood.

Xylocarpus (Meliaceae) J. König
Gk. *xylon* = wood; *karpon* = a fruit. The fruit capsules are large and very woody.

Xylopia (Annonaceae) L.
Gk. *xylon* = wood; *pikros* = bitter. The wood is bitter.

Xylosma (Salicaceae) G.Forst.
Gk. *xylon* = wood; *osme* = scent, fragrance; referring to the plant's aromatic wood.

Xylotheca (Achariaceae) Hochst.

A open friut showing the outer woody nature of the capsule (EJM)

Gk. *xylon* = wood; *theke* = a case. The fruit is a woody capsule.

Xymalos (Monimiaceae) Baill.
Anagram of *Xylosma,* the genus in which these trees were originally placed, from *xylon* = wood; *osme* = scent. *Xymalos* is an unrelated genus.

Xyris (Xyridaceae) L.

(EJM)

Gk. *xyron* = razor; *iris* (q.v.). Greek name for a plant with sharpened leaves. A name used by Dioscorides for *Iris foetidissima*.

Xysmalobium (Apocynaceae) R.Br.
Gk. *xysma* = filings, shavings; *lobion* = dim. of *lobos* = pod, capsule; referring to the fragmented lobes of the corona.

Y

Ypsilopus is a genus of flowering plants from the orchid family, Orchidaceae, native to Africa.

Benny Bytebier (1961–) hails from Sint-Niklaas in Belgium and has a biology degree from the Free University Brussels. He has been working and living in Africa since 1990. He was employed at the University of Nairobi in Kenya and later at the East African Herbarium of the National Museums of Kenya. He moved to Stellenbosch, South Africa in 2001 for a PhD on the large African orchid genus *Disa*. His main interest is systematics, evolution, biogeography and conservation of African orchids. He co-authored *Orchids of South Africa: A Field Guide* (2015). He is currently the curator of the Bews Herbarium at the University of KwaZulu-Natal's Pietermaritzburg campus in South Africa.

Ypsilopus (HS)

Youngia* (Asteraceae) Cass.
For Edward Young (1683–1765), English poet and dramatist best remembered for his poem 'Night-Thoughts'. He obtained a doctorate from Oxford University in cannon law and wrote a number of highly acclaimed satires. In 1728, he became a royal chaplin. And for Thomas Young (1773–1829), English polymath, who at aged 14 had a familiarity with 14 languages. He studied medicine at the universities of London and Edinburgh, and obtained a doctorate in physics at Göttingen. During his career he was a doctor, professor of natural philosophy, a member of the French and Swedish Academy of Sciences, and made discoveries in the fields of wave theory of light, elasticity, vision and colour theory, surface tension, medicine and Egyptian hieroglyphics. The author Cassine wrote: [Named after] '*deux Anglais célèbres, l'un comme poète, l'autre comme physicien,*' (two English celebrities, the one a poet, the other a physician) (Youngia Cassini, Ann. Sci. Nat. (Paris). 23: 88. 1831).

Edward Young

Thomas Young

Ypsilopus (Orchidaceae) Summerh.
Gk. letter *ypsilon* = upsilon (letter Y); *pous* = foot; referring to the diverging or Y-shaped arms at the apex of the stipe of the pollen masses. The pollinia are attached to Y-shaped stipes.

Z

Ziziphus: a genus of about 40 species of spiny shrubs and small trees

Bart Wursten (1953–), originally from the Netherlands, is a self-taught field botanist in Africa. He has particular experience in the floras of Mozambique and Zimbabwe, where he has lived and worked for many years and regularly participates in botanical surveys. Since 2004 he has been one of the main authors, editor and photographer of the online floras of Zimbabwe and Mozambique. His photography has been used in many scientific publications as well as in popular nature books and field guides.

Zaleya (Aizoaceae) Burm.
An ancient Greek name (also called *Zaleia, Zalia*) for Butcher's broom, according to the *Dictionnaire des Sciences Naturelles, Dans Lequel on Traite ...*, Volume 59 by Frédéric Cuvier. Wullei-Tsjarne, botanist, states: 'Burmann's son received this plant from the coast of Coromandel, which Burmann named *Zaleya decandra*, is the same plant that Linnaeus in his Mantissa put under *Trianthema*.'

Zaluzianskya (Scrophulariaceae) F.W.Schmidt

(CM)

For Adam Zalusiansky von Zaluzian (1558–1613), Bohemian botanist and physician, lecturer and administrator at Charles University in Prague, author of *Methodus Herbariae Libri Tres* (1592). He was the first man to argue for the separation of botany from medicine, and for a universal classification of plants years before Linnaeus. He stated (in translation): 'It is customary to connect medicine with botany, yet scientific treatment demands that we should consider each separately. For the fact is that in every art, theory must be disconnected and separated from practice, and the two must be dealt with singly and individually in their proper order before they are united. And for that reason, in order that botany (which is, as it were, a special branch of physics) may form a unit by itself before it can be brought into connection with other sciences, it must be divided and unyoked from medicine.' Quotation from *Herbals: Their Origin and Evolution* (Agnes Arbe).

Zamia = Encephalartos (Zamiaceae) L.
La. *azaniae* = opening while on the tree (the cones, probably from Gk. *azainein* = to dry up). This name is derived from a corrupt form of the word *azania*, used by Pliny the Elder for a pine cone.

Zamioculcas (Araceae) Schott
Zamia (q.v.); *culcas* = Arabic name for *Colocasia antiquorus* (Araceae).

Zanha (Sapindaceae) Hiern
For Karl Hermann Zahn (1865–1940), German botanist, high school teacher and professor of geometry, chemistry and material technology at Karlsruhe University of Applied Sciences. Between 1896 and 1914, he collected a massive number of plants. He became a world authority on *Hieracium*, and wrote, *inter alia*, *Die Hieracien der Schweiz* (1906), *Hieracia Florae Mosquensis* (1911), *Les Hieracium des Alpes Maritimes* (1916), and a monograph between 1921 and 1923 on *Hieracium*, part of a very large genus of flowering plants in the sunflower family (Asteraceae). He recognised 177 subspecies. His work is reflected in Adolf Engler's *Das Pflanzenreich* (*The Plant Kingdom*) (1957) section on *Hieracium*. His collection is housed in the Botanical Museum Berlin-Dahlem (25 000 specimens, 2 800 types). His name was misspelled by Hiern.

Zannichellia (Potamogetonaceae) L.
For Giovanni Gerolamo Zannichelli (1662–1729), Italian botanist, physician and pharmacist with an interest in palaeontology and mineralogy. In 1686, he started a successful apothecary business in Venice. His Santa Fosca pill (1701), an effective laxative, still sold in 1950. He published *Promptuarium Remediorum Chimicorum* (a handbook of chemical medicines). After obtaining his medical degree (1702), he made many trips to explore the natural history around the Adriatic Sea, the mountains of Vicenza and Verona, and elsewhere, building an important natural history cabinet in Venice. His *Istoria delle Piante che Nascono ne' lidi Intorno a Venezia* (*History of the Plants ... around Venice*) was published by his son, posthumously, in 1735.

Zantedeschia (Araceae) Spreng.
For Giovanni Zantedeschi (1773–1846), an Italian physician, pharmacist and botanist. He studied medicine and surgery at the universities of Verona and Padua. His botanical interests centred on the flora of Brescia, northern Italy, where he discovered and described several new genera such as *Laserpitium nitidum*, family Apiaceae. He authored *Descrizione dei Funghi della Provincia Bresciana* (1820) and other works. He corresponded with the German botanist Kurt Sprengel (1766–1833), who named the plant *Zantedeschia* after both Giovanni and his son

(GAN)

Arum lilies, sometimes also called 'Pig lilies', are favoured by procupines. (GAN)

Francesco Zantedeschi (1797–1873), professor of physics and philosophy at the University of Padua, who carried out experiments in electrical currents and magnetism.

Zanthoxylum (Rutaceae) L.
Gk. *xanthos* = yellow; *xylon* = wood. Several species of this genus have yellow heartwood. The roots of some species contain a yellow dye (Jackson).

Zehneria (Cucurbitaceae) Endl.
For Joseph Zehner, Austrian natural history artist and skilful observer who lived in Vienna. No further details found.

Zephyranthes (Amaryllidaceae) Herb.
Gk. *zephyr* = personification of the west wind; *anthos* = flower. The flowers thrive in the west wind, which often brought rain.

Zeuktophyllum (Aizoaceae) N.E.Br.
Gk. *zeuktos* = connected, joint; *phyllon* = leaf (Hans Herre). Liddell & Scott give *zeugos* = a yoke, team or pair; *zeukteria* = a fastening, band.

Zeuxine (Orchidaceae) Lindl.
Gk. *zeuxis* = a yoke or bridge; referring to the partial union of the lip and column, and possibly also to the growing together of the pollinia.

Zeyherella = Englerophytum (Sapotaceae) (Pierre ex Baill.) Aubrév. & Pellegr.
For Carl (Karl) Ludwig Philipp Zeyher (1799–1858), German botanist, botanical collector and adventurer, who arrived in the Cape in 1822 and collected in South Africa for 22 years. He was dogged by misfortune – he was not paid his share of his Cape collection with FW Sieber (1822–1824). His Uitenhage-Clanwilliam collection (1825–1825), sent to his uncle, who died, was procured by the Baden government. A large part of his 1829–1833 (Eastern Cape) collection with Ecklon was destroyed by fire. The ship carrying his 1838 collection (Uitenhage) was wrecked and the cargo lost. He is remembered best for his co-authorship with Christian Friedrich Ecklon of *Enumeratio Plantarum Africae Australis.* His last years were lived in poverty and he died of smallpox.

Zingiber* (Zingiberaceae) Boehm
From a Sanskrit word meaning 'shaped like a horn', or derived from the Greek word *ziniberi*, which probably evolved from an ancient Indian name for ginger.

Zinnia* (Asteraceae) L.
For Johann Gottfried Zinn (1727–1759), a German anatomist, ophthalmologist and botanist. He obtained his doctorate in 1749 from the

University of Göttingen, being one of Albrecht von Haller's (1798–1777) best students. In 1753, he became an extraordinary professor at the University of Göttingen, director of the botanical gardens at Göttingen, and also wrote *A Description of the Flora around Göttingen*. Two years later, in 1755, he became professor of the medical faculty at the University of Göttingen, during which time he wrote *Descriptio Anatomica Oculi Humani*, which contained the first accurate description of the

human eyeball, including details of the vessels and nerves of the eye cavity. Yet his interest in botany remained and, in 1757, just before his early death, he described the genus *Epipactis*, belonging to the Orchidaceae family. He was a member of the Berlin Academy.

Ziziphus (Rhamnaceae) Mill.
Arabic *zizouf* = the name for the lotus or 'jujube tree', *Z. lotus* (syn. *Z. jujuba*). The tree has dark red edible fruits, from which the Victorian sweet 'jujube' originated (Jackson).

Zornia (Fabaceae) J.F.Gmel.
For Johannes Zorn (1739–1799), German pharmacist and botanist, who travelled all over Europe searching for medicinal plants. After studying pharmacy, he became an apothecary in his home town of Kempten, Bavaria. In Nurnberg, between 1779 and 1784, he published *Icones Plantarum Medicinalium* (*Images of Medicinal Plants*), illustrated with 500 hand-coloured engravings, expanded to 600 engravings in his second edition (1799). He was also passionate about the plants of the New World, and published a flora of America in 1786 called *Drey Hundert Auserlesene Amerikanische Gewachse* (*300 Select American Plants*), based on a very rare book (there were only three coloured copies) by NJ Jacquin, *Selectarum Stirpium Americanarum Historia* (1763).

Zostera (Zosteraceae) L.
Gk. *zoster* = belt, girdle; referring to the ribbon-shaped leaves.

Zoutpansbergia (Asteraceae) Hutch.
For the Zoutpansberg mountain range; alluding to the locality where it was first found.

Zygia = Albizia (Fabaceae) Walp.
Gk. *zygon* = a yoke; referring to the stamens being united in a long tube.

Zygoon = Coptosperma (Rubiaceae) Hiern
Gk. *zygon* = a yoke; *oon* = an egg; apparently the name refers to the several ovules crowded on a small placenta (only one of these forms a seed). The word is also used as a specific epithet.

Zygophlebia* (Polypodiaceae) E.Bishop
Gk. *zygon* = a yoke; *phleps, phlebos* = vein. Named for the strong tendency for the fronds to show areolate venation.

Zygophyllum (Zygophyllaceae) L.
Gk. *zygon* = a yoke; *phyllum* = leaf. The leaves are usually bifoliolate – the two leaflets are as if 'yoked together'.

Zyrphelis** (Asteraceae) Cass.
Possible misspelling of Zephyr, the mythological god of the west wind and lover of Flora.

Bryophytes

Bryophytes (mosses and liverworts)

Why are bryophytes put in a different section of this dictionary?

The simple answer is because they are so very different to flowers, trees, grasses, and even ferns. Although they are 'land plants', bryophytes do not possess any vascular tissue – *xylem* (tubes and pipes), which conducts water and nutrients up from the roots, and *phloem* (different tubes and pipes), which distributes food from the leaves/fronds to other parts of the plant.

The term 'bryophyte' is derived from the Latin *bryon* meaning moss and '*phyte*' meaning plant.

So how do bryophytes, which do not have proper, well-developed roots, stems and/or leaves, acquire nutrition?

Unlike vascular plants, bryophytes have threadlike rhizoids – slender, root-like filaments that anchor them to their substrate – that are similar in structure and function to the root hairs of vascular land plants. However, these rhizoids do not extract water and minerals from the substrate like the roots of vascular plants.

Bryophytes receive their water from rainfall, fog, mist and dew that is deposited on their photosynthetic organs – be they leaf-like or comprising a flat thallus. Because bryophytes are prostrate, having much of their surface in contact with their substratum, they also absorb moisture and dissolved minerals that fall on the ground or on their 'leaves'.

A common misconception is that to find bryophytes you need to be in a damp, shaded place like on the banks of a stream. In fact, bryophytes can be found in great variety throughout the year in habitats ranging from arid areas to rainforests and on roof-tops from sea-level to Alpine regions. They occur most abundantly in relatively unpolluted areas because they are most sensitive to air-borne pollutants. Some species have specific habitat preferences while others are found in a variety of habitats. They can be found growing on all sorts of surfaces (or substrates) – soil, rock, tree trunks, leaves, rotting wood, bones, old discarded shoes, gloves, brick walls and even on house roofs – to name a few possibilities.

Bryophytes have neither pollen nor flowers and rely on water to carry the male gametes (the sperm) to the female gametes (the eggs). The spore capsules are produced after the sperm have fertilised the eggs. Hence the spores are part of the sexual reproductive cycle.

Bryophytes vary in size from plants only slightly over a millimetre tall to trailings that grow to strands well over a metre long.

How bryophytes evolved is not certain, and the answer lies in their evolutionary history, which has been traced back to the Carboniferous period, about 320 million years ago, and it is thought that they may be the forerunners of plants emerging from sea to land.

A

Abietinella (Thuidiaceae) Müll.Hal.
La. *abietinus* = resembling the fir tree; *-ella* (diminutive). The leaves somewhat resemble those of fir trees.

Acanthocladiella = Rhacopilopsis (Sematophyllaceae) M.Fleisch.
Gk. *akanthos* = thorn; *kladion* = club-like; *-iella* (diminutive); referring to the small, thorny branches.

Acanthocladium = Wijkia (Sematophyllaceae) Mitt.
Gk. *akanthos* = thorn; *clados* = branch; *-ium* = diminutive; referring to the spines at branch ends.

Acanthocoleus (Lejeuneaceae) R.M.Schust.
Gk. *akanthos* = thorn; *coleus* = sheath.

Acaulon = Astomum (Pottiaceae) Müll.Hal.
Gk. *a-* = without; *kaulon* = stalk; referring to the extremely short stems of the gametophyte.

Acaulonopsis** (Pottiaceae) R.H.Zander & Hedd.
La. *Acaulon* (q.v.); Gk. *-opsis* = resembling (*Acaulon*).

Acrobolbus (Acrobolbaceae) Nees
Gk. *akros* = at the tip, end; *bolbos* = bulb; alluding to the location of the marsupium at the tip of the shoot.

Acromastigum (Lepidoziaceae) A.Evans
Gk. *akro* = at the tip, end; *mastix* = flagellum or whip; alluding to the flagella and their place of origin. The flagella arise from the postical segment of apical cells.

Aerobryopsis (Meteoriaceae) M.Fleisch.
Gk. *aer* = air; *bryon* = moss; *-opsis* = resembling (the genus *Aerobyron*). An 'air plant', this moss dangles in skeins from the branches of trees.

Algaria** (Pottiaceae) Hedd. & R.H.Zander
La. *alga* = seaweed; *aria* = pertaining to.

Alobiella = Alobielliopsis (Cephaloziaceae) (Spruce) Schiffn.
Gk. *alob* = without lobes; *-iella* (diminutive). The leaves are not lobed, unlike typical *Cephalozia*.

Alobiellopsis (Cephaloziaceae) R.M.Schust.
Gk. *Alobiella* (q.v.); *-opsis* = resembling.

Aloina (Pottiaceae) Kindb.
Gk. *aloē* = bitter; *ina* = resembling (diminutive); referring to the fleshy leaves that are similar to those of the *Aloe* (Crum & Anderson, 1981).

Amblystegium = Leptodictyum (Amblystegiaceae) Bruch, Schimp. & W.Gümbel
Gk. *amblys* = blunt; *stegium* = a roof or covering, or *stege* = a small lid or cap (operculum); alluding to the obtuse operculum.

Amphidium (Dicranaceae) Schimp.
Gk. from *amphora* = urn; *-idion* = diminutive; referring to the capsule shape.

Anacolia (Bartramiaceae) Schimp.
Gk. *anakōlos* = short; alluding to the short pedicel.

Anastrophyllum (Anastrophyllaceae) (Spruce) Steph.
Gk. *an* = without; *astro* = star, hence *anastro* = not star-like, underdeveloped; *phyllum* = leaves. The leaves of some plants are small and look underformed.

Andreaea (Andreaeaceae) Hedw.
For Johann Gerhard Reinhard Andreä (1724–1793), German apothecary and mineralogist. He trained in a Frankfurt pharmacy, and studied in Leyden and England, returning to Hannover to take over his father's pharmacy. A natural scientist and writer, he took a scientific journey through Switzerland in 1763, visiting the most important alpine areas, where he recorded observations about the climate,  fossils and minerals, salt works, thermal baths and glaciers. He was appointed by the royal court of Hannover to determine the best soil types for certain kinds of agriculture. In 1769, he published (in German) *A Treatise on a Considerable Number of Soil Types ... and the Use of Them for the Farmer*. Besides natural history, he read great literature in various languages, especially loved the English poets, and was a fine pianist.

Aneura (Aneuraceae) Dumort.
Gk. *a-* = without; *neuron* = nerve, vein; alluding to the lack of a midrib on the thallus.

Anisothecium = Dicranella (Dicranaceae) Mitt.
Gk. *anisos, aniso-* = unequal; *theka* = case, capsule; probably referring to the capsules, which are different to those of *Dicranum*, from which *Anisothecium* was separated (David Meagher).

Anoectangium = Sekra (Pottiaceae) Schwägr.
Gk. *anoiktos* = opened, pitiless, ruthless; *angion* =

vessel, receptacle, container; alluding to the wide-mouthed capsule (Dixon), but more likely alluding to the lack of a peristome (David Meagher).

Anomalolejeunea* =Cheilolejeunea** (Lejeuneaceae) (Spruce) Schiffn.
Gk. *anomalos-* = unusual, irregular; *Lejeunea* (q.v.).

Anomobryum (Bryaceae) Schimp.
Gk. *anoma-* = irregular; *bryon* = lichen, moss; referring to the true *Bryum*.

Anomodon = Leucodon (Leucodontaceae) Hook. & Taylor
Gk. *anom-* = irregular; *odon* = toothed; initially referring to the ciliate inner peristome teeth, but David Meagher states this was proven false and maybe refers to 'various species which are now classified in other genera' (Iwatsuki, 1963).

Anthelia (Antheliaceae) Dumort.
Gk. *anti-* = against; *helios* = sun. This species likes a cold climate – arctic, alpine and subalpine conditions. It can also be found at high altitudes and in snow beds.

Anthoceros = Phaeoceros (Anthocerotaceae) L.
Gk. *anthos* = flower; *keras* = horn, hence 'horn-like flowers'; from the horn-like sporophyte, or for the conspicuously elongated, dark brown, bivalved capsules.

Aongstroemia = Oncophorus (Dicranaceae) Bruch & Schimp.
For Johan Ångström (1813–1879), Swedish bryologist, pterodologist and physician, who became an expert on the mosses. With his Finnish colleague Fredrik Nylander (1820–1880), a botanist and doctor, he explored the east of Finland, the Russian White Sea area surrounded by Karelia, and Lapland, Norway and northwestern Russia in the summer of 1843, and he went on a long voyage to Brazil (1851–1853). In total, Nylander described 39 taxa, consisting of 11 species, eight subspecies and 20 varieties. Johan was the older brother of Anders Jonas (1814–1874), a world-famous physicist and astronomer, professor of physics at Uppsala in 1858, generally regarded as the father of spectrometry. His symbol, Å, is internationally recognised to indicate length.

Aongstroemiopsis (Dicranaceae) M.Fleisch.
Aongstroemia (q.v.); Gk. *-iopsis* = resembling, likeness.

Aphanolejeunea = Cololejeunea (Lejeuneaceae) A.Evans

Gk. *aphanes* = invisible, unseen; *Lejeunea* (q.v.); alluding to the conspicuous character of the species.

Arachniopsis (Lepidoziaceae) Spruce
Gk. *arachnis* = the spider or its web; *-iopsis* = resembling, likeness.

Archidiella = Archidium (Archidiaceae) Irmsch.
Archidium (q.v.); Gk. *-iella* = diminutive.

Archidium** (Archidiaceae) Brid.
Gk. *archidion* = primitive, original, first; *-idion* = diminutive; alluding to the author's view that these were the earliest of all mosses.

Archilejeunea (Lejeuneaceae) (Spruce) Schiffn.
Gk. *arche* = prime, original (Spruce): *Lejeunea* (q.v.). Also *arche,* stated to mean power, sovereignty. The plants are large when compared to other species included in *Lejeunea*.

Asterella = Fimbraria (Aytoniaceae) P.Beauv.
Gk. *astron* = a star; *-ella* = diminutive; referring to the star-like shape of the archegoniophores when seen from above.

Astomiopsis (Ditrichaceae) Müll.Hal.
Gk. *a-* = without; *stoma* = mouth; *iopsis* = resembling *Astomum* (q.v.).

Astomum = Archidium (Pottiaceae) Hampe
Gk. *a-* = without; *stoma* = mouth; alluding to the cleistocarpus capsule.

Athalamia** (Cleveaceae) Falc.
Gk. *Athamas* = of or from Mount Athamas in Sicily; *lamium* = name for dead nettle.

Atractylocarpus (Dicranaceae) Mitt.
Atractylis (q.v.); *karpos* = fruit; perhaps referring to sharp-pointed fruit.

Atrichum (Polytrichaceae) P.Beauv.
Gk. *a-* = without; *thrichos* = hair; alluding to calyptra.

Aulacopilum (Erpodiaceae) Wilson
Gk. *aulax, aulaco* = a furrow or groove; *pilos* = a brimless felt cap; La. *pilum* = javelin. The meaning is obscure.

B

Barbula (Pottiaceae) Hedw.
La. *barbulus* = short beard; alluding to the hairy appearance of the peristome.

Bartramia (Bartramiaceae) Hedw.
For John Bartram (1699–1777), US naturalist, explorer and plant collector, considered the father of US botany. Largely self-educated, he was the first North American experimenter to hybridise flowering plants, founded the 12-acre Bartram Botanical Gardens near Philadelphia that became internationally famous, and was appointed royal botanist for the American colonies by George III (1765). Thanks to Peter Collinson, a Fellow Quaker and London naturalist, Bartram and his son, William, undertook many seed- and specimen-collecting expeditions, some partly financed by Europeans. He wrote journals of his travels. Bartram was a co-founder of the American Philosophical Society. Linnaeus called him the greatest natural scientist in the world.

Bartramidula = **Philonotis** (Bartramiaceae) Bruch & Schimp.
Bartramia (q.v.); *idula* (diminutive).

Bazzania (Lepidoziaceae) Gray
For Matteo Bazzani (1674–1749), Italian physician, naturalist and professor of anatomy at the University of Bologna and the Academy of Sciences. Bazzani studied and taught medicine at Bologna, where most of his research was on bone growth. He published a paper with Giuseppe Pozzi entitled *De Ambigue Prolatis in Judicium Criminationibus Consultationes Physico-Medicae Nonnullae* (1742), which discusses the legal and forensic aspects of four different cases of infanticide. He was a patron of Italian botanist Pier Antonio Micheli (1679–1737), and made a notable oration to honour the awarding of a doctoral degree in philosophy to physicist Laura Bassi, the first woman to join the faculty of a European university.

Brachiolejeunea (Lejeuneaceae) (Spruce) Schiffn.
Gk. *brachys* = short; *Lejeunea* (q.v.); possibly referring to the perianth, which has a short but distinct beak.

Brachymenium (Bryaceae) Schwägr.
Gk. *brachys* = short, stout; *hymen* = membrane; alluding to the short basal membrane.

Brachythecium (Brachytheciaceae) Schimp.
Gk. *brachys* = short; *theka* = case, capsule; alluding to the short, fat capsule (David Meagher).

Braunia (Hedwigiaceae) Bruch & Schimp.
For Alexander Karl (Carl) Heinrich Braun (1805–1877), German botanist from Bavaria. He studied botany at Heidelberg, Paris and Munich, became professor of botany at Freiburg (from 1846) and Giessen (from 1850), and then at the University of Berlin, where he was also director of the Berlin Botanical Garden. He is largely known for his research involving plant morphology, and was author of *Botanical and Physiological Memoirs* (1853) and many other works. He made important contributions in the field of cell theory. He became a foreign member of the Royal Swedish Academy of Sciences in 1852.

Breutelia (Bartramiaceae) (Bruch, Schimp. & W.Gümbel) Schimp.
For Johann Christian Breutel (1788–1875), German bishop of the Moravian church (from 1853), a member of the board of directors, and plant collector, mainly of cryptogams. He collected some 310 mosses from the West Indies (the islands of St Thomas, St Croix, St John, St Kitts and Antigua) and South Africa. He visited the Cape from 1853–1854 and toured with CR Kölbing, visiting many areas on the Garden Route from Cape Town to King William's Town (near East London). His moss collection was deposited with the British Museum. Other collections are with the Municipal Natural History Museums of Bremen and Lübeck, the Botanical Museum at Breslau, and the Herbarium Hieronymus (now in Berlin).

Brothera (Dicranaceae) Müll.Hal.
For Viktor Ferdinand Brotherus (1849–1929), Finnish bryologist and natural history and mathematics teacher. He travelled extensively through eastern Europe and central Asia, and corresponded with many botanists, receiving collections from Turkmenistan, Africa, Australia, Brazil, New Guinea and elsewhere. His unique achievement was his synthesis of moss taxonomy. He identified and classified an estimated 20 000 species of mosses then known to him. He authored, *inter alia,* the 'Musci' in Adolf Engler and Karl von Prantl's *Die Naturlichen Pflanzenfamilien*, and a section on Chinese mosses in *Symbolae Sinicae*. His herbarium of 120 000 specimens was purchased by the University of Helsinki Herbarium, which made him an honorary professor in 1927.

Bruchia (Bruchiaceae) Schwägr.
For Philipp Bruch (1781–1847), German

pharmacist and bryologist, co-author with Wilhelm Philipp Schimper (1808–1880) of the epic six-volume *Bryologia Europaea,* which was published between 1836 and 1855, and which described every species of European moss known at the time. It is still considered one of the classics of its day.

Bryobartramia (Bryobartramiaceae) Sainsbury
For Edwin Bunting Bartram (1878–1964), US botanist and bryologist and great-great-great-great-grandson of John Bartram (1699–1777), whom Linnaeus referred to as the 'greatest natural botanist in the world'. He had a particular interest in mosses, and described many new species and authored many papers and books, including *Honduran Mosses, Mosses of Dominica, British West Indies, Mosses of the Ecuadorian Andes, Costa Rican Mosses* and *Mosses of the Philippines.* He made many botanical trips with the highly regarded US botanist Merritt Lyndon Fernald. He was a member of the Torrey Botanical Club, New England Botanical Club, Sullivan Moss Society, and the Academy of Natural Science, and president of the Philadelphia Botanical Club.

Bryodesma = Selaginella (Selaginellaceae) Soják.
Gk. *bryon* = moss; *desma* = (Burman term for poison). These spike mosses are toxic.

Bryoerythrophyllum (Pottiaceae) P.C.Chen
Gk. *bryon* = moss or maybe *bryo* = to blossom, come into bud; *erythros* = red; *phyllon* = leaf; referring to the brick-red colour of the most widespread species, *B. recurvirostrum* (Crum & Anderson, 1981).

Bryum (Bryaceae) Hedw.
Gk. *bryo* = to swell, derived apparently from *bruein* = to swell, sprout or burgeon. An ancient name for an unidentified bryophyte.

C

Callicostella (Pilotrichaceae) (Müll.Hal.) Mitt.
Gk. *kallos* = beautiful; *costa* = a nerve; *-ella* = diminutive; referring to the beautiful leaves, which have prominent nerves.

Calymperes (Calymperaceae) Sw. ex F.Weber
Gk. *kalumma, kalymma* = veil; *peiro* = to pierce through; alluding to the fissured calyptra.

Calypogeia (Calypogeiaceae) Raddi
Gk. *kalyx* = cup, hence calyx; *hypo* = below; *geia* = ground; alluding to much of the plant's development occurring underground.

Calyptothecium (Pterobryaceae) Mitt.
Gk. *calyptos* = enveloping, hidden, covered; *thekion* = little vessel, container, case; referring to the almost sessile capsule immersed in the perichaetial leaves.

Calyptrochaeta (Daltoniaceae) Desv.
Gk. *kalyptros* = veil, hood; *chaite* = long hair, a mane, bristle; referring to the generally long-haired calyptra.

Campyliadelphus (Amblystegiaceae) (Schimp.) R.S.Chopra
Gk. *kampulos* = curved or bent; *adelphus* = brother. Speculatively, some members of the subfamilies ('brothers') such as Campylioideae and Amblystegioideae are closely related and can hence be confused.

Campylopus (Dicranaceae) Brid.
Gk. *kampulos* = bent, curved; *pous* = foot; referring to the curved seta. *Candelaria* (q.v.); *-iella* = diminutive.

Canomaculina (Parmeliaceae) Elix & Hale.
La. *canus* = grey-white; *macul-* = spot, blotch; *-ina* = a little; grey-white, usually applied to hair covering.

Canoparmelia (Parmeliaceae) Elix & Hale.
Gk. *canus* = hoary or grey-white; *parme* = fruit bowl; *eileo* = closed; referring to the pale upper surface of thalli.

Carbonea (Lecanoraceae) (Hertel) Hertel
Gk. *carbō, carbōn-* = charcoal; named for shiny black carbonaceous excipulums.

Cardotiella (Orthotrichaceae) Vitt
For Jules Cardot (1860–1934), French botanist, bryologist and authority on mosses. He studied flora in the Antarctic, Alaska, Japan, West Africa (Cameroon), Mexico, Madagascar and Taiwan, and authored many books and publications, including *The Mosses of Alaska* (1902–1904, six editions), *La Flore Bryologique des Terres Magellaniques* (1908), *Natural History Moss Plants* (1913) (with Ferdinand Renauld), and *Mousses de Madagascar* (1916). During World War I, his herbarium collection of specimens was severely damaged. He sold the remnants to the Paris Natural History Museum. From 1919 to 1931, he worked for the economic bureau of the Indo-Chinese government. During his career, Cardot named 40 genera and 1 200 species. In 1923, he became a knight of the Legion of Honour for his contributions to science.

Carrpos = **Monocarpus** (Monocarpaceae) Prosk.
For Denis John Carr (1915–2008), English botanist and professor in the department of developmental biology at the Research School of Biological Sciences, Australian National University (1968–1980). He obtained his PhD from Manchester University and taught there (1958–1960), and was professor of botany at Queen's University, Belfast (1960–1967). He and his wife, Stella Grace M (Maisie) (1912–1988), an ecologist, wrote a number of books together, including *People and Plants in Australia* (1981) and *Plants and Man in Australia* (1983). They were noted for their work in the alpine regions of New South Wales and Victoria, and named several plant species, including the bloodwood, *Eucalyptus dampieri,* in 1987.

Catagonium (Catagoniaceae) Müll.Hall. ex Broth.
Gk. *kata* = down, against, under; *gonos, gonium* = seed. Perhaps in reference to the long, exerted sporangia, and to set it apart from *Cryptogonium*, which used to be in the same family.

Caudalejeunea (Lejeuneaceae) (Steph.) Schiffn.
La. *cauda* = tail; *Lejeunea* (q.v.); so named for its caudiform gemmiparous shoots.

Cephalozia (Cephaloziaceae) (Dumort.) Dumort.
Gk. *kephale* = a head; *ozis, ozos* = branch, bud; referring to the head-like shape formed by the enlarged bracts and bracteoles.

Cephaloziella (Cephaloziellaceae) (Spruce) Schiffn.
Cephaloza (q.v.); Gk. *-iella* = diminutive.

Ceratodon (Ditrichaceae) Brid.
Gk. *kerato* = horned; *-odon* = toothed; referring to the forked peristome teeth that resemble the horns of goats.

Chamaebryum** (Gigaspermaceae) Thér. & Dixon
Gk. *khamai* = on the ground, low, small, dwarf; *byron* = lichen, moss; referring to the size of the moss.

Chandonanthus = **Plicanthus** (Jungermanniaceae) Mitt.
La. *chandon* = with mouth wide open, yawning; *anthos* = flower; alluding to the wide mouth of the type species.

Cheilolejeunea (Lejeuneaceae) (Spruce) Schiffn.
Gk. *keilo, cheilos* = lip, edge; *Lejeunea* (q.v.); referring to the fact that the perianth often becomes two-lipped upon the extrusion of the capsule at maturity.

Cheilothela (Ditrichaceae) Lindb. ex Broth.
La. *cheilothela* = thick nipple; referring to the thickened beak on the capsule lid.

Chenia (Pottiaceae) R.H.Zander
For Chén Bāngjié (Chen Pan Chieh) (1907–1970), Chinese bryologist who described the Asiatic species of the family *Pottiaceae*, author of *Studien über die Ostasiatischen Arten der Pottiaceae* (*Studies of the East Asian Species of Pottiaceae*) (1941), and author or editor of *Genera Muscorum Sinicorum* (1963, 1978).

Chiloscyphus (Lophocoleaceae) Corda
Gk. *kheilos-* = lip, edge; *scyphus* = a deep drinking cup; perhaps referring to the frequently triangular-prismatic cup-shaped perianths thought to be a *Lophocolea* characteristic. The two genera, *Chiloscyphus* and *Lophocolea*, have characteristics that merge (Xiaolan He-Nygrén & Sinikka Piippo).

Chorisodontium (Dicranaceae) (Mitt.) Broth.
Gk. *chori-* = separate, apart; *odontos* = toothed. The peristome teeth are separated almost in half, with the striolate below.

Chrysoblastella (Ditrichaceae) R.S.Williams
Gk. *chrysos* = golden; *blastos* = outbreak of (bud, sprout, embryo); *-ella* = diminutive; referring to the yellowish green gametophytes, especially in the young parts.

Chryso-hypnum (Hypnaceae) Hampe
Gk. *chrysos* = gold; *hypnon* = ancient name for a bryophyte, probably a moss; alluding to its appearance. It has dense, bright golden-green tufts or mats.

Cladophascum (Bruchiaceae) Dixon
Gk. *klados* = branch, twig; *phascum* = tuft of moss. The plants are mostly small, forming loose to dense tufts, upright or hanging on branches.

Clasmatocolea (Geocalycaceae) Spruce
Gk. *klasmatos* = fragmented; *koleos* = sheath; alluding to the perianth, which 'is so fragile that the slightest touch breaks off the short unequal lobes at the wide mouth' (Spruce).

Liverwort showing gemma cups (EJM)

Clevea = Athalamia (Cleveaceae) Lindenb. For Per Theodor Cleve (1804–1905), Swedish chemist and geologist; expert in agricultural chemistry, inorganic and organic chemistries, geology, mineralogy and oceanography, and member of the Royal Swedish Academy of Sciences. The mineral cleveite, the first known terrestrial source of helium, was named for him. He was a professor at Uppsala University, and father of the botanist and chemist Astrid Cleve.

Codonoblepharon (Orthotrichaceae) Schwägr. Gk. *kodon* = a bell; *blepharon* = eyelid. Meaning obscure.

Cololejeunea (Lejeuneaceae) (Spruce) Schiffn. Gk. *kolos* = with parts missing, maimed; *Lejeunea* (q.v.); alluding to the lack of under leaves.

Colura (Lejeuneaceae) (Dumort.) Dumort. Gk. *kolouros* = dock-tailed (*kolos* = cut off; *oura* = tail); alluding to the appearance of the perianth of the type species.

Conostomum (Bartramiaceae) Sw. in F.Weber & Mohr

Gk. *konos* = cone; *stoma* = mouth; referring to the perforated cone formed by the fused tips of the peristome teeth.

Cratoneuron (Amblystegiaceae) (Sul) Spruce Gk. *kratos* = strong; *neuron* = nerve: alluding to the strong costa: 'costa stout, subcontinous' Sullivant 1876 (David Meagher).

Crossidium (Pottiaceae) Jur. Gk. *krossos* = fringe; *-idion* = little; alluding to tassel-like fringe on adaxial surface of costa (*Flora of North America*) or the dense fringe provided by filaments covering the costa in the upper part of the leaf (D Meagher).

Cryphaea (Cryphaeaceae) Mohr in F.Weber Gk. *cryph* = hidden; *phae* = dark; referring to the immersed sporophytes.

Cryptochila = Zyzigiella (Jungermanniaceae) R.M.Schust. Gk. *kryptos* = hidden, concealed; *chilos* = lip. Meaning unclear.

Cylindrocolea (Cephaloziellaceae) R.M.Schust. Gk. *kylindros* = cylinder; *koleos* = sheath.

Cryptomitrium (Aytoniaceae) Austin ex Underw.

Gk. *kryptos* = hidden, concealed; *mitrion* = cap, headband. Meaning unclear.

Cyclodictyon (Pilotrichaceae) Mitt.
Gk. *kyklos* = circle, circular; *diktyon* = net, network; alluding to the large, nearly circular laminal cells.

Cygnicollum** (Funariaceae) Fife & Magill
La. *cygnus* = swan; *collum* = neck; referring to the curved sterile lower portion of the capsule.

D

Desmatodon = Tortula (Pottiaceae) Brid.
Gk. *desmos* = band, bundle; *odon* = toothed.

Dicranella (Dicranaceae) (Müll.Hal.) Schimp.
Dicranum (q.v.); Gk. *-ella* = diminutive; alluding to the forking pattern of the frond.

Dicranolejeunea (Lejeuneaceae) Steph.
Dicranum (q.v.); *Lejeunea* (q.v.); possibly referring to the split teeth in the peristome.

Dicranoloma (Dicranaceae) (Renauld) Renauld
Dicranum (q.v.); *loma* = edge, fringe, border; referring to the similarity with many *Dicranum* species and the bordered leaves.

Dicranoweisia (Dicranaceae) Lindb. ex Milde
Dicranum (q.v.); *Weisia* (q.v.).

Dicranum (Dicranaceae) Hedw.
Gk. *dikranos* = pitchfork, two branched; referring to the stems, which may fork but do not branch.

Didymodon (Pottiaceae) Hedw.
Gk. *didymos* = twin; *-odon* = toothed; referring to the peristome teeth grouped in pairs.

Dimerodontium (Fabroniaceae) Mitt.
Gk. *di-* = two; *meros* = part, portion; *odontos* = tooth. Two species of *Dimerodontium* were recognised by Mitten, hence, two parts.

Diplasiolejeunea (Lejeuneaceae) (Spruce) Schiffn.
Gk. *di-* = two; *plasion* = oblong; *Lejeunea* (q.v.); referring to a double sori.

Distichium (Ditrichaceae) Bruch & Schimp.
La. *distichos*, Gk. *distichos* = two-ranked; alluding to the leaves in two rows.

Distichophyllum (Daltoniaceae) Dozy & Molk.
Gk. *distichos* = two-ranked; *phyllos* = leaf; alluding to the differentiated dorsal and ventral leaves found in most species.

Ditrichum (Ditrichaceae) Hampe
Gk. *di-* = two; *trichos* = hair, hairy; alluding to peristome split longitudinally into two segments.

Drepanocladus (Amblystegiaceae) (Müll.Hal.) G.Roth
Gk. *drepanon* = sickle; *clados* = a branch (q.v.); referring to the somewhat sickle-shaped leaves.

Drepanolejeunea (Lejeuneaceae) (Spruce) Schiffn.
Gk. *drepane* = sickle; *Lejeunea* (q.v.); referring to the sickle-shaped leaves.

E

Eccremidium (Ditrichaceae) Hook.f. & Wilson
Gk. *ekkremes* = pendulous, hanging; *-idium* = diminutive; referring to the pendulous capsule.

Ectropothecium (Hypnaceae) Mitt.
Gk. *ektrope-* = turned aside; *thekion* = a container capsule, case. The capsules are small, somewhat spherically shaped.

Encalypta (Encalyptaceae) Hedw.
Gk. *en-* = in; *kalyptos* = enveloping; alluding to the remarkably large calyptra, which covers and extends below the capsule.

Entodon (Entodontaceae) Müll.Hal.l.
Gk. *entos* = inside, within; *odon* = tooth; alluding to the insertion of the outer peristome teeth inside the capsule mouth.

Entodontopsis (Stereophyllaceae) Broth.
Gk. *entos* = inner; *odon* = teeth; *opsis* = resembling.

Entosthodon (Funariaceae) Schwägr.
Gk. *entosthen-* = inside; *odon* = tooth; alluding to the position of the inner peristome, well below (and thus inside) the mouth of the capsule.

Ephemerum (Ephemeraceae) Hampe
Gk. *ephemera* = temporary, short-lived; referring to the short life cycle of the plant.

Eriopus = Calyptrochaeta (Hookeriaceae) Brid.
Gk. *erion* = wool; *-pus* = foot (*pous*, a foot); *eriopus* = woolly foot; possibly referring to the woolly stalks.

Erpodium (Erpodiaceae) (Brid.) Brid.
Gk. *erpo* = creeping; referring to the growth habit of most species.

Erythrodontium (Entodontaceae) Hampe
Gk. *erythro* = red; *odontium* = little teeth; pertaining to the exostome teeth.

Eulejeunea (Lejeuneaceae) Steph.
Gk. *eu-* = good, well, completely; *Lejeunea* (q.v.).

Eustichia (Eustichiaceae) (Brid.) Brid.
Gk. *eu-* = good, well; *stichos* = row. The flattened plant leaves are spirally arranged in two rows.

Exormotheca* (Exormothecaceae) Mitt.
Gk. *exormo-* = out, outside; *theca* = box, case. Meaning unclear.

F

Fabronia (Fabroniaceae) Raddi
For Giovanni Valentino Mattia Fabbroni (1752–1822), Italian naturalist, economist, agronomist and chemist. With the Italian physicist Felice Fontana (1730–1805), he cofounded the natural history museum in Florence (Museo di Fisica e Storia Naturale di Firenze) in 1775, of which he was vice director. He also wrote *Reflexions Sur l'état Actuel de l'Agricolture* (1777–1778), which had an impact on farming methods and agrarian reform. He was involved in economic matters, and was instrumental in the development of the metric system in Italy. As a chemist, he did work in electrochemistry and wrote a work on anthracite, *Dell'Antracite o Carbone di Cava Detto Volgarmente Carbone Fossile* (1790). He became a member of the Accademia dei Georgofili in 1783.

Felipponea = ***Pterogoniadelphus***
(Leucodontaceae) Broth.
For Dr Florentino Silvestre Felippone (1849–1939), botanist, bryologist, naturalist and plant collector from Uruguay. He was interested in the flora and fauna of Uruguay. He was the editor of *Contribution á la Flore Bryologique de l'Uruguay*.

Fissidens (Fissidentaceae) Hedw.
La. *fissus* = a split or crack; *dens* = tooth; alluding to the split peristome teeth.

Floribundaria (Meteoriaceae) M.Fleisch.
La. *floribundus* = profusely flowering; *aria* = pertaining to; apparently alluding to the plant often being found fertile.

Fontinalis (Fontinalaceae) Hedw.
La. *fons, fontanus, fontianalis* = pertaining to or found growing in or by springs; referring to the preferred habitat of the aquatic moss.

Forsstroemia (Leptodontaceae) Lindb.
For Johan Erik Forsström (1775–1824), Swedish pastor, botanist and plant collector. He studied at the University of Uppsala, where one of his instructors was the naturalist Carl Peter Thunberg (1743–1828). In 1800, he accompanied Göran Wahlenberg (1780–1851), the successor to Thunberg, on an expedition through Fennoscandia (Scandinavian Peninsula, Finland, Karelia, and the Kola Peninsula), where he performed entomological and botanical investigations. On the trip, he maintained a diary of his trip to northern Sweden, Finnmark and Moberg, published in 1817 as *Norrlandsstäder och Lapplandsbygd År* (1800). From 1802 to 1815, Forsström was a pastor on Saint Barthélemy of the Leeward Islands, where he collected botanical specimens.

Fossombronia (Fossombroniaceae) Hazsl.
For Conte Vittorio Fossombroni (1754–1844), Italian statesman, mathematician, economist and engineer. He studied mathematics and hydraulics at the University of Pisa and worked in Tuscany as minister to the dukes Pietro Leopoldo and Ferdinand III, where he was distinguished by his efforts to improve the agriculture of Tuscany by drainage and irrigation, especially of the marshy Valdicana Valley, about which he published the treatise *Memorie Idraulico-Storiche Sopra la Valdi-Chiana* (1789). He was made foreign affairs minister, but fled to Sicily when the French occupied Tuscany in 1799. With the fall of Napoleon, he was appointed prime minister of the restored Tuscany under Grand Duke Leopold II.

Frullania (Jubulaceae) Raddi
For Leonardo Frullani (1756–1824), an Italian statesman. He held a law degree from the University of Pisa, became a civil servant in 1790, vice governor of Livorno in 1797, was appointed judge of the criminal court of Florence in 1808, president of the court in 1810, and finance minister of the Duchy of Tuscany under Ferdinand III in 1814.

Frullanoides (Lejeuneaceae) Raddi
Frullania (q.v.); Gk. *-oides* = resembling.

Funaria (Funariaceae) Hedw.
La. *funis* = rope; *-aria* = pertaining to; alluding to the twisted cord-like seta of *F. hygrometrica*.

G

Gammiella (Sematophyllaceae) Broth.
For George Alexander Gammie (1864–1935), Indian botanist, author, an authority especially of Indian flora, and a moss collector in Sikkim. He was employed by the cinchona department of Bengal as deputy superintendent of one of the

cinchona plantations. Later, he became a lecturer on botany at the College of Science at Poona. A commercial problem for the government was how to extract quinine from the rapidly increasing stock of crown and yellow bark. The government quinologist, CR Wood, suggested a method of extraction, which Gammie put into practice. This method was so successful that all the government hospitals and dispensaries in India had all the quinine they required (up to 2.7 tonnes per year).

Gigaspermum (Gigaspermaceae) Lindb.
Gk. *gigas* = giant; *spermum* = seed. The seeds (spores) are very large.

Glyphomitrium = Ptychomitrium (Ptychomitriaceae) Brid.
Gk. *glyphos* = carved, from *glyphein* = to carve, engrave; *mitrion* = cap, headband. Meaning unclear.

Gongylanthus (Arnelliaceae) Nees
Gk. *gongylus* = round; *anthus* = flower; referring to the opposite, rounded, entire leaves that bend upward.

Goniomitrium (Funariaceae) Hook.f. & Wilson
Gk. *gonio* = angular, an angle; *mitria* = cap, headdress; referring to the pleated calyptra.

Gottschea = Schistochila (Schistochilaceae) Nees ex Mont.
For Carl Moritz Gottsche (1808–1892), German physician and bryologist. He studied medicine at the University of Berlin, and practised medicine with botany as a hobby. Later he became director of the Botanical Gardens at Hamburg. A great student of liverworts, he authored, with Johann Bernhard Wilhelm Lindenberg (1781–1851) and Christian Gottfried Nees von Esenbeck (1776–1858), *Synopsis Hepaticarum* (1844–1847), providing many illustrations and producing a major work in the field of hepaticology containing more than 1 600 descriptions. He also worked with Lindenberg on the publication *Species Hepaticarum* (1839–1851). In 1881, he received an honorary doctorate in philosophy from the University of Kiel.

Grimmia (Grimmiaceae) Hedw.
For Johann Friedrich Karl (Carl) Grimm (1737–1821), German botanist, physician and medical researcher in Gotha, personal physician to the Duke of Saxe-Gotha. He studied medicine and natural sciences in Göttingen under Albrecht von Haller in 1758. He became a member of the Leopoldina Society in 1801 and an honorary member of the Botanists of Regensburg. He was involved initially with phanerogams (seed-bearing plants), but later devoted himself especially to cryptogams (plants that do not reproduce by seeds but spores). He took field trips in Thuringia with his friend Bridel, who introduced him to moss collecting. He translated the works of Hippocrates, *Hippocrates Werke, aus dem mit Griechischeen Übersetzt Erläuterungen*, in four volumes (1792).

Gymnocoleopsis (Jungermanniaceae) (R.M.Schust.) R.M.Schust.
Gk. *gymnos* = naked; *koleos* = sheath; *-opsis* = resembling. The genus is similar to the earlier genus *Gymnocolea* (Dum.), in which it was formerly included.

Gymnostomum (Pottiaceae) Nees & Hornsch.
Gk. *gymnos* = naked; *stoma* = mouth; alluding to the lack of a perostome.

Gyroweisia (Pottiaceae) Schimp.
Gk. *gyros* = a ring, circle; *Weissia* (q.v.); alluding to resemblance and well-developed, persistent annulus.

H

Haplocladium (Thuidiaceae) Müll.Hal.
Gk. *haplo-* = single; *klados* = branch, stem; (*cladium* = branchlet); alluding to only once-pinnate branching.

Haplohymenium (Anomodontaceae) Schwägr.
Gk. *haplo-* = single; *hymen* = thin skin, membrane; *-ium* = diminutive; possibly referring to the single peristome (endostome's delicate membrane).

Haplomitrium (Haplomitriaceae) Nees
Gk. *haplo-* = single, simple; *mitrion* = cap; alluding to the translucent sheath covering the developing capsule.

Hedwigia (Hedwigiaceae) P.Beauv.
For Johann (Johannes, Joannis) Hedwig (1730–1799), German botanist, physician, and  expert microscopist, sometimes referred to as the father of bryology because of his study of mosses. He studied at the University of Leipzig, receiving an MD in 1759, and practised medicine for 20 years. He was a professor of medicine and botany at the University of Leipzig, director of the Leipzig Botanical Garden, author of *Fundamentum Historiae Naturalis Muscorum Frondosorum* in

two volumes (1782–1783), author of *Species Muscorum Frondosorum* (1801), a Fellow of the Royal Society, and a foreign member of the Royal Swedish Academy of Sciences.

Hedwigidium (Hedwigiaceae) Bruch & Schimp.
Hedwigia (q.v.); Gk. *-idion* = diminutive. Similar to the genus above.

Helicodontium (Myriniaceae) (Mitt.) A.Jaeger
Gk. *heliko* = spirally twisted; *odontos* = toothed; referring to the peristome, curling inwards to dry.

Hennediella (Pottiaceae) Paris
For Roger Hennedy (1809–1876), Scottish phycologist and professor of botany at Anderson's University in Glasgow from 1863–1876, teacher and friend of Scottish botanist and paleobotanist Robert Brown (1773–1858). Previously, he was a block cutter for calico printers, and later a designer for calico printing before he became a botany professor, a post he occupied until his death. He published *The Clydesdale Flora: A Description of the Flowering Plants and Ferns of the Clyde District*, which ran through four editions, the last revised copy by Thomas King being published in 1891.

Herbertus (Herbertaceae) Gray
For Thomas Herbert (c 1656–1733), eighth Earl of Pembroke and fifth Earl of Montgomery, British politician during the reigns of King William III and Queen Anne, president of the Royal Society, and a patron of the noted Italian botanist Pier Antonio Micheli

Herpetineuron = Anomodon
(Anomodontaceae) (Müll.Hal.) Cardot
Gk. *herpeton* = creeping animal, reptile; *neuron* = nerve or vein; alluding to the characteristic snaking of the costa in the upper part of the leaf.

Heterophyllium (Sematophyllaceae) (Schimp.) Kindb. in M.Fleisch.
Gk. *heteros* = dissimilar; *phyllon* = leaves; *-ium* = little. The leaves are variable in shape.

Heteroscyphus (Geocalycaceae) Schiffn.
Gk. *hetero-* = different; *scyphus* = cup; alluding to the difference in the position of the androecia and in the shape of the perigonal leaves compared with the stem leaves.

Holomitrium (Dicranaceae) Brid.
Gk. *holos* = entire; *mitrion* = cap, headdress; alluding to the entire calyptra.

Homalothecium (Brachytheciaceae) Schimp.
Gk. *homalos* = smooth, flat, (fairly) even; *theka* = case, capsule; referring to the shape of the capsule.

Hookeriopsis (Pilotrichaceae) (Besch.) A. Jaeger
For Sir William Jackson Hooker (1785–1865), father of JD Hooker, professor of botany at Glasgow University, close friend of Sir Joseph Banks, and first director of Kew Gardens. He made botanical trips to Iceland (1809), during which he nearly died in a fire that destroyed his samples; and to France, Switzerland and northern Italy (1814). He authored *Tour in Iceland* (1809), *Muscologia* (1818), *Musci Exotici* (two volumes, 1818–1820), *Flora Scotica* (1821), *British Flora* (1830), *British Flora Cryptogamia* (1833), and many other books, including accounts of botanical expeditions of Sir William Edward Parry, Sir John Franklin, and Frederick William Beechey. He helped establish the Royal Botanic Institution of Glasgow and the Glasgow Botanical Gardens.

Husnotiella = Didymodon (Pottiaceae) Cardot.
For Pierre Tranquille Husnot (1840–1929), French botanist, specialist in mosses, founder of *Revue Bryologique* in 1874, and its editor until 1927. He studied at the Ecole d'Agriculture der Grignon and the Université de Caen, and later he attended lectures by Adolphe Chatin, French mycologist, at the Muséum National d'Histoire Naturelle. Between 1863 and 1886, he undertook collecting trips to Britain, the Pyrenees, the Alps, New Grenada, the French Antilles, and the Canary Islands. His major works include *Hepatologica Gallica* (1881) and *Muscologica Gallica* (1894), which earned him the Prix Montagne of the Académies des Sciences. He was awarded the Legion d'Honneur for his nearly 60 years' service as socialist mayor of Caen.

Hyalolepidozia*** (Lepidoziaceae) S.W.Arnell ex Grolle
Gk. *hualos* = of glass, transparent; *lepis* = scale, flake; *ozoz* = bud. Meaning unclear.

Hygroamblystegium (Amblystegiaceae) Loeske
Gk. *hugros* = damp, moist; *amblys* = blunt;

stegium = a roof or covering; referring to the plant's favoured environment – an aquatic habitat.

Hymenostomum (Pottiaceae) R.Br.
Gk. *hymen* = membrane; *stoma* = mouth; referring to the mouth of the capsule, which is completely covered by a horizontal web derived from the outer membrane of the capsule.

Hymenostylium (Pottiaceae) Brid.
Gk. *hymen* = membrane; *stylos* = pillar; alluding to the stylus capsule.

Hyophila (Pottiaceae) Brid.
Gk. *hyo* = water, rain; *philos* = loving; alluding to its habitat – it is found close to or in water.

Hypnum (Hypnaceae) Hedw.
Gk. *hypnos* = sleep; alluding to the ancient practice of using this moss as a filler for cushions and pillows.

Hypodontium (Pottiaceae) Müll.Hal.
Gk. *hypo-* = beneath; *odontos* = toothed; referring to the narrow exostome teeth that remain attached to the inside of the operculum of *Hypodontium* and fall with it.

Hypopterygium (Hypopterygiaceae) Brid.
Gk. *hypo-* = below; *pterugion* (diminutive of *pterux*, meaning wing); referring to the amphigastria.

I

Inflatolejeunea = Lejeunea (Lejeuneaceae) S.W.Arnell
La. *inflatior* = swollen; *Lejeunea* (q.v.).

Ischyrodon (Fabroniaceae) Müll.Hal.
Gk. *iskhyros* = strong; *odon* = tooth; alluding to the strong teeth of the single peristome.

Isopterygium (Hypnaceae) Mitt.
Gk. *isos* = equal; *pterugion* (diminutive of *pterux*, meaning wing); referring to the delicate, wing-like arrangement of the leaves equally spaced and opposite each other in some species.

Isotachis (Balantiopsaceae) Mitt.
Gk. *isos* = equal; *taxis* = arrangement; alluding to the regular arrangement of the equal leaves and underside of the leaves (stipules).

J

Jaegerina (Pterobryaceae) Müll.Hal.
For August (Augusto) Jaeger (1842–1877), Swiss bryologist and author of *Genera et Species Muscorum,* who had a herbarium of 12 500 specimens of mosses. This became a core component of the William C Steere Bryophyte Herbarium at the New York Botanical Garden, and which was the basis for a series of articles by Jaeger and Frederick William Sauerbeck (1810–1880) that reviewed all known moss species.

Jamesoniella (Jungermanniaceae) (Spruce) Schiffn.

For William Jameson (1796–1873), Scottish botanist, chemist, explorer, and plant collector. (Gk. *-iella* = diminutive). The plant resembles a miniature version of the fern *Jamesonia*. See *Jamesonia* (Pteridaceae) for biography.

Jensenia (Pallaviciniaceae) Lindb.
For Thomas Jensen (1824–1877), Danish bryologist and teacher. He graduated in 1842 and became a candidate of theology in 1849, but earned his living as a teacher of natural history from 1858 at the seminary in Ranum, Denmark. He authored *Bryologia Danica eller de Danske Bladmosser* (1856), and collected in Denmark and Norway. He was recognised as an author of chess problems.

Jungermannia (Jungermanniaceae) L.
For Ludwig Jungermann (1572–1653), German physician and botanist. He studied medicine and botany at Nuremberg. In 1609, he laid out the 1 200 m² historic botanical garden in grounds donated by Louis V, Landgrave of Hesse-Darmstadt, for the newly established university and for the cultivation of *hortus medicus*. From 1614–1625, he taught anatomy and botany at the newly established University at Giessen, and from 1625 at Altdorf bel Nürnberg, where he created another beautiful garden. Jungermann is credited with writing the major part of the text of *Hortus Eystettensis* (1613), with Basilius Bester (1561–1629), a beautifully illustrated book featuring 1 084 species from the garden of Johan Konrad von Gemmingen, prince bishop of Eichätt in Bavaria.

Juratzkaea (Stereophyllaceae) Lorentz
For Jacob Juratzka (1821–1878), Austrian

engineer, botanist and bryologist, author of numerous papers on mosses and books, mainly on mosses in southeastern Austria, including *Zur Moosflora der Obersteiermark* (*The Moss Flora of Upper Styria*) (1871) and *Nachtrag Moosflora der zur Obersteiermark* (*Adding to the Moss Flora of Upper Styria*) (1871). His major work, *Die Laubmoosflora von Oesterreich-Ungarn* (*The Moss Flora of Austria-Hungary*), published after his death in 1882, contains the description of most mosses from Austria and Hungary.

K

Kindbergia (Brachytheciaceae) Ochyra
For Nils Conrad Kindberg (1832–1910), Swedish bryologist. He obtained his PhD from Uppsala University (1857), became senior schoolteacher in nature science and mathematics in Linköping (1860–1910), made several bryological journeys to Norway, Switzerland and Italy, and between 1888 and 1910, published more than 50 papers relating to mosses of North America, many sent to him by John Macoun (1831–1920), a Canadian naturalist. Kindberg authored *Species of European and Northamerican Bryineae* (mosses) (1896), *Genera of European and Northamerican Bryineae* (1897), and *New Canadian Mosses* (1889) (with Macoun). Many of his 'new' species proved to be founded on insignificant variations of earlier described mosses.

Kurzia (Lepidoziaceae) G.Martens
For Wilhelm Sulpiz Kurz (1834–1878), German botanist. He studied at the University of Munich and worked briefly as an apothecary. Following a fallout with his family, he joined the medical services of the Dutch East Indies Army (Koninklijk Nederlands Indisch Leger) in 1856, going under the false, more Dutch-sounding name of Johann Amann. While stationed in Batavia, he collected plants widely and, in 1859, was offered a position as an assistant in the botanical gardens at Bogor. In 1863, he left for India, where he worked as herbarium curator at the Sipbur Botanical Gardens. Between 1863 and 1871, he explored the flora of India, Indonesia, Burma, Malaya and Singapore. He published more than 60 reports and papers during his lifetime. His publications include *Forest Flora of British Burma* in two volumes (1877), *Report on the Vegetation of the Andaman Islands* (1870), *Preliminary Report on the Forest and Other Vegetation of Pegu* (1875), and *Bamboo and Its Use* ... (1876). His knowledge of trees seems extensive. One report states: 'In lower Burma alone, the enumeration of the trees made by Sulpiz Kurz in his *Forest Flora of British Burma* (1877) includes some 1,500 species.'

L

Lejeunea (Lejeuneaceae) Lib.
For Alexandre Louis Simon Lejeune (1779–1858), Belgian pharmacist and botanist, sometimes referred to as the father of Belgian botany. He studied pharmacology and botany at Liege, then enrolled in medical school in 1801 at Paris. Although his studies were disrupted by military conscription, he later became a civilian doctor in Ensival and Verviers, but his passion was botany. He researched and authored *Methodique of Regne Végétal Tableau du Département de l'Ourthe* (1806), *La Flore de Spa* (1811), *Revue de la Flore des Environs de Spa* (1824), and the three-volume *Compendium Florae Belgicae* (1831) with Belgian botanist Richard Joseph Courtois. He was a member of the Belgian Royal Academy of Sciences and Letters, and the Linnaean Society of Paris.

Lepicolea (Lepicoleaceae) Dumort.
Gk. *lepis* = scale; *koleos* = sheath. Meaning unclear.

Lepidopilidium (Pilotrichaceae) (Müll.Hal.) Broth.
Gk. *lepidotos* = scaly; *-idion* = diminutive. Meaning unclear.

Lepidozia (Lepidoziaceae) Dumort.
Gk. *lepis* = scale; *ozoz* = bud; alluding to the scale-like bracts and bracteoles surrounding the perianth.

Leptocolea (Lejeuneaceae) (Spruce) A.Evans
Gk. *leptos* = slender, thin; *koleos* = sheath; perhaps referring to the slender leaves, thin and flat, and the pilose sheaths.

Leptodictyum (Amblystegiaceae) (Schimp.) Warnst.
Gk. *leptos* = slender, thin; *diktyon* = net; alluding to the areolation of the leaves.

Leptodon (Leptodontaceae) D.Mohr.
Gk. *leptos* = slender, thin; *-odon* = teeth; alluding to the narrow peristome teeth.

Leptodontium (Pottiaceae) (Müll.Hal.) Hampe ex Lindb.
Gk. *leptos* = slender, thin; *odontos* = toothed; *-ium* = diminutive; alluding to narrow peristome teeth.

Leptoischyrodon (Sematophyllaceae) Dixon
Gk. *leptos* = slender, thin; *iskhuros* = strong; *-odon* = toothed. Meaning unclear.

Leptophascum (Pottiaceae) (Müll.Hal.) J.Guerra & Cano.
Gk. *leptos* = slender, thin; *phascum* = tuft of moss; perhaps referring to the texture of the moss.

Leptopterigynandrum (Thuidaceae) Müll. al.
Gk. *leptos* = slender, thin; *pterus* = wing; *gyna* = ovary; *andrum* = stamen. Meaning unclear.

Leptoscyphus (Geocalycaceae) Mitt.
Gk. *leptos* = slender, thin; *skyphos* = cup; presumably alluding to the flattened perianth.

Leptotheca (Rhizogoniaceae) Schwägr.
Gk. *leptos* = slender, thin; *theka* = case, capsule; alluding to the narrow capsule.

Leptotrichella (Dicranaceae) (Müll.Hal.) Lindb.
Gk. *leptos* = slender, thin; *trichos* = hair; *-ella* = diminutive; referring to the fine peristome teeth.

Leskeella (Leskeaceae) (Limpr.) Loeske
For Nathanael Gottfried Leske (1751–1786), German naturalist, economist and geologist. He  studied at the Bergakademie of Freiberg, Saxony, under Abraham Gottlob Werner (1749–1817) and the Franckeschen Stiftungen in Halle. He became professor at the University of Leipzig, teaching natural history in 1775 and economics from 1777–1786, when he accepted the chair of financial science and economics at Marburg University, but died shortly thereafter in an accident. His large mineral and natural history collection was sold to the Dublin Society in 1792 and is now housed in the National Museum of Ireland. He wrote on diverse topics and was a co-editor of *Leipziger Magazin zur Naturkunde, Mathematik und Oekonomie* from 1781.

Leskeodon = Distichophyllum
(Hookeriaceae) Broth.
Leskeelle (q.v.); *-odon* = tooth.

Lethocolea (Acrobolbaceae) Mitt.
Gk. *lethe* = missing, overlooked; *koleos* = sheath; alluding to the marsupium that is usually so deeply buried in the soil it is torn off when the plant is collected, or simply overlooked.

Leucobryum (Leucobryaceae) Hampe
Gk. *leukos* = white; *bryon* = lichen, moss; alluding to its colour, although this is variable, from white to greyish to bluish-green.

Leucodon (Leucodontaceae) Schwägr.
Gk. *leukos* = white; *-odon* = toothed; alluding to the pale peristome teeth.

Leucolejeunea (Lejeuneaceae) A.Evans
Gk. *leukos* = white; *Lejeunea* (q.v.); alluding to the pale colour of the plants.

Leucoloma (Dicranaceae) Brid.
Gk. *leukos* = white; *loma* = edge or fringe; alluding to the pale border of the leaves.

Leucoperichaetium** (Grimmiaceae) Magill
Gk. *leukos* = white; *peri* = round, about; *chaite* = bristle.

Levierella (Fabroniaceae) Müll.Hal.
For Émile (Emilio) Levier (1839–1911), Swiss physician, botanist, algologist, bryologist, pteridologist and mycologist.  In 1865, he moved to Florence and collected in the Italian Alps. He visited Spain and Portugal between 1878 and 1879 with Louis François Jules Rodolphe Leresche (1808–1885) to collect plants. They published *A Voyage Botanique en Espagne* (1878), and *Deux Excursions Dans le Nord Bota de l'Espagne et le Portugal* (1880) with Pierre Edmond Boissier (1810–1885), who described some 6 000 species. In 1890, he travelled to Sicily, Turkey and Greece with Stefano Sommier (1848–1922), publishing *A Travers le Caucase: Notes et Impressions d'un Botaniste* (1894). His collection when he died totalled 47 000 specimens – some from western Asia, New Guinea, Madagascar and Central America.

Lindbergia (Leskeaceae) Kindb.
For Sextus Otto Lindberg (1835–1889), Swedish-born physician, naturalist and bryologist. He was educated at Uppsala in Sweden, obtaining a BA in 1856, lectured in pharmacology and did various other studies. Most of Lindberg's time from 1853–1880 was spent making microscopic studies of mosses and hepatics. He collected specimens in Finland, Sweden, Norway, England and Ireland. He became recognised as an eminent European authority on mosses and hepatics, and had the respect of botanists from around the world. In

Liverwort thallus with gemma cups (EJM)

1865, he succeeded William Nylander to the chair in botany at the Botanical Museum in Helsingfors (later University of Helsinki), and was also appointed director of the botanical gardens. Later, he became the dean of the physics-mathematics faculty. He wrote more than 150 papers on bryology in prestigious scientific journals such as the *Journal of the Linnaean Society* and *Journal of Botany*. From 1866 until he died, he was president of the Societas Pro Fauna et Flora Fennica (Finnish Society of Fauna and Flora).

Lophocolea (Geocalycaceae) (Dumort.) Dumort.
Gk. *lophos* = crest, tuft; *koleos* = sheath; referring to the crested or toothed ridges of the perianth of certain species.

Lopholejeunea (Lejeuneaceae) (Spruce) Schiffn.
Gk. *lophos* = crest, tuft; *Lejeunea* (q.v.). The perianth keels have horns or crests.

Lophozia (Jungermanniaceae) (Dumort.) Dumort.
Gk. *lophos* = crest, tuft; *ozoz* = bud; alluding to the typically dentate mouth of the perianth.

Lopidium (Hypopterygiaceae) Hook.f. & Wilson
Gk. *lopis* = scale; *-idion* = diminutive; alluding to the overlapping leaves like scales of a fish.

Ludorugbya (Pottiaceae) Hedd. & R.H.Zander
La. *ludo* = play; *rugbya* = rugby. At the time of discovery (2007), the species was named *springbokorum* – the 'Boks' won the Rugby World Cup.

M

Macrocoma (Orthotrichaceae) (Hornsch. ex Müll. Hal.) Grout
Gk. *makros* = large; *kome* = hair, tufts of hair; possibly alluding to long hairs on calyptra of some species.

Macromitrium (Orthotrichaceae) Brid.
Gk. *makros* = large; *mitra* = mitre, head-band; referring to the shape of the calyptra.

Madotheca (Porellaceae) Dumort.
Gk. *mador* = moisture, wetness(?); *theka* = a case, capsule. Possibly because the capsules are on short fruit stalks in wet environments.

Mannia (Aytoniaceae) Opiz
Named by the Czech Philipp (Filip) Maximilian Opiz, possibly for the Czech Bohemian lichenologist Wenzeslaus (Wenzel) Blasius Mann (1799–1839). Not to be confused with *Mannia* (Simaroubaceae, a different family), which was named for Gustav Mann (1836–1916), German botanist.

Marchantia (Marchantiaceae) L.
For Nicholas Marchant (?–1678), French botanist and director of the ducal gardens at Blois. He obtained a degree in medicine from the University of Paris and became an apothecary to Gaston, duc d'Orléans, the brother of Louis XIII. When Gaston died in 1660, Marchant entered into service of the king, capacity unknown, but in 1774 he became '*concierge et directeur de la culture des plantes du Jardin Royal*'. Marchant was a founding member of the Académie Royale des Sciences, and remained the only botanist in the organisation until the election of Denis Dodart in 1673. He worked with his son Jean, on *Histoire des Plantes* (unpublished), and collaborated in editing Dodart's *Mémoires pour Servir à l'Histoire des Plantes* (1676).

Marsupella (Gymnomitriaceae) Dumort.
Gk. *marsupion* = pouch; *-elle* = diminutive; alluding to the appearance of the perianth that, in some species, encloses the developing capsule.

Marsupidium (Acrobolbaceae) Mitt.
Gk. *marsupion* = pouch; *-idion* = diminutive. As above.

Meiothecium (Sematophyllaceae) Mitt.
La. *meion* = smaller; *thekion* = case, capsule; alluding to the small capsules.

Metacalypogeia (Calypogeiaceae) (S.Hatt.) Inouye
Gk. *meta* = changed; *Calypogeia* (q.v.); possibly referring to a circumboreal (northern regions) genus in addition to *Calypogeia*.

Metzleria = Atractylocarpus (Dicranaceae) Schimp. ex Milde
For Jacob Adolf Metzler (1812–1883), German landowner of Frankfurt, lichenologist, bryologist, a  'man of independent means', and a collector of mosses from southern France, northern Italy and the Alps. In 1869, Metzler began work in the herbarium of the Senckenberg Naturmuseum in Frankfurt, Germany. Theodor Geyler (1834–1889), the curator of the museum, decided to rearrange the herbarium. He put Metzler in charge of the cryptogams section while he took charge of the phanerogams section. Together they added many specimens to the collections and in 1870 a new Botanical Hall in the museum was opened to the public. After Metzler's death, his personal lichen collection was given to the Senckenberg Herbarium.

Microcampylopus (Dicranaceae) (Müll. Hal.) M.Fleisch.
Gk. *mikros* = very small; *kampylos* = curved or bent; *-pus* = foot (*pous*, a foot). Meaning unclear.

Microcrossidium (Pottiaceae) J.Guerra & Cano
Gk. *mikros* = very small; *krossos* = fringe or tassel; *-idion* = diminutive; alluding to the very small tassel-like fringe on the adaxial surface of costa.

Microdus (Dicranaceae) Schimp. ex Besch.
Gk. *mikros* = very small; *-dus* = diminutive. Many species are very small.

Microlejeunea (Lejeuneaceae) Steph.
Gk. *mikros* = very small; *Lejeunea* (q.v.); alluding to the small indusium.

Microlepidozia (Lepidoziaceae) Spruce
Gk. *mikros* = very small; *Lepidoza* (q.v.). A smaller genus in the same family.

Micromitrium (Ephemeraceae) Austin
Gk. *mikros* = very small; *mitra* = mitre; referring to the small calyptra.

Micropoma (Funariaceae) Lindb.
Gk. *mikros* = very small; *poma* = lid. Meaning unclear.

Microthamnium (Hypnaceae) Mitt.
Gk. *mikros* = very small; *thamnos* = shrub. Shrub-like in appearance.

Mielichhoferia (Mniaceae) Nees & Hornsch.
For Mathias Mielichhofer (1772–1847), Austrian botanist, bryologist and mining engineer in Slazburgischen until 1843. He  became interested in flora as a result of his walks in the Alps and in Saxony. In 1813, he became married and shortly thereafter he developed an interest in mosses. It was in this year that he discovered two new mosses, *Mielichhoferia elongate* and *M. mielichoferiana*, which led to the genus *Mielichhoferia* being named after him. The Austrian botanist, mycologist and doctor Anton Eleutherius Sauter (1800–1881), archbishop of Salzburg, called him a happy man as 'every new plant, each rare mineral filled him with great joy'. He had a valuable collection of minerals. His moss herbarium, which was acquired by a landowner, Ratzesberger, is considered lost.

Mikrosbryum (Pottiaceae) Schimp.
Gk. *mikros* = very small; *bryon* = lichen, moss; referring to its small size.

Mittenothamnium (Hypnaceae) Henn.
For William Mitten (1819–1906), British pharmaceutical chemist and bryophyte collector  considered the premier bryologist of the second half of the 19th century, according to the New York Botanical Garden. He began his bryological career under the tutelage of William Borrer and William Jackson Hooker, but declined curatorship of the Kew Gardens herbarium as he had a chemist business in Hurstpierpont, Sussex, which enabled him to better support his family. His collection of bryophytes, the largest in the world in private hands, consisted of some 50 000 specimens, and was purchased at the time of his death by the New York Botanical Garden. He was the father-in-law of Alfred Russel Wallace.

Mnioloma (Calypogeiaceae) Herzog
Gk. *mnion* = moss, seaweed; *loma* = edge, border; alluding to the border of hyaline cells on the leaf, typical of the genus.

Mnium (Mniaceae) Hedw.
Gk. *mnium* = moss.

Molendoa (Pottiaceae) Lindb.
For Ludwig Molendo (1833–1902), German botanist and bryologist, newspaper editor and writer. He studied medicine and botany in Munich, and researched in the Alps and the Fichtelgebirge mountains, Bavaria, with his brother-in-law, Alexander Walter (1813–1890). They published a basic description of the bryoflora of Upper Franconia, *Die Laubmoose Oberfrankens* (1868), listing some 384 species. Molendo worked briefly for the German botanist Carl Friedrich Philipp von Martius and Swiss botanist KW von Nägeli, then turned to journalism, working in Bayreuth, Passau, Regensburg and Munich. He continued collecting for Paul (Pablo) Günter Lorentz (1835–1881) and others, but sold much of his herbarium because he lived under financially difficult circumstances.

Monocarpus** (Monocarpaceae) D.J.Carr
Gk. *mono-* = one; *karpos* = fruit; one fruit.

N

Nanobryum (Fissidentaceae) Dixon
Gk. *nanos* = dwarf; *bryon* = lichen, moss; referring to the small size of the plants.

Neckera (Neckeraceae) Hedw.
For Noel Martin Joseph de Necker (or Natalis Joseph de Necker) (1729–1793), Belgian physician

and botanist of French descent, although some sources say he was German. He was physician to the ruler (elector palatinate), and from 1768, was a regular member of the Palatine Academy of Sciences in Mannheim. He had a special interest in mosses, and especially reproduction in moss-like organisations. His *Traité sur la Mycitologie* (1774) recorded what was known about fungi in his time. He invented the word *achene* = open.

Notoscyphus (Jungermanniaceae) Mitt.
Gk. *notos* = south; *scyphus* = cup, Speculatively 'south cup', possibly referring to this being a Southern Hemisphere bryophyte.

O

Octoblepharum (Calymperaceae) Hedw.
La. *octos* = eight; *blepharis* = eyelash; referring to the eight-toothed peristome.

Odontoschisma (Cephaloziaceae) (Dumort.) Dumort.
Gk. *odontos-* = toothed; *schisma* = cleft, division. Meaning unclear.

Oedipodiella (Gigaspermaceae) Dixon
Gk. *oideos, oedi* = swollen; *pous, pod* = foot; *-iella* = diminutive. Meaning unclear.

Oligotrichum (Polytrichaceae) DC.
Gk. *oligos* = few; *triche* = hair; alluding to the spare hairs on the calyptra.

Oreoweisia (Dicranaceae) (Bruch & Schimp.) De Not.
Gk. *oreo-* = mountain; *Weissia* (q.v.).

Orthodontium (Orthodontiaceae) Schwägr.
Gk. *ortho* = straight, upright; *odontos* = toothed; alluding to the upright peristome teeth.

Orthostichella (Neckeraceae) Müll.Hal.
Orthotrichum (q.v.); Gk. *-ella* = diminutive.

Orthostichopsis (Pterobryaceae) Broth.
Orthotrichum (q.v.); Gk. *-opsis* = resembling.

Orthotrichum (Orthotrichaceae) Hedw.
Gk. *ortho* = straight, upright; *trichos* = hair; alluding to the more or less erect hairs on the capsules of most species.

Oxymitra (Oxymitraceae) Bisch.
Gk. *oxys* = sharp; *mitre* = headband. Meaning unclear.

Oxyrrhynchium (Brachytheciaceae) (Schimp.) Warnst.
Gk. *oxys* = sharp, pungent; *rhynchos* = beak, bill; referring to the operculum of the capsule, which has a long appendage, somewhat beak-like.

Oxystegus = Trichostomum (Pottiaceae) (Limpr.) Hilp.
Gk. *oxys* = sharp, pungent; *stego* = covering.

P

Palamocladium (Brachytheciaceae) Müll.Hal.
Gk. *palame* = palm of the hand; *klados* = branch, stem; alluding to the fasciculate branching.

Pallavicinia (Pallaviciniaceae) Gray
Derivation uncertain. Probably named for Lazarus Opizio Pallavicini (f 1719–1785), botanist and archbishop of Genoa, but it might also be for the Italian naturalist Ignazio Alessandro Pallavicini (1800–1871) or for the Marquis Adalberto

Pallavicini delle Frabose, first president of the Societa Agraria in Turin.

Papillaria (Meteoriaceae) (Müll.Hal.) Lorentz
La. *papillatus* = nipple-like; *papillae* = soft protuberances; *aris* = resembling; alluding to the papillae (nipple-like bumps or projections) on the surface of cells.

Paracromastigum (Lepidoziaceae) Fulford & J.Taylor
Gk. *para* = next, near to; *akro-* = at the tip; *mastix* = flagellum, whip; alluding to the occurrence of *Acromastigum*-type ventral branching, one of the characteristics that separates it from *Lepidozia*.

Pelekium (Thuidiaceae) Mitt.
Gk. *pelekus* = axe; *-ium* = characteristic or resemblance to. The sporophyte and its seta resemble a long-handled axe.

Phaeoceros (Anthocerotaceae) Prosk.
Gk. *phaios* = dusky, brownish grey; *keras* = horn; alluding to the typical colour of the mature sporangium and the hornwort group.

Phasconica = Weissia (Pottiaceae) Müll.Hal.
Gk. *phascum* = tuft of moss; *konikos* = conical.

Phascum (Pottiaceae) Hedw.
Gk. *phascum* = tuft of moss. An ancient Greek word used by Theophrastus. Hedwig used this word to describe all mosses lacking an operculum.

Philonotis (Bartramiaceae) Brid.
Gk. *philo* = loving; *notis* = moisture; referring to its natural habitat.

Phragmilejeunea (Lejeuneaceae) R.M.Schust.
Gk. *phragma* = a hedge; *Lejeunea* (q.v.); the culms being used for fencing.

Physcomitrellopsis (Funariaceae) Broth. & Wager.
Physcomitrium (q.v.); Gk. *-ella* = diminutive; *opsis* = resembling.

Physcomitrium (Funariaceae) (Brid.) Brid.
Gk. *physce* = bladder; *mitrion* = little cap; probably alluding to the calyptra.

Pilotrichella (Meteoriaceae) (Müll.Hal.) Besch.
La. *pilus* = hair; *triche* = hair; *-ella* = diminutive; referring to the axillary hairs usually being reddish throughout.

Pilularia (Marsileaceae) L.
La. *pilularia* = a little ball; referring to this aquatic or semi-aquatic fern's small, globose sporocarps.

Pinnatella (Neckeraceae) M.Fleisch.
La. *pinnata* = referring to a feather; *-ella* = diminutive; alluding to the small, regularly pinnate secondary shoots arising from a creeping primary stem.

Plagiobryum (Bryaceae) Lindb.
Gk. *plagios* = oblique (angle), slanting; *bryon* = lichen, moss; referring to the incurved capsule.

Plagiochasma (Aytoniaceae) Lehm. & Lindenb.
Gk. *plagios* = oblique (angle), slanting; *chasme* = gaping; alluding to the fissure in the side of the carpocephalon in which the archegonia lie.

Plagiochila (Plagiochilaceae) (Dumort.) Dumort.
Gk. *plagios* = oblique, slanting; *cheilos* = lip, edge.

Plagiomnium (Mniaceae) T.J.Kop.
Gk. *plagios* = oblique; *mnion* = moss, seaweed; alluding to the presence of plagiotopic stolons in most species, hence differentiating it from *Mnium*.

Plagiopus (Bartramiaceae) Brid.
Gk. *plagios* = oblique; *-pus, pous* = foot; referring to the curved seta.

Plagiothecium (Plagiotheciaceae) Bruch & Schimp.
Gk. *plagios* = oblique (angle), slanting; *thekion* = small case, capsule; referring to the capsule.

Platyhypnidium (Brachytheciaceae) M.Fleisch.
Gk. *platys* = broad, flat; *hypnon* = moss; *-idion* = diminutive; alluding to the prostrate spreading habit.

Platyneuron (Dicranaceae) (Cardot) Broth.
Gk. *platys* = broad, flat; *neuron* = nerve, vein; alluding to the broad veins or nerves.

Plaubelia (Pottiaceae) Brid.
For Julius August Plaubel (f 1828–1834), German mycologist and homoeopathist of Gotha in Thuringa. He was a contributor to the 50th anniversary of the founding of homeopathy by German physician Christian Friedrich Samuel Hahnemann (1755–1843).

Pleuridium (Ditrichaceae) Rabenh.
Gk. *pleuridion* = on one side; the sporophytes often emerge laterally from the perichaetium.

Pleuropus = Homalothecium
(Brachytheciaceae) Griff.
Gk. *pleuron* = side, rib; *-pus* = foot (*pous* = a foot) referring to the deeply plicate leaves. Meaning unclear.

Plicanthus (Jungermanniaceae) (F.Weber) R.M.Schust.
La. *plica* = fold, pleated; *anthos* = flower; alluding to the strongly plicate perianth.

Pogonatum (Polytrichaceae) P.Beauv.
Gk. *pogonatos* = bearded; *arthron* = joint; alluding to the densely hairy calyptra.

Pohlia (Bryaceae) Hedw.
For Johann Ehrenfried Pohl (1746–1800), German botanist and pathologist, son of Johann Christoph Pohl (1720–1780), professor of pathology at Leipzig University. Pohl (jnr) studied medicine at Leipzig from 1763 to 1769, qualifying as a physician, became associate professor of botany in 1773, and subsequently also acting director of the Leipzig Botanical Gardens in 1789. In that year, he was made associate professor of pathology, and of therapy in 1796. He was a member of the Economic Society of Leipzig, and a member of the Roman Imperial Academy of Sciences.

Polytrichastrum (Polytrichaceae) G.Sm.
Polytrichum (q.v.); *-astrum* = incomplete resemblance.

Polytrichum (Polytrichaceae) Hedw.
Gk. *polys* = many; *trichum* = hair; alluding to the hairy calyptra.

Porella (Porellaceae) L.
Gk. *poros* = pore, pathway; *-ella* = diminutive; referring to a liverwort with pores in the side.

Porothamnium (Neckeraceae) M.Fleisch.
Gk. a combination of *Porotrichum* (q.v.) and *Thamnium* (q.v.).

Porotrichum (Neckeraceae) (Brid.) Hampe
Gk. *poros* = perforation, passage, tube; *triche* = hair, hairy; alluding to the perforated processes of the inner peristome.

Pottia (Pottiaceae) Ehrh. ex Fürnr.
For Johann Friedrich Pott (1738–1805), German botanist, a professor of botany in Braunschweig, Germany, personal physician to the Duke of Brunswick, and 'correspondent with [Carl] Linnaeus (one letter, 1776)'. In 1800, he was made dean of Furstl Medical College. He maintained an extensive herbarium of vascular plants, which was purchased by the Botanical Museum of St Petersburg (currently the Komarov Botanical Research Institute) in 1826.

Pseudocrossidium (Pottiaceae) R.S.Williams
Gk. *pseudo-* = false; resembling but not the real *Crossidium* (q.v.).

Pseudoleskea (Leskeaceae) Bruch & Schimp.
Gk. *pseudo-* = false; resembling but not the real genus *Leskea*. For Nathanael Gottfried Leske  (1751–1786), German naturalist, economist and geologist. He studied at the Bergakademie of Freiberg, Saxony, under Abraham Gottlob Werner (1749–1817), and the Franckeschen Stiftungen in Halle. He became professor at the University of Leipzig, teaching natural history in 1775 and economics from 1777–1786, when he accepted the chair of financial science and economics at Marburg University, but died shortly thereafter in an accident. His large mineral and natural history collection was sold after his death to the Dublin Society in 1792, and is now housed in the National Museum of Ireland. He wrote on diverse topics, and was a co-editor of *Leipziger Magazin zur Naturkunde, Mathematik und Oekonomie* from 1781.

Pseudoleskeopsis (Leskeaceae) Broth.
Gk. *Pseudoleskea* (q.v.); *-opsis* = resembling.

Pseudoscleropodium (Brachytheciaceae) (Limpr.) M.Fleisch.
Gk. *pseudo-* = false or resembling; *sclero*, from *skleros* = hard; *podion* = diminutive of *pous* = foot, stalk or pedicel; resembling but not the real genus *Scleropodium*.

Pseudosymblepharis (Pottiaceae) Broth.
Gk. *pseudo-* = false or resembling; *syn* = together, united; *blepharis* = eyelash. False, the genus *Symblepharis*; referring to the peristome teeth.

Psiloclada (Lepidoziaceae) Mitt.
Gk. *psilos* = bare; *klados* = branch or shoot; referring to the naked branches or shoots.

Psilopilum (Polytrichaceae) Brid.
Gk. *psilos* = bare; *pilos* = felt hat; alluding to the lack of hairs on the calyptra.

Pterobryopsis (Pterobryaceae) M.Fleisch.
Gk. *pteron* = wing, feather; *bryon* = moss, liverwort; *-opsis* = resembling; referring to a moss that grows on trees that look like ferns.

Pterogoniadelphus (Leucodontaceae) M.Fleisch.
Gk. *pteron* = a wing; *gonia* = angle, corner; *adelphus* = brother; referring to the moss's close relationship to *Pterogonum*.

Hornwort (EJM)

Pterogonium (Leucodontaceae) Sw.
Gk. *pteron* = a wing; *gonos, gonium* = seed; referring to the winged seed.

Pterygoneurum (Pottiaceae) Jur.
Gk. *pteros* = wing; *neuron* = nerve; referring to the ridge-like lamellae of the adaxial surface of the costa.

Ptychanthus (Lejeuneaceae) Nees
Gk. *ptychos* = folded or pleated; *anthos* = flower; referring to the deeply and multiply pleated (keeled) perianth.

Ptychomitriopsis (Ptychomitriaceae) Dixon
Ptychomitrium (q.v.); Gk. *-opsis* = resembling.

Ptychomitrium (Ptychomitriaceae) Fürnr.
Gk. *ptychos* = pleat, fold; *mitrion* = little cap, headband; alluding to the plicate calyptrata.

Pyrrhobryum (Rhizogoniaceae) Mitt.
Gk. *pyrros* = flaming, fire-coloured, red; *bryon* = lichen, moss; presumably alluding to the peristome.

Q

Quathlamba (Bartramiaceae) Magill
African language Sotho name for Drakensberg, referring to a mountain range in South Africa where this moss was found.

R

Racomitrium (Grimmiaceae) Brid.
Gk. *rhakos* = rough, ragged, frayed; *mitrion* = little cap; alluding to the calyptra, which in some species is split all around the base.

Racopilum (Racopilaceae) P.Beauv.
Gk. *rhakos* = rough, ragged, frayed; *pilos* = felt hat; alluding to the ragged and hairy cap especially in those species with a mitrate, lobed calyptra.

Radula (Radulaceae) (L.) Dumort.
La. *radula* = a scraper; referring to the flattened and truncate perianth, very much like a scraper for removing paint.

Rauia (Thuidiaceae) Austin
For Ambrosius Rau (1784–1830), German botanist, mineralogist, naturalist and plant collector. He was professor of natural history, forestry and rural economy at the University of Würzburg from 1809 until his death, and author of *Enumeratio Rosarum circa Wirceburgum* (*List of Roses around Wirceburgum*) (1816), *Lehrbuch der Mineralogie* (*Textbook of Mineralogy*) (1818–1826), and *Bemerkungen über das Naturhistorische Mineral-System des Herrn Friederich Mohs* (*Remarks on the Natural History Mineral System of Friederich Mohs*) (1821).

Rauiella (Thuidiaceae) Reimers
For Eugene Abraham Rau (1848–1932), US pharmacist and bryologist, authority on mosses and fungi. He assembled large collections for the American Museum of Natural History, New York. He co-authored with Alpheus Baker Hervey *A Catalogue of North American Musci* (1880). He was elected a correspondent of the Academy of Natural sciences in 1883.

Rectolejeunea (Lejeuneaceae) A.Evans
La. *rectus* = fragile; *Lejeunea* (q.v.); alluding to the leaves of the species, which are easily broken off.

Renauldia = ***Calyptothecium***
(Pterobryaceae) Müll.Hal. in Renauld
For Ferdinand François Gabriel Renauld (1837–1910), French bryologist and army officer who joined the cavalry 1856, and became commander of the Palace of Monaco in 1888. He learned botany through exchanging plants with other botanists, and began a 40-year-long study of bryology in 1870, corresponding with leading botanists such as WP Schimper and PT Husnot. Among his research, he examined the geographical distribution of bryophytes in Forcalquier and Chain Lure (1876), studied the bryological flora of Madagascar and its neighbouring islands (1877), catalogued the vascular plants and mosses that grow spontaneously in the Haute-Saone (1883), and authored *Mousses Nouvelles de l'Amérique du Nord* (1888–1895) and *Histoire Naturelle des Plantes Mousses* (1898–1915) with Jules Cardot (1860–1934).

Rhabdoweisia (Dicranaceae) Bruch & Schimp.
Gk. *rhabdos* = a rod, staff; *Weissia* (q.v.); presumably referring to the ribbed capsule, and the pottiaceous genus *Weissia*, which it resembles.

Rhachithecium (Rhachitheciaceae) Broth. ex Le'Jol.
Gk. *rhakhis* = ridge; *theka* = case, capsule; referring to the ribbed capsule.

Rhacocarpus (Hedwigiaceae) Lindb.
Gk. *rhakos* = a scrap of clothing, frayed, ragged; *karpos* = fruit; alluding to the raggedly split base of the calyptra.

Rhacopilopsis (Hypnaceae) Renauld & Cardot
Gk. *rhakos* = frayed, ragged; *pilos* = felt hair; *-opsis* = resembling; referring to the ragged, hairy calyptra.

Rhaphidorrhynchium (Sematophyllaceae) Besch. ex M.Fleisch.
Gk. *rhaphis, rhaphid* = needle-like; *rhynchos* = nose, beak; alluding to the long, narrow rostrum on the operculum.

Rhizogonium = ***Pyrrhobryum***
(Rhizogoniaceae) Brid.
Gk. *rhiza* = root; *gonium* = in this instance, the sporophytes; alluding to the fact that the sporophytes appear to arise from the 'root' of the plant.

Rhodobryum (Bryaceae) (Schimp.) Limpr.
Gk. *rhodon* = rose; *bryon* = lichen, moss; referring to the terminal rosette of leaves in most of the species.

Rhynchostegiella (Brachytheciaceae) (Schimp.) Limpr.
Gk. *Rhynchostegium* (q.v.); *-iella* = diminutive.

Rhynchostegium (Brachytheciaceae) Bruch & Schimp.
Gk. *rhynchos* = beak; *stegium* = a roof or covering; alluding to the long beaked operculum.

Riccardia (Aneuraceae) Gray
For one or more members of the Riccardi family, Ottavio Riccardi, Francesco Riccardi (1648–1718), Cassandra Capponi Riccardi, Cosimo Riccardi (1671–1751) and Vincenzio Riccardi (1704–1752), all of whom are mentioned in Pier (Pietro) Antonio Micheli's *Nova Plantarum Genera* (1729) as having supported the work. Some sources suggest the genus was named after Vincento Riccardi, a resident of Florence during the early 18th century, while Samuel Frederick Gray suggests it was most likely Octavio Riccardi.

Riccia (Ricciaceae) L.
For Pietro Francisco Ricci, Florentine botanist and 18th-century Italian senator, a member of the Botanical Society in Florence (1729) who left some of his works to the Academy of Florence.

Ricciocarpos (Ricciaceae) Corda
Riccia (q.v.); *karpos* = fruit; alluding to the sporophytes, which resemble those of *Riccia*.

Riella (Riellaceae) Mont.
For Michel-Charles Durieu de Maisonneuve (1796–1878), French soldier and botanist (*Riella* = diminutive form of 'Durieu'). He studied at the Military School of Saint-Cyr and received the rank of sub-lieutenant. He worked in the French army

until 1848 taking part in the battles of Trocado against Spain (1823), and of Smala against the Algerians (1843). In the mid-1820s, he developed an interest in natural history and botany, and went on the Morea expedition to Greece with naturalist Jean Baptiste Bory de Saint-Vincent (1778–1828). He also visited France, Spain and northern Portugal, and studied cryptogams with the top botanists of the day. In 1858, he succeeded Jean François Latterrade (1784–1858) as director of the botanical garden at Bordeux, and was a professor of botany at the University of Bordeaux (1867–1877).

Rigodium = Heterocladium (Thuidiaceae) Kunze ex Schwägr.
Gk. *rigos* = rigid; referring to the extremely wiry and stiff stems and branches.

S

Saelania (Ditrichaceae) Lindb.
For Anders Thiodolf Saelan (1834–1921), Finnish psychiatrist and botanist. He studied physical mathematics at the University of Helsincki (1856) and obtained his PhD with a dissertation entitled *Suicide in Finland in Statistical and Forensic Purposes* (1865). He was an assistant from 1859 to 1866 at the Helsinki Museum. He compiled a flora of eastern Finland, cotyledons and ferns, drew up a list of the botanical collections at the museum with William Nylander (1859), and with Elias Lönnrot compiled the first Finnish flora in the Finnish language (1866). From 1868 to 1904, he was physician at Lapinlahti Central Institution for the insane, and was the nation's leading authority on the subject. In 1916, he published a complete bibliography of Finland's botanical literature up to 1900.

Sanionia (Amblystegiaceae) Loeske
For Carl Gustav Sanio (1832–1891), German botanist. He commenced his studies in 1852 and received his doctorate in medicine and botany at the University of Königsberg in 1858. He lectured at the University of Königsberg from 1858 to 1866, and from 1866 was an independent scholar in Lyck. He made the first scientific description of compression wood, then was an independent researcher in Lyck, where he worked on floristics and wood anatomy.

Scapania (Scapaniaceae) (Dumort.) Dumort.
Gk. *scapanion* = spade or hoe; alluding to the flattened, truncate perianth.

Schiffneriolejeunea (Lejeuneaceae) Verd.
Probably for Victor Félix Schiffner (1862–1944), French botanist, in particular bryologist. He studied at the University of Prague, collected liverwort specimens in Java and Sumatra (1893–1894), thereafter becoming professor of botany at the University of Prague (1895). He collected bryophytes in Brazil in 1901, and on his return was appointed a professor at the University of Vienna (1902–1932). He authored *Hepaticae* in Adolf Engler and Karl von Prantl's *Die Natürlichen Pflanzenfamilien* (1893), and *Die Hepaticae der Flora von Buitenzorg* (*Liverwort Flora of Java*) (1900), but gave up working on exotic liverworts after the first instalments of *Species Hepaticarum* appeared and turned to the European flora. An unfinished manuscript on the liverworts of Brazil, based on his travels in 1901 in the framework of the *Botanischen Expedition der kaiserlichen Akademie der Wissenschaften nach Südbrasilien*, was completed by Sigrid Arnell (in 1964).

Schisma (Herbertaceae) Dumort.
Gk. *schisma* = splitting, cleft, division. Probably relating to the bifidness of the bryophyte's leaves.

Schistidium (Grimmiaceae) Brid.
Gk. *schistos* = divided; *-idion* = diminutive; alluding to the splitting of the calyptra at its base.

Schistochila (Schistochilaceae) Dumort.
Gk. *schistos* = split, cut, divided; *cheilos* or *chilos* = lip; alluding to the laciniate mouth of the perianth.

Schistomitrium = Leucobryum
(Dicranaceae) Dozy & Molk.
Gk. *schistos* = split, cut, divided; *mitra* = headband.

Schizomitrium = Callicostella
(Hookeriaceae) Bruch, Schimp. & W.Gümbel
Gk. *schizo* = split; *mitra* = headband.

Schlotheimia (Orthotrichaceae) Brid.
For Ernst Friedrich von Schlotheim (1764–1832), German paleobotanist. He studied at Göttingen and Freiberg (Saxony), a student of Abraham Werner, one of the founding fathers of geology. He gathered an extensive collection of the remains of carboniferous plant fossils, which he described in his illustrated book *Ein Beitrag zur Flora der Vorwelt* (1804). His major work, *Die Petrefactenkunde* (1820), supplemented by a folio atlas (1822), illustrates the 'petrified and fossil remains of the animal and vegetable kingdom of a former world'. He held various posts: privy councillor; president of the chamber at the Court

Moss on a stone wall (EJM)

of Saxony; and from 1822, curator of the library, art, and natural history collections of the duke of Saxony in Gotha.

Schoenobryum (Cryphaeaceae) Dozy & Molk.
Gk. *schoinos* = a rush, reed; *bryon* = moss; alluding to the cord-like appearance of the stems, especially when dry (David Meagher).

Schwetschkea (Fabroniaceae) Müll.Hal.
For Carl Gustav Schwetschke (1804–1881), German book publisher and politician. He attended the Classical Gymnasium in Halle and studied Latin and Greek at the universities of Heidelberg and Halle. He first worked in his family's book and publishing house, edited the newspaper *Der Hallische Courier* from 1828–1843, and was politically active between the 1840s and 1870s. He was a cofounder of the journal *Nature* (1852), published among other scholarly works the journal *Allgemeine Monatsschrift für Literatur*, and wrote a Latin epic in honour of Otto von Bismarck, *Bismarkias*, (1867). He was a dedicated city councillor in Halle (Saale), and a member of the Frankfort National Assembly.

Sciaromium = Vittia (Amblystegiaceae) (Mitt.) Mitt.
Gk. *skia* = shadow, pertaining to shadow or darkness; *aroma* = fragrance, scent.

Sematophyllum (Sematophyllaceae) Mitt.
Gk. *semato-* = marked; *phyllon* = a leaf; referring to the distinctive alar cells that are typically found at the leaf base in some mosses, and are a distinguishing feature of this genus.

Semibarbula = Barbula (Pottiaceae) Herzog ex Hilp.
La. *semi* = partly; *barbus* = beard; *-ula* = diminutive. Some of the mosses have small, bushy growth 'beards'.

Sendtnera (Herbertaceae) Endl.
For Otto Sendtner (1813–1859), German bryologist and vegetation scientist. He studied under Karl Friederich Schimper (1803–1867), a pioneer in the field of plant morphology, at the University of Munich. Among his achievements, he was appointed curator of the Leuchtenbergsche Naturalienkabinett in Eichstätt (1841); went on a botanical expedition with Mutius von Tommasini (1794–1879) through Istria and Tyrol (1843); was one of the first botanists to study the flora of Bosnia, collecting monocotyledons, algae, mosses and ferns (1857); and conducted phytogeographic studies in Bavaria. He became an associate professor (1854) then second chair of botany, as well as first curator of the herbarium at the University of Munich (1857).

Solenostoma (Jungermanniaceae) Mitt.
Gk. *solen* = pipe, tube; *stoma* = mouth; referring to the perianth's mouth which forms a tubular beak.

Sphaerocarpos*** (Sphaerocarpaceae) Boehm
La. *sphaer-, sphaero-* = spherical, globe-shaped; *carpos* = fruit; referring to the shape of the fruit.

Sphaerothecium (Dicranaceae) Hampe
La. *sphaer-, sphaero-* = spherical, globe-shaped; *theka* = case, capsule. The capsules are ovoid (egg-shaped) rather than sub-globose (almost round).

Sphagnum = Campylopus (Sphagnaceae) L.
Gk. *sphagnos* = a variety of peat or bog moss. Originally a Greek name for an unknown plant.

Splachnobryum (Pottiaceae) Müll.Hal.
Gk. *splachnos* = viscera (soft, sap-like); *bryon* = moss; alluding to the rugose appearance of the apophysis when dry.

Sphenolobus = Anastrophyllum (Anastrophyllaceae) Berggr.
Gk. *spheno-* = wedge-shaped, sphenoid; *lobos* = lobe or pod. The leaf lobes are somewhat wedge-shaped.

Squamidium (Meteoriaceae) (Müll.Hal.) Broth.
Gk. *squama* = a scale or structure resembling a scale; *-idion* = diminutive. Possibly for its scaly leaves.

Stephaniella (Gymnomitriaceae) Jack
For Franz Stephani (1842–1927), German salesman and self-taught bryologist specialising in liverworts. How his interest in bryology was sparked is unknown as he had no formal training in this field, but may have had help from a friend, JB Jack, a botanist from southern Germany. He began publishing papers on the subject of liverworts in 1876. In 1880, he travelled through Europe and North America to improve his commercial and linguistic skills. His major work, the six-volume *Species Hepaticarum*, (1898–1925), lists some 10 000 species of liverworts and hornworts, with more than 4 000 new ones described by Stephani. He published more than 200 articles describing species from six continents. After his death, critics have 'condemned' his work for its poor quality, validity and taxonomic inaccuracies, perhaps as a result of a progressive brain disease that affected him in his later years.

Stephanocoma*** Berkheya (Asteraceae) L.
Gk. *stephanos* = crown; *coma* = head of hair; alluding to the crown-like pappus (William Henry Harvey, *Flora Capensis*).

Stereophyllum (Stereophyllaceae) Mitt.
Gk. *stereos* = stiff, solid; *phyllon* = a leaf; presumably alluding to the stiffness of the leaves.

Stoneobryum (Orthotrichaceae) D.H.Norris & H.Rob.
For Ilma Grace Stone (1913–2001), Australian botanist and bryologist. She studied botany at the University of Melbourne, completing her MSc in 1933. In 1936, she married and spent the next 20 years away from research, raising her family. From 1957, she worked as a part-time demonstrator and researcher at the University of Melbourne. She obtained her PhD in 1963, and was offered a full-time position. She retired in 1978, but was made a senior associate and later associate professor of the university's department of botany. She published 70 papers and authored *The Mosses of Southern Australia* (1976) with Dr George Scott. She was made an honorary member of the British Bryological Society in 1982.

Strepsilejeunea (Lejeuneaceae) Steph.
Gk. *strephein* = to turn; *streptos* = twisted; Lejeunea (q.v.).

Streptocalypta (Pottiaceae) Müll.Hal.
Gk. *streptos* = bent, twisted; *calypta* = veil, lid, hood; veiled with a twisted calypta.

Streptopogon (Pottiaceae) (Taylor) Wilson ex Mitt.
Gk. *streptos* = twisted; *pogon* = bearded. The flowers have twisted petals.

Stylolejeunea (Lejeuneaceae) Sim
Gk. *stylos* = style; Lejeunea (q.v.).

Symphyogyna (Pallaviciniaceae) Nees & Mont.
Gk. *symphusis* = growing together; *gyna* = female, woman (pistil). Female plants may grow in colonies together, although male and female plants may also be intermixed.

Symphyomitra (Acrobolbaceae) Spruce
Gk. *symphusis* = growing together; *mitra* = mitre, headband. Meaning unclear.

Syntrichia (Pottiaceae) Brid.
Gk. *syn-* = together, joined; *trichos* = hair; referring to the attachment of the bases of the hair-like peristome teeth in the inner membrane.

Syrrhopodon (Calymperaceae) Schwägr.
Gk. *syrrhopos* = close together; *odon* = tooth; alluding to the connivent teeth of the peristome.

T

Targionia (Targioniaceae) L.
Named for Dr Cyprian Targioni (f 1720s), Florentine physician (in *Nova Plantarum Genera juxta Tournefortii Methodum Disposita* (1729) by Pier Micheli (1679–1737)), professor of botany at the University of Padua, for Targioni's contribution to Italian natural history, knowledge of herbal medicines, and especially for his museum. This name was subsequently reconfirmed by Linnaeus in his *Species Plantarum* (1753). This genus was not, as is often claimed, named after Italian physician and botanist Giovanni Targioni-Tozzetti (1712–1783), a pupil of Pier Micheli and later curator of the botanical garden and professor of botany at the University of Florence; nor two other Florentine botanists, John Anthony Targioni (dates unknown) and John Targioni Tozzetti, who published in 1734 a work showing the benefits of botanical lectures when studying medicine.

Taxilejeunea (Lejeuneaceae) (Spruce) Schiffn.
Gk. *taxis* = arrangement or *taxus* = yew; *Lejeunea* (q.v.).

Tayloria (Splachnaceae) Hook.
For Thomas Taylor (1775–1848), Irish physician, professor of botany and natural history at the Royal Cork Scientific institution. He qualified with an MB and MD (1814) from Trinity College, Dublin, was elected a Fellow of the King and Queen's College of Physicians and worked at Sir Patrick Dun's Hospital, Dublin. A keen bryologist, he prepared *Muscologia Britannica* (1818) with William Jackson Hooker, the lichens section for James Townsend Mackay's *Flora Hibernica* (1836), and much of the cryptogamic material in Hooker's *Flora Antarctica* (1844), in addition to many other publications. He became a Fellow of the Linnaean Society (1814) and honorary member of the Royal Irish Academy. His 8 000-sheet herbarium was purchased for the Boston Society of Natural History.

Telaranea (Lepidoziaceae) Spruce
La. *tela* = web; *aranea* = spider or spider's web; alluding to the delicate, cobweb-like appearance of the plants.

Tetrapterum (Pottiaceae) Hampe ex A.Jaeger
Gk. *tetra* = four; *pteron* = wing, feather. The spore capsule was regarded as quadrangular rather than rounded in section.

Thamnium (Thamnobryaceae) Schimp.
Gk. *thamnion* = bush, shrub.

Thuidium (Thuidiaceae) Bruch & Schimp.
Gk. *Thuja* (q.v.); *-idion* = diminutive; alluding to the resemblance of the feathery, branched fronds to the foliage of those trees.

Thysanomitrion = Campylopus (Dicranaceae) Schwägr.
Gk. *thysanos* = tasselled or tufted; *mitrios* = cap. Meaning unclear.

Timmiella (Pottiaceae) (De Not.) Limpr.
For Joachim Christian Timm (1734–1805), German botanist, bryologist and burgermeister of Malchin, Mecklenberg, Germany. He studied pharmacy for five years under John Frederick and became his assistant for one year. He became council pharmacist in Malchin in 1760, and became involved in local government matters, being elected a senator in 1771 and mayor in 1790. As a pharmacist, he had a keen interest in botany and collected plants of all kinds, especially cryptogams, primarily in the area around Malchin. He wrote *Florae Megapolitanae Prodromus* (1788), and is regarded as a pioneer of modern botany in Germany.

Tortella (Pottiaceae) (Lindb.) Limpr.
La. *torta* = twisted; *-ella* = diminutive; referring to the characteristic twisting of the long peristome teeth.

Tortula (Pottiaceae) Hedw.
La. *tortula* = slightly twisted; referring to the spiral twisting of the peristome teeth.

Trachyphyllum (Balantiopsidaceae) A.Gepp
Gk. *trachys* = rough; *phyllon* = a leaf; alluding to the papillose surface of the leaves, with papillae at each end of the cells.

Trachypodopsis (Meteoriaceae) M.Fleisch.
Gk. *trachys* = rough; *pous*, *pod* = foot; *-opsis* = resembling.

Trachypus (Meteoriaceae) Reinw. & Hornsch.
Gk. *trachy-* = rough; *pous* = foot; alluding to the markedly papillose seta.

Trematodon (Bruchiaceae) Michx.
Gk. *tremato* = perforated, opening, hole; *odon* = toothed; alluding to the perforated peristome teeth.

Trichosteleum (Sematophyllaceae) Mitt.
Gk. *tricho-* = hair; *stele* = pillar, column; alluding apparently to the long, slender seta.

Trichostomopsis* = *Didymodon (Pottiaceae) Cardot.
Gk. *tricho-* = hair; *stoma* = mouth; *opsis* = resembling.

Trichostomum (Pottiaceae) Bruch
Gk. *tricho-* = hair; *stoma* = mouth; alluding to the peristome of filiform teeth.

Triquetrella (Pottiaceae) Müll.Hal.
Gk. *triquetris* = triangular; *-ella* = diminutive. The leaves are in three distinct rows on the stem – either straight or weakly spiralling in either direction.

Tristichium (Ditrichaceae) Müll.Hal.
Gk. *tri-* = three; *stichos* = row, rank; referring to the leaves of this moss, which are in three rows: two lateral rows of larger leaves, and a third of smaller leaves.

Tritomaria (Jungermanniaceae) Schiffn. ex Loeske
Gk. *Tritoma* (q.v.); *aria* = related to. Leaves usually three-lobed, with (two to four) lobes acute to sub-acute.

Tylimanthus (Acrobolbaceae) Mitt.
Gk. *tylos* = lump, swelling; *anthos* = flower. Members of this liverwort family have their leaves in two rows and produce sporophytes from a fleshy pouch (lump, swelling) called a marsupium.

U

Ulota (Orthotrichaceae) D.Mohr.
Gk. *oulote* = something curled; referring to some species' strongly curled leaves when dry.

V

Vesicularia (Hypnaceae) (Müll.Hal.) Müll.Hal.
La. *versicula* = small blister; *aria* = pertaining to; alluding to the lax areolation of the leaf, consisting of short, broad cells suggestive of inflated vesicles (or bladders).

Vittia (Amblystegiaceae) Ochyra
For Dale Hadley Vitt (1944–), US bryologist, plant collector and peatland expert. He obtained a PhD from the University of Michigan (1970) and became a biology professor at the University of Alberta (1970–2000). He was director of the Devonian Botanical Garden from 1990 to 2000, and was chair of the department and professor of plant biology at Southern Illinois University from 2000 to 2011. Currently, he is professor emeritus and research professor at Southern Illinois University. He specialises, *inter alia*, in moss systematics, community dynamics, and the development of peatlands. He has authored four books and more than 250 papers, and carried out plant collections in many countries. He was editor-in-chief of *The Bryologist* from 1994–2004, and received the Outstanding Scholar Award from Southern Illinois University in 2010.

Vrolijkheidia (Pottiaceae) Hedd. & R.H.Zander
For the Vrolijkheid Nature Reserve in the Little Karoo near Robertson, presumably where this plant was discovered.

W

Webera* = *Pohlia (Bryaceae) Hedw.
For Georg Heinrich Weber (1752–1828), German physician and botanist. He obtained his medical doctorate in 1777 and, in 1780, became professor of medicine and botany at the University of Kiel (1780). He started a polyclinic, which became the Institute of Christian-Albrechts-University (1802), developed the university's botanical garden and, in 1810, was appointed director of the Schleswig-Holstein Medical College in charge of the combined hospitals and the botanical garden. He published extensively on botany, entomology and medicine, including lichens as a life form, the healing-scientific use of mosses and ferns, the Flora of Göttingen, and African medicine. For his research, publications and humanitarian achievements, he was awarded an honorary PhD by the Göttingen Philosophical Faculty in 1824.

Weisiopsis (Pottiaceae) Broth.
For *Weissia* (q.v.); *-iopsis* = resembling. Weiss was originally also spelled 'Weis'.

Weissia (Pottiaceae) Hedw.
For Friedrich Wilhelm Weiss (1744–1826), German physician and botanist. He studied medicine at the University of Göttingen and obtained his doctorate in 1769, his dissertation was titled *Plantae Florae Crypotogamae Gottingensis* (1770), which he significantly expanded. From 1769–1784, he was physician to the count and landgrave of Hessen-Rothenburg, and a councillor in Rotenburg. His main interest was bryology and lichenology. He wrote the first volume of a draft book on *Entwurf einer Forstbotanik* (*Forest Botany*) in 1775. He was a correspondent of Albrecht von Haller

Wijkia (Sematophyllaceae) H.A.Crum.
For Roelof van der Wijk (1895–1981), Dutch bryologist, collector, and professor of plant systematics and plant morphology at the University of Gronigen (1947–1965). He had a lifelong interest in bryophytes, especially mosses of the Malesian region. In 1952 and 1972, he collected mosses and liverworts in west and east Java and Sumatra, and described many new species. His major publication was *Index Muscorum* (five volumes) (1959–1969), with the assistance of WD Margadant and PA Florschutz (1959, 1962). Other works include *A Preliminary Key to the Moss Genera* (1958), *A Preliminary Key to the Genera of Indian Mosses* (1966), *Glossary of Terms Relevant to the Morphology* (1981) and dozens of other papers.

Z

Zygodon (Orthotrichaceae) Hook. & Taylor
Gk. *zygos* = yoke; *odon* = toothed; alluding to the peristome teeth, some 16, in pairs, as if yoked together.

Lichens

How come lichens are included in this book of 'plant' names?*

Hugh Clarke answers this question:

Most readers of this book will have a keen interest in trees and shrubs (woody plants). Others may have an interest in flowers, grasses and ferns (herbaceous plants), and still others are fascinated by the tiny mosses and lichens.

Lichens? Why are they in this book, and are they plants? A definition of a plant from Wikipedia is: 'A plant is a living organism of the kind exemplified by trees, shrubs, herbs, grasses, ferns, and mosses, typically growing in a permanent site, absorbing water and inorganic substances through its roots (and rhizoids in Bryophytes), and synthesizing nutrients in its leaves (thallus in Bryophytes) by photosynthesis using the green pigment chlorophyll.' So the simple answer is that lichens do not fit into this category. However, in the first database I was given they were there, and at that time I did not realise they should be excluded – so this is a bonus for those readers, of which there are many, who have an interest in lichens.

Fruticose

Foliose

Flakes

The disqualifier for lichens being plants is that they do not have roots that absorb water and nutrients as plants do, nor do they have stems and leaves. When one examines a lichen you may see what look like tiny leafless branches (fruticose), flat leaf-like structures (foliose) or crusts that lie on the surface like paint (crustose) – but these are in fact the amalgam of an intimate partnership of a fungus and an alga. Examples of these are shown in the images on this page.

Although there are about 20 000 lichens in the world, I could find virtually no source that explains the scientific meaning of lichen names and, more challengingly, what this name implies. But thanks to publications written by David Meagher (Australia), Gerould Wilhelm & Laura Rericha (USA) and personal correspondence with Matthias Schultz (Germany), Shirley Tucker (USA) and Alan Fryday (USA), I was able to establish many of the answers. Any errors in the text or wrongful interpretations are attributable entirely to the authors.

For those readers who have a special interest in lichens and did not expect to find lichens in a plant names book, we hope this is a surprise addition to the book's content.

---

* The meanings of lichens are exceptionally difficult to determine (especially for non-lichenologists). Where a first name is replaced by a second name, the change is a result of synonymy, and the second name seems to be in current use. Any updated corrections to gascoyne@mweb.co.za would be welcomed.

A

Acarospora (Acarosporaceae) A.Massal.
La. *acaro* = mite (-sized); *spora* = spore; referring to the hundreds of spores in each ascus, compared to eight in most lichens.

Acolium = Cyphelium (Physciaceae) (Ach.) Gray
Gk. *a* = without; *cola* = to inhabit, dweller. Perhaps a non-endemic species.

Acrorixis = Thelenella (Thelenellaceae) Trevis.
Gk. *akro* = at the tip, end; *rhexis* = rupture. Meaning unclear. Possibly because the thick thallus becomes cracked-areolate and warted.

Acroscyphus (Physciaceae) Lév.
Gk. *akro* = at the tip, end; *scyphus* = cup; referring to the position of the ascomata.

Actinoplaca (Gomphillaceae) Müll.Arg.
Gk. *actis, actinos* = rayed, star-like, radiating from a centre; *plakos* = a flat, round plate, or disc (can also mean 'spot'); referring to the star-like radiating margin of the otherwise tightly adpressed, spot-like lichen body (thallus) (Matthias Schultz).

Agyrophora = Umbilicaria (Umbilicariaceae) (Nyl.) Nyl.
Gk. *a-* = without; *gyros* = marked with a sinuous line; *phora* = carrying, bearing. The lichen has non-gyrose ascocarps as opposed to the closely related *Gyrophora*, which does. Both genera are now included in *Umbilicaria*.

Alectoria (Parmeliaceae) Ach.
Gk. *alektor, alektruon* = a cock, chicken; from resemblance of the lichen thallus to the tail feathers of a rooster.

Allarthothelium (Arthoniaceae) (Vain.) Zahlbr.
Gk. *allos* = different; *arthos* = for a fleck or speckles; *thelium* = little nipples, warts; referring to the spot-like ascocarps.

Almbornia (Parmeliaceae) Essl.
For Dr Ove Almborn (1914–1992), Swedish botanist, lichenologist, and plant collector in Europe (Denmark, France and Sweden) and Africa (Tunisia, Mozambique, Namibia and South Africa). He was senior lecturer in the department of systematic botany at the University of Lund, and director of the Botanical Museum in Lund (1966–1979). He spent considerable time researching in South Africa, including five months in 1953. He published many papers relating to Southern African lichens, including *Revision of Some Lichen Genera* (1966), *Lichens at High Altitudes* (1987), *Distribution Patterns in the Lichen Flora* (1988) and *Revision of the Lichen Genus Teloschistes in Central and Southern Africa* (1989). He also made a substantial contribution to European lichen research and literature.

Amandinea (Caliciaceae) Choisy ex Scheid. & H.Mayrhofer
For Madame Amandine Manière, possibly born 1937, a friend of the original genus author, French mycologist and lichenologist Maurice Gustave Benoit Choisy (1897–1966). GB Frige states, 'after French botanist St Amand (one half century)'. We could not find a French botanist by that name, but there is a botanical garden in France named Botanical Gardens Saint-Amand-sur-Sèvre, Deux-Sèvres, France.

Amphiloma = Caloplaca (Teloschistaceae) Körb.
Gk. *amphi* = around; *loma* = edge or fringe. Meaning unclear.

Amphoridium (Verrucariaceae) A. Massal.
Gk. *amphora* = a two-handled jar with a narrow neck; *-ium* = diminutive; referring to the pitcher-shaped fruiting body warts.

Anaptychia (Physciaceae) Körb.
Gk. *ana* = throughout; *ptycho* = to fold; probably from the interwoven hyphae of the algal and medullary layers of the upper cortex.

Anthracothecium (Pyrenulaceae) Hampe ex A.Massal.
Gk. *anthrax* = coal; *theke* = a cup; referring to the black, carbonised structure of the thecium, and to the ascomatal wall of the fruit bodies.

Antrocarpum = Thelotrema (Graphidaceae) G.Mey.
Gk. *antron* = cave, cavern or cavity; *karpos* = fruit. Speculatively, the apothecia resemble small volcanoes or barnacles with cavities where the disk would normally be.

Arctomia (Arctomiaceae) Th.Fr.M.
La. *articus,* from La. *arcticus,* from Gk. *arkticos* = of the north, literally, 'of the (constellation) Bear', from *arktos* = bear. The arctomiaceae family is found in arctic and subarctic habitats.

Arthonia (Arthoniaceae) Ach.
Gk. *arthron* = a joint or *arthos* = fleck, speckled; referring to the speckled appearance of the thallus created by the dispersed apothecia.

Arthopyrenia (Arthopyreniaceae) A.Massal.
Gk. *arthros* = a fleck of speckles; *pyren* = grain; referring to the spot-like ascocarps.

Arthothelium (Arthoniaceae) A.Massal.
Gk. *arthros* = a fleck of speckles; *thelium* = little nipples; referring to the spot-like ascocarps.

Arthrorhaphis (Arthrorhaphidaceae) Th.Fr.

Arthrorhaphis citrinella

Gk. *arthron* = a limb or articulation; *rhaphis* = a needle; referring to the long, acicular, transversely septate ascospores of the type species.

Aspicilia (Megasporaceae) A.Massal.
La. *aspicilia* = eyes of the viper; possibly referring to the lidless 'eyes' of the apotheca.

Asterothyrium (Asterothyriaceae) Müll.Arg.
Gk. *astron* = a star; *thoros* = sperm; possibly referring to the star-like appearance of apotheicia, which is somewhat shield-shaped.

Astrothelium (Trypetheliaceae) Eschw.
Gk. *aster* = a star; *thele* = a nipple, female; referring to the star-like arrangement and shape of the ascomata.

Aulaxina (Gomphillaceae) Fée
Gk. *aulax* = a furrow or groove; *-ina* = resembling.

B

Bacidia (Ramalinaceae) De Not.
La. *baculum* = rod; Gk. *eidos* = shape, hence *bacidia* = rod-shaped; referring to elongated spores.

Bacidina (Ramalinaceae) Vezsda
Bacidia (q.v.); *-ina* = diminutive.

Baeomyces (Baeomycetaceae) Pers.
Gk. *baios* = little, small, scanty or few; *mykes* = a fungus or mushroom; referring to the small, mushroom-shaped fruit body (asocarp).

Bathelium = Trypethium (Trypetheliaceae) Trevis.
Gk. *bathos* = deep; *thele* = a nipple; referring to the shape of the ascomata.

Biatora (Ramalinaceae) Fr.
Gk. *biator* = small; referring to the tiny spores.

Biatorella (Biatorellaceae) Th.Fr.
Biatora (q.v.); *-ella* = diminutive; referring to the numerous minute spores.

Biatorina = Catinaria (Ramalinaceae) A.Massal.
La. *Biatora* (q.v.); *-inus* = indicates possession or likeness; resembling the lichen genus *Biatora* in having apothecium without thalline margin, and with soft, almost colourless excipulum.

Blastenia = Caloplaca (Teloschistaceae) A.Massal.
Gk. *blastos* = bud, sprout, embryo; referring to its rapid growth rate.

Blasteniospora = Xanthoria (Teloschistaceae) Trevis.
Gk. *blastos* = bud, sprout, embryo; *spora* = seed; referring to the spouting spores.

Blennothallia (Collemataceae) Trevis.
Gk. *blenno* = slimy; *thallos* = green shoot or twig, from *thallein* = to bloom; possibly referring to its appearance.

Bombyliospora = Megalospora (Megalosporaceae) De Not.
Gk. *bombyx* = bumblebee; *sporos* = spore; referring to the thick spores (Freige).

Borrera = Teloschistes (Teloschistaceae) Ach.
For William J Borrer, the Elder (1781–1862), British botanist and mycologist, horticulturist, plant collector, Fellow of the Royal Society and the Linnaean Society, who contributed many materials to English botany and was the co-author with Dawson Turner of *Lichenographia Britannica* (1830). He was a friend of Sir Joseph Banks and Sir William Hooker, and was widely considered as the father of British lichenology.

Bottaria = Mycroporum (Mycoporaceae) A.Massal.
For Bartholomew Bottari (1732–1789), Italian naturalist. He lived in Chioggia, a town situated on a small island at the southern entrance to the lagoon of Venice, about 25km south of Venice. He

belonged to the Chioggia School of Naturalists. This school flourished in the late 18th and early 19th centuries. Members included Joseph Valentino Vianelli, Stefano Chiereghin, Fortunato Naccari and Giandomenico Nardo. They studied various aspects of the lagoon, its marine life and marine fishermen. Bottaria gathered a collection of 1 200 specimens of plants, many of which were on the coast of Chioggia.

Brigantiaea (Brigantiaeaceae) Trevis.
For Maria II da Gloria Brigantiae et Borboniae, Portugaliae et Algarbiarum Reginae (1819–1853), also known as Maria da Glória de Bragança, Queen II of Portugal. She was a member of the house of Braganza (an important Portuguese noble family whose dynasty was to rule Portugal from 1640–1910). She married twice, first August Herzog von Leuchtenberg (1835) and later Ferdinand II August Prinz von Sachsen-Coburg-Saalfeld (1836). Doctors warned her of the dangers of giving birth nearly every year. However, she ignored the risks that had killed her mother, and said: 'If I die, I die at my post' – and she did in 1853, aged 35, while giving birth to her 11th child, Prince Eugene, who also died. Some sources say after Francesco Briganti (1802–1865), Italian lichenologist, but this is incorrect.

Bryopogon = Alectoria (Parmeliaceae) Th.Fr.
Gk. *bryo* = to blossom, come into bud; *pogon* = beard; referring to the lichen's fine hairs up to one metre long.

Bryoria (Parmeliaceae) Brodo & D.Hawksw.
This name is a combination of two genera – *Bry*opogon and Alect*oria*.

Buellia (Physciaceae) De Not.
For Esperanzo Buelli, a friend of the Italian lichenologist Giuseppe de Notaris (1805–1877). No further details found.

Bulbothrix (Parmeliaceae) Hale.
Gk. *bolbos* = bulb; *thrix* = hair; referring to the swollen (bulbate) cilia. The bulbous hairs along the margins are diagnostic.

Bullatina (Gomphillaceae) Vezda & Poelt
La. *bullatus* = inflated; *-ina* = diminutive. The lichen's surface is covered with irregular bubble-like or blister-like swellings.

Bunodea = Thelidium (Verrucariaceae)
A.Massal.
Gk. *bounos* = a heap or mound; *-odes* = resembling, of the nature of.

Bunodophoron (Sphaerophoraceae)
A.Massal.
Probably Gk. *bounos* = a heap or mound; *-phorum* = bearing or carrying; referring to the fertile branches that carry the sooty black mazaedia.

Byssolecania (Pilocarpaceae) Vain.
Gk. *byssos* = flax, fine thread; *lekane, lecano* = dish, basin, bow; referring to the byssoid – whispy, like teased wool – margin of the apothecia and the resemblance to the genus *Lecania*.

Byssoloma (Pilocarpaceae) Trevis.
Gk. *byssos* = flax; *loma* = edge or fringe. Many species have a fringe-like margin on the apothecia.

Byssopsora = Bacidia (Ramalinaceae)
A.Massal.
Gk. *byssos* = flax; *psora, psoraleos* = affected with leprosy, scabby; referring to the blackish glandular points on the calyx.

C

Calicium (Physciaceae) Pers.
Gk. *kalukion* = a little cup; referring to the fruiting bodies' shape.

Callopisma (Teloschistaceae) De Not.
Gk. *kallos* = beauty. Named by author Giuseppe De Notaris as such 'on account of its beauty': '*Nomene graeco, ob pulchritudinem*'.

Calopadia (Pilocarpaceae) Vezsda
Gk. *kalos* = beautiful; alluding to the beauty of this lichen.

Caloplaca (Teloschistaceae) Th.Fr.
Gk. *kalos* = beautiful; *plakos* = a flat, round plate or disc; referring to the attractive, round, yellow apotheca, resembling plates in some species.

Candelaria (Candelariaceae) A.Massal.
La. *candelaria* = candle, candle-like, possibly *candela* = tallow candle; *aria* = belonging to; referring to the bright-yellow colouring of the thallus of some species.

Candelariella (Candelariaceae) Müll.Arg.
La. *Candelaria* (q.v.); *-iella* = diminutive.

A lichen field, characteristic of the stony patches of the Namib north of Swakopmund where the sea mist provides the moisture in an area where the average rainfall is <1 mm/year. With a close-up look lichens can be found everywhere. (EJM)

Canomaculina (Parmeliaceae) Elix & Hale
La. *canus* = grey-white; *macul-* = spot, blotch; *-ina* = a little; grey-white, usually applied to hair covering.

Canoparmelia (Parmeliaceae) Elix & Hale
Gk. *canus* = hoary or grey-white; *parme* = fruit bowl; *eileo* = closed; referring to the pale upper surface of thalli.

Carbonea (Lecanoraceae) (Hertel) Hertel
Gk. *carbō, carbōn-* = charcoal; named for shiny black carbonaceous excipulums.

Catapyrenium (Verrucariaceae) Flot.
Gk. *kata* = downward; *pyren* = kernal, nucleus; referring to the sunken perithecia.

Catarrhospora (Lecideaceae) Brusse
La. *catarrhus* = slimy secretion, Gk. *katarrhein* = to flow down; *spora* = seed; referring to the gelatinous coat covering the ascospore.

Catillaria (Catillariaceae) A.Massal.
La. *catillus* = a small bowl, dish or cup; *aria* = resembling; referring to the cup-shaped apothecia.

Catinaria (Ramalinaceae) Vain.
Possibly La. *catinus* = deep bowl; referring to the bowl shape of the fruit body.

Catocarpus (Rhizocarpaceae) (Körb.) Arnold.
Gk. *kata* = downward; *karpos* = fruit; bearing chain-like suspended fruit.

Catolechia (Rhizocarpaceae) Flot.
Gk. *kata* = downward; *lechia* = rushes or sedges. Meaning obscure.

Celidium = Arthonia (Arthoniaceae) Tul.
Gk. *kēlis* = spot; *-idion* = little. Meaning unknown.

Cenomyce = Cladonia (Cladoniaceae) Ach.
Gk. *kenos* = empty; *myce* = fungus; alluding to the hollowness of the fungus-like receptacles.

Cetraria (Parmeliaceae) Ach.
La. *caetra* = shield, a leather shield; referring to the shape of the shape and texture of the thallus.

Chapsa (Graphidaceae) A.Massal.
The name is probably a spelling error. Should be La. *capsa* = a case, box; referring to the development of the ascomata.

Chiodecton (Roccellaceae) Ach.
Gk. *khion* = snow; *dekta* = the same; alluding to the whitish thallus of the type species.

Chroodiscus (Thelotremataceae) (Müll. Arg.) Müll.Arg.
Gk. *chroa* = colour; *diskos* = disk, flat plate; referring to the red apothecial disks of the type species.

Chrysothrix (Chrysothricaceae) Mont.
Gk. *chryso-* = gold, golden; *thrix* = hair; referring to the appearance of the thallus comprising mostly yellow fungal hyphae.

Circinaria (Lecanoraceae) M.Choisy
Gk. *kirkos* = a ring; *-aria* = pertaining to; possibly referring to the dark ring or wall around the disc section.

Cladia (Cladoniaceae) Nyl.
Gk. *klados* = branch, twig; referring to the upright branched stalks comprising the thallus or body.

Cladonia (Cladoniaceae) P.Browne
Gk. *klados* = branch, shoot; referring to the branched podetia of some species.

Clathrina (Cladoniaceae) Müll.Arg.
Gk. *clathratus* = latticed or pierced; *-ina* = diminutive, a little. The surface is pierced and contorted by numerous round-to-elongated holes.

Clathroporina (Pertusariaceae) Müll.Arg.
Gk. *clathratus* = latticed or pierced; *porina* = little pores; referring probably to the small ostiole (hole) of the perithecia through which spores are released.

Cliostomum (Ramalinaceae) Fr.
Gk. *kleio* = I shut; *stoma* = mouth; possibly referring to the staminodes. Meaning unclear.

Clypeococcum (Dacampiaceae) D.Hawksw.
Gk. *clypea* = shield; *kokkos* = berry.

Clypeosphaeria (Clypeosphaeriaceae) Fuckel
Gk. *clypea* = shield; *sphaeria* = spherical, round; referring to the shape of the fruit.

Coccocarpia (Coccocarpiaceae) Pers.
Gk. *coccos* = lice; *carpus, carpon*; = fruit; referring to resemblance of the apothecia to lice of the genus *Coccus*.

Coelocaulon (Parmeliaceae) Link
Gk. *koilos* = hollow; *kaulos* = stem, stalk; referring to the hollow thallus branches.

Coelopogon (Parmeliaceae) Brusse & Kärnef.
Gk. *koilos* = hollow; *pogon* = bearded. These lichens are 'bearded' with a white or cream fungus.

Coenogonium (Coenogoniaceae) Ehrenb.
Gk. *coenos* = common; *gonos, gonium* = seed. These lichen species are characterised by very fluffy or webby thalli in which the algal component is dominant, so the name derivation is obscure.

Collema (Collemataceae) F.H.Wigg.
Gk. *kollēma* = that which is glued; referring to the strongly hardened gelatinous thallus (when dry) of some species.

Collemodiopsis (Collemataceae) (Vain.) B.de Lesd.
Collemodium (q.v.); Gk. *-iopsis* = resembling, likeness.

Collemodium (Collemataceae) Nyl.
Collema (q.v.); *-odion* = diminutive.

Combea (Roccellaceae) De Not.
For Francesco Comba (?–1892), Italian naturalist and taxidermist of the Royal Zoological Museum of the University of Turin. He assisted Vincenzo Griseri, the first person to undertake the rearing of the Bombyx Cynthia silkworm moth on leaves of the castor-oil plant, and the first who introduced it into France. He was also responsible for mounting fossil skeletons and was director of the royal zoo, and 'head of the royal hunts' and painter of scenes from them. He became second preparator of zoology in 1833, and first preparator between 1853 and 1859. With Professor Giuseppe Gené and Vittore Ghiliani, he participated in zoological research on the island of Sardinia (1833–1838). Plants collected during this research in 1838 were used by Giuseppe Giacinto Moris and his pupil and friend Professor De Notaris in a work called *Florula Caprariae* (1839). De Notaris named Combea for his friend.

Coniarthonia (Arthoniaceae) Grube.
Gk. *konis* = dust; *arthos* = fleck, speckled; alluding to the powdery appearance of the ascomata, and *arthos* for fleck-like, sprinkled.

Coniocarpon (Arthoniaceae) DC.
Gk. *konikos* = conical; *karpos* = fruit; referring to the shape of the fruit.

Conotrema (Stictidaceae) Tuck.
Gk. *konos* = cone; *trema* = aperture, hole, especially the female pudendum; referring to the concave, immersed apothecia.

Coprinus (Coprinaceae) Pers.
Gk. *kopros* = dung; *-inus* = belonging to. The moss, *Coprinus,* 'lives' on dung.

Coriscium* = *Omphalina (Tricholomataceae)
Vain.
Gk. *koris* = bedbug; *-cium* = connected with. This is a rejected name, now *Lichenomphalia*.

Cornicularia (Parmeliaceae) (Schreb.) Hoffm.
Gk. *corniculatus* = having a small horn or spur-like appendages; *aria* = possessing; referring to the shape of the thallus tips.

Coronoplectrum (Parmeliaceae) Brusse
Gk. *korone* = crowned; *plectron* = cock's spur.

Corynecystis (Heppiaceae) Brusse
Gk. *koryne* = a club; *cystis* = bladder, pouch.

Crocodia (Lobariaceae) Link
Gk. *krokos* = saffron; probably alluding to the yellowish-brown colour of the thallus.

Crocynia (Crocyniaceae) (Ach.) A.Massal.
Perhaps Gk. *crocys* = the flock or nap on woollen cloth; possibly referring to the surface of the thallus.

Cryptothecia (Arthoniaceae) Stirt.
Gk. *kryptos* = to conceal, hidden; *theke* = a container or sheath; referring to the immersed asci (the reproductive structures are 'hidden' under the surface in this lichen).

Cyanisticta* = *Pseudocyphellaria
(Lobariaceae) Gyeln.
La. *cyaneus* = greenish-blue (Gk. *kyanos*); *sticta* = spotted; possibly referring to the lichen's appearance.

Cyphelium (Physciaceae) Ach.
Gk. *kyphelon* = small bowl or cup according to Acharius (blister in modern Greek); referring to the hymenium sunken into the warty thallus, resembling a cup.

D

Dactylina (Parmeliaceae) Nyl.
Gk. *daktylos* = finger; *-ina* = resembling; referring to the thallus form.

Dermatiscum (Physciaceae) Nyl.
Gk. *derma* = skin; *-iskos* = diminutive suffix; some crustose lichens grow with bark.

Dermatocarpon (Verrucariaceae) Eschw.
Gk. *derma* = skin; *karpos* = fruit; referring to the leathery-looking thallus with its inspersed perithecia.

Dichodium* = *Physma (Collemataceae) Nyl.
Gk. *dicha-* = in two, split, divided in two; *-ium* = diminutive.

Dictyolus* = *Arrhenia (Tricholomataceae) Quél.
Gk. *dictyolus* = a little netted, latticed.

Didymella (Didymellaceae) Sacc.
Gk. *didymos* = in pairs; *-ella* = diminutive.

Didymosphaeria (Didymosphaeriaceae) Fuckel
Gk. *didymos* = twin, in pairs; *sphaeria* = spherical, round.

Digitothyrea (Lichinaceae) P.P.Moreno & Egea
La. *digitus* = a finger; *thyreos* = a large, oblong, door-shaped shield; similar to the genus Thyrea but with elongate lobes.

Dimelaena (Physciaceae) Norman
Gk. *di-* = two, double; *melaina* = black; referring to the dark, two-celled ascospores of the type species, or from the presence of black apothecia and a black margin on the squamules.

Dimerella (Coenogoniaceae) Trevis.
Gk. *di-* = two; *meros* = part, portion; *-ella* = diminutive; referring to two-celled spores.

Diploicia (Physciaceae) A.Massal.

Diploicia subcanescens

Gk. *diplo-* = double, two-fold; *oikia* = house; referring to the two-celled ascospores.

Diploschistella* = *Gyalideopsis
(Gomphillaceae) Vain.
Gk. *diplo-* = double; *schistos* = split, divided; *-ella* = diminutive; referring to the double split between the excipiple (proper margin) and the hymenium in the fruit bodies.

Diploschistes (Thelotremataceae) Norman
Gk. *Diplo-* = double; *schistos* = split, divided; referring to the double split between exciple (proper margin) and hymenium in the fruit bodies.

Diplotomma (Physciaceae) Flot.
Gk. *diplo-* = double; *omma* = eyed; referring to the apothecia, which in this genus has the appearance of dark 'eyes'.

Dirina (Roccellaceae) Fr.
Gk. *deirē* = ridge of a hill, neck; *-ina* denoting likeness. Meaning unclear.

Dirinaria (Physciaceae) (Tuck.) Clem.
From the crustose genus *Dirina; arius* = connection with; perhaps referring to the superficial resemblance to *Dirina*.

E

Echidnocymbium (Ramalinaceae) Brusse
Gk. *echidna* = snake, viper; *kymbe* = boat (La. *cymba*). Meaning unclear.

Echinoplaca (Gomphillaceae) Fée
Gk. *echinos* = prickly, spiny (hedgehog, sea urchin); *placos* = flat surface, crust. The 'hyphophoren' on the flat thallus looks similar to hedgehog spines.

Enchylium = Pterygiopsis (Lichinaceae) A.Massal.
Gk. *enchlyos* = fleshy, succulent; *-ium* = diminutive; referring to its appearance.

Endocarpon (Verrucariaceae) Hedw.
Gk. *endon, endo-* = within; *karpos* = fruit; referring to the immersed perithecia of this genus.

Endopyrenium = Catapyrenium (Verrucariaceae) Flot.
Gk. *endo-* = within; *pyren* = grain; *-ium* = diminutive.

Engizostoma = Valsa (Valsaceae) A.Gray
Gk. *engizo* = to bring near, draw nearer; *stoma* = mouth, a minute opening in leaves and stems. Possibly alluding to stomata, which are numerous and congregated.

Enterographa (Roccellaceae) Fée
Gk. *entero-* = internal; *graphos* = writing, drawing; referring to the black apothecia, variable in shape but generally elongated, thread-like, forming wavy lines.

Enterostigma = Thelotrema (Thelotremataceae) Müll.Arg.
Gk. *entero-* = internal; *enteron* = intestine; stigma.

Ephebe (Lichinaceae) Fr.
Gk. *epi-* = upon, fitting together; Gk. *hebe* = youth, hairy, puberty, early manhood; referring to the threadlike appearance of this lichen.

Ephebeia = Ephebe (Lichinaceae) Nyl.
Relating to *Ephebe* (q.v.).

Eremastrella (Psoraceae) S.Vogel
Gk. *eremos* = uninhabited, solitude, solitary; *-ella* = diminutive.

Erioderma (Pannariaceae) Fée
Gk. *erion* = wool; *derma* = skin; referring to the thick layer of hairs over the thallus surface.

Eucollema = Collema (Collemataceae) (Cromb.) Horw.
Gk. *eu-* = real, typical (of the genus); *collema* = that which is glued; referring to the strongly hardened gelatinous thallus (when dry) of some species.

Eumitria = Usnea (Parmeliaceae) Stirt.
Gk. *eu-* = good, well; *mitre* = cap, headdress or turban.

F

Fellhanera (Pilocarpaceae) Vezsda
Near anagram of *Hafellner* (*Hafellia* q.v.).

Fistulariella (Ramalinaceae) Bowler & Rundel.
La. *fistula* = hollow, tube-like; *-iella* = diminutive; referring to the thallus which is hollow, like a pipe.

Flavoparmelia (Parmeliaceae) Hale
La. *flavus* = yellow; *Parmelia* (q.v.); referring to the colour of the upper surface of this segregate of *Parmelia*.

Flavopunctelia (Parmeliaceae) (Krog) Hale
La. *flavus* = yellow; *punctum* = small spot or dot; possibly referring to the straw-yellow thallus and long, spotty conidia.

G

Gabura (Collemataceae) Adans.
Origin unknown, possibly a town in Bangladesh.

Gasparrinia = Caloplaca (Teloschistaceae) Tornab.

Another form of Frutiose lichen (EJM)

For Guglielmo Gasparrini (1804–1865), Italian botanist and mycologist, professor of plant anatomy and morphology at the University of Pavia (1857–1861), professor of botany at Naples, and director of the Botanical Garden of Naples (1861–1865), during which time he upgraded some areas of the garden such as the arboretum, the botanical museum and the herbarium. He was the author of numerous scholarly publications, and devoted much of his career to morphology, histology and plant systematics. He promoted the use of silkworms in the production of Chinese textiles in the Naples area.

Gloeoheppia (Gloeoheppiaceae) Gyeln.
Gk. *gloios* = sticky; referring to the thallus texture. For Johann Adam Philipp Hepp (1797–1867), German medical doctor, botanist and lichenologist. He studied medicine in Würzburg and worked in a mental hospital in Frankenthal before becoming a general practitioner in 1826. In 1832, he participated in the exploration of Hambahskom, Russia. In 1840, he became co-founder with Carl Heinrich Schultz and others of the Palatinate scientific association Pollichia, and its first chairman. Politically active from the 1830s, he had to flee to Switzerland in 1849 as political refugee, and was sentenced to death in 1851 *in absentia*, but was granted amnesty in 1865. His collection of 2 400 specimens of lichens, 1 650 algae and 770 samples of moss is stored in the herbarium of the botany department of the Natural History Museum in London. He was editor of the exsiccate collection *The Lichens of Europe* (Zurich, 1853–1867) and described several new species.

Glyphis (Graphidaceae) Ach.
Gk. *glyphos* = carved, engraved, from *glyphein* = to carve; referring to the apparently incised lirellae.

Glyphopeltis (Psoraceae) Brusse
Gk. *glyphos* = carved, engraved, from *glyphein* = to carve; *peltis* = small, round shield; referring to the apothecium structure.

Gomphillus (Gomphillaceae) Nyl.
Gk. *gomphos* = a peg, club; *-illus* = diminutive; referring to the secondary thallus form (similar to podetium).

Gonohymenia = Lichenella (Lichinaceae) J.Steetz
Gk. *gonos* = seed, progeny; *hymen* = membrane; referring to the small gelatinous-membranous thallus with the spore-bearing portion included therein.

Gonolecania = Byssolecania (Pilocarpaceae) Zahlbr.
Gk. *gonia* = an angle; *lekane, lecano* = a dish, a basin, bowl; alluding to the form of shields.

Graphina (Graphidaceae) Müll.Arg.
Gk. *graphos* = a letter; *ina* = little; referring to the resemblance of the black lirellae to handwritten script.

Graphis (Graphidaceae) Adans.
Gk. *graphios* = line drawing; referring to the resemblance of the black lirellae to handwritten script or perhaps to the elongate, often branched apothecia that resemble written markings.

Gyalecta (Gyalectaceae) Ach.
Gk. *gyalos* = hollow; *lecton* = container; referring to the cup-shaped apothecia.

Gyalectidium (Gomphillaceae) Müll. Arg.
Gyalecta (q.v.); *-idium* = little; referring to the urn-shaped apothecia.

Gyrostomum (Graphidaceae) Fr.
Gk. *gyros* = a ring, circle; *stoma* = mouth; referring to the cup-shaped ascocarps crowded into a raised circular stroma.

H

Haematomma (Haematommataceae) A.Massal.
Gk. *haimato-* from *haima, haema* or *haemat* = blood; *-omma* = an eye; referring to the red apothecial disks of the genus.

Hafellia (Physciaceae) Kalb, H.Mayrhofer & Scheid.
For Josef Hafellner (1951–), Austrian mycologist at the Institute of Plant Sciences at the University of Graz who obtained a doctorate at Graz (1978)

and did postdoctoral studies at Queensland University (1986). He has lectured at Graz since 1983 in systematic botany and geobotany. His main interests include the taxonomy of lichenicolous fungi, canopy lichens, the taxonomy of vegetatively reproducing lichens, and lichen diversity in Austria, the Alps, Macaronesia, the Sonoran Desert, North America and other areas. He authored world monographs for the genera *Karschia, Letroutia* and *Brigantiaea,* and *Diversity and Ecology of Lichens in Polar and Mountain Ecosystems* (2010) with I Kärnefelt, and V Wirth. He was guest professor at the Ecole Normale Supérieure de St Cloud (1980, 1982).

Hagenia* = *Anaptychia (Physciaceae) Eschw.
The lichen was first described in 1824, so possibly named for Karl Gottfried Hagen (1749–1829). See *Hagenia* (Rosaceae).

Helminthocarpon (Arthoniaceae) Fée
Gk. *helminthian* = to be infested with worms; *karpon* = fruit; referring to the elongated, wrinkled fruits.

Helotium* = *Cudoniella (Helotiaceae) Pers.
Gk. (*h*)*ēlōtos* = nail-shaped; referring to the fungus's shape.

Hemithecium (Graphidaceae) Trevis.
Gk. *hemi* = half; *theka* = case, capsule; possibly in reference to the sessile lirellae.

Heppia (Heppiaceae) Nägeli ex A.Massal.
For Johann Adam Philipp Hepp (1797–1867), German medical doctor, botanist and lichenologist. He studied medicine in Würzburg and worked in a mental hospital in Frankenthal before becoming a general practitioner in 1826. In 1840, he became cofounder (with Carl Heinrich Schultz, 1805–1867) of the Palatinate scientific association, and its first chairman. Politically active from the 1830s (although he failed to get elected to the national assembly), he had to flee to Switzerland in 1849, as a political refugee, and was sentenced to death in 1851 *in absentia*, then granted amnesty in 1865. He died while visiting his daughter in Frankfurt in 1867. He was editor of the exsiccate collection *The Lichens of Europe* (Zurich, 1853–1867), and described several new species.

Heterodermia (Physciaceae) Trevis.
Gk. *hetero-* = different, unlike; *derma* = skin or hide; referring to the presence or absence of a lower cortex.

Huea* = *Caloplaca (Teloschistaceae)
C.W.Dodge & G.E.Baker

For Abbé Auguste-Marie Hue (1840–1917), French botanist and lichen specialist, author of *Lichens de Canisy et des Environs*. He studied lichens collected both by the scientific expedition of Tunisia and the French Antarctic expeditions of 1904–1907 and 1908–1910.

Hyperphyscia (Physciaceae) Müll.Arg.
Gk. *hyper* = above; *Physcia* (q.v.); possibly referring to the substrate (bark) with the lichen above and the substrate below (Mathias Schultz).

Hypogymnia (Parmeliaceae) (Ny) Nyl.
Gk. *hypo-* = below, under; *gymnos* = naked; referring to the smooth, rhizine-free lower surfaces of the cortex and near-naked underside.

Hypotrachyna (Rhizocarpaceae) (Vain.) Hale
Gk. *hypo-* = below, underneath; *trachyno* = rough or shaggy; referring to the somewhat roughened appearance of the densely squarros-rhizinate lower surface.

Hysterina* = *Opegrapha (Roccellaceae) (Ach.) Gray
Gk. *hysteron* = later; *hysteros* = womb; *ina* = small. Meaning obscure.

I

Imbricaria* = *Anaptychia (Physciaceae) (Schreb.) Michx.
La. *imbricaria* = covering with broad overlapping scales, derived from *imbrex* = a tile, roofing tile; referring to the lichen's appearance.

Immersaria (Lecideaceae) Rambold & Pietschm.
La. *immerse* = immersed, covered; referring to the sunken fruiting bodies (apothecia).

Imshaugia (Parmeliaceae) S.F.Mey.

Imshangia aleurtis

For Henry Andrew Imshaug (1925–2010), US-based curator of the cryptogamic herbarium at Michigan State University from 1956–2000. He increased the herbarium collection from 40 000 to more than 145 000 specimens, obtaining specimens from the collections of mainly European lichenologists and through his extraordinary efforts to collect 75 000 specimens from western North America, the Great Lakes regions, the Caribbean, and sub-Antarctic areas from the 1950s to the 1970s. His knowledge of lichens was unsurpassed, and in addition to his curatorial duties, he was a superb teacher who supervised a number of graduate students, many of whom became distinguished lichenologists, including Irwin M Brodo, Richard C Harris and Clifford M Wetmore.

K

Karoowia (Parmeliaceae) Hale
Named for the Great Karoo in the Cape province, South Africa.

Kroswia (Pannariaceae) P.M.Jörg.
For Hildur Krog (1922–), Norwegian lichenologist and collector of fungi and lichens, curator of the Botanical Museum of Oslo, and professor at the University of Oslo; AND Thomas Douglas Victor ('Dougal') Swinscow (1917–1992), on staff at the British Medical Journal, founder of the British Lichen Society and its journal *The Lichenologist*, co-author with Dr Krog of *The Macrolichens of East Africa* (1988).

L

Lagerheimina = Diploschistes (Thelotremataceae) Kuntze
For Nils Gustaf (von, de) Lagerheim (1860–1926), Swedish botanist. He was curator at the Natural History Museum in Lisbon, Portugal, in 1889, and professor in cryptogam systematics and a bacteriologist with a special interest in algae and fungi at the University of Quito, Ecuador, the same year. He left this appointment for health reasons in 1892 to become the curator at the museum in Tromsö, Norway. In 1895, he became a professor in botany at the University of Stockholm. He was mainly a mycologist and phycologist, but was also one of the founders of pollen analysis. He wrote *Ein Beitrag zur Schneeflora Spitzberg* (*A Contribution to the Snow Flora of Spitsbergen*) (1894).

Lasallia (Umbilicariaceae) Mérat
For a 19th-century gardener-botanist named Lasalle who worked at Fountainebleau and the Botanical Garden of Corsica. No further information found.

Lathagrium = Collema (Collemataceae) (Ach.) Gray
La. *letharia* = lethal or poisonous. The lichen is poisonous.

Laurera (Trypetheliaceae) Rchb.
For Johann Friedrich Laurer (1798–1873), German botanist, pharmacologist, anatomist and lichenologist. After training as pharmacist, he studied medicine and botany. He was a private lecturer of anatomy from 1836. In 1863, he became a professor of medical pharmacology at the University of Greifswald and visited South Africa. He collected in Germany, Sweden (Gotland), Switzerland, Austria and Czechoslovakia. He discovered Laurer's canal – a muscular duct in some trematode worms that arises from the oviduct between the ovary and the omphalomesenteric duct and passes to the dorsal surface – as described in *Disquisitiones Anatomicai de Amphistomo Conico* (1830).

Lecanactis (Roccellaceae) Körb.
Gk. *lekane*, lecano = a dish, bowl; *actis* = ray (seemingly of no meaning); referring to the shape of the apothecia.

Lecania (Ramalinaceae) A.Massal.
Gk. *lekane*, La. *lecano* = a dish, bowl; referring to the shape of the apothecia.

Lecanidion = Patellaria (Patellariaceae) Endl.
Gk. *lekane, lecano* = a dish, bowl; *-idion* = diminutive; perhaps referring to the shape of a little apothecia.

Lecanographa (Roccellaceae) Egea & Torrente.
Gk. *Lecanactis* (q.v.); *Opegrapha* (q.v.). The name is a combination of Lecanactis and Opegrapha because of the similarity.

Lecanora (Lecanoraceae) Ach.
Gk. *lekanon* = a small cup, bowl; *-ora* = form, beauty; referring to the cup-shaped apothecia and reproductive structures.

Lecidea (Lecideaceae) Ach.
Gk. *lekis* = small shield; *eidos* = form, resemblance; referring to the shape of the apothecia.

Crustose lichen: Example of the spore-bearing bodies shown as tiny craters on the surface (EJM)

Lecidella (Lecanoraceae) Körb.
Gk. *Lecidea* (q.v.); *-ella* = diminutive. Perhaps similar to *Lecidea* but smaller or inferior to it.

Leciographa = Opegrapha (Roccellaceae) A.Massal.
Gk. *lexis* = small shield; *graphos* = writing; referring to the apothecia and the 'lines' on it.

Leightonia (Trypetheliaceae) Trevis.
For William Allport Leighton (1805–1889), British botanist, lichenologist and curator. He attended  school with Charles Darwin, was articled to become a solicitor (1822), and obtained a BA from Cambridge (1833), being profoundly influenced by Reverend John Henslow, professor of botany. He 'postponed' his ordination for 10 years to compile his *Flora of Shropshire* (1851). He was curator of St Giles's church in Shrewsbury from 1846 to 1848, but then resigned to devote himself again to botany, principally to cryptogams and especially lichens. His publications include *Catalogue of the Cellularies or Flowerless Plants of Great Britain* (1837), and *Lichen Flora of Great Britain, Ireland and the Channel Isles* (1871). He became a Fellow of the Linnaean Society in 1865.

Leiophloea = Arthropyrena (Arthopyreniaceae) Trevis.
Gk. *leios* = smooth; *phloios* = bark; referring to the preferred habitat of the lichen.

Lempholemma (Lichinaceae) Körb.
Gk. *lemphos* = slime; *lemma* = scale; referring to the pulpy texture of these lichens when moistened.

Lenormandia = Normandina (Verrucariaceae) Delise.
For Sébastien-René Lenormand (1796–1871), French lawyer, botanist, algologist and plant  collector. He was a practising lawyer until 1835 in Vire, and thereafter devoted himself to botany during his retirement in Lénaudières and put together a sizeable herbarium. He was one of the leading botanists of his time, also a microscopist, with a network of people with whom he corresponded and from whom he acquired most of his specimens, mainly cryptogamic plants in Europe (France, Greece, the Russian federation), Africa (Guinea, Senegal, South Africa), Madagascar, Australia, New Caledonia, and North America (United States). He also had a particular interest in Pacific flora.

Lepadolemma = Haematomma (Haematommataceae) Trevis.
Gk. *lepidos* = scaly; *lemma* = scales. Meaning obscure.

Lepidoma = Rhizocarpon (Psoraceae) (Ach.) Gray
Gk. *lepis* = scale; *oma* = tumour; referring to the appearance of the lichen.

Lepraria (Stereocaulaceae) Ach.
Gk. *lepra* = a skin disease, leprosy; *-aria* = pertaining to; referring to the scruffy appearance of the thalli.

Leproloma = Lepraria (Stereocaulaceae) Nyl. ex Cromb.
Gk. *lepros* = scurfy; *loma* = edge or fringe; referring to the scurfy appearance of the thalli.

Leproncus = Pertusaria (Pertusariaceae) Vent.
Gk. *leprosus* = scaly; *onkos* = knob; referring to the lichen's knob-like insidia?

Leptogiomyces = Leptogium (Collemataceae) E.A.Thomas ex Cif. & Tomas.
Gk. *leptos* = slender, thin; *myces* = fungus; referring to the needle-like spore.

Leptogiopsis*** = Mastodia (Mastodiaceae) Nyl.
Gk. *leptos* = slender, thin; *-iopsis* = resembling, likeness. The lichen resembles Leptogium.

Leptogium (Collemataceae) (Ach.) Gray
Gk. *leptos* = peeled, slender, thin, weak; *ge* = the earth, land; perhaps referring to thallus lobes that appear as thin shavings on the ground in terricolus species.

Leptotrema (Thelotremataceae) Mont. & Bosch.
Gk. *leptos* = slender, thin; *trema* = aperture, hole,

opening; referring to the pore-like opening of the ascomata.

Lethagrium (Collemataceae) A.Massal.
Gk. *lethagria* = forgetful, morbid or sickly state. The 'sickly state' may allude to the toxicity of some species of this genus; historically, this lichen has been used to poison wolves and foxes.

Leucodecton (Thelotremataceae) A.Massal.
Gk. *leukos* = white; *dekta*= the same outside; referring to the whitish thallus of the type species.

Leucodecton (Parmeliaceae) L.
Gk. *leichen,* originally 'what eats around itself', probably from *leichein* = 'to lick'.

Lichenoides (Physciaceae) Hoffm.
Lichen (q.v.); Gk. *-oides* = resembling.

Lichinella (Lichinaceae) Nyl.
Lichen (q.v.); *-ella* = diminutive.

Lithocia = Verrucaria (Verrucariaceae) Gray
Gk. *lithos* = stone, rock; perhaps referring to its appearance.

Lithoglypha (Acarosporaceae) Brusse
Gk. *lithos* = stone, rock; *glyphos* = carved, engraving, lettering in stone, from *glyphein*, to carve; referring to the script-like apothecia and the lichens habitat (rock).

Lithographa (Trapeliaceae) Nyl.
Gk. *lithos* = stone, rock; *graphos* = writing; referring to the script-like apothecia that resemble those of the genus *Graphis*, and the lichen's habitat on rocks

Lobaria (Lobariaceae) (Schreb.) Hoffm.
Gk. *lobos* = lobe, *arius* = belonging to; referring to the large lobate thallus of the genus.

Lobarina (Lobariaceae) Nyl.
Gk. *lobos* = lobe, pod; *-ina* = resembling.

Lopadium (Pilocarpaceae) Körb.
Gk. *lopas* = flat plate, but defined by the author as 'cup'; alluding to a deep dish or bowl of fruit.

M

Mallotium = Leptogium (Collemataceae) (Ach.) Gray
Gk. *mallotus* = lined in wool, fleecy; referring to the seed cap.

Mazosia (Roccellaceae) A.Massal.
Gk. *mazos* = breast; probably refers to the appearance of the young apothecia, which initially resemble perithecia (globose structures).

Megalospora (Megalosporaceae) Meyen
Gk. *megalo* = large; *spora* = spores. The spores are large in this genus.

Melampylidium = Bactrospora (Roccellaceae) Stirt. ex Müll.Arg.
Gk. *melas* = black; *pylac* = bottleneck; *-idion* = diminutive referring to the black ostiole.

Melanomma (Melanommataceae) Nitschke ex Fuckel
Gk. *melas* = black; *omma* = appearance, or eye; possibly referring to the black 'eyes' of the apothecia.

Melanotheca = Thelidium (Verrucariaceae) Fée
Gk. *melas, melano-* = black; *theka* = a case, capsule; referring to the black-walled perithecia of this genus.

Melaspilea (Melaspileaceae) Nyl.
Gk. *melas* = black; *pilea* = cap, hat; referring to the apothecia, which are abundant, black, minute.

Metasphaeria = Saccothecium (Dothioraceae) Sacc.
Gk. *meta* = after, beyond; *sphaeria* = spherical, round. The globose perithecia embedded in leaf tissues, spherical or slightly depressed.

Micarea (Micareaceae) Fr.
La. *mica* = crumb, morsel; *area* = space; referring to the fine-grained texture of the thallus.

Microphiale = Coenogonium (Coenogoniaceae) (Stizenb.) Zahlbr.
Gk. *mikros* = very small; *fiale, phiale* = a shell; referring to the shape of the fruiting body (apothecia).

Microthelia = Anisomeridium (Monoblastiaceae) Körb.
Gk. *mikros* = very small; *thelius* = a little nipple (warts); referring to the warty ornamentation of the ascospores.

Morchella (Morchellaceae) Dil ex Pers.
German *morchel* = mushroom. The main feature of this mushroom is that it is edible.

Muellerella (Verrucariaceae) Hepp
For Jean (Johannes) Mueller (Müller), called Argoviensis (1828–1896), Swiss botanist with a

Muellerella lichenicola

special interest in tropical lichens and owner of a lichenological herbarium. He studied botany at the University of Zurich, obtaining his PhD in 1857. He was the curator of Augustin Pyramus de Candolle's herbarium (1851–1869), curator of the B Delessert herbarium (1869–1896), and professor of botany at the University of Geneva (1871–1889), during which time he was also director of the Geneva Botanical Gardens (1870–1874). He was the author of *Monographie de la famille Résédacées* (1857), *Principes des Classification des Lichens et Énumération des Lichens des Environs de Genève* (1862) and *Lichenologie Beiträge in Flora Annis 1874–1891 Editi*. He also wrote monographs for the Apocynaceae, Rubiaceae and Euphorbiaceae in Augustin Pyramus de Candolle's *Prodromus* and Carl Friedrich Philipp von Martius's *Flora Brasiliensis* (1862).

Mycarthonia = Arthonia (Arthoniaceae) Reinke.
Gk. *mykes* = fungus; *Arthonia* (q.v.).

Mycobilimbia (Lecideaceae) Rehm
Gk. *mykes* = fungus; the genus *Bilimbia*, a segregate of *Bacidia*.

Mycomicrothelia (Trypetheliaceae) Keissl.
Gk. *mykes* = fungus; *mikros* = small, tiny; *thelius* = a little nipple (wart). The fungus is covered with tiny warts.

Myelochroa (Parmeliaceae) (Asahina) Elix & Hale
Gk. *myelos* = marrow; *chroa* = coloured; referring to the yellow-tinted medulla.

Myriotrema (Graphidaceae) Fée
Gk. *myrios* = innumerable; *trema* = aperture, hole; referring to the numerous ascomata of some species.

N

Namakwa (Parmeliaceae) Hale
An ethnic name, from Namaqualand.

Nemaria = Roccella (Roccellaceae) Navàs
Gk. *nema* = a thread, filament; *aria* = pertaining to; alluding to a branched threadlike structure that grows from a moss spore and eventually develops into the moss.

Neofuscelia (Parmeliaceae) Essl.
La. *neo-* = new; *fusca* = dark brown, dusky; relating to the thallus colour.

Nesolechia = Phacopsis (Parmeliaceae) A.Massal.
La. *neso-* = pertaining to islands; *lechia* = rushes or sedges; pertaining to the habitat.

Normandina (Verrucariaceae) Nyl.
Possibly for Sébastien-René Lenormand (1796–1871), French lawyer and botanist, algologist and plant collector who amassed a substantial herbarium. He had a particular interest in the Pacific flora and published a *Catalogue des Plantes Recueillies a Cayenne* (1859). He collected a large number of botanical specimens from naturalists who travelled to all parts of the world. This is, perhaps, an overstatement as Hippolyte Francois Jaubert of the French Botanical Society wrote an 11-page document, *Notice sur la Vie et les Travaux de Sebastien-René Lenormand* (1872) that seems to rebut this conception.

O

Ocellularia (Graphidaceae) G.Mey.
La. *ocellus* = an eye; *-ula* = diminutive; *aria* = indicating possession; referring to the appearance of the ascomata.

Ochrolechia (Ochrolechiaceae) A.Massal.
Gk. *ochros* = pale yellow, dull red or reddish brown; *lechos* = bed; referring to the yellowish appearance of the apothecia.

Omphalodium = Xanthoparmelua (Parmeliaceae) Meyen & Flot.
Gk. *omphalos* = navel, hub, central part of the flower; *odes, odium* = resembling; referring to the lichen's habit, which is foliose (leaf-like), attached by a central holdfast.

Opegrapha (Roccellaceae) Ach.
La. *oper* = enclose; Gk. *graphos* = writing; or Gk. *ope* = a hole, chink, opening; *graphis* =

Combination of fruticose and foliose lichens on the bark of a tree (EJM)

line drawings; referring to the elongate, script-like ascomata; or referring to the partly open apothecia, rather than closed, as in *Graphis*.

Ophioparma (Ophioparmaceae) Norman
Gk. *ophis* = serpent, snake(-like); La. *parma* = small round shield; referring to the shape of the roundish, disc-like fruit bodies.

P

Pachyphiale (Gyalectaceae) Lönnr.
Gk. *pachys* = thick; *fiale, phiale* = bowl used for offerings; referring to the thick-edged, cup-like apotheca.

Pannaria (Pannariaceae) Delise ex Bory
La. *pannosus* = woolly, felted or *pannus* = cloth, lobe; *arius* = pertaining to; referring to the numerous rhizohyphae on the lower side of the thalli.

Paraparmelia (Parmeliaceae) Elix & J.Johnst.
Gk. *para* = near, beside; *Parmelia* (q.v.).

Parathelium = Theledium (Verrucariaceae) Nyl.
Gk. *para* = near, besides; *thelium* = little nipples; referring to the perithecia with lateral ostioles that form a surface crust.

Parmelia (Parmeliaceae) Ach.
Gk. *parma* = a small round shield; *eileo* = enclosed or wrapped around; referring to the apothecia (*parma*) and thalline exciple of the apothecia (*eileo*).

Parmeliella (Pannariaceae) Müll.Arg.
Parmelia (q.v.); *-ella*= diminutive.

Parmelina (Parmeliaceae) Hale
Parmelia (q.v.); *-inae, ina* = diminutive.

Parmelinella (Parmeliaceae) Elix & Hale
Gk. *Parmelina* (q.v.); *-ella* = diminutive.

Parmelinopsis (Parmeliaceae) Elix & Hale
Gk. *Parmelina* (q.v.); *-opsis* = resembling.

Parmeliopsis (Parmeliaceae) (Nyl. ex Stizenb.). Nyl.
Gk. *Parmelia* (q.v.); *-opsis* = resembling. A segregate of *Parmelia*.

Parmentaria = Theldium (Verrucariaceae) Fée
For Antoine-Augustin Parmentier (1737–1813), French pharmacist, nutritionist and agronomist.

During the Seven Years' War (1756–1763), Parmentier was imprisoned by the Prussians and survived by eating potatoes. After the war, he studied nutritional chemistry and, in 1772, persuaded the Paris Faculty of Medicine that potatoes, banned by the French parliament in 1748 on the grounds that they caused leprosy, were edible. He pioneered the extraction of sugar from sugar beets, studied methods of conserving food, including refrigeration, and published many papers on agronomic matters. As inspector general of the health service (1805), he was responsible for the first mandatory smallpox vaccination campaign.

Parmosticta = Pseudocyphellaria (Lobariaceae) Nyl.
Gk. *parma* = a shield; *stictos* = spotted, prickled. The genus is named as such because there are similarities between *Parmelia* and *Sticta*.

Parmotrema (Parmeliaceae) A.Massal.
La. *parma* = a small, round shield; *trema* = a hole, (especially the female pudendum); probably referring to the perforated apothecia of *Parmotrema perforatum*.

Parmotrema (Patellariaceae) Fr.
La. *patella* = a small, shallow pan; *aria* = pertaining to. Pattelaris refers to a disk.

Peccania (Lichinaceae) A.Massal. ex Arnold.
La. *peccania* = the sinner. Perhaps Abramo Bartolomeo Massalongo felt like a sinner when he realised the 'mistake' or 'sin' he had committed when he named a new lichen genus

Corynophorus in 1856. This name already belonged to the grass genus *Corynephorus* (P.Beauv. 1812). Because homonyms are not allowed in botanical nomenclature, he changed his illegitimate *Corynophorus* name to *Peccania* (Matthias Schultz). (Just possibly named after an Algonquian, native American word = nut.)

Peltigera (Peltigeraceae) Willd.
La. *pelta* = small shield; Gk. *pelte*; *-gera* = bearing, carrying; referring to the shield-shaped thallus of these species.

Peltiphylla = **Psora** (Psoraceae) M.Choisy
Gk. *pelte* = shield; *phyllon* = leaf. Possibly because the shield-shape thallus looks somewhat leaf-like.

Peltula (Peltulaceae) Nyl.
Gk. *pelte* (La. *pelta*) = shield; *-ula* = diminutive, hence small shield; referring to the appearance of the squamules of the type species.

Peridiothelia = *Peridiothelia* (Pleomassariaceae) D.Hawksw.
Gk. *pēridion*, diminutive of *pērā* = leather pouch; *thelius* = a little nipple. Peridium, the outer enveloping coat of the fruit body in many fungi.

Pertusaria (Pertusariaceae) DC.
Gk. *pertusa* = perforated, punctured, thrust through; *aria* = pertaining to; referring to the punctured appearance of the thallus caused by the osteolate warts.

Phaeographina (Graphidaceae) Müll.Arg.
Gk. *phaios* = dusky, brownish grey; *Graphis* (q.v.); *-ina* = pertaining to; possibly referring to the dusky thallus.

Phaeographis (Graphidaceae) Müll.Arg.
Gk. *phaios* = dusky, brownish grey; *Graphis* (q.v.); referring to the brown, *Graphis*-like spores.

Phaeophyscia (Physciaceae) Moberg
Gk. *phaios* = dusky, brownish grey; *Physicia* (q.v.); referring to 'the dark *Physicia*', with brown spores and usually a dark thallus surface.

Phaeotrema (Thelotremataceae) Müll.Arg.
Gk. *phaios* = dusky, brownish grey; *trema* = opening, aperture; referring to the ascomata.

Phloeopeccania (Lichinaceae) J.Steiner
Gk. *phloeos* = bark; *Peccania* (q.v.). The word *phloeos* refers to a (misinterpreted) special layer of the cortex surrounding the lichen body. It has nothing to do with the bark of a tree.

Phlyctella =*Phlyctis* (Phlyctidaceae) Kremp.
Gk. *phlyctis* = blister burn (ulcer); *-ella* = diminutive; referring to the appearance of the ascocarps.

Phlyctis (Phlyctidaceae) (Wallr.) Flot.
Gk. *phlyktaina* = a blister burn, from *phlyein* = to boil over; referring to the white-to-grey crust of the thallus.

Phylliscum (Lichinaceae) Nyl.
Gk. *phyllon* = leaf; *-iskos* = diminutive suffix. The small lobed thallus looks somewhat leaf-like.

Phylloporis = **Strigula** (Strigulaceae) Clem.
Gk. *phyllon* = leaf; *poros* = pore, passage. Possibly referring to *phyllon,* the leaves the lichen is living on, and *poros,* the pore-like opening of the preithecial fruit bodies such as in Porina or Poroscyphus (Matthias Schultz).

Phyllopsora (Ramalinaceae) Müll.Arg.
Gk. *phyllon* = leaf; *psoros* = a scab, scurvy; referring to the squamules occurring in scab-like patches on a basal prothallus.

Physcia (Physciaceae) (Schreb.) Michx.
Gk. *physke* = a blister, wart, sausage, bladder; referring to the well-developed thalline apothecia.

Physciella = *Phaeophysica* (Physciaceae) Essl.
Physcia (q.v.); Gk. *-iella* = diminutive. This genus is smaller than many of the *Physcia*.

Physciopsis = *Hyperphysica* (Physciaceae) M.Choisy
Physcia (q.v.); *opsis* = resembling.

Physconia (Physciaceae) Poelt
Physcia (q.v.) -onia = bladder (inflating), referring to the enlarged thalline apothecia.

Physma (Collemataceae) A.Massal.
Probably Gk. *physao* = to blow up, distend; and *-ma* = to indicate result of an action; referring to the lobes, which usually swell strongly when moistened.

Pilocarpon = **Bysoloma** (Pilocarpaceae) Vain.
Gk. *pilos* = cap, hair; *karpos* = fruit; with cap-shaped fruit or bearing hairy fruit.

Placidium (Verrucariaceae) A.Massal.
Gk. *plax, plakos* = flat, flat-top; *idium* = little; referring to the suppressed squamae (A.Massal. *Symm. Lich.* 1855).

Placodium = **Xanthoria** (Teloschistaceae) F.H.Wigg.

Gk. *plakos* = spot; *-idion* = little; referring to the spot-like thallus.

Placolecanora* = *Lecanora (Lecanoraceae) Räsänen
Gk. *plakos* = flat; *Lecanora* (q.v.); referring to the thallus structure.

Placopsis (Trapeliaceae) (Ny) Linds.
Gk. *plakos* = flat; *-opsis* = resembling; referring in this instance to a plate. Its appearance is similar to *Placodium*.

Placynthiopsis (Placynthiaceae) Zahlbr.
The genus *Placynthium*; *opsis* = resembling; resembling the earlier genus *Placynthium* (Ach.), Gray.

Platismatia (Parmeliaceae) W.Culb. & C.F.Culb.
Gk. *platys* = broad, wide, flat; referring to the broad, flat thallus.

Platygramme (Graphidaceae) Fée
Gk. *platys* = broad, flat; *gramma, gramme* = a line; referring to the flattened, open lirellae of the type species.

Platysma* = *Ramalina (Ramalinaceae) Hill
Gk. *platys* = broad, flat; referring to the broad, flat lobes of this lichen.

Plectocarpon* = *Lithothelium (Roccellaceae) Fée
La. *plecto* = to punish; Gk. *plektos* = twisted, plaited; *carpos* = fruit.

Pleurotrema* = *Plagiocarpa* = *Lithothelium (Pyrenulaceae) Müll.Arg.
Gk. *pleuron* = side, rib; *trema* = aperture, hole, opening. Meaning obscure.

Polyblastia (Verrucariaceae) A.Massal.
Gk. *poly-* = many; *blastos* = bud, sprout, germ; referring to the richly fertile thalli of the type species.

Polyblastiopsis* = *Polyblastia
(Verrucariaceae) Zahlbr.
Gk. *Polyblastia* (q.v.); *iopsis* = resembling.

Polycauliona* = *Caloplaca (Teloschistaceae) Hue
Gk. *poly* = many; *caulos* = stem.

Polycoccum (Dacampiaceae) Saut. ex Körb.
Gk. *poly* = many; *coccus* = seeds or berries.

Porina (Pertusariaceae) Müll.Arg.
Gk. *poro* = pore, pathway, La. *porus* = passage or hole; *ina* = diminutive; presumably referring to the apical ostiole of the perithecia through which ascospores are released.

Porophora* = *Trypethelium (Trypetheliaceae) Zenker ex Göbelez
Gk. *poros* = pore, hole; *phorein* = bearing.

Porpidia (Lecideaceae) Körb.
Gk. *porpe* = buckle, pin, brooch, but defined by the author as 'ring'; *-idion* = diminutive; referring to the apothecia, evocative of little pins or brooches; another source states: because of its white thallus formed by the so-called housing ring. (The latter is a translation, perhaps incorrect, from German.)

Protoblastenia (Psoraceae) (Zahlbr.) J.Steetz
Gk. *protos* = first, primary, earliest; *blastos* = bud, shoot; *enos* = pertaining to; referring to the simple spores.

Psalliota* = *Agaricus (Agaricaceae) (Fr.) P.Kumm.
Gk. *psalion* = ring. The *Agaricus* spp. are responsible for the production of 'fairy rings' in grass.

Psathyrophlyctis (Phlyctidaceae) Brusse
Gk. *psathyros* = weak, loose, tender; La. *phlyctis* = a blister burn, from *phlyein* = to boil over; possibly referring to the blistered appearance of the Thallus.

Pseudobuellia* =*Rinodina (Physciaceae) B.de Lesd.
Gk. *pseudo-* = false, resembling but not the real *Buellia* (q.v.).

Pseudocyphellaria (Lobariaceae) Vain.
Gk. *pseudo-* = false, resembling but not the real genus *Cyphellaria;* referring to the pseudocyphellae on the lower surface of the thallus (sometimes also on the upper surface), structures anatomically and developmentally different from the cyphellae found on the lower surface of species of *Sticta*.

Pseudoparmelia (Parmeliaceae) Lynge.
Gk. *pseudo-* = false, resembling but not the real *Parmelia* (q.v.).

Pseudopyrenula (Trypetheliaceae) Müll.Arg.
Gk. *pseudo-* = false, resembling but not the real *Pyrenula* (q.v.); referring to the closed ascomatal wall.

Psiloparmelia (Parmeliaceae) Hale
Gk. *psilos* = bare; *Parmelia* (q.v.).

Psora (Psoraceae) Hoffm.
Gk. *psora* = the itch, scurvy; referring to the scurfy or scab-like thalli.

Psoroma (Pannariaceae) Ach. ex Michx.
Gk. *psora* = mange. The fungus looks scabby.

Psorotichia (Lichinaceae) A.Massal.
Gk. *psora* = the itch; *teichos* = wall around a city; probably referring to the lichen being found on concrete, walls and rails.

Punctelia (Parmeliaceae) Krog
La. *punctum* = a prick, puncture, small spot or dot; *elio* = to roll up or collect; referring to the numerous pseudocyphellae.

Pyrenastrum = Pyrenula (Pyrenulaceae) Eschw.
Gk. *pyren* = fruit stone, kernel; *-astrum* = partially resembling or inferior to *Prenula* (q.v.).

Pyrenopsis (Lichinaceae) Nyl.
Gk. *pyren* = fruit stone, kernel; *-opsis* = resembling.

Pyrenotea = Lecanatis (Roccellaceae) Fr.
Gk. *pyren* = fruit stone, kernel; *ot-* = small; the lichen looks like a bundle of small stones.

Pyrenotrichum (Pilocarpaceae) Mont.
Gk. *pyren* = fruit stone, kernel; *trichum* = hairy; perhaps alluding to the hairy capsule.

Pyrenowilmsia (Pyrenulaceae) R.C.Harris & Aptroot.
Gk. *pyren* = fruit stone; For Friedrich Wilms (1848–1919), German apothecary, botanical collector and traveller. He came to South Africa in 1883 and initially collected in Pietermaritzburg, but later further north to Lydenburg, which he used as his base for the next 13 years. Among his expeditions, he ventured through the malaria-infested coastal flats of Portuguese East Africa (now Mozambique) in winter, when there were fewer mosquitoes about. He returned to Germany (1896) with an extensive collection of mosses, lichens, ferns and phanerogams, and described his experiences in *Ein Botanischer Ausflug ins Boerenland* (1898). Later, he was appointed as an assistant, mainly in the moss section, of the Botanical Museum Berlin-Dahlem.

Pyrenula (Pyrenulaceae) Ach.
Gk. *pyren* = a nut or hard seed; *-ula* = diminutive; possibly referring to the hard, carbonised ascomatal wall or perhaps referring to the perithecia that are thought to resemble small kernels or grains.

Pyrrhospora (Lecanoraceae) Körb.
Gk. *pyrro* = flaming red; *spora* = spore, seeds; referring to the reddish apothecia of the type species.

Pyxine (Physciaceae) Fr.
La. *pyxis* = a cylindrical box; referring to the dark exciple that encloses the apothecium.

R

Ramalina (Ramalinaceae) Ach.
La. *ramus* = branch, *ramalia* = twigs; *-ina* = diminutive or La. *ramus* = branch; *linum* = thread, rope; referring to the branched thallus of many species, or the cordlike or lined appearance of the surface of the thallus branches.

Ramboldia (Lecanoraceae) Kantvilas & Elix.
For Gerhard Walter Rambold (1956–), German lichenologist, professor of mycology, and head of the DNA Analytics Laboratory at the University of Bayreuth, Germany. He studied at the University of Munich, where he obtained his PhD in 1989. From 1982–2006, he mainly collected lichenised and non-lichenised *Ascomycetes* in central Europe, the Canary Islands, Australia, Southern Africa and the United States (Arizona). He is author of *Bibliotheca Lichenologica*, volumes 34 (1989) and 48 (1992) with D Triebel; has authored or co-authored well over 100 papers on mycology, lichenology and eco-informatics; and is lead editor of the lichen information system LIAS (www.lias.net).

Relicina (Parmeliaceae) (Hale & Kurok.) Hale
Gk. *relicinus* = bent or curled backward or upward, possibly referring to its appearance.

Rhizocarpon (Rhizocarpaceae) Ramond ex DC.
Gk. *rhiza* = root; *karpos* = fruit. The meaning as to what feature of the ascocarp is referred to is unclear.

Ricasolia = Lobaria (Lobariaceae) A.Massal.
For Ricasoli (village) of Arezzo, Tuscany, Italy.

Rimelia (Parmeliaceae) Hale & A.Fletcher
La. *rima* = fissure; *-elia* = a generic ending; referring to the reticulate cracks in the upper surface.

Rimeliella = Canomaculina (Parmeliaceae) Kurok.
Rimelia (q.v.); *-elia* = diminutive.

Rinodina (Physciaceae) (Ach.) Gray
Gk. *rinos* = a shield; *dinos* = rotation, a large round goblet or round cup; possibly referring to

Fruticose lichen (dendritic form) a large tree dripping with lichen from canopy branches. (EJM)

the shape of the apothecia and their often dry or rough-looking discs.

Roccella (Roccellaceae) DC.
La. *rocca* = little rock; referring to this lichen's favoured habitat, maritime rocks.

Roccellina (Roccellaceae) Darb.
Roccella (q.v.); *-ina* = associated with rock. This lichen grows on maritime rocks along the southern coast of South Africa.

S

Saccardoa = Pseudocyphellaria (Lobariaceae) Trevis.
For Pier Andrea Saccardo (1845–1920), Italian botanist, professor of natural history at the University of Padua (1869), and later promoted to professor of botany and director of the botanical gardens (1879). He obtained his PhD from the University of Padua in 1867, having worked there as an assistant from 1866. His only work is *Sylloge*, a list of all the names that had ever been used for fungi, with an entry for each, identifying the currently correct name for that fungus. This consists of 160 000 pages of entries, with only 120 pages of *errata*. Even with the help of Giovanni Battista de Toni, Paul Sydow, Augusto Napoleone Berlese, and his son Daniel Saccardo, this took the last 35 years of his life.

Sagedia = Pertusaria (Pertusariaceae) A.Massal.
Gk. *sago* = large starch grains; perhaps alluding to this lichen's large seed grains.

Santessonia (Physciaceae) Hale & Vobis
For Rolf Santesson (1916–2013), Swedish lichenologist with strong taxonomical interests. He wrote his first scientific paper aged 18. In 1939, he went to the southernmost region of South America, staying two years because of World War II. On his return, he revised several genera, and his work on *Menegazzia* at the Natural History Museum in Stockholm received acclaim as a landmark in South American lichenology. In 1946, he moved to Uppsala, where he worked on foliicolous lichens, and published his major work, *Foliicolous Lichens*, in 1952. Subsequently, he spent much time teaching and supervising, becoming professor and head of the botanical section of the Natural History Museum in Stockholm from 1973–1981. He was elected a member of the Academy of Sciences in 1974, and honoured with the Acharius Medal in 1992.

Sarcographa (Graphidaceae) Fée
Gk. *sarx, sarkos* (*sarco*) = flesh, fleshy; *graphos* = a letter; referring to the swollen stromata composed of many narrow lirellae.

Sarcogyne (Acarosporaceae) Flot.
Gk. *sarx, sarkos* (*sarco*) = flesh, fleshy; *gyne* = woman, female; probably referring to the tendency of a moistened hymenium to turn red (flesh coloured).

Schaereria (Agyriaceae) Körb.
Probably for Ludwig Emanuel Schaerer (1785–1853), Swiss botanist, lichenologist, schoolteacher and clergyman. He studied theology and taught at a primary school in Bern (Switzerland) from 1806–1808. In that year, he was ordained. From 1813, he taught in a high school and became municipal director of an orphanage in 1819. He served as a priest in Lauperswil (Canton of Bern) from 1826, and then Belp (Canton of Zurich) until 1852. He authored several works on lichens, including *Cladoniae Europeae ad Species, Varieties and Forms a Dispositae* (1849), *Enumeratio Critica Lichenum Europaeorum, quos ex Nova Methodo Digerit Ludov* (1850), and *Lichenum Helveticorum Spicilegium. Auctore-ludov. Eman* (1853).

Schismatomma (Roccellaceae) Flot. & Körb. ex A.Massal.
Gk. *schisma* = cleft, division; *omma* = appearance, eye.

Schistoplaca* = *Lecanora (Lecanoraceae) Brusse
Gk. *schistos* = split, cut, divided; *plakos* = *placa* = anything flat, broad, such as a plate.

Schizodiscus (Lecidaceae) Brusse
Gk. *schizein* = to split; *diskos* = disk; referring to the spongy, biconvex, lenticular disc.

Secoliga (Gyalectaceae) Norman
Gk. *sekos* = enclosure, nest; *oligos* = few, small. Meaning unclear.

Siphula (Icmadophilaceae) Fr.
Gk. *siphon* = tube; *-ula* = diminutive; referring to the tube-like shape of the thallus.

Solorina (Peltigeraceae) Ach.
Gk. *solos* = disk; *rinos* = shield; referring to the shape of the apothecia.

Spermatodium (Naetrocymbaceae) Fée ex Trevis.
Gk. *sperma* = a seed; *-odium* = diminutive.

Sphaeria* = *Hypoxylon (Xylariaceae) Haller
La. *sphaer-*, *sphaero-* = spherical, globe shaped; referring to the shape of the fungal fruit bodies (apothecia).

Sphaeromphale* = *Pertusaria (Pertusariaceae) Rchb.
La. *sphaer-*, *sphaero-* = spherical, globe-shaped; *omphalos* = naval. Possibly referring to the shape of the fungal fruit bodies.

Sphaerophorus (Sphaerophoraceae) Pers.
Gk. *sphair-* = spherical, globe-shaped; *phorus* = bearer; refers to the globose apothecia.

Sphinctrina (Sphinctrinaceae) Fr.
Gk. *sphinkter* = band, lace, anything that binds tight; *-ina* = resembling. The apothecia (fruit bodies) have a very narrow opening.

Spiloma* = *Xylographa (Trapeliaceae) Ach.
Gk. *spilos* = a spot; *-oma* = tumour.

Sporodictyon (Pilocarpaceae) A.Massal.
Gk. *sporos* = seed; *diktyon* = net. Meaning obscure.

Sporopodium (Philocarpaceae) Mont.
Gk. *sporos* = seed; *podium* = little foot; referring to the muriform ascospores, the surface of which appears net-like.

Squamaria* = *Cetraria (Parmeliaceae) Hoffm.
Gk. *squama* = a scale or structure resembling a scale; *-aria* = pertaining to. Possibly because the lichen is covered with scales.

Squamarina (Stereocaulaceae) Poelt
Gk. *squama* = a scale or structure resembling a scale; *-ina* = resembling; referring to the appearance of a Squamarina lichen's body (thallus), which is composed of scale-like elements.

Staurothele (Verrucariaceae) Norman
Gk. *stauros* = a cross or upright stake; *thele* = a teat or nipple; referring to the cruciate ostiole of the type species, or an allusion to the muriform spores.

Stegobolus* = *Ocellularia (Graphidaceae) Mont.
Gk. *stego*, *steganos* = covered; *bolus* = bulb.

Stephanophorus* = *Leptogium (Collemataceae) Flot.
Gk. *stephanos* = crown; *phorus* = bearing, carrying. Possibly because in antiquity, Stephanophorus was the chief priest of Pallas, who presided over the rest.

Stereocaulon (Stereocaulaceae) Hoffm.
Gk. *stereos* = stiff, solid; *kaulon* = stalk, stem; referring to the solid podetia of this lichen.

Sticta (Lobariaceae) (Schreb.) Ach.
Gk. *stiktos* = spotted; referring to the spot-like cyphellae of the lower surface.

Stictina* = *Pseudocyphellaria (Lobariaceae) Nyl.
Gk. *Sticta* (q.v.); *-ina* = resembling.

Stigmatidium* = *Enterographa (Roccellaceae) G.Mey.
Gk. *stigmatos* = tattoo mark, spot; *-idion* = diminutive.

Stigmatomma (Verrucariaceae) (Wahlenb.) Trevis.
Gk. *stigmatos* = tattoo mark, spot; *omma* = eye.

Strigula (Strigulaceae) Fr.
La. *strigula* = a scraper, flesh brush; or La. *strigis*, derived from La. *strix* = furrow, channel, groove; *-ula* = diminutive; perhaps referring to the referring to the brush-like appearance of the hymenium or referring to the furrows around the perithecia of the type species.

Synalissa (Lichinaceae) Fr.
Gk. *synalizo* = bring together. The thallus consists of cylindrical to club shaped, deeply branched lobes that hold together only at the base of the cushion and form some kind of fascicle.

Synechoblastus = Collema (Collemataceae) Trevis.
Gk. *synecho* = to hold together; *blastos* = sprout, bud; referring to the repeated septate ascospores.

T

Tapellaria (Philocarpaceae) Müll.Arg.
Anagram of *Patellaria*.

Teloschistes (Teloschistaceae) Norman
Gk. *telos* = end or point; *schistos* = split, divided; referring to the polaribilocular spores, the two end cells divided and rendered remote by an isthmus.

Tephromela (Mycoblastaceae) M.Choisy
Gk. *tephros* = ash-coloured; *melas* = jet-black; referring to the colours of the thallus and the apothecial disk, respectively.

Thamnonoma = Caloplaca (Teloschistacea) (Tuck.) Gyeln.
Gk. *thamno-* = a bush; *noma* = regular, habitat, pasture. The lichen is like a small shrub.

Thelotrema (Graphidaceae) Ach.
Gk. *thele, thelo* = nipple; *trema* = aperture, hole, opening; referring to the morphology of the ascomata in the type species.

Thyrea (Lichinaceae) A.Massal.
Gk. *thyreos* = a large, oblong, door-shaped shield; perhaps referring to the apothecia that are sunken and open by a small pore.

Tichothecium = Verrucaria (Verrucariaceae) Flot.
Gk. *teichos* = wall; *theka* = case, capsule; relating to the walls of a part or cavity of the ascus.

Tomasellia (Naetrocymbaceae) A.Massal.
For Giuseppe (Joseph) Tomaselli (1733–1818), Italian scientist, naturalist and scholar, originally intended to be a priest, then studied chemistry, agriculture, geology and meteorology. He became interested in botany and was associated with the Botanical Garden of Verona, wrote and published a great deal on many subjects, and researched fossil ichthyology. (From an Italian website about the community of Soave.)

Tomaselliella (Arthoniaceae) Cif.
Tomasellia (q.v.); Gk. *-iella* = diminutive.

Toninia (Ramalinaceae) A.Massal.
For Carlo Tonini (1803–1877), chemist and botanist of Verona, who had a herbarium with a well-classified collection of lichens. He was a friend of Massalongo. No further information found.

Topeliopsis (Thelotremataceae) Kantvilas & Vezsda
For Josef Poelt (1924–1995), Austrian lichenologist, curator and head of the department of cryptogams at the Botanical State Collection in Munich (1954–1965), and chair of systematic botany and plant geography at the University of Berlin (1965–1972).  He then moved to the University of Graz, Austria, where he was responsible for the department of botany and the management of the Botanical Gardens of Graz (1972–1990). His first publication was in 1950. He wrote some 200 lichenological publications on floristics, taxonomy, morphology, evolution and biology of lichens, and some 100 papers on bryophytes, non-lichenised fungi, and vascular plants. He authored the highly respected book *Bestimmungsschlüssel Europäischer Flechten* (*Identification Keys for Caucasian Lichens*) (1969). He received the Acharius Medal in 1992, and was an honorary member of a number of distinguished botanical societies.

Tornabenia (Physciaceae) A.Massal.
For Francesco Tornabene Roccaforte (1813–1897), Benedictine friar, founder of the botanical gardens in Catania, and professor of botany at the University of Catania.  He trained at the School of Cosentini, and began his cultural and religious education in the Benedictine monastery of San Nicolò l'Arena. He was a particularly important figure in the history of botany in Catania. He lobbied for the founding of the botanical gardens in Catania in 1843, but it was opened only in 1858. He authored a number of books, including *Lichenographia Sicula* (1849), *Flora Fossile dell'Etna* (1859), *Flora Sicula juxta Methodum Naturalem Vegetabilium Exposita* (1887), and *Flora Aetnea, seu, Descriptio Plantarum in Monte Aetna sponte Nascentium* (1889–1892). He was a member of the Academy Goenia.

Trachylia = Arthonia (Arthoniaceae) Tuck.
Gk. *trachys* = rough, uneven; possibly referring to the crustaceous thallus.

Trapelia (Trapeliaceae) M. Choisy
Gr. *trapelos* = easily turned, changeable; or *trapeza* = table; perhaps referring to the variable, irregular morphology of the exciple. Or could mean referring to the often angulose fruit.

Trapeliopsis (Trapeliaceae) Hertel & Gotth. Schneid.
La. *trapelia* (q.v.); Gk. *-opsis* = resembling. A segregate of Trapelia.

Tricharia (Gomphillaceae) Fée
Gk. *thrichos* = hair; *aria* = like, referring to the long hair-like outgrowths – called hyphophores – which are on the upper surface of the thallus of this lichenised fungi.

Trichoramalina* = *Ramalina (Ramalinaceae) Rundel & Bowler
Gk. *tricho-* = hair, wool, bristle; *Ramalina* (q.v.); referring to this lichen's 'hairy' appearance, a characteristic of *Ramlina*.

Trichotrema* = *Lithothelium (Pyrenulaceae) Clem.
Gk. *tricho-* = hair; *trema* = aperture, hole, opening; possibly referring to the hairy, pore-like opening of the ascomata.

Trypethelium (Trypetheliaceae) Spreng.
Gk. *trype* = hole; *thele* = nipple; possibly referring to the perithecia imbedded in the pseudostroma, or to the many locules and the shape of the ascomata apothecia, which is like an udder with holes (Fee).

Trypethelium (Umbilicariaceae) Hoffm.
La. *umbilicus* = naval-like structure; *aria* = possessing; referring to the navel-like holdfast attaching the thallus to the substratum.

Tylophoron (Arthroniaceae) Nyl. ex Stizenb.
Gk. *tylos* = lump, knob; *phora* = bearing, carrying; referring to the stalks with terminal knoblike reproductive structures.

U

Usnea (Parmeliaceae) Dil ex Adans.
Arabic *ushnah* = moss, although it is a lichen. It has only a superficial resemblance to moss. All species of *Usnea* contain usnic acid, which has antibiotic properties.

V

Valsa (Valsaceae) Fr.
Gr. *walzen* = to dance; Old High Gr. *walzan* = to turn. Meaning obscure.

Verrucaria (Verrucariaceae) Schrad.
La. *verruca* = wart; *aria* = pertaining to; from the wart-like appearance of the ascoma.

X

Xanthodactylon (Teloschistaceae) P.A.Duvign.
Gk. *xanthos* = yellow; *daktylos* = a finger; possibly referring to digitate spikes or divided inflorescence.

Xanthomaculina* = *Xanthoparmelia (Parmeliaceae) Hale
Gk. *xanthos* = yellow; *maculina* = with little spots.

Xanthomendoza (Teloschistaceae) S.Y.Kondr. & Kärnef.
Gk. *xanthos* = yellow; *mendoza* = Mendoza, Argentina. The type species, formerly *Xanthoria mendozae*, was described by the Mendoza Departementos, Argentina.

Xanthoparmelia (Parmeliaceae) (Vain.) Hale
Gk. *xanthos* = yellow; *Parmelia* (q.v.); referring to a segregate of *Parmelia*, which has strong tints of yellow.

Xanthoria (Teloschistaceae) (Fr.) Th.Fr.
Gk. *xanthos* = yellow; La. *orius* = a place suitable for something. The apotheciorum has a thin yellow disk.

Z

Zeora* = *Lecanora (Lecanoraceae) Fr.
The meaning of this name could not be found.

Zschackea* = *Verrucaria (Verrucariaceae) M. Choisy & Werner
For Georg Hermann Zschack (1867–1937), German botanist and teacher in the Anhaltian town of Bernburg, Germany, who devoted his life to the study of lichens. Besides regional studies on the Harz mountains in Germany, Romania and the French island of Corse, he published taxonomic studies of *Verrucariacea*

Glossary of botanical terms used in this book

This glossary briefly covers most of the descriptive terms that you will find in this book. Although the majority of plant names in this book relate to vascular plants (flowers, trees, shrubs, ferns, etc.) there are also a limited number of names relating to bryology (the study of bryophytes – mosses, liverworts and hornworts) and lichenology (the study of lichens). Because of the technical nature of many botanical terms, glossaries are often of limited value. Should you want to know more about the plant, it may be necessary to go to a suitable source to find more elaborate detail. This glossary will, however, give you a headstart in understanding the meanings of words you may not have seen before. To get a better visual understanding of plant terminology, a book that we found useful is *Plant Identification Terminology: An Illustrated Glossary* by James G Harris and Melinda Woolf Harris. It does not cover bryology or lichenology.

Achenes – a small dry indehiscent one-seeded fruit (as of a sunflower).

Acuminate – gradually tapering to a sharp point.

Adaxial – the side next to the axis; the upper surface of something such as a leaf (the opposite of abaxial = away from the axis).

Adpressed – lying close to another organ but not fused to it.

Albumen – the starchy nutritious tissue in a seed.

Algal – referring to green algae.

Amphigastria – stipule-like accessory leaves, different in shape and size to normal leaves.

Anastomosis, anastomosising – the rejoining of veins after branching to form a network.

Androecium – a collective name for the stamens which are the male parts of the flower (see gynoecium).

Androphore – a stalk or column supporting the androecium.

Angulose (fruit) – fruit that is flattened by being compressed with many others.

Annulus – the ring-like structure sometimes found on the stipe (stalk) of some species of mushrooms.

Anther – that part of the stamen bearing the pollen.

Anthesis – the flowering period of a flower from the opening of the bud flower until it wilts.

Apical cells – cells that divide repeatedly to form new cells at a shoot tip.

Apices – (plural of **apex**) – the tip or distal end.

Apophysis (or **hypothysis**) – a swelling of the seta (stalk) immediately below the capsule.

Apotheca – a cuplike, spore-bearing structure in many lichens containing sacs in which sexual spores are developed.

Apothecial disks – the exposed upper surface of the hymenium (q.v.) in a central apothecium ascocarp.

Archegonia – a female reproductive organ containing a single egg cell.

Archegoniophores – a structure that bears the archegonia in some liverworts.

Areolate – marked with areola (pl **areolae**, **areolas**) – see below.

Areolation – a small hollow or cracks in the surface of leaves.

Areole – an open space formed by anastomosing veins; a small pit or raised spot often with a tuft of hairs or spines.

Aril – fleshy thickening on the seed coat.

Aristate (glumes) – the glumes end in a stiff, bristle-shaped point.

Aristulate – having a minute awn or bristle at the tip.

Ascocarp (also called **ascoma**, pl **ascomata**) – the fruiting body of an ascomycetous fungus (term in lichenology).

Ascomata – the fruiting body of an ascomycete fungus.

Ascomycete – an old word now called ascomycota.

Ascus, pl **asci** – a saclike structure that contains and releases the sexual spores (ascospores) in fungi of the ascomycota.

Auriculate – ear-shaped.

Awns – a narrow, bristle-like appendage. Usually found towards the tip of the spikelets of many grasses.

Axil – the point between the stem and any leaf stalk emerging from it.

Barbate – having tufts of long hairs or spines; bearded.

Basal – situated or attached to the base.

Bifid – deeply cut or divided into two parts for about half its length (such as a leaf).

Bilabiate – having two lips, relating to the corollas of many flowers.

Bilamellate – made up of two plates with raised edges.

Bipartite (capsule) – the capsule, a dried fruit that opens at maturity, splits open in two parts.

Bipinnate – twice pinnate, as with the pinnate leaf, with the divisions again pinnately divided.

Bisaccate lip – having two little bags, sacs, or pouches with a conspicuous lip or outgrowth.

Bivalve shell – a shell that is in two halves.

Bracteate thyrses – a cylindrical or egg-shaped inflorescence with bracts (reduced leaf-like structures).

Bracteoles – small bracts borne singly or in pairs on the pedicel or calyx.

Bracts – a usually small leaf-like structure, in the axil of which arises a flower or a branch of an inflorescence.

Bryophyte – a green, seedless, non-vascular plant which includes mosses, hornworts, or liverworts.

Bulbate – swollen, inflated.

Bursicula – a small structure shaped like a pouch or purse.

Calcarate – having a calcar, a spur.

Calyptra – a thin veil, hood or lid.

Calyx, pl **calyces** – the outer whorl of a flower, a collective term for the sepals.

Capsule – a dry dehiscent fruit formed from two or more united carpels which split apart to release seed.

Carpel – the female reproductive organ of a flower, consisting of an ovary, a stigma, and usually a style.

Cataphyll – a leaf form or shoot which does not perform the usual functions of photosynthetic leaves.

Caudiform – having the shape of a tail.

Cilia – hair-like protrusions, black hairs found in some lichens, hair-like threads of inner peristome, a marginal fringe of hairs.

Circumboreal – referring to the northern regions of the world.

Cladode – modified, flattened petiole or branchlet that resembles a leaf and having their function.

Cleft lemma – referring to the lower most bracts (lemma and palea) of a grass spikelet, cut or split about half-way to the base.

Cleistocarpus (capsule) – opening irregularly, not by a lid or valves.

Clinandrium – the portion of an orchid column in which the anther is concealed.

Coalescing – plants come together to form one mass or whole.

Columella – a small column; the central axis of the cone or fruit.

Companulate – bell-shaped (flowers).

Conidia – an asexual spore formed by special structures that arise from the hyphae (in lichenology).

Connate – fusion of similar parts, such as the staminal filaments into a tube.

Corm – a fleshy, swollen underground stem usually with fibrous or scaly outer layer.

Corolla – the collective name for all the petals of a flower, making the second whorl or inner whorl of a flower, a collective term for petals.

Glossary of botanical terms

Corolla spurs – outgrowth from the base of a perianth segment.

Cortex – the layer of tissue inside the corky or epidermal tissues of a vascular plant.

Corymbose – having flat-topped or rounded-topped inflorescence.

Costa – a rib, mid-rib or prominent mid-vein.

Cotyledons – primary leaf or leaves of an embryo, becoming the first seed leaf or leaves.

Crescentic – crescent-shaped.

Cucullate – hooded or hood-shaped.

Cutinized – a wax-like, water-repellent material (cutin) present in the walls of some plant cells.

Cyme – a flat- or round-topped inflorescence in which the terminal flowers open first.

Cypselae, Cypselas – a dry, single-seed, indehiscent fruit.

Decussate – leaves arranged in pairs along the stem, with each pair at right angles to the pair above and/or below.

Dentate leaves – toothed at the leaf edges, the teeth directed outwards rather than forward.

Desiccated – deeply divided; cut into many segments.

Diaphanous – translucent.

Dichotomous – branched or forked into two more or less equal divisions or ways.

Dioecious – male (staminate) and female (pistillate) flowers are borne on different plants.

Discoid – resembling a disc or plate.

Distal – situated away from the point of attachment.

Distichous – the leaves or flowers are in two rows on opposite sides of a stem.

Divaricate, divaricately forked – widely spread out or diverging.

Domatia – conspicuous tufts of hair or small pits; usually found at the junction of the midrib and main lateral vein.

Dorsal – the back surface or outward surface of an organ (see abaxial).

Dorsiventral – having structurally different upper and lower surfaces, e.g. as with some leaves.

Ellipsoid – a solid body that is oval in the long section and circular in the cross section.

Emarginate – having a notched margin or tip, as have some leaves or petals.

Emollient – having the quality of softening or soothing the skin.

Endostome – the inner part of the peristome of a moss.

Epapillose – without having a nipple-like bumps or projections.

Epidermis – the outer layer of tissue in a plant.

Epiphytic (habit) – a plant grows upon another but does not draw food or water from it, e.g. an air-plant.

Exciple, Excipulum – tissue forming margins or walls of an ascoma.

Exindusiate – lacking an indusium.

Exostome – the small aperture or foramen in the outer coat of the ovule of a plant.

Exserted – projecting beyond the surrounding plants, eg the stamens beyond the corolla.

Farinaceaous – mealy or starchy in texture.

Fenestrated (window) – a window-like perforation, opening or translucent area.

Festucine – the colour of yellow straw.

Filament – a thread-like structure such as the stalk of the stamen which supports the anther.

Filiform – thread-like.

Fimbriate – fringed, usually with hairs or hair-like structures.

Fissured – split or cracked, often referring to fissured bark.

Flagella – a whip-like appendage that protrudes from the cell body of certain eukaryotic and eukaryotic cells.

Flagellate – with long, slender runners.

Fronds – a leaf of a fern, cycad or palm.

Fructification – the fruiting process of the plant.

Fruiting capitula (fruiting heads) – plants at the fruiting stage.

Fungal hypha, pl **hyphae** – the long, branching, microscopic filamentous structure of a fungus.

Fusiform – spindle-shape, narrowing gradually from the middle towards each end.

Galeate – helmet-shaped.

Gametophyte – plant that bears gametes; a cell or nucleus that fuses with another of the opposite sex in sexual reproduction.

Gelatinous – containing gelatine, which has the nature of or resembles jelly.

Gemma, pl **Gemmea** (borne on thallus) – a small bud-like structure or cluster of cells, which propagate offspring plants.

Gemmiparous – reproducing by gemmation: asexual reproduction by the production of gemmea; budding.

Gibbaeum – the name of a genus, gibba meaning 'hump'.

Glabrous – smooth, hairless.

Glands – an appendage, protuberance which secretes sticky, oily or sugary substances.

Glandular-hairy – hairs which terminate in a gland, often sticky to the touch.

Globose – globe-shaped, near spherical.

Glomerule – a dense cluster of flowers caused by the condensation of a cyme (a flat-topped or rounded inflorescence).

Glume – one of the paired bracts at the base of a grass spikelet (= flower).

Gymnostemium – a structure formed from the fusion of the style and stamens in the Orchidaceae.

Gynaecium – the female reproductive organs of a flower, a collective term for the carpels or pistils.

Gynophore – the elongated stalk that raises the pistil above other flower parts.

Hemiparastic – a parasitic plant, such as the mistletoe, which obtains a portion of its 'food' from a host plant.

Hermaphroditic – bearing both male and female reproductive organs, having pistils and stamens in the same flower.

Heteromericarpic – having mericarps of more than one kind or form.

Heterostylic – flowers of the same species have styles of different length.

Homogamous – having all flowers alike, of the same sex.

Hyaline (cells) – colourless and partially transparent.

Hymenium – a layer of cells containing the spore-bearing cells.

Hypanthium – a cup-shaped extension of the floral axis formed by the fusion of the calyx, corolla and androecium.

Hyphae – the fine, branching tubes which make up the body (or mycelium) of a multi-cellular fungus.

Hyphophore – simple or complex whip-like outgrowths on the upper surface of the thallus.

Hypogynous – having the floral parts, petals, sepals, and stamens attached below the ovary.

Incised – cut sharply, deeply and irregularly.

Indehiscent – (of a pod or fruit) not splitting open to release the seeds when ripe.

Indumentum (La. literally: 'garment') – an outer covering on a plant, such as hairs, down or bristles.

Indusium, pl **Indusia** – a thin membranous covering, like a 'shield', covering a sorus on a fern frond.

Inflorescence – the complete flower head of a plant including stems, stalks, bracts, and flowers.

Internodes – the portion of the stem between two of the nodes from which leaves emerge.

Involucral bracts (phyllaries) – a leaf-like plant, usually located just below a flower, flower stalk, or inflorescence.

Involucre – a group of bracts surrounding the base of a flowerhead.

Labellum (La. = 'lip') – the lip or lower petal of an orchid, often enlarged or variously shaped.

Laciniate – divided into deep narrow irregular segments.

Lamina – the expanded upper part of a petal, sepal or bract, also leaf blade.

Leaf margins – the boundary area extending along the edge of the leaf which can have many shapes.

Lemma – the lower of two bracts enclosing a grass flower.

Lenticular – lens-like, biconvex in cross-section, spherically shaped.

Lirella, pl **lirellae** – an ascoma with a long, narrow disc often branching somewhat star-like (in lichenology).

Glossary of botanical terms

Lobate – having lobes, e.g. lobate oak leaves.

Lobulate – with lobules, small lobes or lobe-like subdivisions of lobes.

Mazaedia – a fruiting body of fungi of the phylum ascomycota in which there is a powdery mass of spores (in lichenology)

Medial nerve – a large vein situated in the middle of a leaf.

Medulla – loosely packed cells or hyphae.

Mericarps – a carpel with one seed; one of a pair split apart at maturity.

Mesocarp – the middle layer of the pericarp of a fruit.

Moniliform – has a beaded necklace-like appearance; cylindrical and constricted at regular intervals.

Multi-cellular – having or consisting of many cells.

Muriform – resembling courses of bricks or stones in square and regular arrangement.

Nectaries – a nectar-secreting glandular organ in a flower (floral) or on a leaf or stem (extra-floral).

Nodes – the part of a plant stem from which one or more leaves emerge.

Non-resupinate – the opposite of resupinate, which is upside down due to the twisting of the pedicel as in some orchids, and facing upwards rather than downwards.

Operculum – the lid which closes the capsule and permits the spores to escape when it falls.

Osteolate – a pericarp with hole on it.

Ostiole – in some small algae and fungi, a small pore through which spores are discharged (in lichenology).

Ovary – the hollow base of the carpel of a flower, containing one or more ovules.

Palea, pl **paleae** – the upper bract of the floret of a grass.

Panicles – a much-branched cluster of flowers.

Papilla, pl **papillae** – a small pimple-like projection on the surface of cells.

Papillose – having the surface roughened by small protuberances like little warts.

Pappus – the scales or bristles on the fruit of many members of the Asteraceae family.

Parasite – a plant that obtains all of its food from another living plant (the host) to which it is attached.

Pectinate – comb-shaped.

Pedicel – a small stalk bearing an individual flower in an inflorescence.

Pedunculate – having the stalk (peduncle) of a solitary flower or inflorescence.

Peltate – shield-shaped: a flat structure, usually circular, born on a stalk attached to the lower surface.

Perennial – a plant that lives for several years, often flowering annually.

Perianth – the outer, sterile parts of a flower comprising the calyx and corolla.

Pericarp – the part of a fruit formed from the wall of the ripened ovary.

Pericarpium – refers to the wall of a fruit, the peel or rind.

Perigone – the perianth of a flower, the perigonium.

Peristome – the ring of teeth or fringe surrounding the capsule.

Perithecia, pl **perithecium** – a flask-shaped hollow fruiting body in various ascomycetous fungi.

Persistent – something, such as an annulus on a mushroom, which continues exist or occur over a prolonged period.

Petiole – a leaf stalk.

Phyllaries – bracts found in the involucre head of a composite plant.

Pinna, pl **pinnae** – one of the primary divisions or leaflets of a pinnate leaf.

Pinnules – any of the lobes of a leaflet of a pinnate compound leaf, which is itself pinnately divided.

Pistil – the female reproductive organ of a flower.

Pistillate (flower) – bearing a pistil or pistils, but lacking in stamens.

Podetia, pl **podetium** – a stalk on which the ascocarp is borne in various lichens (as of the genus *Cladonia*).

Pollinium, pl **pollinia** – a mass of waxy pollen grain stuck together and transferred as a whole in pollination, often by insects.

Prostrate (habit) – lying flat on the ground.

Prothallus – a thallus-like growth germinating from a spore under the main body of the lichen.

Pseudobulb – a bulbous thickening on the stems of many epiphytic orchids.

Pubscent – covered with short soft hairs.

Pustulate – covered with small blisters or pustules.

Pyxidium – a seed capsule that splits open at the top like a lid.

Quillwort – any fernlike, aquatic marsh plant of the genus Isoëtes, characterised by clustered, quill-like leaves.

Raceme – an unbranched, elongated inflorescence with flowers maturing upwards.

Rachilla – the central stalk of a grass spikelet, to which the glumes and florets are attached.

Rachis – the main stem of a compound leaf or inflorescence.

Radiate, radiating – ridges radiate out, star-like on the bottom some fruit which indicates the number of fruit segments inside.

Receptacle – the end of the pedicel (or peduncle) upon which the flowers are born.

Reticulation (vein) – refers to the network-vein structure.

Rhizohyphae – clustered hyphae, developing from the lower medulla, usually black, bluish or whitish (especially in Pannariaceae) (in lichenology).

Rhizome – a rootstock; a horizontal underground stem.

Rostellum – a small beak, from the upper edge of the stigma in orchids.

Rugose (appearance) – wrinkled, usually transversely.

Samara, pl **samarae** – a dry, indehiscent, winged fruit whose shape allows the seed to be widely dispersed by the wind.

Scales – any thin, flat scarious structure.

Scarious – referring to surface texture which is thin, dry, membranous and not green.

Schizocarp – a dry, dehiscent fruit that at maturity splits into two or more one-seeded carpels.

Sclerophyllous – with stiff, rigid leaves.

Seed pod – a case that holds a plant's seeds.

Segregate – separated from others of the same kind.

Sepals – a segment of the calyx.

Seta – in a moss or liverwort, the stalk supporting the capsule.

Silicles – a dry, dehiscent fruit of the Brassicaceae family.

Skein – a length of thread or yarn, loosely coiled and knotted.

Sorus, pl **sori** is a cluster of sporangia (structures producing and containing spores) in ferns and fungi.

Spadix – a spike with small flowers crowded on a thicken axis, e.g. arum lily.

Spathaceous – spathe bearing.

Spathe – a large leaf or petal-like bract enclosing the entire or partial flower cluster, e.g. an arum lily.

Spathella – an archaic term referring to the lemma of a grass flower, occasionally to the glume of a grass spikelet.

Spatheoles – a small spathe enclosing part of an inflorescence typically surrounding a single raceme or a pair of racemes.

Spermatophytes – plants reproducing by seed.

Spicate – arranged in a spike.

Spicules – a short, pointed epidermal projection.

Spines – a stiff, sharp-pointed structure arising from below the epidermis.

Spinose – bearing spines.

Sporangium, pl **sporangia** – a spore-bearing case or sac.

Spores – reproductive cell.

Sporophyll – a sporangium-bearing leaf, often modified in structure.

Sporophytes – the spore-bearing generation consisting of seta and capsule (only in ferns and bryophytes)

Squamules – the lodicule (paired rudimentary scales of the ovary) of a grass flower.

Squarros – abruptly recurved.

Stamen – male reproductive organ comprising a stalk (filament) and an anther.

Staminal – of or pertaining to the stamens.

Staminodes – a modified stamen which is sterile, producing no pollen.

Stellate – star-shaped.

Stigmatic – belonging to the stigma or having characteristics of the stigma.

Stipe – a stalk that supports some other structure.

Stipule – a leaf-like appendage found at the base of the plant, usually found in pairs.

Stolon – a lengthy, horizontal stem creeping along the ground, rooting at the nodes, thus giving rise to new plants.

Stria, pl **striae** – a fine line or groove.

Stroboli – a cone or an inflorescence that is cone-shaped.

Stroma – a mass or matrix of vegetative hyphae (usually black) with or without tissue in or on which spores are produced.

Sub-capitulate clusters – inflorescence which are nearly 'head-shaped' clusters, but not quite.

Suborbicular – almost circular in outline.

Subtend – to stretch beneath, as in a bract which extends under flowers so as to support and enfold them.

Sulcate – with longitudinal grooves.

Synsepal – with united sepals.

Terminal – at the tip or apex.

Terricolus, **terricolous** – referring to a plant, especially a lichen, growing on soil or on the ground.

Testa, pl **testae** – the protective covering of a seed; the seed coat.

Testiculate (tubers) – resembling testicles.

Thallus – the plant body of algae, fungi, lichens, and some liverworts which lack roots, stems, leaves as found in vascular plants.

Thecium – the hymenium (q.v.): the internal hyphal layers of an apothecium.

Transverse – at a right angle to the longitudinal axis of a structure.

Trichome – a hair or hair-like outgrowth of the epidermis.

Triquetrous nut – having three corners, triangle like, each corner projecting outwards.

Truncated – with the apex or the base squared at the end as if cut off.

Tuber – a swollen stem or root that stores food or water underground, bearing buds from which new plants arise.

Tunic – the outer coating or covering of a seed or bulb.

Umbel – a flat-topped to slightly rounded inflorescence with pedicels arising from more or less a common point.

Umbelliferous – bearing umbels.

Umbillicate – with a depression in the middle, like a naval.

Unguent – a soft greasy or viscous substance used as ointment or for lubrication.

Unilocular – with a single locule or compartment, as in some ovaries.

Uniseriate – arranged in a single row or series.

Urceolate – urn-shaped; large below and contracted toward the mouth.

Ventral – facing towards the stem of a plant (in particular denoting the upper surface of a leaf).

Ventricose – inflated or swollen on one side only, as in some corollas.

Vesicles – an air-filled swelling in a plant.

Virgate (stem) – erect, straight, wand-like, and slender.

Viscous (sap) – thick, sticky, gooey, slow flowing sap.

Xeric – an environment or habitat which receives very little moisture resulting in a dry area.

Bibliography

The following list is a representative indication of the types of books and journals as well as select online sources we accessed in compiling this book – we have over 1 500 records of websites visited.

AusGrass 2: Grasses of Australia. Ausgrass2.myspecies.info

Bailey, H. 1949. *Manual of Cultivated Plants*. Piacnillan Company, New York.
Bean, A and Johns, A. 2005. *Stellenbosch to Hermanus: South African Wild Flower Guide 5*. Botanical Society of South Africa, Cape Town
Borror, DJ. 1971. *Dictionary of Word Roots and Combining Forms*. Palo Alto, California, Mayfield Publishing Company
Botanary, the Botanical Dictionary. Davesgarden.com/guides/botanary.
Britten, J and Boulger, GS. 1893. *A Biographical Index of British and Irish Botanists*. London: West, Newman & Co.
Brummitt, RK and Powell, CE. 1992 (editors). *Authors of Plant Names*. Royal Botanic Gardens, Kew

Charters, M. 2008. The Eponym Dictionary of Southern African Plants. Calflora.net/southafrica/plantnames.html
Clifford, HT and Bostock, PD. 2007. *Etymological Dictionary of Grasses*. Springer-Verlag Berlin Heidelberg New York
Coombs, AJ. 2012. *The A to Z of Plant Names*. Coombs, AJ. 2012. The A to Z of Plant Names. Timber Press, Portland USA
Coombs, AJ. 1994. *Dictionary of Plant Names*. Timber Press, Portland USA

Desmond, R and Ellwood, C. 1994. *Dictionary of British and Irish Botanists and Horticulturists*. Taylor & Francis and the Natural History Museum, London
Dyer, RA. 1976. *Genera of Southern African Flowering Plants*. (2 vols) Department of Argricultural Technical Service. Pretoria

Eggli, U and Newton, L. 2004. *Etymological Dictionary of Succulent Plant Names*. Springer-Verlag Berlin Heidelberg New York

Feige, GB. 1998. *Etymologie der Wissenschaftlichen Gattungsnamen der Flechten*. Published by the author, Essen.
Fiddlehead Forum. The American Fern Society. Amerfernsoc.org/ffa.

Germishuizen, G and Meyer, N (eds). 2003. *Plants of Southern Africa: An Annotated Checklist*. Strelitzia 14. National Botanical Institute, Pretoria.
Gledhill, D. 2008. *The Names of Plants*. Cambridge University Press
Glen, HF and Germishuizen, G. 2010. *Botanical Exploration of Southern Africa – Edition 2*. Strelitzia 26. South African National Biodiversity Institute Pretoria
Glen H.F. 2002. *Cultivated plants* of *Southern Africa : botanical names, common names, origins, literature*. Jacana Johannesburg
Glen, HF. 2004. Sappi, *What's In A Name? The Meaning of the Botanical Names of Trees*. Jacana Johannesbug.
Goldblatt, P and Manning, J. 2000. *Cape Plants: A Conspectus of the Cape Flora of South Africa*. National Botanical Institute of South Africa & M.G. Press, Missouri U.S.A
Gunn, M and Codd, LE. 1981. *Botanical Exploration of Southern Africa*. A.A. Balkema Cape Town
Gunn, M and Codd, LE. 1984. 'Additional Biographical Notes on Plant Collectors in Southern Africa.' Bothalia 15(3/4).

Hardy Fern Library. Hardyfernlibrary.com/ferns.
Harris, JG and Woolf Harris, M. 2004. *Plant Identification Terminology : an illustrated glossary*. Spring Lake Publishing, Payson, USA
Harrison, L. 2012. *Latin for Gardeners: Over 3 000 Plant Names Explained and Explored*. Mitchell Beazley, Great Britain
Harvey, WH. 1838. *The Genera of South African Plants*. A.S. Robertson, Cape Town
Harvey, HH and Sonder, OW. 1894–1925. *Flora Capensis*. Thistleton-Dyer ed. Reeve London
Haynes, J. Etymological Compendium of Cycad Names. Cycad.org/documents/etymological_compendium.pdf
Herre, H. 1971. *Genera of the Mesembryanthemaceae*. Tafelberg-Uitgewers Beperk, Cape Town
Hyam, R and Pankhurst, R. 1995. *Plants and Their Names: A Concise Dictionary*. Oxford University Press

Index Fungorum. Indexfungorum.org.

Jackson, WPU. 1990. *Origins and Meanings of Names of South African Plant Genera.* UCT Ecolab c/o Botany Department, UCT Printing Department, Cape Town

Johnson, AT and Smith, HA. (Orig. 1931) 2008 *Plant Names Simplified.* Old Pond, UK

Kesting, D and Clarke, H. 2008. (3rd ed.) *Botanical Names: What They Mean.* Flora Documentation Programe, Friends of Silvermine Nature Area (FOSNA). Ink and Print, Port Elizabeth

Laurence, JD. 1997. *Plant Collectors in Madagascar and the Comoro Islands.* Department of Botany, MRC-166, National Museum of Natural History, Smithsonian Institution, Washington, U.S.A.

Manning, J. 2007. *Field Guide to Fynbos.* Struik Publishers, Pretoria
Manning, J. 2009. *Field Guide to Wild Flowers of South Africa.* Struik Nature, Cape Town
Meager, G. 2008. An Etymology of Australian Bryophyte Genera 1 – (Liverworts and Hornworts).
http://www.bioone.org/doi/abs/10.3158/0015-0746-47.1.257
Meager, G. 2011. An Etymology of Australian Bryophyte Genera 2 – Mosses.
https://www.rbg.vic.gov.au/documents/Muelleria_29(1),_Meagher.pdf

Neal, B. 1993. *Gardener's Latin – a Lexicon.* Robert Hale Ltd, London

Paxton, J. 1849. *Pocket Botanical Dictionary.* Bradbury & Evans, London
Perrin, D. 1990. *Dictionary of Botanical Names : Australian Plant Names.* Kippa Ring Qld. Australia
Plants of Southern Africa: an online checklist. Sanbi. Posa.sanbi.org/searchspp.php
Pooley, E. 1998. *A Field Guide to Wild Flowers: KwaZulu-Natal and the Eastern Region.* Natal Flora Publications Trust, Durban.
Pooley, E. 2003. *Mountain Flowers: a field guide to the flora of the Drakensberg and Lesotho.* The Flora Publications Trust, Durban.

Quattrocchi, U. 1999. *CRC World Dictionary of Plant Names.* CRC Press. USA
Quattrocchi, U. 2006. *CRC World Dictionary of Grasses.* CRC Press. USA

Red List of South African Plants. Sanbi. Redlist.sanbi.org

Schmidt, E et al. 2002. Trees and Shrubs of Mpumalanga and Kruger National Park. Jacana
Smith, AW. 1963, 1997. A Gardener's Handbook of Plant Names: *Their Meanings and Origins.* Dover Publications Mineola, New York
Stearn, WT. 1966. (3rd ed.) Botanical Latin. Redwood Burn Ltd, Trowbridge, Wilts, Devon for David & Charles Inc , Vermont USA

The Plant List. Royal Botanic Gardens Kew and Missouri Botanical Garden. Theplantlist.org.
Tropicos. Missouri Botanical Garden. Tropicos.org.

Van Wyk, B and Van Wyk, P. 2007. *How to Identify Trees in Southern Africa.* Struik Publishers, Cape Town

Wikipedia. List of botanists by author abbreviation. https://en.wikipedia.org/wiki/List_of_botanists_by_author_abbreviation
Wikipedia. List of South African plant botanical authors. En.wikipedia.org/wiki/List_of_South_African_plant_botanical_authors
Wilhelm, G and Rericha, L. 2003. Lichens of the Chicago region.
www.conservationresearchinstitute.org/assets/lichenflorachicago2.pdf

Zimmer, GF. 1912 (reprint 1946). *A Popular Dictionary of Botanical Names and Terms with their English equivalents.* George Routledge & Sons, London

Sources used to compile the biographies

The short biographies are based on information found on Wikipedia, the second edition of Mary Gunn and LE Codd's *Botanical Exploration of Southern Africa* (*BESA*), Michael Charters' *The Eponym Dictionary of Southern African Plants* (*EDSAP*), many other books, journals and lexicons, personal communications by email and our own research. Often we took fragments from books and articles. For space reasons, only the major sources are shown. Wikipedia content is licensed under the Creative Commons (CC) licences CC BY-SA 3.0 or 4.0.

A

Abelia – **Clarke Abel (1780–1826)** En.wikipedia.org/w/index.php?title= Clarke_Abel&oldid=632170088l. Image: public domain. Source: En.wikipedia.org/w/index.php?title=Clarke_Abel&oldid=632170088. CC BY-SA 3.0

Abildgaardia – **Peter Christian Abildgaard (1740–1801)** Da.wikipedia.org/w/index.php?title=Peter_Christian_Abildgaard&oldid=7282966; Minrec.org/libdetail.asp?id=2. Image: public domain. Source: Dvjb.kvl.dk/upload/dvjb/ill/abildgaard-b.jpg. CC BY-SA 3.0

Acharia – **Erik Acharius (1757–1819)** En.wikipedia.org/w/index.php?title=Erik_Acharius&oldid=621603543& 1911encyclopedia.org/Erik_Acharius. Image: public domain. Source: www-hotel.uu.se/evolmuseum/fytotek/acharius.html. CC BY-SA 3.0. Acharius Herbarium and text. Source: Luomus.fi/en/botanical-and-mycological-collections, Natural History Museum of Helsinki.

Adansonia – **Michel Adanson (1727–1806)** En.wikipedia.org/w/index.php?title=Michel_Adanson&oldid=630087551; Abebooks.co.uk/search/sortby/3/an/Adanson+/tn/+Histoire+Naturelle+Du+Senegal. Image: public domain. Source: Jules Pizzetta – Galerie des naturalistes de J Pizzetta, Paris: Ed. Hennuyer, 1893. CC BY-SA 3.0

Aerisilvaea – **Herbert Kenneth Airy Shaw (1902–1985)** Anbg.gov.au/biography/airy-shaw-herbert-kenneth.html. Image: extracted from *Kew Bulletin* 42(1), 1987; photographer's name unknown.

Afzelia – **Adam Afzelius (1750–1837)** En.wikipedia.org/w/index.php?title=Adam_Afzelius&oldid=599681441 Image: public domain. Artist/source: Oil painting by Carl Fredrik von Breda (1759–1818) – Carl von Linee project. CC BY-SA 3.0.

Aitonia – **William Aiton (1731–1793)** En.wikisource.org/w/index.php?title=Aiton_William_(1731–1793)_(DNB00)& oldid=2335094. CC BY-SA 3.0

Alberta – **Albertus Magnus (c 1200–1280)** En.wikipedia.org/w/index.php?title=Albertus_Magnus&oldid=632284994& encyclopedia.com/topic/Saint_Albertus_Magnus.aspx#1-1G2:2830900073-full. Image: public domain. Artist/source: Tommaso da Modena (1326–1379) First uploaded by sv:Användare:Lamré to Swedish Wikipedia as sv:Bild:AlbertusMagnus.jpg; Kapitelsaal des ehemaligen Dominikanerklosters San Niccolò in Treviso. CC BY-SA 3.0.

Albertisia – **Luigi D'Albertis (1841–1901)** En.wikipedia.org/w/index.php?title=Luigi_D%27Albertis&oldid= 632011011&adbonline.anu.edu.au/biogs/A040002b.htm. Image: public domain. Source: Papuaweb.org/dlib/bk2/dalbertis/index.html. CC BY-SA 3.0.

Alchornea – **Stanesby Alchorne (1727–1800)** Plants.jstor.org/person/bm000392821&chelseaphysicgarden.co.uk/aboutus/former.html; Brisrain.webcentral.com.au/01_cms/details_pop.asp?ID=18.

Aldrovanda – **Ulisse Aldrovandi (1522–1605)** En.wikipedia.org/w/index.php?title=Ulisse_Aldrovandi&oldid= 628777658biology.ualberta.ca/courses.hp/ent207/1999/biolcontrol-1.htm& ius.edu/journalism/J210F03/Carner/PAGE3.html. Image: public domain.
Agostino Carracci (1557–1602) Source: Photographer unknown: Museopalazzopoggi.unibo.it//poggi_eng/palazzo/foto/prot. CC BY-SA 3.0.

Alinula – **Aline Marie Raynal-Roques (1937–)** Babelio.com/auteur/Aline-Raynal-Roques/39696: Image: photographer unknown.

Allenia – **Robert Allen Dyer (1900–1987)** En.wikipedia.org/wiki/Robert_Allen_Dyer: Source: Gunn & Codd, *BESA*. Image: Sanbi

Almbornia – **Dr Ove Almborn (1914–1992)** Plants.jstor.org/person/bm000051095&floraseries.landcareresearch.co.nz/pages/ReferenceList.aspx?id=_dca8a477-b60d-4bde-9079-68c052569741&fileName=Lichen+Ed2+Pan-Z.xml& biomedsearch.com/nih/How-many-species-fungi-are/18490969.html.

Alonsoa – **Cenón (or Zenón) Alonso Acosta Zorilla y Dávila (1756–?)** Translate.google.co.za/translate?hl=en&sl= es&u=ghart27.iespana.es/definiciones/Aindice.html&ei=Kc78TPjKFMueOqm-xdQK&sa=X&oi=translate&ct=result& resnum=6&ved=0CDkQ7gEwBQ&prev=/search%3Fq%3DAlonzo%2BZanoni,%2BSpanish%2Bofficial%2Bat% 2BBogota%26hl%3Den%26rlz%3D1B3GGIC_enZA317ZA317%26prmd%3Dbo.

Alpinia – **Prospero Alpino (1553–1617)** En.wikipedia.org/w/index.php?title=Prospero_Alpini&oldid=601916986. Image: public domain. Artist: Leandro Bassano. Source/photographer: Scan par Valérie75. CC BY-SA 3.0.

Alstonia – **Charles Alston (1683 –1760)** Google.co.za/#q=tCharles+Alston+(1683+%E2%80%931760&nahste.ac.uk/ isaar/GB_0237_NAHSTE_P0250.html; Archives.lib.ed.ac.uk/catalogue/pers/a/250.

Alstroemeria – **Klas von Alstroemer (1736–1796)**
Es.wikipedia.org/wiki/Clas_Alströmer. Public domain. Artist unknown (Swedish painter). Photographer Andersson et al. (2005):ebc.uu.se/undervisning/AlumnCvL/uppsatser/Larjungar/MattsonMarie.pdf CC BY-SA 3.0

Althenia – **Jean Althen (1709–1774)** En.wikipedia.org/w/index.php?title=Jean_Althen&oldid=617817821; Unmondedebrut.pagesperso-orange.fr/HISTOIRE/COLON_ALGERIE/REFERENCE/Garanceexplication.htm. Image: public domain. Author: User Vmenko. Source: own photo CC BY-SA 3.0.

Alvesia – **Bento Antonio Alves (f 1800s)** Source: Charters, *EDSAP*.

Amandinea – **Amandine Manière (1937***)* Sources uncertain.

Ammannia – **Paul Ammann (1634–1691)** En.wikisource.org/w/index.php?title=1911_Encyclop%C3%A6dia_ Britannica/Amman,_Paul&oldid=3552260 Source: 1911 *Encyclopedia Britannica*/Amman Public domain. CC BY-SA 3.0.

Amperea – **André-Marie Ampère (1775–1836)**
En.wikipedia.org/w/index.php?title=Andr%C3%A9-Marie_Amp%C3%A8re&oldid=632557163. Image: public domain. Author: Ambrose Tardieu. Source: The Dibner collection at the Smithsonian Institution. CC BY-SA 3.0.

Amsinckia – **Wilhelm Amsinck (1752–1831)** De.wikipedia.org/w/index.php?title=Wilhelm_Amsinck_(1752%E2%80% 931831)&oldid=127622577. Image: public domain. Author: unknown. Source: *Hamburg Biography*, Vol. 5, p 20 (Orig. Lt. Photo credits in the Hamburg State Archives). CC BY-SA 3.0.

Amsonia – **Charles Amson (1698–1763)** Swcoloradowildflowers.com/biographies%20of%20naturalists.htm; Personal communication with website compiler Al Schneider.

Anderbergia – **Arne Alfred Anderberg (1954–)** Text and image: Personal communication with Arne Anderberg, professor and head of botany, Swedish Museum of Natural History.

Andradia – **Alfredo Augusto Freire de Andrade (1859–1929)** Pt.wikipedia.org/w/index.php?title=Alfredo_Augusto_ Freire_de_Andrade&oldid=40404977; four other websites not shown here. Image: author unknown. Copyright: unknown. Source: Delagoabayworld.wordpress.com/2012/03/10/alfredo-augusto-freire-de-andrade. CC BY-SA 3.0.

Andreaea – **Johann Gerhard Reinhard Andreä (1724–1793)** En.wikipedia.org/w/index.php?title=Johann_Gerhard_ Reinhard_Andreae&oldid=599394934. Image: public domain. Author: Johann Georg Ziesenis. Source: Museum-digital.de. Permissions: Gleimhaus Halberstadt. CC BY-SA 3.0.

Ansellia – **John Ansell (?–1847)** En.wikipedia.org/wiki/Niger_expedition_of_1841.

Antonia – **Archduke Anton Victor of Austria (1779–1835)** En.wikipedia.org/w/index.php?title=Archduke_Anton_ Victor_of_Austria&oldid=629480852. Image: public domain. Author: unknown. Source: Austrian National Library / Österreichische Nationalbibliothek. CC BY-SA 3.0.

Antunesia – **José Maria Antunes (1856–1928)** Plants.jstor.org/person/bm000010248?history=true.

Aongstroemia – **Johan Ångström (1813–1879)** En.wikipedia.org/w/index.php?title=Johan_%C3%85ngstr%C3% B6m&oldid=616319866; Uni-goettingen.de/en/187051.html; Plants.jstor.org/person/bm000372651; three other websites; CC BY-SA 3.0.

Araujia – **Antonio de Araujo de Azevedo (1754–1817)** En.wikipedia.org/w/index.php?title=Ant%C3%B3nio_de_Ara% C3%BAjo_e_Azevedo,_1st_Count_of_Barca&oldid=616189946; Translate.google.co.za/translate?hl=en&sl=pt& u=leiturasdahistoria.uol.com.br/ESLH/Edicoes/17/imprime125466.asp&ei=2w0JTcHdK4SEOoXB8fkE&sa=X&oi= translate&ct=result&resnum=6&ved=0CEEQ7gEwBQ&prev=/search%3Fq%3Danto nio%2Bde%2BAraujo%2Bde% 2BAzevedo%2B%281754-1817%29%26hl%3Den%26sa%3DG%26rlz%3D1B3GGIC_enZA317ZA317%26prmd% 3Divo; three other websites. Image: public domain. Author: unknown. Source: Sgmf.pt/NR/rdonlyres/D17716EB-1A0C-4DB0-BA91-399149F89AB9/2979/Ant%C3%B3niodeAra%C3%BAjoeAzevedo1.pdf. CC BY-SA 3.0.

Arrowsmithia – **John Arrowsmith (1790–1873)**
En.wikipedia.org/w/index.php?title=John_Arrowsmith_(cartographer)&oldid=629904199. Image: public domain. Author: Henry William Pickersgill. Source: Nmm.ac.uk/collections/displayRepro.cfm?reproID=BHC2519&picture= 1#content. CC BY-SA 3.0.

Assonia – **Jordán de Asso (1742–1814)** Es.wikipedia.org/w/index.php?title=Jord%C3%A1n_de_Asso&oldid= 74858397; Translate.google.co.za/translate?hl=en&sl=es&u=pasapues.es/ignaciodeasso/index.php&ei=8_xfTY7_ LcGw8gPrz5HoCw&sa=X&oi=translate&ct=result&resnum=6&ved=0CDYQ7gEwBTgK&prev=/search%3Fq% 3D%2522asso%2522%2BAND%2B%2522assonia%2522%26start%3D10%26hl%3Den%26sa%3DN%26rlz% 3D1B3GGIC_enZA317ZA317%26prmd%3Divnsb. Image: public domain. Source: Estatua dedicada a Jordán de Asso en el Paraninfo de la Universidad de Zaragoza. Author: Ajzh2074 – Trabajo propio. CC BY-SA 3.0.

Astridia – **Astrid Elise Schwantes (née Wilberg) (1887–1960)** Quattrocchi, *CRC World Dictionary of Plant Names 1999*.

Audouinia – **Jean Victoire Audouin (1797–1841)** En.wikipedia.org/w/index.php?title=Jean_Victoire_Audouin&oldid= 607335825&absoluteastronomy.com/topics/Jean_Victoire_Audouin. Image: public domain. Author: unknown. CC BY-SA 3.0.

Augea – **Johann Andreas Auge (1711–1805)** Gunn & Codd, *BESA*; Books.google.co.za/books?id=mkfdQ3l1YQUC&

pg=PA83&lpg=PA83&dq=Johann+Andreas+Auge+%281711-1805%29,&source=bl&ots=S_4_KoFn01&sig=
JyPpettyaizydVuBjE2uNfbyCNk&hl=en&sa=X&ei=q_llVOL9EMHpatyfgagl&ved=0CCkQ6AEwBA#v=onepage&q=
Johann%20Andreas%20Auge%20%281711-1805%29%2C&f=false. Image: Sanbi

Avicennia – **Ibn Sina Avicenna (980–1037)** En.wikipedia.org/w/index.php?title=Avicenna&oldid=633774805 Image: public domain. Source: Omma.ua/uploads/article/2012/02/23/11/ac1ece6b5b3eb78b6316f4a3f0498ac1b61f22f6.jpg. Author: unknown. CC BY-SA 3.0.

Azanza – **Miguel José de Azanza, Duke of Santa Fe (1746–1826)** En.wikipedia.org/w/index.php?title=Miguel_Jos%
C3%A9_de_Azanza,_Duke_of_Santa_Fe&oldid=634711144. Image: public domain. Author: unknown. Source: Viceroys of New Spain. CC BY-SA 3.0.

B

Bachmannia – **Franz Ewald Bachmann (1856–c1916)** Glen & Germishuizen, *BESA* (edition 2), p.82. Image: Sanbi

Baikiaea – **William Balfour Baikie (1825–1864)** En.wikipedia.org/w/index.php?title=William_Balfour_Baikie&oldid=
604285528; Answers.com/topic/william-balfour-baikie. Image: public domain. Author: unknown. Source: *Illustrated London News*, 28 Jan 1865. CC BY-SA 3.0.

Baillonella – **Henri Ernest Baillon (1827–1895)** En.wikipedia.org/w/index.php?title=Henri_Ernest_Baillon&oldid=
608140568&arboretum.sav.sk/aktuality/henri-ernest-baillon-1827-1895. Image: public domain. Author: unknown. Source: Bium. CC BY-SA 3.0.

Baissea – **Nicolas Sarrabat (1698–1739)** En.wikipedia.org/w/index.php?title=Nicolas_Sarrabat&oldid=609627883&
microsofttranslator.com/bv.aspx?ref=SERP&br=ro&mkt=en-ww&dl=en&lp=FR_EN&a=http%3a%2f%2ffr.wikipedia.
org%2fwiki%2fNicolas_Sarrabaton. CC BY-SA 3.0.

Ballya – **Peter René Oscar Bally (1895–1980)** Es.wikipedia.org/w/index.php?title=Peter_Ren%C3%A9_Oscar_
Bally&oldid=74537468&huntbot.andrew.cmu.edu/hibd/departments/Archives/Archives-AG/Bally.shtml and Plants.
jstor.org/person/bm000000397. Image: Bally, Peter René Oscar. Source: courtesy Hunt Institute for Botanical Documentation, Carnegie Mellon University, Pittsburgh. Photograph: Loutfy Boulos. CC BY-SA 3.0

Banksia – **Joseph Banks (1743–1829)** En.wikipedia.org/w/index.php?title=Joseph_Banks&oldid=633273099 Image: public domain. Author: Joshua Reynolds (1723–1792). Source: Csupomona.edu/~larryblakely/whoname/who_banks.
htm. CC BY-SA 3.0.

Barbacenia – **Luís António Furtado de Castro do Rio de Mendonça e Faro (1754–1830)** Pt.wikipedia.org/w/index.
php?title=Lu%C3%ADs_Ant%C3%B3nio_Furtado_de_Castro_do_Rio_de_Mendon%C3%A7a_e_Faro&oldid=
38303058 CC BY-SA 3.0.

Barbarea – **Saint Barbara (Third Century)** En.wikipedia.org/w/index.php?title=Saint_Barbara&oldid=632437996.

Barbaretta – **Mary Elizabeth Barber (1818–1899)** Glen & Germishuizen, *BESA* (edition 2), p.88; En.wikipedia.org/w/
index.php?title=James_Henry_Bowker&oldid=632669085. Public domain. Image: Sanbi CC BY-SA 3.0.

Barleria – **Jacques Barrelier (1606–1673)** Es.wikipedia.org/w/index.php?title=Jacques_Barrelier&oldid=64925867 Image: public domain. Author: unknown. Source unknown. CC BY-SA 3.0

Barnardiella – **Thomas Theodore Barnard (1898–1983)** Glen & Germishuizen, *BESA* (edition 2), p.90.Image: Image: Sanbi CC BY-SA 3.0.

Barringtonia – **Daines Barrington (1727–1800)** En.wikipedia.org/w/index.php?title=Daines_Barrington&oldid=
623453928; Encyclopedia.jrank.org/BAR_BEC/BARRINGTON_DAINES_1727_1800_.html and museumwales.
ac.uk/en/1349. Image: public domain. Author: unknown. Source: Superstock.co.uk/stock-photos-. Image: s/4069-
1437. CC BY-SA 3.0.

Barrowia – **Sir John Barrow, first baronet (1764–1848)**
En.wikipedia.org/w/index.php?title=Sir_John_Barrow,_1st_Baronet&oldid=630031750&libweb5.princeton.edu/
visual_materials/maps/websites/africa/barrow/barrow.html; Encyclopedia.jrank.org/BAR_BEC/BARROW_SIR_
JOHN_1764_1848_.html. Image: public domain. Author: Sir John Barrow. Source: Chemical Heritage Foundation.
CC BY-SA 3.0.

Bartholina – **Thomas Bartholin (1616–1680)**
En.wikipedia.org/w/index.php?title=Thomas_Bartholin&oldid=621263565; historical.hsl.virginia.edu/treasures/
bartholin.html. Image: public domain. Author: Nico. Source: Da.wikipedia.org/wiki/Billede:Thomas_bartholin.jpg. CC BY-SA 3.0.

Bartramia – **John Bartram (1699–1777)**
En.wikipedia.org/w/index.php?title=John_Bartram&oldid=616522429; Answers.com/topic/john-bartram Image: public domain. Author: unknown. Source: Amazon.com/Historic-Print-John-Bartram-1699-1777/dp/B003HXCSCA. No contact address.

Bartsia – **Johann Bartsch (1709–1738)** Translate.google.co.za/translate?hl=en&sl=de&u=wapedia.mobi/de/Johann_
Bartsch&ei=R0AsTeKSLInsOdG9sccK&sa=X&oi=translate&ct=result&resnum=7&ved=0CEwQ7gEwBg&prev=
/search%3Fq%3DJohann%2BBartsch%2B%2B%281709-1738%29,%26hl%3Den%26rlz%3D1B3GGIC_enZ
A317ZA317%26prmd%3Divnso; Linnaeus.nrm.se/flora/di/scrophularia/barts/welcome.html&fleurssauvages.ca/
scrophulariaceae/bartsia_alpina.html.

Bassia – **Ferdinando Bassi (1710–1774)** Translate.google.co.za/translate?hl=en&sl=it&u=sma.unibo.it/
ortobotanico/endirectors.html&ei=MVksTdj6lYWVOrOrzeYK&sa=X&oi=translate&ct=result&resnum=1&ved=
0CB0Q7gEwADgU&prev=/search%3Fq%3DFerdinando%2BBassi%2B%281710-1774%29,%26start%3D20%
26hl%3Den%26sa%3DN%26rlz%3D1B3GGIC_enZA317ZA317%26prmd%3Divnso; Translate.google.co.za/
translate?hl=en&sl=es&u=Es.wikipedia.org/wiki/Bassi&ei=GlosTfn8DceUOs_AhMkK&sa=X&oi=translate&ct=

result&resnum=4&ved=0CDcQ7gEwAzge&prev=/search%3Fq%3DFerdinando%2BBassi%2B%281710-1774%
29,%26start%3D30%26hl%3Den%26sa%3DN%26rlz%3D1B3GGIC_enZA317ZA317%26prmd%3Divnso.

Basteria – **Job Baster (1711–1775)** En.wikipedia.org/w/index.php?title=Job_Baster&oldid=624851240; Books.
google.co.za/books?id=wm6bva_zkacC&pg=PR43&lpg=PR43&dq=Job+Baster+%281711-75%29,&source=bl&
ots=q8UWB8_-Rl&sig=MDMtLV563nMGAibpthe5y4JoXzs&hl=en&ei=4V4sTdmsHcqdOt_s7fQJ&sa=X&oi=book_
result&ct=result&resnum=2&ved=0CBwQ6AEwAQ#v=onepage&q=Job%20Baster%20%281711-75%29%2C&f=
false Image: public domain. Author: Lymantria. Source: Own work; Commons.wikimedia.org/wiki/File:Beeld_Job_
Baster.JPG. CC BY-SA 3.0.

Bauhinia – **Gaspard Bauhin (1560–1624) and Johan Bauhin (1541–1613)** En.wikipedia.org/w/index.php?title=
Gaspard_Bauhin&oldid=631377208. Image: public domain. Author: unknown. Source unknown. Uploaded by
Valérie75 and Johann Bauhin; En.wikipedia.org/w/index.php?title=Johann_Bauhin&oldid=601777128. Image: public
domain. Author: unknown. Source: National Library of Medicine. CC BY-SA 3.0.

Bazzania – **Matteo Bazzani (1674–1749)** Charters, *EDSAP*.

Beaumontia – **Diana Beaumont (née Wordworth) (1765–1831)**
Euppublishing.com/doi/abs/10.3366/anh.2003.30.1.180?journalCode=anh;
Smgrowers.com/products/plants/plantdisplay.asp?plant_id=238.; isearch.avg.com/pages/abt/cs/homepage.
aspx?pid=avg&sg=&cid={6a172397-41f1-4d62-9f39-b09091cbb4ee}&mid=0c41a18a87c847d396ae810f1b444fa7-
1b8d6d5cbfe511841adf26efacf647e2a6799f00&ds=AVG&v=14.2.0.1&lang=en&pr=fr&d=2013-03-20%2016%
3A58%3A07&sap=hp&tc=test16.

Beckeropsis – **Johannes Becker (1769–1833)** Es.wikipedia.org/w/index.php?title=Johannes_Becker&oldid=
68275446&senckenberg.de/files/content/forschung/abteilung/botanik/phanerogamen1/historybotanyfr.pdf. Image:
public domain. Photo supplied by Stefan Dressler curator phanerogams 2 Herbarium Senckenbergianum (c)
Senckenberg Gesellschaft für Naturforschung Frankfurt/M.

Begonia – **Michel Bégon (1638–1710)** En.wikipedia.org/w/index.php?title=Michel_B%C3%A9gon_(1638%E2%80%
931710)&oldid=618251958&williamreesecompany.com/catalogs/cat276.pdf. Image: public domain. Source: Alienor.
org/publications/planrelief/ville.htm.

Behnia – **Wilhelm Friedrich Georg Behn (1808–1878)** En.wikipedia.org/w/index.php?title=Wilhelm_Friedrich_Georg_
Behn&oldid=615585933. Image: public domain. Source: 'zoolmuseum@email.uni-kiel.de'.

Beilschmiedia – **Carl Traugott Beilschmied (1793–1848)**
Archive.org/stream/taxonomicliterature00staf/taxonomicliterature00staf_djvu.txt;
find-a-book.com/db/book1133_9A5AD001U79O.html;
Translate.google.co.za/translate?hl=en&sl=de&u=En.wikilingue.com/es/Carl_Traugott_Beilschmied&ei=
PwAvTZLbMYaAOqO_jbQK&sa=X&oi=translate&ct=result&resnum=9&ved=0CFEQ7gEwCA&prev=/search%
3Fq%3DCarl%2B%28Karl%29%2BTraugott%2BBeilschmied%2B%281793-1848%29,%26hl%3Den%26rlz%
3D1B3GGIC_enZA317ZA317%26prmd%3Divnso&asherbooks.com/9A5AD001U79O_v.html.

Bellardia – **Carlo Antonio Lodovico Bellardi (1741–1826)**
Es.wikipedia.org/w/index.php?title=Carlo_Antonio_Lodovico_Bellardi&oldid=71416154archive.org/stream/
taxonomicliterature00staf/taxonomicliterature00staf_djvu.txt; Plants.jstor.org/person/bm000390060?history=true. CC
BY-SA 3.0.

Bequaertiodendron – **Joseph Charles Corneille Bequaert (1886–1982)**
Antwiki.org/wiki/Bequaert,_Joseph_Charles_Corneille_%281886-1982%29. Image: Antwiki.org/wiki/File:Bequaert.
jpg. CC BY-SA 3.0.

Berardia – **Etienne Bérard (1764–1839)** Books.google.co.za/books?id=rFClzHWN8hlC&pg=PA41&lpg=PA41&
dq=%22M.+Berard%22+AND+%22Professor+of+Chemistry%22+AND+%22Montpellier%22&source=bl&ots=
cVKn7RQtcA&sig=VNE5qSnDAUt6TGSFRNLNVAJzQUY&hl=en&ei=mvgxTayJDcTsOdPw-bQC&sa=X&oi=
book_result&ct=result&resnum=1&ved=0CBgQ6AEwAA#v=onepage&q=%22M.%20Berard%22%20AND%20%
22Professor%20of%20Chemistry%22%20AND%20%22Montpellier%22&f=false.

Berchemia – **Jacob Peter Berthout van Berchem (1763–1832)**
Books.google.co.za/books?id=VwMNAAAAIAAJ&pg=PA45&lpg=PA45&dq=Berchem+%281763-1832%29&
source=bl&ots=s2e3oY-iWL&sig=nSIp_4IhPGGiWivFOEDRv1oXsEg&hl=en&ei=RAMyTbjVIMGcOqyx9bYC&sa=
X&oi=book_result&ct=result&resnum=1&ved=0CBgQ6AEwAA#v=onepage&q=Berchem%20%281763-1832%29&
f=false
Translate.google.co.za/translate?hl=en&sl=fr&u=www3.unil.ch/viatImage: s/index.php%3Fajax%3Dtrue%
26module%3Dpersonne%26IDpers_ass%3D103&ei=4gUyTeW8I8iaOsDykLYC&sa=X&oi=translate&ct=result&
resnum=2&ved=0CCUQ7gEwATgK&prev=/search%3Fq%3DBerchem%2B%281763-1832%29%26start%3D10%
26hl%3Den%26sa%3DN%26rlz%3D1B3GGIC_enZA317ZA317%26prmd%3Divns
archive.org/stream/indexanimaliumsi00sher/indexanimaliumsi00sher_djvu.txt.

Bergenia – **Karl August von Bergen (1704–1759)**
En.wikipedia.org/w/index.php?title=Karl_August_von_Bergen&oldid=587847877.

Bergeranthus – **Alwin Berger (1871–1931)**
En.wikipedia.org/w/index.php?title=Alwin_Berger&oldid=609368907. Image: public domain. Author: unknown; Peter
A. Mansfeld for the filtered image: Source: Alwin-Berger-Archiv, Möschlitz. CC BY-SA 3.0.

Bergia – **Peter Jonas Bergius (1730–1790)**
Sv.wikipedia.org/w/index.php?title=Peter_Jonas_Bergius&oldid=2802644; words.fromoldbooks.org/Chalmers-
Biography/b/bergius-peter-jonas.html; Translate.google.co.za/translate?hl=en&sl=fr&u=fr.wikipedia.org/wiki/Peter_
Jonas_Bergius&ei=tP0yTf24IIubOpLb9LQC&sa=X&oi=translate&ct=result&resnum=2&ved=0CCQQ7gEwAQ&

prev=/search%3Fq%3DPeter%2BJonas%2BBergius%2B%281730-1790%29,%26hl%3Den%26rlz%3D1B3GGIC_ enZA317ZA317%26prmd%3Divnso. Image: public domain. Author – Lithography by unknown person. Source: Libris. kb.se/bib/1579474. CC BY-SA 3.0.

Berkheya – **Johannesle Francq van Berkhey (1729–1812)** Nl.wikipedia.org/w/index.php?title=Johannes_le_Francq_ van_Berkhey&oldid=42273338; Translate.google.co.za/translate?hl=en&sl=nl&u=xs4all.nl/~carlkop/berkhey.html& ei=6QkzTdjEFcqfOrml3LYC&sa=X&oi=translate&ct=result&resnum=1&ved=0CBwQ7gEwAA&prev=/search% 3Fq%3DJohannes%2Ble%2BFrancq%2Bvan%2BBerkhey%26hl%3Den%26rlz%3D1B3GGIC_enZA317ZA317% 26prmd%3Divnso. Image: public domain. Author: H. Pothoven, J. Houbraken. Source: *De geschiedenis wordt vanaf de Romeinen tot aan 1774 verteld*. CC BY-SA 3.0.

Berlinia – **Andreas Berlin (1746–1773)** En.wikipedia.org/w/index.php?title=Apostles_of_Linnaeus&oldid=617364106& st-andrews.ac.uk/~gdk/stabg_new/news/Linnaeus_7.html; Jarvis, *Order out of Chaos: Linnaean Plant Names and their Types*. The Linnaean Society in association with The Natural History Museum. CC BY-SA 3.0.

Bernhardia – **Johann Jakob Bernhardi (1774–1850)** En.wikipedia.org/w/index.php?title=Johann_Jakob_Bernhardi& oldid=600202857&minrec.org/sitemap5.asp&cas.muohio.edu/~meicenrd/Anatomy/Ch0_History/history.html. CC BY-SA 3.0.

Berrisfordia – **Francis ('Frank') Berrisford (1898–1973)**
Source: Personal communication with Berrisford family. Image: Berrisford family.

Bertilia – **Bertil Nordenstam (1936–)** Books.google.co.za/books/about/Bertil_Nordenstam.html?id=XLIYywAACAAJ& redir_esc=y: brief review that was modified by Bertil Noredenstam. He also provided a photograph of himself for this book.

Berzelia – **Jöns Jacob Berzelius (1779–1848)**
En.wikipedia.org/w/index.php?title=J%C3%B6ns_Jacob_Berzelius&oldid=631294730 Image: public domain. Author: unknown. Source: chemistry.msu.edu/Portraits/ Image: S/Berzelius3c.jpg. CC BY-SA 3.0

Beschorneria – **Friedrich Wilhelm Christian Beschorner (1806–1873)** Pl.wikipedia.org/w/index.php?title=Friedrich_ Wilhelm_Christian_Beschorner&oldid=24090575. CC BY-SA 3.0.

Bewsia – **John William Bews (1884–1938)**
Gunn; Codd, *Exploration of Southern Africa*, p.93–94. Image: Sanbi CC BY-SA 3.0.

Bignonia – **Abbé Jean-Paul Bignon (1662–1743)**
En.wikipedia.org/w/index.php?title=Jean-Paul_Bignon&oldid=629078201. Image: public domain. Author: Edelinck. Source: author(s): Alice Stroup. Royal Funding of the Parisian Académie Royale des Sciences during the 1690s. Source: *Transactions of the American Philosophical Society*, New Series, Vol. 77, No. 4 (1987), pp.i-x; 1–167. CC BY-SA 3.0.

Bilderdykia – **Willem Bilderdyk (1756–1831)** En.wikipedia.org/w/index.php?title=Willem_Bilderdijk&oldid= 601208543&1911encyclopedia.org/Willem_Bilderdijk&dare.ubvu.vu.nl/bitstream/1871/15772/1/Bilderdijkbetween2. pdf. Image: public domain. Author: Charles Howard Hodges (1764–1837). Source: Rijksmuseum, Amsterdam. CC BY-SA 3.0.

Bischofia – **Gottlieb Wilhelm Bischoff (1797–1854)** De.wikipedia.org/w/index.php?title=Gottlieb_Wilhelm_Bischoff& oldid=133939819Image: public domain. Author: unknown. Source: Heidelberg University Library. CC BY-SA 3.0.

Blackiella – **John McConnell Black (1855–1951)**
En.wikipedia.org/wiki/John_McConnell_Black; En.wikipedia.org/wiki/File:John_McConnell_Black.jpeg Image: Author: unknown. Source: Cropped from collections.slsa.sa.gov.au/resource/B+6074.

Blackwellia – **Elizabeth Blackwell (1707–1758)**
Es.wikipedia.org/w/index.php?title=Elizabeth_Blackwell_(ilustradora)&oldid=79359399. Image: public domain. Author: unknown. Source: Transferred from De.wikipedia; transferred to Commons by User:Ireas using CommonsHelper. CC BY-SA 3.0.

Blaeria – **Patrick Blair (1666–1728)**
En.wikipedia.org/w/index.php?title=Patrick_Blair_(surgeon)&oldid=604782363; alternative-perth.co.uk/patrickblair. htm.

Blainvillea – **Henri Marie Ducrotay de Blainville (1777–1850)**
En.wikipedia.org/w/index.php?title=Henri_Marie_Ducrotay_de_Blainville&oldid=627283280; Encyclopedia.jrank.org/ BER_BLA/BLAINVILLE_HENRI_MARIE_DUCROTAY.html Image: public domain. Author: unknown. Source: hbs. bishopmuseum.org/dipterists/dipt-b.html CC BY-SA 3.0.

Blighia – **William Bligh (1754–1817)**
En.wikipedia.org/w/index.php?title=William_Bligh&oldid=633139991; Adbonline.anu.edu.au/biogs/A010111b.htm. Image: public domain. Author: User Hephaestos on En.wikipedia. Source: National Library of Australia. CC BY-SA 3.0.

Blindia – **Jean-Jacques Blind (1806–1867)** Short biography. Charters, *EDSAP*.

Blotiella – **Marie-Laure Tardieu-Blot (1902–1998)** Fr.wikipedia.org/wiki/Marie-Laure_Tardieu-Blot. CC BY-SA 3.0.

Blumea – **Karl Ludwig von Blume (1796–1862)** En.wikipedia.org/w/index.php?title=Carl_Ludwig_Blume& oldid=579662603. Image: public domain. Author: Carl Ludwig Blume (1796–1862) Source: Botanicus.org/ item/31753002794367. CC BY-SA 3.0.

Bobartia – **Jacob Bobart the Elder (1599–1680)**
En.wikipedia.org/w/index.php?title=Jacob_Bobart_the_Elder&oldid=631879668. Image: public domain. Author: James Granger. Source: Granger, *A Biographical History of England: from Egbert the Great to the Revolution*. CC BY-SA 3.0.

Bobgunnia – **Charles Robert 'Bob' Gunn (1927–)** Amazon.com/Fabaceae-Charles-Weitzman-Dallwitz-Kirkbride/ dp/1887905251.

Boeckeleria – **Johann Otto Boeckeler (1803–1899)**
En.wikipedia.org/w/index.php?title=Johann_Otto_Boeckeler&oldid=614633433. CC BY-SA 3.0.
Boeckhia – **Philipp August Böckh (1785–1867)** En.wikipedia.org/w/index.php?title=Philipp_August_B%C3%B6ckh& oldid=631580555. Image: public domain. Author: Franz Krüger (1797–1857). Source: Alfred Gudeman: Imagines philologorum. Berlin/Leipzig: Teubner 1911, S. 21. Digitale Sammlungen der HU Berlin, ID 7188. CC BY-SA 3.0.
Boehmeria – **Georg(e) Rudolf Boehmer (1723–1803)**
En.wikipedia.org/w/index.php?title=Georg_Rudolf_Boehmer&oldid=588091076. CC BY-SA 3.0.
Boerhavia – **Herman Boerhaave (1668–1739)**
En.wikipedia.org/w/index.php?title=Herman_Boerhaave&oldid=634455638. Image: public domain. Author: J. Chapman, artist. Source: Ihm.nlm.nih.gov/Image: s/B29694. CC BY-SA 3.0.
Boivinella – **Louis Hyacinthe Boivin (1808–1852)** Translate.google.co.za/translate?hl=en&sl=es&u=esacademic. com/dic.nsf/eswiki/180917&ei=Zlc8Tf2KJcehOq6HgZML&sa=X&oi=translate&ct=result&resnum=1&ved= 0CB0Q7gEwADg8&prev=/search%3Fq%3DLouis%2BHyacinthe%2BBoivin%2B%281808-1852%29,%26start% 3D60%26hl%3Den%26sa%3DN%26rlz%3D1B3GGIC_enZA317ZA317%26prmd%3Divnso.
Bojeria – **Wenceslas Bojer (1797–1856)**
Glen & Germishuizen, *BESA* (edition 2), p.100; Plants.jstor.org/person/bm000000841. Image: Sanbi
Bolusafra – **Harry Bolus (1834–1911)**
En.wikipedia.org/w/index.php?title=Harry_Bolus&oldid=633086363 Public domain. Image: Sanbi CC BY-SA 3.0.
Bolusanthemum – **Harriet Margaret Louisa Bolus (1877–1970)** Glen & Germishuizen, *BESA* (edition 2), p.101. Image: Sanbi
Bonamia – **François Bonamy (1710–1786)** Translate.google.co.za/translate?hl=en&sl=fr&u=dictionnaire. sensagent.com/bonamy/fr-fr/&ei=2XRATdTsCtGYOozS6Ycl&sa=X&oi=translate&ct=result&resnum=10&ved= 0CGAQ7gEwCQ&prev=/search%3Fq%3DFran%25C3%25A7ois%2BBonamy%2B%281710-1786%29,%26hl% 3Den%26rlz%3D1B3GGIC_enZA317ZA317%26prmd%3Divnso; Google book references.
Bonatea – **Guiseppe Antonio Bonato (1753–1836)**
Miosjournal.org/journal/2009/08/BonateaSpeciosaStudy.html Image: public domain. Source: www2.biusante. parisdescartes.fr/img/?p=6&mod=a&orig=anm.
Bonnaya – **Charles François, Marquis de Bonnay (1750–1825)**
En.wikipedia.org/w/index.php?title=Charles_Fran%C3%A7ois,_Marquis_de_Bonnay&oldid=623090130. Image: public domain. Author: unknown. Source: Château de Versailles. CC BY-SA 3.0.
Borbonia – **Gaston, Duke of Orléans (1608–1660)**
En.wikipedia.org/w/index.php?title=Gaston,_Duke_of_Orl%C3%A9ans&oldid=632947637; Charters, *EDSAP*. Image: public domain. Artist: Anthony van Dyck (1599–1641). Source: Musée Condé. CC BY-SA 3.0.
Borrera – **William J Borrer (1781–1862)**
En.wikipedia.org/w/index.php?title=William_Borrer&oldid=626828721; Quattrocchi, *CRC World Dictionary of Plant Names*. CC BY-SA 3.0.
Boschia – **Johannes van den Bosch (1780–1844)**, https://en.wikipedia.org/wiki/Johannes_van_den_Bosch Image: Public domain. Author: Cornelis Kruseman (1797-1857). Source: Rijksmuseum Amsterdam.
Boscia – **Louis Augustin Guillaume Bosc (1759–1828)**
En.wikipedia.org/w/index.php?title=Louis_Augustin_Guillaume_Bosc&oldid=628652112&1911encyclopedia. org/Louis_Augustin_Guillaume_Bosc. Image: public domain. Author: unknown. Source: hbs.bishopmuseum.org/ dipterists/Image: s/bosc.jpg CC BY-SA 3.0
Bottaria – **Bartholomew Bottari (1732–1789)**
Translate.google.co.za/translate?hl=en&sl=it&u=tuttochioggia.it/personaggi.htm&ei=pywsT6atGoXAhAebr4TaCg& sa=X&oi=translate&ct=result&resnum=5&sqi=2&ved=0CD8Q7gEwBA&prev=/search%3Fq%3DVenetian%2B% 2522Bartolomeo%2BBottari%2522%2B%281732-1789%29%2Bnaturalist%26hl%3Den%26prmdo%3D1%26biw% 3D1226%26bih%3D756%26prmd%3Dimvnso; other websites, no longer functional.
Bouchea – **Peter Carl (Karl) Bouché (1783–1856) & Peter Friedrich Bouché (1785–1756)** Charters, *EDSAP*.
Bougainvillea – **Louis-Antoinede Bougainville (1729–1811)** En.wikipedia.org/w/index.php?title=Louis_Antoine_de_ Bougainville&oldid=632007669 Image: public domain. Author: Jean-Pierre Franque (1774–1860) Permission. Florian FDP. CC BY-SA 3.0.
Bowiea – **James Bowie (1789–1869)** Glen & Germishuizen, *BESA* (edition 2), p.105; Oxforddnb.com/view/ printable/3163.
Bowkeria – **James Henry Bowker (1822–1900)**
En.wikipedia.org/w/index.php?title=James_Henry_Bowker&oldid=632669085; Glen & Germishuizen, *BESA* (edition 2), p.105. Image: public domain. Author: unknown. Source: Bowker.info/BowkerChildrenJamesHenryBowker.htm. CC BY-SA 3.0.
Brackenridgea – **William Dunlop Brackenridge (1810–1893)**
Commons.wikimedia.org/wiki/File:William_Dunlop_Brackenridge_(young).jpg; Charters, *EDSAP*. Image: public domain. Author: unknown. Source: Smithsonian Institution Archives.
Brasenia-**Christoph Brasen (1738–1774)** Google.co.za/search?q=Christoph+Brasen+%281738-1774% 29&ie=utf-8&oe=utf-8&aq=t&rls=org.mozilla:en-US:official&client=firefox-a&channel=nts&gfe_rd=cr&ei= 3Yh1VJ3OCYyp8wf9i4Ew& other sources.
Braunia – **Alexander Braun (1805–1877)** En.wikipedia.org/w/index.php?title=Alexander_Braun&oldid=616444136 Image: public domain. Author: Artist not listed. Source: Frontispiece from Leben nach seinen handschriftlichen Nachlass (Berlin 1882). CC BY-SA 3.0.

Braunsia – **Hans Heinrich Justus Carl Ernst Brauns (1857–1929)** Biodiversityexplorer.org/people/brauns-hhjce.htm; Van Son, 1933 unpublished account of Brauns; Nature.com/nature/journal/v123/n3100/abs/123499a0.html.
Brayulinea – **William L. Bray (1865–1953) & Edwin Burton Uline (1867–1933)** En.wikipedia.org/w/index.php?title= William_L._Bray&oldid=615068618&people.wku.edu/charles.smith/chronob/BRAY1865.htm; Charters, *EDSAP*. Image: William Bray, Smithsonian Institution Archives.
Breonadia – **Jean Nicolas Bréon (1785–1864)** Charters, *EDSAP*; Dorr, *Plant Collectors in Madagascar and Comoro Islands*, p.64–5.
Breutelia – **Johann Christian Breutel (1788–1875)** Uni-goettingen.de/de/187053.html&jstor.org/pss/2479419; Archive.org/stream/floraofislandofs17mill/floraofislandofs17mill_djvu.txt; Glen & Germishuizen, *BESA* (edition 2), p.109.
Breweria – **Samuel Brewer (1670–1743)** En.wikipedia.org/w/index.php?title=Samuel_Brewer&oldid=604271912.
Brianhuntleya – **Brian Huntley (1944–)**
    Glen & Germishuizen, *BESA* (edition 2), p.221; other sources not recorded. Image: Sanbi.
Bridelia – **Samuel Elisée von Bridel (1761–1828)** Charters, *EDSAP*; Quattrochi, *CRC World Dictionary of Plant Names*.
Brigantiaea – **Maria II of Portugal (1819–1853)** En.wikipedia.org/wiki/Maria_II_of_Portugal; Thepeerage.com/p10450. htm#i104492. Image: public domain. Author: unknown. Source: Royalcollection.org.uk/egallery/object.asp?category= 277&pagesize=60&object=420393&row=678&detail=magnify. CC BY-SA 3.0.
Brothera – **Viktor Ferdinand Brotherus (1849–1929)** En.wikipedia.org/w/index.php?title=Viktor_Ferdinand_ Brotherus&oldid=628391241&uni-goettingen.de/de/187053.html. Image: unknown.
Broussonetia – **Pierre Marie Auguste (1761–1807)** En.wikipedia.org/w/index.php?title=Pierre_Marie_Auguste_ Broussonet&oldid=60372867. Image: public domain. Author: Alcide-Louis-Joseph Railliet et L. Moulé. Source: Bium CC BY-SA 3.0.
Browallia – **Johannes Browallius (1707–1755)** En.wikipedia.org/w/index.php?title=Johannes_Browallius&oldid= 634979490. Image: public domain. Author: Margareta Capsia. Source: Transferred from En.wikipedia. (Original text: Helsinki Museum). CC BY-SA 3.0.
Brownanthus – **Nicholas Edward Brown (1849–1934)** Glen & Germishuizen, *BESA* (edition 2), p.111. Image: public domain; En.wikipedia.org/wiki/N._E._Brown. Author: unknown. Source: En.wikipedia.org, first uploaded by en:user:Paul Venter.
Brownleea – **John Brownlee (1781–1847)** Glen & Germishuizen, *BESA* (edition 2), p.111. Image: Sanbi
Bruchia – **Philipp Bruch (1781–1847)** En.wikipedia.org/w/index.php?title=Philipp_Bruch&oldid=588096472; short biography. CC BY-SA 3.0.
Bruguiera – **Jean Guillaume Bruguière (1750–1798)** En.wikipedia.org/w/index.php?title=Jean_Guillaume_Brugui% C3%A8re&oldid=621887580. Image: public domain. Source: Cotebleue.org.
Brunia – **Alexander Brown (f 1692–1698)** Glen & Germishuizen, *BESA* (edition 2), p.110; short biography.
Brunsvigia – **Charles I, Duke of Brunswick-Wolfenbüttel (1713–1780)** En.wikipedia.org/w/index.php?title=Charles_ I,_Duke_of_Brunswick-Wolfenb%C3%BCttel&oldid=540346093. Image: public domain. Artist: unknown. Source: Bildindex.de. CC BY-SA 3.0.
Bryobartramia – **Edwin Bunting Bartram (1878–1964)** Charters, *EDSAP*.
Buchenroedera – **Wilhelm Ludwig von Buchenröder (1783–1841)** Glen & Germishuizen, *BESA* (edition 2), p.115.; short biography.
Buchnera – **Andreas Elias Buchner (1701–1769)** De.wikipedia.org/w/index.php?title=Andreas_Elias_B%C3% BCchner&oldid=133297726. Image: public domain. Author: Johann Jakob Haid. Source: Dspace.utlib.ee/dspace/ handle/10062/6693. CC BY-SA 3.0.
Buddleja – **Adam Buddle (1660–1715)** En.wikipedia.org/w/index.php?title=Adam_Buddle&oldid=595834625. CC BY-SA 3.0.
Buellia – **Esperanzo Buelli (Dates unknown)** Wsl.ch/wsl/info/mitarbeitende/scheideg/pdf/108.pdf; very short biography.
Bulliarda – **Jean Baptiste François Bulliard (1752–1793)** Botlib.huh.harvard.edu/libraries/mycology/Myco_books9. htm; Books.google.co.za/books?id=Qf0qFZlrjpEC&pg=PT117&lpg=PT117&dq=Jean+Baptiste+Fran%C3% A7ois+Bulliard+%281752-1793%29&source=bl&ots=WAwsnOF2Cx&sig=TvjkIIhU3xXZeqLM-2Wvx9gBZvc&hl= en&sa=X&ei=Ztt2VOioE4nb7AbqpYG4DA&ved=0CDsQ6AEwCg#v=onepage&q=Jean%20Baptiste%20Fran% C3%A7ois%20Bulliard%20%281752-1793%29&f=false (Williams, *Botanophilia in Eighteenth-Century France*).
Bunburya – **Sir Charles Bunbury, Eighth Baronet (1809–1886)** En.wikipedia.org/w/index.php?title=Sir_Charles_ Bunbury,_8th_Baronet&oldid=630031551. Image: public domain. Author: Eden Upton Eddis (1812–1901) Source: Katherine M. Lyell (Hrsg.): *The Life of Sir Charles J.F. Bunbury*, Bart. John Murray, London 1906, (volume 1); Glen & Germishuizen, *BESA* (edition 2), p.112. CC BY-SA 3.0.
Burchellia – **William John Burchell (1782–1863)** En.wikipedia.org/w/index.php?title=William_John_Burchell&oldid= 621603794; Glen & Germishuizen, *BESA* (edition 2), p.117. Image: Sanbi CC BY-SA 3.0.
Burkea – **Joseph Burke (1812–1873)** Glen & Germishuizen, *BESA* (edition 2), p.118–120.
Burmannia – **Johannes Burman (1707–1780)** En.wikipedia.org/w/index.php?title=Johannes_Burman&oldid= 608367291. Image: public domain. Author: JM Quinkhard. Source: Thesaurus zeylanicus: exhibens plantas in insula Zeylana nascentes, inter quas plurimae novae species; genera inveniuntur, omnia iconibus illustrata, ac descripta / cura; studio Joannis Burmanni. CC BY-SA 3.0.
Burnatia – **Emile Burnat (1828–1920)** En.wikipedia.org/w/index.php?title=%C3%89mile_Burnat&oldid=588098659; very short biography. CC BY-SA 3.0.
Buttonia – **Edward Button (1836–1900)** Glen & Germishuizen, *BESA* (edition 2), p.123. Image: Sanbi.

C

Caesalpinia – **Andrea Cesalpino (1519–1603)** En.wikipedia.org/w/index.php?title=Andrea_Cesalpino&oldid=610756483. Image: public domain. Artist: Dessin d'fr:G. Zocchi, gravé par fr:F. Allegrini. Source: Gravure de 1765 d'Après un Portrait du Muséum du Jardin Botanique de Pise tirée d'*Herbals* d'Agnes Arber (1912). CC BY-SA 3.0.
Caesia – **Federico Cesi (1585–1630)** En.wikipedia.org/w/index.php?title=Federico_Cesi&oldid=628925866. Image: public domain. Author: Pietro Fachetti. Source: Comitatinazionali.librari.beniculturali.it – material of the national committee for the celebrations of IV the centenarian of the foundation of the Academy of the Members of the Accademia dei Lincei. CC BY-SA 3.0.
Cailliea – **René Caillie (1799–1838)** En.wikipedia.org/w/index.php?title=Ren%C3%A9_Cailli%C3%A9&oldid=616893088; Nndb.com/people/416/000114074. Image: public domain. Artist: Amélie Legrand de Saint-Aubin (1798–?) Source/photographer: Viguier, Pierre (2008), *Sur les Traces de René Caillié: Le Mali de 1828 Revisité* (in French), Versailles, France: Quae, ISBN 978-2-7592-0271-3. CC BY-SA 3.0.
Calandrinia – **Jean Louis Calandrini (1703–1758)** En.wikipedia.org/w/index.php?title=Jean-Louis_Calandrini&oldid=57797698; Charters, *EDSAP*. Image: public domain. Artist: Robert Gardelle. Source: Digital photograph of the original, probably by Queen Adelaide. CC BY-SA 3.0.
Caldesia – **Ludovico Caldesi (1822–1884)** It.wikipedia.org/w/index.php?title=Lodovico_Caldesi&oldid=61810156. CC BY-SA 3.0.
Calomeria – **Joséphine de Beauharnais (1763–1814)** En.wikipedia.org/w/index.php?title=Jos%C3%A9phine_de_Beauharnais&oldid=632858315. Image: public domain. Artist: François Gérard (1779–1837). Source: Musée national du Château de Fontainebleau. CC BY-SA 3.0.
Calpurnia – **Calpurnius (1st century AD)** Charters, *EDSAP*.
Camellia – **Georg Joseph Kamel (1661–1706)** En.wikipedia.org/w/index.php?title=Georg_Joseph_Kamel&oldid=632065239. CC BY-SA 3.0.
Caperonia – **Noël Caperon or Capperon (died 1572)** Raven, *English Naturalists from Neckam to Ray*.
Cardotiella – **Jules Cardot (1860–1934)** En.wikipedia.org/w/index.php?title=Jules_Cardot&oldid=607641939& Plants.jstor.org/person/bm000001274. CC BY-SA 3.0.
Carrpos – **Denis John Carr** (**1915–2008**) Eoas.info/biogs/P001501b.htm; Anbg.gov.au/biography/carr-denis-john.html; Trove.nla.gov.au/people/616210?c=people. Image: Author: unknown. Source: Australian National University Archives: university photographs, ANUA 225, photograph of Professor Denis John Carr, 1968.
Casearia – **Johannes Casearius (1642–1678)** Quattrocchi, *CRC World Dictionary of Plant Names*; short biography.
Casimiroa – **Casimiro Gomez (1740–1818)** Personajes Ilustres | Gobierno del Estado de Hidalgohidalgo.gob.mx/?p=121& Pers com. from David Hollombe.
Cassebeera – **Johann Heinrich Cassebeer (1784–1850)** De.wikipedia.org/w/index.php?title=Johann_Heinrich_Cassebeer&oldid=133963032. CC BY-SA 3.0.
Cassinia – **Henri Cassini (1781–1832)** En.wikipedia.org/w/index.php?title=Henri_Cassini&oldid=634254169&catnaps.org/cassini/cass5.htm. Image: public domain. Author: Engraved by Ambroise Tardieu in Paris. Source: Theo de Boer books database. CC BY-SA 3.0.
Cautleya – **Proby Cautley (1802–1871)** En.wikipedia.org/w/index.php?title=Proby_Cautley&oldid=623265931 Image: public domain. Artist: unknown. Source: Brown, *A Memoir of Colonel Sir Proby Cautley, 1802–1871, Engineer and Paleontologist*. CC BY-SA 3.0.
Cavacoa – **Alberto Judice Leote Cavaco (1916–?)** Es.wikipedia.org/w/index.php?title=Alberto_Judice_Leote_Cavaco&oldid=78306958 CC BY-SA 3.0.
Celsia – **Olof Celsius (1670–1756).** En.wikipedia.org/w/index.php?title=Olof_Celsius&oldid=61180195; *The Linnaean Correspondence*, Linnaeus.c18.net/Doc/lbio.php. Image: public domain. Author: unknown. Source: Plantexplorers.com. CC BY-SA 3.0.
Chapmanolirion – **James Chapman (1831–1872)** En.wikipedia.org/w/index.php?title=James_Chapman_(explorer)&oldid=605566901; Glen & Germishuizen, *BESA* (edition 2), p.126. CC BY-SA 3.0.
Chenia – **Chén Bāngjié (Chen Pan Chieh) (1907–1970)** Tropicos.org/Name/35001700?projectid=21. Source: Archiv, Botanischer Garten und Botanisches Museum Berlin-Dahlem, Freie Universität Berlin.
Chomelia – **Pierre Jean Baptiste Chomel (1671–1740)** Fr.wikipedia.org/wiki/Pierre-Jean-Baptiste_Chomel Image: public domain. Author: Gerardgiraud. Source: Tavail staff. CC BY-SA 3.0.
Christella – **Konrad H. Christ (1833–1933)** Fr.wikipedia.org/wiki/Konrad_Hermann_Heinrich_Christ; Plants.jstor.org/person/bm000001496. CC BY-SA 3.0.
Cienfuegosia – **Bernardo de Cienfuegos (c 1580–1640)** Es.wikipedia.org/w/index.php?title=Bernardo_de_Cienfuegos&oldid=71913717. CC BY-SA 3.0.
Clausena – **Peder Claussen Friis (1545–1614)** No.wikipedia.org/w/index.php?title=Peder_Clauss%C3%B8n_Friis&oldid=12965449. Image: public domain. Author: Kolbjørn Andersen. Source: Peder_claussen_friis.jpg. CC BY-SA 3.0.
Clevea – **Per Theodor Cleve (1804–1905)** En.wikipedia.org/w/index.php?title=Per_Teodor_Cleve&oldid=604798557; Charters, *EDSAP*. Image: public domain. Author: Osti, Heinrich, 1826–1914. Source: App.ub.uu.se/epub/bildsok/bibrecord.cfm?bibid=5467. CC BY-SA 3.0.
Cliffortia – **George Clifford (1685–1760)** Nl.wikipedia.org/w/index.php?title=George_Clifford_(1685-1760)&oldid=37290851. Image: public domain. Author: unknown. Source: George Clifford Herbarium (1685–1760). CC BY-SA 3.0.
Clivia – **Charlotte Percy, Duchess of Northumberland (1787–1866)** En.wikipedia.org/w/index.php?title=Charlotte_Percy,_Duchess_of_Northumberland& oldid= 635498467; Cliviasociety.org/history/ Image: public domain. Author:

William Oakley Burgess. Source: Npg.org.uk/collections/search/portraitLarge/mw40147/Charlotte-Florentia-Percy-ne-Clive-Duchess-of-Northumberland?LinkID= mp51003& role= sit& rNo= 1. CC BY-SA 3.0.
Clutia – Theodorus Augerius Clutius (Outgers Cluyt) (1577–1636) Charters, *EDSAP*; Quattrocchi, *CRC World Dictionary of Plant Names*.
Coddia – Leslie Edward Wastell Codd (1908–1999) Glen & Germishuiz en, *BESA* (edition 2), p.132. Image: Sanbi
Coldenia – Cadwallader Colden (1688–1776) En.wikipedia.org/w/index.php?title= Cadwallader_ Colden& oldid= 607074803& Answers.com/topic/colden-cadwallader. Image: public domain. Author: Nonenmac at En.wikipedia. Source: Transferred from En.wikipedia by SreeBot Original caption: '*From Harper's Encyclopedia of United States History, Harper; Brothers, 1905.*' CC BY-SA 3.0.
Columellea – Columella (4–c 70 BCE) En.wikipedia.org/w/index.php?title= Columella& oldid= 634204109 Image: public domain. Author: Jean de Tournes. Source: Original: Jean de Tournes, *Insignium aliquot virorum icones*. Lugduni: Apud Ioan. Tornaesium 1559; immediate source: Villa.culture.fr/accessible/en/uc/01_ 01_ 01. CC BY-SA 3.0.
Columnea – Fabio Colonna (1567–1640) En.wikipedia.org/w/index.php?title= Fabio_ Colonna& oldid= 621279709. Image: public domain. Source: Gravure de 1606 tiré de *Ekphrasis de Colonna*, tirée d'Herbals d'Agnes Arber (1912). CC BY-SA 3.0.
Combea – Francesco Comba (?–1892) Charters, *EDSAP*.
Commelina – Jan Commelin (1629–1692) En.wikipedia.org/w/index.php?title= Jan_ Commelin& oldid= 622122200; Doaks.org/library-archives/library/library-exhibitions/botany-of-empire/gardens/horti-medici-amstelodamensis; Brynmawr.edu/library/exhibits/darwin/Commelin.html. Image: public domain. Artist: Gerard Hoet (1688–1733) Source: Ahm.adlibsoft.com/ahmonline/dispatcher.aspx?action= search& database= ChoiceCollect& search= priref= 39423, allowed by Amsterdam Museum.
Comptonanthus – Robert Harold Compton (1886–1979) Glen & Germishuiz en, *BESA* (edition 2), p.134. Image: Sanbi.
Corbichonia – Jean Corbichon (or Corbechon) (c 1350) Chaudon, *Universal Dictionary* (1785); Biblicalcyclopedia.com/C/corbichon-% 28or-corbechon% 29-jean.html (from McClintock & Strong, *Biblical Cyclopedia*).
Cordia – Euricius and Valerius Cordus (1486–1535) De.wikipedia.org/w/index.php?title= Euricius_ Cordus& oldid= 134080814; En.wikipedia.org/w/index.php?title= Valerius_ Cordus& oldid= 560059252. Image: public domain. Author: Original uploader was Ringbang at En.wikipedia. Source: Originally from En.wikipedia. CC BY-SA 3.0.
Coulteria – Thomas Coulter (1793–1846) En.wikipedia.org/w/index.php?title= Thomas_ Coulter& oldid= 588114276; Sbcity.org/about/history/thomas_ coulters_ 1835_ map.asp. Image: public domain. Author: unknown. Source: Tcd.ie/Botany/herbarium/history.php.
Courbonia – Alfred Courbon (1829–1895) Charters, *EDSAP*. The original sources mislaid.
Courtoisia – Richard Joseph Courtois (1806–1835) Wiki.arts.kuleuven.be/wiki/index.php?title= Courtois,_ Richard-Joseph_ (1806–1835).
Crabbea – George Crabbe (1754–1832) En.wikipedia.org/w/index.php?title= George_ Crabbe& oldid= 609139426& ourcivilisation.com/smartboard/shop/anecdtes/c18/crabbe.htm. Image: Author: unknown. Source: National Portrait Gallery, London: NPG 1495 CC BY-SA 3.0.
Craibia – William Grant Craib (1882–1933) En.wikipedia.org/w/index.php?title= William_ Grant_ Craib& oldid= 634375743; Journals.cambridge.org/action/displayAbstract?fromPage= online& aid= 8372744.
Cullen – William Cullen (1710–1790) En.wikipedia.org/w/index.php?title= William_ Cullen& oldid= 634479760. Image: public domain. Author: unknown. Source: Universitystory.gla.ac.uk/Image: /?id= UGSP00091 CC BY-SA 3.0
Cullumia – Sir John Cullum, sixth Baronet (1733–1785) En.wikipedia.org/w/index.php?title= Sir_ John_ Cullum,_ 6th_ Baronet& oldid= 600853513. Image: public domain. Author**:** Painting by Angelica Kauffmann. Source: Vads.ac.uk/large.php?uid= 86174. CC BY-SA 3.0.
Cunonia – Johann Christian Cuno (1708–1780) En.wikipedia.org/w/index.php?title= Johann_ Christian_ Cuno& oldid= 60069922. Image: public domain. Author: Carel Augusti Arstenius (painter), GF Marstaller (engraver). Source: Portraitindex.de/db/apsisa.dll/ete?action= displayDetails/1& sstate= eJydlf9T2z YUwHFZ BxWELwJaQte15NquW2mlw9hxdHcbTeBgF1rWrD_ sLnc-xX6JNWz Jk-Sw7K-fJDsmjIT19ksSvfd5el_ 1Mu_ g3xGGIKn-mYIYVqUiLCAiWHE2Z ypfJiAkZ 6-c-jdoSQ6Ix7t_ eAkLPMbxA7deO6jv7e_ tb80evW9WZ IV0-d3MgfOz 88v86JKtJ2jNXB1xfpkmxd14Z mvmlXOwXllBD45_ O26k2onAjrvpVF6iOS055VLhMvEVHcCbgCbVHlc8JqKbin41AM09Qsuaa9I7jyyH7z GuFU8RLhRNKkDrOMNz PS6utGcNPEdPCuCERgrEKQVBhB8OWz CAKIujjFYL6o. CC BY-SA 3.0.
Curroria – Andrew Beveridge Curror (1811–1843) Charters, *EDSAP*; Floradeguinea.com/static/pdf/Bothalia39_ 2_ -_ Pteridophytes_ _ Figueiredo_ Gascoigne_ and_ Roux.pdf.
Curtisia – William Curtis (1746–1799) En.wikipedia.org/w/index.php?title= William_ Curtis& oldid= 621603876. Image: public domain. Author: unknown. Source: *Curtis's Botanical Magaz ine* at online scan Botanicus.org/page/468612. CC BY-SA 3.0.
Cussonia – Pierre Cusson (1727–1783) Charters, *EDSAP*.
Cuviera – Georges Cuvier (1769–1832) En.wikipedia.org/w/index.php?title=Georges_Cuvier&oldid=629936377. Image: public domain. Author: unknown. Source: Georges_Cuvier.jpg. CC BY-SA 3.0.

D

Dahlgrenodendron – Rolf Martin Theodor Dahlgren (1932–1987) En.wikipedia.org/w/index.php?title=Rolf_Dahlgren&oldid=619565592; Glen & Germishuiz en, *BESA* (edition 2), p.140 Image. bihrmann.com/caudiciforms/DIV/hist5.aspBi Photographer unknown.

Dahlia – **Andreas (Anders) Dahl (1751–1789)** En.wikipedia.org/w/index.php?title=Anders_Dahl&oldid=635733166
Dalbergia – **Carl Gustav Dahlberg (1721–1781)**
   Index to the Linnaean Herbarium. Sources.archive.org/stream/indextolinneanhe00jack/indextolinneanhe00jack_djvu.txt Short biography.
Dalechampia – **Jacques Daléchamps (1513–1588)** Charters, *EDSAP*. Image: public domain. Portrait de Daléchamps. Artist unknown. Source: Gravure du début du XVIIe siècle.
Danthonia – **Étienne Danthoine (1739–1794)** Es.wikipedia.org/wiki/Etienne_Danthoine.
Darea – **George Dare (f 1680s–1690s)** Single line entry.
Daubenya – **Charles Daubeny (1795–1867)** En.wikipedia.org/w/index.php?title=Charles_Daubeny&oldid=615805301; Charters, *EDSAP*. Image: public domain. Author: Hills (studio), unknown. Source: The Royal Society ? CC BY-SA 3.0.
Davallia – **Edmund Davall (1763–1798)** En.wikipedia.org/w/index.php?title=Edmund_Davall&oldid=619646235. CC BY-SA 3.0
Decorsea – **Gaston-Jules Decorse (1873–1907)**
   The Eponym Dictionary of Birds By Bo Beolens, Michael Watkins, Michael Grayson; The Eponym Dictionary of Reptiles By Bo Beolens, Michael Watkins, Michael Grayson.
Deinbollia – **Peter Vogelius Deinboll (1783–1874)** no.wikipedia.org/w/index.php?title=Peter_Vogelius_Deinboll&oldid=12098542. Image: public domain. Author: unknown. Source: Reprinted from OddMagnarSyversen: In light and shadow: the schools in Whitstable through the ages.Whitstable1979. CC BY-SA 3.0.
Delairea – **Eugene Delaire (1810–1856)** Charters, *EDSAP*.
Denekia – **Carl Henry Deneke (1735–1803)** Charters, *EDSAP*.
Derenbergia – **Julius Derenberg (1735–1803)** Single line biography.
Deroemeria – **Rudolf Benno von Römer (Roemer) (1803–1870)** Charters, *EDSAP*; l-iz.de/Bildung/Bücher/2013/11/Roemers-Garten-Buechersammlung-Rudolph-Benno-von-Roemer-52453.html.
Deschampsia – **Louis Auguste Deschamps (1765–1842)** nationaalherbarium.nl/FMCollectors/D/DeschampsLA.htm& Plants.jstor.org/stable/10.5555/al.ap.person.bm000002008 and others.
Descurainia – **François Déscourain (1658–1740)** Diccionario de Epónimos: DéscourainTranslate.google.co.za/translate?hl=en&sl=fr&u=doc.rero.ch/record/27750/files/32-13.pdf&prev=search.
Desmazeria – **Jean Baptiste Henri Joseph Desmazieres (1786–1862)**
ru.wikipedia.org/?oldid=661036 and other sources. Image: public domain. Author; Source: unstated. CC BY-SA 3.0.
Devia – **Miriam Phoebe de Vos (1912–2005)** Glen & Germishuizen, *BESA* (edition 2), p.145. Image: Sanbi.
Dewinteria – **Bernard de Winter (1924–)** Glen & Germishuizen, *BESA* (edition 2), p.146. Image: Sanbi.
Dicksonia – **James Dickson (1738–1822)** En.wikipedia.org/w/index.php?title=James_Dickson_(botanist)&oldid=619934889&sites.google.com/site/scottishfungi/biographies. Image: public domain. Author: Engraving after Henry Perronet Briggs. Source: Ihm.nlm.nih.gov/luna/servlet/detail/NLMNLM~1~1~101413659~173725:James-Dickson-F-L-S-H-P--Briggs-esq. CC BY-SA 3.0.
Dielsia – **Friedrich Ludwig Emil Diels (1874–1945)** Glen & Germishuizen, *BESA* (edition 2), p.146. Image: Sanbi
Dietrichia – **Friedrich Gottlieb Dietrich (1765–1850)** De.wikipedia.org/w/index.php?title=Friedrich_Gottlieb_Dietrich&oldid=131496553; Charters, *EDSAP*. Image: public domain. Author: unknown. Source: lgplus.ch/download/derWegderAlpen.pdf. CC BY-SA 3.0.
Dintera – **Moritz Kurt Dinter (1868–1945)** En.wikipedia.org/w/index.php?title=Kurt_Dinter&oldid=588096136; Glen & Germishuizen, *BESA* (edition 2), p.148–501. Image: Sanbi CC BY-SA 3.0.
Dioscorea – **Pedanius Dioscorides (c 40–90)** En.wikipedia.org/w/index.php?title=Pedanius_Dioscorides&oldid=627509372; short biography. Image: public domain. Author: unknown. Source: Huntbot.andrew.cmu.edu. CC BY-SA 3.0.
Dirichletia – **Gustav Lejeune Dirichlet (1805–1859)** *En.wikiped*ia.org/w/index.php?title=Peter_Gustav_Lejeune_Dirichlet&oldid=623298053. Image: public domain. Author and source: unknown. CC BY-SA 3.0.
Dittrichia – **Manfred Dittrich Manfred Dittrich (1934–)** Charters, *EDSAP*.
Dobrowskya – **Joseph Dobrowsky (1753–1829)** En.wikipedia.org/w/index.php?title=Josef_Dobrovsk%C3%BD&oldid=628714519. Image: public domain. Author: Jan Vilimek (1860–1938). Source: České album, sbírka podobizen předních českých velikánů, mužů i žen práce, kteří život svůj zasvětili povznesení národa svého. CC BY-SA 3.0.
Dodonaea – **Rembert Dodoens (1517–1585)** En.wikipedia.org/w/index.php?title=Rembert_Dodoens&oldid=636310167. Image: public domain. Author and source: unknown. CC BY-SA 3.0.
Doellia – **Johann Christoph Döll (1808–1885)** Wunschmann, 'Doell, Johann Christoph' in *General German Biography*, published by the Historical Commission of the Bavarian Academy of Sciences, Volume 47 (1903), p.740, digital full text De.wikisource.org/w/index.php?title=ADB:D%C3%B6ll,_Johann_Christoph&oldid=1986334.
Dombeya – **Joseph Dombey (1742–1794)** En.wikipedia.org/w/index.php?title=Joseph_Dombey&oldid=62964693; Charters, *EDSAP*. CC BY-SA 3.0.
Doodia – **Samuel Doody (1656–1706)** En.wikipedia.org/w/index.php?title=Samuel_Doody&oldid=588079197 (1656–1706); Rbg-web2.rbge.org.uk/bbs/learning/bryohistory/Bygone%20Bryologists/SAMUEL%20DOODY.pdf. CC BY-SA 3.0.
Doria – **Andrea Doria (1468–1560)** En.wikipedia.org/w/index.php?title=Andrea_Doria&oldid=631187720. Image: public domain. Artist: Sebastiano Del Piombo. Source: Web Gallery of Art. CC BY-SA 3.0.
Dorotheanthus – **Dorothea Schwantes (1849–?)** Plantzafrica.com/plantcd/dorotheanthusbell.htm
Dortmannia – **Jan Dortmann (dates unknown)** Short biography; Don, *A General History of the Dichlamydeous Plants (1831–1828)*.
Dovea – **Heinrich Wilhelm Dove (1803–1879)** En.wikipedia.org/w/index.php?title=Heinrich_Wilhelm_Dove&oldid=

614634786; Charters, *EDSAP*. Image: public domain. Author: unknown. Source: Sil.si.edu/digitalcollections/hst/scientific-identity/explore.htm. CC BY-SA 3.0.
Dregea – Johann Franz Drège (1794–1881) Glen & Germishuizen, *BESA* (edition 2), p.154–156. Image: Sanbi
Droguetia – Marc-Julien Droguet (1769–1836) Short biography on Openlibrary.org/search?author_key=OL7031135A&first_publish_year=1806; Droguet, Marc-Julien (1769–1836), *Propositions sur les Préceptes d'Hygiène Navale à Observer sous les Tropiques du Cancer, du Capricorne*.
Duchesnea – Antoine Nicholas Duchesne (1747–1827) En.wikipedia.org/w/index.php?title=Antoine_Nicolas_Duchesne&oldid=624864978; other sources. Image: public domain. Author: copper engraving by JD Nargeot. Source: Acknowledgements to the Bibliothèque Nordique (department of the Bibliothèque Sainte-Geneviève). CC BY-SA 3.0.
Dufourea – Léon Jean Marie Dufour (1780–1865) Charters, *EDSAP;* En.wikipedia.org/w/index.php?title=L%C3%A9on_Jean_Marie_Dufour&oldid=59350095. Image: public domain. Author: unknown. Source: Hbs.bishopmuseum.org/dipterists/dipt-d.html. CC BY-SA 3.0.
Dumasia – Jean Baptiste Dumas (1800–1884) Charters, *EDSAP*; En.wikipedia.org/w/index.php?title=Jean-Baptiste_Dumas&oldid=630629049. Image: public domain. Author: Magnus Manske. Source: Web4.si.edu/sil/scientific-identity/display_results.cfm?alpha_sort=d. CC BY-SA 3.0.
Dumortiera – Barthélemy Charles Joseph Dumortier (1797–1878) En.wikipedia.org/w/index.php?title=Barth%C3%A9lemy_Charles_Joseph_Dumortier&oldid=632778747; other sources. Image: public domain. Author: Louis Gailait (1810–1887). Source: Kikirpa.be/www2/cgi-bin/wwwopac.exe?DATABASE=object&LANGUAGE=2&OPAC_URL=&%250=20036758&LIMIT=50 – Musées Royaux des Beaux-Arts de Belgique. CC BY-SA 3.0.
Duranta – Castor Durantes (1529–1590) En.wikipedia.org/w/index.php?title=Castore_Durante&oldid=598476755&taccuinistorici.it/ita/news/moderna/bletteratura/Herbario-novo-Castor-Durante.html. Image: public domain. Author: unknown. Source: AstiLibri: Frontispiece to one of his works. CC BY-SA 3.0.
Duthiastrum – Augusta Vera Duthie (1881–1963) Glen & Germishuizen, *BESA* (edition 2), p.160. Image: Sanbi
Duvalia – Henri Auguste Duval (1774 1814) De.wikipedia.org/w/index.php?title=Henri-Auguste_Duval&oldid=116556259.
Duvernoia – Johann Georg Duvernoy (1692–1759) Charters, *EDSAP.*
Dyerophytum – Sir William Thiselton-Dyer (1843–1928) En.wikipedia.org/w/index.php?title=William_Turner_Thiselton-Dyer&oldid=621604565. Image: public domain. Author: unknown. Source: *The Gardener's Chronicle*, (1899), Vol. 25, Series 3, London, p.8. CC BY-SA 3.0.
Dymondia – Margaret E. Dryden-Dymond (1909–1952) (1774–1814) Plantzafrica.com/plantcd/dymondiamarg.htm.

E

Eberlanzia – **Friedrich Gustav Eberlanz (1879–1966)** Taken directly from Glen & Germishuizen, *BESA* (edition 2), p.160.
Ecklonea – **Christian Friedrich Ecklon (1795–1868)** En.wikipedia.org/w/index.php?title=Christian_Friedrich_Ecklon&oldid=607369260; Glen & Germishuizen, *BESA* (edition 2), p.162–164. CC BY-SA 3.0.
Edmondia – **Edmund Davall (1763–1798)** Charters, *EDSAP*; personal communication with Henry J Noltie, botanical historian, Royal Botanical Garden Edinburgh.
Eenia – **Ture Johan Gustaf Een (1837–1883)** Glen & Germishuizen, *BESA* (edition 2), p.101. Image: Sanbi
Ehretia – **George Dionysius Ehret. (1708–1770)** En.wikipedia.org/w/index.php?title=Georg_Dionysius_Ehret&oldid=600699247&rousseaustudies.free.fr/articlebotanical.pdf. Image: public domain. Artist: Johann Jakob Haid (1704–1767). Source: University of Strassbourg. CC BY-SA 3.0.
Ehrharta – **Jakob Friedrich Ehrhart (1708–1770)** De.wikipedia.org/w/index.php?title=Jakob_Friedrich_Ehrhart&oldid=135589792. Image: public domain. Author: unknown. Source: Herba.msu.ru/botanists/ehrhart/ehrhart_en.html. CC BY-SA 3.0.
Eichhornia – **Johann Albrecht Friedrich von Eichorn (1779–1856)** De.wikipedia.org/w/index.php?title=Friedrich_Eichhorn&oldid=134257911. Image: public domain. Author: Machahn. Source: scan. CC BY-SA 3.0.
Ekebergia – **Carl Gustaf Ekeberg (1716–1784)** En.wikipedia.org/w/index.php?title=Carl_Gustaf_Ekeberg&oldid=568762455; Glen & Germishuizen, *BESA* (edition 2), p.166. Image: Sanbi CC BY-SA 3.0.
Elsiea – **Elsie Elizabeth Esterhuysen (1912–2006)** Glen & Esterhuizen, *BESA* (edition 2), p.169. Image: Sanbi
Emelianthe – **Emily Pauline Reitz Ferguson (1872–?)** Charters, *EDSAP.*
Eminia – **Mehmed Emin Pasha (Isaak Eduard Schnitzer) (1840–1892)** En.wikipedia.org/w/index.php?title=Emin_Pasha&oldid=660792367. Image: public domain. Author: Emin Pasha (1840–1892). Source: Pasha, *Emin Pasha, his life and work*, Volume 1, 1898.
Emorya – **William H. Emory (1811–1887)** En.wikipedia.org/w/index.php?title=William_H._Emory& oldid= 632563700. Image: public domain. Authors: Mathew Brady (1822–1896), Levin Corbin Handy (1855–1932). Source: Library of Congress Prints and Photographs Division; Brady-Handy Photograph Collection; Hdl.loc.gov/loc.pnp/cwpbh.04689. CC BY-SA 3.0.
Engleria – **Adolf Engler (1844–1930)** Glen & Germishuizen, *BESA* (edition 2), p.168. Image: Sanbi.
Escallonia – **Antonio Escallón y Flórez (1739–1819)** Charters, *EDSAP.*
Eschscholzia – **Johann Friedrich von Eschscholtz (1793–1831)** En.wikipedia.org/w/index.php?title=Johann_Friedrich_von_Eschscholtz & oldid= 631357335. Image: public domain. Author: unknown. Source: *Zoologischer Atlas*, enthaltend Abbildungen und Beschreibungen neuer Thierarten, waehrend des Flottcapitains von Kotz ebue z weiter Reise um die Welt, auf der Russisch – Kaiserlichen Kriegsschlupp Predpriaetie in den Jahren 1823–1826 – Berlin, 1829. CC BY-SA 3.0.

Esterhuysenia – **Elizabeth ('Elsie') Esterhuysen** See Elsiea.
Eugenia – **Prince Eugène of Savoy (1663–1736)** En.wikipedia.org/w/index.php?title=Prince_Eugene_of_Savoy&oldid=637169201. Image: public domain. Artist: Jacob van Schuppen (1670–1751). Source: Geheugenvannederland.nl/?/en/items/RIJK01:SK-A-373. CC BY-SA 3.0.
Eulalia – **Eulalie Delile (1800–1840)** Short biography on Es.wikipedia.org/w/index.php?title=Eulalie_Delile&oldid=59292171&ausgrass2.myspecies.info/content/eulalia-0. CC BY-SA 3.0.
Euphorbia – **Euphorbus (1250 B.C.E)** Charters, *EDSAP*.

F

Fabronia – **Giovanni Valentino Mattia Fabbroni (1752–1822)** En.wikipedia.org/w/index.php?title=Giovanni_Fabbroni&oldid=619928879; En.wikipedia.org/wiki/Giovanni_Fabbroni. Image: public domain. Source: Biografisch Portaal van Nederland. CC BY-SA 3.0.
Fagelia – **Hendrik Fagel (1706–1790)** Kalden.home.xs4all.nl/verm/fagel-dublin-lecturesENG.html.
Fagonia – **Guy-Crescent Fagon (1638–1718)** En.wikipedia.org/w/index.php?title=Guy-Crescent_Fagon&oldid=588082445. Image: public domain. Author and source unknown. CC BY-SA 3.0.
Faidherbia – **Louis Faidherbe (1818–1889)** Charters, *EDSAP;* En.wikipedia.org/w/index.php?title=Louis_Faidherbe&oldid=627474378. Image: public domain. Artist: Marie-Madeleine Rignot-Dubaux (1857–1887). Source: Own photograph at Musée de l'Armée. CC BY-SA 3.0.
Falkia – **Johan Peter Falk (1733–1774)** Charters, *EDSAP*.
Fallopia – **Gabriele Falloppio (1523–1562)** En.wikipedia.org/w/index.php?title=Gabriele_Falloppio&oldid=628083059. Image: public domain. Author: Wiccan Quagga at En.wikipedia. Source: Peoples.ru/science/professor/gabriello. CC BY-SA 3.0.
Fanninia – **George Fox Fannin (1832–1865)** Charters, *EDSAP*.
Faurea – **William Caldwell Faure (1822–1844)** Glen & Germishuizen, *BESA* (edition 2), p.173.
Felicia – **Fortunato Felice (1723–1789)** En.wikipedia.org/w/index.php?title=Fortunato_Felice&oldid=629817969; Classiques-garnier.com/numerique-en/index.php?option=com_content&view=article&id=65%3Ayverdon-encyclopedia&catid=33%3Acatalogue-bases-dicenc&Itemid=30. Image: public domain. Author : Savoia (talk) Source: Musee d'Yverdon, Nr Lausanne, Switzerland. CC BY-SA 3.0.
Felipponea – **Dr Florentino Silvestre Felippone (1849–1939)** Short biography. Charters, *EDSAP*.
Fellhanera – near anagram of Josef Hafellner. See Hafellia.
Feretia – **Pierre Victor Adolphe Ferret (1814–1882)** Beolens *et al.*, *The Eponym Dictionary of Birds*; Youscribe.com/catalogue/livres/savoirs/sciences-humaines-et-sociales/origine-du-nom-de-famille-ferret-202079; Trove.nla.gov.au/result?q=ferret&l-decade=184.
Fernandoa – **Ferdinand 11 of Portugal (1816–1885)** En.wikipedia.org/w/index.php?title=Ferdinand_II_of_Portugal&oldid=631568106 Image: public domain. Author: Adolphe Pincon (1847–1884). Source: Purl.pt/5292/1. CC BY-SA 3.0.
Ferraria – **Giovanni Baptista Ferrari (1584–1655)** En.wikipedia.org/w/index.php?title=Giovanni_Baptista_Ferrari&oldid=608504895; Illustratedgarden.org/mobot/rarebooks/author.asp?creator=Ferrari,%20Giovanni%20Battista&creatorID=36. CC BY-SA 3.0.
Feuilleea – **Louis Feuillée (1660–1732)** En.wikipedia.org/w/index.php?title=Louis_Feuill%C3%A9e&oldid=616901783. Image: public domain. Author: unknown. Source: www-obs.cnrs-mrs.fr/tricent/astronomes/feuillee.ht. CC BY-SA 3.0.
Ficinia – **Heinrich David August Ficinus (1782–1857)** De.wikipedia.org/w/index.php?title=Heinrich_David_August_Ficinus&oldid=12877279. CC BY-SA 3.0.
Finckea – **August Fincke (1805–1873)** Short biography in Charters, *EDSAP*.
Fingerhuthia – **Carl Anton Fingerhuth (1774–1863)** Short biography in Charters, *EDSAP*.
Fintelmannia – **Ferdinand Fintelmann (1774–1863)** De.wikipedia.org/w/index.php?title=Ferdinand_Fintelmann&oldid=126195858. CC BY-SA 3.0.
Firmiana – **Karl Joseph von Firmian (1716–1782)** It.wikipedia.org/w/index.php?title=Carlo_Giuseppe_di_Firmian&oldid=66665117. Image: public domain. Artist: Giuseppe Franchi (1731–1806). Source: Based on a public domain text from *Meyers Encyclopedia* (edition 4) (1888–1890). CC BY-SA 3.0.
Flacourtia – **Étienne de Flacourt (1607–1660)** En.wikipedia.org/w/index.php?title=%C3%89tienne_de_Flacourt&oldid=609199756. Image: public domain. Author: unknown. Source: Unknown. CC BY-SA 3.0.
Flanagania – **Henry George Flanagan (1861–1919)** Glen & Germishuizen, *BESA* (edition 2), p.175. Image: Sanbi.
Flemingia – **John Fleming (1747–1829)** Charters, *EDSAP*.
Fleurya – **Camile Fleury (F 1817–1820)** Charters, *EDSAP*.
Flueggea – **Johannes Flüggé (1775–1816)** En.wikipedia.org/w/index.php?title=Johannes_Fl%C3%BCgg%C3%A9&oldid=623629458.
Fockea – **Gustav Woldemar Focke (1810–1877)** De.wikipedia.org/w/index.php?title=Gustav_Woldemar_Focke&oldid=131082373. Image: public domain. Author: unknown. Source: Hollanders, *Oberneuland – Bilder aus alten Truhen*, Döll Verlag, Bremen, 2005. CC BY-SA 3.0.
Forbesia – **John Forbes (1799–1823)** Histsem.unibas.ch/forschung/abschlussarbeiten/detailseite/?tx_x4equalificationgeneral_pi1[showUid]=5785&cHash=431906cf00c79eb86494fc823037cef6; Glen & Germishuizen, *BESA* (edition 2), p.176.
Forsskaolea – **Pehr Forsskål (1732–1763)** En.wikipedia.org/w/index.php?title=Peter_Forssk%C3%A5l&oldid=

635438794. Image: public domain. Author: unknown. Source: Svt.se/content/1/c6/30/17/44/forrskal_180.jpg at Svt.se/svt/jsp/Crosslink.jsp?d=25699&a=301744. CC BY-SA 3.0.
Forsstroemia – **Johan Erik Forsström (1775–1824)** En.wikipedia.org/w/index.php?title=Johan_Erik_Forsstr%C3%B6m&oldid=600201864. CC BY-SA 3.0.
Fossombronia – **Vittorio Fossombroni (1754–1844)** En.wikipedia.org/w/index.php?title=Vittorio_Fossombroni&oldid=633111215. Image: public domain. Author: unknown. Source: Transferred from It.wikipedia. CC BY-SA 3.0.
Frankenia – **Johan Frankenius (1590–1661)** Charters, *EDSAP*.
Freesia – **Friedrich Heinrich Theodor Freese (1795–1876)** Charters, *EDSAP*.
Fresenia – **Georg Fresenius (1808–1866)** En.wikipedia.org/w/index.php?title=Georg_Fresenius&oldid=615590775; Es.wikipedia.org/wiki/Johann_Baptist_Georg_Wolfgang_Fresenius. Image: public domain. Author: unknown. Source: Transferred from De.wikipedia; transferred to Commons by User: Ireas using CommonsHelper (Original text: National Weather Service Weather Forecast Office). CC BY-SA 3.0.
Freylinia – **Pietro Lorenzo, count de Freylino (1754–1820)** En.wikipedia.org/wiki/Freylinia_tropica.
Fridericia – **Frederick William III of Prussia. (1770– 1840)** En.wikipedia.org/w/index.php?title=Frederick_William_III_of_Prussia&oldid=625455044. Image: public domain. Author: unknown. Source: Macht und Dienst. Zur Darstellung des Brandenburgisch-preußischen Herrscherhauses in Gemälde und Graphik 1650–1900. CC BY-SA 3.0.
Friedrichsthalia – **Emanuel von Friedrichstal (1809–1842)** En.wikipedia.org/w/index.php?title=Emanuel_von_Friedrichsthal&oldid=631108744.
Friesodielsia – **Elias Magnus Fries (1794–1878) and Friedrich Ludwig Emil Diels (1874–1945)** En.wikipedia.org/w/index.php?title=Elias_Magnus_Fries&oldid=618833490; En.wikipedia.org/w/index.php?title=Ludwig_Diels&oldid=588096252. Image: public domain. No author or source.
Frithia – **Frank Frith (1872–1954)** Glen & Germishuizen, *BESA* (edition 2), p.179.
Frulliana – **Leonardo Frullani (1756–1824)** De.wikipedia.org/w/index.php?title=Leonardo_Frullani&oldid=69662497. CC BY-S.A.3.0
Fuchsia – **Leonhart Fuchs (1501–1566)** En.wikipedia.org/w/index.php?title=Leonhart_Fuchs&oldid=620077299. Image: public domain. Artist: Heinrich Füllmaurer (tätig um 1530/40). Source: Württembergisches Landesmuseu, Stuttgart. CC BY-SA 3.0.
Fuirena – **Jørgen Fuiren (1581–1628)** *Danish Biographical Encyclopedia*; Runeberg.org/dbl/5/0492.html.
Furcraea – **Antoine François, Comte de Fourcroy (1755–1809)** En.wikipedia.org/w/index.php?title=Antoine_Fran%C3%A7ois,_comte_de_Fourcroy&oldid=633733215. Image: public domain. Author: François-Séraphin Delpech (1778–1825). Source: Sil.si.edu. CC BY-SA 3.0.

G

Gagea – **Sir Thomas Gage (1781 – 1820)** Short biography in Charters, *EDSAP*.
Gaillardia – **Antoine René Gaillard de Charentonneau (1719–1789)** Charters, *EDSAP*.
Gaillonia – **Benjamin Gaillon (1782–1839)** Fr.wikipedia.org/w/index.php?title=Benjamin_Gaillon&oldid=106739898. Image: public domain. Author: Portrait drawn by Antoine Chazal and engraved by Ambroise Tardieu (1827). Source: unknown. CC BY-SA 3.0.
Galenia – **Claudius Galen (130–201 C.E.)** En.wikipedia.org/w/index.php?title=Galen&oldid=63608645. Image: public domain. Author: Georg Paul Busch (engraver). Source: *The Lancet*. CC BY-SA 3.0.
Galinsoga – **Ignacio Mariano Martinez de Galinsoga (1766–1797)** Jrpermar.com/Heraldica/martinde.htm; En.wikipedia.org/w/index.php?title=Ignacio_Mariano_Martinez_de_Galinsoga&oldid=567178580.
Galpinia – **Ernest Edward Galpin (1858–1941)** Glen & Germishuizen, *BESA* (edition 2), p.181–184. Image: Sanbi.
Galtonia – **Francis Galton (1822–1911)** En.wikipedia.org/w/index.php?title=Francis_Galton&oldid=637160359. Image: public domain. Author: unknown. Source: scanned from Pearson, *The Life, Letters, and Labors of Francis Galton*. CC BY-SA 3.0.
Gammiella – **George Alexander Gammie (1864–1935)** *A Sketch of the History of Indian Botany*; Ces.iisc.ernet.in/biodiversity/sahyadri/wgbis_info/botanical.htm. Image: public domain. Author: unknown. Source: Records.ancestry.com.
Garcinia – **Laurent Garcin (1683–1751)** *Dictionary Journalists*, Dictionnaire-journalistes.gazettes18e.fr/journaliste/330-laurent-garcin. This dictionary is available to the public courtesy Voltaire Foundation. CC BY-SA 3.0.
Gardenia – **Alexander Gardener (1730–1791)** En.wikipedia.org/w/index.php?title=Alexander_Garden_(naturalist)&oldid=617491881.
Gasparrinia – **Guglielmo Gasparrini (1804–1865)** It.wikipedia.org/w/index.php?title=Guglielmo_Gasparrini&oldid=65962551. Image: public domain. Author: unknown. Source: Basilicata.cc. CC BY-SA 3.0.
Gazania – **Theodorus Gaza (1398–1478)** En.wikipedia.org/w/index.php?title=Theodorus_Gaza&oldid=634979194. Image: public domain. Author: anonymous. Source: Österreichischen Nationalbibliothek. CC BY-SA 3.0.
Geigeria – **Philipp Lorenz Geiger (1785–1836)** Charters, *EDSAP*. Image: public domain. Author: unknown. Source: Images from the History of Medicine. CC BY-SA 3.0.
Genlisea – **Stéphanie Félicité, Comtesse de Genlis (1746–1830)** En.wikipedia.org/w/index.php?title=St%C3%A9phanie_F%C3%A9licit%C3%A9,_comtesse_de_Genlis&oldid=629236325. Image: public domain. Artist: Jacques-Antoine-Marie Lemoine (1751–1824). Source: Artexpertswebsite.ca/pages/artists/lemoine.php. CC BY-SA 3.0.
Georgeantha – **Alex George (1939–)** En.wikipedia.org/w/index.php?title=Alex_George&oldid=60581744. Image: public domain. Author and source: Cas Liber. CC BY-SA 3.0.

Gerardia – **John Gerard (1545–1607)** En.wikipedia.org/w/index.php?title=John_Gerard&oldid=623973580; Special. lib.gla.ac.uk/exhibns/printing/ Image: public domain. Artist unknown. Frontispice of *The Herball* (1636). CC BY-SA 3.0.
Gerbera – **Traugott Gerber (1710–1743)** De.wikipedia.org/w/index.php?title=Traugott_Gerber&oldid=127098714. CC BY-SA 3.0.
Germanea – **Jean-Joseph de Saint-Germain (1719–1791)** Loiseleur-Deslongchamps, *Herbier Général de l'Amateur, Contenant la Description, l'Histoire...* (volume 5); Masterart.com/Jean-Joseph-Saint-Germain-%E2%80%9CMa%C3%AEtre-fondeur-ciseleur%E2%80%9D-1748-Pierre-Roy-PortalDefault.aspx?tabid=53&dealerID=309&objectID=470625.
Gerrardanthus – **William Tyrer Gerrard (c 1831–1866)** Glen & Germishuizen, *BESA* (edition 2), p.187. Image: Sanbi
Girardinia – **Jean Pierre Louis Girardin (1803–1884)** Fr.wikipedia.org/w/index.php?title=Jean_Pierre_Louis_Girardin&oldid=108861284; Historicalautographs.co.uk/catalogue.asp?content=General%20Science%20and%20Invention. CC BY-SA 3.0.
Gisekia – **Paul Dietrich Giseke (1741–1796)** En.wikipedia.org/w/index.php?title=Paul_Dietrich_Giseke&oldid=595835038. CC BY-SA 3.0.
Gleditsia – **Johann Gottlieb Gleditsch (1714–1786)** De.wikipedia.org/w/index.php?title=Johann_Gottlieb_Gleditsch_(Botaniker)&oldid=134184151. Image: public domain. Author: unknown. Source: Hanstedt, *Adam Schwappach. Ein Forstwissenschaftler und sein Erbe*', 2001, p. 241. CC BY-SA 3.0.
Gleichenia – **Wilhelm Friedrich von Gleichen (1717–1783)** Deutsche-biographie.de/sfz21166.html; Worldcat.org/title/auserlesene-mikroskopische-entdeckungen-bey-den-pflanzen-blumen-und-bluethen-insekten-und-andern-merkwuerigkeiten/oclc/003155817.
Glekia – **Georg Ludwig Engelhard Krebs (1792–1844)** En.wikipedia.org/w/index.php?title=Georg_Ludwig_Engelhard_Krebs&oldid=595900683; Glen & Germishuizen, *BESA* (edition 2), p.248–249. Image: Sanbi CC BY-SA 3.0.
Gloeoheppia – **Adam Philipp Hepp (1797–1867)** Es.wikipedia.org/w/index.php?title=Johann_Adam_Philipp_Hepp&oldid=73333612; Frahm & Eggers, *Lexikon Deutschsprachiger Bryologen* (volume 1).
Gloveria – **Ruth Glover (f 1908–1925)** *Annals of the South African Museum* (volume IX) (1911–18) Archive.org/stream/annalsofsouthafr09sout/annalsofsouthafr09sout_djvu.txt; Glen & Esterhuizen, *BESA* (edition 2), p.176. Image: Sanbi
Gmelina – **Johann Georg Gmelin (1709–1755)** En.wikipedia.org/w/index.php?title=Johann_Georg_Gmelin&oldid=624717342. Image: public domain. Artist Johann Jakob Haid (1704–1767). Source: St Petersburg, Archiv der Akademie der Wissenschaften. CC BY-SA 3.0.
Gorskia – **Stanislaw Batys Górski (1802–1864)** Encyklopedia.puszcza-bialowieska.eu/index.php?dzial=haslo&id=622; Translate.google.com/translate?hl=en&sl=pl&u=encyklopedia.puszcza-bialowieska.eu/index.php%3Fdzial%3Dhaslo%26id%3D622&prev=search. Image: public domain. Source: Encyklopedia.puszcza-bialowieska.
Gorteria – **David de Gorter (1717–1783) & Johannes de Gorter (1689–1762)** En.wikipedia.org/w/index.php?title=David_de_Gorter&oldid=589268200. Image: public domain. Author: David de Gorter. Source: *Flora VII Provinciarum Belgii Doederati Indigena*; De.wikipedia.org/w/index.php?title=Johannes_de_Gorter&oldid=134951950. Image: public domain. Artist: JM Qunkhard. Source: Ihm.nlm.nih.gov/luna/servlet/detail/NLMNLM~1~1~101417063~178578: Johannes-de-Gorter-J-M--Quinkhard – US National Library of Medicine. CC BY-SA 3.0.
Gottschea – **Carl Moritz Gottsche (1808–1892)** En.wikipedia.org/w/index.php?title=Carl_Moritz_Gottsche&oldid=588087; 'Gottsche, Karl Moritz' in *General German Biography*, Historical Commission of the Bavarian Academy of Sciences, Volume 49, (1904), p.491–493. CC BY-SA 3.0.
Grangea – **Nicholas Tourtechot-Granger (c 1680–1734)** Charters, *EDSAP*.
Greenwayodendron – **Percy James Greenway (1897–1980)** Glen & Germishuizen, *BESA* (edition 2), p.197–8. Image: Sanbi.
Grevea – **H Grevé (?–1895)** Source: Dorr, *Plant Collectors in Madagascar and Comoro Islands*, 1997 p.192.
Grevillea – **Charles Francis Greville (1749–1809)** En.wikipedia.org/w/index.php?title=Charles_Francis_Greville&oldid=611658423. Image: public domain. Author: George Romney (1734–1802). Source: Thepeerage.com/p41634.htm. CC BY-SA 3.0.
Grewia – **Nehemiah Grew (1641–1712)** En.wikipedia.org/w/index.php?title=Nehemiah_Grew&oldid=620876293. Image: public domain. Author: White, *Makers of British Botany*, Plate 4, scan from Archive.org/details/makersofbritishb00oliv.
Greyia – **Sir George Grey (1812–1898)** En.wikipedia.org/w/index.php?title=George_Grey&oldid=631100939; Adbonline.anu.edu.au/biogs/A010439b.htm; Glen & Germishuizen, *BESA* (edition 2). Image: public domain. Author: Mundy, Daniel Louis (1826/7–1881) (photographer). Source: cropped version of original at Alexander Turnbull Library, Wellington, New Zealand. Reference number: G-623.
Grimmia – **Johann Friedrich Karl (Carl) Grimm (1737–1821)** Es.wikipedia.org/w/index.php?title=Johann_Friedrich_Carl_Grimm&oldid=65857661; Frahm & Eggers, *Lexikon Deutschsprachiger Bryologen* (volume 1).
Grisebachia – **August Heinrich Rudolph Grisebach (1814–1879)** Charters, *EDSAP;* De.wikipedia.org/wiki/August_Grisebach#/media/File:August_Heinrich_Rudolf_Grisebach.jpg. Image: public domain. Author: unknown. Source: Lsa.umich.edu/herb/malpigh/Intro/Authors/Gris.
Grossera – **Wilhelm Carl Heinrich Grosser (1869–1942)** Very short biography. Source: International Plant Names Index.
Grubbia – **Michael (Mikael) Grubb (af Grubbens) (1728–1808)**
Glen & Germishuizen, *BESA* (edition 2), p.197–8. Image: Sanbi.

Guatteria – **Giovanni Battista Guatteri (1739–1793)** It.wikipedia.org/w/index.php?title=Giambattista_Guatteri&oldid= 64387920; Scricciolo.com/Nuovo_Neornithes/Guatteri_Giambattista.htm. Image: public domain. Author: unknown. Source: 2.bp.blogspot.com/_Q8JBk7Z3KHk/ScbPZCUpyxI/AAAAAAAAAR0/mUbYOE0G9H0/s400/Guatteri.jpg. CC BY-SA 3.0.
Guettarda – **Jean-Étienne Guettard (1715–86)** En.wikipedia.org/w/index.php?title=Jean-%C3%89tienne_Guettard& oldid=632921269; Answers.com/topic/guettard-jean-tienne#ixzz1Llb2deAi. Image: public domain. Author: Théodore Charpentier. Source: photo by Bernard Gineste, in *Corpus Etampois*. CC BY-SA 3.0.
Guibourtia – **Nicholas Jean Baptiste Gaston Guibourt (1790–1861)** Fr.wikipedia.org/w/index.php?title=Nicolas_ Jean-Baptiste_Gaston_Guibourt&action=history; Guibourt, *An Illustrated Quinonologue* (1790–1867); Biusante. parisdescartes.fr/guibourt/exposition_guibourt_2.htm. Image: public domain. Author: unknown. Source: courtesy BIU Santé (Paris), Biusante.parisdescartes.fr/guibourt/exposition_guibourt_accueil.htm.
Guilandina – **Guilandinus Melchior (c 1520–1589)** Encyclopedia.com/doc/1G2-2830904644.html; Words. fromoldbooks.org/Chalmers-Biography/g/guilandinus-melchior.html.
Guilleminea – **Jean Baptiste Antoine Guillemin (1796–1842)** Charters, *EDSAP*.
Guizotia – **François Guizot (1787–1874)** En.wikipedia.org/wiki/Fran%C3%A7ois_Guizot&bookrags.com/biography/ francois-pierre-guillaume-guizot/#gsc.tab=0; Image: public domain. Author: Jehan Georges Verbert (1840–1902). Source/photographer: Versailles, Châteaux de Versailles et de Trianon; En.wikipedia.org/wiki/Fran%C3%A7ois_Guiz ot#/media/File:Guizot,_Fran%C3%A7ois_-_2.
Gunillaea – **Gunilla Thulin (1948–)** Text and image: Personal communication with Mats Thulin, professor of systematic botany at Uppsala University, Sweden.
Gunnera – **Johan Ernst Gunnerus (1718–1773)** En.wikipedia.org/w/index.php?title=Johan_Ernst_Gunnerus&oldid= 634996514&gullmedalje.com/medaljer/1102.htm. Image: public domain. Author: unknown. Source: Ub.ntnu.no/ formidl/utgivelser/til_opplysning/to_nr13.html. CC BY-SA 3.0.
Gussonea – **Giovanni Gussone (1787–1866)** En.wikipedia.org/w/index.php?title=Giovanni_Gussone&oldid= 612799622; Plants.jstor.org/stable/10.5555/al.ap.person.bm000392058. Image: public domain. No source details. CC BY-SA 3.0.
Gutenbergia – **Johann Gutenberg (1400–1468)** En.wikipedia.org/w/index.php?title=Johannes_Gutenberg&oldid= 637800196; Ideafinder.com/history/inventors/gutenberg.htm; Britannica.com/EBchecked/topic/249878/Johannes-Gutenberg. Image: public domain. Author: Shizhao, User: Liberal Freemason; Source: Sru.edu/depts/cisba/compsci/ dailey/217students/sgm8660/Final. CC BY-SA 3.0.
Guthriea – **Francis Guthrie (1831–1899)** Glen & Germishuizen, *BESA* (edition 2), p.201. Image: Sanbi.

H

Hackel – **Eduard Hackel (1850–1926)**
En.wikipedia.org/w/index.php?title=Eduard_Hackel&oldid=588106442. Image: public domain. Author: Photo by G Haslinger. Source: From the Biblioteca Digital de Botânica, Universidade de Coimbra, Bibdigital.bot.uc.pt. CC BY-SA 3.0.
Haenelia – **Eduard Gustav Haenel (1804–1856)**
Dictionary.sensagent.com/Eduard%20H%C3%A4nel/de-de/ (based on *Meyers Encyclopedia* (edition 4), 1888–1890); Uni-magdeburg.de/mbl/Biografien/1232.htm.
Hafellia – **Josef Hafellner (1951–)** Charters, *EDSAP;* personal communication with Joseph Hafellner, Institute of Plant Science, Karl Franzens University, Austria.
 En.wikipedia.org/w/index.php?title=Karl_Gottfried_Hagen&oldid=637638812; Manchester.edu/kant/bio/FullBio/ HagenKG.html. Image: public domain. Author: Neumann-Meding. Source: own work. CC BY-SA 3.0.
Hainardia – **Pierre Hainard (1936–)** Charters, *EDSAP*. Image & author: unknown. Source: Rts.ch/la-1ere/ programmes/les-dicodeurs/2206796-les-dicodeurs-du-28-09-2009.html.
Hakea – **Christian Ludwig von Hake (1745–1818)** Charters, *EDSAP;* En.wikipedia.org/wiki/Bremen-Verden.
Halesia – **Stephen Hales (1677–1761)**
En.wikipedia.org/w/index.php?title=Stephen_Hales&oldid=640414619. Image: public domain. Author: Mezzotint by J McArdell after T Hudson. Source: WellcomeImage:s.org/indexplus/Image:/V0002504.html. CC BY-SA 3.0.
Hallackia – **Russell Hallack (1824–1903)** Glen & Germishuizen, *BESA* (edition 2), p.203.
Halleria – **Albrecht von Haller (1708–1777)** En.wikipedia.org/w/index.php?title=Albrecht_von_Haller&oldid= 638077483; Answers.com/topic/albrecht-von-haller. Image: public domain. Artist: Johann Rudolph Huber (1668–1748). Source: Burgerbibliothek Bern, Negativnummer 2453WP. CC BY-SA 3.0.
Hallia – **Birger Mårten Hall (1741–1815)** Sv.wikipedia.org/w/index.php?title=Birger_Martin_Hall&oldid=27900550. Image: public domain. Author: BanWisco. Source: own work. CC BY-SA 3.0.
Hallianthus – **Harry Hall (1906–1986)** Glen & Germishuizen, *BESA* (edition 2), p.203. Image: Sanbi
Hammeria – **Steven Allen Hammer (1951–)** Charters, *EDSAP;* Cssainc.org/index.php?searchword= Steven+Hammer&option=com_search&Itemid. Image source: Little Sphaeroid Press.
Hardwickia – **Thomas Hardwicke (1755–1833)** En.wikipedia.org/w/index.php?title=Thomas_Hardwicke&oldid= 607478511. Image: public domain. Author: J Lucas; Lithograph Louis Haghe. Source: Archive.org/stream/ IllustrationsOfIndianZoology1/Hardwicke1#page/n7/mode/1u. CC BY-SA 3.0.
Harrisia – **William Harris (1860–1920)** Botanicgardens.ie/herb/books/inetok.htm; Archive.org/stream/ bulletinofbritis04britlond/bulletinofbritis04britlond_djvu.txt.
Hartmanthus – **Heidrun Elsbeth Klara Hartmann (1942–)** Glen & Germishuizen, *BESA* (edition 2), p.205.

Hartogia – **Johannes (Jan, Johan) Hartog (1663–1722)** Glen & Germishuizen, *BESA* (edition 2), p.205–6.
Hartwegia – **Andreas Johann Hartweg (1777–1831)** Crocoll.net/Dahlien/infos/Hartweg.php5/orchidologists/karl-theodore-hartweg.shtm. En.wikipedia.org/wiki/Karl_Theodor_Hartweg.
Harveya – **William Henry Harvey (1811–1866)** Glen & Germishuizen, *BESA* (edition 2), p.207–8. Image: Sanbi.
Haumaniastrum – **Lucien Leon Hauman (1880–1965)** Pt.wikipedia.org/w/index.php?title=Lucien_Hauman&oldid= 35169482; Charters, *EDSAP.*
Haworthia – **Adrian Hardy Haworth (1768–1833)** En.wikipedia.org/w/index.php?title=Adrian_Hardy_Haworth&oldid= 624329190. Image: public domain. Author: unknown. Source: Image: S18.fotki.com/v345/photos/6/642761/3958513/ AdrianHardyHaworth17671833-vi.jpg. CC BY-SA 3.0.
Hebenstretia – **Johann Christian Hebenstreit (1720–1791)** De.wikipedia.org/w/index.php?title=Johann_Christian_ Hebenstreit_(Botaniker)&oldid=124347808 (text virtually a copy).
Hedwigia – **Johann Hedwig (1730–1799)** En.wikipedia.org/w/index.php?title=Johann_Hedwig&oldid=600838515. Image: public domain. Author: unknown. Source: courtesy Ashmolean Museum, Oxford. CC BY-SA 3.0.
Heeria – **Oswald von Heer (1809–1883)** En.wikipedia.org/w/index.php?title=Oswald_Heer&oldid=637548134; Equisetites.de/palbot/science_history/palaeobotanists.html. Image: public domain. Author and source: unknown.
Heimia – **Ernst Ludwig Heim (1747–1834)** En.wikipedia.org/wiki/Ernst_Ludwig_Heim; Ncbi.nlm.nih.gov/pubmed/3318181; Archive.org/stream/ anintroductiont00garrgoog/anintroductiont00garrgoog_djvu.txtl. Image: public domain. Artist: Julius Hübner (1806–1882). Source: Gallery: Alte Nationalgalerie Berlin.
Heinsia – **Daniel Heinsius (1580–1655)** En.wikipedia.org/w/index.php?title=Daniel_Heinsius&oldid=621250576; Mywire.com/a/Oxford-Dictionary-Renaissance/Heinsius-Daniel/9511706. Image: public domain. Author: unknown. Source: Telemachos.hu-berlin.de/bilder/gudeman/gudeman.html.
Hellmuthia – **Hellmuth Steudel (1816–1886)** Averbeck et al., *Von der Kaltwasserkur bis zur Physikalischen Therapie,* 2013.
Hennediella – **Roger Hennedy (1809–1877)** *Dictionary of National Biography* (volume 25), 1885–1900; The Friends of Glasgow Necropolis. Image: public domain. Author: unknown. Source: The Friends of Glasgow Necropolis.
Heppia – **Johann Adam Philipp Hepp (1797–1867)** Frahm & Eggers, *Encyclopedia of German Bryologen* (volume 1, 1995).
Herbertus – **Thomas Herbert (c 1656–1733)** En.wikipedia.org/wiki/Thomas_Herbert,_8th_Earl_of_Pembroke. Image: public domain; Commons.wikimedia.org/wiki/File:Thomas_Herbert,_8th_Earl_of_Pembroke. Author: John Greenhill (died 1676). Source: National Portrait Gallery: NPG 5237.
Hermannia – **Paul Hermann (1646–1695)** Glen & Germishuizen, *BESA* (edition 2), p.210–211.
Hermbstaedtia – **Sigismund Friedrich Hermbstaedt (1760–1833)** De.wikipedia.org/w/index.php?title=Sigismund_ Friedrich_Hermbst%C3%A4dt&oldid=129093939; Deutsche-biographie.de/sfz30190.htmlSigismund Friedrich Hermbstaedt. Image: public domain. Source: Spektrum.de/lexikon/biologie/hermbstaedt-sigismund-friedrich/31455; wellcomeimages.org/indexplus/obf_images/dd/c0/516f20747bf1b0e1c257d648e383.jpg. CC BY-SA 3.0.
Herrea – **Adolar Gottlieb Julius (Hans) Herre (1895–1979)** Glen & Germishuizen, *BESA* (edition 2), p.212. Image: Sanbi CC BY-SA 3.0.
Herschelia – **John Frederick William Herschel (1792–1871)** Glen & Germishuizen, *BESA* (edition 2), p.212–213. Image: Author: Julia Margaret Cameron. Source: John Herschel (Metropolitan Museum of Art copy, restored) CC BY-SA 3.0.
Hertia – **Johann Casimir Hertius (1679–1748)** Uni-giessen.de/cms/fbz/fb11/dekanat/geschichte/Facultas% 20Medica%20Giessen/Vom%20Anteil%20der%20Facultas%20M.
Hessea – **Christian Henrich Friedrich Hesse (1772–1837)** Glen & Germishuizen, *BESA* (edition 2), p.213.
Heudelotia – **Jean P Heudelot (1802–1837)** *Biographical Etymology of Marine Organism Names*, Bemon.loven.gu.se/ petymol.h.html. Image: Berhaut, *Flore illustrée du Sénégal.*
Hewittia – **Hewett Cottrell Watson (1804–1881)** Personal communication with Henry Noltie, botanical historian, Royal Botanical Garden Edinburgh; En.wikipedia.org/w/index.php?title=Hewett_Watson&oldid=632348687. Image: public domain. Author and source: unknown.
Heywoodia – **Arthur William Heywood (1853–1918)** Hsrc.ac.za/Document-1546.phtml; *From Forestry to Soil Conservation: British Trees*; Sro.sussex.ac.uk/12056/1/Showers.pdf.
Hibbertia – **George Hibbert (1757–1837)** En.wikipedia.org/w/index.php?title=George_Hibbert&oldid=628786488; Ucl. ac.uk/lbs/person/view/16791; Commons.wikimedia.org/wiki/File:George_Hibbert_by_Thomas_Lawrence,_1811. JPG. Image: public domain. Author: Stephencdickson. Source: own work. CC BY-SA 4.0.
Hiernia – **William Philip Hiern (1839–1925)** Es.wikipedia.org/w/index.php?title=William_Philip_Hiern&oldid= 74668402; Rammuseum.org.uk/collections/collectors/william-philip-hiern-18391925.
Hilleria – **Matthaeus Hiller (1646–1725)** Johann Reinhard Hedinger (1664–1704) by Wolfgang Schöllkop (1999); Books.google.co.za/books?id=RkY0I_4e9ZgC&pg=PA153&dq=Matthaeus+Hiller+%281646-1725%29,&hl=en&ei= p0DZTYrACoW6hAez153JBg&sa=X&oi=book_result&ct=result&resnum=6&ved=0CEMQ6AEwBQ#v=onepage& q&f=falsef; Wikimedia Commons. Image: public domain. Author: unknown. Source: Tübinger Professorengalerie.
Hilliardia – **Olive Mary Hilliard (1925–)** En.wikipedia.org/w/index.php ?title=Olive_Mary_Hilliard&oldid=59583528; Glen & Germishuizen, *BESA* (edition 2), p.215. Image: Sanbi CC BY-SA 3.0.
Hippocratea – **Hippocrates of Kos (Cos) (c 460–370 BCE)** Charters, *EDSAP.* En.wikipedia.org/w/index.php?title= Hippocrates&oldid=641476564. Image: public domain. Author: Engraving by Peter Paul Rubens, 1638. Source: courtesy National Library of Medicine.
Hirschfeldia – **Christian Cay Lorenz Hirschfeld (1742–1792)** De.wikipedia.org/w/index.php?title=Christian_Cay_

Lorenz_Hirschfeld&oldid=134748765; Books.simsreed.com/architecture/architecture_source.php?category= architecture%20-%20source&stk=24286. Image: public domain. Artist Christian Cay Lorenz Hirschfeld (1742–1792) Portraitindex.de/dokumente/html/obj34014182. CC BY-SA 3.0.

Hoarea – Sir Richard Hoare, second Baronet (1758–1838) En.wikipedia.org/w/index.php?title=Sir_Richard_Hoare,_2nd_Baronet&oldid=637208126. Image: Regencyhistory. net/2014/05/sir-richard-colt-hoare-2nd-baronet-1758.html.

Hoffmannia – Georg Franz Hoffmann (1760–1826) De.wikipedia.org/w/index.php?title=Georg_Franz_Hoffmann& oldid=134702327. Image: public domain. Author: unknown. Source: Botsad.msu.ru/story.htm. CC BY-SA 3.0.

Hoffmannseggia – Johann Centurius Hoffmansegg (1766–1849) En.wikipedia.org/w/index.php?title=Johann_ Centurius_Hoffmannsegg&oldid=599983728. Image: public domain. Source: Michael T Stieber and FA Stafleu, *(Illustration): HI-IAPT Portraits of Botanists No. 88. Johann Centurius, Graf von Hoffmannsegg,* Taxon, Vol. 27, No. 4 (Aug., 1978). CC BY-SA 3.0.

Holmskioldia – Johan Theodor Holmskjold (1732-1794) Da.wikipedia.org/w/index.php?title=Theodor_Holmskiold& oldid=7800978. Image: public domain. Artist: Jens Juel (1745–1802) Source: Museum of National History at Feederiksborg Castle. CC BY-SA 3.0.

Holubia – Emil Holub (1847–1902) En.wikipedia.org/w/index.php?title=Emil_Holub&oldid=607815247; Britannica. com/EBchecked/topic/269714/Emil-Holub. Image: public domain. Author: Penarc. Source: Czeck wikipedist: Mikeshk. CC BY-SA 3.0.

Hoodia – William Chamberlain Hood (1790–1879) Archiver.rootsweb.ancestry.com/th/read/ENG-SURREY/2005-01/1106474018; *The British Medical Journal*, Bmj.com/content/bmj/1/997/local/admin.pdf; David Hollombe personal communication.

Hookeriopsis – William Hooker (1785–1865) En.wikipedia.org/w/index.php?title=William_Hooker_(botanist)&oldid= 640788312. Image: public domain. Artist: Spiridione Giambardella. Source unknown. CC BY-SA 3.0.

Hoslundia – Ole Haaslund-Smith (Schmidt) (died 1801) Plants.jstor.org/stable/10.5555/al.ap.person.bm000127486; Hopkins, *Peter Thonning and Denmark's Guinea Commission: A study in Nineteenth-Century African Colonial Geography*, 2013.

Hottonia – Petrus Houttuyn (1648–1709) Translate.google.co.za/translate?hl=en&sl=cs; u=botany.cz/cs/hotton/&ei= XVfdTeHnOoWq-gaky-m4Dw&sa=X&oi=translate&ct=result&resnum=5&ved=0CDwQ7gEwBA&prev=/search% 3Fq%3DPetrus%2BHouttuyn%2B%28Pieter%2BHotton%29%2B%281648-1709%29,%26hl%3Den%26rlz% 3D1B3GGIC_enZA317ZA317%26prmd%3Divnsob (Homo Botanicus:Hotton, Pieter by Jana Möllerová).

Hoya – Thomas Hoy (c 1750 – 1822) En.wikipedia.org/w/index.php?title=Thomas_Hoy_(botanist)&oldid=621604156.

Huea – Abbé Auguste-Marie Hue (1840–1917) Charters, *EDSAP*.

Huernia – Justus Heurnius (1587–1652) Charters, *EDSAP*.

Hugonia – Augustus Johannes Hugo (?-1753) Charters, *EDSAP*.

Humea – Lady Amelia Hume (17511809) En.wikipedia.org/wiki/Charles_Long,_1st_Baron_Farnborough; History of European Botanical Discoveries in China (1898) By Emil Bretschneider; other snippet sources. CC BY-SA

Huperzia – Johann Peter Huperz (1771–1816) Quattrocchi, *CRC World Dictionary of Medicinal and Poisonous Plants: Common Names*; Books.google.co.za/books?id=YC_IAgAAQBAJ&pg=PA2015&lpg=PA2015&dq= Johann+Peter+Huperz+%281771-1816%29&source=bl&ots=cUZeCkgeVD&sig=AXUwvL-HaMsg3w8gl8jWuViX-ao&hl=en&sa=X&ei=w5uvVIG4I8PyUMeqgvAI&redir_esc=y#v=onepage&q=Johann%20Peter%20Huperz%20% 281771-1816%29&f=false.

Husnotiella – Pierre Tranquille Husnot (1840–1929) Plants.jstor.org/person/bm000003894.

Huttonaea – Caroline Hutton (née Atherstone) (1826–1908) Glen & Germishuizen, *BESA* (edition 2), p.226.

I

Ibbetsonia – Agnes Ibbetson (1757–1827) En.wikipedia.org/w/index.php?title=Agnes_Ibbetson&oldid=599830012.

Ihlenfeldtia – Hans-Dieter Ihlenfeldt (1932–) Glen & Germishuizen, *BESA* (edition 2), p.197–8. Image: Sanbi

Illigera – Johann Karl Wilhelm Illiger (1775–1813) En.wikipedia.org/w/index.php?title=Johann_Karl_Wilhelm_Illiger& oldid=634760311.

Imperata – Ferrante Imperato (1550–1625) En.wikipedia.org/w/index.php?title=Ferrante_Imperato&oldid=621278874; Antiquariaatjunk.com/item.php?item=6309; Summagallicana.it/lessico/i/Imperato%20Ferrante.htm. Image, author & source: unknown. CC BY-SA 3.0.

Imshaugia – Henry Andrew Imshaug (1925–2010) Jstor.org/pss/3244780; SpeciEs.wikimedia.org/wiki/Henry_ Andrew_Imshaug. Image: Source: Herbarium.msu.edu/SSP/SSP_Introduction.html.

Inezia – Inez Clare Verdoorn (1896–1989) Glen & Germishuizen, *BESA* (edition 2), p.441. Image: Sanbi.

Ingenhoussia – Jan Ingenhousz or Ingen-Housz (1730–1799) Charters, *EDSAP*; En.wikipedia.org/w/index. php?title=Jan_Ingenhousz&oldid=631326927. Image: public domain. Author: unknown. Source: Britannica.com/eb/ art-11958/Ingenhousz-detail-of-an-engraving?articleTypeId=1.

J

Jacksonago – Benjamin Daydon Jackson (1846–1927) En.wikisource.org/wiki/The_Times/1927/Obituary/Benjamin_Daydon_Jackson. Image: Npg.org.uk/collections/ search/portrait/mw194756/Benjamin-Daydon-Jackson. Source: National Portrait Gallery.

Jacobsenia – Hermann Johannes Heinrich Jacobsen (1898–1978) Es.wikipedia.org/w/index.php?title=Hermann_

Johannes_Heinrich_Jacobsen&oldid=77243339. Image: Uni-kiel.de/nickol/Garten/Hermann-Jacobsen.html.
Jacquemontia – **Victor Jacquemont (1801–1832)** Fr.wikipedia.org/w/index.php?title=Victor_Jacquemont&action= history. Image: public domain. Author: Vivant Beauce. Source: Édouard Charton (dir.), *Le Magasin pittoresque*, Paris, 1851. CC BY-SA 3.0.
Jacquesfelixia – **Henri Jacques-Félix (1907–2008)** Es.wikipedia.org/w/index.php?title=Henri_Jacques-F%C3% A9lix&oldid=78004726; Quattrocchi, *CRC World Dictionary of Grasses: Common Names, Scientific Names...* (volume 1).
Jaegerina – **August (Augusto) Jaeger (1842–1877)** Charters, *EDSAP*.
Jamesbrittenia – **James Britten (1846–1924)** En.wikipedia.org/w/index.php?title=James_Britten&oldid=607591119. Image: Author: unknown. Source: website of the Catholic Truth Society. CC BY-SA 3.0.
Jamesonia – **William Jameson (1796–1873)** En.wikipedia.org/w/index.php?title=William_Jameson&oldid= 618583708; En.wikisource.org/wiki/Jameson,_William_%281796-1873%29_%28DNB00%29 bbc.co.uk/arts/ yourpaintings/paintings/william-jameson-17961873-87590. Image: public domain. Author: unknown. Source: Royal Botanic Gardens, Kew: Herbarium, Library, Art; Archives.
Jamesoniella – See Jamesonia above.
Jaumea – **Jean Henri Jaume Saint-Hilaire (1772–1845)** En.wikipedia.org/w/index.php?title=Jean_Henri_Jaume_ Saint-Hilaire&oldid=600149179.
Jensenia – **Thomas Jensen (1824–1877)** Denstoredanske.dk/ Image: public domain. Author: unknown. Source: Sorensen, SA: Nordisk, Skaktidende, Kjobenhavn, Wilhelm Priors Hofbokhandel, 1878, p. xiv.
Jensenobotrya – **Emil Jensen (1889–1963)** Author investigation. Source: Sam Cohen Library, Scientific Society, Swakopmund, Namibia.
Jordaaniella – **Pieter Gerhardus Jordaan (1913–1987)** Glen & Germishuizen, *BESA* (edition 2), p.234. Image: Sanbi
Julbernardia – **Marie Joseph Jules Pierre Bernard (1876–?)** Short biography in Charters, *EDSAP*.
Jumellea – **Henri Lucien Jumelle (1866–1935)** Plants.jstor.org/stable/history/10.5555/al.ap.person.bm000004162. Image: courtesy Hunt Institution of Botanical Documentation, Carnegie Mellon University.
Jungermannia – **Ludwig Jungermann (1572–1653)** De.wikipedia.org/w/index.php?title=Ludwig_Jungermann&oldid= 132917095; En.wikipedia.org/wiki/Botanischer_Garten_Gie%C3%9Fen. Image: public domain. Author: Bartholomew Kilian II (1630–1696). Source: 'The University of Giessen 1607–1907: Volume 1 page 176'. CC BY-SA 3.0.
Juratzkaea – **Jacob Juratzka (1821–1878)** Uni-goettingen.de/de/187061.html; Es.wikipedia.org/w/index.php?title= Jacob_Juratzka&oldid=78430881.
Justicia – **James Justice (1698–1763)** Freepages.genealogy.rootsweb.ancestry.com/~brookefamily/justicejamessir. html; En.wikipedia.org/wiki/James_Justice. CC BY-SA 3.0.
Juttadinteria – **Helena Jutta Dinter (née Schilde) (1868–1945)** Glen & Germishuizen, *BESA* (edition 2), p.148–1508. Image: Sanbi.

K

Kaempferia – **Engelbert Kaempfer (1651–1716)** En.wikipedia.org/w/index.php?title=Engelbert_Kaempfer&oldid= 635092270; Nl.wikipedia.org/w/index.php?title=Engelbert_Kaempfer&oldid=39233550; Jhmas.oxfordjournals.org/ content/XXI/3/237.extract; Kwanten.home.xs4all.nl/kaempfer.htm. Image: public domain. Author: original uploader was Fuelbottle at En.wikipedia. Source: En.wikipedia. CC BY-SA 3.0.
Kalmia – **Pehr Kalm (1716–1779)** Fi.wikipedia.org/w/index.php?title=Pehr_Kalm&oldid=14376758. Image: public domain. Author: JG Geitel. Source: En.wikipedia, were first loaded by en:User:Hephaestos on 5 February 2004. CC BY-SA 3.0.
Kaulfussia – **Georg Friedrich Kaulfuss (1786–1830)** Severens.net/Auteurs/BiografieAuteurs/AuteursG/Georg_ Friedrich_Kaulfuss.html.
Keetia – **Johan Diederik Möhr Keet (1882–1976)** Glen & Germishuizen, *BESA* (edition 2), p.237. Image: Sanbi.
Kennedia – **John Kennedy (1759–1842)** En.wikipedia.org/w/index.php?title=Lee_and_Kennedy&oldid=639157999; Charters, *EDSAP*; personal communication with David Hollombe, botanical history researcher.
K*ensitia* – **Harriet Margaret Louisa Bolus (née Kensit) (1877–1970)** Glen & Germishuizen, *BESA* (edition 2), p.197–8.Image: Sanbi.
Kickxia – **Jean Kickx Sr (1775–1831) & Jean Kickx Jr (1803–1864)** En.wikipedia.org/w/index.php?title=Jean_Kickx_ (1775%E2%80%931831)&oldid=620390985; En.wikipedia.org/w/index.php?title=Jean_Kickx&oldid=606878442; From the *History of the Ghent Botany* Plantkunderug.wordpress.com/uit-de-historiek-van-de-gentse-plantkunde. Image: public domain. Author: unknown. Source: Meemelink.com/portraits%20pages/prints-portraits%20A-J.htm. CC BY-SA 3.0.
Kiggelaria – **Franz Kiggelaer (1648–1722)** En.wikipedia.org/w/index.php?title=Franz_Kiggelaer&oldid=568223414. CC BY-SA 3.0.
Killickia – **Donald Joseph Boomer Killick (1926–2008)** Glen & Germishuizen, *BESA* (edition 2), p.239. Image: Sanbi.
Kindbergia – **Nils Conrad Kindberg (1832–1910)**
Biographi.ca/009004-119.01-e.php?BioId=41064&query=;
Linnaeus.nrm.se/botany/fbo/hand/kindberg.html.en; Biodiversitylibrary.org/creator/7. Image: public domain. Author: unknown, signed by himself. Source: Botaniska Notiser för År 1912. Lund: C. W. K. Gleerups, 1912, p. 119. CC BY-SA 3.0.
Kirkia – **John Kirk (1832–1922)** Glen & Germishuizen, *BESA* (edition 2), p.240–241. Image: Sanbi.

Klattia – **Friedrich Wilhelm Klatt (1825–1897)** Charters, *EDSAP.*
Kleinia – **Jacob Theodor Klein (1685–1759)** En.wikipedia.org/w/index.php?title=Jacob_Theodor_Klein&oldid= 639682009; Geology-books.com/servlet/the-12/Conchology-book,--Jacob/Detail. Image: public domain. Author: Jacob Wessel (1710–1780). Source: Image: taken by User: Mathiasrex Maciej Szczepańczyk. CC BY-SA 3.0.
Klenzea – **Possibly for Leo von Klenze (1784–1864)** En.wikipedia.org/w/index.php?title=Leo_von_Klenze&oldid= 631696100. Image: public domain. Author: Franz Hanfstaengl (1804–1877). Source: scan. CC BY-SA 3.0.
Klingia – **Erich Kling (1854–1892)** De.wikipedia.org/w/index.php?title=Erich_Kling&oldid=137646238; Ftp.scd.univ-metz.fr/pub/Theses/1990/De_Souza.Kodzo_Nyamadi.LMZ908.pdf. Image: public domain. Author: unknown. Source: Schmidt, *Deutschlands Kolonien* (Bd. 2), Verlag des Vereins der Bücherfreunde Schall; Grund, Berlin 1898, S. 192. CC BY-SA 3.0.
Knightia – **Thomas Andrew Knight (1759–1838)** En.wikipedia.org/w/index.php?title=Thomas_Andrew_Knight& oldid=622653979. Image: public domain. Author: Painted by Solomon Cole (1806–1893). Source: Bbc.co.uk/arts/yourpaintings/paintings/thomas-andrew-knight-17581838-frs-fls-prhs-87576. CC BY-SA 3.0.
Kniphofia – **Johann(es) Hieronymus Kniphof (1704–1763)** En.wikipedia.org/w/index.php?title=Johann_Hieronymus_ Kniphof&oldid=59583759&. Image: public domain. Author: unknown. Source: Academictree.org/chemistry/peopleinfo.php?pid=66945. CC BY-SA 3.0.
Knowltonia – **Thomas Knowlton (1691–1781)** Charters, *EDSAP*; Digitalcollections.nypl.org/items/510d47e2-a660-a3d9-e040-e00a18064a99. Image: Pub.com. Author: digital collections. Source: and publisher: The New York Public Library, Astor, Lennox, and Tilden Foundation.
Kobresia – **Paul von Kobres (1747–1823)** Charters, *EDSAP.*
Kochia – **Wilhelm Daniel Joseph Koch (1771–1849)** En.wikipedia.org/w/index.php?title=Wilhelm_Daniel_Joseph_ Koch&oldid=600127312. Image: public domain. No author or source information. CC BY-SA 3.0.
Koeleria – **Georg Ludwig Koeler (1765–1807)** Charters, *EDSAP.*
Kohautia – **Francisci (Franz) Kohaut (?–1822)** Charters, *EDSAP.*
Kohleria – **Johann Michael Kohler (1810/1816–c 1884)** Charters, *EDSAP.*
Kohlrauschia – **Henriette Kohlrausch (née Eichmann) (1781–1841)** Charters, *EDSAP.*
Kolleria – **Franz von Koller (1767–1826)** De.wikipedia.org/w/index.php?title=Franz_von_Koller&oldid=133830823; Primaplana.cz/news/rakousky-general-franz-baron-von-koller. Image: public domain. (Illustration taken from obristvi.cz) Author and source: unconfirmed. CC BY-SA 3.0.
Kosteletzkya – **Vincenz Franz Kosteletzky (1801–1887)** Charters, *EDSAP.*
Kotschya – **Georg Theodor Kotschy (1813–1866)** En.wikipedia.org/w/index.php?title=Theodor_Kotschy&oldid= 616372775. Image: public domain. Author: unknown. Source: Schweinfurth 1868. CC BY-SA 3.0.
Krauseola – **Ernst Hans Ludwig Krause (1859–1942)**
Frahm & Eggers, *Lexikon Deutschsprachiger Bryologen* (volume 2, 2001). Books.google.co.za/books?id= 8VEUIH15wAYC&pg=PA255&dq=Ernst+Hans+Ludwig+Krause+%281859-1942%29,&hl=en&ei= Nxj2TbbvJMODOsLppakH&sa=X&oi=book_result&ct=result&resnum=2&ved=0CC8Q6AEwATgK#v=onepage&q= Ernst%20Hans%20Ludwig%20Krause%20%281859-1942%29%2C&f=false. Image: From Commons.wikimedia.org/wiki. Author: unknown. Source: Herbarium Hamburgense Ernst Schäfer (1910–1992) Source: German Federal Archives. (Archiv Porträtsammlung). CC BY_SA 3.0.
Kraussia – **Christian Ferdinand Friedrich von Krauss (1812–1890)** Glen & Germishuizen, *BESA* (edition 2), p.245–248. Image: Sanbi.
Krebsia – **Georg Ludwig Engelhard Krebs (1792–1844)** Glen & Germishuizen, *BESA* (edition 2), p.246–249. Image: Sanbi.
Kroswia – **Hildur Krog (1922–2014)** En.wikipedia.org/w/index.php?title=Hildur_Krog&oldid=642079022; Translate.google.com/translate?hl=en&sl=es&u=dicci-eponimos.blogspot.com/2009/11/swinscow-dougal.html&prev=search. CC BY-SA 3.0.
Kunzea – **Gustav Kunze (1793–1851)**
Charters, *EDSAP*; Sv.wikipedia.org/w/index.php?title=Gustav_Kunze&oldid=29247683. Image: public domain. Author: unknown. Source: Orchids-flowers.com/category/famous-botanists/page/3. CC BY-SA 3.0.
Kurzia – **Wilhelm Sulpiz Kurz (1834–1878)** Es.wikipedia.org/w/index.php?title=Wilhelm_Sulpiz_Kurz&oldid= 75785315&archive.org/stream/taxonomicliteratur00staf/taxonomicliteratur00staf_djvu.txt; Gutenberg.org/cache/epub/19846/pg19846.txt.
Bioone.org/doi/abs/10.3158/0015-0746-47.1.257?journalCode=fbot. CC BY-SA 3.0.
Kyllinga – **Peder Lauridsen Kylling (c 1640–1696)** En.wikipedia.org/w/index.php?title=Peder_Lauridsen_Kylling& oldid=554635073. CC BY-SA 3.0.

L

Labourdonnaisia – **Count Bertrand François Mahé de La Bourdonnais (1699–1753)** En.wikipedia.org/w/index.php?title=Bertrand-Fran%C3%A7ois_Mah%C3%A9_de_La_Bourdonnais&oldid=606128438. Image: public domain. Artist: Antoine Graincourt (1748–1823). Source: Palace of Versailles, Uploaded by Der Bischof mit der E-Gitarre. CC BY-SA 3.0.
Lachenalia – **Werner de Lachenal (1736–1800)** Books.google.co.za/books?id=g3gCAAAAQAAJ&pg=PA438&dq= Werner+de+Lachenal&hl=en&ei=ZNn4TaXDDoig8QOMw5zfCw&sa=X&oi=book_result&ct=result&resnum= 1&ved=0CCoQ6AEwAA#v=onepage&q=Werner%20de%20Lachenal&f=false; Wright, *The Royal Dictionary-cyclopedia*; Books.google.co.za/books?id=mdAa4GBVJ5gC&pg=PA34&dq=Werner+de+Lachenal&hl=en&ei=

9t74Tc2xOsao8QPJmbziCw&sa=X&oi=book_result&ct=result&resnum=1&ved=0CCkQ6AEwADgo#v=onepage& q=Werner%20de%20Lachenal&f=false; Thurmann, *Fragment Pour Servir à l'Histoire*. Image: public domain. Author: unknown. Source: Regionatur.ch/Natur und Landschaft der Region Basel.

***Lafoensia* – João Carlos de Bragança (1719–1806)** Pt.wikipedia.org/w/index.php?title=Jo%C3%A3o_Carlos_de_ Bragan%C3%A7a_e_Ligne_de_Sousa_Tavares_Mascarenhas_da_Silva&oldid=3777600. Image: public domain. Author & source: unknown. CC BY-SA 3.0.

***Lagerheimina* – Nils Gustaf von Lagerheim (1860–1926)** Es.wikipedia.org/w/index.php?title=Nils_Gustaf_von_ Lagerheim&oldid=78564827&uni-goettingen.de/de/187063.html. Image: Author: Photographer: Ferd. Flodin, Norrmalmstorg. Source: Genealogi.se. CC BY-SA 3.0.

***Lagerstroemia* – Carl Magnus von Lagerström (1691–1759)** De.wikipedia.org/w/index.php?title=Magnus_Lagerstr% C3%B6m_(1691%E2%80%931759)&oldid=130944106. Image: public domain. Source: Svenskt Porträttarkiv, Nationalmuseum. CC BY-SA 3.0.

***Laggera* – Dr Franz Josef Lagger (1802–1870)** Short biography in Quattrocchi, *CRC World Dictionary of Plant Names*.

***Lagunaria* – Andrés Laguna de Segovia (1499–1559)** En.wikipedia.org/w/index.php?title=Andr%C3%A9s_Laguna& oldid=601652042. Image: public domain. Author: unknown. Source: Gobiernodecanarias.org/educacion/fundoro/ ciencia_es.html. CC BY-SA 3.0.

***Lamarckia* – Jean-Baptiste Antoine Pierre de Monet Lamarck (1744–1829)** En.wikipedia.org/w/index.php?title= Jean-Baptiste_Lamarck&oldid=64268187. Artist: Charles Thévenin (1764–1838). Source: unknown. CC BY-SA 3.0.

***Lambertia* – Aylmer Bourke Lambert (1761–1842)** En.wikipedia.org/w/index.php?title=Aylmer_Bourke_Lambert& oldid=638156426; Illustratedgarden.org/mobot/rarebooks/author.asp?creator=Lambert,+Aylmer+Bourke&creatorID= 2. Image: public domain. Author: unknown. Source: courtesy Royal Society. CC BY-SA 3.0.

***Lancisia* – Giovanni Maria Lancisi (1654–1720)** En.wikipedia.org/w/index.php?title=Giovanni_Maria_Lancisi&oldid= 628714323. Image: public domain. Author & source: unknown. CC BY-SA 3.0.

***Landolphia* – Jean-François Landolphe (1747–1825)** Fr.wikipedia.org/w/index.php?title=Jean-Fran%C3%A7ois_ Landolphe&action=history. Image: public domain. Author: Bertrand et pillet. Source: Librairie.maritime.com. CC BY-SA 3.0.

***Landtia* – Jørgen Landt (c 1753–1804)** De.wikipedia.org/w/index.php?title=J%C3%B8rgen_Landt&oldid=131107734. CC BY-SA 3.0.

***Lapeirousia* – Philippe-Isidore Picot de Lapeyrouse (Lapeirouse) (1744–1818)** Catalyst.library.jhu.edu/catalog/bib_ 1846735. Image: public domain. Author: Archaeodontosaurus. Source: Museum of Toulouse.

***Laportea* – François Louis Nompar de Caumat de Laporte (1810–1880)** En.wikipedia.org/w/index.php?title= Francis_de_Laporte_de_Castelnau&oldid=633300528; Glen & Germishuizen, *BESA* (edition 2), p.125–126; Plants. jstor.org/person/bm000069773. Image: public domain. Author: unknown. Source: Papers of Gregory M Mathews (1900–1949) located at National Library of Australia Manuscript collection MS 1465. CC BY-SA 3.0.

***Larochea* – Daniel Delaroche (de la Roche) (1743–1813)** De.wikipedia.org/w/index.php?title=Daniel_Delaroche& oldid=134029089. CC BY-SA 3.0.

***Larryleachia* – Leslie Charles (Larry) Leach (1909–1996)** Glen & Germishuizen, *BESA* (edition 2), p.256–257. Image: Sanbi.

***Lasallia* – Lasalle (c 1800)** Single line biography in Charters, *EDSAP*.

***Lastrea* – Charles Jean Louis Delastre (1792–1859)** Charters, *EDSAP*.

***Launaea* – Jean Claude Mien Mordant de Launay (c 1750–1816)** Charters, *EDSAP*.

***Laurembergia* – Peter Lauremberg (1585–1639)** De.wikipedia.org/w/index.php?title=Peter_Lauremberg&oldid= 127765841. CC BY-SA 3.0.

***Laurentia* – Laurent Garcin (1683–1751)** Jstor.org/pss/2481531; Books.google.co.za/books?id=3wtSuKSCke8C&pg= PA436&dq=Marco+Antonio+Laurenti&hl=en&ei=2isGTsy0MsWy8gOxrsTbDQ&sa=X&oi=book_result&ct=result& resnum=3&ved=0CDQQ6AEwAjge#v=onepage&q=Marco%20Antonio%20Laurenti&f=false; Mazzetti, *Memorie Storiche Sopra l'Università e l'Istituto delle Scienze di Bologna e...*; Books.google.co.za/books?id=XtXaAAAAMAAJ& q=Marco+Antonio+Laurenti&dq=Marco+Antonio+Laurenti&hl=en&ei=KzAGTojJH4rX8gPhI9S-DQ&sa=X&oi=book_ result&ct=result&resnum=4&ved=0CDYQ6AEwAzgo; Seligardi, *Lavoisier in Italia: la Comunità Scientifica Italiana e la Rivoluzione Chimica*.

***Laurera* – Johann Friedrich Laurer (1798–1873)** Books.google.com/books?id=8VEUIH15wAYC&pg= PA272&lpg=PA272&dq=Johann+Friedrich+Laurer+(1798–1873)&source=bl&ots=y3Y8aPC0KF&sig= MSW6cbHHXcG2iO66xpIgLX9ma1g; Frahm & Eggers, *Encyclopedia of German Bryologen* (volume 1); Uni-goettingen.de/de/l/187063.html; Merriam-webster.com/medical/laurer%27s%20canal.

***Lavatera* – Lavater brothers, Johann Heinrich (1611–1691) & Johann Jacob (1594–1636)** De.wikipedia.org/w/ index.php?title=Johann_Heinrich_Lavater&oldid=128563797.

***Lavrania* – John Jacob Lavranos (1926–)** Glen & Germishuizen, *BESA* (edition 2), p.255. Image: Sanbi

***Leachia* – Leslie Charles (Larry) Leach (1909–1996)**
Glen & Germishuizen, *BESA* (edition 2), p.256. Image: Sanbi.

***Lebeckia* – Heinrich Julius Lebeck (1772–1800)** Glen & Germishuizen, *BESA* (edition 2), p.257. Image: Sanbi

***Lecomtedoxa* – Paul Henri Lecomte (1856–1934)** En.wikipedia.org/w/index.php?title=Paul_Henri_Lecomte&oldid= 612235383. Image: Author & source: unknown. CC BY-SA 3.0.

***Ledebouria* – Carl Friedrich von Ledebour (Ledebur) (1785–1851)** De.wikipedia.org/w/index.php?title=Carl_ Friedrich_von_Ledebour&oldid=135348438. Image: public domain. Author: original uploader was Dobschütz at En.wikipedia. Source: Original text: Ledebur.de. CC BY-SA 3.0.

***Ledermanniella* – Carl Ludwig Ledermann (1875–1958)** Translate.google.co.za/translate?hl=en&sl=de&u=bgbm.fu-

berlin.de/bgbm/museum/expo/1998/lederman.htm&ei=QiQLTpHFJojG-QaqqZzPAg&sa=X&oi=translate&ct=result& resnum=8&ved=0CEkQ7gEwBw&prev=/search%3Fq%3DCarl%2BLudwig%2BLedermann%2B%281875-1958% 29,%26hl%3Den%26rlz%3D1B3GGIC_enZA317ZA317%26prmd%3Divnsob. Image: Author: unknown. Source: Bgbm.fu-berlin.de/de/node/1338.
Leersia – **Johann Daniel Leers (1727–1774)** Books.google.co.za/books?id=y_HDxSH2P9kC&pg=PA19&dq= Leers+%281727-1774%29,&hl=en&ei=flILTsz3EYiv8QO7pI2TAQ&sa=X&oi=book_result&ct=result&resnum=5& ved=0CDsQ6AEwBDgU#v=onepage&q=Leers%20%281727-1774%29%2C&f=false; Wessinghage, *Die Hohe Schule zu Herborn und Ihre Medizinische Fakultät, 1584–1817*.
Lefebvrea – **Charl Charlemagne Théophile Lefebvre (1811–1860)** Charters, *EDSAP.*
Leightonia – **William Allport Leighton (1805–1889)** Fenscore.man.ac.uk/cgi-bin/fensearch?lichen&website.lineone. net/~margaret_cole/SFG7/botanical%20recording%20in%20shropshire.htm&darwincountry.org/explore/000407. html. Image: public domain. Author: unknown. Source: Photograph. Shrewsbury Museums Service.
Leipoldtia – **Christian Frederik Louis Leipoldt (1880–1947)** Glen & Germishuizen, *BESA* (edition 2), p.259–260. Image: Sanbi.
Lejeunea – **Alexandre Louis Simon Lejeune (1779–1858)** En.wikipedia.org/w/index.php?title=Alexandre_Louis_ Simon_Lejeune&oldid=600406307. Image: public domain. Author: unknown. Source: Membres.multimania.fr/ jbeaujean/robermt/bio_c1.htm. CC BY-SA 3.0.
Lellingeria – **David Bruce Lellinger (1937–)** Es.wikipedia.org/w/index.php?title=David_Bruce_Lellinger&oldid= 76644084. CC BY-SA 3.0.
Lenormandia – **Sébastien-René Lenormand (1796–1871)** Plants.jstor.org/stable/10.5555/al.ap.person. bm000079067.
Leobordea – **Leon Emmanuel de Laborde (1807–1869)** Fr.wikipedia.org/w/index.php?title=L%C3%A9on_de_ Laborde&oldid=111664429; Jordanjubilee.com/history/laborde-linant.htm. Image: public domain. Author: Charles Reutlinger (1816–c 1880). Source: unknown. CC BY-SA 3.0.
Leskeella – **Nathanael Gottfried Leske (1751–1786)** De.wikipedia.org/w/index.php?title=Nathanael_Gottfried_ Leske&oldid=129354917; Scricciolo.com/linnaeus_notes.htm. Image: public domain. Author & source: unknown. CC BY-SA 3.0.
Lespedeza – **Vicente Manuel de Céspedes y Velasco (1721–1794)** En.wikipedia.org/w/index.php?title=Vicente_ Manuel_de_C%C3%A9spedes&oldid=645610856. CC BY-SA 3.0.
Lessertia – **Jules Paul Benjamin Delessert (1773–1847)** En.wikipedia.org/w/index.php?title=Jules_Paul_Benjamin_ Delessert&oldid=625061185. Image: public domain. Author & source: unknown. CC BY-SA 3.0.
Letestuella – **Georges Marie Patrice Charles Le Testu (1877–1967)** Plants.jstor.org/stable/10.5555/al.ap.person. bm000038253; Quattrocchi, *CRC World Dictionary of Grasses: Common Names, Scientific Names* (volume 1).
Letrouitia – **Marie-Agnès Letrouit-Galinou (1931–)** International Association for Lichenology, Lichenology.org/index. html?/Awards/AchariusMedallists.html. Image: Author: unknown. Source: Ouest-france.fr.
Levierella – **Émile (Emilio) Levier (1839–1911)** Plants.jstor.org/stable/10.5555/al.ap.person.bm000004973; Es.wikipedia.org/w/index.php?title=Emilio_Levier&oldid=69384543. Image: public domain. Author: unknown. Source: Hunt Institute for Botanical Documentation, Carnegie Mellon University, Pittsburgh, PA. CC BY-SA 3.0.
Leysera – **Friedrich Wilhelm von Leysser (1731–1815)** En.wikipedia.org/w/index.php?title=Friedrich_Wilhelm_von_ Leysser&oldid=588089829; Minrec.org/libdetail.asp?id=860. CC BY-SA 3.0.
Lichtensteinia – **Martin Heinrich Karl von Lichtenstein (1780–1857)** Glen & Germishuizen, *BESA* (edition 2), p.125–126. Image: Sanbi.
Lidbeckia – **Erik Gustavus Lidbeck (1724–1803)** Botaniskatradgarden.se/in-english/history.html; Ub.lu.se/en/the-correspondence-of-eric-gustaf-lidbeck. Image: public domain. Author: unknown. Source: National Library of Sweden, Kb.se/samlingarna/digitala/Linnes-natverks/Lidback-.
Lightfootia – **John Lightfoot (1735–1788)** Nationmaster.com/encyclopedia/John-Lightfoot-FRS; Sites.google.com/ site/scottishfungi/biographies.
Lindbergia – **Sextus Otto Lindberg (1835–1889)** En.wikipedia.org/w/index.php?title=Sextus_Otto_Lindberg&oldid= 616142121; Sok.riksarkivet.se/sbl/Presentation.aspx?id=10440. Image: public domain. Author: unknown. Source: Geni.com. CC BY-SA 3.0.
Lindernia – **Franz Balthazar von Lindern (1682–1755)** De.wikipedia.org/w/index.php?title=Franz_Balthazar_von_ Lindern&oldid=132391357. Image: public domain. Author or copyright holder: JM Weis. CC BY-SA 3.0.
Lindneria – **Otto Lindner (1852–1915)** Glen & Germishuizen, *BESA* (edition 2), p.271; Charters, *EDSAP.*
Lindsaea – **John Lindsay (c 1750–1803)** Ou.edu/cas/botany-micro/ben/ben120.html; 'In Search of Fern Seed'.
Linociera – **Geoffrey Linocier (f 1500s)** Christies.com/LotFinder/lot_details.aspx?; Abebooks.fr/servlet/ SearchResults?sortby=1&vci=114319&prl=30.
Lintonia – **Andrew Linton (died 1951)** Charters, *EDSAP.*
Linzia – **Johann Michael Linz (1771–1855)** Charters, *EDSAP.*
Lippia – **Augustin Lippi (1678–1705)** De.wikipedia.org/w/index.php?title=Augustin_Lippi&oldid=133868599. CC BY-SA 3.0.
Listia – **Friedrich Ludwig List (c 1837)** Charters, *EDSAP.*
Littonia – **Samuel Litton (1781–1847)** Botanicgardens.ie/glasra/ns3_1.htm; Catalogue.nli.ie/Record/vtls000144383. Image: public domain. Catalogue.nli.ie/Record/vtls000186773.
Lloydia – **Edward Lhuyd (1660–1709)** En.wikipedia.org/w/index.php?title=Edward_Lhuyd&oldid=607538366. Image: public domain. Author: unknown. Source: Delwedd.llgc.org.uk/delweddau/xbc/Lhyud.jpg. CC BY-SA 3.0.
Lobelia – **Mathias de L'Obel (Lobel, Lobelius) (1538–1616)** En.wikipedia.org/w/index.php?title=Matthias_de_l%

27Obel&oldid=638884611; Galileo.rice.edu/Catalog/NewFiles/lobel.html. Image: public domain. Author: Francis Delaram. Source: Summagallicana.it/lessico/l/Lobel%20Mathias.htm. CC BY-SA 3.0.
Lochnera – **Michael Friedrich Lochner von Hummelstein (1662–1720)** Books.google.com/books?isbn=1615393439; The Lochner Family Chronicle, Volume 2 edited by Johann Karl Lochner. Image: public domain. Author: unknown. Source: courtesy Welcome Library, London. CC-BY-SA 4.0.
Loddigesia – **Joachim Conrad Loddiges (1738–1826)** En.wikipedia.org/w/index.php?title=Loddiges&oldid= 640357601. Image: public domain. Author: John Renton, son of the headgardener at Loddiges' nursery (c 1820); photo of Conrad Loddiges painting taken by Jon Agar. Source: Commons.wikimedia.org/wiki/File:Joachim_Conrad_ Loddiges_%281738-1826%29.jpg. CC BY-SA 3.0.
Loeseneriella – **Ludwig Eduard Theodor Loesener (1865–1941)** Books.google.co.za/books?id=8VEUIH15wAYC& pg=PA289&dq=Ludwig+Eduard+Theodor+Loesener+%281865-1941%29,&hl=en&ei=syEXTsy8OMKp8QP5kOgZ &sa=X&oi=book_result&ct=result&resnum=10&ved=0CFYQ6AEwCQ#v=onepage&q=Ludwig%20Eduard% 20Theodor%20Loesener%20%281865-1941%29%2C&f=false; Frahm & Eggers, *Lexikon Deutschsprachiger Bryologen* (volume 2).
Loethainia – **Rudolf Benno von Römer (1803–1870)** Charters, *EDSAP.*
Lonicera – **Adam Lonitzer (Lonicer, Lonicerus) (1528–1586)** En.wikipedia.org/w/index.php?title=Adam_Lonicer& oldid=622561394&minrec.org/libdetail.asp?id=890&marelibri.com/topic/23392-main/books/AUTHOR../47750. Image: public domain. Author: Adam Lonitzer (1528–1586). Source: *Kreuterbuch: künstliche Conterfeytunge der Bäume, Stauden, Hecken.* CC BY-SA 3.0.
Loudetia – **Eduard Loudet (1811–1867)** Clifford & Bostock, *Etymological Dictionary of Grasses.*
Luckhoffia – **James Lückhoff (dates unknown) and Carl August Lückhoff (1914–1960)** Glen & Germishuizen, *BESA* (edition 2), p.275. Image: Sanbi.
Lüderitzia – **August Lüderitz (Luederitz) (1838–1922)** Glen & Germishuizen, *BESA* (edition 2), p.276–8. Image: Sanbi.
Ludwigia – **Christian Gottlieb Ludwig (1709–1773)** Phil-hum-ren.uni-muenchen.de/GermLat/Acta/Jonsson.htm; Christies.com/LotFinder/lot_details.aspx?intObjectID=1422755. Image: public domain. Authors: Johan Jacob Haid (1704–1767), Elias Gottlob Haussman (1695–1764). Source: Commons.wikimedia.org/wiki/ File:ChristianGottliebLudwig.jpg.
Lumnitzera – **István (Stefani, Stefan) Lumnitzer (1750–1806)** Quattrocchi, *CRC World Dictionary of Plant Names.*

M

Macadamia – **John Macadam (1827–1865)** En.wikipedia.org/w/index.php?title=John_Macadam&oldid=639760487; Adb.anu.edu.au/biography/macadam-john-4054. Image: public domain. Author: John Macadam. Source: University of Melbourne Medical School Jubilee, 1914. CC BY-SA 3.0.
Macfadyena – **James Macfadyen (1799–1850)** En.wikipedia.org/w/index.php?title=James_Macfadyen&oldid= 621604253. Image: public domain. Author: unknown. Source: Snipview.com. CC BY-SA 3.0.
Mackaya – **James Townsend Mackay (1775–1862)** En.wikipedia.org/w/index.php?title=James_Townsend_Mackay& oldid=603196842; Botanicgardens.ie/herb/floras/cavan/cavan01.pdf. CC BY-SA 3.0.
Macledium – **William Sharp Macleay (1792–1865)** En.wikipedia.org/w/index.php?title=William_Sharp_Macleay& oldid=639956638. Image: public domain. Author: unknown. Source: Anbg.gov.au/bot-biog/bot-biog-M.html, originally Mitchell Library, Government Printing Office. CC BY-SA 3.0.
Maclura – **William Maclure (1763–1840)** En.wikipedia.org/w/index.php?title=William_Maclure&oldid=615810100& faculty.evansville.edu/ck6/bstud/maclure.html. Image: public domain. Author: unknown. Source: Infed.org/thinkers/ maclure.htm. CC BY-SA 3.0.
Macowania – **Dr Peter MacOwan (1830–1909)** Glen & Germishuizen, *BESA* (edition 2), p.281. Image: Sanbi
Magnolia – **Pierre Magnol (1638–1715)** En.wikipedia.org/w/index.php?title=Pierre_Magnol&oldid=614086129. Image: public domain. Author and source: unknown.
Mairia – **Louis Maire (f 1815–1833)** Glen & Germishuizen, *BESA* (edition 2), p.284.
Maltebrunia – **Konrad Bruun (Conrad Malte-Brun) (1755–1826)** Online encyclopedia originally appearing in Volume V17, Page 515 of the 1911 *Encyclopedia Britannica.*
Mandevilla – **Henry John Mandeville (1773–1861)** Oxford DNB article: Mandeville, John Henry by Maxine Hanaon.
Mannia – **Wenzeslaus (Wenzel) Blasius Mann (1799–1839)** Charters, *EDSAP.*
Marattia – **Giovanni Francesco Maratti (1704–1777)** Es.wikipedia.org/w/index.php?title=Giovanni_Francesco_ Maratti&oldid=65003015. CC BY-SA 3.0.
Marchantia – **Nicholas Marchant (?–1678)** Adapted from Encylopedia.com relating to Nicholas Marchant; *Complete Dictionary of Scientific Biography* 2008; Encyclopedia.com/doc/1G2-2830902810.html. CC BY-SA 3.0.
Markhamia – **Clements Robert Markham (1830–1916)** En.wikipedia.org/w/index.php?title=Clements_Markham& oldid=648729150. Image: public domain. Author: George Grantham Bain, Collection (Library of Congress). CC BY-SA 3.0.
Marlothia – **Hermann Wilhelm Rudolf Marloth (1855–1931)** Glen & Germishuizen, *BESA* (edition 2), p.285. Image: Sanbi.
Maronea – **Nicolai Marogna or Nicholas Maronae (f 1600s)** Diccionario de Epónimos dicci-eponimos.blogspot. com/2010/01/maronae-nicholas.html.
Marsdenia – **William Marsden (1754–1836)** En.wikipedia.org/w/index.php?title=William_Marsden_(orientalist)&oldid= 619614586. Image: public domain. Author: Mary Dawson Turner (née Palgrave) (?–1850), after Thomas Phillips

etching. Source: NPG D22585 William Marsden (1815). CC BY-SA 3.0.
Marsilea – **Luigi Ferdinando Marsigli (1658–1730)** De.wikipedia.org/w/index.php?title=Luigi_Ferdinando_Marsigli& oldid=13886394. Image: public domain. Author: unknown. Source: Scricciolo.com/Nuovo_Neornithes/cronologia.htm. CC BY-SA 3.0.
Martynia – **John Martyn (1699–1768)** En.wikipedia.org/w/index.php?title=John_Martyn_(botanist)&oldid=610118664. Image: public domain. Author: unknown. Source: courtesy National Portrait Gallery, London.
Mascarenhasia – **Don Pedro Mascarenhas (1470–1555)** En.wikipedia.org/w/index.php?title=Pedro_Mascarenhas& oldid=639276776. Image: public domain. Author: Faria e Sousa. Source: Faria e Sousa. Ásia Portuguesa. Tomo II. Antonio Craesbeeck de Mello, 1674. CC BY-SA 3.0.
Massonia – **Francis Masson (1741–1805)** Glen & Germishuizen, *BESA* (edition 2), p.287-290. Image: Sanbi
Mastersiella – **Maxwell Tylden Masters (1833–1907)** En.wikipedia.org/w/index.php?title=Maxwell_T._Masters&oldid= 636822795; Ncbi.nlm.nih.gov/pmc/articles/PMC2357616/?page=1Image: public domain. Author: unknown. Source: *Revue de l'horticulture Belge et Étrangère* by Frédéric Burvenich, Édouard Pynaert, Émile Rodigas, August van Geert; HJ van Hulle (editors). CC BY-SA 3.0.
Matthiola – **Pietro Andrea Gregorio Mattioli (1501–1577)** En.wikipedia.org/w/index.php?title=Pietro_Andrea_ Mattioli&oldid=645921920. Image: public domain. Author: unknown. Source: Ihm.nlm.nih.gov. Image: s/B18950. CC BY-SA 3.0.
Maughaniella – **Herbert Maughan-Brown (1883–1940)** *Diccionario de Epónimos*, Dicci-eponimos.blogspot. com/2010/01/maughan-brown-herbert.html.
Mauhlia – **Johannis Mauhle (dates unknown)** Stöver, *The life of Sir Charles Linnæus...*, Books.google.co.za/ books?id=AoidX8bJK-oC&pg=PA257&dq=%22Mauhle&hl=en&ei=oQEjTpfcK8-YOtTGlfEO&sa=X&oi=book_ result&ct=result&resnum=1&ved=0CDIQ6AEwAA#v=onepage&q=%22Mauhle&f=false; *Journal of Botany, British and foreign*, Vol. 51, Books.google.co.za/books?id=vs9EAAAAYAAJ&q=%22Mauhle&dq=%22Mauhle&hl=en&ei= rwgjTtKOBMmCOuH4mfEO&sa=X&oi=book_result&ct=result&resnum=3&ved=0CDQQ6AEwAjgU.
Maurocenia – **Giovanni Francesco Morosini (1658–1739)** *Diccionario de Epónimos*, Dicci-eponimos.blogspot. com/2010/01/morosini-giovanni-francesco.html.
Merciera – **Marie Philippe Mercier (1781–1831)** Plants.jstor.org/person/bm000024076.
Mercurialis – **Hieronymus Mercurialis (real name 'Girolamo Mercuriale') (1530–1606)** En.wikipedia.org/w/index. php?title=Girolamo_Mercuriale&oldid=638876158. Image: public domain. Author: Lavinia Fontana (1552–1614). Source: Walters Art Museum. CC BY-SA 3.0.
Merremia – **Basius Merrem (1761–1824)** En.wikipedia.org/w/index.php?title=Blasius_Merrem&oldid=599881374. Image: public domain. Author and source: unknown.
Mertensia – **Franz Karl (Carl) Mertens (1764–1831)** En.wikipedia.org/w/index.php?title=Franz_Carl_Mertens&oldid= 595834952. Image: public domain. Artist: Karl Mertens (son of subject Franz Karl Mertens), printed by C Hullmandel, c 1820. Source: HIBD via German Wikipedia.
Merwilla – **Frederick Ziervogel van der Merwe (1894–1968)** Glen & Germishuizen, *BESA* (edition 2), p.431.
Merxmuellera – **Hermann Merxmüller (1920–1988)** Glen & Germishuizen, *BESA* (edition 2), p.298–298. Image: Sanbi.
Metzgeria – **Johann Baptiste Metzger (1771–1844)** De.wikipedia.org/w/index.php?title=Johann_Baptist_Metzger& oldid=131641150. Image: public domain. Artist and printmaker: Julius Schnorr von Carolsfeld (1794–1872). CC BY-SA 3.0.
Metzleria – **Jacob Adolf Metzler (1812–1883)** Frahm & Eggers, *Lexikon Deutschsprachiger Bryologen* (volume 1); Senckenberg.de/files/content/forschung/abteilung/botanik/phanerogamen1/historybotanyfr.pdf.
Meyerophytum – **Louis Gottlieb Meyer (1867–1958)** Glen & Esterhuizen, *BESA* (edition 2), p.197–8. Image: Sanbi.
Mezleria – **Johann Georg Metzler (1761–1833)** En.wikipedia.org/w/index.php?title=Karl_Ludwig_Giesecke&oldid= 618338925. Image: public domain. Author: Sir Henry Raeburn. Source: Uibk.ac.at/mineralogie/oemg/bd_146/146_ 451-479.pdf. CC BY-SA 3.0.
Mielichhoferia – **Mathias Mielichhofer (1772–1847)** Frahm & Eggers, *Lexikon Deutschsprachiger Bryologen* (volume 2, 2001). Image: public domain. Source: Biographien.ac.at/oebl/oebl_M/Mielichhofer_Mathias_1772_1847.xml.
Mikania – **Josef (Joseph) Gottfried Mikan (1743–1814)** En.wikipedia.org/w/index.php?title=Joseph_Gottfried_ Mikan&oldid=595835452.
Milicia – **Senhor Milicia (Dates unknown)** Source**:** Glen, *Sappi tree spotting: What's in a name?*
Millettia – **Charles Millet (1792–1873)** Charters, *EDSAP.*
Mittenothamnium – **William Mitten (1819–1906)** En.wikipedia.org/w/index.php?title=William_Mitten&oldid= 588080750. Image: public domain. Author: unknown. Source: *William Mitten. A Sketch with Bibliography*; The Bryologist 10; American Bryological and Lichenological Society. CC BY-SA 3.0.
Moenchia – **Conrad (Konrad) Moench (Mönch) (1744–1805)** Encyclopedia.com/doc/1G2-2830902997.html. CC BY-SA 3.0
Moesslera – **Johann Christoph Mössler (Moessler) (1770–1840)** Charters, *EDSAP.*
Mohria – **Daniel Matthias Heinrich Mohr (1780–1808)** De.wikipedia.org/w/index.php?title=Daniel_Matthias_Heinrich_ Mohr&oldid=134441551. CC BY-SA 3.0.
Molendoa – **Ludwig Molendo (1833–1902)** Jan-peter-frahm.de/Archive/Guide.pdf; Senckenberg.de/files/content/ forschung/abteilung/botanik/index_collectorum.pdf.
Monsonia – **Lady Ann Monson (née Vane) (1714–1776)** Glen & Germishuizen, *BESA* (edition 2), p.197–8.
Montanoa – **Luis José Montaña Carrascó (1755–1820)** Mexico-tenoch.com/enmarca.php?de=mexico-tenoch.com/ lamedicinaenmexico/luisjosemontana.html; ejournal.unam.mx/rfm/no48-4/RFM48411.pdf.

Montbretia – **Antoine François Ernest Coquebert de Montbret (1781–1801)** Charters, *EDSAP.* ***Montinia*** – **Lars Jonasson Montin (1723–1785)** Sv.wikipedia.org/w/index.php?title=Lars_Montin&oldid=29256918. CC BY-SA 3.0.
Moquinia – **Christian Horace Bénédict Alfred Moquin-Tandon (1804–1863)** En.wikipedia.org/w/index.php?title= Alfred_Moquin-Tandon&oldid=615014837. Image: public domain. Author and source: unknown. CC BY-SA 3.0.
Moraea – **Sara Elizabeth Moraea (1716–1806)** Sv.wikipedia.org/w/index.php?title=Sara_Elisabeth_Moraea& oldid=29537058. Image: public domain. Artist JohanHenrik Scheffel, autumn 1739. Source: Blogg.hd.se/ slaktforskning/2007/05/04/sara-lisa-von-linne-och-hennes-liv.
Morysia – **Charles Bourgevin of Vialart, Count of Saint Morys (1772–1817)** Appl-lachaise.net/appl/article.php3?id_ article=1898. Image: public domain. Author: unknown. Source: Appl-lachaise.net. CC BY-SA 3.0.
Mossia – **Charles Edward Moss (1870–1930)** Glen & Germishuizen, *BESA* (edition 2), p.197–8. Image: Sanbi. CC BY-SA 3.0.
Muellerella – **Jean (Johannes) Mueller (Müller), called Argoviensis, (1828–1896)** Google.com/search?q= Jean+(Johannes)+Mueller+(M%C3%BCller),+called+Argoviensis+(1828%E2%80%931896),+&ie=utf-8&oe=utf-8& client=firefox-b&gfe_rd=cr&ei=YJOEV5bPG_Co8weghbqgDg. CC BY-SA 3.0.
Muiria – **Dr John Muir (1874–1947)** Glen & Germishuizen, *BESA* (edition 2), p.306–307. Image: Sanbi.
Mundia – **Johannes Ludwig Leopold Mund (1791–1831)** Glen & Germishuizen, *BESA* (edition 2), p.309–3107. Image: Sanbi CC BY-SA 3.0.
Muraltia – **Johannes von Muralt (1645–1733)** De.wikipedia.org/w/index.php?title=Johannes_von_Muralt_(Mediz iner)&oldid=116319367. Image: public domain. Author and source: unknown. CC BY-SA 3.0.
Murdannia – **Munshi Murdan Ali (f 1840s)** Charters, *EDSAP.*
Muriea – **James Murie (1832–1925)**
Nature.com/nature/journal/v129/n3264/abs/129752c0.html; Archives.wellcomelibrary.

N

Nachtigalia – **Gustav Hermann Nachtigal (1834–1885)** En.wikipedia.org/w/index.php?title=Gustav_Nachtigal&oldid= 649565236. Image: public domain. Glen & Germishuizen, *BESA* (edition 2), p.311. Image: Sanbi.
Nathusia – **Hermann Engelhard von Nathusius (1809–1879)** En.wikipedia.org/w/index.php?title=Hermann_von_ Nathusius&oldid=633565400. Image: public domain. Author: H Schnäbeli. Source: Scan from photo. CC BY-SA 3.0
Nebelia – **Daniel Nebel (1663–1733) and William Bernhard Nebel (1699–1748)** *Heidelberger Gelehrtenlexikon*; Dicci-eponimos.blogspot.com/2010/01/nebel-daniel.html. Image: public domain. Source: Academictree.org.
Neckera – **Noel Martin Joseph de Necker (or Natalis Joseph de Necker) (1729–1793)** De.wikipedia.org/w/index. php?title=No%C3%ABl_Martin_Joseph_de_Necker&oldid=130411839. Image: public domain. Author: Anton Karcher. Source: Bibliorto.cab.unipd.it/iconoteca-dei-botanici/iconoteca-database/PUV1117315.CC BY-SA 3.0.
Neesenbeckia – **Christian Gottfried Daniel Nees von Esenbeck (1776–1858)** En.wikipedia.org/w/index.php?title= Christian_Gottfried_Daniel_Nees_von_Esenbeck&oldid=633550110 Image: public domain. Author: Fotographie von Johann Robert Weigelt, 1855. Source: Botanik.uni-bonn.de/system/Nees-Homepage/nees.html.
Negria – **Cristoforo Negri (1809–1896)** En.wikipedia.org/w/index.php?title=Cristoforo_Negri&oldid=626130739. Image: public domain. Author: Alexandre Quinet. Source: Bibliothèque Nationale de France. CC BY-SA 3.0.
Nelia – **Gert Cornelius Nel (1885–1950)** Glen & Germishuizen, *BESA* (edition 2), p.311. Image: Sanbi.
Nelsia – **Louis Nels (1855–1910)** Fr.wikipedia.org/w/index.php?title=Louis_Nels&action=history. CC BY-SA 3.0.
Nelsonia – **David Nelson (?–1789)** En.wikipedia.org/w/index.php?title=David_Nelson_(botanical_collector)&oldid= 649315415.
Neobakeria – **John Gilbert Baker (1834–1920)** En.wikipedia.org/w/index.php?title=John_Gilbert_Baker&oldid= 637880000. Image: public domain. Author: Berthold Carl Seemann (1825–1871). Source: Plantillustrations.org/ illustration.php?id_illustration=176522. CC BY-SA 3.0.
Neoboivinella – **Louis Hyacinthe Boivin (1808–1852)** Es.wikipedia.org/w/index.php?title=Louis_Hyacinthe_Boivin& oldid=77922879. CC BY-SA 3.0.
Neobolusia – **Harry Bolus (1834–1911)** Glen & Germishuizen, *BESA* (edition 2), p.102–103. Image: Sanbi CC BY-SA 3.0.
Neodregea – **Isaac Louis Drège (1853–1921)** Glen & Germishuizen, *BESA* (edition 2), p.154–156.
Neoglaziovia – **Auguste François Marie Glaziou (1828–1906)** Nrm.se/english/researchandcollections/botany/ botanicalhistory/augustefrancoismarieglaziou.13418_en.html. Image: public domain. Author: unknown. Source: Casaruibarbosa.gov.br. CC BY-SA 3.0.
Neohenricia – **Dr Marguerite Gertrude Anna Henrici (1892–1971)** Glen & Germishuizen, *BESA* (edition 2), p.210. Image: Sanbi.
Neoluederitzia – **Franz Adolph Eduard Lüderitz (1834–1886)** Glen & Germishuizen, *BESA* (edition 2), p.275–276. Image: Sanbi CC BY-SA 3.0.
Neopatersonia – **Florence Mary Paterson (née Hallack) (1869–1936)** Glen & Germishuizen, *BESA* (edition 2), p.275–276. Image: Sanbi.
Neorautanenia – **Martti Rautanen (1845–1926)** Glen & Germishuizen, *BESA* (edition 2), p.354; Fi.wikipedia.org/wiki/ Martti_Rautanen. Image: public domain. Author: unknown. Source: Inkeri.ru/wp-content/uploads/%D0%98%D0% BD%D0%BA%D0%B5%D1%80%D0%B8-%E2%84%96-2-075.pdf. CC BY-SA 3.0.
Neorosea – **Valentin Rose 'the Younger' (1762–1807)** En.wikipedia.org/w/index.php?title=Valentin_Rose_ (pharmacologist)&oldid=625771752. Image: public domain. Author: unknown. Source: Diariolahora@gmail.com. CC BY-SA 3.0.

Nestlera – **Christian Gottfried Nestler (1778–1832)** De.wikipedia.org/w/index.php?title=Christian_Gottfried_Nestler&oldid=135103453. CC BY-SA 3.0.
Neumannia – **Joseph Henri François Neumann (1800–1858)** Es.wikipedia.org/w/index.php?title=Joseph_Henri_Fran%C3%A7ois_Neumann&oldid=75134395; En.wikisource.org/wiki/Popular_Science_Monthly/Volume_15/September_1879/The_Vanilla-Plant. CC BY-SA 3.0.
Nevillea – **Neville Stuart Pillans (1884–1964)** Glen & Germishuizen, *BESA* (edition 2), p.341. Image: Sanbi.
Newtonia – **Isaac Newton (1642–1726)**
En.wikipedia.org/wiki/Isaac_Newton. Image: public domain. Author: Sir Godfrey Kneller (1646–1723). Portrait of Isaac Newton, source: Newton.cam.ac.uk/art/portrait.
Newtonia – **Francisco Xavier Oakley de Aguiar Newton (1864–1909)** Charters, *EDSAP*.
Nicandra – **Nikander of Colophon (Nikandros Kolophonios) (c 100–150 CE)** Charters, *EDSAP*.
Nicolasia – **Nicholas Edward Brown (1849–1934)**
Glen & Germishuizen, *BESA* (edition 2), p.341; En.wikipedia.org/w/index.php?title=N._E._Brown&oldid=641031838. Image author: unknown. Source: En.wikipedia, first uploaded by en: user: Paul venter, 2007-03-01. CC BY-SA 3.0.
Nicolsonia – **Jean Barthélemy Maximilien Nicolson (1734–1773)**
Mclellan III, *Colonialism and Science: Saint Domingue and the Old Regime*, p.114, University of Chicago Press, 2010.
Nicotiana – **Jean Nicot (1530–1600)** En.wikipedia.org/w/index.php?title=Jean_Nicot&oldid=649819060. Image: public domain. Author: Gravure publiée par l'éditeur Albin Michel. Source: Nemausensis.ifrance.com/gard/Image: mois/jeannicot/jeannicot.htm. CC BY-SA 3.0.
Niebuhria – **Carsten Niebuhr (1733–1815)** En.wikipedia.org/w/index.php?title=Carsten_Niebuhr&oldid=639434310. Image: public domain. Author: unknown. Source: Carsten_niebuhr.jpg. CC BY-SA 3.0.
Niemeyera – **Felix von Niemeyer (1820–1871)** En.wikipedia.org/w/index.php?title=Felix_von_Niemeyer&oldid=625616975. Image: public domain. Author: unknown. Source: *Biographisches Lexikon Hervorragender Ärzte des Neunzehnten Jahrhunderts*, Berlin, Wien 1901, p.1207–1209. CC BY-SA 3.0.
Nierembergia – **Juan Eusebio Nieremberg (1595–1658)** En.wikipedia.org/w/index.php?title=Juan_Eusebio_Nieremberg&oldid=608337956. Image: public domain. Author source unknown. CC BY-SA 3.0.
Nivenia – **James Niven (1776–1827)** Glen & Germishuizen, *BESA* (edition 2), p.315.
Nolletia – **Jean Antoine Nollet (1700–1770)** En.wikipedia.org/w/index.php?title=Jean-Antoine_Nollet&oldid=648454243. Image: public domain. Author: unknown. Source: En.wikipedia. CC BY-SA 3.0.
Noltea – **Ernst Ferdinand Nolte (1791–1875)** En.wikipedia.org/w/index.php?title=Ernst_Ferdinand_Nolte&oldid=595837325. CC BY-SA 3.0.
Norlindhia – **Nils Tycho Norlindh (1906–1995)** Glen & Germishuizen, *BESA* (edition 2), p.316. Image: Sanbi
Normandina – **Sébastien-René Lenormand (1796–1871)** En.wikipedia.org/w/index.php?title=S%C3%A9bastien_Ren%C3%A9_Lenormand&oldid=646797874. CC BY-SA 3.0.
Notonia – **Benjamin Noton (1812–1835)** Robert Wight; Walker-Arnott in Prodromus Florae Peninsulae Indiae Orientalis (1834); Kew.org/science-conservation/research-data/resources/legumes-of-the-world/genus/neonotonia.
Nurmonia – **William Munro (1818–1880)** En.wikipedia.org/w/index.php?title=William_Munro&oldid=618393309. Image: public domain. Author: unknown. Source: Robswebstek.com. CC BY-SA 3.0.
Nuxia – **Jean Baptiste François de Lanux (la Nux) (1702–1772)** Gw5.geneanet.org/index.php3?b=robillard1&lang=fr;p=jean+baptiste+francois;n=de+lanux.
Nylandtia – **Petrus (Peter, Pierre) Nylandt (c 1635–1710)** Translate.google.co.za/translate?hl=en&sl=es&u=apicultura.wikia.com/wiki/Petrus_Nylandt&ei=9b5GTrLmBYbzsgbwqdmtBw&sa=X&oi=translate&ct=result&resnum=10&ved=0CFgQ7gEwCQ&prev=/search%3Fq%3DPetrus%2BNylandt,%26hl%3Den%26rlz%3D1B3GGIC_enZA317ZA317%26prmd%3Divnsb.
Nymania – **Carl Fredrik Nyman (1820–1893)** Sok.riksarkivet.se/sbl/Presentation.aspx?id=8490. Image: public domain. Author: unknown. Source: Riksarkivet (National Archives).

O

Obetia – **Louis Jean Marie Obet (1777–1856)**
Oedera – **Georg Christian Oeder (1728–1791)** En.wikipedia.org/w/index.php?title=Georg_Christian_Oeder&oldid=648236371. Image: public domain. Author: unknown. Source: Transferred from En.wikipedia; transferred to Commons by User: Magnus Manske using CommonsHelper. CC BY-SA 3.0.
Ohlendorffia – **Johann Heinrich Ohlendorff (1788–1857)** De.wikipedia.org/w/index.php?title=Johann_Heinrich_Ohlendorff&oldid=120759840. CC BY-SA 3.0.
Oldenburgia – **Franz (Frantz) Pehr Oldenburg (1740–1773)** Glen & Germishuizen, *BESA* (edition 2), p.319.
Oldenlandia – **Heinrich Bernhard Oldenland (c 1663–1697)** Glen & Germishuizen, *BESA* (edition 2), p.319–321.
Olfersia – **Ignaz Franz Werner Maria von Olfers (1793–1871)** En.wikipedia.org/w/index.php?title=Ignaz_von_Olfers&oldid=598339634; De.wikisource.org/wiki/ADB:Olfers,_Ignaz_von. Image: public domain. Author: Zeichnung von Kaulbach. Source: Margarete von Olfers, Elisabeth Staegemann, Leipzig 1937 (Buch).
Olinia – **Johan Hendric (Hendrik) Olin (1764–1824)** Quattrocchi, *CRC World Dictionary of Plant Names*, Books.google.co.za/books?id=YC_IAgAAQBAJ&pg=PA2684&lpg=PA2684&dq=Dissertatio+arnica+in+1799.&source=bl&ots=cU0kBgkeUy&sig=aigEIHj6DQe-UA1maRUpLR9-zbM&hl=en&sa=X&redir_esc=y#v=onepage&q=Dissertatio%20arnica%20in%201799.&f=false.
Oliverella – **Daniel Oliver (1830–1916)** En.wikipedia.org/w/index.php?title=Daniel_Oliver&oldid=588069507.

Ottosonderia – **Otto Wilhelm Sonder (1812–1881)** Glen & Germishuizen, *BESA* (edition 2), p.319–400. Image: Sanbi CC BY-SA 3.0.

P

Pagella – **Mary Maud Page (1867–1925)** Glen & Germishuizen, *BESA* (edition 2), p.325.
Pahudia – **Charles Ferdinand Pahud (1803–1873)** Nl.wikipedia.org/w/index.php?title=Charles_Ferdinand_Pahud&oldid=43098298. Image: public domain. Artist: Jacob Spoel (1820–1868). Source: Rijksmuseum, Amsterdam. CC BY-SA 3.0.
Palhinhaea – **Ruy Teles Palhinha (1871–1957)** Pt.wikipedia.org/w/index.php?title=Rui_Teles_Palhinha&oldid=37829933. Image: author and source unknown. CC BY-SA 3.0.
Palmstruckia – **Wilhelm Palmstruch (1770–1811)** En.wikipedia.org/wiki/Johan_Wilhelm_Palmstruch. Image: Author: unknown. Source: Biodiversity Heritage Library, Biodiversitylibrary.org/page/14885373.
Pancovia – **Thomas Panckow (1622–1665)** Abebooks.it/ricerca-libro/autore/panckow-thomas-pancovius. Image: public domain. Author: unknown. Source: Benl.ebay.be.
Pappea – **Carl (Karl) Wilhelm Ludwig Pappe (1803–1862)** Glen & Germishuizen, *BESA* (edition 2), p.319–400. Image:
Parkinsonia – **John Parkinson (1567–1650)** En.wikipedia.org/w/index.php?title=John_Parkinson_(botanist)&oldid=65155294. Image: public domain. Author: unknown. Source: From Parkinson, John (1640) *Theatrum Botanicum*. CC BY-SA 3.0.
Parmentaria – **Antoine-Augustin Parmentier (1737–1813)** En.wikipedia.org/w/index.php?title=Antoine-Augustin_Parmentier&oldid=647967355. Image: public domain. Author and source: unknown. CC BY-SA 3.0.
Patersonia – **William Paterson (1755–1810)** Glen & Germishuizen, *BESA* (edition 2), p.329–331. Image: Sanbi
Pauletia – **Jean Jacques Paulet (1740–1526)** En.wikipedia.org/w/index.php?title=Jean-Jacques_Paulet&oldid=595836725. Image: public domain. Author: JB Bailliere. Lithography by Becquet. Source: Lhm.nlm.nih.gov/Image: s/B21091. CC BY-SA 3.0.
Paullinia – **Simon Paulli (1603–1680)** En.wikipedia.org/w/index.php?title=Simon_Paulli&oldid=613631094. Image: public domain. Author: unknown. Source: Ihm.nlm.nih.gov/Image: s/B21088. CC BY-SA 3.0.
Paulownia – **Anna Pavlovna (1795–1865)** En.wikipedia.org/w/index.php?title=Anna_Pavlovna_of_Russia&oldid=636860179. Image: public domain. Author: Jan Baptist van der Hulst (1790–1862). Source unknown. CC BY-SA 3.0.
Pavonia – **José Antonio Pavón y Jiménez (1754–1840)** Biologia-en-internet.com/fteixido/s-xviii/jose-antonio-pavon-jimenez-1754-1840; Encyclopedia.com/doc/1G2-2830904926.html.
Pearsonia – **Professor Henry Harold Welch Pearson (1870–1916)** Glen & Germishuizen, *BESA* (edition 2), p.332–333. Image: Sanbi.
Pechuel-Loeschea – **Moritz Eduard Pechuël-Loesche (1840–1913)** Glen & Germishuizen, *BESA* (edition 2), p. 333.
Peddiea – **John Peddie (?–1840)** Glen & Germishuizen, *BESA* (edition 2), p. 333. CC BY-SA 3.0.
Peersia – **Victor Stanley Peers (1874–1940)** Glen & Germishuizen, *BESA* (edition 2), p.334. Image: Sanbi.
Peglera – **Alice Marguerite Pegler (1861–1929)** Glen & Germishuizen, *BESA* (edition 2), p.334. Image: Sanbi.
Pegolettia – **Francesco Balducci Pegoletti (f 1310–1347)** En.wikipedia.org/w/index.php?title=Francesco_Balducci_Pegolotti&oldid=557216030. CC BY-SA 3.0.
Penaea – **Pierre Pena (1535–1605)** Es.wikipedia.org/w/index.php?title=Pierre_Pena&oldid=64468676. CC BY-SA 3.0.
Pentheriella – **Arnold Penther (1865–1931)** Glen & Germishuizen, *BESA* (edition 2), p.335. Image: Sanbi.
Pentzia – **Carolus Johannes Pentz (f 1700s)** Charters, *EDSAP*.
Pereskia – **Nicholas-Claude Fabri (Fabry) de Peiresc (1580–1637)** Fr.wikipedia.org/wiki/Nicolas-Claude_Fabri_de_Peiresc. Image: public domain. Author and source: unknown. CC BY-SA 3.0.
Perlebia – **Karl (Carl) Julius Perleb (1794–1845)** Ms.wikipedia.org/w/index.php?title=Karl_Julius_Perleb&oldid=2821384. CC BY-SA 3.0.
Peyrousea – **Jean François de Galaup, Comte de la Pèrouse (1741–1788)** En.wikipedia.org/w/index.php?title=Jean-Fran%C3%A7ois_de_Galaup,_comte_de_Lap%C3%A9rouse&oldid=651618120. Image: public domain. Artist: Thomas Woolnoth (1785–1857). Source: Nla.gov.au/pub/nlanews/2001/oct01/french_books.html. CC BY-SA 3.0.
Phaenohoffmannia – **Heinrich Karl Hermann Hoffmann (1819–1891)** En.wikipedia.org/w/index.php?title=Hermann_Hoffmann&oldid=62607391. CC BY-SA 3.0.
Pharnaceum – **Pharnaces II (63–47 BCE)** En.wikipedia.org/w/index.php?title=Pharnaces_II_of_Pontus&oldid=649232040. CC BY-SA 3.0.
Phelypaea – **Louis Phélypeaux (1643–1727)** En.wikipedia.org/w/index.php?title=Louis_Ph%C3%A9lypeaux,_comte_de_Pontchartrain&oldid=618216573. Image: public domain. Author: Robert Nanteuil (1623–1678). Source: Princeton Digital Collections, Robert Nanteuil Collection. CC BY-SA 3.0.
Philippia – **Rodolfo Amando Philippi (1808–1904)** Plants.jstor.org/person/bm000006508. Image: public domain. Author: unknown. Source: Commons.wikimedia.org/wiki/File:Philippi_Rodolfo_Amando_1808-1904.png. CC BY-SA 3.0.
Pillansia – **Neville Stuart Pillans (1884–1964)** Glen & Germishuizen, *BESA* (edition 2), p.341.Image: Sanbi.
Pisonia – **Willem Piso (c 1611–1678)** Charters, *EDSAP*. Image: public domain. Author: Pies, Eike, Artist: Jan de Baen. Source: Commons.wikimedia.org/wiki/File:Willem_gemalt_V.jpg.
Pistorinia – **Santiago Pistorini (?–1776)** Díez, *El Real Tribunal del Protomedicato Castellano* (siglos XIV–XIX).
Pitcairnia – **William Pitcairn (1712–1791)** En.wikipedia.org/w/index.php?title=William_Pitcairn&oldid=619301337. Image: public domain. Author: Sir J Reynolds. Engraved by J Jones. Source: Grosvenorprints.com/stock_detail.php?ref=12782. CC BY-SA 3.0.

Plaubelia – **Julius August Plaubel (f 1828–1834)** Homeoint.org/seror/biograph/plaubel.htm.
Pluchia – **Noël-Antoine Pluche (1688–1761)** Oll.libertyfund.org/people/noel-antoine-pluche; The Online Library of Liberty. Image: public domain. Author: Nicolas Blakey. Source: Allposters.co.uk/-sp/L-abbe-Noel-Antoine-Pluche-1688-1761-Posters_i7218245_.htm.
Plukenetia – **Leonard Plukenet (1641–1706)** En.wikipedia.org/wiki/Leonard_Plukenet; Plantzafrica.com/plantefg/ericaplukenetii.htm. Image: public domain. Author: unknown. Source: Huntbot.andrew.cmu.edu/HIBD/Exhibitions/OrderFromChaos/OFC-Pages/02Linnaeus/sources/Plukenet-Leonard.shtm. CC BY-SA 3.0.
Plumeria – **Charles Plumier (1646–1704)** En.wikipedia.org/w/index.php?title=Charles_Plumier&oldid=61690221. Image: public domain. Author: unknown. Source: Upload.wikimedia.org/wikipedia/en/2/28/Plumier_Charles.jpg. CC BY-SA 3.0.
Poellnitzia – **Karl von Poellnitz (1896–1945)** Eden-plants.com/shared/introduction_haworthia_photographs.html; De.wikipedia.org/w/index.php?title=Karl_von_Poellnitz&oldid=129491416. Image: Author: Alfred Zantner. Source: Sukkulentenkunde, Zurich 1948, p.63. CC BY-SA 3.0.
Pohlia – **Johann Ehrenfried Pohl (1746–1800)** De.wikipedia.org/w/index.php?title=Johann_Ehrenfried_Pohl&oldid=124708437. CC BY-SA 3.0.
Poinciana – **Phillippe de Longvilliers de Poincy (1583–1660)** En.wikipedia.org/w/index.php?title=Phillippe_de_Longvilliers_de_Poincy&oldid=637552758. CC BY-SA 3.0.
Poinsettia – **Joel Roberts Poinsett (1775–1851)** En.wikipedia.org/w/index.php?title=Phillippe_de_Longvilliers_de_Poincy&oldid=637552758. Image: public domain. Author: unknown. Source: Findagrave.com/cgi-bin/fg.cgi?page=pv&GRid=6912237. CC BY-SA 3.0.
Poivrea – **Pierre Poivre (1719–1786)** En.wikipedia.org/w/index.php?title=Pierre_Poivre&oldid=646225132. Image: public domain. Author: unknown. Source: Ahistoryblog.com/2013/04/25/pierre-poivre-1719-1786-nutmeg-and-spice. CC BY-SA 3.0.
Polemannia – **Peter Heinrich Poleman (Polemann, Pohlmann) (c 1780–1839)** Glen & Germishuizen, *BESA* (edition 2), p.346.
Polevansia – **Illtyd Buller Pole Evans (1879–1968)** Glen & Germishuizen, *BESA* (edition 2), p.343.Image: Sanbi.
Polhillia – **Roger Marcus Polhill (1937–)** Charters, *EDSAP*. CC BY-SA 3.0.
Pollichia – **Johann Adam Pollich (1741–1780)** En.wikipedia.org/w/index.php?title=Johan_Adam_Pollich&oldid=588094822. CC BY-SA 3.0.
Pontederia – **Giulio (Julius) Pontedera (1688–1757)** De.wikipedia.org/w/index.php?title=Giulio_Pontedera&oldid=135943863. Image: public domain. Author: F Masetti (designer). Source: Bibliotecabertoliana.it. CC BY-SA 3.0.
Popowia – **Johannes Siegmund Valentin Popowitsch (1705–1774)** De.wikipedia.org/w/index.php?title=Johann_Siegmund_Popowitsch&oldid=136298166. Image: public domain. Author: engraving by E Mansfeld. Transferred from Bar.wikipedia to Commons by User: Magnus Manske using CommonsHelper. CC BY-SA 3.0.
Pottia – **Johann Friedrich Pott (1738–1805)** Charters, *EDSAP*.
Pouzolzia – **Pierre Marie Casimir de Pouzolz (1785–1858)** Charters, *EDSAP*.
Priestleya – **Joseph Priestley (1733–1804)** En.wikipedia.org/w/index.php?title=Joseph_Priestley&oldid=653832981. Image: public domain. Author: Ellen Sharples (1769–1849). Source: Original is housed at the National Portrait Gallery, London. CC BY-SA 3.0.
Printzia – **Jacob Gabriel Printz (1740–1779)** Charters, *EDSAP*.
Pseudobaeckea – **Dr Abraham Baeck (Baekea, Bäck) (1713–1795)** De.wikipedia.org/w/index.php?title=Abraham_B%C3%A4ck&oldid=139143033; Thomas, *The Universal Dictionary of Biography and Mythology* (volume 1, part 2). Image: public domain. Author: unknown. Source: Svenskt biografiskt handlexikon. CC BY-SA 3.0.
Pseudoleskea – **Nathanael Gottfried Leske (1751–1786)** En.wikipedia.org/w/index.php?title=Nathanael_Gottfried_Leske&oldid=627236876. Image: public domain. Author and source unknown. CC BY-SA 3.0.
Puccinellia – **Benedetto Luigi Puccinelli (1808–1850)** Es.wikipedia.org/w/index.php?title=Benedetto_Puccinelli&oldid=76693319. Image: public domain. Author: unknown. Source: Loschermo.it. CC BY-SA 3.0.
Pueraria – **Marc Nicolas Puerari (1766–1845)** Es.wikipedia.org/w/index.php?title=Marc_Nicolas_Puerari&oldid=76697237; www2.kb.dk/elib/mss/hcabio//note2.htm. CC BY-SA 3.0.
Putterlickia – **Aloys Putterlick (1810–1845)** Ru.wikipedia.org/?oldid=66279749.
Pyrenowilmsia – **Friedrich Wilms (1848–1919)** Glen & Germishuizen, *BESA* (edition 2), p.462.

Q

Quartinia – **Léon Richard Quartin-Dillon (1814–1841)** Charters, *EDSAP*; Uhlig (ed.), *Encyclopaedia Aethiopica*: He-N, 2007.

R

Rabiea – **William (Bill) Abbot Rabie (1869–1936)** Quattrocchi, *CRC World Dictionary of Plant Names;* personal communication with Chris Rabie, John Krige and others. Image: courtesy Dianne Coetzee, vice-chairwoman, Rabie/Rabe Family Bond.
Radyera – **Robert Allen Dyer (1900–1987)** Glen & Germishuizen, *BESA* (edition 2), p.160–161. Image: Sanbi
Rafnia – **Carl Gottlob Rafn (1769–1808)** En.wikipedia.org/w/index.php?title=Carl_Gottlob_Rafn&oldid=605574730. Image: public domain. Source: courtesy The Royal Library, National Library of Denmark and Copenhagen University Library.

Ramboldia – **Gerhard Walter Rambold (1956–)** Personal communication with Gerhard Walter Rambold, German lichenologist, professor of mycology and head of the DNA analytics laboratory at the University of Bayreuth, Germany.
Randia – **Isaac Rand (1674–1743)** En.wikipedia.org/w/index.php?title=Isaac_Rand&oldid=621936179; Jstor.org/pss/1587324. CC BY-SA 3.0.
Raspalia – **François Vincent Raspail (1794–1878)** Translate.google.com/translate?hl=en&sl=fr&u=appl-lachaise.net/appl/article.php3%3Fid_article%3D758&ei=qxZdTPSyGsyFsAaGoI2zBw&sa=X&oi=translate&ct=result&resnum=3&ved=0CCEQ7gEwAg&prev=/search%3Fq%3DFran%25C3%25A7ois%2BVincent%2BRaspail%2; En.wikipedia.org/w/index.php?title=Fran%C3%A7ois-Vincent_Raspail&oldid=607693023. Image: public domain. Author and source: unknown.
Rauia – **Ambrosius Rau (1784–1830)** Minrec.org/libdetail.asp?id=1177.
Rauiella – **Eugene Abraham Rau (1848–1932)** Ansp.org/research/library/archives/0900-0999/coll0906; Academy of Natural Sciences of Drexel University. Image: Author: unknown. Source: Find-a-grave.com.
Rauvolfia – **Leonhard Rauwolf (Rauwolff) (1535–1596)** En.wikipedia.org/w/index.php?title=Leonhard_Rauwolf&oldid=650983568. CC BY-SA 3.0.
Rawsonia – **Rawson William Rawson (1812–1899)** Glen & Germishuizen, *BESA* (edition 2), p.354. Image: Sanbi
Rehmannia – **Matthew Augustine Joseph Rehmann (1779–1831)** Translate.google.com/translate?hl=en&sl=de&u=drw.saw-leipzig.de/30447.html&prev=search. Image: public domain. Author: unknown. Source: Drw.saw-leipzig.de/30447.html.
Reichardia – **Johann Jakob (Jacob) Reichard (1743–1782)** Quattrocchi, *CRC Dictionary of Plant Names*.
Relhania – **Rev. Richard Relhan (1754–1823)**
En.wikipedia.org/w/index.php?title=Richard_Relhan&oldid=621604393. CC BY-SA 3.0.
Renauldia – **Ferdinand François Gabriel Renauld (1837–1910)** Archive.org/stream/mobot31753003541452/mobot31753003541452_djvu.txt; botanicus.org/ViewTitles.aspx; personal communication with Denis Lamy. Image: public domain. Author and source: unknown. CC BY-SA 3.0.
Rendlia – **Alfred Barton Rendle (1865–1938)** En.wikipedia.org/w/index.php?title=Alfred_Barton_Rendle&oldid=618862774. Image: Author: Walter Stoneman. Source: courtesy National Portrait Gallery. CC BY-SA 3.0.
Renealmia – **Paul Reneaulme (1560–1624)** De.wikipedia.org/w/index.php?title=Paul_Reneaulme&oldid=124943609. Image: public domain. Author: unknown. Source: *Diccionario de Epónimos*, Dicci-eponimos.blogspot.com/2010/01/reneaulme-paul.html. CC BY-SA 3.0.
Rennera – **Otto Renner (1883–1960)** Sysbot.biologie.uni-muenchen.de/botsyst/ic/ic-pha-r.htm. Image: Author: unknown. Source: Dicciomed.eusal.es/creadores.php?idcre=484.
Requienia – **Esprit Requien (1788–1851)** Image: public domain. Author: unknown. Source: Landrucimetieres.fr/spip/spip.php?article2608.
Retzia – **Anders Jahan Retzius (1742–1821)** En.wikipedia.org/wiki/Anders_Jahan_Retzius. Image: public domain. Author: unknown. Source: *Observationes Botanicae: Sex Fasciculis Comprehensae/Andreae Johannis Retzii; quibus accedunt Ioannis Gerhardi Koenig…*
Reyemia – **Heinrich Meyer (f 1861–1886)** Glen & Germishuizen, *BESA* (edition 2), p.197–8. Sanbi.
Reynoutria – **Karel van Sint Omaars (Omaers) (c 1532/1533–1569)** Botany.cz/cs/reynoutre; Homo Botanicus: Sint Omaars, Karel van (van Reynoutre). Author John Möllerová | 5 3rd 2009; Euppublishing.com/doi/abs/10.3366/anh.1997.24.3.423. The albums of Karel van Sint Omaars (1533–1569) (Libri picturati A 16–31, in the Jagiellon Library in Krakow).
Rhynea – **Willem ten Rhyne (Wilhelmi ten Rhijne) (1647–1700)** En.wikipedia.org/w/index.php?title=Willem_ten_Rhijne&oldid=627177075; Abebooks.com/Dissertatio-acupunctura-Wilhelmi-ten-Rhyne-M.D/1090763851/bd. Image: public domain. Author: John Sturt (1658–1730), englischer Kupferstecher. Source: *Porträt in Ten Rhijnes Buch Dissertatio de Arthritide: Mantissa Schematica: De Acupunctura: Et Orationes Tres*, R Chiswell, London 1683.
Riccardia – **the Riccardi family (various dates)** Riccardia S Gray 1821, corr. Trev. 1874 nominum conservum; Perold, 2002, Studies in the liverwort family Aneuraceae from Southern Africa 4 Riccardia obtusa. *Bothalia* 32(2): 181–184.
Riccia – **Pietro Francisco Ricci (f 1729)** Loudon *et al.*, *An Encyclopedia of Plants: Comprising the Description, Specific Character...*, 1829.
Richardia – **Richard Richardson (1663–1741)** Quattrocchi, *CRC World Dictionary of Plant Names: Common Names, Scientific Names, Eponyms...*
Richardia – **Louis Claude Richard (1754–1821)** En.wikipedia.org/w/index.php?title=Louis_Claude_Richard&oldid=627842244. Image: public domain. Author: unknown. Source: Eattheweeds.com/hydrilla. CC BY-SA 3.0.
Riella – **Michel-Charles Durieu de Maisonneuve (1796–1878)** En.wikipedia.org/w/index.php?title=Michel_Charles_Durieu_de_Maisonneuve&oldid=588084499. CC BY-SA 3.0.
Rikliella – **Martin Albert Rikli (1868–1951)** Ngzh.ch/archiv/1951_96/96_4/96_39.pdf. An obituary of Martin Rikli.
Riocreuxia – **Alfred Riocreux (1820–1912)** En.wikipedia.org/w/index.php?title=Alfred_Riocreux&oldid=647469112; Antiquariaatjunk.com/item.php?item=5873. CC BY-SA 3.0.
Ritchiea – **Joseph Ritchie (1792–1819)** En.wikipedia.org/w/index.php?title=Joseph_Ritchie&oldid=653833095; Libweb5.princeton.edu/visual_materials/maps/websites/africa/lyon/lyon.html. CC BY-SA 3.0.
Rivina – **Augustus Quirinus Rivinus (1652–1723)** En.wikipedia.org/w/index.php?title=Augustus_Quirinus_Rivinus&oldid=614439921. Image: public domain. Author: Martin Bernigeroth (1670–1733). Source: De.wikipedia.org/w/index.php?title=Datei:WP_August_Quirinus_Rivinus.jpg. CC BY-SA 3.0.
Robinia – **Jean Robin (1550–1629)** Fr.wikipedia.org/w/index.php?title=Jean_Robin_(botaniste)&action=history; De.academic.ru/dic.nsf/dewiki/687642; Translate.google.com/translate?hl=en&sl=de&u=de.academic.ru/dic.nsf/

dewiki/687642. Image: public domain. Author: unknown. Source: BIU Santé (Paris).
Robsonodendron – **Norman Keith Bonner Robson (1928–)** En.wikipedia.org/w/index.php?title=Norman_Robson&oldid=595900951. CC BY-SA 3.0.
Rochea – **Daniel de la Roche (Delaroche) (1743–1813)** De.wikipedia.org/w/index.php?title=Daniel_Delaroche&oldid=134029089; Es.wikipedia.org/w/index.php?title=Daniel_Delaroche&oldid=70867004. CC BY-SA 3.0.
Rochelia – **Anton Rochel (1770–1847)** Sk.wikipedia.org/w/index.php?title=Anton_Rochel&oldid=5789355. CC BY-SA 3.0.
Roella – **Willem (Wilhelm) Röell (1700–1775)** Nl.wikipedia.org/w/index.php?title=Willem_R%C3%B6ell_(1700-1775)&oldid=40352084. Image: public domain. Author: August Johannes Le Gras (1864–1915). Source: Uva.nl. CC BY-SA 3.0.
Rogeria – **Jacques-François Roger (1787–1849)** Fr.wikipedia.org/w/index.php?title=Jacques-Fran%C3%A7ois_Roger&action=history. Image: Author: unknown. CC BY-SA 3.0.
Rohria – **Julius Philip Benjamin von Röhr (1737–1793)** En.wikipedia.org/w/index.php?title=Julius_von_R%C3%B6hr&oldid=608588737; Jstor.org/discover/10.2307/1218471?uid=3739368&uid=2&uid=4&sid=21106072796761. CC BY-SA 3.0.
Rondeletia – **Guillaume Rondelet (1507–1566)** En.wikipedia.org/w/index.php?title=Guillaume_Rondelet&oldid=653991194; *Encyclopaedia Londinensis* (volume 22). Image: public domain. Author: unknown. Source: Commons.wikimedia.org/wiki/File:Guillaume-Rondelet-1507-1566.jpg. CC BY-SA 3.0.
Roodia – **Petrusa Benjamina Rood (1861–1946)** Glen & Germishuizen, *BESA* (edition 2), p.365. Image: Author: unknown. Source: Geni.com. CC BY-SA 3.0.
Rosenia – **Eberhard Rosén (Rosenblad) (1714–1796) and Nils Rosén (1706–1773)** Sv.wikipedia.org/w/index.php?title=Eberhard_Rosenblad&oldid=29751036. Image: public domain. Author: unknown. Source: Libris.kb.se/bib/1579474; Sv.wikipedia.org/w/index.php?title=Nils_Ros%C3%A9n_von_Rosenstein&oldid=29751888. Image: public domain. Author: Lorentz Pasch the Younger (1733–1805); Svenskhistoria.se/arkiv/797.htm. CC BY-SA 3.0.
Rothia – **Albrecht Wilhelm Roth (1757–1834)** En.wikipedia.org/w/index.php?title=Albrecht_Wilhelm_Roth&oldid=599830034. Image: public domain. Author: unknown. Source: Alamy.com. CC BY-SA 3.0.
Rothmannia – **Jöran (Georg) Johansson Rothman (1739–1778)** De.wikipedia.org/w/index.php?title=G%C3%B6ran_Rothman&oldid=134940588. CC BY-SA 3.0.
Rottboellia – **Christen Friis Rottböll (1727–1797)** En.wikipedia.org/w/index.php?title=Christen_Friis_Rottb%C3%B8ll&oldid=631389543. Image: public domain. Author and source: unknown. CC BY-SA 3.0.
Roubieva – **Guillaume Joseph Roubieu (1757–1834)** Charters, *EDSAP.*
Royena – **Adriaan van Royen (1704–1779)** En.wikipedia.org/w/index.php?title=Adriaan_van_Royen&oldid=635317270; Hieronymus van der Mij. Image: public domain. Author: Hieronymus van der Mij. Source: Collectie Icones Leidenses 157; Socrates.leidenuniv.nl. CC BY-SA 3.0.
Ruckeria – **Johann Friedrich Rücker (f 1600s)** Glen & Germishuizen, *BESA* (edition 2), p.370.
Ruellia – **Jean Ruel (Jean de la Ruelle) (1474–1537)** Meyer et al., *De Historia Stirpium Commentarii Insignes*, Books.google.com/books?id=qCGdBd4TaswC&pg=PA789&lpg=PA789&dq=Jean+Ruel+(Jean+de+la+Ruelle)+(1474-1537)&source=bl&ots=aTpcWuYYZ-&sig=3UHsPZP4R6j8wyebZOywv3Zqhiw&hl=en&ei=eBt8TI2kJZGWswali5myDQ&sa=X&oi=book_result&ct=result&resnum=4&ved=0CCEQ6AEwAzgK#.
Rumohra – **Karl Friedrich Felix von Rumohr (1785–1843)** En.wikipedia.org/w/index.php?title=Carl_Friedrich_von_Rumohr&oldid=650692466. Image: public domain. Artist: Friedrich Nerly. Source: Alte Nationalgallerie. CC BY-SA 3.0.
Rungia – **Friedlieb Ferdinand Runge (1795–1867)** En.wikipedia.org/w/index.php?title=Friedlieb_Ferdinand_Runge&oldid=621154179. Image: public domain. Author: Original uploader was Manuel at De.wikipedia. Source: Transferred from De.wikipedia; transferred to Commons by User:Kilom691 using CommonsHelper. CC BY-SA 3.0.
Ruppia – **Heinrich Bernhard Ruppius (Rupp, Ruppee) (1688–1719)** De.wikipedia.org/w/index.php?title=Heinrich_Bernhard_Rupp&oldid=123892019. CC BY-SA 3.0.
Ruschia – **Ernst Julius Rusch (1867–1957)** Glen & Germishuizen, *BESA* (edition 2), p.371–372. Image: Sanbi CC BY-SA 3.0.
Ruschianthemum – **Ernst Franz Theodor Rusch (1897–1964)** Glen & Germishuizen, *BESA* (edition 2), p.371–372. Image: Sanbi.
Ruspolia – **Prince Eugenio Ruspoli (1866–1893)** En.wikipedia.org/w/index.php?title=Emanuele_Ruspoli,_1st_Prince_of_Poggio_Suasa&oldid=654501967. Image: public domain. Author: unknown. Source: Strangebehaviors.wordpress.com. CC BY-SA 3.0.
Russelia – **Dr Alexander Russell (c 1715–1768)** Electricscotland.com/history/nation/russell.
Ruthea – **Johann Friedrich Ruthe (Ruthé or von Ruthe) (1788–1859)** En.wikipedia.org/w/index.php?title=Johann_Friedrich_Ruthe&oldid=617281618. Image: public domain. Author: unknown. Source: Myheritage.de.
Ruttya – **John Rutty (1697–1775)** En.wikipedia.org/w/index.php?title=John_Rutty&oldid=586374171; Botanicgardens.ie/herb/books/inltos.htm. Image: public domain. Author: unknown. Source: Snipview.com.

S

Saccardoa – **Pier Andrea Saccardo (1845–1920)** En.wikipedia.org/w/index.php?title=Pier_Andrea_Saccardo&oldid=540660483; Mushroomthejournal.com/greatlakesdata/Authors/Saccardo24.html. Image: Author Curtis Gates Lloyd (1898–1925). Source: Commons.wikimedia.org/wiki/File:Pier_Andrea_Saccardo_1845-1920.jpg.

Saelania – **Anders Thiodolf Saelan (1834–1921)** Sv.wikipedia.org/w/index.php?title=Thiodolf_S%C3%A6lan&oldid=28304184. Image and author: unknown. Source: Geni.com.
Saintpaulia – **Walter von Saint Paul-Illaire (1860–1940)** De.wikipedia.org/w/index.php?title=Walter_von_Saint_Paul-Illaire&oldid=137202875. Image: Author: Christian Wilhelm Allers (1857–1915). Source: Allers, *Unser Bismarck*, 1895, Seite 113.
Saltera – **Terence Macleane Salter (1883–1969)** Glen & Germishuizen, *BESA* (edition 2), p.374. Image: Sanbi.
Salvadora – **Juan (Joan) Salvador y Bosca (1598–1681)** Mcnbiografias.com.
Salvinia – **Anton Maria Salvini (1653–1729)** It.wikipedia.org/w/index.php?title=Anton_Maria_Salvini&oldid=71210768. Image: public domain. Author: C Magalli. Source: *Sonnets de Anton Maria Salvini, Academician of Bran*, Florence. CC BY-SA 3.0.
Sandersonia – **John Sanderson (1820–1881)** Glen & Esterhuizen, *BESA* (edition 2), p.374–3758. CC BY-SA 3.0.
Sanionia – **Carl Gustav Sanio (1832–1891)** Frahm & Eggers, *Lexikon Deutschsprachiger Bryologen* (volume 1, 1999). Books.google.com/books?id=8VEUIH15wAYC&pg=PA431&lpg=PA431&dq=Carl+Gustav+Sanio+(1832–1891)&source=bl&ots=y3YgdQAUKO&sig=SjkSIJ1Ce8-3gXW77cdZnGU-tYk.
Sansevieria – **Raimondo di Sangro (1710–1771)** En.wikipedia.org/w/index.php?title=Raimondo_di_Sangro&oldid=647372928. Image: public domain. Author: unknown. Source: Italian Wikipedia, scanned from a work published in 1836, drawing of Raimondo di Sangro. CC BY-SA 3.0.
Santessonia – **Rolf Santesson (1916–2013)** Sv.wikipedia.org/w/index.php?title=Rolf_Santesson&oldid=24858380. Image: Author: unknown. Source: Journals@cambridge.org. *The Lichenologist* 46(2) 135–139 (2014). CC BY-SA 3.0.
Savia – **Gaetano Savi (1769–1844)** En.wikipedia.org/w/index.php?title=Gaetano_Savi&oldid=600077502. Image: Author: unknown. Source: Brunelleschi.imss.fi.it/genscheda.asp?appl=LST&xsl=biografia&lingua=ENG&chiave=300516. CC BY-SA 3.0.
Scaevola – **Gaius Mucius Scaevola (6 BCE)** En.wikipedia.org/w/index.php?title=Gaius_Mucius_Scaevola&oldid=655103624. Image: public domain. Photographer: Jastrow. Source: Louvre Museum.
Schaereria – **Ludwig Emanuel Schaerer (1785–1853)** Fr.wikipedia.org/w/index.php?title=Ludwig_Emanuel_Schaerer&action=history. Image: public domain. Author: unknown. Source: Sac-albis.ch.
Schefflera – **Johann Peter Ernst von Scheffler (c 1739–1808)** En.wikipedia.org/w/index.php?title=Schefflera&oldid=651399728.
Schiffneriolejeunea – **Victor Félix Schiffner (1862–1944)** En.wikipedia.org/w/index.php?title=Victor_F%C3%A9lix_Schiffner&oldid=614152330.
Schinziophyton – **Hans Schinz (1858–1941)** Glen & Germishuizen, *BESA* (edition 2), p.390. Image: Sanbi.
Schkuhria – **Christian Schkuhr (1741–1811)** Library.umass.edu/spcoll/exhibits/herbal/schkur.htm; Es.wikipedia.org/w/index.php?title=Christian_Schkuhr&oldid=71290645.
Schlechteranthus – **Max (Maximilian) Schlechter (1874–1960)** Glen & Germishuizen, *BESA* (edition 2), p.383. Image: Sanbi.
Schlechteria – **Friedrich Richard Rudolf Schlecter (1872–1925)** Glen & Germishuizen, *BESA* (edition 2), p.381–383. Image: Sanbi.
Schlotheimia – **Ernst Friedrich von Schlotheim (1764–1832)** En.wikipedia.org/w/index.php?title=Ernst_Friedrich,_Baron_von_Schlotheim&oldid=619592644. Image: public domain. Author: unknown. Source: as above.
Schlumbergera – **Frédéric Schlumberger (1823–1893)** Charters, *EDSAP*.
Schmidelia – **Casimir Christoph Schmidel (1718–1792)** Minrec.org/artwork.asp?artistid=43&cat=1; Encyclopedia.com/doc/1G2-2830903885.html. CC BY-SA 3.0.
Schmidtia – **Johann Anton Schmidt (1823–1905)** Uni-goettingen.de/de/187070.html; Plants.jstor.org/stable/10.5555/al.ap.person.bm000007528.
Schoenefeldia – **Wladimir de Schoenefeld (1816–1875)** En.wikipedia.org/w/index.php?title=Wladimir_de_Schoenefeld&oldid=628735777.
Schonlandia – **Selmar Schönland (1860–1940)** Glen & Germishuizen, *BESA* (edition 2), p.386. Image: Sanbi CC BY-SA 30.
Schotia – **Richard van der Schot (c 1730–1790)** Plants.jstor.org/person/bm000004014.
Schrebera – **Johann Christian Daniel von Schreber (1739–1810)** En.wikipedia.org/w/index.php?title=Johann_Christian_Daniel_von_Schreber&oldid=650925547. Image: public domain. Author: unknown. Source: J Dörfler (1906–1907) *Botaniker Porträts* (secondary source: Illinois Mycological Association). CC BY-SA 30.
Schwabea – **Samuel Heinrich Schwabe (1789–1875)** En.wikipedia.org/w/index.php?title=Heinrich_Schwabe&oldid=657392966. Image: public domain. Author: unknown. Source: Saburchill.com. CC BY-SA 30.
Schwantesia – **Martin Heinrich Gustav (Georg) Schwantes (1881–1960)** De.wikipedia.org/w/index.php?title=Gustav_Schwantes&oldid=138004773. Image: Author: unknown. Source: *Kakteenkunde*, Jg. 1936, Heft 2; S. 27 CC BY-SA 3.0.
Schweiggera – **August Friedrich Schweigger (1783–1821)** De.wikipedia.org/w/index.php?title=August_Friedrich_Schweigger&oldid=132160369. Image: public domain. Author: Knorre del. Fr. Bolt sc. Source: Ihm.nlm.nih.gov. Image: s/B22986. CC BY-SA 3.0.
Schwenckia – **Martin Wilhelm Schwenke (1707–1785)** Snippet information only. Short Biography.
Schwetschkea – **Carl Gustav Schwetschke (1804–1881)** De.wikipedia.org/w/index.php?title=Carl_Gustav_Schwetschke&oldid=136948569. CC BY-SA 3.0.
Searsia – **Paul Bigelow Sears (1891–1990)** En.wikipedia.org/w/index.php?title=Paul_Sears&oldid=660339815. Image: courtesy Botanical Society of America. CC BY-SA 3.0.
Sebaea – **Albertus Seba (1665–1736)** En.wikipedia.org/w/index.php?title=Albertus_Seba&oldid=608682808. Image:

public domain. Artists: Jacobus Houbraken (1698–1780) after Jan Maurits Quinkhard (1688–1672). Source: Taschen. com/pages/en/catalogue/classics/reading_room/24.albertus_sebas_collection_of_natural_specimens_and_ its_ pictorial_inventory.6htm. CC BY-SA 3.0.

Seemannaralia – **Berthold Carl Seemann (1825–1871)** En.wikipedia.org/w/index.php?title=Berthold_Carl_Seemann& oldid=660437388. Image: public domain. Source: Commons.wikimedia.org/wiki/File:Berthold_Carl_Seemann.jpg. CC BY-SA 3.0.

Seetzenia – **Ulrich Jasper Seetzen (1767–1811)** En.wikipedia.org/w/index.php?title=Ulrich_Jasper_Seetzen&oldid= 660485173. Image: public domain. Authors: Frederik Christiaan Bierweiler (1783–c 1831) after Eberhard Christian Dunker (1735–1817). Source: Bibelwissenschaft.de. CC BY-SA 3.0.

Seidelia – **Christopher Friedrich Seidel (f 1869–c 1893)** Atlanticcoastcamelliasociety.org/journal/ACCS-0592-OCR-OPT.pdf; Kiki.huh.harvard.edu/databases/botanist_search.php?start=1&name=Eid&id=&remarks=&specialty=& country=&individual=on.

Selliguea – **Alexandre François Gilles (Selligue Paris) (1784–1845)** Gbamici.sns.it/eng/biografia/biografia_pt5. htm; Letemps.ch/Page/Uuid/67dd5318-5b0d-11e0-b9d5-f41fea4c20d4/Alexander_Selligue_premier_distillateur_ industriel_de_schistes.

Semonvillea – **Charles-Louis Huguet, Marquis de Semonville (1759–1839)** En.wikipedia.org/w/index.php?title= Charles_Louis_Huguet,_marquis_de_S%C3%A9monville&oldid=660408131. Image: public domain. Author: Gérard René Le Villain (1740–1836). Source: Bibliothèque Nationale de France. CC BY-SA 3.0.

Sendtnera – **Otto Sendtner (1813–1859)** En.wikipedia.org/w/index.php?title=Otto_Sendtner&oldid=659612867. CC BY-SA 3.0.

Senebiera – **Jean Senebier (1742–1809)** En.wikipedia.org/w/index.php?title=Jean_Senebier&oldid=660850895. Public domain. Author: unknown. Source: Unige.ch/450/photo450-1/archives.html. CC BY-SA 3.0.

Serruria – **Joseph (Josephus) Serrurier (1663–1742)** De.wikipedia.org/w/index.php?title=Josephus_Serrurier& oldid=122941528. Image: public domain. Author: unknown. Source: Profs.library.uu.nl/index.php/profrec/ getprofdata/1893/35/39/0. CC BY-SA 3.0.

Shantzia – **Homer Leroy Shantz (1876–1958)** Glen & Germishuizen, *BESA* (edition 2), p.390–391. Image: Sanbi

Sheilanthera – **Sheila (dates unknown), wife of Ion James Muirhead Williams (1912–2001)** Glen & Germishuizen, *BESA* (edition 2), p.451. Image: Sanbi CC BY-SA 3.0.

Sherardia – **William Sherard (1659–1727)** En.wikipedia.org/w/index.php?title=William_Sherard&oldid=660333228. CC BY-SA 3.0.

Shutereia – **James Shuter (1775–1826)** The London *Literary Gazette and Journal of Belles Lettres, Arts, Sciences* (1826); Books.google.com/books?id=24tHAAAAYAAJ&pg=PA111&lpg=PA111&dq=Dr.+James+Shuter+(% 3F-1826)+botanist.&source=bl&ots=g2gdK0Tcpk&sig=ZLyJnsKRQi3Qf-8d-hDNp6feT3U&hl=en&ei= JyiOTPawCIOLswaA3LjeAQ&sa=X&oi=book_result&ct=result&resnum=1&ved=0CBEQ6AEwADgK#v=onepage& q=Dr.%20James%20Shuter%20(%3F-1826)%20botanist.&f=false.

Sickmannia – **Johann Rudolph Sickmann (1779–1849)** Snippet information only.

Sigesbeckia – **Johann Georg Siegesbeck (1686–1755)** Scricciolo.com/linnaeus_polemic.htm.

Smelophyllum – **Timofei Andreevich Smielowski (1769–1815)** Phaidra.univie.ac.at/detail_object/o:175436.

Smithia – **Edward Smith (1759–1828)** En.wikipedia.org/w/index.php?title=James_Edward_Smith&oldid=660822595. Image: public domain. Author: RA Russel, painter to their majesties. Source: Hbs.bishopmuseum.org/dipterists/dipt-s. html. CC BY-SA 3.0.

Soliva – **Salvador Soliva (1750–1793)** Raco.cat/index.php/AnnalsMedicina/article/download/283323/371222.

Sonderina – **Otto Wilhelm Sonder (1812–1881)** Glen & Germishuizen, *BESA* (edition 2), p.400. Image: Sanbi.

Sowerbaea – **James Sowerby (1757–1822)** En.wikipedia.org/w/index.php?title=James_Sowerby&oldid=653805806. Author: Thomas Heaphy. Source: *Mineralogical Record*, vol. 26, July–August, 1995. CC BY-SA 3.0.

Sparrmannia – **Anders Erikson Sparrman (1748–1820)** Glen & Germishuizen, *BESA* (edition 2), p.401. Image: Sanbi.

Spetaea – **Franz Speta (1941–?)** De.wikipedia.org/w/index.php?title=Franz_Speta&oldid=123712539. Image: Source: Zobodat.at/personen.php?id=277#. CC BY-SA 3.0.

Spielmannia – **Jakob Reinbold Spielmann (1722–1783)** De.wikipedia.org/w/index.php?title=Jacob_Reinbold_ Spielmann&oldid=123761976. Image: public domain. Author: Christophe Guérin. Source: US National Library of Medicine. CC BY-SA 3.0.

Sponia – **Jacob (Jacques) Spon (1647–1685)** El.wikipedia.org/w/index.php?title=%CE%96%CE%B1%CE%BA% CF%8C%CE%BC%CF%80_%CE%A3%CF%80%CE%BF%CE%BD&oldid=4440630. Image: public domain. Author: unknown. Source: Huguenots-france.org/france/lyon/celebrites_lyon/spon.htm. CC BY-SA 3.0.

Sporledera – **Friedrich Wilhelm Sporleder (1787–1875)** De.wikisource.org/wiki/ADB:Sporleder,_Friedrich_Wilhelm; De.wikipedia.org/w/index.php?title=Friedrich_Sporleder&oldid=119266649. CC BY-SA 3.0.

Sprekelia – **Johann Heinrich von Spreckelsen (1691–1764)** De.wikipedia.org/w/index.php?title=Johann_Heinrich_ von_Spreckelsen&oldid=136707607. CC BY-SA 3.0.

Staavia – **Martin Staaf (1731–1788)** Scricciolo.com/linnaeus_notes1.htm.

Staberoha – **Johann Heinrich Julius Staberoh (1785–1857)** De.wikipedia.org/wiki/Johann_Gottfried_Hempel; Archive.org/stream/dasapothekenwese00bere/dasapothekenwese00bere_djvu.txt; Archive.org/details/ preussischepharm00link. CC BY-SA 3.0.

Stadmannia – **Jean Frédéric Stadtmann (1762–1807)** Ru.wikipedia.org/?oldid=66275072. Image: public domain. Author: unknown. Source: Royal Society of Arts and Sciences, Reduit, Mauritius. CC BY-SA 3.0.

Staehelina – **Benedict Staehelin (Staehelin) (1695–1750)** Unigeschichte.unibas.ch/materialien/rektoren/ benedict-staehelin.html. *The Universal Dictionary of Biography and Mythology* (volume IV), p.2054 Books. google.com/books?id=LP2C-ORXkLkC&pg=PA2054&dq=Benedict+Staehelin,+Swiss+botanist&hl=en&ei=9z GnTNvxM9O6jAev4ay-DA&sa=X&oi=book_result&ct=result&resnum=3&ved=0CC4Q6AEwAg#v=onepage&q= Benedict%20Staehelin%2C%20Swiss%20botanist&f=false.
Stangeria – **William Stanger (1811–1854)** Glen & Germishuizen, *BESA* (edition 2), p.403. Image: Sanbi CC BY-SA 3.0.
Stapelia – **Johannes Bodaeus van Stapel (1602–1636)** Glen & Germishuizen, *BESA* (edition 2), p.213. (under Heurnius). Image: Sanbi CC BY-SA 3.0.
Staudtia – **Alois Staudt (?–1897)** Charters, *EDSAP*.
Stayneria – **Frank James Stayner (1907–1981)** Glen & Germishuizen, *BESA* (edition 2), p.197–8. Image: Sanbi.
Stephaniella – **Franz Stephani (1842–1927)** En.wikipedia.org/w/index.php?title=Franz_Stephani&oldid=659733353. Image: Source: Brill.com.
Steudelia – **Ernst Gottlieb von Steudel (1783–1856)** Botanicus.org/creator/558.
Stirtonia – **Charles Howard Stirton (1946–)** Glen & Germishuizen, *BESA* (edition 2), p.197–8. Image: Sanbi.
Stobaea – **Kilian Stobaeus (1690–1742)** Sv.wikipedia.org/w/index.php?title=Kilian_Stob%C3%A6us&oldid= 29750978. Image: public domain. Author: Carl Peter Mörth. Source: Åberg, Alf: Ramlösa. En hälsobrunns historia under 250 år (1957). CC BY-SA 3.0.
Stoeberia – **Ernst Stöber (1889–?)** Glen & Germishuizen, *BESA* (edition 2), p.197–198. Image: Sanbi.
Stokoeanthus – **Thomas Pearson Stokoe (1868–1959)** Glen & Germishuizen, *BESA* (edition 2), p.197–8. Image: Sanbi CC BY-SA 3.0.
Stoneobryum – **Ilma Stone (1913–2001)** Trove.nla.gov.au/people/1475669?q=&c=people. Image: Source: Vibedo. com.
Strelitzia – **Charlotte of Mecklenburg-Strelitz (1744–1818)** En.wikipedia.org/w/index.php?title=Charlotte_of_ Mecklenburg-Strelitz&oldid=661117245. Image: public domain. Artist: Johann Georg Ziesenis. Source: Destee.com/index.php?threads/interesting-portraits-of-the-european-nobility-and-kings.73489. CC BY-SA 3.0.
Sturmia – **Jacob Sturm (1771–1848)** Sil.si.edu/DigitalCollections/NHRareBooks/Sturm/sturm-introduction.htm. Image: public domain. Author: unknown. Source: Hbs.bishopmuseum.org.
Suessenguthiella – **Karl Suessenguth (1893–1955)** De.wikipedia.org/w/index.php?title=Karl_Suessenguth&oldid= 129272889. Image: Source: courtesy Botanic Garden and Botanical Museum BerlinDahlem. CC BY-SA 3.0.
Susanna – **Susan Phillips (née Kriel) (dates unknown)** Glen & Germishuizen, *BESA* (edition 2), p.339. Image: Sanbi CC BY-SA 3.0.
Sutera – **Johann Rudolf Suter (1766–1827)** En.wikipedia.org/w/index.php?title=Johann_Rudolf_Suter&oldid= 660539122; hls-dhs-dss.ch/textes/d/D26186.php Historisches Lexikon der Schweiz.
Sutherlandia – **James Sutherland (1639–1719)** Glen & Germishuizen, *BESA* (edition 2), p.411. Image: Sanbi
Swartzia – **Olof Peter Swartz (1760–1818)** En.wikipedia.org/w/index.php?title=Olof_Swartz&oldid=661376706. Image: public domain. Author: unknown. Source: Swedish Museum of Natural History. CC BY-SA 3.0.
Swertia – **Emanuel Sweert (1552–1612)** En.wikipedia.org/w/index.php?title=Emanuel_Sweert&oldid=662692851. Image: public domain. Source: Picturingplants.com. CC BY-SA 3.0.
Synnotia – **Walter Synnot (1773–1851)** Glen & Germishuizen, *BESA* (edition 2), p.413.

T

Tabernaemontana – **Jakob Theodor von Bergzaben (Jacobus Theodorus) (1522–1590)** En.wikipedia.org/w/index.php?title=Jacobus_Theodorus_Tabernaemontanus&oldid=624067831h; De.wikipedia.org/w/index.php?title=Tabernaemontanus&oldid=142189600. Image: public domain. Author: unknown. Source: Boissard, Jean-Jacques: Bibliotheca Chalcographica pt 9. CC BY-SA 3.0.
Tacca – **Pietro Tacca (1577–1640)** En.wikipedia.org/w/index.php?title=Pietro_Tacca& oldid= 662700114. CC BY-SA 3.0.
Talbotia – **William Henry Fox Talbot (1800–1877)** En.wikipedia.org/w/index.php?title=Henry_Fox_Talbot&oldid=660895197. Image: public domain. Author: John Moffat. Source: Fox Talbot, Lifelines 38, Shire publications Ltd, Princes Risborough, 3rd Edition 1997.
Targionia – **Dr Cyprian Targioni (f 1720–1748)** Rees, *The Cyclopaedia; Or, Universal Dictionary of Arts, Sciences and Literature* (volume 35), 1819.
Tavaresia – **Joaquim da Silva Tavares (1866–1931)** Translate.google.co.za/translate?hl=en&sl=pt&u=culturacentro. pt/museuit.asp%3Fid%3D110&ei=EzqqTMDbDMKeOvH-gaUM&sa=X&oi=translate&ct=result&resnum=2&ved= 0CB0Q7gEwAQ&prev=/search%3Fq%3DJoaquim%2Bda%2BSilva%2BTavares%2B%281866-1931%29%26hl% 3Den%26sa%3DG%26rlz%3D1B3GGIC_enZA317ZA317%26prmd%3Dbo; Translate.google.co.za/translate?hl= en&sl=pt&u=fmsoares.pt/aeb/crono/biografias%3Fregisto%3DJoaquim%2520da%2520Silva%2520Tavares&ei= EzqqTMDbDMKeOvH-gaUM&sa=X&oi=translate&ct=result&resnum=4&ved=0CCUQ7gEwAw&prev=/search% 3Fq%3DJoaquim%2Bda%2BSilva%2BTavares%2B%281866-1931%29%26hl%3Den%26sa%3DG%26rlz% 3D1B3GGIC_enZA317ZA317%26prmd%3Dbo. Image: Pt.wikipedia.org/wiki/Ficheiro:Joaquim_da_Silva_Tavares. png. Author and source: Brotéria. CC BY-SA 3.0.
Tayloria – **Thomas Taylor (1775–1848)** En.wikipedia.org/w/index.php?title=Thomas_Taylor_(botanist)&oldid= 659401442. CC BY-SA 3.0.
Teclea – **Mara Takla Haymanot (c 1137)** En.wikipedia.org/w/index.php?title=Mara_Takla_Haymanot&oldid= 644568743. CC BY-SA 3.0.

Tedingea – **Edward 'Ted' George Hudson Oliver (1938–) and Inge Oliver (born Nitzsche) (1947–2003)** Glen & Germishuizen, *BESA* (edition 2), p.197–8. Image: Sanbi.
Teedia – **Johann Georg Teede (f 1700s)** Books.google.co.za/books?id=S35MAAAAYAAJ&pg=PA288& lpg=PA288&dq=Journal+f%C3%BCr+die+Botanik+Teede&source=bl&ots=oKRxjPMcwL&sig= WivPGWhwhvKe68naU8kc4f1ayxA&hl=en&ei=-k25TsWqFMa6hAeit5CgBw&sa=X&oi=book_result&ct=result& resnum=4&ved=0CDAQ6AEwAw#v=onepage&q&f=false; *Journal für die Botanik* (volume 2), 1799.
Thalia – **Johannes Thal (1542–1583)** De.wikipedia.org/w/index.php?title=Johannes_Thal&oldid=138092473. Image: public domain. Source: Datuopinion.com/johannes-thal CC BY-SA 3.0.
Theilera – **Sir Arnold Theiler (1867–1936)** Glen & Germishuizen, *BESA* (edition 2), p.416. Image: Author. National Photo Company. Source: *National Photo Company Collection*. CC BY-SA 3.0.
Theodora – **Charles Theodore Elector and Duke of Bavaria (1724–1799)** En.wikipedia.org/w/index.php?title= Charles_Theodore,_Elector_of_Bavaria&oldid=651712025. Image: public domain. Artist: Anna Dorothea Lisiewska-Therbusch (1721–1782). Source: Reiss-Engelhorn-Museen, Mannheim.
Thodaya – **David Thoday (1883–1964)** Glen & Germishuizen, *BESA* (edition 2), p.417. Image: Sanbi.
Thorncroftia – **George Thorncroft (1857–1934)** Glen & Germishuizen, *BESA* (edition 2), p.420–421. Image: Sanbi.
Thunbergia – **Carl Pehr (Peter) Thunberg (1743–1828)** Glen & Germishuizen, *BESA* (edition 2), p.422–424.Image: Sanbi.
Tieghemia – **Philippe Édouard Léon van Tieghem (1839–1914)** En.wikipedia.org/w/index.php?title=Philippe_%C3% 89douard_L%C3%A9on_Van_Tieghem&oldid=660033038. Image: public domain, photographie: Eugène Pirou. Source: *Le Miroir*, 10 mai 1914. CC BY-SA 3.0.
Tillaea – **Michelangelo Tilli (Michele Angelo Tilli) (1655–1740)** En.wikipedia.org/w/index.php?title=Michelangelo_ Tilli&oldid=659791017. Image: public domain. Author: unknown. Source: Navigationdusavoir.net/PortalPisa/ seafarers/moderna_carriere1.html. CC BY-SA 3.0.
Tillandsia – **Elias Tillandz (1640–1693)** Biocity.turku.fi/about-us/elias-tillandz-prize/elias-tillandz.
Timmiella – **Joachim Christian Timm (1734–1805)** En.wikipedia.org/w/index.php?title=Joachim_Christian_ Timm&oldid=640129176. Image: public domain. Author: FG Herzog. Source: Transferred from De.wikipedia to Commons by Leyo using CommonsHelper. CC BY-SA 3.0.
Tinnea – **Alexandria Tinne (1835–1869)** En.wikipedia.org/w/index.php?title=Alexandrine_Tinn%C3%A9&oldid= 663191001. Image: public domain. Author: unknown. Source: Digitaal Vrouwenlexicon van Nederland. CC BY-SA 3.0.
Tittmannia – **Johann August Tittman (1774–1840)** Charters, *EDSAP;* Don, *A General System of Gardening and Botany*, 1832.
Todea – **Heinrich Julius Tode (1733–1797)** De.wikipedia.org/w/index.php?title=Heinrich_Julius_Tode&oldid= 128091614. CC BY-SA 3.0.
Tomasellia – **Giuseppe (Joseph) Tomaselli (1733–1818)** Charters, *EDSAP.*
Toninia – **Carlo Tonini (1803–1877)** Charters, *EDSAP.* Image: public domain. Source: Dicci-eponimos. blogspot.com.
Topeliopsis – **Josef Poelt (1924–1995)** Plants.jstor.org/stable/history/10.5555/al.ap.person.bm000006593; Deutsche-biographie.de/sfz96530.html. Image: journals.cambridge.org. Source: *The Lichenologist* (volume 28 issue 2).
Torenia – **Rev. Olof Torén (1718–1753)** Webcache.googleusercontent.com/search?q=cache:EtxKGRVW8VAJ:goran. waldeck.se/Ento3E.htm+Rev.+Olof+Tor%C3%A9n+%281718-1753%29,+Swedish+clergyman,+traveller,+botanist& cd=3&hl=en&ct=clnk LINNAEUS' DISCIPLES AND APOSTLES; Quattrocchi, *CRC World Dictionary of Plant Names: Common Names, Scientific Names, Eponyms…*
Tornabenia – **Francesco Tornabene Roccaforte (1813–1897)** Dipbot.unict.it/orto-botanico/tornabene.htm (Hortus Botanicus Catinensis). Image: public domain: Source: Nuovosoldo.it.
Tournefortia – **Joseph Pitton de Tournefort (1656–1708)** En.wikipedia.org/w/index.php?title=Joseph_Pitton_de_ Tournefort&oldid=660068704. Image: public domain. Source: Commons.wikimedia.org/wiki/File:Tournefort_Joseph_ Pitton_de_1656-1708.jpg. CC BY-SA 3.0.
Tradescantia – **John Tradescant (aka 'Treadeskant') (c 1570–1638)** Practicallyedible.com/edible.nsf/pages/ johntradescant#ixzz11m2D3W8A; En.wikipedia.org/wiki/John_Tradescant_the_elder; En.wikipedia.org/wiki/John_Tradescant_the_Younger. Image: public domain. Tradescant the elder. Source: Thecultureconcept.com/; Tradescent the younger. Source: Mexicolore.co.uk. CC BY-SA 3.0.
Tragia – **Hieronymus Tragus (1498–1554)** Es.wikipedia.org/w/index.php?title=Hieronymus_Tragus&oldid=73224003. Image: public domain. Author: David Kandel (1546). Source: Kreütter Büch (1546) a Herbal. CC BY-SA 3.0.
Treichelia – **Alexander Johann August Treichel (1837–1901)** Source: Polen, Deutsche und Kaschuben**:** *Alltag, Brauchtum und Volkkultur auf dem Gut Hochpaleschken in Westpreussen um 1900.* Bernhard Lauer, Hanna Nogossek, Brüder Grimm-Gesellschaft, 1997 – 111 pages.
Trieenea – **Elizabeth ('Elsie') Esterhuysen (1912–2006)** Glen & Germishuizen, *BESA* (edition 2), p.169. Image: Sanbi.
Triumfetta – **Giovanni Battista Trionfetti (1658–1708)** Antiquariaatjunk.com/item.php?item=7783; Translate.google. co.za/translate?hl=en&sl=it&u=it.wikipedia.org/wiki/Giovan_Battista_Trionfetti&ei=wGC1TI25MpH44gbrstWgDQ& sa=X&oi=translate&ct=result&resnum=9&ved=0CDQQ7gEwCA&prev=/search%3Fq%3Dgiovanni%2Bbattista% 2Btrionfetti%26hl%3Den%26rlz%3D1B3GGIC_enZA317ZA317%26prmd%3Dbo.
Tulbaghia – **Ryk Tulbagh (Rijk Tulbagh) (1699–1771)** Glen & Germishuizen, *BESA* (edition 2), p.428. Image: Sanbi
Turnera – **William Turner (c 1508–1568)** En.wikipedia.org/w/index.php?title=William_Turner_(naturalist)&oldid= 660787772. Image: public domain. Lhccshtd.org. CC BY-SA 3.0.

Turraea – **Giorgio della Torre (Turre, Turra) (1607–1688)** Etymological explanation de'nomi generic plant By Alexandre Etienne Guillaume Baron de Theis (1815).
Tysonia – **William Tyson (1851–1920)** Glen & Germishuizen, *BESA* (edition 2), p.197–8. Image: Sanbi.

U

Ugena – **Manuel Munõz de Ugena (1747–1807)** Dicci-eponimos.blogspot.com/2009/11/ugena-manuel-munoz-de.html; Museodelprado.es/investigacion/biblioteca/fondo-antiguo/obras-destacadas/obra/browse/3/volver/72/actualidad/munoz-de-ugena-manuel-1747-1807.
Ursinia – **Johann Heinrich Ursinus (1608–1666)** En.wikipedia.org/w/index.php?title=Johannes_Heinrich_Ursinus&oldid=548758296. Image: public domain. Georg Christoph Eimart (painter) (1603–1685), Jacob von Sandrart (engraver)(1630–1708). Source: Flickr.com/photos/63794459@N07/6395051203/sizes/o/in/photostream.

V

Vachellia – **Rev. George Harvey Vachell (1798–1839)** *Alumni Cantabrigienses: A Biographical List of All Known Students ...*, Volume 2 By John Venn; Geni.com/people/George-Vachell/6000000023294759917.
Vahlia – **Martin Henrichsen Vahl (1749–1804)** En.wikipedia.org/w/index.php?title=Martin_Vahl&oldid=659872054. Image: public domain. Source: Commons.wikimedia.org/wiki/File:Martin_Vahl.jpg. CC BY-SA 3.0.
Vallisneria – **Antonio Vallisnieri (Vallisneri) (1661–1730)** En.wikipedia.org/w/index.php?title=Antonio_Vallisneri&oldid=661434510. Image: public domain. Source: *Observaciones y disertaciones sobre la física, la medicina y la historia natural*. CC BY-SA 3.0.
Vallota – **Antoine Vallot (1594–1671)** Fr.wikipedia.org/wiki/Antoine_Vallot.
Vandenboschia – **Benjamin Roelof van den Bosch (1810–1862)** Nl.wikipedia.org/w/index.php?title=Roelof_Benjamin_van_den_Bosch&oldid=42452612; Biografischportaal.nl/persoon/57574780; *Nieuw Nederlandsch Biografisch Woordenboek*.
Vanheerdea – **Pieter van Heerde (1893–1979)** Glen & Germishuizen, *BESA* (edition 2), p.435. Image: Sanbi.
Vanwykia – **Piet van Wyk (1931– 2006)** Glen & Germishuizen, *BESA* (edition 2), p.438–439. Image: Sanbi.
Vanzijlia – **Dorothy Constantia van Zijl (1886–1938)** Plants.jstor.org/stable/10.5555/al.ap.person.bm000123808.
Vascoa – **Vasco da Gama (1460 or 1469–1524)** En.wikipedia.org/w/index.php?title=Vasco_da_Gama&oldid=663327116. Author: unknown. Source: National Museum of Ancient Art. CC BY-SA 3.0.
Veltheimia – **August Ferdinand von Veltheim (1741–1801)** De.wikipedia.org/w/index.php?title=August_von_Veltheim&oldid=136747730; Minrec.org/libdetail.asp?id=1418. Image: public domain. Author: unknown. Source: Uni-magdeburg.de/mbl/Biografien/0416.htm, Herzog August Library Wolfenbüttel. CC BY-SA 3.0.
Verea – **James Vere (1739–1803)** Charters, *EDSAP*.
Vernonia – **William Vernon (1666–1711)** Plants.jstor.org/stable/10.5555/al.ap.person.bm000359038; Plantsystematics.org/reveal/pbio/usda/fnach7.html.
Vicoa – **Giovanni Battista (Giambattista) Vico (1688–1744)** En.wikipedia.org/w/index.php?title=Giambattista_Vico&oldid=660880567; Reference.com/browse/Vico. Image: public domain. Author: Francesco Solimena (1657–1747) Source: Universalis.fr/encyclopedie/S182531/VICO_G.htm. CC BY-SA 3.0.
Vieusseuxia – **Gaspard Vieusseux (1746–1814)** Fr.wikipedia.org/w/index.php?title=Gaspard_Vieusseux&oldid=112615805. Image: public domain. Source: Gefor.4t.com/arte/pintura/infectio37.html. CC BY-SA 3.0.
Vigna – **Domenico Vigni, (Domenicus Vigna) (1577–1647)** Nature.jardin.free.fr/1107/vigna_caracalla.html; Kew.org/science-conservation/research-data/resources/legumes-of-the-world/genus/vigna.
Villarsia – **Dominique Villars (1745–1814)** Bibliotheque-dauphinoise.com/histoire_plantes_dauphine.html. Image: public domain. Source: Commons.wikimedia.org/wiki/File:Villars_Antoine_1745-1814.jpg. CC BY-SA 3.0.
Virgilia – **Publius Vergilius Maro (70–19 BCE)** En.wikipedia.org/w/index.php?title=Virgil&oldid=663317517; Online-literature.com/virgil/(The Literature Network). Image: public domain. Source: Online-literature.com/virgil. CC BY-SA 3.0.
Vittia – **Dale Hadley Vitt (1944–)** Plants.jstor.org/stable/10.5555/al.ap.person.bm000009820.
Vlokia – **Johannes Hendrik Jacobus Vlok (1957–)** Glen & Germishuizen, *BESA* (edition 2), p.444. Image: Sanbi Eeu.org.za/downloads/touwsrivier-documents/Appendix%202.2_Specialist%20CV.pdf.
Vogelia – **Benedict Christian Vogel (1745–1825)** Ru.wikipedia.org/?oldid=67994032.
Volkiella – **Otto Heinrich Volk (1903–2000)** Plants.jstor.org/stable/10.5555/al.ap.person.bm000037504; Glen & Germishuizen, *BESA* (edition 2), p.445. Image: Sanbi.
Vossia – **Johann Heinrich Voss (1751–1826)** En.wikipedia.org/w/index.php?title=Johann_Heinrich_Voss&oldid=663296857. Image: public domain. Artist: Georg Friedrich Adolph Schöner (1774–1841). Source: *Das Jahrhundert der Freundschaft. Johann Wilhelm Ludwig Gleim und seine Zeitgenossen*, ed. Ute Pott (Göttingen: Wallstein, 2004), p.104. (James Steakley). CC BY-SA 3.0.
Vulpia – **Johann Samuel Vulpius (1760–1846)** En.wikipedia.org/w/index.php?title=Vulpia&oldid=645127242.

W

Wachendorfia – **Evert Jacob van Wachendorff (1702–1758)** Plants.jstor.org/stable/10.5555/al.ap.person.bm000339153.
Wahlenbergia – **Georg Göran Wahlenberg (1780–1851)** Sv.wikipedia.org/w/index.php?title=G%C3%

B6ran_ Wahlenberg&oldid=28593902. Image: public domain. Author: Johan Elias Cardon. Source: Uppsala Universitetsbibliotek.

Walafrida – **Walahfrid Strabo (or Walafrid Strabus) (c 809–849)** De.wikipedia.org/w/index.php?title= Walahfrid_ Strabo& oldid= 141412536. CC BY-SA 3.0.

Walleria – **Horace Waller (1833–1896)** En.wikipedia.org/w/index.php?title= Horace_ Waller_ (activist)& oldid= 663502476; Glen & Germishuizen, *BESA* (edition 2), p.450–451. Image: Sanbi CC BY-SA 3.0.

Wallinia – **Georg/Jöran Wallin the Younger (1686–1760)** Sv.wikipedia.org/w/index.php?title= Georg_ Wallin_ den_ yngre& oldid= 29170779. Image: public domain. Author: unknown. Source: *Svenskt Biografiskt Handlexikon*. CC BY-SA 3.0.

Waltheria – **Augustin Friedrich Walther (1688–1746)** En.wikipedia.org/w/index.php?title= Augustin_ Friedrich_ Walther& oldid= 659738269. Image: public domain. Author: unknown. Source: Aus der Bildersammlung der Bibliothek des Evangelischen Predigerseminars in der Lutherstadt Wittenberg. CC BY-SA 3.0.

Warburgia – **Otto Warburg (1859–1938)** En.wikipedia.org/w/index.php?title= Otto_ Warburg_ (botanist)& oldid= 656935369. Author: Kluger Z oltan. Source: National Photo Collection, Serial # 003616, photo code D22-123. CC BY-SA 3.0.

Wardia – **Nathaniel Bagshaw Ward (1791–1868)** En.wikipedia.org/w/index.php?title= Nathaniel_ Bagshaw_ Ward& oldid= 660631763. Image: public domain. Source: Scan aus einem buch; Original: Bodleian Library, EW Godwin, Artistic Conservatories, Shelfmark 19186. CC BY-SA 3.0.

Warneckea – **Otto Warnecke (f 1899–1908)** *Botanisches Z entralblatt: BBC Beihefte* (volume 34, 35, part 1), Association Internationale de Botanistes, 1917.

Watsonia – **William Watson (1715–1787)** En.wikipedia.org/w/index.php?title= William_ Watson_ (scientist)& oldid= 661282594. Image: public domain. Author: Magnus Manske on En.wikipedia. CC BY-SA 3.0.

Webbia – **Philip Barker Webb (1793–1854)** En.wikipedia.org/w/index.php?title= Philip_ Barker-Webb& oldid= 660713263. Image: public domain. Source: Commons.wikimedia.org/wiki/File:Philip_ Barker_ Webb_ 1793-1854.jpg. CC BY-SA 3.0.

Webera – **Georg Heinrich Weber (1752–1828)** De.wikipedia.org/w/index.php?title= Georg_ Heinrich_ Weber& oldid= 132513219. Image: public domain. Author: Joachim Johann Friedrich Bünsow. Source: Voigt & Lohff: *A Home for Surgery*, 1803–1986. CC BY-SA 3.0.

Websteria – **George. W. Webster (1833–1914)** Bulletin of the Torrey Botanical Club 14(7): 135. 1887.

Wedelia – **Georg Wolfgang Wedel (1645–1731)** Galileo.rice.edu/Catalog/NewFiles/wedel.html (The Galileo Project). Image: De.wikipedia.org/wiki/Datei:Fotothek_ df_ tg_ 0009002_ Mediz in_ ^_ Porträt.jpg Public domain. Author: Georg Wolfgang Wedel. Source: Saxon State Library, Dresden.

Weihea – **Carl Ernst August Weihe (1779–1834)** En.wikipedia.org/w/index.php?title= Carl_ Ernst_ August_ Weihe& oldid= 659406762.

Weinmannia – **Johann Wilhelm Weinmann (1683–1741)** Illustratedgarden.org/mobot/rarebooks/author.asp?creator= Weinmann,+Johann+Wilhelm& creatorID= 60. Source: Extract from article © 1995–2012 Missouri Botanical Garden.

Weissia – **Friedrich Wilhelm Weiss (1744–1826)** Books.google.com/books?id= 2an5IcZ DwC4C& pg= PA193& lpg= PA193& dq= Friedrich+Wilhelm+Weiss+(1744-1826),& source= bl& ots= OBx9PX3JTN& sig= PdevADPjbjLmx17uZ yM3VjXTsys. Göttinger Biologen 1737–1945: Eine Biographisch-Bibliographische Liste by Gerhard Wagenitz .

Wellstedia – **James Raymond Wellsted (1805–1842)** En.wikipedia.org/w/index.php?title= James_ Raymond_ Wellsted& oldid= 659462219.

Welwitschia – **Friedrich Martin Joseph Welwitsch (1806–1872)** En.wikipedia.org/w/index.php?title= Friedrich_ Welwitsch& oldid= 660733341. Image: Author: unknown. Source: Reinikka, *History of the Orchid*, Timber Press.

Whiteheadia – **Henry Whitehead (1817–1884)** Glen & Germishuizen, *BESA* (edition 2), p.459. Image: Sanbi.

Wiborgia – **Eric Nissen Viborg (1759–1822)** En.wikipedia.org/w/index.php?title= Erik_ Viborg& oldid= 659730446. Image: public domain. Author: unknown. Source: bibliotek.science.ku.dk/life150/forskere/viborg. CC BY-SA 3.0.

Widdringtonia – **Samuel Edward Widdrington (formerly Cook) (1787–1856)** En.wikipedia.org/w/index.php?title= Samuel_ Edward_ Cook& oldid= 660093881.

Wiesneria – **Julius Ritter von Wiesner (1838–1916)** En.wikipedia.org/w/index.php?title= Julius_ Wiesner& oldid= 663984750. Image: Author: unknown. Source: Ihm.nlm.nih.gov/Image: s/B25835. CC BY-SA 3.0.

Wigandia – **Johann Wigand (1523–1587)** De.wikipedia.org/w/index.php?title= Johannes_ Wigand& oldid= 139744073. Image: public domain. Author: unknown. Source: Museen in Thüringen/FSU Jena Kunstodie. CC BY-SA 3.0.

Wijkia – **Roelof van der Wijk (1895–1981)** Charters, *EDSAP*.

Willdenowia – **Karl Ludwig Willdenow (1765–1812)** En.wikipedia.org/w/index.php?title= Carl_ Ludwig_ Willdenow& oldid= 662780682. Image: public domain. Author: Laurens Geulp. Source: Bpun.unine.ch/IconoNeuch/Portraits/A-Z /W.htm. CC BY-SA 3.0.

Willkommia – **Heinrich Moriz Willkomm (1821–1895)** De.wikipedia.org/w/index.php?title= Heinrich_ Moritz _ Willkomm& oldid= 140167375; Saebi.isgv.de/biografie/Moritz _ Willkomm_ % 281821-1895% 29. Image: public domain. Source unknown. CC BY-SA 3.0.

Wimmerella – **Elfrieda Franz Wimmer (1881–1961)** Translate.google.com/translate?hl= en& sl= de& u= data.onb.ac.at/ nlv_ lex/perslex/nlv_ perslexikon.htm.

Wisteria – **Caspar Wistar (1761–1818)** En.wikipedia.org/w/index.php?title= Caspar_ Wistar_ (physician)& oldid= 660436327. Image: public domain. Author: Magnus Manske on En.wikipedia. Source: Originally from En.wikipedia. CC BY-SA 3.0.

Withania – **Henry Thomas Maire Witham (1779–1844)** En.wikipedia.org/w/index.php?title= Henry_ Witham& oldid= 659530856. Image: public domain. Author: unknown. Source: Makers of British Botany; this scan obtained from Archive. org/details/makersofbritishb00oliv. CC BY-SA 3.0.

Witsenia – **Nicolaas (Nicolaes) Witsen (1641–1717)** En.wikipedia.org/w/index.php?title= Nicolaes_ Witsen& oldid=

660706705. Image: public domain. Author: Peter Schenk the Elder (1660–1711). Source: Sporen van de Compagnie (ISBN 90-6707-169-2). CC BY-SA 3.0.
Wolffia – **Johann Friedrich Wolff (1778–1806)** En.wikipedia.org/w/index.php?title= Johann_ Friedrich_ Wolff& oldid= 600294761. Image: public domain. Source: Myheritage.de/photo-1001218_ 70396851_ 70396851/johann-friedrich-wolff. CC BY-SA 3.0.
Woodia – **John Medley Wood (1827–1915)** Glen & Germishuiz en, *BESA* (edition 2), p.464–466. Image: Sanbi.
Woodsia – **Joseph Woods (1776–1864)** En.wikipedia.org/w/index.php?title= Joseph_ Woods& oldid= 661056167. CC BY-SA 3.0.
Woodwardia = **Thomas Jenkinson Woodward (1745–1820)** En.wikipedia.org/w/index.php?title= Thomas_ Jenkinson_ Woodward& oldid= 595837472. CC BY-SA 3.0.
Wooleya – **Charles Hugh Frederick Wooley (1894–1969)** Charters, *EDSAP*; *Journal of South African Botany* (volumes 25–27), 1959.
Wormskioldia – **Morten Wormskjold (1783–1845)** En.wikipedia.org/w/index.php?title= Morten_ Wormskjold& oldid= 65948708. Image: public domain. Author: Bertha1949. Source: own work. CC BY-SA 3.0.
Wrightia – **William Wright (1735–1819)** En.wikipedia.org/w/index.php?title= William_ Wright_ (botanist)& oldid= 660000781. Image: public domain. Author: Thomas Phillips. Source: Comrie, *History of Scottish Medicine* (volume 2), 1932. Wellcome Historical Medical Museum, London.
Wulfhorstia – **August Wulfhorst (1861–1936)** Glen & Germishuiz en, *BESA* (edition 2), p.467.
Wurmbea – **Christoph Carl Friedrich von Wurmb (1742–1782)** De.wikipedia.org/w/index.php?title=Friedrich_ von_ Wurmb&oldid=134798621.

X

Ximenesia – **José (Joseph) Salvador Ximénez (Ximenes) Peset (1713–1803)** En.wikipedia.org/wiki/Francisco_ Xim%C3%A9nez
Translate.google.co.za/translate?hl=en&sl=es&u=biografiasyvidas.com/biografia/x/ximenez_francisco.htm&ei=7H_ vTPq_BImdOpbw0JcK&sa=X&oi=translate&ct=result&resnum=11&ved=0CEoQ7gEwCg&prev=/search%3Fq% 3DFrancisco%2BXim%25C3%25A9nez%26hl%3Den%26rlz%3D1B3GGIC_enZA317ZA317%26prmd%3Dio.
Ximenia – **Francisco Ximénez (1666–1729)** En.wikipedia.org/w/index.php?title=Francisco_ Xim%C3%A9nez&oldid= 648091477. CC BY-SA 3.0.

Y

Youngia – **Edward Young (1683–1765) and Thomas Young (1773–1829)** En.wikipedia.org/w/index.php?title= Edward_Young&oldid=665910032; En.wikipedia.org/w/index.php?title=Thomas_Young_(scientist)&oldid= 663286021. Image: public domain. Edward Young: Source: detail of illustration from Johnson, *Works of the English Poets with Prefaces, Biographical and Critical* (68 volumes) (volume 15). London: J. Nichols, 1779. Thomas: Source: National Portrait Gallery NPG 1899D7714. CC BY-SA 3.0.

Z

Zaluzianskya – **Adam Zalusiansky von Zaluzian (1558–1613)** Fernkloof.com/herbs/herbs-sept-2009.pdf.
Zanha – **Karl Hermann Zahn (1865–1940)** De.wikipedia.org/w/index.php?title=Karl_Hermann_Zahn&oldid= 139389880. Image: courtesy Botanic Garden and Botanical Museum Berlin-Dahlem. CC BY-SA 3.0.
Zannichellia – **Giovanni Gerolamo Zannichelli (1662–1729)** Unipd.it/vallisneri/en/herbariums/index.html. CC BY-SA 3.0.
Zantedeschia – **Giovanni Zantedeschi (1773–1846) and his son Francesco Zantedeschi (1797–1873)** En.wikipedia.org/w/index.php?title=Giovanni_Zantedeschi&oldid=600191790. CC BY-SA 3.0.
Zehneria – **Joseph Zehner (dates unknown)**
Zeyherella – **Carl (Karl) Ludwig Philipp Zeyher (1799–1858)** Glen & Germishuizen, *BESA* (edition 2), p.197–8. CC BY-SA 3.0.
Zinnia – **Johann Gottfried Zinn (1727–1759)** En.wikipedia.org/w/index.php?title=Johann_Gottfried_Zinn&oldid= 660167420; Whonamedit.com/doctor.cfm/3203.html. Image: public domain. Source: Commons.wikimedia.org/wiki/ File:Johann_Gottfried_Zinn.jpg.
Zornia – **Johannes Zorn (1739–1799)** En.wikipedia.org/w/index.php?title=Johannes_ Zorn&oldid=608367383.
Zschackea – **Georg Hermann Zschack (1867–1937)** Bgbm.org/Sipman/zschackia. Image: courtesy Botanic Garden and Botanical Museum Berlin-Dahlem.

Authors' acknowledgements

This publication would not have been possible without the help of a number of people, known and/or unknown to us, who gave of their knowledge, time and expertise to bring it to fruition. We are grateful for their support in many ways, such as helping us trace some of the more obscure names, providing historical information and Greek and Latin translations, giving permission for use of photographs and other such contributions.

Alice Notten, interpretive officer: Kirstenbosch National Botanical Garden, who generously and speedily answered our many questions
Al Schneider, compiler of Swcoloradowildflowers.com for information about John Amson (*Amsonia*)
Alan Fryday, Herbarium, department of plant biology, Michigan State University, for permission to use a photograph of HA Imshaug (*Imshaugia*)
Anders Hagborg, associate, department of science and education, Field Museum, Chicago
Andre Rabie, professor, faculty of law, University of Stellenbosch for documentation and photographs of WA Rabie (*Rabiea*)
Andreas Beck, PhD, curator and research scientist, Botanische Staatssammlung, München, Germany, who obtained a copy of the much needed book *Etymologie der Wissenschaftlichen Gattungsnamen der Flechten*, and its author's widow, Ilse Feige, for permission to use the content
Anita Stridvall, Lichen Gallery (online), for permission to use her late husband, Leif Stridvall's lichen photographs, *Arthrorhaphis, Diploicia, Imshaugia* and *Muerrerella*.
Arne Anderberg, professor, head of botany, Swedish Museum of Natural History, Stockholm for details of his career and photograph
Bertil Nordenstam, professor emeritus at the Swedish Museum of Natural History in the department of phanerogamic botany (recently retired) for help with many queries
Ben-Erik van Wyk, professor of botany and biotechnology at the University of Johannesburg
Bruce Bayer, author of *All You Want to Know about Haworthias and Gasterias*, etc, curator of Karoo Botanical Garden, Worcester
Bryan Simon and Yucely Alfonso, authors of *Grasses of Australia*, for use of information from their website Ausgrass2.myspecies.info
Cameron MacMaster, wildflower photographer, plant collector and grower
Celeste McNamara, visiting assistant professor of history at the College of William and Mary
Charles H Smith, professor and science librarian, department of library public services, Western Kentucky University and author of *Some Biogeographers, Evolutionists and Ecologists: Chrono-Biographical Sketches* – for a photograph of William Bray (*Brayulinea*)
Christoph Scheidegger, professor of biodiversity and conservation biology, Swiss Federal Institute for Forest, Snow and Landscape Research – for permission to use the image of Ludwig Emanuel Schaerer (*Schaereria*)
Christopher Cupido, research scientist, South Africa National Biodiversity Institute
Claude Durieux, Librairie.maritime.com, for permission to use a photograph of Jean-François Landolphe (*Landolphia*)
Colin Paterson-Jones, wildflower photographer
Cor Kwant, The Ginkgo Pages, for helping us locate an image of E Kaempfer (*Kaempferia*)
Crestina Forcina, picture researcher, historical collection, Wellcome Trust, London, for permission to use the image of Michael Friedrich Lochner von Hummelstein (*Lochnera*)
Dale Hadley Vitt, professor of plant biology at Southern Illinois University

Authors' acknowledgements

David Cumming, succulent plant enthusiast
David Hollombe, independent researcher, who has an unerring ability to ferret out obscure information and has unfailingly and generously helped us with difficult etymological problems
David Meagher, school of botany, University of Melbourne, for use of bryophyte and liverwort papers
David Victor, International Cultivar Registrar for Geranium and Erodium, president of the Saxifrage Society and the geranium group of the Hardy Plant Society and former chairman of the British Clematis Society and of the peony group of the Hardy Plant Society
Denis Lamy, chief editor, department of systematics and evolution, Paris, for *Renauldia* information;
Deon Kesting, emeritus professor of library science at the University of Cape Town
Deon Viljoen of the Karoo Botanical Gardens, Worcester
Dethardt Goetze, PhD, curator of the Botanical Garden, University of Rostock, Germany, for *Krauseola* information
Dianne Coetzee, vice-chairwoman, Rabie/Rabe Family Association for providing information about WA Rabie and a photograph (*Rabiea*)
Dr Kobus Venter, South African botanist
Dr Matt Buys, South African National Biodiversity Institute
Dr Randall J Bayer, professor and chair, department of biological sciences, University of Memphis, Memphis, Tennessee
Dr Ted Oliver, on staff at Sanbi (retired), world authority on Ericaceae
Elizma Fouché, Sanbi publishing, for the provision of a number of images that appear in editions 1 and 2 of Botanical Collectors in Southern Africa
Ernst van Jaarsveld, botanist and former horticulturist in charge of the succulent collection at the Botanical Society Conservatory, Kirstenbosch and now at Babylonstoren
Eugene Moll, extraordinary professor in the department of biodiversity and conservation biology at the University of the Western Cape, Cape Town (see 'The history of this book')
Fanie Venter, adjunct researcher at Australian Tropical Herbarium and director at Botanical and Environmental Consultancy
Geoffrey Andrew, author of *Fernkloof Plant Names Explained*
Gerhard Walter Rambold, German lichenologist, professor of mycology and DNA analytics, University of Bayreuth, Germany – for *Ramboldia* information
Gerould Wilhelm, PhD, botanist, research taxonomist and director of research for the Conservation Research Institute – for use of lichen name descriptions in his and Laura Rericha's *Lichens of the Chicago Region*
Gideon Smith, chief director of research and scientific services with Sanbi, honorary professor to the John PH Acocks Chair in the botany department at the University of Pretoria, co-author of *Guide to the Aloes of South Africa*
Greig Russell, independent researcher, for information relating to Franz Pehr Oldenburg (*Oldenburgia*)
Guy Cobolet, director of the BIU Santé (Paris) for providing a photograph of Nicolas Jean-Baptiste Gaston Guibourt (*Guibourtia*)
Hamish G Robertson – director of natural history collections, Iziko Museums of South Africa – for documentation relating to HHJCE Brauns (*Braunsia*)
Hannelie Snyman, researcher at Sanbi
Helen Hopper, archivist, Noel Butlin Archives Centre, Canberra University, Australia – for permission to use a photograph of DJ Carr (*Carrpos*)
Henrik Dupont, research librarian and map curator, The Royal Library, National Library and Copenhagen University – for permission to use a photograph of CG Rafn (*Rafnia*)
Henry J Noltie, PhD, botanical historian, Royal Botanical Garden Edinburgh – who, among other things, helped us find after whom *Hewittia* was probably named
Dr Herbert Schwabl, president of the board of directors of PADMA AG, for information about Joseph Rehmann (*Rehmannia*)
Hugh Glen, PhD, senior researcher, Sanbi – for support, photos, connections and information in the book he co-authored with Germishuizen, *Botanical Exploration of Southern Africa* (edition 2), and his book *Sappi Tree Spotting: What's in a Name?*
Ingo Breuer, nurseryman, *Haworthia* grower and authority and plant collector
J Dustin Williams, archivist and research scholar, Hunt Institute for Botanical Documentation, Carnegie Mellon University – for permission to use a picture of PRO Bally (*Ballya*)
Jacques van Rooy, Sanbi bryophyte researcher

Jan Meemelink, Antiquariaat Jan Meemelink, the Netherlands, for permission to use a photograph of MT Masters (*Mastersiella*)

Jan Mokre, author of Fürstliche Forstwirtschaft und Bürgerliche Freiheit. Die Botanische Studien- und Sammelreise des Joseph van der Schot nach Nordamerika, in Elisabeth Zeilinger, *Osterreich und die Neue Welt: Symposion in der Osterreichischen Nationalbibliothek, Wien*, Osterreichische Nationalbibliothek, 1993, pp.107–120

Joanna McManus, picture curator, The Royal Society, for permission to use a picture of AB Lambert (*Lambertia*)

Johan Krige, retired consulting engineer, who provided previously unrecorded details concerning WA Rabie (*Rabiea*)

John Manning, specialist scientist at National Botanical Institute, Kirstenbosch, author of many South African plant guides

Josef Hafellner, Institut für Pflanzenwissenschaften, Karl-Franzens-Universitat, Graz, Austria, who provided information relating to *Brigantiaea, Hafellia* and *Fellhanera*

Julia Buckley, illustrations assistant, library, art; archives, Royal Botanic Gardens, Kew for the images of HK Airy Shaw (*Aerisilvaea*) and W Jameson (*Jamesonia*);

Justin Wynns, PhD Student, University of Copenhagen

Karel Sacek (Obristvi.cz/osobnosti_koller_frantisek.htm) for permission to use a picture of Franz von Koller (Kolleria);

Karin Zimmermann, Heidelberg University library, for permission to use a picture of GW Bischoff (*Bischofia*)

Kevin Balkwill, University of the Witwatersrand

Knud Høgsberg, information specialist, librarian, faculty library of natural and health sciences, University of Copenhagen for permission to use an image of EN Viborg (*Wiborgia*) (The genus name was misspelled by Thunberg)

Koos Roux, Sanbi.

Laurence 'Larry' Dorr, research botanist and associate curator, US National Herbarium, Smithsonian Institution, for photographs

Len Newton, professor of botany, Kenyatta University and co-author of *Etymological Dictionary of Succulent Plant Names*

Leslie Powrie, who provided a comprehensive Southern African plant genera list

Marinda Koekemoer, author of *Grasses of Southern Africa* and curator at Sanbi

Martin Nickol, garden curator and deputy director, Botanical Garden of the University of Kiel – for permission to use a photograph of HJH Jacobsen (*Jacobsenia*)

Mats Thulin, professor of systematic botany at Uppsala University, Sweden – for information on *Afroqueta* and *Gunillaea*

Mathias Böhm, Österreichische Nationalbibliothek, Bildarchiv und Grafiksammlung – for permission to use a photograph of Jacob Juratzka (*Juratzkaea*)

Matthias Schultz, PhD and research associate in biodiversity, evolution and ecology of plants at the Herbarium Hamburgense, University of Hamburg, Germany – who gave help with lichen names, especially *Peccania*

Michael Baars, information specialist, Huygens Institute for the History of the Netherlands, Royal Netherlands Academy of Arts and Sciences – for permission to use the Fabbroni-portrait (*Fabronia*)

Mienkie Welman, from the National Herbarium, National Botanical Institute, Pretoria

National Geographic Society (with IBG) for the image of Rawson W Rawson (*Rawsonia*)

Nikki Clarke, daughter of Hugh Clarke (author), for hours spent doing a provisional type layout and proof of the draft publication

Norbert Kilian, Archive, Berlin-Dahlem Botanical Garden and Botanical Museum, Free University, Berlin for use of photographs

Peter Bruyns, Bolus Herbarium, University of Cape Town

Peter Goldblatt, Missouri Botanical Garden

Pieter Winter of Sanbi for help and advice

R Stangl, librarian of the University of Vienna botany department

Rachel Knowles, author and historian (Regencyhistory.net), and Andrew Knowles, copyright holder, for permission to use a photograph of Sir Richard Colt Hoare (*Hoarea*)

Rachel Ingold, curator, history of medicine collections, Rubenstein Rare Book and Manuscript Library, Duke University, for permission to use a picture of J de Gorter (*Gorteria*)

Authors' acknowledgements

Ralph S Stewart, David M Johnson and John T Hickel for the names and meanings of ferns in the
   Pteridophyte genera, in *Fiddlehead Forum, Bulletin of the American Fern Society*, Vol. 10, Nos. 4; 5
Reuben Roberts, who provided Hugh Clarke with the original Sanbi generic names spreadsheet
Rodney Moffett, professor at the University of the Free State, for a picture of WA Rabie's tombstone
Russell Wagner, Little Sphaeroid Press, for a photograph of Steven Hammer (*Hammeria*)
Sanbi publishing, Sanbi, for the use of photos
Shaun Russell, director, Wales Environment Research Hub – who helped with botanical etymology
Shirley Tucker, PhD, ex Louisia State University and one of America's foremost lichenologists – who
   helped with the meanings of lichen names
Stefan Dressler, curator phanerogams, 2 Herbarium Senckenbergianum, Frankfurt am Main – for
   permission to use an image of Johannes Becker (*Beckeropsis*)
Steven Hammer, author of *Lithops, Treasures of the Veld*
Susan Marsh, Jotello F Soga Library, faculty of veterinary science – for permission to use a photograph of
   Sir Arnold Theiler (*Theilera*)
The late Koos Roux, curator of the Compton Herbarium, who made available the latest APGII plant list
The Berrisford family for a photograph and information about Francis 'Frank' Berrisford (*Berrisfordia*)
Tony Dold, botany department, Rhodes University
Tony Rebelo, who provided the electronic version of the late WPU Jackson's book
Trudi Stols, librarian, Sam Cohen Library, Scientific Society, Swakopmund, Namibia – for information on
   Emil Jensen (*Jensenobotrya*) and Ernst Stöber (*Stoeberia*);
Umberto Quattrocchi, author of *CRC World Dictionary of Plant Names* and publisher, for permission to
   use some of his information;
Urs Eggli, curator of the municipal succulent collection, Zurich, author of *Glossary of Botanical Terms* and
   editor of *Illustrated Handbook of Succulent Plants: Dicotyledons*
Vincent DeVries, *Haworthia* grower
Welland Cowley, Cape Flora Nursery, Port Elizabeth, South Africa
William R Buck, Institute of Systematic Botany, New York Botanical Garden.

The authors would also like to acknowledge any other people whose names may have been accidentally
   omitted.

Contributors & photographers

A Eugene Moll (EM)
B Michael Charters (MC)
C Darrel Plowes
D AB (Tony) Cunningham (AC)
E EGH (Ted) Oliver (EGHO)
F Sizwe Cawe (SC)
G Marie Jordaan
H David Gwynne-Evans (DG)
I Brian Schrire (BS)
J John Manning
K Himansu Baijnath (HB)
L Ernst van Jaarsveld
M John Burrows (JB)
N René Glen
O Hugh Glen (HG)
P John Rourke
Q Richard James Poynton
R Peter Linder (PL)
S Rodney Moffet
T Peter Goldblatt (PG)
U Braam van Wyk (BvW)
V Tony Dold (TD)
W Ernst van Jaarsveld/ Uschi Pond (UP)
X Ted Woods (TW)
Y Benny Bytebier
Z Bart Wursten (BW)
   Hugh Clarke (HC)

Addendum: guest photographers

Our special thanks go to the following botanists for providing superb illustrations at short notice.

Adam Harrower (AH) has been a horticulturist at Kirstenbosch Gardens for 15 years. He curates the woody plant collection and arboretum. Discovering the hidden wonders of the natural world through the eye of a camera lens has contributed to his passion for plants.

Corinne Merry (CM) has a PhD in theoretical physics. Now retired, she spends her days walking in the mountains with dog and camera and photographing the flowers she loves with her husband, Charles, at her side. She is involved in the flora documentation programme of the Friends of Silvermine nature area and co-authored *Common Wild Flowers of Table Mountain and Silvermine*.

Geoff Nichols (GN) worked in the Durban Parks for 21 years and has diplomas in agriculture, horticulture and parks; recreation administration. He pioneered growing indigenous medicinal plants at Silverglen. His interest in indigenous gardening in the urban environment and is consultant specialising in the development of the 'habitat gardening concept'. He has published books on plant propagation, medicinal plants and wild gardening.

In 2015 he was awarded the Marloth Medal by Botanical Society of South Africa.

Graham Grieve (GG) – after retirement his primary natural history interest was birding, but a tertiary-level research project broadened his interest to include keeping a photographic record of plants. His main focus is on the plants of the Pondoland Centre of Endemism.

Greg Nicolson (GAN) was born in Cape Town. He developed an interest in fynbos through spending time hiking in the Cape Mountains. He now works as a botanist and photographer based in Cape Town, and his unique photographs have been widely used in many publications.

Herbert Stärker is from Vienna, Austria. He is a retired banker who initially visited South Africa in 1976. His first dedicated orchid trip to South Africa was in 2006, followed by numerous subsequent trips to photograph orchids. After meeting Bill Liltved he started to contribute pictures to the Cape Orchids project. From 2012–2015 he photographed plants in the field for *Orchids of South Africa: A Field Guide*.

Neil Crouch (NC) has a passion for hiking and photographing the mountains of Central and Southern Africa. His botanical interests are broad, fed by exposure to the exciting fields of ethnobotany and biosystematics. Ferns, bulbs and succulents are of particular interest, and are reflected in many of his photographs and related publications.

Nick Helme (NH) is the sole proprietor of Nick Helme Botanical Surveys, established in 2001. He undertakes botanical and ecological assessments throughout the Greater Cape Floristic Region (CFR), is a lead author of CFR assessments for the Sanbi Red List of SA Plants and was co-author of the *Fynbos Forum Ecosystem Guidelines* and of the fynbos chapter in the *SA Vegetation Map*. He has discovered more than 50 previously undescribed plant species in the Cape region alone, some of which he has helped describe.

Richard Boon (RB) is the author of *Trees of Eastern South Africa: A Complete Guide*. He works for the eThekwini Municipality's environmental planning and climate protection department and heads up the biodiversity planning branch. Richard is a practical botanist with over 25 years of field experience.

Editor's endnote

As you read this dictionary, you will have learned much about how and why many of our plant genera were named; and you will discover that ±1 000 of the ±5 000 genera have been named after people. This is because some have either been great botanists and/or plant collectors and are so honoured for all time while others funded botanical exploration or were good friends and/or respected colleagues of those who named new genera, etc. Having a plant genus (or even a species) named after you is indeed a great tribute. So the dictionary reflects much of our rich botanical history. Being the editor gave me important insights into the content and scope of this dictionary. As a Southern African botanist with a life-long passion for biodiversity conservation and people, I have learned that life treats people differently. Through my involvement with students at many levels, both professionally as well as voluntarily, I have learned much about people – their strengths and weaknesses. From the dictionary I have noticed that getting a genus named after you is a little hit-and-miss, and in my view many deserving botanists have not been accorded this honour or even given recognition.

Because I have had time to ponder this matter I have taken the liberty of doing something about it here. Since this book has *gravitas* I have taken it upon myself to nominate a few other botanical greats that have not had a genera named after them. I have achieved this by:
- Carefully selecting the people invited to author the A–Z pages – these are some of the top botanists in South Africa; many are household names.
- Choosing the photographs to illustrate additional genera in the dictionary that the authors and/or the A–Z people were unable to supply.
- Compiling a short list of people – all of whom have lived in the last 150 years, who are now deceased, that made significant contributions to botany and whose names and achievements also deserve to be remembered. This is my way of acknowledging a few of our other botanical 'masters'.

To assist in compiling my list of great botanists I consulted Glen and Germishuizen's *Botanical Exploration of Southern Africa* (Strelitzia 26, 2010), which records short biographies of all plant collectors from the region who have specimens housed in herbaria. In doing so I also used personal knowledge to keep the list short, and one of the criteria I used that showed their commitment was that they collected well in excess of 6 000 herbarium specimens or made a substantial contribution to the contemporary botanical landscape. If you wish to know more about anyone on my list (as I have only provided an extremely brief biography for each) then please consult Glen and Germishuizen for more details.

Editor's endnote

My list of additional famous botanists

Amy Jacot Guillarmod (1911–1992), great teacher and researcher and author of *The Flora of Lesotho*
AW Exell (1901–1993), a founder of AETFAT (Association pour l'Étude Taxonomique de la Flore d'Afrique Tropicale) and writer/editor of many of the floras of the region

'Bertie' Mogg (1889–1980), dynamo plant collector (±40 000 plants collected) and teacher
'Bob' Drummond (1924–2008), taxonomic whizz-kid at Southern Rhodesia Government Herbarium and much more
'Buddy' Barker (1907–1994), curator of the Compton Herbarium and Cape taxonomist

DS Hardy (1931–1998), collector and horticultural succulent expert

EACLE 'Ted' Schelpe (1924–1985), professor, and fern and global orchid expert
EP Phillips (1884–1967), author of *The Genera of South African Flowering Plants*, among others

Frank White (1927–1994), best known for his *Vegetation of Africa* map and companion tome

GW Reynolds (1895–1967), the African and Madagascan *Aloe* expert of the time, despite being an amateur

HG Fourcade (1865–1948), forester and land surveyor – pioneered aerial photography
Hiram Wild (1917–1982), Southern Rhodesia Government Herbarium keeper, sunflower expert, starter of *Kirkia* and instrumental in getting *Flora Zambesiaca* initiated
HP van der Schijff (1921–1997), best known for his work in the Kruger Park and for planting indigenous trees on the Pretoria University campus
Hugh Taylor (1925–1999), fynbos botanist *par excellence*

JC Smuts (1870–1950), grass expert and famous for his book *Holism and Evolution*
John Acocks (1911–1979), author of *Veld Types*
John Hutchinson (1884–1972), keeper at Kew, in charge of the African section and phylogenist
Joseph Burtt Davy (1870–1940), founder of what is now South African National Biodiversity Institute and fern expert

Keith Coates Palgrave (1926–1991), author of *Trees of Southern Africa*

Lucy 'Chips' Chippendale (1913–1992), author of the grasses in *The Grasses and Pastures of South Africa*

Margaret Levyns (1890–1975), active in botany at University of Cape Town and among many others publications was author of *Guide to the Flora of the Cape Peninsula*

RGN Young (1904–1979), one of the last great botanical explorers of the region
'Roddy' Ward (1926–2015), KZN botanist *par excellence*
RS Adamson (1885–1965), professor and author, with Salter, of *The Flora of the Cape Peninsula*

Willy Giess (1910–2000), founder of the Windhoek Herbarium and author of *The Vegetation of SWA* (now Namibia)

Doubtless, a few other names could be added, and these show my own bias, but this 'roll of honour' simply demonstrates that there are many other unsung heroes and heroines in the world of botany.

Finally I wish to thank all the donors, sponsors and subscribers for believing in this dictionary. The support I received to my call for assistance with funding was astonishing – my target was reached in less than two months!

I would also like to take this opportunity to thank all those botanists and photographers who contributed to making this a superb book – your generous support and efficient service is appreciated.

Our thanks also to Charles Botha of Botanical Education Trust and Marylynn Grant of the Flora & Fauna Publication Trust for all their support and commitment to the Dictionary.

And our special thanks to Carol Broomhall and her capable team at Jacana (Nadia Goetham, Megan Mance and Shawn Paikin) as well as Joey Kok, who edited the initial draft. We appreciated your dedication, expertise and professionalism.

Thank you all!

Eugene Moll

A postscript from the authors

Our grateful thanks to Eugene Moll, our botanical editor. He went way beyond being a botanical editor, being fund-raiser, colour-images coordinator, and a wonderful supporter of this book throughout the many years it took to come to fruition.

Hugh Clarke & Michael Charters

About the authors and editor

Hugh Gascoyne Clarke has written books in the fields of botany including wild flowers of Table Mountain and wild flowers of Table Mountain and Silvemine as well books in retail, sales management, selling, labour law and children's fiction. He has a BA degree from the University of Witwatersrand, majoring in philosophy and political science, a diploma in marketing from the Institute of Marketing Management, and Master's degree in Business Leadership from UNISA. Although a businessman for most of his life, in his younger years he also did some part-time lecturing at the Cape Technikon and University of Cape Town. His interests have included travel, hiking, bird and game watching, marathon running, classical music especially guitar and most recently botany. Married, he has four children and eleven grandchildren.

Michael Charters is a non-professional botanist and wildflower photographer who has been documenting the flora of Southern California photographically for the past twenty years, taking hundreds of fields trips and some 200 000 photographs. He initially became interested in the flora of the Cape region of South Africa because of the affinity to that of Southern California, and he has visited South Africa four times, spending weeks at a time photographing plants in the field in both the East and Western Cape. His work has appeared in dozens of books and scholarly articles.

Eugene John Moll graduated from UKZN with a PhD in plant ecology and spent 10 years working as a Botanical Survey Officer, 20 years at UCT and 10 years at the University of Queensland before retiring to Cape Town in 2003. He holds an honorary position at UWC in the Department of Biodiversity and Conservation Biology and teaches an Honours module in Fynbos Ecology. He has spent his life being passionate about people-and-biodiversity conservation and has published in journals, written books and articles, and served the botanical community in many ways.